B/NA

$$V_{bi} = \psi_n - \psi_p = \frac{kT}{8} \ln\left(\frac{N_A N_D}{n_i^2}\right) = 0.0259 \ln\left(\frac{10^{19} \times 10^{16}}{(9.65 \times 10^9)^2}\right) = 0.895V$$

Si at 300°K & $N_A = 10^{19}/cc$

$N_D = 10^{16}/cc$

INTERNATIONAL SERIES IN THE
SCIENCE OF THE SOLID STATE
General Editor: B. R. PAMPLIN

VOLUME 14

Electrical Transport in Solids

Other Titles in the
International Series in the Science of the Solid State
(*Editor*: B. R. PAMPLIN)

An important new review journal

Progress in Crystal Growth and Characterization
Editor-in-chief: B. R. PAMPLIN

Free specimen copy available on request

Electrical Transport in Solids

With Particular Reference to Organic Semiconductors

by

KWAN C. KAO

Professor of Electrical Engineering
University of Manitoba, Canada

and

WEI HWANG

Assistant Professor of Electrical Engineering
Columbia University, USA

PERGAMON PRESS

OXFORD · NEW YORK · TORONTO · SYDNEY · PARIS · FRANKFURT

U.K.	Pergamon Press Ltd., Headington Hill Hall, Oxford OX3 0BW, England
U.S.A.	Pergamon Press Inc., Maxwell House, Fairview Park, Elmsford, New York 10523, U.S.A.
CANADA	Pergamon of Canada, Suite 104, 150 Consumers Road, Willowdale, Ontario M2J 1P9, Canada
AUSTRALIA	Pergamon Press (Aust.) Pty. Ltd., P.O. Box 544, Potts Point, N.S.W. 2011, Australia
FRANCE	Pergamon Press SARL, 24 rue des Ecoles, 75240 Paris, Cedex 05, France
FEDERAL REPUBLIC OF GERMANY	Pergamon Press GmbH, 6242 Kronberg-Taunus, Hammerweg 6, Federal Republic of Germany

First edition 1981

British Library Cataloguing in Publication Data

Kao, Kwan Chi
Electrical transport in solids.—
(International series in the science of
the solid state; vol. 14).
1. Organic semiconductors
I. Title   II. Hwang, Wei
III. Series
537.6'22   QC611.8.07   79–40551
ISBN 0–08–023973–0

Printed and bound in Great Britain by
William Clowes (Beccles) Limited, Beccles and London

SD 4/8/81
JW

DEDICATED TO OUR PARENTS

Preface

THIS book attempts to give a clear picture of electrical transport problems and to provide analytical methods for exploring the behaviour of charge carriers in semiconductors and insulators, which is responsible for their macroscopic electrical and optical properties. In recent years electrical transport, including optical and photoelectronic phenomena in solids with a large energy band gap and a small carrier mobility and particularly in organic semiconductors, has become an increasingly important area of research partly because the materials of this category possess a high potential for solid state electronic devices and partly because they can be used as model materials for studying the electronic processes in general molecular solids which are directly related to the properties of organic substances, and as a stepping-stone towards the understanding of the behaviour of biological systems. In materials of this category, traps arising from various imperfections play a major role in the electrical transport and, therefore, special emphasis is given to the effects of traps and electrical contacts and the correlation between theoretical and experimental results with particular reference to organic semiconductors. The theoretical analyses are general and thus applicable to both inorganic and organic solids. However, although most of the electrical transport properties of organic solids are not unlike those of inorganic ones, there are wide departures in many features between them. This book provides the principles to explain such departures.

It is important to mention that the topics of electrical transport treated in the present book, even confined in the area of organic semiconductors, are a vast subject. To deal with such a vast subject, it is almost unavoidable that some topics are deliberately overemphasized and others less discussed or excluded, not because they are of less importance, but simply at least in part it is our selections on account of the limited volume of this book. Since the amount of work done in this field and available in the literature is huge, the literature coverage in this book is representative and selective rather than exhaustive. An extensive bibliography with over 2600 references is included so that the reader, fascinated by some topics, can be referred to them for more information and guided to further studies of the subjects.

We wish to thank our many friends for their encouragement during the writing of this book, the University of Manitoba Research Board for the grant towards the expenses in preparing the manuscript, and the Editor of Pergamon Press for his helpful co-operation and prompt actions.

We are indebted to the publishers and the authors cited for granting permission to reproduce figures and tables appearing in this book. The publishers and the authors own the copyright of those figures and tables published originally by them, and we wish to acknowledge particularly the permission from the publishers listed in alphabetical order below:

Academic Press, Inc., for Figs. 1.19, 3.36, 4.14, 5.7, 5.41, and 5.42.

Akademische Verlagsgesellschaft (*Z. Phys. Chem.*), for Fig. 4.19.

American Chemical Society, for Fig. 1.3.

American Institute of Mining, Metallurgical, and Petroleum Engineers, Inc., for Fig. 2.26.

American Institute of Physics, for Figs. 1.4, 1.5, 1.8, 1.9, 1.10, 2.15, 2.18, 2.23, 2.25, 2.33, 2.34, 3.7, 3.8, 3.9, 3.16, 4.9, 4.10, 4.11, 4.12, 4.16, 4.17, 4.20, 5.3, 5.6, 5.15, 5.18, 5.30, 5.34, 5.35, 5.36, 5.37, 5.38, 5.43, 6.13, 6.14, 6.18, 6.20, 6.23, 6.31, 6.32, 7.6, 7.7, 7.22, 7.23, 7.24, 7.25, 7.26, 7.27, 7.37, 7.38, 7.39, 7.40, 7.41, 7.43, 7.49, and 7.50; and Tables 2.3, 3.5, 3.6, and 6.2.

American Physical Society, for Figs. 1.6, 2.22, 2.28, 2.29, 2.38, 3.6, 3.11, 3.12, 3.27, 5.23, 5.24, 5.25, 5.29, 5.44, 5.45, 5.46, 5.47, 5.48, 6.15, 6.16, 6.17, 7.8, 7.9, 7.10, 7.32, 7.33, 7:34, 7.35, and 7.36; and Tables 1.6, 1.7, 1.8, and 7.2.

Armour Research Foundation of ITT, for Figs. 1.6, 5.24, and 5.25.

Associated Book Publishers Ltd. (London), for Table 1.12.

Australian Chemical Society (*Australian Journal of Chemistry*), for Figs. 5.9, 5.10, 5.11, 5.12, 7.12, 7.13, 7.17, 7.18, 7.19, and 7.20; and Table 5.1.

Chemical Society (Faraday Society, London), for Fig. 5.19.

Chemical Society of Japan, for Figs. 6.25, 6.28, and 6.33.

Elsevier Sequoia, for Fig. 3.24.

Gordon and Breach Science Publishers, Inc., for Figs. 3.10, 4.21, 7.14, 7.15, 7.30, and 7.31; and Tables 1.2 and 1.3.

Institute of Electrical and Electronics Engineers, Inc., for Figs. 5.1 and 5.2.

Institute of Physics (London), for Fig. 1.14.

International Business Machines Corporation, for Table 6.1.

John Wiley and Sons, Inc., for Figs. 2.9, 7.1, and 7.5.

Macmillan (Journals) Ltd., for Fig. 6.34.

Macmillan Publishing Co., Inc., for Figs. 1.6, 5.24, and 5.25.

Mills and Boon Ltd., for Fig. 5.8.

North-Holland Publishing Co., for Figs. 1.11, 5.4, 5.26, 5.28, 5.33, 6.9, 6.10, 7.16, 7.51, and 7.52.

Oxford University Press (Clarendon Press), for Figs. 1.2 and 2.37.

Pergamon Press Ltd., for Figs. 2.19, 2.20, 2.29, 2.31, 2.32, 2.36, 3.14, 3.15, 3.17, 3.18, 3.19, 3.20, 3.21, 3.22, 3.23, 3.30, 3.31, 3.32, 3.33, 3.34, 4.15, 5.5, 5.27, 5.32, 5.39, 5.40, 6.12, 6.19, 7.42, 7.44, 7.45, 7.46, 7.47, and 7.48.

Physica Status Solidi, for Figs. 1.16, 3.13, 5.14, 5.20, and 7.21; and Table 1.9.

Physical Society of Japan, for Figs. 1.13, 6.11, and 6.27.

Plenum Press, for Figs. 2.24, 6.7, and 6.8; and Table 1.10.

Proceedings of the International Symposium, Clausthal-Göttingen (1965), for Fig. 2.14.

Royal Society (London), for Figs. 2.39, 2.40, and 2.41.

Societa Italiana Di Fisica (*Nuovo Cimento*), for Figs. 4.18 and 5.21.

Springer-Verlag (*Z. Phys.*), for Fig. 5.31.

Taylor and Francis Ltd., for Fig. 5.17.

Verlag Chemie GmbH, for Figs. 1.12 and 3.35.
Verlag der Zeitschrift für Naturforschung, Tübingen, for Figs. 7.28 and 7.29.

Winnipeg, Manitoba, Canada   KWAN CHI KAO
December 1979   WEI HWANG

Contents

6. Photoelectronic Processes

7. Luminescence

List of Principal Symbols

Symbol	Definition	Dimensions
A	Richardson's constant	$\text{A m}^{-2}\,^\circ\text{K}^{-2}$
A^*	Effective Richardson's constant	$\text{A m}^{-2}\,^\circ\text{K}^{-2}$
B	Magnetic flux density	Weber m^{-2}
C_n	Capture rate (or simply capture) coefficient or trapping coefficient for electrons	$\text{m}^3\,\text{sec}^{-1}$
C_p	Capture rate (or simply capture) coefficient or trapping coefficient for holes	$\text{m}^3\,\text{sec}^{-1}$
C_r	Band-to-band recombination rate coefficient	$\text{m}^3\,\text{sec}^{-1}$
d	Specimen thickness	m
d_{eff}	Effective specimen thickness	m
D	Carrier diffusion coefficient	$\text{m}^2\,\text{sec}^{-1}$
D_n, D_p	Electron and hole diffusion coefficients, respectively	$\text{m}^2\,\text{sec}^{-1}$
D_s, D_T	Singlet and triplet exciton diffusion coefficients, respectively	$\text{m}^2\,\text{sec}^{-1}$
$D_T(E)$	Quantum mechanical transmission function	—
E	Energy level	eV
E_c, E_v	Energy levels at conduction band and valence band edges, respectively	eV
E_{Dn}, E_{Dp}	Electron and hole demarcation levels, respectively	eV
E_F	Fermi energy level	eV
E_{Fm}, E_{Fs}	Fermi energy levels in the metal and semi-conductor, respectively	eV
E_{Fn}, E_{Fp}	Quasi-Fermi levels for electrons and holes, respectively	eV
E_g	Energy band gap	eV
E_I	Ionization energy of the impurity centre	eV
E_r	Recombination centre energy level	eV
E_t	Trapping centre energy level	eV
E_{tn}, E_{tp}	Electron and hole trapping energy levels, respectively	eV
E_u, E_l	The upper and lower limits of the trapping energy levels, respectively	eV
$f(E)$	Fermi–Dirac distribution function	—

Symbol	Definition	Dimensions
f_n, f_p	Fermi–Dirac distribution functions for trapped electrons and holes, respectively	—
f_r	Probability for a recombination centre to be occupied	—
F	Electric field	$V\ m^{-1}$
F_{av}	Average applied electric field (V/d)	$V\ m^{-1}$
$g(E)$	Density of states	$m^{-3}\ (eV)^{-1}$
g_n, g_p	Degeneracy factors of trap states for electrons and holes, respectively	—
G	Charge carrier generation rate	$m^{-3}\ sec^{-1}$
h	Planck's constant	$J\ sec$
$\hbar$	$h/2\pi$	$J\ sec$
h_n, h_p	Trap density distribution functions for electrons and holes, respectively	$m^{-3}\ (eV)^{-1}$
H_a, H_e	Trap densities for hole traps confined in a single discrete energy level for planar and non-planar electrode geometries, respectively	m^{-3}
H_b, H_f	Trap densities for hole traps distributed exponentially within the forbidden energy gap for planar and non-planar electrode geometries, respectively	m^{-3}
	Gaussianly within the forbidden energy gap	
H_{an}, H_{ap}	Trap densities for electron and hole traps confined in a single discrete energy level, respectively	m^{-3}
H_{bn}, H_{bp}	Trap densities for electron and hole traps distributed exponentially within the forbidden energy gap, respectively	m^{-3}
H_{cn}, H_{cp} (or H_c)	Trap densities for electron and hole traps distributed uniformly within the forbidden energy gap per unit energy interval, respectively	$m^{-3}\ (eV)^{-1}$
H_d	Trap density for hole traps distributed	m^{-3}
i	Light intensity or photon flux	$m^{-2}\ sec^{-1}$
I	Total current	A
J, J_n, J_p	Total, electron, and hole current densities, respectively	$A\ m^{-2}$
k	Boltzmann constant	$eV\ {}^\circ K^{-1}$
l	$= T_c/T$	—
L_a	Ambipolar diffusion length	m
m	Rest electron mass	kg
m_e^*	Effective electron mass	kg
m_h^*	Effective hole mass	kg
n, n_t, n_T	Free, trapped, and total electron densities, respectively	m^{-3}
n_i	Electron (or hole) carrier density of the intrinsic semiconductor	m^{-3}

Symbol	Definition	Dimensions
n_c	Refractive index	—
n_0, n_{t0}	Free and trapped electron densities in thermal equilibrium, respectively	m^{-3}
n_{at}	Trapped electron density for the traps confined in a single discrete energy level	m^{-3}
n_{bt}	Trapped electron density for the traps distributed exponentially within the forbidden energy gap	m^{-3}
n_{ct}	Trapped electron density for the traps distributed uniformly within the forbidden energy gap	m^{-3}
n_r	Density of occupied recombination centres (captured electrons)	m^{-3}
n_{r0}	Density of occupied recombination centres (captured electrons) in thermal equilibrium	m^{-3}
n_{ra}	Density of occupied acceptor-type recombination centres (captured electrons)	m^{-3}
n_{rd}	Density of empty donor-type recombination centres (captured holes)	m^{-3}
N_A	Acceptor dopant concentration	m^{-3}
N_D (or N_d)	Donor dopant concentration	m^{-3}
N_c, N_v	Effective densities of states in the conduction and valence bands, respectively	m^{-3}
N_n	Density of empty electron traps	m^{-3}
N_r	Density of total recombination centres	m^{-3}
N_{ra}	Density of total acceptor-type recombination centres	m^{-3}
N_{rd}	Density of total donor-type recombination centres	m^{-3}
N_t	Density of total traps	m^{-3}
N_{tn}	Density of total electron traps	m^{-3}
N_{tp}	Density of total hole traps	m^{-3}
$N_t(E)$	Energy distribution function of traps	$m^{-3}(eV)^{-1}$
p, p_t, p_T	Free, trapped, and total hole densities, respectively	m^{-3}
p_0, p_{t0}	Free and trapped hole densities in thermal equilibrium, respectively	m^{-3}
p_{at}	Trapped hole density for the traps confined in a single discrete energy level	m^{-3}
p_{bt}	Trapped hole density for the traps distributed exponentially within the forbidden energy gap	m^{-3}
p_{ct}	Trapped hole density for the traps distributed uniformly within the forbidden energy gap	m^{-3}
p_r	Density of captured holes in recombination centres	m^{-3}
p_{r0}	Density of captured holes in recombination centres in thermal equilibrium	m^{-3}

Symbol	Definition	Dimensions
q	Electronic charge	C
R	Electron–hole recombination rate	$\text{m}^{-3}\,\text{sec}^{-1}$
R_a	Recombination rate for acceptor-type recombination centres	$\text{m}^{-3}\,\text{sec}^{-1}$
R_d	Recombination rate for donor-type recombination centres	$\text{m}^{-3}\,\text{sec}^{-1}$
S	Thin film thickness	m
$[S]$, $[S_G]$	Free and trapped singlet exciton densities, respectively	m^{-3}
SCL	Space-charge-limited	A
t_t	Charge carrier transit time	sec
T	Absolute temperature	°K
T_c	Characteristic constant (or temperature) for the exponential distribution of traps	°K
$[T]$, $[T_G]$	Free and trapped triplet exciton densities, respectively	m^{-3}
TSC	Thermally stimulated current	A
v	Thermal velocity of carriers	m sec^{-1}
V	Applied voltage	V
V_d	Contact potential or diffusion potential	V
V_T	$= kT/q$	V
W	Width of space charge region or width of Schottky barrier	m
W_a	Location of the virtual anode	m
W_c	Location of the virtual cathode	m
ε	Permittivity (or dielectric constant) or absorption coefficient	F m^{-1} m^{-1}
ε_0	Permittivity (or dielectric constant) of vacuum	F m^{-1}
ε_r	Relative permittivity	—
η	Quantum yield or quantum efficiency	—
θ	Ratio of free to total carrier densities	—
μ	Carrier mobility	$\text{m}^2\,\text{V}^{-1}\,\text{sec}^{-1}$
μ_n, μ_p	Electron and hole mobilities, respectively	—
v	Frequency of light wave	sec^{-1}
ρ	Space charge density	C m^{-3}
σ	Electric conductivity	$\text{ohm}^{-1}\,\text{m}^{-1}$
σ_n	Electron capture cross-section	m^2
σ_p	Hole capture cross-section	m^2
σ_R	Recombination cross-section	m^2
σ_t	Standard deviation of the Gaussian function	eV
σ_{tn}, σ_{tp}	Standard deviations of the Gaussian functions for electron and hole traps, respectively	eV
τ	Carrier lifetime or relaxation time of a carrier between scatterings	sec
τ_d	Dielectric relaxation time	sec

Symbol	Definition	Dimensions
τ_0	Diffusion length minority carrier lifetime	sec
τ_n, τ_p	Electron and hole lifetimes, respectively	sec
ϕ_B	Potential barrier height from the metal to the non-metal material	eV
ϕ_m, ϕ_s	Work function of the metal and the semiconductor, respectively	eV
χ	Electron affinity of the solid	eV
ω	Angular frequency	rad sec^{-1}
$\langle v\sigma_n \rangle$	$= C_n$	m^3 sec^{-1}
$\langle v\sigma_p \rangle$	$= C_p$	m^3 sec^{-1}
$\langle v\sigma_R \rangle$	$= C_r$	m^3 sec^{-1}

CHAPTER 1

Some Basic Concepts Related to Electrical Transport in Organic Semiconductors

1.1. CONCEPTS OF π-ELECTRONS, σ-ELECTRONS, LOCALIZED AND DELOCALIZED ORBITALS AND STATES

In inorganic semiconductors the structure is generally characterized by their strong covalent or ionic bonding between atoms in the lattice so that the charge transport can easily take place through a strong exchange interaction of overlap atomic orbitals in a close-packed structure; whereas in organic semiconductors the bonding is mainly due to van der Waals or London forces between molecules and is therefore rather weak, and also the overlap of the molecular orbitals and the intermolecular electron exchange are small so that the structure is not favourable for charge transport. To study the electrical transport in a material, two basic questions would be automatically asked and they are: How much energy is required to ionize the molecules (or atoms) in a solid so as to produce free charge carriers?, and How do the free carriers move through the solid? To answer these questions for organic semiconductors we have to know the electronic structure of these materials.

The crystalline organic semiconductors belong to the category of molecular crystals which have the characteristics of wide energy band gap, low carrier mobility, and low melting point. The typical values of these physical parameters for some organic semiconductors are given in Table 1.1. The electric conductivity of these materials is very low; for example, the conductivity of anthracene is about 10^{-20} to 10^{-16} ohm^{-1} cm^{-1}. Most organic semiconductors should really be designated as insulators. They are called semiconductors because their dark electric conductivity increases exponentially with temperature and some other properties are similar to those of inorganic semiconductors. However, the intermolecular separation is large in molecular crystals so that the molecular energy levels are relatively less disturbed. We shall discuss briefly the structure of some typical organic semiconductors before going to the electronic states of the molecules. The structures of naphthalene and anthracene were, among many crystals, the first studied by Bragg [1922] using the X-ray diffraction method. The structures are shown in Fig. 1.1. Their structure is based on a monoclinic unit cell with three unequal axes, two of which are at right angles to each other. The **c**-axis is not perpendicular to the plane of **a** and **b**, but we normally use **c′** axis which is

TABLE 1.1. *Chemical structure, melting point, ionization energy, electron affinity, energy band gap, activation energy, and drift mobility for naphthalene, anthracene, and tetracene*

Organic semiconductor	Chemical structure	Melting point (°C)	Ionization energy, E_i (eV)	Electron affinity, χ (eV)	Energy band gap $E_g = E_i - \chi$ (eV)	Activation energy for conductivity, $\Delta E\sigma$ (eV)	Type of charge carriers	Crystal axis	Mobility at 300°K (cm^2/V-sec)	Temperature dependence n for $\mu \propto T^n$
Naphthalene $C_{10}H_8$		80.2	6.75[a]	1.55[b] (0.9–2.0)[c]	5.2[b]	3.6[f]	Electrons	a	0.51[g]	−0.1[g]
								b	0.63	0.0
								c'	0.68	−0.9
							Holes	a	0.88[g]	−0.1[g]
								b	1.41	−0.8
								c'	0.99	−2.1
Anthracene $C_{14}H_{10}$		216.0	5.65[a]	2.45[b] (2.2–3.1)[c]	3.2[b] (3.7)[d]	2.6[f]	Electrons	a	1.70[h]	−1.5[h][j]
								b	1.00	−1.0
								c'	0.40	>0
							Holes	a	1.00[h]	−1.0[h][j]
								b	2.00	−1.5
								c'	0.80	−1.2
Tetracene $C_{18}H_{12}$		357.0	5.30[a]	2.90[b]	2.4[b] (>2.9)[e]	1.6[f]	Electrons	a	0.50[i]	—
								b		
								c'		
							Holes	a	0.01[i]	−1.0[i]
								b		
								c'		

(a) Lyons and Morris [1960].
(b) Le Blanc [1967].
(c) Gutmann and Lyons [1967].
(d) Experimental value [Silver et al. 1966].
(e) Possible value [Pope et al. 1965].
(f) Kommandeur [1965].

(g) Mey and Hermann [1973].
(h) Baessler et al. [1969], Geacintov et al. [1966], Szymanski and Labes [1969], Kondrasiuk and Szymanski [1972].
(i) Fourny and Delacote [1969], Munn and Siebrand [1970].

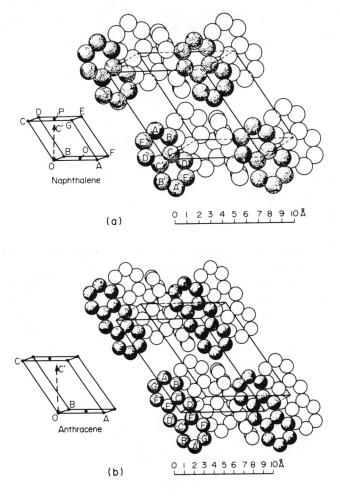

Naphthalene

(a)

Anthracene

(b)

FIG. 1.1. Base-centred monoclinic crystal structures: (a) the arrangement of naphthalene molecules in the unit cell, and (b) the arrangement of anthracene molecules in the unit cell. $OA = \mathbf{a}$, $OB = \mathbf{b}$, $OC = \mathbf{c}$, angle $AOC = \beta$, $c' \perp \mathbf{a} \times \mathbf{b}$ plane.

perpendicular to both $\mathbf{a}$ and $\mathbf{b}$. The values of $\mathbf{a}$, $\mathbf{b}$, $\mathbf{c}$ and the angles between $\mathbf{a}$ and $\mathbf{c}$ for some organic crystals are given in Table 1.2. There are two molecules per unit cell, the structure is close packed, and each molecule has 12 nearest neighbours. Chemically, the semiconductors, naphthalene, anthracene, and tetracene, are conjugated aromatic hydrocarbon compounds and the family of the compounds can be regarded as derivatives of benzene. The term "conjugated" refers to the regular alternation of single and double chemical bonds in the molecules. It is easy to understand the structure of these compounds from a simplest one—benzene. The carbon atom in the benzene molecule has three trigonal sp^2 hybrid orbitals with the interbond angle of 120° on the same molecular plane for σ-bond formation and a lone $2p_z$ orbital with its axis perpendicular to the molecular plane. One of the hybrid orbitals is used to bind the

TABLE 1.2. *Unit cell parameters of various aromatic hydrocarbon crystals [after Singh and Mathur 1974]*

Crystal	a (Å)	b (Å)	c (Å)	β
Naphthalene[a]	8.235	6.003	8.658	122° 55′
Anthracene[b]	8.562	6.038	11.184	124° 42′
β-Perylene[c]	11.27	5.88	9.65	92.1°
Biphenyl[d]	8.12	5.63	9.51	95.1°
p-Terphenyl[e]	8.106	5.613	13.613	92° 1′
p-Quaterphenyl[f]	8.05	5.55	17.81	95.8°

[a] Cruickshank [1957].
[b] Mason [1964].
[c] Tanaka [1963].
[d] Hargreaves and Rigvi [1962].
[e] Rietweld *et al.* [1970].
[f] Pickett [1936].

hydrogen atom and the other two to bind two neighbouring carbon atoms as shown in Fig. 1.2(a); the lone $2p_z$ orbitals overlap to form a π-orbital and the electron charge density in the π-orbitals is symmetrically distributed to form two streamer-type layers stretching right round the ring, one above and the other below the plane of the nuclei (the molecular plane) as shown in Fig. 1.2(b). The σ-orbitals are symmetrical around the bond axis giving localized C–C and C–H bonds. The electrons which occupy the σ-orbitals are called σ-electrons, and they are concentrated mainly along the line joining

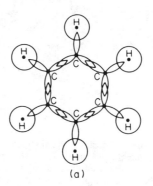

(a)

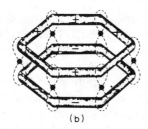

(b)

FIG. 1.2. (a) The σ-hybrids of the carbon atoms of benzene, and (b) the π-molecular orbitals in benzene. [After Coulson 1961.]

two nuclei and are localized. The electrons which occupy the π-orbitals are called π-electrons which do not participate in any bonds and are delocalized over all the carbon atoms so that they move freely inside the molecule and are, therefore, sometimes called the mobile electrons or the unsaturated electrons because they result from unsaturated bonds. It can be imagined that the six delocalized π-electrons form an electric current flowing round the ring; one electron moves from atom 1 to atom 2, but another electron moves from atom 2 to atom 1, constituting equal currents flowing in opposite directions with no net flow of charge. Through such mobile π-electrons, electrical influences are easily propagated from one part of the molecule to another. These π-electrons are also responsible for the propagation of chemical and physical influence along the complete length of a large molecule in a biological system [Coulson 1961].

The delocalized π-orbitals have been treated by means of many approximation methods such as the valence band theory, molecular orbital theory, free electron theory, etc. [Coulson 1961, Salem 1966, Murrell *et al.* 1965, Streitwieser 1961, Kuhn 1949, Kuhn *et al.* 1960]. Le Blanc [1961, 1962, 1963] and Thaxton *et al.* [1962] have applied the tight binding approximation to construct crystal wave functions in order to study the band structure and the transport of charge carriers in the naphthalene–anthracene series of organic crystals. The crystal wave function is based on Hartree's one-electron model and constructed from linear combination of one-electron molecular wave functions $\phi_n(\mathbf{r} - \mathbf{r}_n)$ formed within the Hückel approximations of Slater $2p_z$ atomic orbitals [Slater 1930, Seitz 1940], and it is given by

$$\psi_{\mathbf{k}}(\mathbf{r}) = N^{-1/2} \sum_{n=1}^{N} \exp(j\mathbf{k} \cdot \mathbf{r}_n) \phi_n(\mathbf{r} - \mathbf{r}_n) \tag{1.1}$$

where N is the total number of molecules in the crystal, $\mathbf{r}_n$ is the position vector of the geometrical centre of the nth molecule, and $\mathbf{k}$ is the wave vector. The crystal potential field used in the Hamiltonian, $H = (-\hbar^2/2m)\nabla^2 + V(r)$, is the sum of the Hartree potential of each molecule $V_n(\mathbf{r} - \mathbf{r}_n)$

$$V(\mathbf{r}) = \sum_{n=1}^{N} V_n(\mathbf{r} - \mathbf{r}_n) \tag{1.2}$$

Using a set of vectors $\boldsymbol{\alpha} = (\mathbf{a} + \mathbf{b})/2$, $\boldsymbol{\beta} = (-\mathbf{a} + \mathbf{b})/2$, and $\mathbf{c}$ to connect nearest-neighbour molecules to form a so-called "pseudo unit cell" each containing one molecule, and appropriate simplifications, Thaxton *et al.* [1962] have obtained expressions for the energies of both electrons and holes (the eigenvalue of ψ_k), each having the form

$$E(\mathbf{k}) = \text{constant} + 2\sum_{s} E_s \cos(\mathbf{k} \cdot \mathbf{r}_s) \tag{1.3}$$

where E_s is the resonance integral which is given by

$$E_s = \int \phi_{n+s} V_{n+s} \phi_n \, d\tau \tag{1.4}$$

in which

$$\phi_n = \sum_i C_{ni} \mu_i \tag{1.5}$$

with

$$\mu_i = (\alpha^5/\pi)^{1/2}(\mathbf{n}_i \cdot \mathbf{r}_i)\exp(-\alpha r_i) \tag{1.6}$$

The constant in eq. (1.3) is immaterial to the determination of bandwidths and mobilities. In eqs. (1.5) and (1.6), μ_i is the Slater-type $2p_z$ atomic orbitals, $\mathbf{n}_i$ is the unit vector defining the direction of the $2p_z$ orbital, α (should not be confused with α as one side of the pseudo unit cell) is a parameter equal to 3.08×10^{-8} cm^{-1} used by Slater [1930], and C_{ni} is the Hückel coefficients. Thaxton *et al.* [1962] have calculated C_{ni} from the secular equations using the approximation of Hückel [Hückel 1932, Dandel *et al.* 1959] for four molecular organic crystals, and their values are given in Fig. 1.3. The magnitudes of these coefficients are the same for holes and electrons, and if they differ in sign, the lower sign is associated with the antibonding π-orbital of lowest energy and the upper sign is associated with the bonding π-orbital of highest energy. The squares of these coefficients give the relative weights of charge density distribution of an excess electron or an excess hole in the molecule.

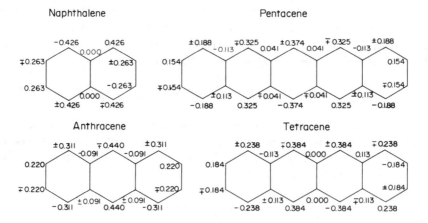

FIG. 1.3. Values of coefficients C_{ni} for molecular wave functions for naphthalene, anthracene, tetracene, and pentacene. The magnitudes are the same for electrons and holes, and if they differ in phase, the upper sign is for holes and the lower for electrons. [After Thaxton, Jarnagin, and Silver 1962.]

The π-electrons are completely delocalized and mobile along the conjugated carbon atoms in any individual molecule, but an injected excess carrier, electron, or hole is not easy to proceed from molecule to molecule in the crystal because the overlap of wave functions of adjacent molecules is small. Since the overlap is small, the time required for resonance transfer of the excess carrier between adjacent molecules will be relatively long and it might be so long that other transport mechanisms may take place more readily [Le Blanc 1961]. In Sections 1.2–1.10 we shall first answer the question about the carrier transport, and partly answer the question about the carrier generation. The photogeneration of carriers will be discussed in Section 6.1.

To avoid any confusion with the four terms: localized orbitals and states and delocalized orbitals and states: we summarize their definitions as follows. The orbitals around the bond axis in the localized C–C and C–H bonds are called the localized orbitals, while the orbitals which do not participate in any bonds and are delocalized over all the carbon atoms within a single molecule are called the delocalized orbitals. The so-called "localized states" are generally referred to as the states in which the electrons are trapped, such as an excess electron interacts with a neutral molecule and is trapped in a self-induced potential well due to polarization. The delocalized states are the states in which the electrons can move freely over several lattice sites such as the states in the electron band or in the hole band.

1.2. THE ENERGY BAND MODEL

The experimental fact of a negative temperature coefficient of carrier drift mobility [Le Blanc 1960, 1963] has led to the development of the Bloch band theory for carrier transport in organic crystals. The assumptions used generally in the band theory [Slater 1959] are: (i) the use of one electron in a periodic potential–one electron approximation, (ii) the neglect of multiplet structure on individual atoms, and (iii) the treatment of the electron–lattice interaction as a small perturbation. The most serious one is assumption (iii) for organic semiconductors. Ioffe [1956, 1959] has pointed out that for materials with carrier mobilities less than $100 \, \text{cm}^2/\text{V-sec}$ and the normal carrier free path less than the wavelength of thermal electrons, the electron–lattice interaction would be so strong that the band description for such materials is inappropriate because of the violation of assumption (iii). In most organic semiconductors the carrier mobilities are usually very low and so the difficulties in the use of the band theory are expected to arise from the uncertainty principle [Ioffe 1956, Fröhlich and Sewell 1959, Bosman and van Daal 1970]. From the uncertainty principle, the bandwidth W should follow the inequality relation

$$W > \hbar/\tau \tag{1.7}$$

where τ is the relaxation time of the carrier between scatterings, which is

$$\tau = \frac{\lambda}{v} \tag{1.8}$$

where λ and v are, respectively, the mean free path and the drift velocity of the carrier. If $W < kT$, the drift mobility of the carrier can be written as [Gutmann and Lyons 1967]

$$\mu = (q/kT) \langle \tau v^2 \rangle \tag{1.9}$$

and $(\langle v^2 \rangle)^{1/2}$ is of the same order as v_{max}, the maximum value of v, which is given by [Glarum 1963, Kepler 1960]

$$v_{max} \simeq Wa/\hbar \tag{1.10}$$

where a is the lattice constant. From eqs. (1.7)–(1.10) we obtain

$$\lambda > a \tag{1.11}$$

$$\mu > \left(\frac{qa^2}{\hbar}\right)(W/kT) \tag{1.12}$$

for $W < kT$. The inequality relations (1.11) and (1.12) imply that the validity of the band theory requires that the carrier mean free path must be larger than the lattice constant and the carrier mobility must be larger than $3W/kT$ if a is assumed to be 5 Å

The basic mechanism of charge transport in a crystal depends on the nature of the electron exchange interactions and the electron–phonon interactions. In inorganic semiconductors the electron exchange interactions are much larger than the electron–phonon interactions, so that the electrons behave as quasi-free particles occasionally scattered by phonons, and therefore the charge transport is coherent. In organic semiconductors the electron exchange interactions are much smaller than in inorganic semiconductors, while the electron–phonon interactions may be much the same, so that there is a possibility that the electron–phonon interactions may dominate in organic semiconductors. Although the band model has been put forward by several investigators [Gutmann and Lyons 1967] to explain the charge transport phenomena, the small mean free path (of the order of a lattice constant) and the small carrier mobility make the assumption of coherent transport barely self-consistent. Furthermore, for the band model to be valid the bandwidth must also be larger than the lattice vibration energy [Glarum 1963]

$$W > \hbar\omega_0 \qquad (1.13)$$

or

$$\mu > (qa^2/\hbar^2)(\hbar\omega_0/kT) \qquad (1.14)$$

where ω_0 is the Debye frequency for acoustic waves.

However, the band model can explain well several important transport phenomena such as the anisotropy of conductivity and mobility, the temperature dependence of mobility, and the anomalous Hall effect. The energy band model for organic solids based on the assumption of narrow bands has been discussed as early as 1954 [Meier 1974]. More systematic calculations of the energy band structure of aromatic hydrocarbon crystals were first made by Le Blanc [1961, 1962], and then followed by Thaxton *et al.* [1962] and modified by several investigators [Katz *et al.* 1963, Silbey *et al.* 1965, Glaeser and Berry 1966, Chojnacki 1968, 1969, Tanaka and Niira 1968, Mathur and Kumar 1973, Singh and Mathur, 1974].

Using linear combinations of single carbon atomic $2p_z$ orbitals (LCAO) described by eqs. (1.1)–(1.6) in the tight-binding approximation, Le Blanc [1961, 1962] has calculated the bandwidth for anthracene, which is $W < kT$ at room temperature (e.g. $W = 0.56kT$ for electron band in anthracene). Later Thaxton *et al.* [1962] have extended Le Blanc's calculation to naphthalene, tetracene, and pentacene. However, Katz *et al.* [1963] have commented Le Blanc's approach on three points. (i) There are two molecules per unit cell in the lattice, implying that there are two bands for electrons and two bands for holes arising from the symmetric and antisymmetric combinations of molecular wave functions in a unit cell. It is inappropriate to replace the actual crystal structure by a pseudo unit cell having only one molecule per unit cell. (ii) It is inappropriate to use the value of 3.08 Å for the orbital exponent α in a Slater-type wave function for the carbon $2p_z$ atomic orbitals from which the molecular wave functions are constructed because this value seriously underestimates the magnitude of the tails of the wave functions. (iii) It is inaccurate to neglect all three-centre integrals because the terms involving these integrals are not small and the neglect of them may result in a change of calculated energy by 25 %. Katz *et al.* [1963] have employed self-consistent field (SCF) atomic orbitals in the LCAO method to calculate the band structures with the aforementioned three points taken into account. Instead of using eq. (1.1), Katz *et al.* used

$$\psi = (N^{-1/2}) \sum_{l=0}^{N-1} (\pm 1)^l \exp(j\mathbf{k} \cdot \mathbf{r}_l) \phi(\mathbf{r} - \mathbf{r}_l) \tag{1.15}$$

where the index *l* labels the molecules of the unit cells so that the molecule at the corner of each cell has an even index, while the one at the centre of each cell has an odd number, *N* is the number of molecules (twice the number of unit cells), $\mathbf{r}_l$ is the vector to the centre of each molecule so that $\phi(\mathbf{r} - \mathbf{r}_l)$, the one-electron molecular wave function, has a different orientation in space depending on whether *l* is even or odd. In the Hamiltonian $H = (-\hbar^2/2m)\nabla^2 + V(r)$, the crystal potential field is the same as that given by eq. (1.2). The energy or the eigenvalue of ψ can be calculated from

$$E_\pm(\mathbf{k}) = \frac{\int \psi_\pm^* H \psi_\pm \, d\tau}{\int \psi_\pm^* \psi_\pm \, d\tau}$$

$$= E_0 + \sum_n E_n + \sum_l (\pm l)^l E_l \cos(\mathbf{k} \cdot \mathbf{r}_l) \tag{1.16}$$

where E_0 is the energy of the isolated negative or positive ion relative to infinite separation of the electron or hole and the neutral molecule, and E_n and E_l are given by

$$E_n = \int \phi^*(\mathbf{r}) V_n(\mathbf{r} - \mathbf{r}_n) \phi(\mathbf{r}) \, d\tau \tag{1.17}$$

$$E_l = \int \phi^*(\mathbf{r} - \mathbf{r}_l) V_l(\mathbf{r} - \mathbf{r}_l) \phi(r) \, d\tau \tag{1.18}$$

It is the intermolecular resonance (or exchange) integral E_l which determines the band structure. To calculate E_l we have to know the molecular orbitals of a positive or negative ion. These orbitals can be approximated by a linear combination of neutral carbon $2p_z$ wave functions μ_i as given by eq. (1.5). But Katz *et al.* did not use eq. (1.6) for μ_i but used the best available carbon atomic wave functions represented in the form of a linear combination of four Slater wave functions [Clementi *et al.* 1962]

$$\mu_i(r) = (\mathbf{n}_i \mathbf{r}) \sum_{i=1}^{4} a_i (\alpha_i^5/\pi)^{1/2} \exp(-\alpha_i r) \tag{1.19}$$

where $\mathbf{n}_i$ is the unit vector defining the direction of the $2p_z$ orbital and the coefficients a_i and orbital exponents α_i are those given by Clementi *et al.* [1962].

    Katz *et al.* have calculated E_l and the excess electron and hole band structures for naphthalene, anthracene, and several polyphenyls. The shapes of the excess electron band and excess hole band for anthracene are shown in Fig. 1.4. The band containing an excess electron in an otherwise neutral crystal is called the "excess electron band", and that containing an excess hole is called the "excess hole band". This nomenclature, rather than the usual "conduction band" and "valence band", is adopted to emphasize the fact that the calculations are for the electronic states of two distinct systems rather

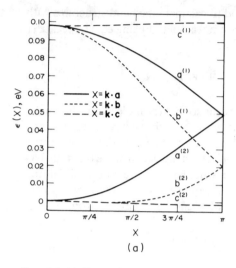 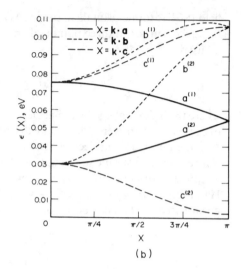

FIG. 1.4. (a) Shape of the excess electron band of anthracene. Band widths in 10^{-4} eV: $a^{(1)} = 487$, $a^{(2)} = 491$, $b^{(1)} = 775$, $b^{(2)} = 209$, $c^{(1)} = 22$, $c^{(2)} = 17$; splitting in c^{-1} direction $= 979 \times 10^{-4}$ eV. (b) Shape of the excess hole band of anthracene. Bandwidths in 10^{-4} eV; $a^{(1)} = 209$, $a^{(2)} = 243$, $b^{(1)} = 329$, $b^{(2)} = 757$, $c^{(1)} = 314$, $c^{(2)} = 272$; splitting in c^{-1} direction $= 452 \times 10^{-4}$ eV. [After Katz, Rice, Choi, and Jortner 1963.]

than two states of the same system [Le Blanc 1961]. It can be seen in Fig. 1.4 that the hole band exhibits either a zero or a small splitting, while the electron band splits appreciably in the c^{-1} direction. The bandwidths for anthracene $W \geq kT$ except the excess electron band in the c^{-1} direction.

It should be noted that Katz *et al.* have not taken into account the effects of molecular vibration and polarization. Silbey *et al.* [1965] have modified the calculations of Katz *et al.* by including the effects of molecular vibrations and adding an exchange potential term to the Hartree potential, and found that the bandwidths are smaller than those calculated by Katz *et al.* and close to those of Thaxton *et al.* Glaeser and Berry [1965] have estimated the effects of electron polarization and their results have shown that the bandwidths would be narrowed considerably if the effects of polarization are included. Recently, Singh and Mathur [1974] have reformulated the band theory by taking into account the configuration interaction [Morris and Yates 1971] and found that the bandwidths of several organic crystals determined by their model are different from those based on the model used by Katz *et al.* Some of their results are given in Table 1.3. They have also reported that for conduction in narrow band, mean free path is constant rather than the mean free time; and that the applicability of band theory generally increases with the increasing size of the molecule except in the c' direction where it decreases. The narrowness of the bandwidths with the delocalized picture retained is physically non-realistic, and this has led to the consideration of other models for electrical transport such as the hopping model and the tunnelling model which will be discussed in the next two sections. However, although the band structure calculations have been criticized for failure to take into account several important effects such as lattice expansion, self-trapping, and the hydrogen atoms, they can predict quite well

TABLE 1.3. *Bandwidths and splittings in* c^{-1} *direction (in eV) for various aromatic hydrocarbon crystals [after Singh and Mathur 1974]*

Crystal	$a^{(1)}$	$a^{(2)}$	$b^{(1)}$	$b^{(2)}$	$c^{(1)}$	$c^{(2)}$	Splitting
	Hole						
Naphthalene	0.03	0.02	0.15	0.16	0.11	0.15	0.01
Anthracene	0.06	0.05	0.24	0.13	0.11	0.13	0.11
β-Perylene	0.03	0.03	0.51	0.45	0.06	0.07	—
Biphenyl	0.08	0.09	0.20	0.04	0.06	0.19	—
p-Terphenyl	0.02	0.03	0.14	0.09	0.08	0.06	—
p-Quaterphenyl	0.001	0.006	0.14	0.13	0.07	0.02	—
	Electron						
Naphthalene	0.10	0.11	0.06	0.15	0.02	0.02	0.18
Anthracene	0.22	0.21	0.09	0.34	0.01	0.02	0.43
β-Perylene	0.11	0.11	0.16	0.07	0.08	0.15	0.15
Biphenyl	0.17	0.17	0.05	0.35	0.08	0.05	0.21
p-Terphenyl	0.15	0.16	0.07	0.35	0.03	0.05	0.22
p-Quaterphenyl	0.17	0.18	0.10	0.42	0.01	0.04	0.29

[1] and [2] denote, respectively, the upper and lower bands in the a^{-1}, b^{-1}, and c^{-1} directions.

many transport phenomena. In the following we shall discuss some of them based on the band model.

1.2.1. Carrier drift mobility

Most organic materials are anisotropic in electrical behaviour as expected from their crystal and band structures. To calculate the mobility in a specific direction, we have to know the velocity component $v_i(\mathbf{k})$ as well as the carrier distribution function that describes the scattering of the charge carriers on their way through the lattice. The mobility tensor is generally calculated in an orthogonal coordinate system whose axes are parallel to **a** and **b** of the unit cell and **c'** perpendicular to the *ab* plane for the naphthalene–anthracene system. According to the band model [Katz *et al.* 1963], the velocity component is given by

$$v_i(k) = (1/\hbar)\nabla_k E_i(k) \tag{1.20}$$

the components of the mobility tensor by

$$\mu_{ij} = (q/kT)\tau \langle v_i v_j \rangle \tag{1.21}$$

for the constant relaxation time (or mean free time) approximation and by

$$\mu_{ij} = (q/kT)\lambda \langle v_i v_j/|v| \rangle \tag{1.22}$$

for the constant mean free path approximation, where the subscripts i, j refer to any directions of **a**, **b**, and **c**. The bracket $\langle \ \rangle$ indicates an average over the Boltzmann distribution of charge carriers in the energy bands. Since the Fermi–Dirac distribution function at thermal equilibrium is given by

$$f_0 = \{\exp[(E - E_F)/kT] + 1\}^{-1}$$
$$\simeq \exp[(E_F - E)/kT] = \exp(E_F/kT)\exp(-E/kT) \tag{1.23}$$

for wide band gap organic semiconductors, we can write $\langle v_i v_j \rangle$ as [Katz *et al.* 1963]

$$\langle v_i v_j \rangle = \frac{\int \left\{ \frac{\partial E_+}{\partial k_i} \frac{\partial E_+}{\partial k_j} \exp\left[-E_+/kT\right] + \frac{\partial E_-}{\partial k_i} \frac{\partial E_-}{\partial k_j} \exp\left[-E_-/kT\right] \right\} dk}{\hbar^2 \int \left\{ \exp\left[-E_+/kT\right] + \exp\left[-E_-/kT\right] \right\} dk} \quad (1.24)$$

If the variation of the Boltzmann factors $\exp\left[-E_\pm/kT\right]$ is neglected, the mean velocity is given by

$$\langle v^2 \rangle = \langle v_a^2 \rangle + \langle v_b^2 \rangle + \langle v_c^2 \rangle \quad (1.25)$$

The anisotropy in the energy bands will contribute greatly to the mobility anisotropy. Using eqs. (1.20)–(1.25), the mobilities in **a**, **b**, or **c′** direction can be computed straightforwardly using numerical methods. Tables 1.4 and 1.5 show the ratios of mobilities in naphthalene and anthracene at room temperature, in which μ_{aa}, μ_{bb}, and $\mu_{c'c'}$ refer to the mobilities in the **a**, **b**, and **c′** directions, respectively. It can be seen that the calculations based on the band model agree at least qualitatively with the experimental values except those involving the drift mobility in the **c′** direction for electrons. This discrepancy may be attributed to the inaccurate calculations of the resonance integrals due to the neglect of the effects of hydrogen atoms, molecular vibrations, etc., and to the inaccurate wave function for the region between molecules; or to the failure of the band model because the bandwidth is too narrow [Glaeser and Berry 1966].

Recently, Kepler and Hoesterey [1974] have reported that both electron and hole drift mobilities in anthracene are independent of the electric field up to fields in which the drift velocity of the carrier is comparable to estimates of their thermal velocity as calculated from the anthracene band structure. The measured mobilities for electrons and holes in the **c′** direction are, respectively, 0.4 and 0.8 cm^2/V-sec. According to the band model we would expect the drift mobility to be field dependent for drift velocities well below their thermal velocity [Conwell 1967]. These results, coupled with the small

TABLE 1.4. *Ratios of drift mobilities in naphthalene at room temperature*

	Hole mobilities			Electron mobilities		
	μ_{aa}/μ_{bb}	$\mu_{bb}/\mu_{c'c'}$	$\mu_{aa}/\mu_{c'c'}$	μ_{aa}/μ_{bb}	$\mu_{bb}/\mu_{c'c'}$	$\mu_{aa}/\mu_{c'c'}$
Band model [Katz et al. 1963]	0.14	5.60	0.76	1.60	25.00	39.00
Band model [Singh and Mathur 1974]	0.09 (const. τ)	4.17 (const. τ)	–	2.41 (const. τ)	5.56 (const. τ)	–
	0.10 (const. λ)	3.70 (const. λ)	–	2.17 (const. λ)	4.00 (const. λ)	–
Hopping model [Glaeser and Berry 1966]	0.36	1.10	0.38	1.00	2.00	2.20
Experimental [Mey and Hermann 1973]	0.62	1.42	0.89	0.81	0.93	0.75

TABLE 1.5. *Ratios of drift mobilities in anthracene at room temperature*

Carrier	Ratio	Theoretical						G	Experimental
		A	B	C	D	E	F		H
Holes	μ_{bb}/μ_{aa}	1.90	1.70	2.90	2.30	2.70	2.30	3.57 (const. τ) / 3.22 (const. λ)	2.00
	$\mu_{cc'}/\mu_{aa}$	0.25	0.20	0.33	0.44	0.40	0.04	1.85 (const. τ) / 1.84 (const. λ)	0.80
Electrons	μ_{bb}/μ_{aa}	0.88	0.85	0.64	0.31	0.40	0.99	0.30 (const. τ) / 0.30 (const. λ)	0.59
	$\mu_{cc'}/\mu_{aa}$	0.004	0.005	0.001	0.003	0.01	0.0025	0.003 (const. τ) / 0.003 (const. λ)	0.23

A Le Blanc [1961, 1967].
B Thaxton et al. [1962].
C Katz et al. [1963].
D Silbey et al. [1965].

E Glaeser and Berry [1966].
F Chojnacki [1967, 1969].
G Singh and Mathur [1974].
H Kepler [1960, 1962].

mean free path (of the order of a lattice constant), are consistent with a localized rather than a delocalized picture of carrier motion. However, there is some evidence that the electron–phonon interaction is strong in organic semiconductors and the carrier transport may be better described by a type of hopping motion with a mean free path of the order of a lattice constant [Glarum 1963, Siebrand 1964, Friedman 1965, Gosar and Choi 1966, Munn and Siebrand 1970].

(A) Temperature dependence of mobility

The temperature dependence of mobility follows the relation

$$\mu = CT^{-n} \tag{1.26}$$

where C and n are constants. The value of n is generally $0 < n < 2.5$. This temperature effect has been observed in many organic materials such as in naphthalene [Mey and Hermann 1973], anthracene [Raman *et al.* 1964, Delacote 1969, Le Blanc 1960, Kepler 1960, 1962, 1964, Castro and Hornig 1965, Pethig and Morgan 1967, Fourny and Delacote 1969, Chojnacki 1968, 1969, Nakada and Ishihara 1964, Pott and Williams 1969], tetracene [Szymanski and Labes 1969], phthalocyanine [Meier and Albrecht 1969, Usov and Benderskii 1970, Heilmeier and Harrison 1963, Fillard and Schott 1966]. Typical experimental mobilities in anthracene are given in Fig. 1.5. There is

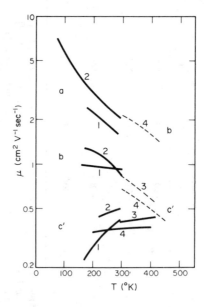

Fig. 1.5. Temperature dependence of carrier mobilities in anthracene. The solid lines are electron mobilities in the directions indicated on the left of the lines. The broken lines are hole mobilities in the directions indicated on the right. [After Munn and Siebrand 1970.] 1, Pott and Williams [1969]; 2, Kepler [1960, 1962, 1964]; 3, Nakada and Ishihara [1964]; 4, Fourny and Delacote [1969].

considerable scatter among the measurements within a narrow temperature range. Nevertheless, it is consistent that the mobilities along all crystal axes decrease with increasing temperature except $\mu_{c/c}$ for electrons, which increases. That the value of n in eq. (1.26) is positive can be considered as evidence for the quasi-free carrier motion in energy bands. The carrier transport in energy bands is governed by (i) the interaction with the periodic lattice potential, the thermal expansion of molecular crystals is large, being about 3 % from 77° K to 300° K, which would cause the mobility to decrease with temperature [Katz *et al.* 1963]; (ii) the weak scattering with phonons, the relaxation time is temperature dependent because the phonon density increases with increasing temperature, thus reducing the mobility [Conwell 1967]; and (iii) the distribution of carriers within the energy bands is sensitive to temperature when $W > kT$. It is possible that the factors (i) and (ii) are predominant in narrow bandwidth molecular crystals.

The positive temperature coefficient of electron mobility in the c' direction can be considered as further evidence of the strong electron–phonon interaction in molecular crystals. Munn and Siebrand [1969, 1970] have suggested three possible modes: (a) hopping in the slow electron limit, (b) hopping in the slow phonon limit, and (c) coherent transport in the slow phonon limit; and concluded that all three modes contribute to electrical transport in these materials. The positive temperature dependence of $\mu_{c/c}$ for electrons may be due to the dominant "slow electron" limit of hopping transport, while the carrier motion in directions other than c' may be due to either the dominant "slow phonon" coherent transport (the band model) or the dominant "slow phonon" hopping transport which have a negative temperature dependence. However, we should not completely ignore the possibility of the influence of shallow electron or hole traps, under which the mobility can increase with increasing temperature.

(B) Isotope effect

Recent measurements of the electron mobility in the c' direction in deuterated anthracene $(C_{14}D_{10})$ yield a value much higher than that in normal anthracene $(C_{14}H_{10})$ at room temperature [Munn *et al.* 1970, Morel and Hermann 1971]. The results of Morel and Hermann give $\mu_{c/c}$ for electrons $= 1.36 \pm 0.30 \text{ cm}^2/\text{V-sec}$ which is about 2.5 fold increase over the value of $0.4 \text{ cm}^2/\text{V-sec}$ for normal anthracene [e.g. Kepler 1960]. Munn and Siebrand [1970] attribute such an increase in $\mu_{c/c}$ to the possibility that the electron transport in the c' direction is intermediate between slow electron and slow phonon hopping. However, Morel and Hermann [1971] have added another possibility for this increase. That is due to strong shallow trapping because of the presence of impurities in the normal anthracene specimens. Hoesterey and Letson [1963] have reported that the presence of only 2 ppm naphthacene (tetracene) in anthracene can lower the room temperature mobility by an order of magnitude. In organic semiconductors there always exists a high concentration of impurities and defects, and the imperfections are not easy to remove. Therefore, the trapping effect should not be ignored in interpreting experimental results. Furthermore, all microscopic theories are based on a perfect or ideal crystal structure, but unfortunately all experimental results are obtained from imperfect crystal specimens. It is quite dangerous to correlate the theory with experiment without careful analysis. The trapping effect will be discussed in detail in Chapters 3–5.

(C) Pressure dependence of mobility

Several investigators have observed the increase of the mobility with pressure. In anthracene the mobility increases steadily with increasing pressure [Kepler 1962, Harada *et al.* 1964, Kajiwara *et al.* 1967]. A similar phenomenon has also been observed in phthalocyanine and diphenyl picryl hydrazyl (DPPH) [Osugi and Hara 1966], in quaterrylene, violanthrone, and others [Boguslavskii and Vannikov 1970]. Typical experimental results on pressure dependence of mobility in anthracene are shown in Fig. 1.6. It is interesting to note that $\mu_{c'c'}$ for electrons in the c' direction is practically independent of pressure, which is consistent with the temperature dependence results. The compressibilities in anthracene crystals along the **a**, **b**, and **c**′ are in the ratio of 3.5 : 1.2 : 1.0 [Danno and Inokuchi 1968, Afanaseva *et al.* 1967, Huntington *et al.* 1969]. This trend is similar to the trend shown in Fig. 1.6. It is possible that the pressure dependence of mobility may be associated with the change of unit cell dimensions with compression. This may, in turn, result in more overlapping of the wave functions of adjacent molecules leading to a decrease in the energy barriers and to an increase in mobility no matter whether a band model or a hopping model is assumed [Wood *et al.* 1966]. To interpret the pressure dependence results, additional information is needed on the dependence of the electronic exchange integrals on the intermolecular separation and relative orientation, and on the pressure dependence of molecular orientation and lattice vibration frequencies [Munn and Siebrand 1970].

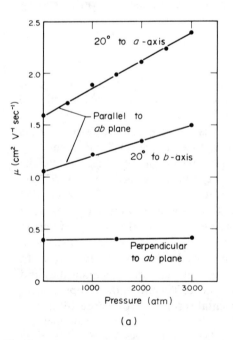

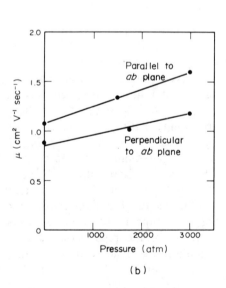

(a)   (b)

FIG. 1.6. Pressure dependence of mobility for (a) electrons and (b) holes in anthracene. [After Kepler 1962.]

1.2.2. Hall effect

As organic semiconductors have a very small conductivity, it is not easy to measure the Hall effect in the dark as well as at low electric fields (in the ohmic region). Even the Hall effect can be measured under such conditions; the results do not yield any information as to whether the carriers are holes or electrons. For this reason the Hall effect measurement in low-conductivity (high resistivity) materials is usually performed by photoinjecting either holes or electrons at one of the electrodes so that we know exactly which type of carriers is responsible for the observed Hall effect. The Hall effect measured by this way is sometimes called the "photo-Hall effect". Theoretical analysis for the effect of magnetic fields on the charge transport based on the band model has shown that the charge carriers in molecular crystals may exhibit an anomalous Hall effect [Le Blanc 1963, Friedman 1964, Munn and Siebrand 1970]. The term anomalous means that the negative Hall effect is observed in the hole band or the positive Hall effect in the electron band, just opposite to the signs normally expected in inorganic semiconductors (e.g. *p*-Ge or *n*-Ge). By solving the Boltzmann's transport equation, the Hall mobility (μ_H) to drift mobility (μ_D) ratio is given by [Jones and Zener 1934, Wilson 1953]

$$\frac{\mu_H}{\mu_D} = \frac{kT \langle v_j^2 M_i^{-1} - v_j v_i M_{ji}^{-1} \rangle}{\langle v_i^2 \rangle \langle v_j^2 \rangle} \tag{1.27}$$

where M_{ij}^{-1} is the symmetric inverse effective mass tensor defined as

$$M_{ij}^{-1} = \frac{1}{\hbar^2} \left(\frac{\partial^2 E}{\partial k_i \partial k_j} \right) \tag{1.28}$$

and v_i and v_j are the group velocities defined by eq. (1.20). Using eqs. (1.27) and (1.28), Spielberg *et al.* [1971] have calculated the ratio of μ_H/μ_D for naphthalene based on the band model for three types of wave functions for which the transfer integrals are evaluated. Their theoretical results are shown in Table 1.6. They have also measured the drift mobilities and Hall mobilities in high purity single crystals of naphthalene with charge carriers photoinjected on the surface of the specimen. Their experimental results are shown in Table 1.7, where the minus sign indicates that the Hall effect is anomalous. By comparing the theoretical μ_H/μ_D of Table 1.6 with the experimental μ_H/μ_D, it can be seen that only those based on the band structure calculations of Katz *et al.* and Silbey *et*

TABLE 1.6. *Ratios of the Hall to the drift mobilities for naphthalene as calculated from three wave functions (1), (2), and (3) of the indicated investigators [after Spielberg et al. 1971]*

Magnetic field parallel to	Holes			Electrons		
	a	*b*	*c'*	*a*	*b*	*c'*
(1)	−15.8	1.1	−15.8	−9.9	1.5	−10.7
(2)	−1.3	0.08	−1.3	0.70	1.3	−1.9
(3)	−0.21	0.75	−0.2	0.82	0.87	−0.095

(1) Thaxton *et al.* [1962].
(2) Katz *et al.* [1963].
(3) Silbey *et al.* [1965].

TABLE 1.7. (A) The measured photo-Hall mobilities, (B) the measured drift mobilities, and (C) the ratios of the Hall mobility to the drift mobility in naphthalene at room temperature for electrons and holes along the a, b, and c' directions. $B_\parallel$ and $I_\parallel$ indicate the axes to which the magnetic field B and the current I are parallel. The mobilities are given in $cm^2/V\text{-}sec$ [after Spielberg et al. 1971]

(A)

$I_\parallel/B_\parallel$	Electrons		
	a	b	c'
a	—	+1.7	−1.8
b	+0.55	—	−1.5
c'	+0.54	+1.3	—
	Holes		
a	—	+0.46	−2.9
b	−4.7	—	−4.4
c'	−1.4	+0.19	—

(B)

	Electrons
μ_a	0.6
μ_b	0.5
$\mu_{c'}$	0.6
	Holes
μ_a	0.8
μ_b	1.2
$\mu_{c'}$	0.5

(C)

$I_\parallel/B_\parallel$	Electrons		
	a	b	c'
a	—	+2.8	−3.0
b	+1.1	—	−2.9
c'	+0.9	+2.2	—
	Holes		
a	—	+0.57	−3.6
b	−3.9	—	−3.7
c'	−2.7	+0.37	—

TABLE 1.8. (A) The measured photo-Hall mobilities, (B) the measured drift mobilities, and (C) the ratios of the Hall mobility to the drift mobility in anthracene at room temperature for holes along the a, b, and c' directions. $B_\parallel$ and $I_\parallel$ indicate the axes to which the magnetic field B and the current I are parallel. The mobilities are given in $cm^2/V\text{-}sec$ [after Korn et al. 1969]

(A)

$I_\parallel/B_\parallel$	a	b	c'
a	—	+7.6	−1.5
b	−8.6	—	−3.1
c'	−3.8	+7.0	—

(B)

μ_a	1.0
μ_b	2.0
$\mu_{c'}$	0.8

(C)

$I_\parallel/B_\parallel$	a	b	c'
a	—	+7.6	−1.5
b	−4.3	—	−1.5
c'	−4.8	+8.8	—

al. yield the correct signs for electrons. Spielberg *et al.* [1971] have suggested that because of the good agreement between the theory and the experiment, particularly for those calculated using the wave functions of Katz *et al.*, the band model is applicable to describe the charge carrier transport in this material. Korn *et al.* [1969] have measured the μ_H/μ_D for holes in anthracene using the same technique used by Spielberg *et al.* [1971]. Their results are shown in Table 1.8. It can be seen that when the current is in the c' direction, the μ_H/μ_D ratio is negative for the magnetic field in the **a** direction and positive in the **b** direction. Since anthracene crystals cleave readily only in the *ab* plane it is important to specify clearly the directions of electric and magnetic fields when measuring the Hall effect. On the basis of the band model the sign of the Hall mobility depends on the relative population of the conducting states. In narrow bands, all levels are almost equally populated, so that the contribution by states with negative effective mass is almost equal to that by states with positive effective mass. Under a magnetic field each effective mass contribution is weighted according to the carrier velocity in that state, which depends on the band structure. If the negative effective mass contribution is dominant, a negative Hall effect may occur. This implies that the carriers are deflected in a direction opposite to the Lorentz force.

It should be noted that the anomalous Hall effect cannot be considered as sufficient evidence for supporting the band model. Munn and Siebrand [1970] have pointed out that for sufficiently narrow bands the hopping model (the random incoherent transport) may be a more proper model for charge transport and that the anomalous Hall effect can also be explained on the basis of the hopping model.

Finally, it is worthy to mention that based on the band structure calculation by Katz *et al.* [1963] the relaxation times for holes and electrons are, respectively, 1.3×10^{-14} and 2.4×10^{-14} sec for anthracene. The values of $\hbar/\tau$ are 0.05 and 0.03 eV for excess holes and electrons, respectively, which are of the order of the calculated bandwidths. From the experimental mobilities the estimated mean free paths are about 3.5 Å for holes and 4.5 Å for electron in the *ab* plane, which are of the same order of magnitude as the lattice constant. Such low mean free paths indicate that the carrier transport mechanism may be intermediate between the weak scattering approximation where the delocalized Bloch states are dominant and the strong scattering approximation where the localized states are dominant. To advance the understanding of this problem, the information about the mechanism and magnitude of the interaction of charge carriers with intramolecular and intermolecular vibration is important.

1.3. THE TUNNELLING MODEL

The tunnelling model has been proposed and discussed by Eley *et al.* [1953, 1955, 1960, 1962, 1963, 1967]. This model assumes that an electron in a π-molecular orbital on one molecule, when excited to a higher energy level (e.g. to a singlet state), can tunnel through a potential barrier to a non-occupied state of a neighbouring molecule with energy conserved in the tunnelling process as shown in Fig. 1.7. An excited state of a molecule can be either a singlet or a triplet, and the energy depends on the spin (cf. Chapter 7). The electron in the excited state may tunnel to its neighbour molecule or return to its ground state but, in general, the probability for the former is much larger than that for the latter. The tunnelling probability would be still higher if the excited electron were in the long life triplet state.

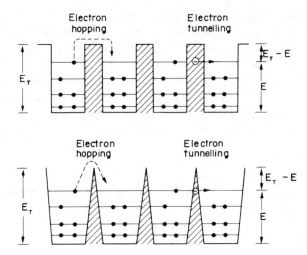

F IG . 1.7. Schematic diagrams illustrating the electron hopping
across and the electron tunnelling through a square
and a triangular potential barrier. The electron hopping or tunnelling in one direction is equivalent to the
hole hopping or tunnelling in the opposite direction.

For a square potential barrier, E_T can be assumed to be the ionization potential of the molecule. In reality the tunnelling electron would experience a potential which is the sum of the approximate Coulomb potential attracting the electron to a positive ion and the potential of electron affinity of the originally neutral molecule [Kemeny and Rosenberg 1970]. These potentials vary smoothly and are better approximated by a triangular than by a square barrier (Fig. 1.7). With the triangular potential barrier the tunnelling model can predict the magnitude of carrier mobility [Eley 1967, Eley and Willis 1961], and explain the compensation law (to be discussed in Section 1.6) and the anisotropy of the conductivity [Hänsel 1970, Kemeny and Rosenberg 1970], possibly because the triangular barrier shape facilitates the intermolecular electron transport since the barrier width becomes smaller for the excited electron at a higher level. The tunnelling model fails to explain the negative temperature dependence of mobility and the difference between electron and hole mobilities [Meier and Albrecht 1969, Keller and Rast 1962, Tredgold 1962].

In the band model there is no explicit mention of potential barrier between molecules. However, it can be imagined that an excited electron may tunnel over a distance of many molecules. Thus the tunnelling model may be considered as a band model when the potential varies periodically and regularly throughout the crystal, and the width of the potential barrier is much less than 10 Å. By considering a two-potential well system with a single potential barrier to be the same as a system containing a great number of potential wells, Keller and Rast [1962] have calculated the bandwidth (the energy level splitting—a single energy level in any one potential well is replaced by a band of levels, and the number of levels in the band equals the number of wells) of anthracene. Their estimate of the bandwidth is 0.029 eV, which is close to the value calculated on the band model by Katz *et al.* [1962].

1.4. THE HOPPING MODEL

We have mentioned in Section 1.2 the difficulties with the band model for electrical transport in materials in which the carrier mobility is very low. It can be imagined that for narrow bands the absorption or emission of single phonons becomes impossible in view of energy and wave vector conservation. If there is phonon scattering in the narrow band it must be a multiple-phonon process and then the electron motion will be random and incoherent. For this reason a hopping model may be preferable to the band model. A carrier can move from one molecule to another by jumping over the barrier via an excited state as shown in Fig. 1.7. The criterion to determine whether the charge transport in molecular crystals takes place coherently according to the band model or by random jumps according to the hopping model depends on the electron–lattice interactions—that is, depends on whether the strongest coupling is with intermolecular (lattice) or intramolecular (nuclear) vibrations, whether it is linear or quadratic in the phonon coordinates, and whether it is strong compared with the intermolecular electron exchange interactions. The vibration periods are typically 10^{-12} sec for intermolecular modes and 10^{-14} sec for intramolecular modes. By denoting the electron relaxation time, intermolecular vibration period and intramolecular vibration period by τ, τ_{vl}, and τ_{vn}, respectively, we have the following two important cases [Gutmann and Lyons 1967].

Case 1. $\tau < \tau_{vn} < \tau_{vl}$. In this case the electron motion is so rapid that the vibration motion can be regarded as stationary and as a perturbation to the motion of the electrons. The electrons can be thought of as waves travelling over several lattice sites before being scattered. The band model is applicable for this case.

Case 2. $\tau_{vn} < \tau < \tau_l$. In this case the molecule vibrates (intramolecular vibration) while the electron remains on a particular lattice site. This implies that during the time when the electron remains on the lattice site, the nuclei of the molecule on this particular lattice site move to new equilibrium positions. This gives rise to the formation of a "polaron". The polaron theory will be discussed in the next section. The interaction of electrons and phonons in the lattice site may lead to self-trapping in which the electrons polarize the molecules and are trapped in self-induced potential wells. This case may lead either to random hopping transport or to coherent band transport. For the former the electron trapped in such a potential well requires an activation energy to surmount a barrier of a height equal to the binding energy of the polaron in order to move to the neighbouring site. We shall discuss this case in more detail as follows.

In general, $\tau_{vn} \simeq 0.01\ \tau_{vl}$ in most organic molecular crystals. It is therefore likely that the interaction between the intermolecular vibrations and the carriers is weak and the motion of acoustic phonons is relatively slow, and that the intramolecular or simply the molecular vibrations or optical phonons interact much more strongly with the carriers. When one-electron energy bands are wider than the phonon energy band, the charge carrier motion through the lattice is fast and the carriers respond rapidly to the lattice motion. But when the charge carrier motion is slow, then the lattice will follow the charge carriers, and as these slow carriers migrate through the lattice they carry a lattice distortion with them. On the assumption that only a linear interaction with intermolecular (lattice) vibrations is considered, the effect of intramolecular vibrations being neglected, and that all lattice modes have approximately the same frequency,

Glarum [1963] has calculated the electron mobilities for organic molecular crystals in terms of the band and hopping models, and concluded that the effect of electron–lattice interactions is to reduce the electron bandwidth and the carrier transport might be intermediate between the coherent and hopping limits. A similar conclusion has also been obtained from a consideration of the interaction of charge carriers with the molecular vibrations [Siebrand 1964]. The electron–phonon interactions have also been treated by Friedman [1965] for the coherent transport, and by Gosar and Choi [1966] for the hopping transport based on a linear interaction between charge carriers and intermolecular vibrations with predominant acoustic lattice modes. All of these studies cannot explain the positive temperature dependence of the electron mobility in the c' direction in anthracene.

However, for the hopping transport the electron–phonon interaction must be strong. Because of strong interaction, the velocity of phonons will determine the velocity of charge carriers. It is, therefore, possible that the predominant electron–phonon coupling is with intramolecular (optical) vibrations, rather than with intermolecular (acoustic) vibrations, since the properties peculiar to molecular crystals can be attributed to the presence of discrete molecules with internal degrees of freedom [Munn and Siebrand 1970]. A model of the hopping transport based on a linear interaction of charge carriers with optical phonons has been put forward by Holstein [1959] to account for carrier drift mobilities in molecular crystals. But his model still fails to explain the temperature dependence of carrier mobilities in anthracene. Rashba [1966] and Munn and Siebrand [1970] have pointed out that a linear interaction has a substantial effect only if it involves an energy greater than a typical phonon energy, whereas a quadratic interaction can have a substantial effect when the energy which it involves is greater than that of intermolecular electronic and vibrational interactions. Furthermore, for the out-of-plane bending modes in aromatic hydrocarbons the changes of all the frequencies for electronic excitation are about 20–30% [Best *et al.* 1948, Munn and Siebrand 1970].

Using a quadratic electron–phonon (optical) coupling in molecular out-of-plane vibrational coordinates, Munn and Siebrand [1969, 1970] have treated the charge carrier transport in aromatic hydrocarbon crystals based on a linear chain model for both the hopping transport limit and the coherent transport limit. In their analysis the electron (or hole) wave function ψ is written in the tight-binding approximation as a linear superposition of localized molecular ion electronic wave functions ϕ:

$$\psi[R, \{x_r\}] = \sum_n a_n\{x_r\}\phi\,(R - na, x_n) \tag{1.29}$$

where R is the position coordinate of the carrier, $\{x_r\}$ denotes the dependence on the set of internuclear coordinates x_r of all the oscillators for each molecule, and n labels the site of the molecular ion. The coefficients $a_n\{x_r\}$ satisfy

$$\begin{aligned}
jh(\partial a_n/\partial t) = &\left\{\sum_r \left[-(\hbar^2/2m_R)\left(\frac{\partial^2}{\partial x_r^2}\right) + \frac{1}{2}m_R\omega_0^2 x_r^2 \right.\right. \\
&\left.\left. + \frac{1}{4}m_R\omega_1^2 x_r(x_{r+1} + x_{r-1}) \right] - \frac{1}{2}m_R\omega_2^2 x_n^2 \right\}a_n \\
&- J(a_{n+1} + a_{n-1})
\end{aligned} \tag{1.30}$$

in which the summation is the vibrational Hamiltonian for a set of oscillators of frequency ω_0 and reduced mass m_R. The single excess electron is assumed to interact only with the intramolecular vibrations, the intermolecular vibrations being suppressed by fixing the centres of mass of the oscillators at sites na. There are three interactions: a mechanical coupling ω_1 between adjacent oscillators, an electronic coupling J between adjacent molecules, and an electron–phonon quadratic interaction ω_2. The electron–phonon interaction is taken to be the strongest of the three. In the hopping model there are two limiting modes of carrier transport. In the slow electron limit the electron exchange energies are small compared with phonon dispersion energies, and the transfer of the electron between adjacent molecules is the rate-determining step. In the slow phonon limit the electron exchange energies are large compared with phonon dispersion energies, and the transfer of the electron between adjacent molecules is limited by the rate of phonon transfer. Munn and Siebrand [1969, 1970] have emphasized that since phonon bandwidths are only a few cm^{-1} compared with computed carrier bandwidths of few tens of cm^{-1} for aromatic hydrocarbon crystals, any hopping transport in these materials is likely to proceed in the slow phonon limit.

For the hopping transport in the slow electron limit, the condition is

$$\hbar\omega_2^4/\omega_0^3 \gg \hbar\omega_1^2/\omega_0 \gg 4J \tag{1.31}$$

Thus, solution of eqs. (1.29) and (1.30) for this model yields the carrier mobility [Munn and Siebrand 1970]

$$\mu_{he}^{(1)} = \frac{2\pi\ qJ^2a^2}{\hbar B\ kT}\left[\frac{\sigma}{1+\sigma}\right] \tag{1.32}$$

where

$$\sigma = \exp(-\hbar\omega_0/kT) \tag{1.33}$$

and B is the phonon bandwidth given by

$$B = \hbar\omega_1^2/\omega_0 \tag{1.34}$$

the subscript h and e denote, respectively, the hopping transport and the slow electron limit, and the superscript 1 refers to the one-oscillator model. For the two-oscillator model, we have

$$\mu_{he}^{(2a)} = \frac{2\pi\ qJ^2a^2}{\hbar kT}\left(\frac{1-\sigma}{1+\sigma}\right)\left(\frac{1-\sigma^\alpha}{1+\sigma^\alpha}\right)$$
$$\times\left[B^{-1}\left(\frac{\sigma}{1-\sigma}\right)+(\gamma B)^{-1}\left(\frac{\sigma^\alpha}{1-\sigma^\alpha}\right)\right] \tag{1.35}$$

in which $\alpha\omega_0$ and γB are, respectively, the frequency and the bandwidth of the second oscillator (α times the frequency and γ times the bandwidth of the first oscillator). Equation (1.35) is based on the assumption that energy is conserved for each oscillator separately. If energy is allowed to redistribute when the two oscillators are degenerate, then the carrier mobility is given by

$$\mu_{he}^{(2b)} = \frac{4\pi q\ J^2a^2}{\hbar B kT}\left[\sigma(1+\sigma^2)/(1+\sigma)^3\right] \tag{1.36}$$

Figure 1.8 shows $\mu_{he}^{(1)}$, $\mu_{he}^{(2a)}$, and $\mu_{he}^{(2b)}$ as functions of T/θ for $\gamma = 1$, where $\theta = \hbar\omega_0/k$. The

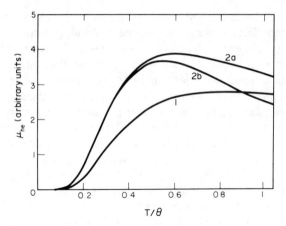

F IG. 1.8. Hopping mobility in the slow electron limit as a
function of temperature. Curve 1 is for the one-
oscillator model, curve 2a for the degenerate two-
oscillator model, and curve 2b for the nearly
degenerate two-oscillator model. [After Munn and
Siebrand 1970.]

value of θ for molecular crystals is of the order of the melting point. At high
temperatures $\mu_{he}^{(2b)}$ approach to $\mu_{he}^{(1)}$.

For the hopping transport in the slow phonon limit, the condition is

$$\hbar\omega_2^4/\omega_0^3 \gg 4J \gg \hbar\omega_1^2/\omega_0 \tag{1.37}$$

Thus, solution of eqs. (1.29) and (1.30) for this model yields the carrier mobility [Munn
and Siebrand 1970]

$$\mu_{hp}^{(1)} = \frac{\pi q B^2 a^2}{16\hbar\, JkT}\left[\sigma(1+\sigma^2)/(1+\sigma)(1+\sigma)^3\right] \tag{1.38}$$

where the subscript p denotes the slow phonon limit. Similarly, for the two-oscillator
model, we have

$$\mu_{hp}^{(2a)} = \frac{\pi q B^2 a^2}{16\hbar\, JkT}\left[\left(\frac{1-\sigma^\alpha}{1+\sigma^\alpha}\right)\frac{\sigma(1+\sigma^2)}{(1-\sigma)(1+\sigma)^3} + \gamma^2\left(\frac{1-\sigma}{1+\sigma}\right)\frac{\sigma^\alpha(1+\sigma^{2\alpha})}{(1-\sigma^\alpha)(1+\sigma^\alpha)^3}\right] \tag{1.39}$$

If energy is allowed to redistribute when the two oscillators are degenerate, then the
carrier mobility is given by

$$\mu_{hp}^{(2b)} = \frac{\pi q B^2 a^2}{8\hbar\, JkT}\left[\sigma/(1-\sigma)^2\right] \tag{1.40}$$

Figure 1.9 shows $\mu_{hp}^{(1)}$, $\mu_{hp}^{(2a)}$, and $\mu_{hp}^{(2b)}$ as functions of T/θ for $\gamma = 1$. It can be seen
that at $\sigma = 1$ there exists a divergence in $\mu_{hp}^{(2b)}$, which is not expected in a realistic
many-oscillator model in which the factors $1 - \sigma^\alpha$ are sufficient to suppress all
divergences [Munn and Siebrand 1970]. In practice the one-oscillator model is broadly
similar to the two- and the many-oscillator model for the temperature range of interest.

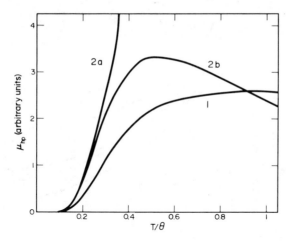

FIG. 1.9. Hopping mobility in the slow phonon limit as a function of temperature. Curve 1 is for the one-oscillator model, curve 2a for the degenerate two-oscillator model, and curve 2b for the nearly degenerate two-oscillator model. [After Munn and Siebrand 1970.]

Munn and Siebrand [1970] have also considered the case for the quasi-free electron coherent transport in the slow phonon limit. The condition for this case is that the electron coupling is much stronger than both the vibrational coupling and the electron–phonon coupling. The carrier mobility for the one-oscillator model is given by

$$\mu_{bp}^{(1)} = (q/kT)(16J^2\omega_0\omega_2^2 a^2/\pi^3\hbar^2\omega_1^4)[(1-\sigma)^3/\sigma^{3/2}] \qquad (1.41)$$

for elastic scattering for which $\eta = J/\hbar\omega_0 < \frac{1}{2}$; and is given by

$$\mu_{bp}^{(1)} = (q/kT)(32J^2\omega_0\omega_2^2 a^2/\pi^3\hbar^2\omega_1^4)[(1-\sigma)^3/\sigma^{1/2}]\phi(T) \qquad (1.42)$$

for inelastic scattering; for $\frac{1}{2} \leq \eta < 1$, $\phi(T)$ is given by

$$\begin{aligned}
\phi(T) = (1-\xi)^{-1}\{&(1+2\sigma)^{-1}[1-\xi^{(1-1/2\eta)}] \\
&+ (2\sigma)^{-1}[\xi^{(1-1/2\eta)}-\xi^{1/2\eta}] \\
&+ (2\sigma+\sigma^2)^{-1}[\xi^{1/2\eta}-\xi]\}
\end{aligned} \qquad (1.43)$$

and for $\eta \geq 1$, $\phi(T)$ is given by

$$\begin{aligned}
\phi(T) = (1-\xi)^{-1}\{&(1+2\sigma)^{-1}[1-\xi^{1/2\eta}] \\
&+ (1+2\sigma+\sigma^2)[\xi^{1/2\eta}-\xi^{(1-1/2\eta)}] \\
&+ (2\sigma+\sigma^2)^{-1}[\xi^{(1-1/2\eta)}-\xi]\}
\end{aligned} \qquad (1.44)$$

where $\xi = \exp(-4J/kT) = \sigma^{4\eta}$, the subscript bp denotes band or quasi-free electron transport in the slow phonon limit. Figure 1.10 shows μ_{bp} as a function of T/θ for various values of η. It is interesting to note that the variation of μ_{bp} with J at low temperatures is quite different from that at high temperatures. The coherent transport, unlike the hopping transport, is sensitive to the number of oscillators considered.

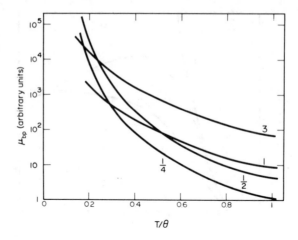

FIG. 1.10. Coherent mobility in the slow phonon limit as a
function of temperature. The curves are labelled
with their value of $\eta = J/\hbar\omega_0$. [After Munn and
Siebrand 1970.]

All the models so far discussed in this subsection are one-dimensional whereas the experimental mobilities are measured in a three-dimensional crystal. Furthermore, the information about the values of ω_0, ω_1, ω_2, and J is incomplete or not available. This makes it difficult to compare the theory with the experiment quantitatively. However, Munn and Siebrand have attempted to interpret the mobility results for anthracene based on their theory. By assuming $\hbar\omega_0 = 350\ \mathrm{cm}^{-1}$, $\hbar\omega_1 = 60\ \mathrm{cm}^{-1}$ corresponding to a phonon bandwidth $B = \hbar\omega_1^2/\omega_0 = 10\ \mathrm{cm}^{-1}$, $\hbar\omega_2 = 250\ \mathrm{cm}^{-1}$ for anthracene, they have computed $\mu_{hp}^{(1)}$, electron mobility in the **a** direction, using eq. (1.38) and $\mu_{he}^{(1)}$, electron mobility in the **c′** direction, using eq. (1.32), and leaving the absolute magnitudes determined by the values of J and a. Their computed results are shown in Fig. 1.11, and they are in qualitative agreement with the available experimental results. They have concluded that as the electron bandwidth in the **c′** direction is much smaller than that in other directions and possibly smaller than the phonon bandwidth in anthracene, the carrier transport may be accurately described by the hopping model in the slow electron limit or in the intermediate between slow electron and slow phonon hopping; and that the characteristics of electron and hole mobilities in other diagonal directions can be interpreted in terms of slow phonon coherent transport or a mode intermediate between hopping and coherent transport in the slow phonon limit.

Mey and Hermann [1973] have also used the models of Munn and Siebrand [1970] to explain their experimental results on the temperature independent electron mobilities in the **a** and **b** directions in naphthalene. The comparison between the theory and the experiment leads to a similar conclusion that the carrier transport mechanism in naphthalene is intermediate between the band and the hopping models.

Before closing this section, we should mention a very simple hopping model applied to the carrier transport in organic solids [Mehl and Wolf 1964, Pohl *et al.* 1962, Boguslavskii and Vannikov 1970]. A similar model has been applied to the motion of ions in ionic crystals [Mott and Gurney 1940, Seitz 1940]. The carrier mobility based

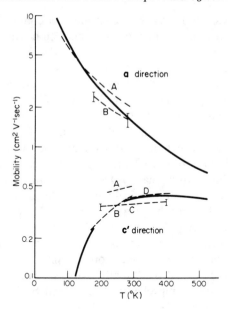

F IG. 1.11. Electron mobilities in anthracene. The solid lines are the theoretical curves. Hopping mobility in the slow phonon limit for the **a** direction and hopping mobility in the slow electron limit for the **c'** direction. The broken lines are the experimental data from A, B, C, and D references. [After Munn and Siebrand 1969.] A, Kepler [1964]; B, Pott and Williams [1969]; C, Nakada and Ishihara [1964]; D, Fourny and Delacote [1969].

on this simple hopping model can be written as

$$\mu = \frac{qa^2 v}{kT} \exp(-\Delta E/kT) \tag{1.45}$$

where a is the width and ΔE is the height of the potential barrier, and v is a frequency factor. Pohl [1962] has also introduced a different hopping model based on the formation of ion–exciton complexes. A transfer of charge can take place by the reaction [Pohl 1962, Gutmann and Lyons 1967]

$$M_1^+ + M_2^- M_3^+ \rightarrow M_1^+ M_2^- + M_3^+$$

where the charge has been transferred from site 1 to site 3, or by

$$M_1^+ + M_2^- M_3^+ \dots M_n^- M_{n+1}^+ \rightarrow M_1^+ M_2^- \dots M_{n-1}^+ M_n^- + M_{n+1}^+$$

where the transfer is from site 1 to site $n+1$. Hopping thus occurs between the ion and the exciton.

1.5. POLARONS IN MOLECULAR CRYSTALS

In the previous section we have pointed out that the interaction between a slow electron and molecular vibrations may be too strong to be considered as a perturbation, and that the electron may polarize the molecule and then be trapped in a self-induced potential well. The polarization of the lattice caused by the electron will act back on the electron itself and reduce its energy. The polarization tends to follow the electron as it moves through the lattice; the combination of the electron and its induced polarization field can be considered as a quasi-particle, which is generally called a polaron [Fröhlich and Seitz 1950, Kuper and Whitefield 1962]. The most important effect of the lattice polarization is the attendant increase in the effective mass of the electron. The size of a polaron is measured by the extent of the region over which the distortion or deformation of lattice due to polarization is introduced. In ionic crystals the electron–phonon coupling arises mainly from the long range strong Coulomb interaction between the electron and optical lattice modes, and therefore the radius of the distorted region is much larger than a lattice constant. The polarons in ionic crystals are sometimes called the "large polarons". In molecular crystals, the electron–phonon coupling is strong but of short range, the distortion may occur predominantly within the order of a lattice constant around the electron. The size of the polarons in this case is small, so such polarons are called the "small polarons". The theory and experiment work on polarons have been comprehensively reviewed by Langreth [1967], Appel [1968], Austin and Mott [1969], Adler [1968], Bogomolov *et al.* [1968], Bosman and van Daal [1970], Allcock [1956], and Fröhlich [1954], with specific emphasis on inorganic materials. The polaron theory has been applied to molecular crystals by Holstein [1959], Siebrand [1962], Kemeny and Rosenberg [1970], and Vilfan [1973].

The self-trapping was first introduced by Landau [1933]. But up to now to what extent self-trapping of electrons in a crystal lattice occurs is still in doubt. The theoretical treatment of a polaron based on perturbation theory or variational principles may not be valid for the actual interaction between an electron and its distorted crystal surroundings where the distortion in shape and extension is a function of the electron velocity. Experimental evidence is similarly in doubt because an ideal crystal lattice is never realistic and lattice imperfections trap carriers, strongly controlling the carrier velocity. However, there is a great deal of theoretical work on polarons. To review this is beyond the scope of this book. The reader who wishes to know more details on this subject is referred to the review articles cited above. But we shall summarize the basic concepts of polarons and the roles they may play in the electrical transport as follows.

For large polarons Fröhlich was the first to calculate the eigenstates of the polarons, and later several investigators also attempted to calculate the same for weak and intermediate electron–phonon coupling [Fröhlich 1954, Gurari 1953, Lee *et al.* 1953]. The energy eigenvalues for low values of k is given by

$$E(k) = E_p(o) + \hbar^2 k^2 / 2m_p^* \qquad (1.46)$$

where $E_p(o)$ is the self-energy and m_p^* the effective mass of the polaron, which are given by

$$E_p(o) = -\alpha\hbar\omega \qquad (1.47)$$

$$m_p^* \simeq m^* (1 + \alpha/6) \qquad (1.48)$$

and α is the coupling constant given by

$$\alpha = \frac{q^2}{\hbar}\left(\frac{1}{\varepsilon_\infty} - \frac{1}{\varepsilon_s}\right)\left(\frac{m^*}{2\hbar\omega}\right)^{1/2} \tag{1.49}$$

m^* is the rigid band effective mass of the electron, including the effect of electronic polarization due to the motion of a slow electron, ω is the angular frequency of the lattice oscillators; and ε_∞ and ε_s are, respectively, the high frequency and low frequency dielectric constants. Equation (1.46) implies that the electron–phonon (lattice vibrations) interaction tends (a) to lower the energy of the electron at rest by the self-energy $-\alpha\hbar\omega$ leading to a shift in the relative positions of the conduction band and the valence band, (b) to change the effective mass of the electron as shown in eq. (1.48), and (c) to give rise to scattering of the electron by phonons. In general, if $\alpha < 1$ the electron–phonon coupling is weak; if $\alpha \geq 10$ the coupling is strong; and if $\alpha \simeq 5$ the coupling can be considered as intermediate coupling. Small values of α, m^*, and m_p^* lead to the formation of large and weakly coupled polarons, while large values of these quantities lead to small and strongly coupled polarons. For a good review of the literatures and the techniques applied to polaron problems the reader is referred to Schultz [1956], Lee and Pines [1953], Feynman [1955], Feynman *et al.* [1962], Kuper and Whitefield [1962], Devreese [1972], and Spear [1974]. For molecular crystals it is most likely the polarons are small and strongly coupled polarons. In the following we shall discuss briefly the small polarons.

For small polarons, the lifetime of a polaron at any site is long because of strong electron–lattice coupling. The polaron can move either as a result of tunnelling between equivalent localized polaron states centred at different sites or by hopping between two non-equivalent localized states, involving emission and absorption of phonons. Tunnelling is analogous to a wave-like motion, which is of the band conduction and in which the vibrational states involve only a few quanta and are well separated. Hopping is a phonon-activated process which is predominant at high temperatures and involves a large number of highly excited vibrational levels so that the polaron motion which is greatly affected by interactions with vibrations becomes random and non-wave like, and cannot be described in terms of a band structure. In general, if the phonon bandwidth is small in comparison with the polaron bandwidth the band-type conduction may be predominant; if the reverse is true the conduction may be mainly by hopping. Holstein [1959] has found that at $T \leq 0.4\ \hbar\omega/k$ a band-type conduction may be assumed, and at $T > 0.5\ \hbar\omega/k$ a hopping-type conduction may be assumed, ω being the optical mode vibrational angular frequency. The polaron bandwidth decreases while the polaron effective mass increases rapidly with increasing temperature. Munn and Siebrand [1970] have derived expressions for the carrier mobilities based on the criteria for the band- or hopping-type of carrier transport (cf. Section 1.4). Many investigators have attempted to modify the theory for a better fit with experimental results [Seager and Emin 1970, Bosman and van Daal 1970, Gosar and Vilfan 1970, Emtage 1971, Jurgis and Silinsh 1972, Vilfan 1973].

However, the general conclusion has been that the carrier transport lies in the intermediate between the coherent and hopping transport when the carrier mobility is of the order of 1 cm^2/V-sec and the corresponding mean free path of the order of the intermolecular spacing. Using the Kubo linear response formalism [Kubo 1952, 1957] and assuming that the electron–phonon interaction is linear in phonon operators,

Vilfan [1973] has derived an expression for carrier mobility based on the small polaron model taking into account the effect of the electron–hole correlation, and it is given by

$$\mu = \frac{q}{\hbar^2 kT} \left[\frac{\pi}{8(B^2 - B_{ij}^2)} \right]^{1/2} \exp\left[-\frac{1}{2} \frac{(E_b - E_{ij})^2}{\hbar^2(B^2 - B_{ij}^2)} \right] \sum_{i-j} \omega_{ij}^2 r_{ij} r_{ij} \qquad (1.50)$$

where ω_{ij} are the electron transfer integrals; r_{ij} is the distance between the centres of nearest-neighbour molecules i and j; and E_b, E_{ij}, B, and B_{ij} are, respectively, given below:

$$E_b = \sum_\lambda \frac{|V_\lambda|^2}{\hbar\omega_\lambda} \qquad (1.51)$$

$$E_{ij} = \frac{96W^2}{Ms} I_3' \qquad (1.52)$$

$$B^2 = \frac{48W^2}{Mr^2 \hbar s/r} \left(\frac{2kT}{\hbar s/r} I_3 + I_4 \right) \qquad (1.53)$$

$$B_{ij}^2 = \frac{48W^2}{Mr^2 \hbar s/r} \left(\frac{2kT}{\hbar s/r} I_3' + I_4' \right) \qquad (1.54)$$

in which $\hbar\omega_\lambda$ is the energy of the phonon in the state λ, where λ represents the wave vector k_q and polarization branch p; M is the molecular mass; s is the velocity of sound in the crystal; r is the distance between two nearest-neighbour molecules; V_λ is the coupling constant between the electron and the phonon in the mode λ; and V_λ, W, I_n, and I_n' are given by

$$|V_\lambda|^2 = \frac{4\alpha^2}{MNs^2} \left(\frac{q}{4\pi\varepsilon_0} \right)^2 \frac{1}{r^8} \frac{\hbar s}{r} \sum_l \frac{\sin^2 K_{q\lambda} r_{il}}{|K_{q\lambda} \cdot r_{il}|} \qquad (1.55)$$

$$W = \frac{\alpha}{2} \left(\frac{q}{4\pi\varepsilon_0} \right)^2 \frac{1}{r^4} \qquad (1.56)$$

$$I_n = \frac{2\pi}{v} \int_0^{Q_m} Q^{n-3} \left(1 - \frac{\sin 2Q}{2Q} \right) dQ \qquad (1.57)$$

$$I_n' = \frac{2\pi}{v} \int_0^{Q_m} Q^{n-3} \left(1 - \frac{\sin 2Q}{2Q} \right) \frac{\sin Q}{Q} dQ \qquad (1.58)$$

where α is the mean molecular polarizability; r_{il} is the distance between the centres of the molecules i and l; the summation over l runs all molecules in the crystal but it gives satisfactory results by taking the directional average of $|V_\lambda|^2$ times the number of nearest neighbours; N is the number of molecules in the crystal; Q_m is the dimensionless radius of the Debye sphere; v is the volume of the sphere; $Q = k_q r$. By taking the radius of the sphere equal to $K_{qm} = \pi/r$, then $Q_m = \pi$.

Equation (1.50) includes both band and hopping transport mechanisms and consists of a $T^{-1.5}$ temperature dependence and a $\exp(-E_a/kT)$ thermally activated process. Vilfan [1973] has shown that the small polaron binding energy E_b is of the order of or somewhat smaller than the electron bandwidth for anthracene (and for most organic molecular crystals), and that the time of formation of the small polarons determined by $\hbar B$ is of the order of the localized electron lifetime determined from the electron bandwidth. Therefore, the electrical transport at room temperature for most organic molecular crystals is intermediate between the band-type transport and the small polaron hopping transport. This implies that the phonon cloud begins to rise around the electron when the electron reaches a certain molecule; but before the cloud is formed to its stationary configuration, the electron leaves the molecule. Vilfan [1973] has calculated the mobilities for anthracene crystals using various transfer integrals and the following parameters: $r = 5.2$ Å, $s = 3.14 \times 10^3$ m sec^{-1}, $Ms^2 = 18$ eV, $\alpha/4\pi\varepsilon = 25.3$ Å^3 for two cases with and without the effect of the electron–hole correlation. Table 1.9 shows some of his computed results using eq. (1.50) and transfer integrals from Silbey *et al.* [1965], including the vibrational overlap factor of 0.35 estimated by the method described by Siebrand [1964]. It can be seen that his computed results are in good agreement with the experimental results of Kepler [1962] except the electron mobility in the c' direction. Vilfan [1973] has attributed this discrepancy to the underestimation of the transfer integrals in the c' direction or to some other electron–phonon interaction mechanisms which would be highly anisotropic [Delacote and Tiberghien 1968].

For anthracene eq. (1.50) can be approximated to $AT^{-1.5} \exp[-E_a/kT]$. Vilfan's theoretical temperature dependent results are in qualitative agreement with the experimental results of Kepler [1960], Pott and Williams [1969], and Fourny and Delacote [1969]. Vilfan [1973] has also reported that the effect of electron–hole correlation is to increase the carrier mobility by about 80%. This is due to the fact that when the energy fluctuations of the lattice sites i and j are correlated, an electron at the lattice site i is exchanged with a hole at the lattice site j, and they do not feel the phonons which produce these fluctuations.

At low temperatures the bandwidth of small polarons becomes so narrow that a band-type carrier transport is possible. Emtage [1971] has reported that at high electric fields an electron may gain from the field an energy comparable with the bandwidth and that charge transport may proceed by hopping from site to site even when the temperature is low, and then the high field conductivity becomes much less than the low field conductivity by non-localized electrons. Such a field-induced transition from band to hopping modes of transport is marked by a region of constant current or negative differential resistance.

Holstein [1973] has reported that on the basis of the small polaron hopping model the sign of the Hall coefficient depends on the number of sites involved in the elementary jump process. The sign of the Hall coefficient for holes is the same as that for electrons when the jump process involves three sites, is opposite to that for electrons when it involves four sites. Kemeny and Rosenberg [1970] have postulated that the tunnelling (band motion) of small polarons in thermally activated energy levels of molecules is responsible for electric conduction in organic and biological semiconductors. It should also be noted that the polarons may also be involved in the carrier transport in metal–insulator–metal systems under carrier injection conditions [Timan 1973], and that excitonic polarons may play a very important role in transport phenomena in molecular crystals [Munn and Siebrand 1970, Chaikin *et al.* 1972]. For more details about the theory of small polarons the reader is referred to the original papers of Holstein [1959], Toyazawa [1961], Sewell *et al.* [1962], Siebrand [1964],

TABLE 1.9. *Components of mobility tensor in cm²/V-sec for anthracene at room temperature [after Vilfan 1973]*

Component		Electron–hole correlation not included			Electron–hole correlation included			Experimental mobilities
		i	ii	iii	i	ii	iii	iv
Electron	aa	6.80	0.83	2.70	12.30	1.50	4.90	1.70
	bb	5.50	0.67	2.14	9.90	1.22	3.80	1.00
	c'c'	0.018	0.0022	0.0041	0.033	0.0040	0.0075	0.40
Hole	aa	4.00	0.49	1.60	7.35	0.90	2.90	1.00
	bb	9.90	1.22	3.60	18.10	2.21	6.50	2.00
	c'c'	3.50	0.43	0.72	6.40	0.78	1.30	0.80

(i) Transfer integral from Silbey et al. [1965], vibrational overlap not included.
(ii) Transfer integral from Silbey et al. [1965], including overlap factor. The value of 0.35 is chosen for the squared vibrational overlap factor.
(iii) Transfer integral from Glaeser and Berry [1966], oscillator strength = 2.0, vibrational overlap not included.
(iv) Experimental values from Kepler [1962].

Lang and Firsov [1963, 1964], Emin [1970, 1971, 1972, 1973], Jurgis and Silinsh [1972], Vilfan [1973], and appropriate references given at the end of this book.

1.6. DARK ELECTRIC CONDUCTION

There has been considerable interest in the electric conductivity of organic semiconductors because the relationship between the electric conductivity and the chemical structure may lead to the understanding of biological systems and to the possibility for a synthesis of organic semiconductors for practical applications. The dark electric conductivity of most organic semiconductors measured using non-injecting (or slightly injecting) electrodes at low fields is given by [Boguslavskii and Vannikov 1970]

$$\sigma = \sigma_0 \exp\left[-\Delta E_\sigma / 2kT\right] \tag{1.59}$$

where σ_0 is called the pre-exponential factor and ΔE_σ is the activation energy. The interpretation of both σ_0 and ΔE_σ is not straightforward, but rather ambiguous since electric conduction may involve various transport processes. The basic equation for σ is

$$\sigma = q(\mu_n n + \mu_p p) \tag{1.60}$$

where n and p can be expressed as

$$n = N_c \exp\left[(E_{Fn} - E_c)/kT\right]$$
$$= 2(2\pi m_e^* kT/h^2)^{3/2} \exp\left[(E_{Fn} - E_c)/kT\right] \tag{1.61}$$
$$p = N_v \exp\left[(E_v - E_{Fp})/kT\right]$$
$$= 2(2\pi m_h^* kT/h^2)^{3/2} \exp\left[(E_v - E_{Fp})/kT\right] \tag{1.62}$$

1.6.1. The activation energy

If $\mu_n N_c$ and $\mu_p N_v$ are independent of temperature in the range of temperatures in which the thermal activation energy ΔE_σ derived from the Arrhenius type of temperature dependence of σ, then ΔE_σ may be associated with the excitation energy for thermal generation of carriers. The excitation mechanisms are as follows.

(a) ΔE_σ corresponds to the energy band gap E_g for intrinsic conduction, or corresponds to the distance between the donor level and the conduction band edge for n-type extrinsic conduction, or corresponds to the distance between the acceptor level and the valence band edge for p-type extrinsic conduction. Extrinsic semiconductors usually exhibit an activation energy corresponding to extrinsic conduction at low temperatures and an activation energy corresponding to intrinsic conduction at high temperatures.

(b) ΔE_σ corresponds to the depth of traps, i.e. $E_c - E_t$ for electron traps and $E_t - E_v$ for hole traps, E_t being the trapping level. This implies that the conduction is dominated by the thermal release of trapped carriers [Helfrich 1967].

(c) ΔE_σ may correspond to the energy required to raise the electrons from the ground state to an excited state so that they can tunnel the potential barrier efficiently to make a major contribution to electric conduction. Such a potential barrier is generally the barrier between molecules [Eley and Parfitt 1955].

(d) ΔE_σ may correspond to the height of the potential barrier between the electrode and the solid specimen which must be overcome for carrier injection from the electrode [Riehl *et al.* 1966]. For details the reader is referred to Chapter 2.

It is clear that the thermal activation energy ΔE_σ, which may be interpreted in terms of one of the above four possible processes, is related to the structure of the solid specimen, and that on the basis of the value of ΔE_σ alone it is difficult to identify the conduction process. However, several investigators [Okamoto and Brenner 1964] have reported that ΔE_σ, which is identified with the energy band gap, decreases with increasing number of π-electrons up to a limiting value as shown in Fig. 1.12, since the excitation process is strongly dependent on the position of the π-electrons within the molecule, and their delocalization. That ΔE_σ tends to remain unchanged with the number of π-electrons for high molecular weight materials may be due to their unfavourable structure with respect to intermolecular overlapping.

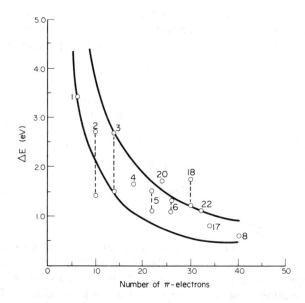

Fig. 1.12. Variation of activation energy with number of π-electrons. The numbers, denote the materials: 1, benzene; 2, naphthalene; 3, anthracene; 4, tetracene; 5, pentacene; 6, hexacene; 8, quaterrylene; 17, isoviolanthrene; 18, pyranthrene; 20, coronene; 22, ovalene. [After Meier 1974.]

In general, the conductivity of organic semiconductors increases and their thermal activation energy decreases with increasing pressure [Samara and Drickamer 1962, Harada *et al.* 1964, Aust *et al.* 1964, Bentley and Drickamer 1965, Schwartz *et al.* 1964]. Some significant results are given in Table 1.10. Generally, the increase in pressure results in an increase of overlapping of the wave functions of adjacent molecules, leading to a decrease in potential barrier (ΔE_σ) or to an increase in carrier mobility [Kawabe *et al.* 1966, Wood *et al.* 1966, Harada *et al.* 1964, Le Blanc 1961].

However, Batley and Lyons [1966] have related the pressure dependence of conductivity to the pressure dependence of polarization. The thermal activation energy

TABLE 1.10. *The effect of pressure on electric conductivity, activation energy, carrier concentration, and mobility of quaterrylene and violanthrone [after Boguslavskii and Vannikov 1970]*

| Substance | Pressure (kg cm^{-2}) | | | | n_{160}/n_{atm} | μ_{160}/μ_{atm} |
| | 1 | | 160 000 | | | |
	σ (Ω^{-1}cm^{-1})	ΔE (eV)	σ (Ω^{-1}cm^{-1})	ΔE (eV)		
Quaterrylene	8.3×10^{-9}	0.61	1.4×10^{-4}	0.16	6×10^3	3
Violanthrone	4.3×10^{-11}	0.78	1.4×10^{-3}	0.20	8×10^4	400

The subscripts 160 and atm refer, respectively, to the pressure of 160,000 kg cm^{-2} and to the atmospheric pressure.

ΔE_σ can be expressed in terms of ionization energy I, electron affinity χ, and polarization energy ΔP [Gutmann and Lyons 1967]

$$\Delta E_\sigma = I - \chi - \Delta P \tag{1.63}$$

The polarization energy increases with increasing pressure, thus ΔE_σ decreases with increasing pressure. This model has been used to calculate the polarization energy of many organic materials based on the assumption that the compressed materials are isotropic and that the increase in conductivity at high pressures is mainly due to the decrease in thermal activation energy [Drickamer 1967, Boguslavskii and Vannikov 1970].

It should be noted that in some organic semiconductors, particularly in polymeric organic semiconductors, the pressure and temperature dependence of conductivity is partly due to the change of carrier mobility. This will be discussed in Subsection 1.6.2.

It is also interesting to note that apart from the normal pressure effect just described, there is also an anomalous pressure effect for which the conductivity decreases with increasing pressure. Such an anomalous effect has been observed in ferrocene [Okamoto *et al.* 1964], and in [(C$_6$H$_5$)$_3$PCH$_3$] (TCNQ)$_2$ and [(C$_6$H$_5$)$_3$AsCH$_3$] (TCNQ)$_2$ [Shirotani *et al.* 1972]. In Li(TCNQ) and Na(TCNQ) the conductivity increases with increasing pressure up to about 5 kbar, decreases to a minimum and then increases again with further increase of pressure. These effects have been attributed to pressure-induced new phase transitions arising from a change in the periodicity in the TCNQ$^-$ column [Sakai *et al.* 1972].

Northrop and Simpson [1956] have reported that there is a close relation between the conductivity, the activation energy, and the triplet excitation energy in pure aromatic hydrocarbons and in solid solutions in which one hydrocarbon is introduced into another as an impurity. Rutkowsky *et al.* [1968] have reported that ΔE_σ is linearly proportional to the triplet excitation energy for polycyclic organic compounds. Wilk [1960] has reported that ΔE_σ is associated with the triplet excitation energy for compounds consisting of only a few rings in which the triplet states are relatively remote from the singlet states, and that there is no relationship between ΔE_σ and the singlet excitation energy, indicating that the triplet levels are important to electric conduction in organic semiconductors. This may suggest that the contributions of triplet states to electric conduction may be due to a singlet–triplet excitation [Kleinermann and McGlynn 1962, Rosenberg 1962].

1.6.2. The compensation effect

If $\mu_n N_c$ and $\mu_p N_v$ are dependent upon temperature, the pre-exponential factor σ_0 in eq. (1.59) is no longer constant. Many *et al.* [1955] were the first to note that a correlation exists between σ_0 and ΔE_σ. The relation of $\ln \sigma_0$ and ΔE_σ take the general linear form [Gutmann and Lyons 1967]

$$\ln \sigma_0 \simeq \alpha \Delta E_\sigma + \beta \tag{1.64}$$

Equation (1.64) is generally referred to as the compensation effect which has been observed in many organic crystals [Eley 1967, Eley *et al.* 1968, Rosenberg *et al.* 1968, Johnston and Lyons 1970]. Equation (1.64) is of interest mainly because of their connection with the carrier mobility and the effective density of states.

The compensation effect is most likely to arise from the method of calculation of σ_0 and ΔE_σ [Johnston and Lyons 1970]. There are two possibilities for eq. (1.64) to hold, and they are:

(i) If α is negative ($\alpha < 0$), σ_0 increases as ΔE_σ decreases. For example, heat-treated polynitriles and bis-(1,2-dicyanoethylene.1,2-dithiolo)metal salts follow this trend [Meier 1974].

(ii) If α is positive ($\alpha > 0$), σ_0 increases as ΔE_σ increases. This trend has been found to hold for many substances, such as in inorganic semiconductors (TiO_2, Fe_2O_3, UO_2, ZnO, SiC, and $BaTiO_3$) [Mooser and Pearson 1960], polyaromatic series [Eley 1962, 1967], amino acids and proteins [Cardew and Eley 1959], monomeric phthalocyanines [Hamann 1967], chloro-substituted copper phthalocyanines [Starke and Hamann 1970], and many other organic materials [Eley and Newman 1970, Meier 1974]. This trend is sometimes referred to as the Meyer–Neldel rule, which was first observed in inorganic semiconductors by Meyer and Neldel [1937].

Equations (1.59) and (1.64) may be combined to the expression in the form

$$\sigma = \sigma_0' \exp(\Delta E_\sigma / 2kT_0) \exp(-\Delta E_\sigma / 2kT) \tag{1.65}$$

where σ_0' and T_0 are constants, which, in fact, are:

$$\sigma_0' = \exp(\beta), \qquad T_0 = (2k\alpha)^{-1} \tag{1.66}$$

and T_0 is called the "characteristic temperature" of the substance. For a single organic substance ΔE_σ can be varied for determining σ_0' and T_0 by any of the following three processes: (a) hydration of the crystals, (b) formation of weak donor–acceptor complexes, and (c) use of different *cis–trans* isomers of the substance when possible [Rosenberg *et al.* 1968]. One method of determining the constants is to plot $\ln \sigma_0$ as a function of ΔE_σ, the slope of the line yields $(2kT_0)^{-1}$, and the intercept yield σ_0'. Eley [1967] has proposed the following mechanisms for the origin and motion of the charge carriers for which $\ln \sigma_0$ increases with ΔE_σ: (a) intrinsic bulk thermal generation and transport by hopping over intermolecular barriers, (b) bulk generation with carrier tunnelling through intermolecular barriers, (c) narrow band transport, and (d) electron (or hole) injection from the electrodes into the conduction band (or the valence band) for large energy band gap substances. But he has not drawn a firm conclusion about these mechanisms for the explanation of this compensation effect. However, Kemeny and Rosenberg [1970] and Hänsel [1970] have used the mechanism (b) proposed by Eley [1967] to explain this effect. This mechanism is tunnelling through intermolecular barriers from activated energy levels of molecules.

In organic semiconductors molecules are surrounded with potential barriers. Electrons have to overcome the barrier height or to tunnel through the barriers before they can travel from molecules to molecules. The former process requires a higher activation energy and the latter requires a long time. Both processes lead to low values of mobility and hence conductivity. Kemeny and Rosenberg [1970] have shown that the triangular potential barrier (Fig. 1.7) assumption gives a qualitative but not quantitative agreement with experiments. Their estimated electron effective mass $100m_0$ is much larger than those theoretically predicted for peptides [Yomosa 1964, Evans and Gergely 1949]. In most biological systems the energy bands are narrow and the small polarons are most likely to be formed in such systems. Later, Kemeny and Rosenberg [1970] proposed another model based on the small polaron theory to explain the compensation effect. Their model is based on the tunnelling (band motion) of small polarons in thermally activated energy levels of molecules and the expressions for electric conductivity derived by them are in the form of eq. (1.65) with the characteristic temperature T_0 related to the Debye temperature.

It should be noted that the relation between $\ln \sigma_0$ and ΔE_σ may be affected by absorbed films which form surface states with the effect of increasing or decreasing σ_0 depending upon whether they act like acceptors or donors [Eley 1967], and that the activation energy ΔE_σ is a function of the delocalization or resonance energy of the π-electrons of molecules [Hänsel 1970]. The larger the delocalization energy characterized by strongly delocalized π-electrons, the smaller is the activation energy.

1.6.3. The conformons

In electric conduction in organic and biological semiconductors there are two rather remarkable properties: (i) the compensation behaviour and (ii) the anomalously large value of the conductivity beyond that predicted by the conventional solid state theory [Kemeny and Goklany 1973]. Property (i) has been discussed in Section 1.6.2. Property (ii) has been observed in some of these materials such as in oxidized cholesterol, 11 *cis* retinal, etc. The pre-exponential factor σ_0 of eq. (1.59) can be expressed as

$$\sigma_0 = qN\mu \tag{1.67}$$

with

$$N = \frac{N_A \rho Z}{M} \tag{1.68}$$

based on the conventional solid state theory, where N_A is the Avogadro's number, M is the molecular weight, ρ is the density, and Z is the number of charge carriers available for excitation per molecule. The value of σ_0 in some cases is too large by as much as 10–12 orders of its magnitude calculated from eqs. (1.67) and (1.68), even by assuming that $Z = 1$ and $\mu = 10^5 \text{ cm}^2 \text{ V}^{-1} \text{ sec}^{-1}$. The nature of the molecular states, intermolecular potential barriers, and the polaron effect govern the carrier mobility. For most organic or biological semiconductors the mobility of charge carriers is much smaller than 10^5 cm^2 V^{-1} sec^{-1} (usually between 10 and 10^{-5} cm^2 V^{-1} sec^{-1}). Thus the large value of σ_0 must be associated with a large density of activated charge carriers. Using the concept of the conformon, which is referred to as the activated charge carrier plus the accompanying conformational changes carrying energy and entropy [Volkenstein 1972, Pullmann 1976], Kemeny and Goklany [1973] have postulated that the effective density of states for activated charge carriers would be greatly increased by the interaction between the activated carriers and the other degrees of the molecules (the vibrational motion), which may cause a change in electronic state and give rise to an activation entropy ΔS because of a change in vibration frequencies probably associated with conformational changes. Such an interaction may lead to the *trans*-conformation of the macromolecules. Kemeny and Goklany [1973] have attributed the conduction mechanism to conformon hopping without involving tunnelling. The small polaron concept does not involve entropy changes while the conformon concept does.

The entropy can take on substantial values because energy can be distributed over many degrees of freedom. In usual intrinsic semiconductors the charge carriers are always created in pairs, the number of electrons being equal to the number of holes. Under an electric field an electron and a hole move in opposite directions, and thus one negatively charged molecule and one positively charged molecule are always present for each electron–hole pair. According to Kemeny and Goklany [1973], one or both of these molecular ions can exist in various conformations and/or coordinations. Because of large entropy of the activated state, the number of activated carriers is not only governed by the activation energy ΔE_σ but also by the activation entropy ΔS, and this accounts for the high conductivity provided that this mechanism does not decrease the carrier mobility. The relation between the conformon concept and the electric conductivity is at present still conjectural. However, this concept has been used to explain both the compensation behaviour and the anomalously large value of σ_0 [Kemeny and Goklany 1973, Kaplan and Mahanti 1975].

1.6.4. The anisotropy

It is well known that the conductivity of aromatic hydrocarbons with the applied electric field parallel to the *ab* plane is larger than that with the field perpendicular to this plane. For aromatic hydrocarbons it is generally believed that the anisotropy is due to the directional dependence of the overlapping of π-orbitals between neighbouring molecules, and therefore this phenomenon may be associated with the carrier mobility rather than the carrier concentration [Inokuchi *et al.* 1961, 1962]. In fact, the carrier mobility in these substances is strongly anisotropic as shown in Fig. 1.6. The association of the π-electron overlapping with the anisotropy is further confirmed by the experimental fact that the photoconductivity in these organic crystals such as in anthracene is strongly dependent on the crystallographic direction [Hasegawa 1964, Compton *et al.* 1957], and that the activation energy of the conductivity is isotropic [Inokuchi *et al.* 1961, Kepler *et al.* 1960, Kronick and Labes 1961, Hänsel 1970]. Most organic crystals exhibit an anisotropic conductivity, such as the triphenylphosphonium

(TCNQ)$_2$ complexes and other DA complexes, the organometallic compounds, and the organic compounds with intermolecular hydrogen bonds, etc. [Meier 1974].

1.6.5. The frequency dependence

Many investigators [Eley and Parfitt 1955, Garrett 1959, Huggins and Sharbaugh 1963, Storbeck and Starke 1965] have reported that the conductivity of some organic semiconductors under a.c. fields may be higher by several orders of magnitude than under d.c. fields. This phenomenon may be associated with the polarization of the molecules. However, in films of polymeric complexes of tetracyanoethylene the conductance increases with increasing frequency up to about 10 MHz, and then becomes independent of frequency from 10 MHz to 200 MHz; and these films possess an extremely high effective dielectric constant which decreases with increasing frequency [Boguslavskii and Vannikov 1970]. A similar phenomenon of frequency dependent conductance has been observed in radiation- and heat-treated polyethylene [Tanaka and Fan 1963, Boguslavskii and Vannikov 1970] and in many non-crystalline solids [Tauc 1974, Mott and Davis 1971]. A study of organic specimens with different specific conductivities has revealed that the lower the specific conductivity the stronger is the frequency dependence of the conductivity [Boguslavskii and Vannikov 1970]. Both the dielectric constant and the specific conductivity can be considered to be associated with the carrier transport.

In general, the conductivity decreases with increasing frequency in the case of band conduction process, while it increases with increasing frequency in the case of hopping conduction process [Pohl 1967, Pollak 1962, Blagodarov *et al.* 1970]. Taking this trend as an indication, the hopping conduction process occurs in many organic semiconductors [Boguslavskii and Vannikov 1970, Brenig *et al.* 1972, Buravov *et al.* 1970]. In some cases, for example, in phthalocyanine films, the band conduction occurs at high temperatures and the hopping transport becomes dominant at low temperatures [Vidadi *et al.* 1969].

In organic polymeric semiconductors there are two stages in the movement of a current carrier—motion within the macromolecule and passage from one macromolecule to another, that is, the intramolecular and intermolecular transfer of the current carrier. The comparatively high conductivity and low activation energy of the intramolecular transfer do not appear in d.c. measurements but may appear in a.c. measurements. If we assume that free carriers appear only within macromolecules and can be detected only in a.c. measurements, and that the carriers are retarded by the barrier between macromolecules, we can predict the frequency dependent conductivity by means of the Maxwell–Wagner two-layer condenser model [von Hippel 1954].

In fact, the available experimental results about the frequency dependence of a.c. conductivity have revealed a considerable similarity of behaviour for a very wide range of materials, ordered and disordered, conducting by electrons, holes, ions, polarons, etc., involving various types of chemical bonds and various electronic energy level structures [Jonscher 1972, 1973]. The frequency dependence of a.c. conductivity can be expressed by the empirical relation

$$\sigma_{a.c.}(\omega) \propto \omega^n \qquad (1.69)$$

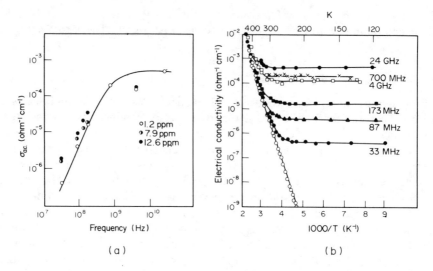

FIG. 1.13. (a) The frequency dependence of electric conductivity of Li-doped NiO in the low
temperature range where $\sigma \simeq \sigma_{\text{a.c.}}$ in eq. (1.70). The solid curve is calculated from
eq. (1.70) with $\tau = 2.2 \times 10^{-10}$ sec. (b) The d.c. and high frequency conductivities
of 1.2 ppm Li-doped NiO. [After Kabashima and Kawakubo 1968.]

where n is not a constant for all substances, but is a function of temperature,
approaching unity at low temperatures and decreasing to 0.5 or less at high
temperatures. Typical results are shown in Fig. 1.13 for lithium-doped NiO crystals and
in Fig. 1.14 for stearic acid films. In Fig. 1.13(a) the frequency dependence for three
different lithium concentrations follows the relation

$$\sigma_{\text{a.c.}}(\omega) = [\sigma(\omega) - \sigma_{\text{d.c.}}] \propto \frac{\omega^2}{\omega^2 + 1/\tau^2} \tag{1.70}$$

where $\sigma(\omega)$ and $\sigma_{\text{d.c.}}$ are, respectively, the total and d.c. conductivities, and τ the
relaxation time. This type of frequency dependence is closely related to the frequency
dependence of the real part of the permittivity

$$\left[\frac{\varepsilon(\omega)}{\varepsilon_0} - \frac{\varepsilon(\infty)}{\varepsilon_0} \right] \propto \frac{1}{\omega^2 + 1/\tau^2} \tag{1.71}$$

where $\varepsilon(\infty)$ is the value of $\varepsilon(\omega)$ at frequencies $\omega \gg 1/\tau$. And in Fig. 1.13(b) it can be seen
that the activation energy is the highest for the d.c. conductivity and decreases with
increasing frequency. This substance gives ω^2 dependence followed by saturation. But
for other substances, for example, organic ferroelectric crystals of triglycene sulphate
give $n = 0.8$ in the frequency range from 10^{-3} Hz to 10^5 Hz, multimolecular layer films
of stearic acid $CH_3(CH_2)_{16}COOH$ give $n < 1$ [Nathoo and Jonscher 1971, Jonscher
1973]. The general experimental facts are:

(a) $\sigma_{\text{a.c.}}(\omega)$ either follows a power law of eq. (1.69) with the exponential $n < 1$ or
     follows a ω^2 dependence followed by saturation according to eq. (1.70). It is

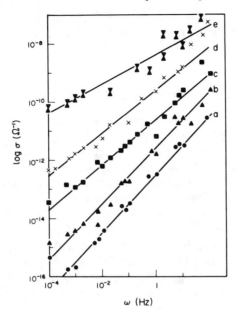

F IG. 1.14. The frequency dependence of electric conductance in multimolecular layers of stearic acid at various temperatures.
a, 105° K; *b*, 295° K; *c*, 338° K; *d*, 362° K; *e*, 373° K. Film
thickness, 270 Å; area, 1 mm². [After Nathoo and Jonscher
1971.]

possible that these two types of frequency dependence are superimposed for
some cases.

(b) There is no correlation between $\sigma_{a.c.}(\omega)$ and $\sigma_{d.c.}$.

(c) $\sigma_{a.c.}(\omega)$ has a weaker temperature dependence than $\sigma_{d.c.}$. Generally, n decreases
with increasing temperature. The ω^2 dependence is not very sensitive to
temperature.

Jonscher [1972, 1973] has proposed a stochastic model of hopping in time and in
space to explain physically the observed frequency dependent behaviour, since the time
and frequency dependences of the electric current are intimately related through the
Fourier transformation. The ω^n dependence of conductivity can be considered as an
indication of the existence of a wide distribution of transition probabilities for charge
carriers, and the ω^2 dependence of conductivity can be considered as evidence of the
presence of molecular dipolar loss mechanisms or two-centre hopping. In the limit of
the high frequency ω^2 dependence, $\sigma_{a.c.}(\omega)$ is practically independent of temperature,
and this may be explained in terms of tunnelling transitions between closely regularly
spaced sites. The weak temperature dependence at low frequencies may be due to the
effect of the increasing thermal activation of the rate-limiting hopping process.
Furthermore, carrier transfer through interfaces either between the specimen and
electrodes or between grain boundaries in the bulk may also contribute to ω^n
dependence. A Maxwell–Wagner system with a very wide distribution of conductive

and capacitive domains (conducting particles dispersed in a non-conducting matrix instead of a two-layer model) to represent a heterogeneous structure can show a similar ω^n dependence [Jonscher 1973].

1.7. THERMOELECTRIC EFFECTS

The measurement of thermoelectric power or Seebeck coefficient as a function of temperature is one of the important methods for investigating electronic properties of solids. It provides information on the concentration of charge carriers and the mechanisms of electric conduction. When a material specimen is subjected to a temperature gradient under a condition in which no current is drawn, a thermoelectromotive force or a potential difference is developed between terminals B and C as shown in Fig. 1.15(a). This phenomenon was first discovered by Seebeck in 1822, and is generally referred to as the Seebeck effect. This effect for n-type materials can be easily understood as a result of the diffusion of electrons from the high temperature end to the low temperature end causing the build-up of positive space charge near the high temperature end and negative space charge near the low temperature end which sets up an electric field (or a potential difference) to make the net current flow zero. A similar

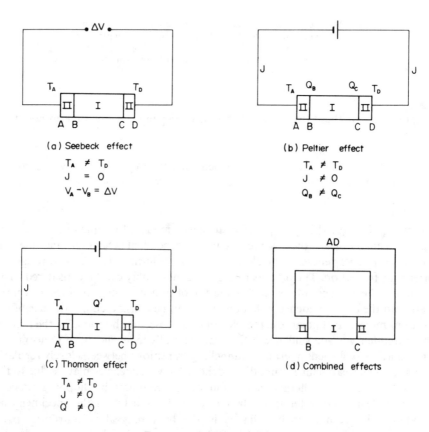

FIG. 1.15. Thermoelectric effects.

process appears in *p*-type materials. The thermoelectric power *S*, which is sometimes called the Seebeck coefficient, is defined by

$$S = \lim_{\Delta T \to 0} \frac{\Delta V}{\Delta T} = \frac{dV}{dT} \tag{1.72}$$

where ΔV is the potential difference produced due to the temperature difference ΔT between two points. To measure ΔV across the specimen *BC*, we need a conducting contact of thickness *AB* at one end anu a conducting contact of thickness *CD* at the other end, as shown in Fig. 1.15. Theoretically, thermoelectric power would be produced in any materials including highly conductive materials if there exists a temperature gradient. However, for good conducting materials with a high thermal conductivity such as copper, gold, etc., the potential differences across *AB* and *CD* are negligibly small as compared with ΔV across the semiconductor specimen *BC*. If we keep the temperature difference between *B* and *C* small enough to maintain a constant temperature gradient across the specimen, then

$$S = \frac{\Delta V}{\Delta T} \bigg]_{\text{between } B \text{ and } C} = \frac{\Delta V}{\Delta T} \bigg]_{\text{between } A \text{ and } D} \tag{1.73}$$

The second thermoelectric effect is the Peltier effect discovered by Peltier in 1834. This effect is in fact the inverse of the Seebeck effect. In this effect, a steady current passing through a semiconductor, whose temperature is initially uniform throughout, will set up a temperature difference between *B* and *C* due to heat evolved at one junction (cold junction) and heat absorbed at the other (hot junction). The Peltier effect is usually expressed as the Peltier coefficient Π which is defined by

$$\Pi = \frac{Q}{J} = ST \tag{1.74}$$

where *Q* is the heat evolved per second at the junction and *J* is the current flowing through the specimen. Π is actually the measure of the total free energy carried by the charge carriers during the current flow and is therefore dependent on the direction of the flow of the majority carriers. Thus, each electron contributed to Π is proportional to its relative contribution to the total conductivity. For two junctions as shown in Fig. 1.15(b) we can write

$$\Pi_{BC} = \Pi_B - \Pi_C = \frac{Q_B - Q_C}{J} \tag{1.75}$$

The third thermoelectric effect is the Thomson effect discovered by Thomson in 1857. This effect is the generation or absorption of heat Q' due to the current flow through a homogeneous semiconductor (or any homogeneous material) in which there is a temperature gradient $\partial T / \partial X$. For two junctions as shown in Fig. 1.15(c), Q' is defined by

$$Q' = \tau J \frac{\partial T}{\partial x} \tag{1.76}$$

where *J* is the current and τ is known as the Thomson coefficient. τ is positive if heat is absorbed when the conventional current and $\partial T/\partial x$ are in the same direction, and it is negative if heat is evolved when J and $\partial T/\partial x$ are in opposing directions.

These three thermoelectric effects have been experimentally established. If we connect the two materials I and II to form a closed circuit as shown in Fig. 1.15(d) with a temperature difference dT between the junction B and C, then we can equate the sum of all types of heat energy generated and absorbed per unit time in the circuit to zero under steady state conditions according to the law of energy conservation. The Peltier heat in watts generated at one junction (say, junction B) due to a current dJ flowing through it is $dQ_B = (\Pi_{II \to I})_B dJ$ while the heat absorbed at the other junction (say, junction C) is $dQ_C = (\Pi_{I \to II})_C dJ$; we have

$$\Pi_{I \to II} dJ = \Pi_{II \to I} dJ - \left(\frac{\partial \Pi_{II \to I}}{\partial T} dT \right) dJ \tag{1.77}$$

The Thomson heat in watts generated in material II is $dQ'_{II} = \tau_{II} dJ dT$ while the heat absorbed in material I is $dQ'_I = \tau_I dJ dT$. The electric power consumed in the circuit due to Seebeck effect is $S dT dJ$. Therefore, the last term in eq. (1.77) can be written as

$$\frac{\partial \Pi_{II \to I}}{\partial T} dT dJ + (\tau_{II} - \tau_I) dJ dT = S dT dJ$$

or

$$\frac{\partial \Pi_{II \to I}}{\partial T} + \tau_{II} - \tau_I = S \tag{1.78}$$

Equation (1.78) expresses a relationship between the coefficients S, Π, and τ. Another relationship between the three coefficients may be derived from the second law of thermodynamics by considering that the thermoelectric processes are reversible. Then at any junction we have the following relationship [Ioffe 1960]:

$$\frac{d\Pi}{dT} - \frac{\Pi}{T} = \tau_{II} - \tau_I \tag{1.79}$$

The combination of eqs. (1.78) and (1.79) leads to

$$S = \frac{\Pi}{T} \tag{1.80}$$

This is the first of Kelvin's relationships to relate the Seebeck and Peltier coefficients. Differentiating eq. (1.80) with respect to T

$$\frac{dS}{dT} = \frac{T(d\Pi/dT) - \Pi}{T^2}$$

$$= \frac{1}{T} \left(\frac{d\Pi}{dT} - \frac{\Pi}{T} \right) \tag{1.81}$$

and substitution of eq. (1.79) into eq. (1.81) yields

$$\frac{dS}{dT} = \frac{\tau_{II} - \tau_I}{T} \tag{1.82}$$

Because of the closed circuit we may write

$$\frac{d}{dT}(S_{II} - S_{I}) = \frac{\tau_{II} - \tau_{I}}{T} \tag{1.83}$$

Thus, eq. (1.83) is the second of Kelvin's relationships to relate the Seebeck and Thompson coefficients, and can be written as

$$\frac{dS}{dT} = \frac{\tau}{T} \tag{1.84}$$

From eqs. (1.80) and (1.84), the Thomson heat generated in a closed circuit is

$$Q' = (\tau_{II} - \tau_{I}) J (T_B - T_C) \tag{1.85}$$

and the Peltier heat is

$$Q = S J T \tag{1.86}$$

The thermoelectric effects involve both charge carrier and energy transport. In metals the change of temperature does not change the concentration of electrons but brings about only a slight redistribution of their thermal velocities, and it is obvious that a large thermoelectromotive force cannot arise. But in semiconductors the change of temperature causes a change in both the concentration and the kinetic energy of mobile charge carriers. Supposing that the temperature at one end is higher than that at the other of an n-type semiconductor, the number of electrons diffusing from the hot end to the cold end is more than that in the opposite direction at the start, but the difference in the electron flow between these two opposite directions gradually decreases to zero as an internal potential difference is gradually set up between the two ends until the electron current due to the diffusion process is equal to the reverse electron current caused by such a potential difference. Thus, under such a dynamic equilibrium, the net flow of electrons per unit time passing through any cross-section of the semiconductor in both directions is zero. However, the velocities or the kinetic energies of electrons from the hot end are higher than those from the cold end through the same cross-section, and this results in a continuous transfer of heat energy in the direction of the temperature gradient without involving actual charge transfer.

In general, measurements of the thermoelectric power provide useful information about the mechanisms of electrical transport. The polarity of the thermal e.m.f. indicates the type of dominant carriers or the type of dominant electric conduction. The thermoelectric power S is negative if the majority carriers are electrons and the polarity of the thermal e.m.f. at the hot end is positive such that

$$S = \frac{dV/dx}{dT/dx} < 0 \quad (n\text{-type})$$

and S is positive if the majority carriers are holes and the polarity of the thermal e.m.f. at the hot end is negative such that

$$S = \frac{dV/dx}{dT/dx} > 0 \quad (p\text{-type})$$

The quantity S may be used to determine the mobility ratio, the concentration of carriers, the position of the Fermi level, etc., and it has been used for transport studies by many investigators [Smith 1964, Johnson 1956, Johnson and Lark-Horowitz 1953, Cutler and Mott 1969, Mott and Davis 1971, Cutler 1972]. Supposing that an *n*-type semiconductor forms a junction with a metal and that the metal work function ϕ_m is larger than the semiconductor work function ϕ_s, then at equilibrium the Fermi level must be the same in both materials and the electrons that flow from the metal to the semiconductor must absorb an amount of energy $E_C - E_F$ at the junction in order to reach the edge of the conduction band of the semiconductor. Moreover, in the semiconductor, the electrons which carry the current have a certain mean kinetic energy $\langle E \rangle$ so that the total energy change at the junction is $(E_C - E_F) + \langle E \rangle$. The Peltier coefficient defined by eq. (1.74) can also be interpreted as the mean energy transported per unit charge. Thus we can write

$$\Pi = -\frac{(E_C - E_F) + \langle E \rangle}{q} \tag{1.87}$$

The mean kinetic energy $\langle E \rangle$ of the charge carriers depends on the precise form of the density of states distribution function and on the nature of the scattering processes. From eq. (1.74) we can write the thermoelectric power as

$$S = \frac{\Pi}{T} = -\frac{(E_C - E_F) + \langle E \rangle}{qT}$$

$$= -\frac{k}{q}\left[\frac{E_C - E_F}{kT} + \frac{5}{2} + \gamma\right] \tag{1.88}$$

for *n*-type semiconductors, and, similarly, we can write

$$S = \frac{k}{q}\left[\frac{E_F - E_V}{kT} + \frac{5}{2} + \gamma\right] \tag{1.89}$$

for *p*-type semiconductors, where

$$\gamma = \frac{d(\ln \tau)}{d(\ln E)} \tag{1.90}$$

τ is the relaxation time between the collisions having a defined mean free path. For intrinsic or non-degenerate semiconductors involving two types of charge carriers, S is given by

$$S = -\frac{k}{q}\left\{\left[\frac{E_c - E_{Fn}}{kT} + \frac{5}{2} + \gamma\right]n\mu_n - \left[\frac{E_{Fp} - E_V}{kT} + \frac{5}{2} + \gamma\right]p\mu_p\right\}/(n\mu_n + p\mu_p) \tag{1.91}$$

In the case of predominantly lattice scattering, eqs. (1.88), (1.89), and (1.91) become

$$S = -\frac{k}{q}\left[\frac{E_C - E_F}{kT} + 2\right] = -\frac{k}{q}\left[\ln\frac{N_c}{n} + 2\right] \tag{1.92}$$

for *n*-type non-degenerate semiconductors,

$$S = \frac{k}{q}\left[\frac{E_F - E_V}{kT} + 2\right] = \frac{k}{q}\left[\ln\frac{N_v}{p} + 2\right] \tag{1.93}$$

for p-type non-degenerate semiconductors, and

$$S = -\frac{k}{q}\left[\frac{(\mu_n/\mu_p)-1}{(\mu_n/\mu_p)+1}\right]\left[\frac{E_g}{2kT}+2\right] \tag{1.94}$$

for intrinsic semiconductors.

In non-degenerate semiconductors, n-type or p-type, S decreases with increasing carrier concentration. S is small for intrinsic and for degenerate semiconductors since in both cases the charge is transported approximately equally by electrons travelling in opposite directions. Although for covalent lattices $\langle E\rangle/kT = \frac{5}{2}+\gamma = A_r = 2$ and for ionic lattices $A_r = 2.5$ to 4, in organic molecular complexes, values of A_r ranging from 0.7 to 17 have been reported [Gutmann and Lyons 1967]. Since the scattering process alone cannot account for values of A_r greater than 4, several investigators [Labes *et al.* 1960, 1962, Gutmann and Lyons 1967] have reported that the large values of A_r may indicate that there exist discrete conductive states localized at lattice sites which charge carriers can occupy by means of a hopping process. In general, S is expected to be temperature dependent. But actually S is proportional to T at low temperatures, is nearly constant at intermediate temperatures, and becomes proportional to $1/T$ only at high temperatures, and such a kind of temperature dependence of S has been observed in many inorganic semiconductors (e.g. V_2O_5, Cu_2O, SiC, etc.) [Ioffe 1957, 1960] and in a number of molecular complexes [Gutmann and Lyons 1967]. It should be noted that a calculation and an accurate measurement of the thermoelectric power of semiconductors are rather difficult because internal polarization[Kallmann and Pope 1958] and space charge layers [Lyons 1963] would affect the results.

If the temperature dependence of the energy gap E_g follows the relation

$$E_g = E_{g0} - aT \tag{1.95}$$

substitution of eq. (1.95) into eq. (1.94) gives

$$E_{g0} = -2q\left(\frac{b+1}{b-1}\right)\frac{dS}{d(1/T)} \tag{1.96}$$

where E_{g0} is the energy gap at $T = 0$, a is the temperature coefficient of the activation energy E_g, and $b = \mu_n/\mu_p$, which is assumed to be independent of temperature. If E_{g0} is known, b can be determined from eq. (1.96). For certain purposes it is convenient to write S as a function of the dark electric conductivity σ. For extrinsic semiconductors S can be written as

$$\begin{aligned}S &= C_{\text{ext}} -\frac{k}{q}\ln\sigma\\ &= C_{\text{ext}} - 86\times 10^{-6}\ln\sigma = C_{\text{ext}} - 2\times 10^{-4}\log\sigma\end{aligned} \tag{1.97}$$

and for intrinsic semiconductors S can be written as

$$\begin{aligned}S &= C_{\text{int}} -\frac{k}{q}\left(\frac{b-1}{b+1}\right)\ln\sigma\\ &= C_{\text{int}} - 2\times 10^{-4}\left(\frac{b-1}{b+1}\right)\log\sigma\end{aligned} \tag{1.98}$$

Brennan *et al.* [1962] have reported that the thermoelectric power in pyrolyzed polyacrylonitrile (PAN) follows eq. (1.98) rather than eq. (1.97), indicating that this material is an intrinsic semiconductor if a band model is applicable. On the basis of eqs. (1.94) and (1.95) they have obtained for PAN-675, $b = 1.03$ and $a = -2.9 \times 10^{-3}$ eV deg^{-1} using $E_{g0} = 0.2$ eV from the thermoelectric power measurements, and have predicted the carrier concentration, carrier mobility, and carrier effective mass from the dark electric conductivity measurements. Hamann [1967] has measured the thermoelectric power in copper phthalocyanine single crystals and some of his results are shown in Fig. 1.16. It can be seen that the value of S depends on temperature as well as on the treatment of the specimens. He has proposed a trapping model to explain his results and deduced from this model the following parameters: the copper phthalocyanine is an intrinsic semiconductor having an activation energy of 2 eV and involving the presence of the order of 10^{13}–10^{14} electron traps per cm^3 located at 0.88 eV below the conduction band.

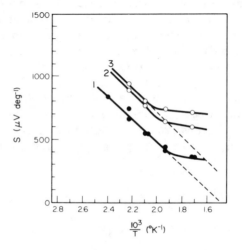

FIG. 1.16. Thermoelectric power in copper phthalocyanine single crystals as a function of temperature. 1, in high-purity nitrogen; 2, in air; 3, in air after annealing in air at 310°C. [After Hamann 1967.]

S generally decreases with pressure and, in some cases, S changes sign as the pressure is increased. This pressure effect may be due to a phase transition or due to the change in carrier concentration or mobility of electrons or holes, or both caused by a specific influence of pressure. In the case of bipolar conduction, S can be expressed as [Meier 1974]

$$S = \frac{(E_p p \mu_p) - (E_n n \mu_n)}{qT(p\mu_p + n\mu_n)}$$

1.99)

where E_p and E_n denote the total energy (potential and kinetic energy) transported by holes and electrons, respectively, from the hot to the cold electrode. Since the potential energy corresponds to the energy difference between the Fermi level and the band edge, and the kinetic energy $A_r kT = \langle E \rangle$, which depends on the scattering processes, E_p and

E_n may be written as

$$E_p = E_F - E_V + A_r\, kT \tag{1.100}$$

$$E_n = E_C - E_F + A_r\, kT \tag{1.101}$$

Polymers with indophenine [Meier 1974], phthalocyanine [Vaisnys and Kirk 1966], and pyrene-1.8 Br_2 complex [Andersen *et al.* 1966] exhibit the pressure effect. At atmospheric pressure the pyrene-1.8 Br_2 shows a dominant *n*-type conduction, but at high pressures the majority carriers become holes changing the conduction from a dominant *n*-type to a dominant *p*-type. This may be due to a pressure-induced orientation of the crystalline, thus weakening the trapping effect [Meier 1974].

The thermoelectric effect can be utilized to make a thermoelectric generator or a thermoelectric cooling device [Ioffe 1960]. The efficiency of the thermoelectric generators or devices depends on (a) the thermo-e.m.f. *S*, (b) the electric conductivity σ, and (c) the thermal conductivity *K*. By introducing the quantity Z_T [Ioffe 1960],

$$Z_T = \frac{\sigma S^2}{K} \tag{1.102}$$

the higher the value of Z_T, the higher is the efficiency. The thermoelectric powers in organic semiconductors reaching 1–3 mV $°K^{-1}$ have been observed, for example, in phthalocyanine [Meier 1974], but their efficiency is small because of low electric conductivity. However, the efficiency may be improved by polymerization or doping [Meier 1974]. Several investigators [Katon 1966, Wildi 1966, Vozzhennikov *et al.* 1969] have reported such possibilities of improving the thermoelectric efficiency.

1.8 CARRIER GENERATION

Excluding the impact ionization process, which may lead to destructive breakdown of the material specimen, the following mechanisms are, in general, responsible for carrier generation which in turn determines the dark and photoconductivity of the material.

1.8.1. Injection of carriers from electrodes

Electrons injected from the cathode or holes injected from the anode result in space charge conduction. The injection may be thermionic, or quantum mechanic tunnelling, or both, or via surface states, and the details of these mechanisms will be given in Chapter 2.

Electrons or holes may be emitted from illuminated electrodes into a solid specimen, and this process is generally referred to as the photoemission from electrodes and may result in an increase in photoconductivity in the same manner as the contribution of carrier injection from electrodes to the dark electric conductivity.

1.8.2. Intrinsic excitation

Electrons and holes may be generated by thermal excitation or photoexcitation from the valence band to the conduction band in organic semiconductors in a similar manner to that occurring in inorganic semiconductors. The dark conductivity of many organic compounds depends on the molecular structure and increases with increasing number of π-electrons in the individual groups of the compounds. Although this indicates indirectly that the carrier generation is intrinsic and related to the band gap, this is not sufficient to rule out other possible conduction mechanisms. The singlet or triplet states, for example, can be populated either thermally or optically, and then such excitons may give rise to carriers by exciton collisions with electrodes or any boundaries. The long wavelength photoconduction observed in many organic compounds has been attributed to either a $\pi \rightarrow \pi^*$ (transition of an electron in a bonding π-molecular orbital to an antibonding π^* molecular orbital) or $n \rightarrow \pi^*$ (transition of a non-bonding n-electron to an antibonding π^* molecular orbital) excitation. According to Rosenberg [1961, 1962], the activation energy for the generation of charge carriers may be associated with the energy for the formation of the triplet state and some vibrational energy levels of a molecule. In many organic compounds the optical activation energy is essentially equal to the thermal activation energy of dark conductivity and agrees with the excitation energy of the $\pi \rightarrow \pi^*$ transition. For this case both the dark and photoelectric conductions may be considered to be intrinsic [Rexer 1966, Terenin 1961, Dyne *et al.* 1965, Baverstock *et al.* 1970]. However, in some organic compounds the optical activation energy agrees with the excitation energy of the $\pi \rightarrow \pi^*$ transition but differs from the thermal activation energy of dark conductivity [Sano and Akamatu 1962, Meier *et al.* 1969, Kulshreshtha and Mookherji 1970]. For this case it is possible that the photoconduction is intrinsic, while the dark conduction is extrinsic. In spite of the agreement between the electron transition energy and the photo- or dark conductivity, the mechanism of triplet participation in the generation of charge carriers has been commented upon by Kleinerman *et al.* [1962].

The intrinsic photogeneration processes can be summarized as follows:

(i) *One-quantum processes*—the one-quantum generation means that only one photon or exciton is involved in a single interaction process. (a) *Direct ionization*. This process implies that the absorbed photons create directly free electron–hole pairs without intermediate steps which involve excitons. In anthracene, photo energy in the range of 4.1–5.0 eV is required to generate directly mobile electrons and holes [Castro and Hornig 1965, Chaiken and Kearns 1966]. (b) *Indirect ionization*. This process implies that the absorption of a quantum produces an exciton, and the carriers are then generated through a reaction of the exciton with or without a defect [Lyons 1957, 1963]. The generation involving collision with impurities or surfaces is not considered as intrinsic generation. The exciton may move between neighbours in a way similar to a charge transfer process, and during the transfer the thermal or field dissociation of the exciton to form free carriers may occur.

(ii) *Two-quantum processes*—the two-quantum generation means that two photons, or two excitons, or one photon and one exciton are involved in a single interaction process. Several possible mechanisms may take place, but the most probable mechanism for the carrier generation in the bulk of molecular crystals is the collision of two singlet excitons [Choi and Rice 1963, Choi 1967, Johnston and Lyons 1970, Kepler 1971, Lavrushko and Benderskii 1971, Silver *et al.* 1963, Silver 1971, Hasegawa and Yoshimura 1965]. However, Kearns [1963] has studied the relative rates of charge carrier generation from two molecular excitons when the excitons are (a) both singlet,

(b) both triplet, and (c) one singlet and one triplet, and shown that (c) gives the higher carrier generation rate.

(iii) *Multi-quantum processes*—the multi-quantum generation means that more than two quanta (each of which can be either a photon or an exciton) are involved in a single interaction process. Carrier generation by a three-quantum process has been reported by several investigators [Singh *et al.* 1965, Kepler 1971] involving the photoionization of singlet exciton states.

1.8.3. Extrinsic excitation

The excitation of doped impurities to contribute the major carriers is well known in inorganic extrinsic (*n*-type or *p*-type) semiconductors. A similar extrinsic conduction occurs in organic semiconductors. Naphthacene (tetracene) doped in an anthracene crystal creates two levels (a donor level and an acceptor level) in the energy band gap [Gutmann and Lyons 1967]. That the doping impurities increase the carrier concentration in poly-*N*-vinylcarbazole, phthalocyanines, and other organic semiconductors [Meier *et al.* 1969, Meier 1974] is a good indication of extrinsic excitation. Trapping centres may be associated with impurities or structural defects. Trapping of charge carriers is a well-known phenomenon (see Chapter 3), and detrapping through thermal or optical excitation is, of course, a source of carriers, but this source would be exhausted unless the traps are refilled continuously.

The collision of excitons with impurities or surfaces will generate carriers. The detrapping of a trapped carrier by a photon or an exciton has been reported by many investigators [Hasegawa and Schneider 1964, Kokado and Schneider 1964, Adolph *et al.* 1964].

1.8.4. Field-assisted generation

The electric field may separate the generated carriers before geminate recombination due to mutual Coulombic interaction (Onsager effect) [Onsager 1938]. The electric field may also modify the potential barrier profile of a trap to make it easy for the trapped carrier to be liberated from the trap (Poole–Frenkel effect) [Ieda *et al.* 1971, Arnett and Klein 1975]. Although avalanche ionization under a high field is unlikely to occur in a low mobility molecular crystal even at its breakdown strength, the mutual enhancement between the electric and thermal conduction processes may cause a rapid increase in the production of thermally generated carriers [Kao 1976, Kao and Rashwan 1978, O'Dwyer 1973].

1.9. FORMATION OF TRAPS

In general, there exist a greater number of trapping levels in the forbidden energy gap of an insulator or a semiconductor. These trapping levels consist of many localized states which may act as traps or recombination centres [Rose 1951, 1963, Milnes 1973]. They are mainly due to imperfections in a crystal caused either by structural defects or impurities or both. Some general considerations have been suggested to relate the discrete trapping levels with chemical impurities introduced into the lattice (chemical traps) and to relate the quasi-continuous trapping level distribution with the imperfection of the crystal structure (structural traps). We now discuss briefly the origin of such imperfections as follows.

1.9.1. Structural defects excluding impurities

Studies of trapping mechanisms based on thermally stimulated conductivity (TSC) measurements [Parkinson *et al.* 1974, Williams 1974], photostimulated conductivity (PSC) measurements [Aris *et al.* 1974], and isothermal decay current (IDC) measurements [Sworakowski 1974] have revealed that the various depths of electron and hole traps are associated with physical rather than chemical defects. It should be noted that physical defects could be produced by the introduction of foreign chemical impurities such as molecules of inert gas used during crystal growth, which do not themselves trap the carriers [Owen *et al.* 1974]. The polarization energy due to a charge carrier localized near an imperfection site in the crystal is different from that in a perfect lattice, thus resulting in trapping of the carrier in the imperfection site [Sworakowski 1970, 1973, 1974, 1976, Silinsh 1970, Munn 1975]. The formation, the depths, and the concentration of traps are directly dependent on the conditions under which the crystals are grown, the structure of the crystals, and subsequent handling of the crystal specimens [Thomas *et al.* 1965, Lupien *et al.* 1972, Weston *et al.* 1973]. Some typical experimental results of carrier traps in anthracene crystals are shown in Table 1.11. It is important to note that as the methods adopted for preparing the crystal specimens play a very significant role in the experimental results, caution is needed when comparing experimental results reported by different investigators. Structural imperfections are present in crystals in the form of fissures, crystalline boundaries, grain boundaries, dislocations, gaps, and point defects [Bullmann 1970, Thomas *et al.* 1972, Crawford and Slifkin 1975]. We shall discuss them as follows:

(A) Point defects

The formation of point defects without impurities is due to the occupation of interstitial sites by the structural units (molecules, atoms, or ions normally located in the lattice) with vacancies left behind in the lattice (Frenkel defects) or due to the migration of the structural units to the surfaces of the crystal with vacancies left behind in the lattice (Schottky defects). Such defects may occur when the stoichiometric ratio of a compound such as Na/Cl in NaCl crystal is disturbed, or the lattice dislocations are present. In molecular crystals Schottky defects rather than Frenkel defects are more likely to occur because of the large size of the molecules. In general, the energy required to create a vacancy is less than that required to create an interstitial occupation since the latter is always accompanied with a considerable lattice distortion. Supposing that only vacancies (point defects) predominate in an ionic crystal, charge neutrality (and stoichiometry, if there are no impurities) requires an equal number in both anion and cation vacancies. A positively charged vacancy may act as a trap and capture a free electron or a negatively charged vacancy act as a trap and capture a free hole to achieve local charge neutrality. This trapped electron or hole may be excited or detrapped by

TABLE 1.11. *Carrier traps in anthracene crystals grown by various methods and under various conditions*

Specimen preparation	Trap density H_b (cm^{-3})	l (from $J \propto V^{l+1}$)	kT_c (eV)	Trapping level E_t (eV)	Trap density N_t (cm^{-3})	Reference
Melt-grown	1.5×10^{19}	1.4	0.035			A. B
	2.73×10^{18}	1.5	0.0375			
	1.5×10^{18}	1.64	0.041			
	3.5×10^{13}	4.8	0.12			
	1.1×10^{17}	1.0	0.25			
Vapour-grown	1.2×10^{17}	2.28	0.057			A. B
	1.1×10^{13}	2.5	0.0625			
	3.6×10^{15}	2.6	0.065			
Solution-grown	5.75×10^{16}	1.9	0.0475			A. B
	1.75×10^{15}	4.2	0.13			
	4.2×10^{14}	5.6	0.14			
Melt-grown (zone refined)				0.53 ± 0.03	1.5×10^{19}	B. C. D
Doped with tetracene				0.43		E
Doped with perylene				0.25		E, F

A. Reucroft and Mullins [1973].
B. Thomas *et al.* [1968].
C. Schadt and Williams [1969].
D, Sworakowski and Pigon [1969].
E, Hoesterey and Letson [1963].
F. Sworakowski and Mager [1969].

the absorption of a photon of approximate energy. Thus, charge carriers trapped at defects may lead to optical absorption. Colour centres (such as *F* centres) in an alkali halide are formed by such a trapping of electrons at negative ion vacancies (a positively charged vacancy means the vacancy with a negative ion missing or called the negative ion vacancy). The coloration (e.g. yellow for NaCl and blue for KCl with *F*-centres) is due to the presence of a characteristic absorption band whose absorption peak lies in or near the visible region (e.g. the colour of the light for which the coloured alkali halides are transparent).

In inorganic crystals with either metallic, ionic, or valence bonds, the basic properties of point defects are well understood [Crawford and Slifkin 1975]. However, in molecular or organic crystals the published work on point defects is scarce. A study of self-diffusion in anthracene crystals was first reported by Sherwood and Thomson [1960]. Since then, a number of studies of lattice defects and defect-controlled properties of both organic and inorganic molecular solids have been reported, and a brief account of these studies has been given by Chadwick and Sherwood [1975]. Using the self-diffusion technique, it has been shown that vacancies are the major point defects in organic crystals [Sherwood 1969]. The self-diffusion can be described by the empirical equation [Shewmon 1963]

$$D = D_0 \exp(-E_{act}/RT) \qquad (1.103)$$

where E_{act} is the activation energy, D_0 is the pre-exponential factor, and R is constant and independent of temperature. In general, solids, which comprise globular molecules having almost free rotation in the crystal lattice, show a high translational mobility or a high plasticity and a low activation energy; while those comprising large asymmetric molecules which form a lattice of low symmetry have an extremely low self-diffusion rate and hence a high activation energy. For most intrinsic organic solids the ratio of the activation energy E_{act} to the sublimation energy L_S is within the range between 1.6 and 2.5 as shown in Table 1.12. The concentration of radiative tracer as a function of distance x from the source at a given temperature after an annealing period t is given by

$$A = [A_0/(\pi Dt)^{1/2}] \exp(-x^2/4Dt) \qquad (1.104)$$

The measurement of A/A_0 enables the determination of D. In general, the observed diffusion coefficient D_{obs} comprises those in the lattice D_l and in the dislocations D_d

$$D_{obs} = (1-g)D_l + gD_d \qquad (1.105)$$

where g is the fraction of the diffusing tracer in the dislocations. The diffusion

TABLE 1.12. *Values of* $D_{(melting\ point)}$, D_0, E_{act}, *and* E_{act}/L_s *of some organic solids [after Sherwood 1972]*

	$D_{(melting\ point)}$ (m^2 sec^{-1})	D_0 (m^2 sec^{-1})	E_{act} (kJ mole^{-1})	E_{act}/L_s
Anthracene	10^{-16}	1×10^6	202	2.3
Naphthalene	10^{-15}	2×10^{11}	179	2.4
Phenanthrene	10^{-15}	3×10^{13}	202	2.4
Biphenyl	10^{-15}	4×10^{10}	169	2.3
Benzene	10^{-13}	1×10^5	95	2.1
Cyclohexane	10^{-11}	4×10^2	69	1.9

coefficient in the grain boundaries D_g is of the same order in magnitude as that in the dislocations. In order to obtain true lattice diffusion coefficients, D_d, D_g, and g should also be measured.

Table 1.12 shows $D_{(melting\ point)}$, D_0, E_{act}, and E_{act}/L_S for some organic solids. The variation in self-diffusion coefficient is reflected more in the pre-exponential factor than in the activation energy. D_0 increases with increasing size of the molecules in the solid because of the increasing difficulty for self-diffusion in lattice vacancies. D_0 decreases with increasing grain boundaries indicating the increasing easiness for self-diffusion in grain boundaries. Many factors such as the crystal growth rate and the annealing would affect the concentration of vacancy defects. The concentration of vacancies can be estimated by the following equation [Mott and Gurney 1940]:

$$n_s = (N - n_s) \exp\left(-\frac{E_s}{kT}\right)$$

$$\simeq N \exp\left(-\frac{E_s}{kT}\right) \tag{1.106}$$

where N is the total number of lattice points (or the total number of molecules per unit volume) and E_s is the energy required to form a vacancy. By assuming E_s to be approximately equal to the lattice energy, E_s can be expressed as [Meier 1974]

$$E_s \simeq L_s - RT + \frac{9k\,\theta_d}{8} \tag{1.107}$$

where θ_d is the Debye temperature. Equations (1.106) and (1.107) imply that the concentration of vacancies in a crystal increases with increasing temperature and the density of a crystal decreases with increasing vacancy concentration, and that rapid cooling from the melt would produce a high concentration of vacancies, while slow growth and annealing of a crystal reduce the concentration of vacancies [Sherwood 1967]. According to Chadwick and Sherwood [1975], the basic point defects in molecular solids are molecular vacancies. The structure, complexity, and aggregation of such point defects vary considerably from one type of molecular solids to another. The variations reflect the nature of the intermolecular forces in the solids and hence the carrier-trapping energy levels when such defects act as carrier traps.

(B) Other structural defects

It can be imagined that any structural defects which would perturb the periodic nature of the lattice will perturb the motion of charge carriers and hence the electric conductivity. The degree of perturbation depends on the type of structural defects and the defects act generally as traps. Calculations of trap depths for realistic models of defects in crystals are very complicated [Sworakowski 1970, 1973, 1974, Munn 1975] and the discussion of the various models for such calculations is beyond the scope of this book. However, defects directly related to the growth and the structure of the crystal are worthy to be briefly described as follows:

(a) DISLOCATIONS

There are two major types of dislocations: the edge dislocation and the screw dislocation. The dislocation refers to the configuration of atoms which exists at the boundary between a slipped and an unslipped area (one part of the crystal slides across a neighbouring part along a direction, called the slip direction, lying in the surface of slip) and spreads across the crystal as the slipped area grows at the expense of the unslipped one. The boundary line between a slipped and an unslipped area is called the dislocation line. The edge dislocation may be envisaged as an additional half-plane of atoms inserted in the middle of a perfect crystal as shown in Fig. 1.17(a). The horizontal plane in which the additional plane of atoms terminates is called the slip plane. Such an

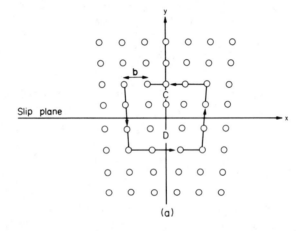

(a)

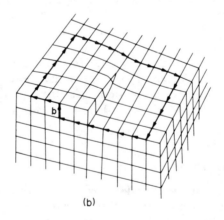

(b)

FIG. 1.17. Schematic diagrams illustrating the edge and screw dislocations. (a) An additional plane of atoms at $x = 0$ and $y > 0$ inserted into the otherwise perfect lattice forms an edge dislocation. Above the slip plane is the compression region and below it the dilation region with the largest strain near the origin. The Burgers vector **b** is perpendicular to the line of the dislocation. (b) Screw dislocation. The Burgers vector **b** is parallel to the line of the dislocation.

additional plane of atoms would create two distinct features governing the electronic behaviour [Matare 1971]:

(i) a lattice deformation or a stress field resulting in a compression zone C just above the dangling bond and a dilation zone D below the free bond;
(ii) the dangling and free bonds tend to form energy states in the forbidden gap.

The screw dislocation may be thought of as a spiral arrangement of atomic planes, with a step up for each revolution. The dislocation may be defined in terms of the Burgers vector. By making a closed circuit in the perfect crystal by means of steps from lattice site to lattice site through Bravais lattice vectors, the sum of the vectors will be zero. The Bravais lattice vector $\mathbf{b}$, by which the endpoint fails to coincide with the starting point, is called the Burgers vector of the location. For the edge dislocation $\mathbf{b}$ is perpendicular to the line of dislocation, while for the screw dislocation $\mathbf{b}$ is parallel to the line of dislocation as shown in Fig. 1.17(b).

Both edge and screw dislocations would not only introduce trapping energy states in the forbidden gap but also change the width of the forbidden gap because of the presence of dangling and free bonds, and local compression and dilation zones. As a result, the electron band structure at the site of a dislocation is rather complicated. If the distance between the individual sites of dislocation is larger than the lattice constant, alternating forbidden-gap widening and narrowing occur [Matare 1971]. The formation of dislocations depends on the method of growing crystals. Crystals grown from the melt contain a very high concentration of dislocations because of the stress created during solidification. For example, anthracene crystals grown by this method contain dislocations of 10^5–10^7 cm^{-2} [Sherwood and White 1967, Williams and Thomas 1967]. But anthracene crystals grown in solution have dislocations of about 10^3 cm^{-2}, and grown in the vapour phase, the dislocation density is further reduced to 10^2 cm^{-2} [Sherwood and White 1967, Williams and Thomas 1967] because less stress is involved during crystal growth for these two methods. Like point defects, the concentration of dislocations can be reduced by an annealing process [Corke *et al.* 1968]. In anthracene Thomas and Williams [1969] have reported that the concentration of traps is proportional to the concentration of dislocations. It is important to reduce the concentration of structural defects in order to study the effects of other factors controlling the electronic properties, such as chemical dopants.

(b) PLANAR DEFECTS

Low angle lineage and grain boundaries between otherwise relatively dislocation free regions often occur in crystals, and they may be considered as a planar array of edge dislocations as shown in Fig. 1.18. In terms of the angle of misfit between the two grains, medium angle grain boundaries are within the range of $1° < \theta < 25°$, the low angle grain boundaries within $0.1° < \theta < 5°$, and the lineage within $0 < \theta < 1°$ [Matare 1971]. In fact, there is no clear-cut demarcation among these definitions. There are also twist boundaries which are formed from a sequence of screw dislocations. In general, low angle grain boundaries are composed of a mixture of both types. Unless very carefully prepared, most real crystals consist of many slightly misaligned grains, separated by low angle grain boundaries.

Fɪɢ. 1.18. Low angle grain boundaries may be thought of as a planar array of dislocations.

Structural defects introduce relatively deep traps or recombination centres which have an exponential distribution of trapping energy levels in the forbidden gap [Sworakowski and Pigon 1969, Sworakowski 1970, 1973, Thomas *et al.* 1968, 1969, 1971, Helfrich *et al.* 1965, Reucroft *et al.* 1973]. The effects of traps on electrical transport are discussed in later chapters.

1.9.2. Defects due to chemical impurities

Chemical impurities, which may be foreign atoms, molecules, or ions, can occupy substitutional or interstitial positions in an otherwise defect free lattice, or be incorporated into a crystal near dislocations to favour the formation of dislocations. The action may also result in the formation of point defects or other structural defects. An impurity is generally different from the host atom or molecule in mass, electronic configuration, and valency. A different mass would affect the lattice vibrations locally, and a different electronic configuration and valency would introduce new electronic energy states in the forbidden gap, thus increasing or decreasing the total conduction carrier concentration in the crystal. It is well known that the substitution of a tetravalent atom by a pentavalent atom (e.g. *P*) in silicon will make the silicon *n*-type conduction, and by a trivalent atom (e.g. *B*) will make the silicon *p*-type conduction. While the above impurities increase the conductivity of the silicon, the doping of gold into silicon leads to the formation of recombination centres, thus reducing the minority carrier lifetime and hence the total conductivity. Similar effects of impurities also occur in ionic and molecular crystals. In general, foreign impurities whose electron affinity is higher than that of the host molecules in the crystal can act as electron traps, while those whose ionization energy is smaller than that of the host molecules can act as hole traps. This may result in the experimental fact that tetracene doped into an anthracene crystal forms shallow discrete traps for both electrons and holes, while anthraquinone and anthrone in anthracene form deep traps for electrons and holes [Hoesterey and Letson 1963].

Impurities may act as trapping or recombination centres and may also interact with host molecules involving in mode of energy transfer through triplet states and other energy levels. The impurity states in molecular crystals have been theoretically treated by several investigators [Merrifield 1963, Craig 1965, Craig and Thirunamachandran 1973, Craig and Philpott 1966, Rashba 1963, Dubovskii and Konobeev 1965, Body and Ross 1965, Rice and Jortner 1967]. A great deal of experimental data about the effects of impurities is available in the literature [Baessler *et al.* 1966, 1969, Sworakowski and Mager 1969, Hamann 1973, Hamann *et al.* 1973, Itoh *et al.* 1973, 1974, Schmillen and Falter 1969, Bogus 1965, 1966, 1967, Northrop and Simpson 1956, 1958, Jones 1968, Morgan and Pethig 1969, Eley *et al.* 1968, Johnston and Lyons 1970, Barbe and Westgate 1970, Sussman 1967, Meier 1974]. Because of limited space, for detailed information about the impurity states and their effects, the reader is referred to the references cited above and also those given at the end of this book.

1.10. LIFETIME AND RELAXATION ELECTRIC CONDUCTION

According to van Roosbroeck [1960, 1961, 1972, 1973], semiconductors can be classified into two distinct types depending on whether the carrier lifetime τ_0 (diffusion length minority carrier lifetime) is greater or smaller than the dielectric relaxation time τ_d. The semiconductors with $\tau_0 > \tau_d$ are generally referred to as the lifetime semiconductors and they have been extensively studied for about four decades, such as germanium and silicon. The semiconductors with $\tau_0 < \tau_d$ are generally referred to as the relaxation semiconductors, examples of which are some high resistivity crystalline solids such as direct gap compound semiconductors (e.g. GaAs), organic semiconductors (e.g. anthracene), and amorphous semiconductors (e.g. chalcogenide glasses).

1.10.1. The concept of the lifetime regime

Before discussing the relaxation regime, it is important to review briefly the physical concept of the lifetime regime of the conventional semiconductors. The condition $\tau_0 > \tau_d$ is often not clearly and precisely defined but, in most cases, it is assumed that the "neutrality condition" prevails in the conventional semiconductors. This implies that the dielectric relaxation is so rapid compared with other time dependent events that it can be assumed to be instantaneous. The relaxation time is given by

$$\tau_d = \frac{\varepsilon}{\sigma} = \rho\varepsilon \tag{1.108}$$

where σ is the electric conductivity which is a measure of the concentration and the mobility of available mobile carriers and ε is the permittivity which is a measure of the strength of the Coulombic interaction inside the semiconductor (the dielectric medium). Thus, the product $\rho\varepsilon$ can be thought of as the RC time constant of the materials. Obviously, the larger the value of τ_d, the longer is the time required to attain equilibrium. In conventional semiconductors τ_d is of the order of 10^{-12} sec and τ_0 is normally larger than 10^{-9} sec. Thus it can be imagined that injection of minority carriers (electrons) of concentration Δn into the semiconductor will result in the change of the quasi-Fermi level following the relation

$$n = n_0 + \Delta n = N_c \exp\left[-(E_c - E_{Fn})/kT\right] \tag{1.109}$$

where n_0 is the concentration of electrons in equilibrium. After the injection, the majority carriers (holes) will quickly respond to neutralize excess Δn so as to maintain the "neutrality condition" as shown in Fig. 1.19(a). The processes to attain equilibrium are relaxation and recombination processes. The field created by the excess minority carrier charge $-q\Delta n$ will attract majority carrier Δp from the vicinity until the excess charge is completely neutralized. The characteristic time for this process is the relaxation time τ_d. After this relaxation process, the semiconductor is in the "neutrality condition". To reach the equilibrium condition, the recombination process is required to reduce Δn with time until the law of mass action is restored. The characteristic time

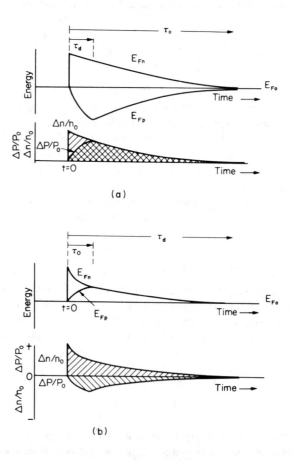

FIG. 1.19. Response of quasi-Fermi levels E_{Fn} and E_{Fp} and of the relative carrier concentrations to an applied pulse of excess minority carriers (electrons) Δn (a) in the lifetime regime $\tau_0 > \tau_d$, and (b) in the relaxation regime $\tau_0 < \tau_d$. [After Fritzsche 1973.]

for this process is the minority carrier lifetime τ_0. In equilibrium E_{Fn} and E_{Fp} are coincided to form one Fermi level E_{F0} which determines the equilibrium electron and hole concentrations n_0 and p_0. In non-equilibrium Δp has to increase locally to neutralize Δn, so E_{Fp} follows a similar relation as eq. (1.109):

$$p = p_0 + \Delta p = N_v \exp\left[-(E_{Fp} - E_v)/kT\right] \tag{1.110}$$

Therefore both Δn and Δp decrease with time, and E_{Fn} and E_{Fp} split rapidly into two levels (within the relaxation time τ_d) and then combine slowly to E_{F0} at equilibrium within the lifetime τ_0 as shown in Fig. 1.19(a).

1.10.2. The concept of the relaxation regime

From eq. (1.108), to satisfy the condition $\tau_0 < \tau_d$ for the occurrence of the relaxation regime, τ_d must be large. Materials with a low carrier mobility and a low concentration of mobile carriers are most likely of this category. For example, a semiconductor of resistivity of 10^8 ohm-cm has a relaxation time of about 10^{-4} sec, which is normally larger than the minority carrier lifetime which is generally lower than 10^{-8} sec for high resistivity materials. Following this classification, gold-doped silicon and germanium may become relaxation semiconductors at extremely low temperatures at which τ_0 becomes smaller than τ_d. In the lifetime regime the neutrality condition is retained long before the excess carriers Δn disappear, while in the relaxation regime the situation is just reversed. After the injection of Δn minority carriers into a relaxation semiconductor, the law of mass action is quickly restored by reducing the local majority carrier concentration within the minority carrier lifetime τ_0. Thus, at the completion of this

TABLE 1.13. *Comparison between lifetime semiconductors and relaxation semiconductors*

Item \ Type	Lifetime semiconductors	Relaxation semiconductors
τ_0, τ_d	$\tau_0 > \tau_d$	$\tau_0 < \tau_d$
I–V characteristics (forward biased)	$I \propto V^n$ or $I \propto \exp(qV/nkT)$ with $n \geqslant 1$ (super-linear)	$I \propto V^n$ with $n < 1$ or $n = 1/2$ (sublinear)
E_{Fn}, E_{Fp} (after carrier injection)	$E_{Fn} \neq E_{Fp}$ Local space charge neutrality, local non-equilibrium	$E_{Fn} = E_{Fp}$ Local space charge enhancement–space charge effect, local equilibrium
The region near the contact of minority carrier injection	Majority carrier enhancement (resistivity decreasing)	Majority carrier depletion (resistivity increasing)
General properties	Low resistivity, high mobility, and narrow energy band gap (e.g. conventional semiconductors, Si and Ge)	High resistivity, low mobility, and wide energy band gap (e.g., organic and amorphous semiconductors or conventional semiconductors at low temperatures)
Electrical transport (electronic, photoelectric and galvanomagnetic properties)	Extensively studied (well understood)	Not yet explored (not fully understood)

process the *np* product follows the relation

$$pn = (n_0 + \Delta n)(p_0 - \Delta p) = n_i^2 = p_0 n_0 \tag{1.111}$$

This leads to

$$\Delta p = p_0 \Delta n / (n_0 + \Delta n) \tag{1.112}$$

If the injection is so high such that $\Delta n > n_0$, then

$$\Delta p \rightarrow p_0 \tag{1.113}$$

This implies that under an extreme condition all mobile majority carriers may have disappeared [Queisser 1972]. It is clear that in the lifetime regime the injection of minority carriers tends to decrease the resistivity of the materials, while in the relaxation regime, the injection of minority carriers tends to increase the resistivity of the material as can be seen from eqs. (1.111)–(1.113) and Fig. 1.19(b). This feature of injection is characteristic for the relaxation regime, and it is sometimes called "recombinative space charge injection". In general, a majority carrier depletion region is adjacent to the minority carrier-injecting contact, followed by a narrow recombination front. It is this majority carrier depletion region which causes a sublinear voltage dependence of currents ($I \propto V^{1/2}$) since an increase in voltage enhances the space charge in the depletion

TABLE 1.14. *Some similarities between relaxation organic semiconductors and relaxation amorphous semiconductors*

Item \ Case	Organic semiconductors	Amorphous semiconductors	References
Typical examples	Anthracene, tetracene	As_2Se_3, GeTe Chalcogenide glasses	A, B
Activation energy	$\sigma = \sigma_0' \exp(\Delta E_\sigma / 2kT_0)$ $\times \exp(-\Delta E_\sigma(2kT)$	$\sigma = \sigma_0 \exp\left(-\dfrac{\Delta E_\sigma}{2kT}\right)$	C, D
AC conductivity $\sigma(\omega) \propto \omega^n$	$n \leq 1$	$0.7 < n < 1.0$	E, F, G
Charge carrier transport	Slow-phonon hopping limit, slow-electron hopping limit	Hopping conduction following $\sigma \propto T^{-4}$	H, I
Hall mobility	Small and anomalous	Small and anomalous	I, J
Drift mobility	≤ 1 cm^2/V-sec	≤ 1 cm^2/V-sec	B, K, L
Switching and memory phenomena	Filamentary conduction	Filamentary conduction	L, M, N, O, P
Optical absorption and the threshold for photoconduction	Depending on exciton and trapping energy levels	Depending on the distribution of localized states	A, Q

A, Gutmann and Lyons [1967].
B, Tauc [1974].
C, Roberts [1972].
D, Rosenberg *et al.* [1972].
E, Pollak and Geballe [1961].
F, Jonscher [1972].
G, Hayward and Pethig [1975].
H, Mott [1972].
I, Munn and Siebrand [1970].

J, Friedman [1972].
K, Kepler [1960].
L, Fritzsche [1972].
M, Hwang and Kao [1974].
N, Elsharkawi and Kao [1976].
O, Saji and Kao [1976].
P, Nakashima and Kao [1978].
Q, Mott and Davis [1972].

region, thus increasing differential resistance and sometimes even creating a negative differential resistance region [Queisser 1972]. After the elapse of τ_0, the space charge then slowly relaxes to retain the equilibrium condition as shown in Fig. 1.19(b). Table 1.13 lists the basic difference between the lifetime and relaxation semiconductors in their electrical behaviour, and Table 1.14 summarizes the similarities between relaxation organic semiconductors and relaxation amorphous semiconductors.

A field domain (or a potential disturbance) created by an injected pulse of minority carriers will move under the influence of an applied electric field. In the relaxation regime the disturbance will move in the same direction as the majority carriers because the excess minority carriers are quickly reduced by the recombination process, while the disturbance in majority carrier concentration moves slowly to relax towards neutralization. In the lifetime regime, the disturbance moves in the direction of minority carriers.

It should be noted that this book deals mainly with the solids in the lifetime regime unless otherwise stated. The electrical transport in the relaxation regime will be discussed in Section 5.7.

CHAPTER 2

Charge Carrier Injection from Contacts

2.1. CONCEPTS OF ELECTRICAL CONTACTS, POTENTIAL BARRIERS, AND ELECTRIC CONDUCTION

An electrical contact is generally referred to as a contact between a metal and a non-metallic material which may be an insulator or a semiconductor, and its function is either to enable or to block carrier injection. Contacts such as metal–electrolyte contacts and electrolyte–insulator or electrolyte–semiconductor contacts are also common electrical contacts. Electrical contacts are heterojunctions but normally exclude those between two different semiconductors or between two different metals or between a semiconductor and an insulator. In this section we shall confine ourselves to the discussion of only those described above.

When two materials with different Fermi levels are brought into contact, free carriers will flow from one material into the other until an equilibrium condition is established, that is, until the Fermi levels of both materials are aligned (the Fermi levels for electrons in both materials are equal at the contact). Such a net carrier flow will set up a positive space charge on one side and a negative space charge on the other side of the interface, forming an electric double layer. This double layer is generally referred to as the potential barrier, and the potential across it is called the "contact potential". The function of this double layer is to set up an electric field to stop any further net flow of free carriers from one material to the other, though thermodynamically the flow of free carriers in both directions always exists but it would be very small and equal in quantity, maintaining a statistically zero net flow under a thermal equilibrium condition.

The Fermi level E_F is sometimes called the "electrochemical potential" or simply the "chemical potential". The Fermi level can be considered as a reference level. A state of energy equal to E_F will have the same probability ($f = 1/2$) for it to be occupied or to be vacant. This implies that the probability for a state at the level ΔE about E_F to be occupied is equal to that at the level ΔE below E_F to be vacant. In the following we shall discuss the features of various types of potential barriers formed by some ideal contacts. It should be noted that although the energy band diagrams might be used to analyse and to predict some consequences, the surface states, created due to the presence of contaminating impurities and crystallographic defects and the lattice mismatching which unavoidably exists in the interface, affect to a great extent the electrical performance of a contact. However, the physics of contacts including the effects of surface states is not yet fully understood, and the science and technology of producing a desired electrical contact for a certain application (e.g. ohmic contact to an organic semiconductor) is not yet explored.

The electrical performance of the contact between an electrolyte and a non-electrolyte solid is mainly based on an electron exchange reaction between the energy levels associated with ions in the electrolyte and the energy levels in the solid. The equilibrium for such a continuous exchange of carriers between the ions of the electrolyte and the surface of the solid is dynamic, and at the dynamic equilibrium the forward and reverse rates of the reaction are equal. This type of electrical contacts can be made to provide or to impede carrier injection by choosing a right electrolyte for a particular solid.

Since the wave functions or orbitals of the outermost electrons of atoms or molecules are overlapped to some extent in a solid because of interatomic or intermolecular bonding, the charge carriers injected from a metallic or an electrolytic electrode into a solid become delocalized inside the solid. Charge carriers produced by thermal or optical excitation inside a solid usually do not alter the electrical neutrality of the solid as a whole (they may cause a net space charge of one sign in a local domain and a net space charge of opposite sign in the other domain due to the difference in mobility and diffusion constant between electrons and holes, but the solid as a whole can always be assumed to be electrically neutral) because such excitations produce either equal numbers of both electrons and holes, or one type of free carrier (mobile) and another type of bound charge (non-mobile) of equal number. But charge carriers injected from a contact will produce a net space charge in the solid, and it is this net space charge that produces the so-called "space-charge-limited" (SCL) current under an applied electric field. In this chapter we shall discuss in some detail the phenomena of charge carrier injection from a contacting electrode to a semiconductor or to an insulator through various potential barriers.

2.1.1. Electrical contacts, work functions, and contact potentials

The nature of a contact is a complex affair. No single crystal without imperfections of any kind has been found in this world, and therefore there always exist structural imperfections (e.g. dislocations in a crystal lattice) and chemical imperfections (e.g. impurities already existing internally in the crystal or due to external contamination) on even a cleavage surface. Imperfections in structure are normally accompanied by imperfections in geometric shape, and imperfections due to impurities always produce protuberances and depressions on the surface. It should be noted that it is very difficult to completely avoid impurity contamination even though the surface is cleaved in a vacuum chamber because a vacuum of 10^{-12} torr still contains about 3×10^4 particles per cm^3. However, a surface with protuberances gently undulating over it may be considered a mathematically smooth surface, and that with protuberances existing as irregular steep and jagged asperities, a rough surface.

When two surfaces are brought into contact, some parts of surfaces may not be in contact and some parts in real contact with mechanical actions and reactions between the surfaces [Harper 1967]. A contact which is prevented to be intimate by an extraneous body (contaminating particle) is referred to as an impeded contact because in such a contact the mechanical forces are transferred from one surface to the other through the intermediary of this extraneous body. In the absence of extraneous bodies, an intimate contact may be formed by long range molecular forces with a finite gap between the two surfaces (close contact), or by much stronger short range molecular forces (true contact). For mathematical analyses we always assume that the contact is a true intimate contact and the gap between the surfaces is so small that it can be considered transparent to quantum mechanical tunnelling of electrons. However, a practical contact is not so ideal: imperfections on the surfaces coupled with the non-homogeneity of the solid lead to the concepts of filamentary carrier injection, which will be discussed in Chapter 5.

The simplest contact between a metal and a non-metallic material is the contact between a metal and a vacuum. When two metallic plates are placed in parallel in a vacuum with a small separation, the current flow is negligibly small if the applied voltage across the two plates is small. The reason is not because there are no free electrons in the metal nor because the electrons, when present in the vacuum, are not mobile in the vacuum; but it is because the electrons in the metal have to surmount a potential barrier before they can leave the metal and enter the vacuum. This potential barrier between the highest energy level of the electrons in the metal termed the "Fermi level" of the metal and the lowest energy level of the electrons in the vacuum termed the "vacuum level" is called the work function of the metal, ϕ_m, as shown in Fig. 2.1(a). It is given by

$$\phi_m = \zeta - E_{Fm} \tag{2.1}$$

in which ζ is the difference in potential energy of the electrons between the inside and the outside of the metal, and depends on the structure of the crystal and the condition of the surface. The higher the cohesive energy of the metal, the higher is the work function, but, on the other hand, the work function may be appreciably altered by the presence of an absorbed or adsorbed layer of foreign atoms or molecules on the surface [Wigner and Bardeen 1935, Bardeen 1936]. Thus, the work function consists of two parts: (i) the energy of binding the electron, and (ii) the energy required to move the electron through an electrostatic double layer at the surface. This implies that ζ must depend partly on the structure of the surface and on the dipole moment of such a double layer.

A metal as a whole is electrically neutral, but at its surface facing a vacuum as a discontinuity the electron distribution may be unsymmetrical with respect to the ion cores, thus resulting in the formation of a double layer in a similar way to that due to a net charge flow. If σ_s is the charge per unit area, ε_0 is the permittivity in vacuum, and t is the thickness of the layer, then the dipole moment per unit area is $\sigma_s t$ and the potential is $\sigma_s t / \varepsilon_0$. This potential may be positive or negative outwardly depending on the double layer with positive or with negative charge on the outside. For clean metal surfaces the magnitude of the dipole moment of such an intrinsic double layer is different for different orientations of the crystal plane and the potential is of the order of 1/2 to 1 V, and for alkali metals it is less than half a volt. But for contaminated metal surfaces the adsorbed layer of foreign atoms, either neutral or ionized, may greatly modify the surface potential barrier and may then raise or lower the work function of the metal by more than 2 eV. For example, atoms of an electronegative gas, such as oxygen, adsorbed on the surface will capture electrons from the metal and form a layer of

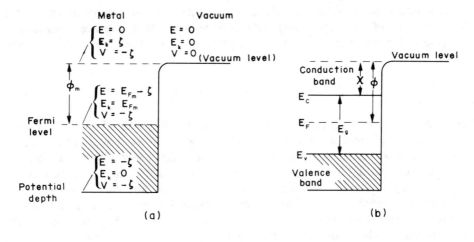

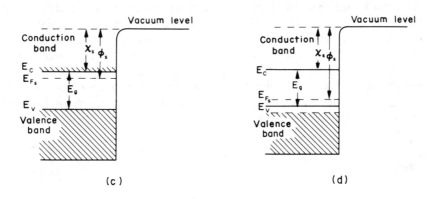

FIG. 2.1. Energy band diagrams showing the work function of (a) metal, ϕ_m; (b) dielectric, ϕ; (c) n-type semiconductor, ϕ_s; and (d) p-type semiconductor, ϕ_s.

negative ions. This layer will in turn induce a layer of positive image charge in the metal. A double layer formed in this way with the negative potential or negative charge outwards always tends to raise the work function. Conversely, caesium, barium, or thorium atoms may give up their outermost electrons to the metal first before they become adsorbed on the metal surface (such as on the tungsten surface) to form a double layer with the positive charge outwards. Such a double layer always tends to lower the work function. The adsorbed layer may be electrically neutral, but it would still be polarized by the field at the surface of the metal to form a double layer. It is obvious that the effect of the layer formed by neutral atoms with or without permanent dipoles is smaller than that formed by ionized atoms.

The work function of a metal surface is therefore mainly determined by the top few layers of atoms, and not by the metal as a whole. As the properties of the double layer are temperature dependent, the work function would also be expected to be temperature dependent.

It should also be noted that even if there is no contamination layer on the surface, the work function of a polycrystalline metal surface may vary from place to place. The domains of different work functions are patches, and thus the average work function of a surface can be expressed as

$$\langle \phi_m \rangle = \sum_i a_i \phi_{mi} \Big/ \sum_i a_i \qquad (2.2)$$

where ϕ_{mi} is the work function of patch i of area a_i. The potential outside the metal surface containing patches of different double layers is not constant, but varies in the same manner as does the work function. The resulting field between different patches is called the patch field. The presence of patches on the surface affects the measured value of the work function and the electron emission [Herring and Nichols 1949].

The work function of the metal can be determined by the following methods: (i) *Contact potential measurements*—contact potential between two materials is defined as the difference in their work functions. If the work function of one material is known and the surface of this material is used as a standard surface, then the unknown ϕ_m of the other material can be determined by measuring the contact potential between the surfaces of these two materials using the Kelvin method [Brattain and Garrett 1959]. It should be noted that the whole system containing the two materials for contact potential measurement must be kept completely isothermal, otherwise errors would be introduced due to thermoelectric effect if a pair of surfaces is at different temperatures. (ii) *Photoelectric emission method*—the photoelectric emission yield is a function of temperature. Fowler was the first to derive an expression for the photoelectric emission yield as a function of work function, photon energy, and temperature [Zworykin and Ramberg 1949], which is given by

$$J = AT^2 f\left(\frac{h\nu - \phi_m}{kT}\right)$$

or

$$\ln\frac{J}{T^2} = B + F(x) \qquad (2.3)$$

where $B = \ln A$ and $F(x) = \ln f\left(\frac{h\nu - \phi_m}{kT}\right)$. By plotting J/T^2 as a function of $h\nu/kT$, known as the "Fowler plot", we can compare the theoretical Fowler plot based on eq. (2.3) with the experimental Fowler plot. A vertical shift of the data to fit $F(x)$ enables the determination of B, while a horizontal shift enables the determination of ϕ_m/kT. This is one of the most accurate methods for determining the work functions of metals because in most cases the theoretical and experimental curves fit very well. (iii) *Thermionic emission method*—this method is based on the common plot of Richardson lines following the equation

$$J = A(1-r)T^2 \exp[-\phi_m/kT]$$

or

$$\ln\frac{J}{T^2} = \ln A(1-r) - \frac{\phi_m}{kT} \qquad (2.4)$$

where r is the surface reflective coefficient which depends on surface conditions. The slope of the Richardson line would be $-\phi_m/k$ and thus enable the determination of ϕ_m.

However, ϕ_m is most likely temperature dependent even for a clean metal surface [Seitz 1940] because interactions between the electrons in the solid have not been taken into account to derive eq. (2.4). If ϕ_m is a function of T, then the apparent work function ϕ_m^* is given by

$$\phi_m^* = \phi_m - T\frac{d\phi_m}{dT} \qquad (2.5)$$

It can be imagined that although the above three methods can be used to estimate the work functions of the metals, some discrepancies between measured ϕ_m are expected because of the uncertainties in measuring ϕ_m thermionically as discussed above, and particularly the measurement involves very high temperatures. Surface preparation for photoelectric emission measurements is extremely important since patchiness of the surface means that ϕ_m is not constant over it and this may well spoil the fit of the Fowler plot. Again, the accuracy of the contact potential method depends so much on the choice of the standard surface of known ϕ_m and the surrounding ambient. However, the Fowler plot satisfactorily explains the temperature dependence of photoelectric emission, so that ϕ_m determined by the photoemission method may be considered as the true ϕ_m at $0°K$.

In general, all the fundamental principles outlined above for metals can also be applied to non-metallic materials. In non-metallic materials the Fermi level is always located within the energy band gap except for the degenerate condition of extrinsic semiconductors. The work function for these materials is defined by

$$\phi = \chi + (E_c - E_F) \qquad (2.6)$$

Figure 2.1 shows the definition of ϕ for metals, insulators, n-type and p-type extrinsic semiconductors. The definition of ϕ for intrinsic semiconductors is the same as for insulators except for the latter the energy band gap is normally much smaller. It should be noted that the conditions for electron emission from a semiconductor or from an insulator differ appreciably from those for metals. There are no electrons at the Fermi level E_F, so the electrons which may be removed from the interior of a non-metallic material must be either in the conductor band, or in the valence band, or in the impurity levels. In a metal the minimum energy required to be imparted to an electron to remove it from the metal at $T = 0$ into vacuum is the energy difference between the Fermi level and the potential energy given by eq. (2.1); and at $T > 0$ the Fermi–Dirac distribution function becomes smeared out over a distance of the order kT, that is, over only a small fraction of an electronvolt. In a semiconductor (or in an insulator) the situation is different; only a very small portion of electrons in the conduction band requires the minimum excess energy (of the order of χ, the electron affinity) to leave the semiconductor to vacuum level, electrons at the impurity levels require higher energy and those in the valence require even much higher energy to do so. After an electron has left the semiconductor, the remaining electrons in the semiconductor restore their statistical distribution. If an electron in the conduction band receives an energy greater than χ and leaves the semiconductor, its place is immediately taken by an electron either from the impurity levels or from the valence band. Since the electron distribution inside the semiconductor before and after the emission of an electron is determined by the energy levels of these electrons with respect to the Fermi level E_F, therefore, the free energy required for an electron emitted from the semiconductor to the vacuum level is the work function defined by eq. (2.6). The work function of a semiconductor depends on the location of E_F, which is a function of temperature, impurity concentration, external pressure, etc. However, in semiconductors and insulators the electron affinity χ is an important quantity defined as the energy required for an electron to be removed from the bottom edge of the conduction band at the surface to a point in vacuum just outside the semiconductor. χ in high resistivity non-degenerate semiconductors or in insulators can be determined by measuring the threshold wavelength for photoemission corresponding to the energy separation between the electron energy in vacuum just outside the surface and the highest energy level that the electrons occupy in the crystal, that is $\chi + E_g$. The yields associated with the threshold wavelengths corresponding to the transitions from the conduction band, from impurity and trapping levels, and from surface states are usually very small. However, the following conditions must be satisfied in order to determine χ accurately using this method: (a) the materials must have a high resistivity and be non-degenerate so that the photoelectric threshold is insensitive to the height of the potential barrier—this means that the band bending over the escape depth of the emitted electrons must be negligibly small, and (b) the concentration of surface states is very small so that the photoelectric yield from the valence band rather than from the surface states is predominant. However, the electron affinity of organic semiconductors is not easy to determine. The method generally used for measuring χ is the electron beam retardation method [Gutmann and Lyons 1967]. This method consists of a collimated electron beam striking in high vacuum the target which may be a silver electrode covered with a thin layer of organic semiconductor under study. The beam current due to electrons accepted by the target can be stopped by a retarding potential. Thus, by plotting the logarithm of the beam current as a function of retarding potential, which is usually linear, and comparing the plot for the target with that without a layer of organic semiconductor, we can measure the difference in retarding potential between the two plots to retard the same beam current; this potential difference is taken as the contact potential between the electrode and the organic semiconductor V_d. If the work function of the electrode is known, χ can then be determined. Nelson [1956, 1957, 1958] has measured χ for several organic dye solids using this method and found the values to agree well with the values of the differences between the threshold wavelength (energy) for photoemission (ionization energy) and that for photoconduction. It should be noted that optical absorption is a complex event in organic materials, which depends upon the energy band structure, the type of optical transition involved, and the nature of the dominant scattering processes; and therefore the interpretation of photoelectric data may not be the same as for inorganic semiconductors. The values of χ and E_g for some common organic semiconductors are given in Table 1.1. Materials which have a positive electron affinity readily form negative ions and are therefore considered to be electronegative, while those which have a negative electron affinity (e.g. alkali elements) readily form positive ions and are electropositive. Electronegativity is defined as the power of an atom in a molecule to attract electrons to itself.

The contact potential is defined as the potential difference created between two dissimilar materials when they are brought into intimate contact, and it is basically equal to the difference in the work functions of the two materials. Taking a metal–n-type semiconductor contact with $\phi_m > \phi_s$ as an example, the contact potential is given by

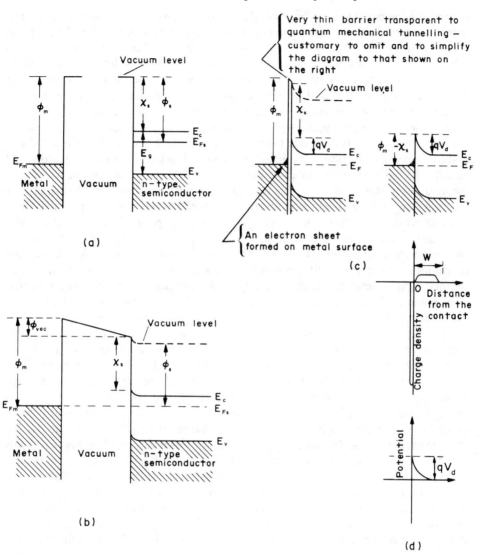

FIG. 2.2. Energy level diagrams for a contact between a metal and an *n*-type semiconductor for $\phi_m > \phi_s$ and without surface states: (a) before contact, (b) in thermal equilibrium, (c) in intimate contact, and (d) charge density distribution and contact potential.

$$V_d = \frac{1}{q}(\phi_m - \phi_s) = \frac{1}{q}[\phi_m - \chi_s - (E_c - E_F)] \tag{2.7}$$

Since $E_c - E_F$ is sensitive to temperature and impurity (particularly donor or acceptor) concentration, V_d depends strongly on temperature and impurity concentration. Figure 2.2(a) shows that before contact the electrons in both isolated metal and semiconductor experience a short range binding force exerted by the lattice of their own crystal (metal or semiconductor). This implies that the electrons have to surmount a very steep potential barrier in order to leave the solid into vacuum. If we now allow the electrons to flow from one solid to the other by making an electrical contact between the two solids (e.g., by connecting a metal wire to the back faces of the solids leaving the front faces not in contact), as shown in Fig. 2.2(b), there will be a net flow of electrons from the n-type semiconductor to the metal because $\phi_s < \phi_m$, and hence a space charge will build up near the two surfaces to hinder the electron flow. The net flow (although the electron flow from both solids always exists due to thermal excitation) will cease when the space charge density builds up to such a level that the Fermi levels in the metal and in the n-type semiconductor are aligned to the same height. This means that a thermal equilibrium has been reached between these two solids. The space charge will establish and electrostatic field and hence a potential difference between the metal and the semiconductor bulks, which is generally referred to as the "contact potential" and sometimes called the "diffusion potential" because in the space charge region the carrier movement is due mainly to a diffusion process since the thickness of this region is normally very large in comparison with the mean free path of the carriers. Under the condition shown in Fig. 2.2(b), in which the separation between the front surfaces of the metal and the semiconductor is still large, the major portion of the contact potential is the potential difference across the vacuum gap, only a small portion being due to the small space charge inside the semiconductor. When the separation between these two solid front surfaces decreases, the positive space charge region in the n-type semiconductor will extend much further into the bulk because electrons are driven further away from the surface due to the proximity effect; so that the portion of the contact potential across the vacuum gap decreases and the portion across the space charge region increases. When the separation reduces to the value of the order of interatomic distance and the contact becomes an intimate contact, the whole contact potential will be practically across the space charge region, the portion across the vacuum gap (atomic scale) being negligibly small as shown in Fig. 2.2(c). In general, this extremely thin barrier between the two surfaces is omitted in the energy level diagram because it is transparent quantum mechanically to electron tunnelling. It should be noted that because the free carrier density in the semiconductor is much smaller than that in the metal, the space charge region extends much further into the bulk of the semiconductor as shown in Fig. 2.2(c); the space charge region in the metal side is very thin and can be thought of as an electric charge sheet at the surface containing charge carriers equal in quantity but opposite in sign to that in the semiconductor. This also implies that a potential barrier is naturally accompanied by a double layer.

The width of the double layer or the space charge region is the width of the potential barrier denoted by W in Fig. 2.2(d). The electrons at the bottom of the conduction band of the semiconductor must have an energy equal to or larger than the height of the potential barrier qV_d before they can leave the semiconductor and pass into the metal side. Similarly, the electrons at the Fermi level of the metal must have an energy equal to or larger than the height of the potential barrier $\phi_m - \chi_s$ before they can be injected from the metal into the semiconductor. Potential barriers of this type, which are formed by a space charge double layer, and whose width and height are dependent on applied voltage, are generally referred to as Schottky barriers. However, if the height of a potential barrier is of the order of or smaller than the thermal energy kT, or if the width of a potential barrier is of the order or smaller than the wavelength of a conduction electron or hole (this means that the carriers may quantum mechanically tunnel through), the barrier does not perform effectively as a barrier.

Before closing this subsection we would like to clarify the concept of the vacuum level shown in Fig. 2.2. The vacuum level serves as a reference for the potential energy of the electrons at a given position. In an isolated homogeneous material it corresponds to the potential energy of an electron at a point where the attractive force between the electron and the surface of this material is negligible. In fact, the vacuum level does not have any absolute meaning. It represents only the relative energy of electrons at rest located just outside the various regions of the material at a position not to be influenced by the attraction force from the surface. However, the potential energy ϕ_{vac} in the vacuum level due to the difference in work function between two different materials [see Fig. 2.2(b)] will create a built-in field in the vacuum gap

$$F_{in} = -\frac{1}{q}\frac{\partial \phi_{vac}}{\partial x} \tag{2.8}$$

even in the absence of an applied voltage. Thus, under such a condition the electrons at the surface would experience a force in the x-direction due to F_{in}.

2.1.2. Types of electrical contacts

To define different types of electrical contacts we choose a metal–semi-conductor–metal system and assume that the semiconductor is intrinsic or lightly doped and the two metallic electrodes are identical. Before an intimate contact is made, we assume that the work function of the metal ϕ_m and that of the semiconductor ϕ are not equal. Therefore after they are brought into contact charge transfer between the electrode and the semiconductor will prevail until the Fermi levels of the electrode and the semiconductor are aligned to the same height. Depending on the values of ϕ_m and ϕ and other conditions, there are many types of electrical contacts and they are now discussed as follows.

(A) Neutral contacts

The word "neutral" implies that the regions adjacent to the contact on both sides are neutral electrically. To satisfy the condition of electrical neutrality, no space charge will exist and no band bending will be present within the semiconductor so that both the conduction and the valence band edges will be flat right up to the interface. A condition like this is sometimes referred to as the flat band condition. The possibilities for neutral contacts are: (i) When $\phi_m = \phi$ the contact is neutral as shown in Fig. 2.3(a) because when they are brought into contact the probability for the electrons to flow from the metal to the semiconductor is equal to the probability for the electrons to flow in the reverse direction, thus there is no net flow and hence no space charge formed near the interface. (ii) When $\phi_m \neq \phi$ at low temperatures or with an electron-trapping level at a distance sufficiently above E_F (or a hole-trapping level below E_F) in wide band gap semiconductors [Simmons 1971], the contact can be neutral because the trapped space charge in the traps will be too small under such conditions to cause significant band bending as shown in Fig. 2.3(b) and (c). A neutral contact is defined as one at which the

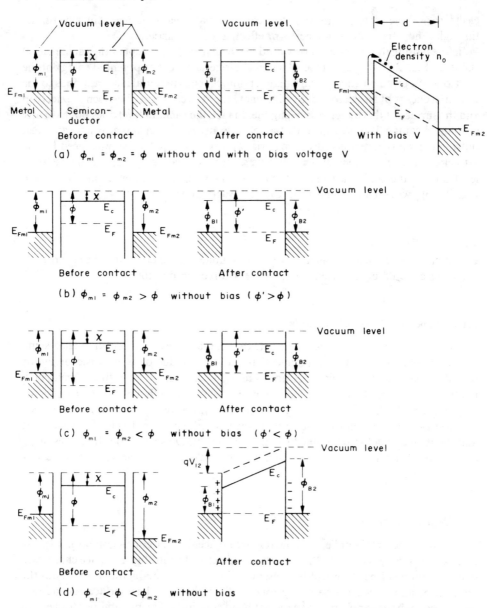

FIG. 2.3. Energy level diagrams for a neutral contact between a metal and an intrinsic semiconductor (or an insulator). $\phi_{B1} = \phi_{m1} - \chi$, $\phi_{B2} = \phi_{m2} - \chi$.

carrier concentration at the contact is equal to that in the bulk of the semiconductor.

We now return to Fig. 2.3(a). If a d.c. voltage V is applied between the two electrodes, and electrode 1 (the cathode) can supply a maximum electron density n_0 through a thermionic emission process to maintain the current flow in the semiconductor, then current recorded at electrode 2 (the anode) is given by

$$J = qn_0\mu \frac{V}{d} = qn_0\mu F \tag{2.9}$$

Equation (2.9) follows Ohm's law, and the contact is ohmic if (i) there is no band bending, so F is constant throughout the semiconductor for a given V, (ii) μ is independent of F—this requires the current to be not too large to cause the change of μ with F through the Joule-heating effect, and (iii) the current drawn through the semiconductor is less than the saturated thermionic emission current from the cathode. J is proportional to F until J is equal to the saturated thermionic emission current, for which F becomes F_0. When F is increased beyond F_0, the thermionic emission current is no longer capable of replacing those drawn out at the anode and under such a condition the contact ceases to be ohmic and tends to become blocking, and the conduction becomes electrode limited.

Supposing that the work functions of electrode 1, electrode 2, and the semiconductor follow $\phi_{m1} < \phi < \phi_{m2}$ and that the contacts are still neutral such as for the case of wide band gap semiconductors at low temperatures; then, as has been discussed in subsection 2.1.1, there will be a potential difference across the semiconductor given by

$$V_{12} = \frac{1}{q}\left[(\phi_{m2} - \chi) - (\phi_{m1} - \chi)\right]$$

$$= \frac{1}{q}(\phi_{m2} - \phi_{m1}) \tag{2.10}$$

as shown in Fig. 2.3(d), and a built-in field in the semiconductor is given by

$$F_{12} = -\frac{dV_{12}}{dx} \simeq -\frac{V_{12}}{d} \tag{2.11}$$

which can be very large for semiconductor thin films for which d is small.

(B) Blocking contacts

The Schottky barrier shown in Fig. 2.2 is formed by an electron-blocking contact for which $\phi_m > \phi_s$. The condition for a contact to be blocking seen by electrons from the metal is $\phi_m > \phi_s$ for a metal–n-type semiconductor junction, or $\phi_m > \phi$ for a metal–intrinsic semiconductor (or metal–insulator) junction. Under such a condition electrons will flow from the semiconductor to the metal, leaving a positive space charge region (or called a depletion region) in the semiconductor as shown in Fig. 2.4 in which W is the width of the depletion region. ϕ_B is the height of the potential barrier which an electron in the metal has to surmount in order to pass into the semiconductor. Such a contact is sometimes referred to as a rectifying contact because under forward bias electrons can flow easily from the semiconductor to the metal, while under reverse bias the flow of electrons from the metal is limited by the electrons available over the Schottky barrier, the density of which is much smaller than that in the bulk of the semiconductor. So a blocking contact can be defined as one which creates a depletion region extended from the interface to the inside of the semiconductor. With this contact the thermionic emission from the metal tends to be saturated. This is why, from the

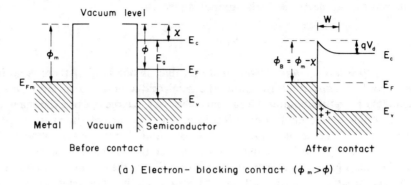

(a) Electron- blocking contact ($\phi_m > \phi$)

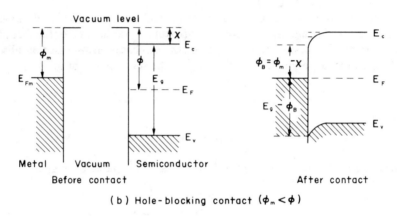

(b) Hole-blocking contact ($\phi_m < \phi$)

FIG. 2.4. Energy level diagrams for a blocking contact between a metal and an intrinsic semiconductor (or an insulator).

current injection point of view, such a contact is called the "blocking contact", and why the conduction is electrode limited under reverse bias. Electron emission from a metal across the blocking contact may be due either to a thermionic process or to a high field tunnelling process, and these will be discussed in Section 2.2.

The condition for a contact to be blocking seen by holes from the metal side (or by electrons from the opposite side) is $\phi_m < \phi_s$ for a metal–p-type semiconductor junction, or $\phi_m < \phi$ for a metal–intrinsic semiconductor (or metal–insulator) junction. Such a contact will block the hole emission from the metal as shown in Fig. 2.4.

(C) Ohmic contacts

An ohmic contact between a metal and a semiconductor is defined as one which has a negligibly small impedance as compared with the series impedance of the bulk of the semiconductor. This implies that the free carrier density at and in the vicinity of the contact is very much greater than that in the bulk of the semiconductor (e.g. thermally generated carriers in the bulk), so that the contact may act as a reservoir of carriers. So an ohmic contact can also be defined as one which creates an accumulation extended from the interface to the inside of the semiconductor. Unfortunately, the terminology "ohmic" is not appropriate in so far as the current–voltage relationship is not linear. With ohmic contacts the current–voltage relationship is non-linear and depends on many factors, and this will be discussed in detail in Sections 3.1 and 3.2. In general, the conduction is ohmic at low fields if the metal does not inject excess carriers (more than the thermally generated carriers in the semiconductor), and becomes non-linear or non-ohmic when the carrier injection from the electrode or the space charge effect becomes predominant.

There are two ways of making ohmic contacts: (a) to choose metals of low work functions such that $\phi_m < \phi_s$ (for metal–n-type semiconductor junctions) or $\phi_m < \phi$ (for metal–intrinsic semiconductor or metal–insulator junctions) for electron injection, or to choose metals of high work functions such that $\phi_m > \phi_s$ (for metal–p-type semiconductor junctions) or $\phi_m > \phi$ (for metal–intrinsic semiconductor or metal–insulator junctions) for hole injection, to lower the potential barrier for efficient thermionic emission so as to make the free carrier density higher (or the impedance smaller) at the contact than that in the bulk of the semiconductor; and (b) to dope the semiconductor surface heavily near the contact to make the potential barrier thin enough for efficient quantum mechanical tunnelling. In general, the resistivity of most organic semiconductors is very large so that the electrical contact impedance is normally considered to be negligibly small as compared with the resistance of the organic semiconductor specimen.

Figure 2.5 shows the energy level diagram for an ohmic contact between a metal and an intrinsic semiconductor for $\phi_m < \phi$ and without surface states. The distributions of charge carriers and potential in the semiconductor are governed by the Poisson's equation

$$\frac{dF}{dx} = \frac{qn}{\varepsilon} \tag{2.12}$$

and the current flow equation

$$J = q\mu n F - qD\frac{dn}{dx} = 0 \tag{2.13}$$

since there is no external applied field. Using the boundary condition (Fig. 2.5)

$$\frac{\psi(x)}{q} - \frac{\phi_m - \chi}{q} = -\int_0^x F\,dx \tag{2.14}$$

and the Einstein relation

$$\frac{D}{\mu} = \frac{kT}{q} \tag{2.15}$$

we obtain

$$n(x) = n_s \exp\left[-(\psi_{(x)} - \phi_m + \chi)/kT\right] \tag{2.16}$$

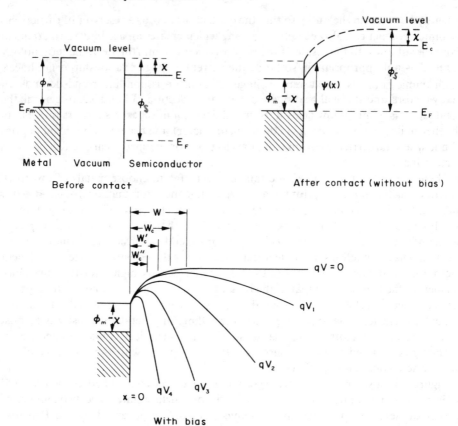

FIG. 2.5. Energy level diagrams for an ohmic contact between a metal and an intrinsic semiconductor (or an insulator). $\phi_m < \phi$ and applied voltages $V_4 > V_3 > V_2 > V_1 > 0$.

where n_s is the electron density at the contact $x = 0$ (where $\psi = \phi_m - \chi$). Thus eq. (2.12) can be written as

$$\frac{d^2\psi}{dX^2} = -\frac{q^2 n_s}{\varepsilon} \exp[-(\psi - \phi_m + \chi)/kT] \qquad (2.17)$$

Using the boundary condition: $d\psi/dx = 0$ when $\psi = \phi - \chi$. The solution of eq. (2.17) yields

$$\left(\frac{d\psi}{dx}\right)^2 = \frac{2q^2 n_s kT}{\varepsilon} \left\{ \exp[-(\psi - \phi_m + \chi)/kT] - \exp[-(\phi - \phi_m)/kT] \right\}$$

or

$$\frac{d\psi}{dx} = \left(\frac{2q^2 n_s kT}{\varepsilon}\right)^{1/2} \left\{ \exp[-(\psi - \phi_m + \chi)/kT] - \exp[-(\phi_s - \phi_m)/kT] \right\}^{1/2} \qquad (2.18)$$

Integration of eq. (2.18) gives the width of the accumulation region

$$W = \left(\frac{2\varepsilon k T}{q^2 N_c}\right)^{1/2} \exp\left[\frac{\phi_s - \chi}{2kT}\right]\left[\frac{\pi}{2} - \sin^{-1}\left\{\exp\left(-\frac{\phi_s - \phi_m}{2kT}\right)\right\}\right] \qquad (2.19)$$

It is clear that when $\phi_s = \phi_m$, $W = 0$, the contact becomes neutral; that when $\phi - \phi_m < 4kT$, W increases with decreasing barrier height $\phi_m - \chi$; and that when $\phi - \phi_m > 4kT$ eq. (2.19) reduces to

$$W \simeq \frac{\pi}{2}\left(\frac{2\varepsilon k T}{q^2 N_c}\right)^{1/2} \exp\left(\frac{\phi_s - \chi}{2kT}\right) \qquad (2.20)$$

Under such a condition W is independent of the barrier height $\phi_m - \chi$ and the electrode work function ϕ_m, but rather, depends on the energy separation between the Fermi level and the bottom edge of the conduction band; or in other words, depends on the free carrier density in the bulk of the semiconductor. This implies that W increases with decreasing free carrier density in the bulk of the semiconductor.

It is important to note that although the semiconductor itself may be intrinsic, the ohmic contact injects free carriers into the semiconductor with a quantity overwhelming those generated thermally inside the semiconductor and thus makes the electric conduction in the intrinsic semiconductor become extrinsic and SCL in nature. In fact, even for an insulator the ohmic contact always tends to inject electrons to it when $\phi_m < \phi$, or to inject holes to it when $\phi_m > \phi$ for raising or lowering the Fermi level in the insulator by an amount of $|\phi - \phi_m|$ in order to make the Fermi levels in both the metal and the insulator aligned at the same level. Since the ohmic contact acts as a reservoir of free charge carriers, the electric conduction is controlled by the impedance of the bulk of the semiconductor (or of the insulator), and is therefore bulk limited.

Since the injected space charge density decreases with increasing distance from $x = 0$ and reaches the value equal to that thermally generated in the bulk of the semiconductor at $x = W$, the internal field created by this accumulated space charge will therefore decrease with increasing distance. It is now interesting to see how an applied electric field affects the band bending. Figure 2.5 shows that by applying an average field F_{av} (applied voltage $V = V_1$ divided by specimen thickness) of low magnitude or of the order of the internal field near $x = W$, the applied field will be equal and opposite to the internal field at $x = W_c$ where the product of the diffusion field and the charge carrier density is equal to the product of the applied field and the charge carrier density in the bulk of the semiconductor [Rose 1963]. So, in general, we call this point as the "virtual cathode" at which $dV/dx = 0$. This situation is very similar to the SCL situation in vacuum diodes due to thermionic emission. Under equilibrium condition the negative potential gradient at $x < W_c$ tends to send back to the contact all the electrons that represent the excess of the SCL current permitted by the semiconductor. Thus, at $x = W_c$, that is, at the virtual cathode, we can assume that the electrons are released without initial velocity. When the applied voltage is increased to $V_2(> V_1)$, this field will balance a higher internal field at $x = W_c'$ ($< W_c$). This also implies that the electron density at $x = W_c'$ is higher than that at $x = W_c$ for supplying a higher SCL current at a higher applied field. In Fig. 2.5 the distances W, W_c, W_c', and W_c'' are greatly exaggerated to illustrate the physical picture.

The higher the applied field, the more close is the virtual cathode to the contact

[Wright 1961]. When the applied field is so high that the virtual cathode coincides with the contact at $x = 0$, the effect of space charge ceases (since there is no accumulated charge region) and the conduction becomes ohmic following Ohm's law. Beyond this, any further increase in applied field will make the conduction change from being bulk limited to electrode limited, because the conduction becomes to be governed by the rate of electron injection from the cathode. However, if $\phi_m - \chi$ is large and the applied field is sufficiently high (but not high enough to cause breakdown), the potential barrier may become so thin that quantum mechanical tunnelling becomes important. The tunnelling process through such a barrier will be discussed in Section 2.2.

The energy level diagrams for an ohmic contact to inject electrons to an n-type semiconductor and an ohmic contact to inject holes to a p-type semiconductor are shown in Fig. 2.6.

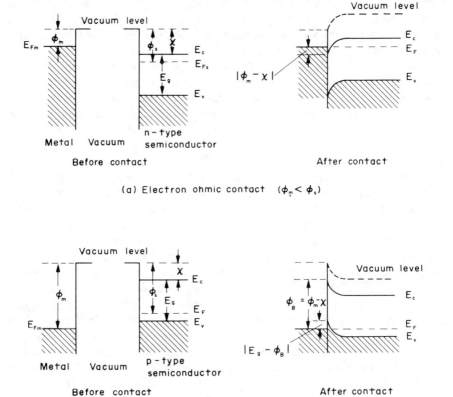

(a) Electron ohmic contact $(\phi_m < \phi_s)$

(b) Hole ohmic contact $(\phi_m > \phi_s)$

FIG. 2.6. Energy level diagrams for an ohmic contact between a metal and an extrinsic semiconductor: (a) n-type, (b) p-type.

It should be noted that the surface states will greatly affect the efficiency of carrier injection and these effects will be discussed in Section 2.1.3. The barrier heights for injection of electrons and holes from various metals into anthracene are given in Table 2.1,

TABLE 2.1. *Barrier heights for carrier injection into anthracene crystals*

Metal	Work function (eV)	Barrier height for hole injection into valence band (eV)	Barrier height for electron injection into 1st conduction band (eV)	Barrier height for electron injection into 2nd conduction band (eV)	References
Au	5.22	1.17	–	–	Williams and Dresner [1967]
Ag	4.31	1.20	–	–	Williams and Dresner [1967]
Al	4.20	1.86	–	–	Williams and Dresner [1967]
Pb	3.97	1.60	1.92	2.60	Williams and Dresner [1967], Baessler *et al.* [1969], Caywood [1970]
Mg	3.68	1.97	1.75	2.30	Williams and Dresner [1967], Baessler *et al.* [1969]
Ce	2.84	2.20	1.52	2.10	Vaubel and Baessler [1968]
Pt	5.30	0.89	–	–	Dresner [1970]
Na	2.28	–	0.90	1.45 / 1.37	Vaubel and Baessler [1968] / Donnini and Abetino [1969]
Cs	1.96	–	0.70	1.25	Vaubel and Baessler [1968]
Ca	2.71	–	1.07	1.65	Baessler *et al.* [1969]
Ba	2.48	–	–	1.63	Baessler [1970], Caywood [1970]
K	2.18	–	–	1.60	Donnini and Abetino [1969]

TABLE 2.2. *Electrode materials for carrier-injecting contacts to anthracene*

Electrodes form	Electron injection materials (cathode)	Hole injection materials (anode)	References
Liquid contacts	A solution of negative anthracene ion Sodium + anthracene + tetrahydrofuran	A solution of positive anthracene ion (1) $kI + I_2$ in water (2) $AlCl_3$ + anthracene + nitromethane	Helfrich and Schneider [1965, 1966]
	Lithium + anthracene + nitromethane	$AlCl_3$ + anthrance + ethylenediamine	Zschokke-Gränacher *et al.* [1967]
Solid contacts	Sodium–potassium alloy	Evaporated gold	Mehl and Funk [1967]
	Sodium + tetrahydrofuran + anthracene	(1) $AlCl_3$ + anthracene + nitromethane (2) Silver paste (3) Gold paste (4) Evaporated silver (5) Evaporated aluminium (6) Conducting glass (SnO_2)	Williams and Schadt [1971]
	(1) n^+–Silicon wafers covered with 20–40 Å SiO_2 (2) A fine grid structure of evaporated aluminium on a glass substrate oxidized to 50 Å of Al_2O_3	(1) Evaporated transparent films of Cu_2O–CuI (2) Evaporated Se–Te alloy (3) Colloidal black platinum paste (4) Iodized copper paste	Dresner and Goodman [1970]
	Carbon fibres	Evaporated indium	Williams *et al.* [1972]

and the electrode materials generally used for making ohmic contacts with anthracene are listed in Table 2.2. It should be noted that if an alkali metal is deposited onto an anthracene crystal, the direct contact with the anthracene is not the metal itself but the alkali–anthracenide charge transfer complexes to enable the charge carrier injection into anthracene. For example, a solution of charge transfer complexes made of sodium and anthracene in tetrahydrofuran is applied to the surface of the anthracene crystal, and a good electron-injecting contact can be obtained when the solvent is evaporated [Pott and Williams 1969]. Great care must be taken in making such contacts. The chamber for making such contacts must be completely free of oxygen, moisture, and other materials which may react chemically with the solution and, after a solid electrode is formed, a layer of epoxy or other protective material must be applied to cover the electrode to protect it from contacting the surrounding atmosphere [Kunkel and Kao 1976].

For the semiconductors or insulators containing shallow traps confined in a discrete energy level above E_F with an electron ohmic contact as shown in Fig. 2.7, the expression for the width of the accumulated region has been derived by Simmons [1971] and is given by

$$W = \frac{\pi}{2}\left(\frac{2\varepsilon k T}{q^2 N_t}\right)^{1/2} \exp\left(\frac{\phi - \chi - E_t}{2kT}\right) \tag{2.21}$$

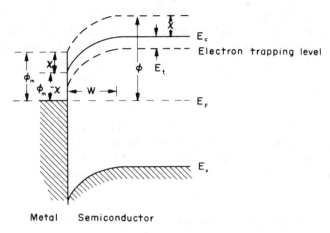

F IG. 2.7. Energy band diagram illustrating an electron ohmic contact between a metal and a semiconductor (or an insulator) with shallow electron traps.

for $\phi - \phi_m > 4kT$, where N_t is the density of shallow traps and E_t is the trapping level measured from the bottom edge of the conduction band. For the effects of deep traps and traps of various distributions in energy the reader is referred to the papers of Simmons [1971].

(D) Metal–electrolyte contacts

The electric conduction in the electrolyte is ionic rather than electronic. When the positive ions (cations) reach the cathode or when the negative ions (anions) reach the anode under an applied field, these ions will pile up at the metal surface until the electric field at the metal surface becomes high enough to raise the electrons on the negative ions to the Fermi level of the metal anode, or to lower the holes on the positive ions to the Fermi level of the metal cathode so that the charges of these ions can be transferred to the metal electrodes. Such a field at the metal surface is of the order of 10^8 V cm^{-1}, or of the same magnitude that exists between ionic planes in an ionic solid [Rose 1963]. In general, chemical reactions set in before the electrons or holes on the electrolyte ions are moved to the Fermi level to provide the necessary energy for the charge transfer from the electrolyte ion to metal electrodes.

(E) Electrolyte–semiconductor (or –insulator) contacts

Similar to the metal–semiconductor contacts, there is a charge transfer taking place, when an electrolyte and a semiconductor are brought into contact, until a potential difference, known as the "Galvani potential", is established at equilibrium. This potential across this region is small because its permittivity is much larger than that of semiconductor V_s, (b) the Helmholtz region formed by a layer of ions in the electrolyte firmly held to the semiconductor at the interface V_H, and (c) a more diffuse mobile space charge region extending into the electrolyte (the Gouy region) V_G. Region (b) is sometimes called the inner Helmholtz layer and region (c) the outer Helmholtz layer. The Gouy region is generally thin because of the high ion concentration, and the potential across this region is small because its permittivity is much larger than that of the semiconductor and of the Helmholtz region. So the Galvani potential is mainly the sum of the potential across the Helmholtz region and that across the space charge region in the semiconductor. The Helmholtz region (in here it refers to inner Helmholtz layer) is a fixed layer of one or two atomic diameter (or molecular diameter) in thickness corresponding to the ions absorbed by the particles in the semiconductor, to which they owe their charges. The diffuse space charges in the Gouy region usually are excess ions of charges opposite to those in space charge region in the semiconductor as shown in Fig. 2.8.

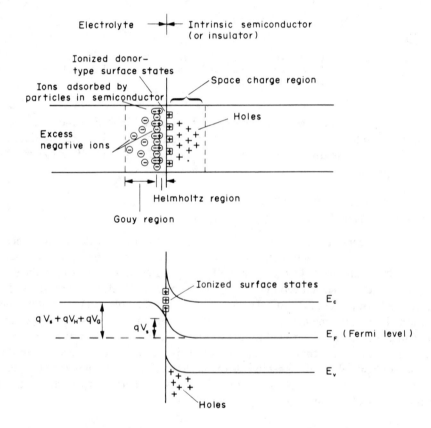

FIG. 2.8. Schematic diagrams showing the energy levels of an electrolyte–intrinsic semiconductor (or insulator) interface and the Helmholtz and Gouy regions in the electrolyte and the space charge region in the semiconductor.

In the electrolyte–semiconductor systems the electrolyte is "reducing" when there is electron transfer from the electrolyte to the semiconductor because "reduction" (Red) means the increase in negative charge or the gain of electrons in the semiconductor. Similarly, the electrolyte is "oxidizing" when there is electron transfer from the semiconductor to oxidized species in the electrolyte because "oxidation" (Ox) means the decrease in negative charge or the loss of electrons in the semiconductor. A redox reaction may simply be expressed by [Mehl 1971]

$$Ox + e^- \underset{\longleftarrow}{\longrightarrow} Red$$

The Red species would be equivalent to an occupied electron level corresponding to a valence band, whereas the Ox species is equivalent to an empty electron level corresponding to a conduction band. The Red and Ox species are separated by an energy which is required for an electron to transfer from the former to the latter. This energy is analogous to the energy band gap in an intrinsic semiconductor.

In general, the electrolyte–semiconductor systems have a rectifying property. The electrons on the negative ions in the electrolyte have an energy level usually located below the conduction band or even below the valence band edge of the semiconductor. These electrons cannot enter the conduction band when the system is biased with positive potential on the semiconductor side. The electrolyte under such a bias (corresponding to a reverse bias in p–n junctions) acts as a blocking contact to electron injection [Rose 1963]. However, when the system is biased with negative potential on the semiconductor side (particularly when the semiconductor is n-type), the electrons can

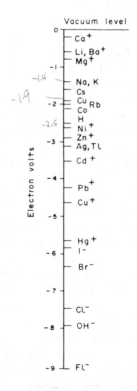

F IG. 2.9. Electronic energy levels of electrons attached on some atoms or ions in solution, derived from heats of hydration. [After Rose 1963.]

flow easily from the conduction band of the semiconductor to the electrolyte and then a large current can be obtained. This condition is equivalent to a forward bias in *p–n* junctions. The electron energy levels of electrons attached on some atoms or ions in solution are given in Fig. 2.9. Of course, by choosing suitable electrolytes with suitable electronic energy levels of their ions, electrolyte can form an ohmic injecting contact for carriers. Kallmann and Pope [1960] and Mark and Helfrich [1962] have used iodine–iodide oxidizing electrolyte as an ohmic contact for injecting holes to anthracene, Helfrich and Schneider [1965, 1966] have used a solution of negative anthracene ions (anthracene and sodium in tetrahydrofuran) as an ohmic contact for injecting electrons to anthracene.

2.1.3. Surface states

A sudden termination of the lattice of a solid crystal at its surface usually gives rise to the distortion of the band structure near the surface, thus causing the bending of energy bands in order to bring the solid to an equilibrium state or, in other words, to make the Fermi level at the surface identical to that in the bulk. If a semiconductor surface is made in intimate contact with a metal or an electrolyte whose structure or interparticle distance is different from that of the semiconductor, localized surface states with energies lying within the forbidden energy gap may result because of the dangling bonds or of the interruption (or the discontinuity) of the periodic lattice structure at the interface. Bardeen [1947] was the first to propose the existence of surface states for the explanation of the experimental fact that the contact potential between a metal and a germanium or silicon is independent of the work function of the metal and of the conductivity of the semiconductor, and Shockley and Pearson [1948] were the first to observe experimentally the existence of surface states on semiconductor thin films.

The surface states partly originate from the discontinuities of the periodic lattice structure at the surface leading to the so-called "Tamm states" [Tamm 1932]— associated with an asymmetric termination of the periodic potential and a large separation or weak interaction between atmos (or between molecules); or to the so-called "Shockley states" [Shockley 1939]—associated with a symmetrical termination of the periodic potential and a small separation or strong interaction between atoms (or between molecules); and partly originate from foreign materials adsorbed on the surface. The latter is caused by the fact that the surface atoms in general are extremely reactive due to the presence of unsaturated bonds, and thus the crystal surface is generally covered with one or more layers of a compound produced by a reaction between the surface atoms (or molecules) and their surroundings. The surface states may also be created by the imperfect structure on the surface such as abrasion on the surface due to grinding, cutting, etching, or polishing, etc., and at the electrolyte–non-electrolyte interfaces, by the chemisorption of reactions or reaction intermediates. It is important to note that apart from the cause by foreign impurities, there is no direct or unambiguous evidence that surface states lying within the forbidden energy gap must arise because of the sudden termination of an otherwise perfectly periodic lattice structure. However, it would seem reasonable to consider the connection of surface states with chemical binding (dangling bonds) at the surface. A dangling bond means a domain from which an electron normally participating in chemical binding in the solid has been removed in creating the surface. For example, a germanium atom at the surface is bonded to two neighbours leaving two of the four covalent bonds unsaturated, then these two unsaturated bonds generally referred to as dangling bonds will act as acceptors tending to capture electrons available around there. It should be noted that the surface density on an atomically clean surface of germanium is about 10^{11}–10^{12} per cm^2 which is much smaller than the value of about 10^{15} per cm^2 expected from the number of dangling bonds if these were not mutually saturated. Hanemann [1961] has pointed out that the layer of surface atoms may become distorted in such a way as to help mutual saturation of the unsaturated bonds of neighbouring atoms, thus reducing the chance for the surface states to be formed by dangling bonds. Such surface layer distortion has been reported by Farnsworth *et al.* [1955] based on their experimental investigation on the structure of atomically clean surfaces. However, Lax [1961] has suggested that the surface states may be associated with flaws or impurities on the surface and that the surface state density may not be related to the intrinsic structure of the solid. But Harper [1967] has suggested that "abandoning the idea of surface states arising from the sudden termination of the lattice very much weakens the idea that they will arise at lattice imperfection, whether surface or interior". It is obvious that as atomically clean surfaces are not easy to achieve, the contamination on the surfaces due to the unavoidable adsorption of or reaction with foreign impurities (e.g. oxygen or oxide) plays the most important role in creating surface states. In general, gas atoms or molecules having an electron affinity greater than the work function of the semiconductor may capture electrons from the surface and their behaviour is like acceptors (acceptor-like surface states) tending to bend the energy bands upward. For example, if oxygen atoms, which are very electronegative, are adsorbed on the surface of a semiconductor to form surface states within the forbidden gap, these surface states would intercept electrons from going to the conduction band and leave some holes in the valence band, thus bending the bands upward. Therefore electronegative atoms or molecules, such as oxygen, adsorbed on the surface act as acceptors tending to produce a depletion region on an n-type semiconductor, and an accumulation region on a p-type semiconductor as shown in Fig. 2.10(a) and (b). Similarly, gas atoms or molecules of electropositivity adsorbed on the surface of a semiconductor, such as chlorpromazine on anthracene surface [Gutmann and Lyons 1967], will form donor-like surface states as shown in Fig. 2.10(c) and (d). In Fig. 2.10 qV_s created by surface states dominates the contact potential and is independent of the work functions if the surface state density is larger. It should be noted that an acceptor-like or a donor-like surface state can be considered to be electrically neutral, but the acceptor-like one would become negatively charged after capturing an electron and the donor-like one positively charged after giving up an electron (or in other words, capturing a hole), since any excess charge in the surface states must be compensated by the change in free carrier concentration beneath the surface in the solid (equivalent to say that the holes are attracted to the negatively charged surface or the electrons to the positively charged surface) in order to maintain charge neutrality, thus forming a double layer. It can be imagined that the effects of surface states are very important in semiconductors and insulators because the space charge region produced by such a double layer usually extends to some depth, but they are not as important in metals since a large quantity of free electrons present can easily compensate any surface charges.

In general, surface states can be divided into fast and slow surface states, depending

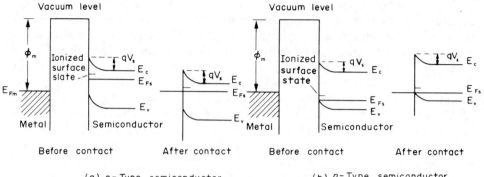

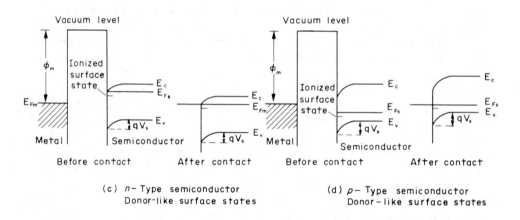

FIG. 2.10. Energy level diagrams showing the effects of surface states on band bending, (a) and (b) for acceptor-like surface states, (c) and (d) for donor-like surface states.

upon the speed with which they interact with the semiconductor space charge region. Fast surface states exist at the semiconductor–insulator (oxide) interface due mainly to inherent structure of the surface; the time required for their interaction with the carriers near the surface to reach an equilibrium state is of the order of nanoseconds or less and their density is about 10^{12} per cm^2, lower than that of slow states. Fast surface states play an important role in carrier recombination processes which affect greatly the electrical and optical properties of semiconductors. Slow surface states have much longer relaxation times which are of the order of milliseconds or longer and their density is of the order of 10^{13}–10^{15} per cm^2 [Brattain and Bardeen 1953] and carry either positive or negative electric charges, depending on the nature of these surface states (that means they may be acceptor-like or donor-like states). Slow surface states generally exist at the outer surface of the oxide layer (i.e. at the oxide–gas interface). The occupation of the slow surface states is affected by the ambient atmosphere, indicating

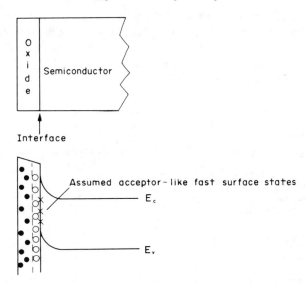

x x Fast surface states (at the interface.)

O O Slow surface states — immobile charges held in traps
                              that either are part of the intrinsic
                              defects in the oxide or are chemisorbed
                              ambient ions.

• • Slow surface states — mobile ions or vacancies in the oxide.

FIG. 2.11. The fast and slow surface states near the semiconductor surface covered with an
         oxide layer.

that they participate or originate in adsorption processes. There are two types of
charges in the oxide layer (Fig. 2.11):

(a) *Immobile atmosphere, charges*—they are either associated with ionic defects in
    the oxide in which the electronic transfer takes place between the defects and the
    semiconductor during the formation of the oxide on the semiconductor surface
    and usually not afterwards, or associated with adsorbed ambient ions located at
    the semiconductor–oxide interface, which modify the distribution of semicon-
    ductor interface states. The immobile (or called "fixed" or "built-in") charges do
    not participate in electron transfer processes across the interfaces.
(b) *Mobile charges*—they are traps within the oxide and they can react electrically
    with the semiconductor with a three-dimensional distribution.

Since both fast and slow surface states act as traps, the density of such traps at the
surface is much higher than the trap density in the bulk. The surface states acting as a
stepping-stone to assist carrier injection from a metal electrode to polyethylene
terephthalate have been reported by Takai *et al.* [1975]. The surface states on
anthracene surface with electrolytic electrode containing iodine or cerium [Kallmann
and Pope 1960, 1962] assist hole injection into anthracene. Adsorbed gases such as
iodine [Labes *et al.* 1962, Bauser 1965] or oxygen [Braun 1968, Johnston and Lyons

1970] also greatly affect carrier injection. Table 2.3 shows the effects of adsorbed impurities on carrier injection into anthracene. Excellent discussions on surface states can be obtained in the following references: Many *et al.* [1965], Kontecky [1960], Levine and Mark [1969], Garrett and Brattain [1955], Eley *et al.* [1968], Stern and Green [1973], Pope *et al.* [1971].

TABLE 2.3. *Effects of adsorbed impurities on carrier injection in anthracene [after Hoesterey 1962]*

Surface condition	Relative hole current	Relative electron current
Nitrobenzene	50	0.1
1,3-Dinitrobenzene	30	0.1
1,3,5-Trinitrobenzene	8	2.5
Iodine	100	0.1
Anthrone	4	0.5
Anthraquinone	4	0.5
Triphenylamine	1	1
UV[a]	30	5
Unblocked[b]	15	2
Clean	1	1[c]

[a] Crystal exposed to 365 mμ in air for 30 min.
[b] Crystal in contact with conducting coating on quartz electrode.
[c] The electron current is 0.1 of the hole current.

2.2. CHARGE CARRIER INJECTION THROUGH POTENTIAL BARRIERS FROM CONTACTS

In 1920–30, when the copper oxide and selenium were the best-known rectifiers, some people thought at that time that the rectification was a bulk effect. Until after 1930, it became generally accepted that the rectification phenomena are associated with the potential barriers near the interface between a metal and a semiconductor. We have mentioned in Section 2.1 that when a depletion layer is formed on the semiconductor side near the interface, the conduction becomes electrode limited because the free carrier concentration in this case is higher in the bulk than what the contact can provide. In Table 2.4 are given the conditions for the formation of depletion, accumulation, or inversion regions in n-type, p-type, and intrinsic semiconductors. In this section we shall consider only those potential barriers leading to an electrode-limited electric conduction. This implies that the current–voltage (J–V) characteristics are controlled by the charge carrier injection from the injecting contacts. There are two possible ways for charge carriers to inject from an electrode to a semiconductor and they are (a) field-enhanced thermionic emission, and (b) quantum mechanical tunnelling, and these are now analysed.

2.2.1. The lowering of the potential barrier and the Schottky effect

The potential barrier at the interface between a metal and a semiconductor prevents the easy injection of electrons from the metal into the semiconductor. In this subsection

TABLE 2.4. *Conditions for the formation of various types of space charge regions*

Space charge region	*n*-type semi-conductor	*p*-type semiconductor	Intrinsic semiconductor or insulator	Remark
Depletion	$n_s < n_0$ $n_s > p_s$	$p_s < p_0$ $p_s > n_s$	Similar to *n*-type if $n_s < n_0, n_s > p_s$	Rectification
	Bands bending upward	Bands bending downward	or Similar to *p*-type if $p_s < p_0, p_s > n_s$	
Accumulation	$n_s > n_0$ $n_s > p_s$	$p_s > p_0$ $p_s > n_s$	similar to *n*-type if $n_s < n_0, n_s > p_s$ or	No rectification but with ohmic action
	Bands bending downward	Bands bending upward	Similar to *p*-type if $p_s > p_0, p_s > n_s$	
Inversion + depletion	$n_s < n_0$ $n_s < p_s$	$p_s < p_0$ $p_s < n_s$	Similar to *n*-type if $n_s < n_0, n_s < p_s$ or	Rectification
	Bands bending upward	Bands bending downward	Similar to *p*-type if $p_s < p_0, p_s < n_s$	

n_s, p_s, n_0, and p_0 are, respectively, the electron and hole densities at the surface and in the bulk.

we shall discuss the factors causing the lowering of such a barrier for neutral and blocking contacts.

(A) Neutral contacts

The phenomenon that the height of the potential barrier is lowered due to the combination of the applied electric field and the image force is called the "Schottky effect". Figure 2.12 shows that the potential barrier height is $\phi_m - \chi$ if the image force is ignored and the applied electric field is zero. When these two parameters are taken into account, the potential barrier height measured from the Fermi level of the metal is given by

$$\psi(x) = \phi_m - \chi - \frac{q^2}{16\,\pi\varepsilon x} - qFx \qquad (2.22)$$

It should be noted that the potential energy due to image force $q^2/(16\pi\varepsilon x)$ is not valid at $x = 0$. To avoid this, we assume that this expression is valid from $x = x_0$ corresponding to $q^2/(16\pi\varepsilon x_0) = \phi_m - \chi$ to $x = \infty$, and that the image force is constant from $x = 0$ to $x = x_0$. We may also assume that the electron sea in the metal at E_F is extended to x_0.

The image force tends to attract the emitted electrons back to the metal, while the driving force due to the applied field tends to drive the emitted electrons away from the metal. There is an optimum point where the net force acting on the electrons is zero and so $\psi(x)$ becomes a minimum. By setting $d\psi(x)/dx = 0$, we obtain $x = x_m$ corresponding to a minimum potential barrier height. Thus we have

$$x_m = \left(\frac{q}{16\pi\varepsilon F}\right)^{1/2} \qquad (2.23)$$

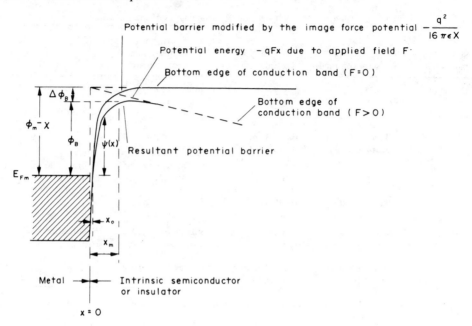

FIG. 2.12. Energy level diagram showing the lowering of potential barrier due to the combination of the image force and the applied uniform field for a neutral contact.

and the total lowering of the potential barrier height

$$\Delta \phi_B = (\phi_m - \chi) - \phi_B$$
$$= \left(\frac{q^3 F}{4\pi\varepsilon}\right)^{1/2} = \beta_{sc} F^{1/2} \tag{2.24}$$

with $$\beta_{sc} = (q^3/4\pi\varepsilon)^{1/2}$$

So that the effective potential barrier height can be written as

$$\phi_B = (\phi_m - \chi) - \left(\frac{q^3}{4\pi\varepsilon}\right)^{1/2} F^{1/2} \tag{2.25}$$

which is field dependent.

(B) Blocking contacts

Following a simple Schottky barrier model and taking a blocking contact between a metal and an *n*-type semiconductor as an example (the same principle can be applied straightforwardly to *p*-type semiconductors or to intrinsic semiconductors with $\phi_m > \phi_s$ with the bands bending up to block the electron injection, or with $\phi_m < \phi_s$ with the bands bending down to block the hole injection), the width of the Schottky barrier is given by [Gossick 1964]

$$W = \left[\frac{2\varepsilon(\phi_m - \phi_s + qV)}{q^2 N_d}\right]^{1/2} \tag{2.26}$$

W is a function of applied voltage across the junction V. The derivation of eq. (2.26) is based on the following assumptions (Schottky model):

(1) The barrier height ϕ_B is large compared with kT.
(2) $N_{d\,(\text{ionized})} = \text{constant}$ for $0 < x < W$, and

$$N_{d\,(\text{ionized})}\, W \gg \int_0^w n\, dx \qquad N_{d\,(\text{ionized})} > p_s = p\,(x = 0)$$

$$N_{d\,(\text{ionized})}\, W \gg \int_0^w p\, dx \qquad N_{d\,(\text{ionized})} > n_s = n\,(x = 0)$$

$$N_{d\,(\text{ionized})} = n\,(x = w)$$

This implies that the barrier is practically depleted of free carriers.
(3) The resistance is much higher in the barrier than outside the barrier, so that the applied voltage can be considered to be completely absorbed across the barrier.
(4) The mean free path of the electrons is small compared with W.

The potential energy of the Schottky barrier measured from the bottom edge of the conduction band at $x = 0$ is

$$-\frac{q^2 N_d}{\varepsilon}\left(Wx - \tfrac{1}{2}x^2\right) \tag{2.27}$$

This potential energy is equal to zero at $x = 0$ and equal to qV_d (contact potential energy $= \phi_m - \phi_s$ when $x = W$) as shown in Fig. 2.13(a). This potential energy will increase with increasing applied voltage V (with negative potential at the metal) as shown in Fig. 2.13(b). Thus the total potential barrier height measured from the Fermi level of the metal is given by

$$\psi(x) = \phi_m - \chi - \frac{q^2}{16\pi\varepsilon x} - \frac{q^2 N_d}{c}\left(Wx - \tfrac{1}{2}x^2\right) \tag{2.28}$$

Obviously, there is an optimum point at $x = x_m$ where $\psi(x) = \phi_B$ is a maximum. By setting $[d\psi(x)]/dx = 0$ we obtain

$$x^3 - Wx^2 + (16\pi N_d)^{-1} = 0 \tag{2.29}$$

since $W \gg X_m$. By neglecting x^3 in eq. (2.29) we have

$$X_m = \frac{1}{4(\pi W N_d)^{1/2}} \tag{2.30}$$

Substituting eq. (2.30) into eq. (2.28) and assuming $W \gg x_m$, we obtain

$$\phi_B = \phi_m - \chi - \left[\frac{q^6(\phi_m - \phi_s + qV)N_d}{2(8\pi)^2 \varepsilon^3}\right]^{1/4} \tag{2.31}$$

which again is field dependent. The total lowering of the potential barrier is [Simmons 1970]

$$\Delta\phi_B = \phi_m - \chi - \phi_B$$
$$= \left[\frac{q^6(\phi_m - \phi_s + qV)N_d}{2(8\pi)^2 \varepsilon^3}\right]^{1/4} \tag{2.32}$$

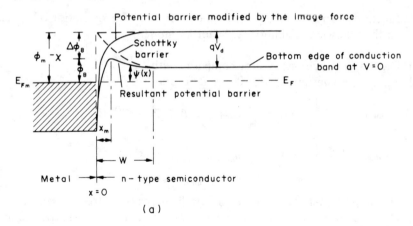

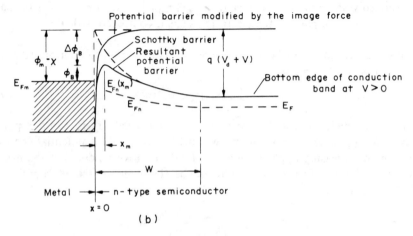

F IG. 2.13. Energy level diagrams showing the lowering of potential barrier due to the combination of the image force and the depletion layer effects for a blocking contact between a metal and an *n*-type semiconductor: (a) without applied voltage, and (b) with applied reverse bias voltage *V*.

It is important to mention that if $\phi_B < kT$ the electrons easily surmount the barrier and inject from the metal to the semiconductor, and so the barrier does not function as a barrier to block the carrier injection. Equally true, if the wavelength of the electron is larger than W though $\phi_B > kT$, the barrier becomes transparent to the electrons and in this case the barrier also does not function as a barrier to block the carrier injection. In the following we shall assume that W is much larger than the wavelength of electrons when considering field-enhanced thermionic emission, so that the tunnelling injection can be neglected; and that when considering tunnelling field emission W and ϕ_B are in such values that the thermionic injection can be neglected. However, it should be noted that under certain conditions both types of carrier injection may be equally important. Since the barrier width becomes much narrower at the energy levels far removed from the Fermi level of the metal, particularly close to the peak of the barrier, we shall also consider the case of thermally assisted tunnelling field emission.

2.2.2. Field-enhanced thermionic emission

If the emitted electrons are not influenced by either space charge or traps or, in other words, if all electrons emitted from the cathode are carried away in the conduction band and collected at the anode, the field-enhanced thermionic emission current is given by [e.g. Simmons 1971]

$$J = A^* T^2 \exp(-\phi_B/kT) \tag{2.33}$$

This equation may be written in the form

$$J = qn\bar{v}$$

with $n = N_c \exp[-\phi_B/kT]$ and $\bar{v} = (\frac{2}{\pi}kT/m)^{1/2}$ and then the expression is self-explanatory, where ϕ_B is given by eq. (2.25) or eq. (2.31) depending on the type of contacts, and A^* is the Richardson constant which is given by

$$A^* = A = \frac{4\pi q k^2 m}{h^3} = 120 \text{ A/cm}^2 \text{ (degree)}^2 \tag{2.34}$$

for thermionic emission into a vacuum. For thermionic emission into a semiconductor or an insulator, A^* is quite different from A. The factors governing the value of A^* are discussed as follows.

(A) The effect of effective mass [Crowell 1965, 1969]

For isotropic materials we can write

$$\frac{A_1^*}{A} = \frac{m^*}{m} \tag{2.35}$$

and for anisotropic materials, eq. (2.35) becomes

$$\frac{A_1^*}{A} = \frac{1}{m}(l_1^2 m_y^* m_z^* + l_2^2 m_z^* m_x^* + l_3^2 m_x^* m_y^*)^{1/2} \tag{2.36}$$

where $l_1, l_2,$ and l_3 are the direction cosines relative to the principal axes of the constant energy ellipsoid and $m_x^*, m_y^*,$ and m_z^* are the corresponding components of the effective mass tensor. The values of A_1^*/A for some inorganic semiconductors [Crowell 1965, 1969, Sze 1969] are summarized in Table 2.5.

TABLE 2.5. *Values of A_1^*/A*

Semiconductor	Silicon	Germanium	GaAs
n-type ⟨111⟩	2.15	1.07	0.072
n-type ⟨100⟩	2.05	1.19	
p-type	0.66	0.36	0.62

Data obtained from Crowell [1965, 1969].

(B) The correction due to the drift and diffusion of carriers in the depletion region

If the mobility of the carriers in the material is low, it will control the thermionic emission current. There are two approaches to this problem depending on the type of contacts.

(a) NEUTRAL CONTACTS

For most metal–insulator contacts we can consider them neutral contacts. Thus we follow the approach of O'Dwyer *et al.* [O'Dwyer 1973, Emtage and O'Dwyer 1966]. The total current density in the insulator is given by

$$J = qn(x)\mu \left[-\frac{d\psi(x)/q}{dx} \right] - qD\frac{dn(x)}{dx} \tag{2.37}$$

Since μ is assumed to be small and constant we can use the Einstein relation

$$\frac{D}{\mu} = \frac{kT}{q} \tag{2.38}$$

Thus eq. (2.37) can be written as

$$J = -n(x)\mu\frac{d\psi(x)}{dx} - kT\mu\frac{dn(x)}{dx} \tag{2.39}$$

The first term on the right-hand side of eq. (2.39) represents the drift current from the metal to the insulator, while the second term represents the diffusion current from the insulator to the metal. $\psi(x)$ in eq. (2.22) and in Fig. 2.12 can be written as

$$\frac{\psi(x)}{kT} = a - (bx + c/x) \tag{2.40}$$

where

$$a = \frac{\phi_m - \chi}{kT} \qquad b = \frac{qF}{kT} \qquad c = \frac{q^2}{16\pi\varepsilon kT} \tag{2.41}$$

Multiplying both sides of eq. (2.39) with $\exp(\psi/kT)$, integrating from $x = x_m$ to $x > x_m$ and using the boundary conditions

$$\psi(x = x_m) = \phi_B$$

$$n(x = x_m) = N_0 = N_c\exp\left(-\frac{\phi_B}{kT}\right) \tag{2.42}$$

we obtain

$$n(x) = \exp\left[-\frac{\psi}{kT} \right]\left[N_0 - \frac{J}{kT\mu}\int_{x_m}^{x} \exp(\psi/kT)\,dx \right] \tag{2.43}$$

With the aid of the following approximations

$$\int_0^x \exp[-(bx+c/x)]\,dx \simeq \int_0^\infty \exp[-(bx+c/x)]\,dx$$

$$-\frac{1}{b}\exp(-bx) \quad \text{if} \quad x \gg \left(\frac{c}{b}\right)^{1/2} \tag{2.44}$$

and

$$\int_0^\infty \exp[-(bx+c/x)]\,dx \simeq \frac{1}{b}, \quad \text{if} \quad (bc)^{1/2} \ll 1 \tag{2.45}$$

$$\simeq \pi^{1/2}\left(\frac{c}{b^3}\right)^{1/4}\exp[-2(bc)^{1/2}] \quad \text{if} \quad (bc)^{1/2} \gg 1 \tag{2.46}$$

we can obtain approximate solutions of eq. (2.43) by assuming x_m to be very small approaching $x \to 0$. First we have to examine the validity of eqs. (2.44)–(2.46) for our use. For example, taking $F = 10^4$ V cm^{-1} and $T = 300°$ K we have $(c/b)^{1/2} = (1.175 \times 10^{-4}/0.62 \times 10^4) = 1.9 \times 10^{-8}$ m, and $(bc)^{1/2} = (1.175 \times 10^{-4} \times 0.62 \times 10^4) = 0.75$. For fields lower than 10^4 V cm^{-1}, we use eqs. (2.44) and (2.45). Substitution of eqs. (2.44) and (2.45) into eq. (2.43) yields

$$n(x) = \frac{J}{q\mu F} + \exp\left[\frac{qFx}{kT}\right]\left[N_0\exp\left(-\frac{\phi_m-\chi}{kT}\right)-\frac{J}{q\mu F}\right] \tag{2.47}$$

To avoid divergence of $n(x)$ for large x, the coefficient of the exponential term $\left[N_0\exp\left(-\frac{\phi_m-\chi}{kT}\right)-\frac{J}{q\mu F}\right]$ must vanish, so we obtain

$$J = q\mu FN_c\exp\left[-\left(\frac{\phi_m-\chi}{kT}\right)\right]\exp\left(-\frac{\phi_B}{kT}\right)$$

$$= A_2^* T^2 \exp\left(-\frac{\phi_B}{kT}\right) \tag{2.48}$$

where

$$A_2^* = \frac{q\mu FN_c\exp\left[-\left(\frac{\phi_m-\chi}{kT}\right)\right]}{T^2} \tag{2.49}$$

Thus the correction factor is

$$\frac{A_2^*}{A_1^*} = \left(\frac{2\pi m^*}{kT}\right)^{1/2}\mu F\exp\left[-\left(\frac{\phi_m-\chi}{kT}\right)\right] \tag{2.50}$$

For fields higher than 10^4 V cm^{-1} we have to use eqs. (2.44) and (2.46). Substitution of eqs. (2.44) and (2.46) into eq. (2.43) yields

$$n(x) = \frac{J}{q\mu F} + \exp\left(\frac{qFx}{kT}\right)\left\{N_0\exp\left[-\left(\frac{\phi_m-\chi}{kT}\right)\right]\right.$$

$$\left. -\frac{J(\pi kT)^{1/2}}{kT\mu(4q\varepsilon F^3)^{1/4}}\exp(-\Delta\phi_B/kT)\right\} \tag{2.51}$$

Again, to avoid divergence of $n(x)$ for large x, the coefficient of the exponential term must vanish. Thus, we obtain

$$
\begin{aligned}
J &= q\mu N_0 \left(\frac{kT}{\pi}\right)^{1/2} \left[4\varepsilon\left(\frac{F}{q}\right)^3\right]^{1/4} \exp\left[-\frac{(\phi_m - \chi) - \Delta\phi_B}{kT}\right] \\
&= q\mu \left(\frac{kT}{\pi}\right)^{1/2} \left[4\varepsilon\left(\frac{F}{q}\right)^3\right]^{1/4} N_c \exp\left(-\frac{\phi_B}{kT}\right) \exp\left(-\frac{\phi_B}{kT}\right) \\
&= A_2^* T^2 \exp\left(-\frac{\phi_B}{kT}\right)
\end{aligned}
\tag{2.52}
$$

For this case

$$
A_2^* = \left\{q\mu\left(\frac{kT}{\pi}\right)^{1/2}\left[4\varepsilon\left(\frac{F}{q}\right)^3\right]^{1/4} N_c \exp\left(-\frac{\phi_B}{kT}\right)\right\}\Big/ T^2
\tag{2.53}
$$

and the correction factor is

$$
\frac{A_2^*}{A_1^*} = \mu(2m^*)^{1/2}\left[4\varepsilon\left(\frac{F}{q}\right)^3\right]^{1/4}\exp\left(-\frac{\phi_B}{kT}\right)
\tag{2.54}
$$

It can be seen that if A_2^*/A_1^* is less than unity the thermionically emitted current will be controlled by diffusion. O'Dwyer [1973] has pointed out that the condition for $A_2^*/A_1^* < 1$ is

$$
\mu \leq 5/F^{3/4}
\tag{2.55}
$$

in which μ is in $cm^2 V^{-1} sec^{-1}$ and F in MV cm^{-1}, and that there is no incontrovertible experimental evidence of diffusion-limited thermionic emission.

(b) BLOCKING CONTACTS

Reffering to Fig. 2.13, the electron density in the region between $x = x_m$ and $x = W$ is given by [Crowell and Sze 1966]

$$
n(x) = N_c \exp\{-[\psi(x) - E_{Fn}]/kT\}
\tag{2.56}
$$

and the current density by

$$
J = -n(x)u\frac{d\psi(x)}{dx} - kT\mu\frac{dn(x)}{dx} = -n(x)\mu\frac{dE_{Fn}}{dx}
\tag{2.57}
$$

where $\psi(x)$ and E_{Fn} are measured from the Fermi level E_{Fm} of the metal as shown in Fig. 2.13. The quasi-Fermi level E_{Fn} is sometimes called imref—the backward spelled Fermi used for describing carrier distribution under non-equilibrium conditions such as under an applied field, to distinguish it from the Fermi level for equilibrium conditions. Equations (2.56) and (2.57) are valid only for $x > x_m$. For $0 < x < x_m$ the density of electrons cannot be described by E_{Fn} nor be associated with N_c because of the rapid change of the potential energy in the distance comparable to the electrons mean path. Crowell and Sze have assumed that the barrier in the region $0 < x < x_m$ acts as a sink of electrons, then the current flow in this region is

$$J = q(N_0 - N_m)v_R \tag{2.58}$$

where v_R is the effective recombination velocity, N_0 is the quasi-equilibrium electron density at x_m

$$N_0 = N_c \exp(-\phi_B/kT) \tag{2.59}$$

and N_m is the electron density at x_m when the current is flowing

$$N_m = N_c \exp\{-[\phi_B - E_{Fn}(x_m)]/kT\} \tag{2.60}$$

Using the boundary condition

$$E_{Fn}(x = w) = -qV \tag{2.61}$$

and from eqs. (2.56)–(2.60) we obtain

$$J = \frac{qN_c v_R}{1 + (v_R/v_D)} \exp\left[-\frac{\phi_B}{kT}\right][1 - \exp(-qV/kT)]$$

$$\simeq \frac{qN_c V_R}{1 + (v_R/v_D)} \exp\left(-\frac{\phi_B}{kT}\right)$$

$$= A_2^* T^2 \exp\left(-\frac{\phi_B}{kT}\right) \quad \text{for large } V \tag{2.62}$$

where

$$v_D = \left[\int_{x_m}^{W} \frac{q}{\mu kT} \exp\left(-\frac{\phi_B - \psi}{kT}\right)dx\right]^{-1} \tag{2.63}$$

is the effective diffusion velocity associated with the diffusion of electrons from $x = W$ to $x = x_m$. Thus

$$A_2^* = \frac{qN_c v_R v_D}{(v_D + v_R)T^2} \tag{2.64}$$

and the correction factor is

$$\frac{A_2^*}{A_1^*} = \frac{qN_c v_R v_D}{(v_D + v_R)T^2 A_1^*}$$

$$= \left(\frac{2\pi m^*}{kT}\right)^{1/2}\left(\frac{v_D v_R}{v_D + v_R}\right) \tag{2.65}$$

In general, if $v_D \gg v_R$ the effect of diffusion is not important, but if $v_R \gg v_D$ the diffusion process is dominant.

(C) The effects of phonon-scattering and quantum mechanical reflection

When an electron crosses the peak of the potential barrier, there is a probability that it will be back-scattered by the scattering between the electron and the optical phonon.

This effect will reduce the net current over the barrier. Crowell and Sze [1966] have proposed that this effect can be viewed as a small perturbation, and that by neglecting the scattering by acoustic phonons the probability of electron emission over the peak of the potential barrier is given by [Sze 1969]

$$f_p \simeq \exp\left(-\frac{x_m}{\lambda}\right) \tag{2.66}$$

where x_m is given by eq. (2.23) and λ is the optical phonon mean free path which is

$$\lambda = \lambda_0 \tanh\left(\frac{E_p}{2kT}\right) \tag{2.67}$$

where E_p is the optical phonon energy and λ_0 is the high energy low temperature asymptotic value of the phonon mean free path. They have also suggested that the effect of optical phonon scattering due to f_p can be taken into account by replacing v_R with a smaller recombination velocity $f_p v_R$ in eqs. (2.62), (2.64), and (2.65) because $f_p < 1$.

Over the Schottky barrier there is quantum mechanical reflection of electrons, and below the peak of the barrier there is tunnelling of electrons through the thinner part of the barrier. Crowell and Sze [1966] have calculated the ratio f_q of the total current flow, taking into account the effects of electron tunelling and quantum mechanical reflection, to the current flow neglecting these effects, as a function of electron energy and it is

$$f_q = \int_{-\infty}^{\infty} \frac{D_q}{kT} \exp(-E/kT)\,dE \tag{2.68}$$

where D_q is the predicted quantum mechanical transmission coefficient, and E is the electron energy which is related to the barrier height and hence the electric field. Therefore the effective recombination velocity is $f_p f_q v_R$.

Taking the two effects into account eq. (2.64) has to be modified as follows:

$$A_2^{**} = \frac{qN_c f_p f_q v_R v_D}{(v_D + f_p f_q v_R)T^2} \tag{2.69}$$

(D) Apart from the effect of contamination at the interface between the metal and the semiconductor (or insulator) the following factors may also be responsible for the discrepancy between the theoretically expected and the experimental value of A^*: (i) the space charge existing in the vicinity of the contact, (ii) the emitting area much less than the actual electrode surface area because of filamentary injection [Kao 1976, Hwang and Kao 1974], (iii) the value of the dielectric constant ε less than the d.c. or static value of ε in the vicinity of the contact [Henisch 1957, Sze et al. 1964, Thornber et al. 1967], (iv) the actual shape of the potential barrier different from the ideal ones shown in Figs. 2.12 and 2.13 because of the imperfections such as surface states, and (v) the implicit assumption that there is no appreciable electron–electron interaction, and thus no interaction terms are included when the Fermi–Dirac distribution is used to derive the J–V characteristics—this assumption may be valid for low currents but may not for high currents.

So far, we have discussed the factors which may affect the value of A^*. In general, if $\phi_B \gg kT$ or if the electron mean free path in the semiconductor or insulator is large compared to the distance over which the potential changes by kT near the top of the barrier, the effects of diffusion and electron collision are unimportant and can be neglected [Bethe 1942, Pollack and Seitchik 1969]. If these effects are unimportant the thermionic emitted current $\ln J$ should be linearly proportional to $V^{1/2}$ at a fixed temperature, and $\ln(J/T)$ should be linearly proportional to $1/T$ at a fixed applied voltage. Figures 2.14 and 2.15 are some typical examples about field-enhanced thermionic emission. In Fig. 2.15, I is the total current and I_T is the temperature independent portion possibly due to quantum mechanical tunnelling (field emission).

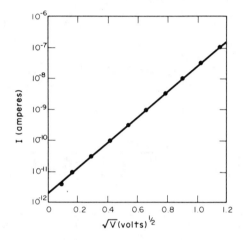

FIG. 2.14. The current–voltage characteristics of a Zn–ZnO–Au system due to Schottky thermionic emission. [After Mead 1966.]

2.2.3. Field emission

Field emission is referred to as the quantum mechanical tunnelling of electrons through a potential barrier from a metal to a semiconductor or an insulator under an intense electric field. At low temperatures most electrons tunnel at the Fermi level of the metal-constituting field emission (F-emission), at intermediate temperatures most electrons tunnel at an energy level E_m (above the Fermi level of the metal) constituting the so-called thermionic field or thermally assisted field emission (T–F emission), and at very high temperatures the main contribution is the thermionic emission. In this subsection we are concerned only with the F-emission. Since the first treatment of this problem by Fowler and Nordheim [1928], many investigators have modified their treatment for emission into a semiconductor or an insulator rather into a vacuum [Duke 1969, Good and Muller 1956, Burstein and Lundqvist 1969, Padovani and Stratton 1966].

The field-emitted current is given by

$$J = q \int D_T v_x n(E)\, dE \qquad (2.70)$$

where D_T is the quantum mechanical transmission function or the transition probability defined as the ratio of the transmitted to the incident current, v_x is the

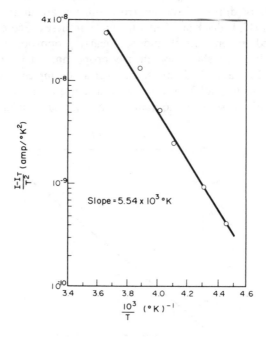

FIG. 2.15. $(I - I_T)/T^2$ as a function of $1/T$ for a Pb–Al$_2$O$_3$–Pb system. [After Pollack 1963.]

electron velocity in the x-direction, and the electron density with energies between E and $E + dE$ is

$$n(E)\,dE = g(E)\,f(E)\,dE$$

$$= \frac{8\pi m \sqrt{2mE}}{h^3} f(E)\,dE$$

$$= \frac{4\pi p^2\,2dp}{h^3} f(E)$$

and $n(E)\,dE$ within $dp_x\,dp_y\,dp_z$ is therefore

$$n(E)\,dE = \frac{4\pi p^2\,2dp}{h^3} f(E) \frac{dp_x\,dp_y\,dp_z}{4\pi p^2\,dp}$$

$$= \frac{2}{h^3} f(E)\,dp_x\,dp_y\,dp_z \tag{2.71}$$

Thus, the net current flow from region 1 to region 2 (cf. Fig. 2.16) is

$$J = \frac{2q}{h^3} \int D_T v_{x1} \left[f_1(E_2) - f_2(E_2) \right] dp_{x1}\,dp_{y1}\,dp_{z1} \tag{2.72}$$

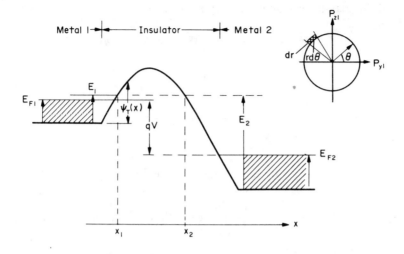

FIG. 2.16. Energy level diagram for tunnelling analysis. The insulator region may represent an insulating film or a barrier formed near the contact between a metal and an insulator (or a semiconductor).

where

$$f_1(E_1) = \left[\exp\{(E_1 - E_{F1})/kT\} + 1\right]^{-1} \tag{2.73}$$

$$f_2(E_2) = \left[\exp\{(E_2 - E_{F2})/kT\} + 1\right]^{-1}$$
$$= \left[\exp\{(E_1 + qV - E_{F1})/kT\} + 1\right]^{-1} \tag{2.74}$$

$$v_{x1} = \frac{\partial \omega}{\partial k} = \frac{\partial E_{x1}}{\partial p_{x1}} \tag{2.75}$$

and

$$D_T = \exp\left\{ -\frac{4\pi}{h} \int_{x_1}^{x_2} \left[2m(\psi_T - E_{x1})\right]^{1/2} dx \right\} \tag{2.76}$$

D_T is based on the WKB (Wentzel–Kramers–Brillouin) approximation and the condition for its validity is that the electron wavelength is small compared to the region over which appreciable variations in the potential energy of the electrons occurs [Stratton 1969], and E_{x1} is the portion of the electron energy E_1 for electron motion in the x-direction. An applied voltage V to favour the electron tunnelling from region 1 to region 2 will lower E_{F2} as shown in Fig. 2.16. Assuming that the current flow is dominant from region 1 to region 2 at E_1 whose energy is close to E_{F1}, and letting

$$\alpha = \frac{4\pi(2m)^{1/2}}{h} \tag{2.77}$$

and

$$\varepsilon_x = E_{F1} - E_{x1} \tag{2.78}$$

then eq. (2.76) may be expanded in Taylor series

$$\ln D_T(E_{1x}) = -\alpha \int_{x_1}^{x_2} (\psi_T - E_{x1})^{1/2} \, dx$$

$$= -\alpha \int_{x_1}^{x_2} (\psi_T - E_{F1} + \varepsilon_x)^{1/2} \, dx$$

$$= -[b_1 + c_1 \varepsilon_x + f_1 \varepsilon_x^2 + \ldots] \tag{2.79}$$

in which

$$b_1 = \alpha \int_{x_1}^{x_2} (\psi_T - E_{F1})^{1/2} \, dx \tag{2.80}$$

$$c_1 = \frac{1}{2}\alpha \int_{x_1}^{x_2} (\psi_T - E_{F1})^{-1/2} \, dx \tag{2.81}$$

and

$$f_1 = \frac{1}{4}\alpha \left\{ \frac{1}{x_2 - x_1} \left(\frac{1}{\psi'_T(x_1)} - \frac{1}{\psi'_T(x_2)} \right) \int_{x_1}^{x_2} \frac{dx}{(\psi_T - E_{F1})^{1/2}} \right.$$
$$\left. - \frac{1}{2} \int_{x_1}^{x_2} \frac{dx}{(\psi_T - E_{F1})^{1/2}} \left[1 - \frac{\psi'_T(x)}{(x_2 - x_1)} \left(\frac{x - x_1}{\psi'_T(x_2)} + \frac{x^2 - x}{\psi'_T(x_1)} \right) \right] \right\} \tag{2.82}$$

b_1, c_1, and f_1 are functions of applied voltage V through the voltage dependence of x_1 and x_2. Substituting eqs. (2.73)–(2.75) into eq. (2.72), we have

$$J = \frac{2q}{h^3} \int_0^\infty [f_1(E_1) - f_2(E_2)] \, dE_{x1} \int D_T(E_1, p_{y1}, p_{z1}) dp_{y1} dp_{z1} \tag{2.83}$$

since

$$D_T(E_{x1}) = D_T\left(E_1 - \frac{p_{y1}^2 + p_{z1}^2}{2m} \right) = D_T(E_1, p_{y1}, p_{z1}) \tag{2.84}$$

and the conservation law

$$\left. \begin{aligned}
p_{y1}^2 + p_{z1}^2 &= p_{y2}^2 + p_{z2}^2 = p_\perp^2 \\
E_1 &= E_2 = E \\
E_1 - E_{x1} &= E_\perp = \frac{p_\perp^2}{2m}
\end{aligned} \right\} \tag{2.85}$$

We can write

$$-dE_{x1} = \frac{p_\perp dp_\perp}{m} = \frac{rdr}{m} \tag{2.86}$$

$$dp_{y1} dp_{z1} = rdrd\theta \tag{2.87}$$

with

$$r^2 = p_{y1}^2 + p_{z1}^2 \tag{2.88}$$

as shown in Fig. 2.16. Thus, we obtain

$$J = \frac{4\pi mq}{h^3} \int_0^\infty [f_1(E_1) - f_2(E_2)] dE_{x1} \int_0^{E_{x1}} D_T(E_{x1}) dE_{x1} \qquad (2.89)$$

Integration of eq. (2.89) by parts yields

$$J = \frac{4\pi qmkT}{h^3} \int_0^\infty D_T(E_{x1}) \left\{ \frac{1 + \exp[(E_{F1} - E_{x1})/kT]}{1 + \exp[(E_{F1} - E_{x1} - qV)/kT]} \right\} dE_{x1} \qquad (2.90)$$

Substituting eq. (2.79) into eq. (2.90) and integrating it, we obtain

$$J = \frac{A^* T^2 \exp(-b_1)}{(c_1 kT)^2} \frac{\pi c_1 kT}{\sin(\pi c_1 kT)} [1 - \exp(-c_1 V)] \qquad (2.91)$$

where $A^* = (4\pi qm^* k^2)/h^3$.

We consider only two terms in eq. (2.79) and this is justified because

$$\frac{1}{kT} - c_1 > (2f_1)^{1/2} \qquad (2.92)$$

in order to satisfy the condition for **WKB** approximation. This also implies that $c_1 kT < 1$ and that no singularity would be involved in eq. (2.91).

The temperature dependence of J can be easily found from eq. (2.91)

$$\frac{J(T)}{J(T=0)} = \frac{\pi c_1 kT}{\sin(\pi c_1 kT)}$$

$$\simeq 1 + \frac{1}{6}(\pi c_1 kT)^2 + \ldots \qquad (2.93)$$

and $c_1 kT < 1$.

Considering a Schottky barrier shown in Fig. 2.17(b) and neglecting the effect of image force for simplicity, we have for low temperatures

$$x_1 = 0$$

$$x_2 = \frac{\varepsilon \phi_B}{q^2 N_d W}$$

and

$$\psi_T(x) = E_{Fm} + \phi_B - \frac{q^2 N_d}{\varepsilon} \left(Wx - \frac{1}{2}x^2 \right) \qquad (2.94)$$

$$E_{F1} = E_{Fm}$$

Using these parameters, integration of eqs. (2.80) and (2.81) gives

$$b_1 = \frac{1}{E_{00}} \left\{ \phi_B^{1/2} (\phi_B + qV)^{1/2} + qV \ln \left[\frac{(\phi_B + qV)^{1/2} + \phi_B^{1/2}}{(qV)^{1/2}} \right] \right\} \qquad (2.95)$$

and

$$c_1 = \frac{1}{E_{00}} \ln \left[\frac{(\phi_B + qV)^{1/2} + \phi_B^{1/2}}{(qV)^{1/2}} \right] \qquad (2.96)$$

where

$$E_{00} = \frac{2q}{\alpha} \left(\frac{N_d}{2\varepsilon} \right)^{1/2} \qquad (2.97)$$

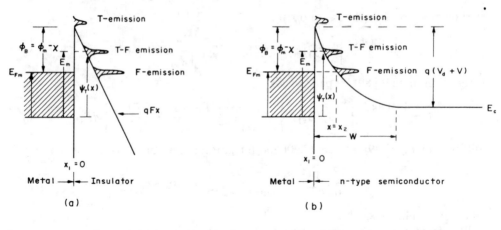

FIG. 2.17. Schematic energy level diagrams showing the thermionic (T) emission, thermionic field ($T-F$) emission, and field (F) emission for (a) neutral contacts, and (b) blocking contacts.

If $|qV| > \phi_B$, then eqs. (2.95) and (2.96) become

$$b_1 = 2\phi_B^{3/2}/3E_{00}(\phi_B + qV)^{1/2} \tag{2.98}$$

$$c_1 = \phi_B^{1/2}/E_{00}(\phi_B + qV)^{1/2} \tag{2.99}$$

and eq. (2.91) becomes

$$J = \frac{A^* T^2 \pi E_{00} \exp[-2\phi_B^{3/2}/3E_{00}(\phi_B + qV)^{1/2}]}{kT[\phi_B/(\phi_B + qV)]^{1/2} \sin\{\pi kT[\phi_B/(\phi_B + qV)]^{1/2}/E_{00}\}} \tag{2.100}$$

For low temperatures, eq. (2.100) reduces to

$$J = A^* T^2 \left(\frac{E_{00}}{kT}\right)^2 \frac{(\phi_B + qV)}{\phi_B} \exp\left[-\frac{2\phi_B^{3/2}}{3E_{00}(\phi_B + qV)^{1/2}}\right] \tag{2.101}$$

Considering a triangular barrier as shown in Fig. 2.17(a) then we have

$$\left.\begin{array}{c} x_1 = 0 \\[4pt] x_2 = \dfrac{\phi_B}{qF} \\[4pt] \psi_T(x) = E_{Fm} + \phi_B - qFx \\[4pt] E_{F1} = E_{Fm} \end{array}\right\} \tag{2.102}$$

and

Using these parameters, integration of eqs. (2.80) and (2.81) gives

$$b_1 = \alpha\frac{2\phi_B^{3/2}}{3qF} \tag{2.103}$$

$$c_1 = \alpha\frac{\phi_B^{1/2}}{qF} \tag{2.104}$$

Thus the field emission current is

$$J = \frac{A^* T^2 \pi \exp\left(-2\alpha\phi_B{}^{3/2}/3qF\right)}{(\alpha\phi_B{}^{1/2}kT/qF)\sin\left(\pi\alpha\phi_B{}^{1/2}kT/qF\right)} \qquad (2.105)$$

For low temperatures

$$J = \frac{A^* T^2}{\phi_B}\left(\frac{qF}{\alpha kT}\right)^2 \exp\left[-\frac{2\alpha\phi_B{}^{3/2}}{3qF}\right] \qquad (2.106)$$

Equations (2.101) and (2.106) are essentially equivalent to the Fowler–Nordheim equation, since for the Schottky barrier the field at the interface is proportional to the square root of the effective barrier height.

If the effect of image force is taken into account, the integral for b_1 and c_1 may be written in terms of complete elliptic integrals of the first and second kinds. The field emission current taking into account this effect has been derived [O'Dwyer 1973, Murphy and Good 1956], but it should be noted that up to the present there is no incontrovertible experimental evidence as to which expression is more accurate to describe field emission. It is most likely that the field emission is filamentary [Kao 1976, Hwang and Kao 1974] and the current is SCL under high fields. If this is so, all expressions for field emission have to be modified. However, the experimental results

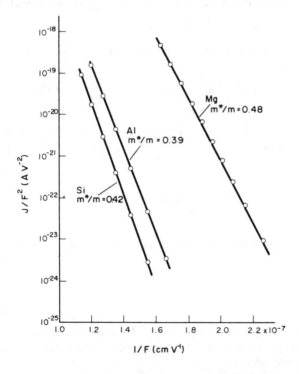

FIG. 2.18. Fowler–Nordheim plot of current density as a function of electric field for field emission from various metal electrodes into SiO_2 at room temperature. m^*/m is the ratio of electron effective mass to rest mass. [After Lenzlinger and Snow 1969.]

showing a linear relation between $\ln J/F^2$ and $1/F$ such as shown in Fig. 2.18 may be taken as an indication of field emission.

2.2.4. Thermionic field emission

In the intermediate temperature range (when $c_1 kT > 1$) most of electrons tunnel at an energy level E_m which is lower than $\phi_B + E_{Fm}$ but higher than E_{Fm} as shown in Fig. 2.17. For a Schottky barrier Stratton [1962] has derived an expression for J as a function of V using a procedure similar to that given in Section 2.2.3. The thermionic field emission current density is given by [Padovani and Stratton 1966]

$$J = J_s \exp(qV/E') \tag{2.107}$$

where

$$J_s = A^* T^2 \left(\frac{\pi E_{00}}{k^2 T^2}\right)^{1/2} \left[qV + \frac{\phi_B}{\cosh^2(E_{00}/kT)}\right]^{1/2} \exp\left(-\frac{\phi_B}{E_0}\right) \tag{2.108}$$

$$E' = E_{00}[E_{00}/kT - \tanh(E_{00}/kT)]^{-1} \tag{2.109}$$

and

$$E_0 = E_{00} \coth(E_{00}/kT) \tag{2.110}$$

It can easily be shown that for T–F emission the energy level E_m, which represents the peak of the energy distribution of the emitted electrons, will be such that

$$c_1(E_m)kT = 1 \tag{2.111}$$

and that the energy distribution of the emitted electrons is a Gaussian distribution with half-width [Stratton 1964]

$$\Delta = 2(\ln 2)^{1/2} E_{00}^{1/2} [qV + \phi_B/\cosh^2(E_{00}/kT)]^{1/2} \tag{2.112}$$

where

$$c_1(E_m) = \frac{1}{E_{00}} \ln\left[\frac{(\phi_B + qV)^{1/2} + (\phi_B + E_{Fm} - E_m)^{1/2}}{(E_m + qV - E_{Fm})^{1/2}}\right] \tag{2.113}$$

and

$$E_m = E_{Fm} + \frac{\phi_B - qV \sinh^2(E_{00}/kT)}{\cosh^2(E_{00}/kT)} \tag{2.114}$$

The above results are only valid in a certain temperature range given by the two conditions

$$c_1 kT > 1$$
$$D_T(E_m) < 1/e \tag{2.115}$$

This implies that $E_m < E_{Fm} + \phi_B$. When $E_m \geq E_{Fm} + \phi_B$, the thermionic field emission will change to thermionic emission. The minimum voltage to be applied for thermionic field emission is

$$qV > \phi_B + \frac{3E_{00}}{2} \frac{\cosh^2(E_{00}/kT)}{\sinh^3(E_{00}/kT)} \tag{2.116}$$

The plot of eq. (2.116) for a typical Schottky barrier—gold–gallium–arsenide barrier—

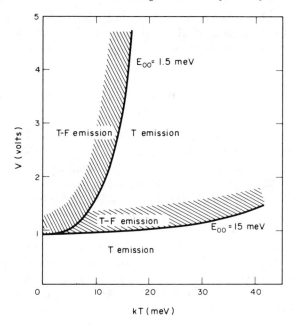

FIG. 2.19. Minimum applied reverse bias (V) for thermionic field emission as a function of temperature for Au–GaAs Schottky barrier for two values of the parameter E_{00}. $T-F$ and T denote, respectively, the thermionic field emission and thermionic emission. [After Padovani and Stratton 1966.]

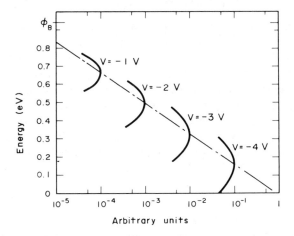

FIG. 2.20. The energy distribution of the thermionic field emitted electrons (density in arbitrary units) from Au electrode to GaAs at 140°C as a function of applied bias V. [After Padovani and Stratton 1966.]

of barrier height of 0.95 eV is shown in Fig. 2.19 and the typical energy distribution of electrons emitted from the metal (Au) into the semiconductor (GaAs) as a function of applied voltage for the temperature of 140° C is shown in Fig. 2.20.

2.3. TUNNELLING THROUGH THIN INSULATING FILMS

If an insulator is sufficiently thin or contains a large number of imperfections, or both, electrons may tunnel directly from one electrode to another and constitute a measurable current without involving the movement of carriers in the conduction band or in the valence band of the insulator. The tunnelling current–voltage characteristics depend on the thin film fabrication procedures and, particularly, the properties of the interfaces between the insulating film and the electrodes, and in addition, on the intrinsic properties of the insulator and the work functions of the metallic electrodes. It is most likely that the tunnelling is filamentary and the current density is not uniform over the electrode surface even with a planar symmetrical electrode geometry [Kao 1976, Hwang and Kao 1974]. A great deal of work, both theoretical and experimental, has been published with the aim of providing a quantitative explanation of tunnelling phenomena and information of fabrication techniques for device applications. Duke [1969], Burstein and Lundqvist [1969], and Simmons [1971] have given a quite comprehensive review on this subject. Since Frenkel [1930] reported his approximate analysis of electron tunnelling through a thin insulating film, several investigators [Sommerfeld and Bethe 1933, Holm 1951, Murphy and Good 1956, Stratton 1962, Simmons 1963, 1964] have extended and elaborated his work. The following discussion is mainly based on the work of Stratton [1962] and Simmons [1963, 1964].

2.3.1. A generalized potential barrier

A great many features of tunnelling phenomena in insulating films are essentially of a one-dimensional nature. If the potential barrier for tunnelling extends in the x-direction, the momentum components of the electrons in the y- and z-directions, which are normal to the direction of the current flow, can be considered to be merely fixed parameters. Thus, the probability that an electron at the energy level E_x can penetrate a potential barrier of height $\psi_T(x)$ and of width $S_2 - S_1$ as shown in Fig. 2.21 can be calculated by the well-known WKB or WKBJ (Wentzel–Kramers–Brillouin–Jeffreys) approximation method. Using this method, it can be shown that the number of electrons tunnelling per second from electrode 1 to electrode 2 as shown in Fig. 2.21 is given by

$$N_1 = \frac{4\pi m}{h^3} \int_0^{E_m} D_T(E_x) dE_x \int_0^\infty f(E) dE_r \qquad (2.117)$$

and that the number of electrons tunnelling per second from electrode 2 to electrode 1 is given by

$$N_2 = \frac{4\pi m}{h^3} \int_0^{E_m} D_T(E_x) dE_x \int_0^\infty f(E + qV) dE_r \qquad (2.118)$$

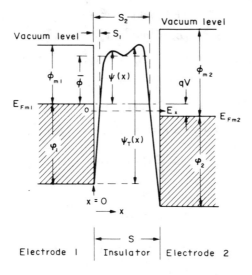

Fᴵɢ. 2.21. General potential barrier for a metal–insulating film–metal system. The width of the barrier for tunnelling at the energy level E_x is $S_2 - S_1$.

The current density due to the net flow of electrons from electrode 1 to electrode 2 through the forbidden energy gap of the insulating film is given by [Simmons 1963]

$$J = q(N_1 - N_2)$$
$$= \int_0^{E_m} D_T(E_x)\xi d E_x \qquad (2.119)$$

where E_m is the maximum energy of the electrons in the electrode, $D_T(E_x)$ is the probability that an electron at the energy level E_x can penetrate a potential barrier of height $\psi_T(x)$ and of width $S_2 - S_1$, which is derived on the basis of the WKBJ approximation and given by

$$D_T(E_x) = \exp\left(-\frac{4\pi}{h} \int_{S_1}^{S_2} \{2m[\psi_T(x) - E_x]\}^{1/2} dx \right) \qquad (2.120)$$

ξ is generally called the supply function, and $\xi d E_x$ represents the difference between the number of electrons having energy in the range from E_x to $E_x + d E_x$ incident on one side per second per unit area and those incident on the opposite side of the barrier. ξ is given by

$$\xi = \xi_1 - \xi_2 \qquad (2.121)$$

$$\xi_1 = \frac{4\pi m q}{h^3} \int_0^\infty f(E) d E_r \qquad (2.122)$$

$$\xi_2 = \frac{4\pi m q}{h^3} \int_0^\infty f(E + qV) d E_r \qquad (2.123)$$

and
$$E = \tfrac{1}{2}m(v_x^2 + v_y^2 + v_z^2)$$
$$E_r = \tfrac{1}{2}m(v_y^2 + v_z^2)$$
$$\tag{2.124}$$

For a generalized barrier, the barrier height can be written as

$$\psi_T(x) = \varphi_1 + \psi(x) \tag{2.125}$$

Substituting eq. (2.125) into eq. (2.120) and simplifying it, we obtain [Simmons 1963]

$$D_T(E_x) \simeq \exp\left[-C(\varphi_1 + \bar{\phi} - E_x)^{1/2}\right] \tag{2.126}$$

where
$$C = \frac{4\pi}{h}(2m)^{1/2}\beta|S_2 - S_1| \tag{2.127}$$

$$\bar{\phi} = \frac{1}{|S_2 - S_1|}\int_{S_1}^{S_2}\psi(x)dx \tag{2.128}$$

$$\beta = 1 - \frac{1}{8\bar{g}^2|S_2 - S_1|}\int_{S_1}^{S_2}[g - \bar{g}]^2 dx \tag{2.129}$$

and
$$\bar{g} = \frac{1}{|S_2 - S_1|}\int_{S_1}^{S_2}g(x)dx$$
$$g = [\psi_T(x) - E_x]^{1/2} \tag{2.130}$$

(a) FOR $T = 0$

For this case eqs. (2.121)–(2.123) become

$$\xi_1 = \frac{4\pi m q}{h^3}(\varphi_1 - E_x) \tag{2.131}$$

$$\xi_2 = \frac{4\pi m q}{h^3}(\varphi_1 - E_x - qV) \tag{2.132}$$

$$\xi^* = \xi_1 - \xi_2 = \frac{4\pi m q}{h^3}(qV), \quad 0 < E_x < \varphi_1 - qV$$
$$= \frac{4\pi m q}{h^3}(\varphi_1 - E_x), \quad \varphi_1 - qV < E_x < \varphi_1$$
$$= 0, \quad E_x > \varphi_1 \tag{2.133}$$

Substitution of eqs. (2.126) and (2.133) into eq. (2.119) yields

$$J \simeq J_0\{\bar{\phi}\exp(-C\bar{\phi}^{1/2}) - (\bar{\phi} + qV)\exp[-C(\bar{\phi} + qV)^{1/2}] \tag{2.134}$$

where
$$J_0 = \frac{q}{2\pi h}(\beta|S_2 - S_1|)^{-2} \tag{2.135}$$

Equation (2.134) can be applied to any shape of potential barrier provided that the mean barrier height $\bar{\phi}$ is known. Of course, if the $J - V$ characteristic is known, then $\bar{\phi}$

can be determined. The first term on the right-hand side of eq. (2.134) can be interpreted as the current flow from electrode 1 to electrode 2 and the second term as the current flow from electrode 2 to electrode 1. For low applied voltages, $qV \to 0$, $\beta \simeq 1$, and $\overline{\phi} \gg qV$, then eq. (2.134) becomes

$$J \simeq J_0 qV [C\overline{\phi}^{1/2}/2 - 1] \exp(-C\overline{\phi}^{1/2}) \tag{2.136}$$

J is a linear function of V for very low voltages. At $T = 0$ or at a fixed temperature it is not possible to determine whether the current flow is a tunnelling process or not, at low voltages from the J–V characteristics alone.

(b) FOR $T > 0$

For this case, eqs. (2.122) and (2.123) become

$$\xi_1 = \frac{4\pi m q k T}{h^3} \ln \{1 + \exp [(E_{Fm1} - E_x)/kT]\} \tag{2.137}$$

and

$$\xi_2 = \frac{4\pi m q k T}{h^3} \ln\{1 + \exp[(E_{Fm1} - E_x - qV)/kT]\} \tag{2.138}$$

Assuming that the current flow is predominantly electrons with energies close to E_{Fm1} then the integral in eq. (2.120) can be expanded with respect to Θ_x which is

$$\Theta_x = E_{Fm1} - E_x = \varphi_1 - E_x \tag{2.139}$$

By carrying out a Taylor expansion, eq. (2.120) becomes

$$\ln D_T(E_x) = -[b_1 + c_1 \theta_x + f_1 \theta_x^2 + \ldots] \tag{2.140}$$

where

$$b_1 = \frac{4\pi (2m)^{1/2}}{h} \int_{S_1}^{S_2} [\varphi_1 + \psi(x) - E_x]^{1/2} \, dx \tag{2.141}$$

$$c_1 = \frac{2\pi (2m)^{1/2}}{h} \int_{S_1}^{S_2} [\varphi_1 + \psi(x) - E_x]^{-1/2} \, dx \tag{2.142}$$

which are functions of V through $\psi(x)$ which is a function of V. For most practical cases [Murphy and Good 1956]

$$1 - c_1 kT > kT(2f_1)^{1/2} \tag{2.143}$$

the term with the quadratic and high orders can be neglected. Thus, applying eq. (2.119), we obtain

$$J = \frac{4\pi m q k T}{h^3} \int_0^{E_m} \exp\left\{ -[b_1 + c_1(\varphi_1 - E_x)] \right\} \ln\left\{ \frac{1 + \exp[(\varphi_1 - E_x)/kT]}{1 + \exp[(\varphi_1 - E_x - qV)/kT]} \right\} dE_x \tag{2.144}$$

After integration [Stratton 1962] we have

$$J = \frac{4\pi m q}{h^3 c_1} \exp(-b_1)[1 - \exp(-c_1 V)] \frac{\pi c_1 kT}{\sin \pi c_1 kT} \tag{2.145}$$

Thus, at a given applied voltage, the ratio of the tunnelling current through a thin insulating film for $T > 0$ to that for $T = 0$ is

$$\frac{J(T > 0)}{J(T = 0)} = \frac{\pi c_1 kT}{\sin \pi c_1 kT} \simeq \frac{\pi c_1 kT}{\pi c_1 kT - (\pi c_1 kT)^3/3! + \ldots} \simeq 1 + \frac{1}{6}(\pi c_1 kT)^2 \quad (2.146)$$

Equations (2.145) and (2.146) are identical to eqs. (2.91) and (2.93) indicating that the basic tunnelling process in field emission and that in tunnelling through a thin film are identical. The only difference between these two cases is the values of b_1 and c_1, which depend on the profile of the potential barrier. However, the slight quadratic dependence of the current on temperature is characteristic of the tunnelling process. Figure 2.22 shows the good agreement between the experimental results and eq. (2.146) for temperature dependence of tunnelling current through Al_2O_3 films. It should be noted that at $T = 0$, eq. (2.145) is more general than eq. (2.134) and can be used for any shape of the potential barrier without the need of approximating the shape of the barrier to a rectangular one before calculations.

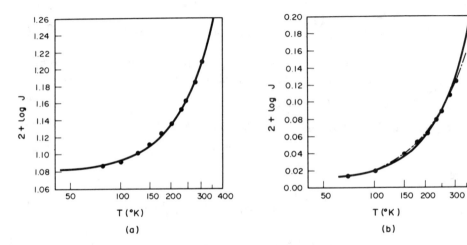

FIG. 2.22. Temperature dependence of the tunnelling current density in an $Al–Al_2O_3–Al$ system with the oxide thickness of 30 Å: (a) 0.5 V, and (b) 0.05 V. The points are experimental and the solid curves are theoretical results based on eq. (2.146). [After Hartman and Chivian 1964.]

2.3.2. A rectangular potential barrier

(A) Between dissimilar electrodes

We shall outline Simmons' treatment of this problem as follows: if the electrodes are dissimilar, an intrinsic field F_i will exist within the insulating film due to the contact potential difference as shown in Fig. 2.23. This intrinsic field is given by

$$F_i = \frac{\phi_{m2} - \phi_{m1}}{qS} = \frac{\phi_2 - \phi_1}{qS} = \frac{\Delta\phi}{qS} \quad (2.147)$$

The effect of this field is to produce an asymmetric potential barrier

$$\phi_2 = \phi_1 + \Delta\phi \quad (2.148)$$

We can easily use eqs. (2.126)–(2.130), (2.134), and (2.135) to deduce the following cases.

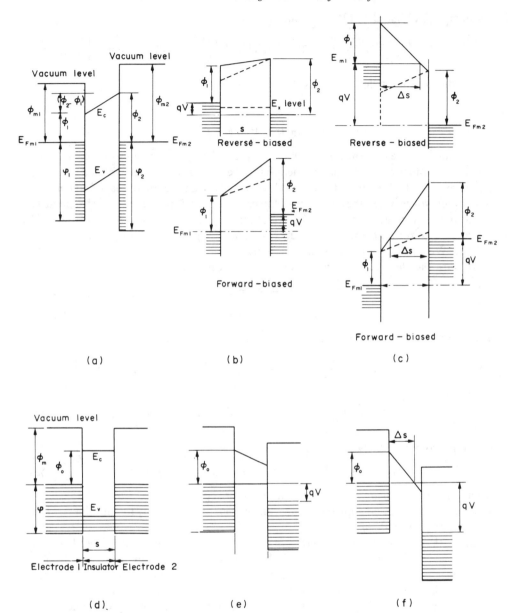

FIG. 2.23. Rectangular potential barriers in an insulating film between dissimilar electrodes for (a) $V = 0$, (b) $V < \phi_1/q$, and (c) $V > \phi_1/q$; and between two similar electrodes for (d) $V = 0$, (e) $V < \phi_0/q$, and (f) $V > \phi_0/q$. [After Simmons 1963.]

(a) LOW VOLTAGE RANGE: $V \simeq 0$

For this case we have

$$|S_2 - S_1| = S \qquad (2.149)$$

$$\overline{\phi} = (\phi_1 + \phi_2)/2 \qquad (2.150)$$

From eqs. (2.127) and (2.136) we obtain

$$J = \left(\frac{q^2}{Sh^2}\right)[m(\phi_1 + \phi_2)]^{1/2} V \exp\left[-\left(\frac{4\pi S}{h}\right)m^{1/2}(\phi_1 + \phi_2)^{1/2}\right] \quad (2.151)$$

This indicates that at low voltages the J–V relation is linear.

(b) INTERMEDIATE VOLTAGE RANGE: $0 < V < \phi_1/q$

For this case the current flow depends on the polarity of the applied voltage at the electrodes. Simmons [1963] has defined that the electrodes are reverse biased when the electrons are tunnelling from the electrode of lower work function to that of higher one, or in other words, when the electrode of lower work function is at the negative polarity of the applied voltage as shown in Fig. 2.23. Similarly, the electrodes are forward biased when the electrode of lower work function is at the positive polarity of the applied voltage. We shall summarize Simmons' work as follows:

(i) *With the electrodes reverse biased.* For this case we have

$$|S_2 - S_1| = S \quad (2.152)$$

$$\overline{\phi} = (\phi_1 + \phi_2 - qV)/2 \quad (2.153)$$

Substitution of eqs. (2.152) and (2.153) into eq. (2.134) gives

$$\begin{aligned}
J_1 = \frac{q}{4\pi h(\beta S)^2}\Bigg\{&(\phi_1 + \phi_2 - qV)\exp\left[-\left(\frac{4\pi\beta S}{h}\right)m^{1/2}(\phi_1 + \phi_2 - qV)^{1/2}\right] \\
&- (\phi_1 + \phi_2 + qV)\exp\left[-\left(\frac{4\pi\beta S}{h}\right)m^{1/2}(\phi_1 + \phi_2 + qV)^{1/2}\right]\Bigg\}
\end{aligned} \quad (2.154)$$

where β, from eq. (2.129), is given by

$$\beta = 1 - (qV)^2/\{96[\tfrac{1}{2}(\phi_1 + \phi_2) + \varphi_1 - E_x - \tfrac{1}{2}qV]^2\} \quad (2.155)$$

where the values of β are between 0.96 and 1 for most practical cases. In eq. (2.154) β, for simplicity and to a good approximation, can be assumed to be unity for all values of V in the range $0 < V < \phi_2/q$.

(ii) *With the electrodes forward biased.* For this case eqs. (2.152)–(2.154) are applicable except that the validity of these equations is within the voltage range $0 \le V \le \phi_1/q$. Thus, within the voltage range $0 \le V \le \phi_1/q$, $J_1 = J_2$, where J_2 is the forward-biased current; implying that the J–V characteristics within this range are independent of bias polarity. It should be noted that in this section ϕ_2 is assumed to be larger than ϕ_1.

(c) HIGH VOLTAGE RANGE: $V > \phi_1/q$

Again, we shall consider two cases as follows:

(i) *With the electrodes reverse biased.* For this case we have

$$\Delta S = |S_2 - S_1| = \frac{S\phi_1}{qV - \Delta\phi} \quad (2.156)$$

$$\overline{\phi} = \phi_1/2 \tag{2.157}$$

Substitution of eqs. (2.156) and (2.157) into eq. (2.134) gives

$$J_1 = \frac{q(qV - \Delta\phi)^2}{4\pi h\phi_1(\beta S)^2} \left\{ \exp\left[-\frac{4\pi m^{1/2}\beta S\phi_1^{3/2}}{h(qV - \Delta\phi)} \right] - \left(1 + \frac{2qV}{\phi_1}\right) \right.$$
$$\left. \times \exp\left[-\left(\frac{4\pi m^{1/2}\beta}{h}\right)\left(\frac{S\phi_1^{3/2}[1 + (2qV/\phi_1)]^{1/2}}{qV - \Delta\phi}\right) \right] \right\} \tag{2.158}$$

where β, from eq. (2.129), is equal to 23/24.

(ii) *With the electrodes forward biased.* For this case we have

$$|S_2 - S_1| = \frac{S\phi_2}{qV + \Delta\phi} \tag{2.159}$$

$$\overline{\phi} = \frac{\phi_2}{2} \tag{2.160}$$

$$\beta = \frac{23}{24} \tag{2.161}$$

Substitution of eqs. (2.159)–(2.161) into eq. (2.134) gives

$$J_2 = \frac{1.1q(qV + \Delta\phi)^2}{4\pi h\phi_2 S^2} \left\{ \exp\left[-\frac{23\pi m^{1/2}S\phi_2^{3/2}}{6h(qV + \Delta\phi)} \right] - \left(1 + \frac{2qV}{\phi_2}\right) \right.$$
$$\left. \times \exp\left[\left(-\frac{23\pi m^{1/2}}{6h}\right)\left(\frac{S\phi_2^{3/2}[1 + (2qV/\phi_2)]^{1/2}}{qV + \Delta\phi}\right) \right] \right\} \tag{2.162}$$

It can be seen that the second terms in eqs. (2.158) and (2.162) can be neglected for $V \gg \Delta\phi/q$ and these equations then reduce to the form of the Fowler–Nordheim equation for field emission which is $J = aF^2 \exp(-b/F)$ with $F = V/S$.

(d) THE EFFECT OF THE IMAGE FORCE

The effect of the image force is to reduce the width and the height of the potential barrier by rounding off the corners [Geppert 1963, Simmons 1963]. The image potential can be readily derived [Simmons 1963] and is given by

$$V_i = -\frac{q^2}{4\pi\varepsilon}\left\{ \frac{1}{2x} + \sum_{n=1}^{\infty}\left[\frac{nS}{(nS)^2 - x^2} - \frac{1}{nS} \right] \right\} \tag{2.163}$$

This expression can be approximated to a form more convenient to handle as follows [Simmons 1963]:

$$V_i = -\frac{0.795q^2 S}{8\pi\varepsilon x(S - x)} \tag{2.164}$$

When the effect of the image force is taken into account, the potential of the barrier will be modified as shown in Fig. 2.24. In the following we shall give ϕ with this effect for electrodes reverse or forward biased.

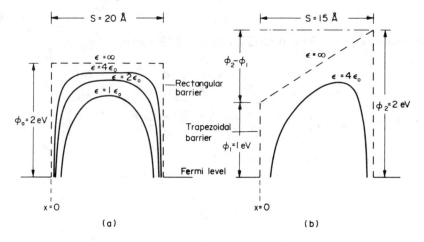

FIG. 2.24. The effect of image force on the profile of the potential barrier: (a) an insulating
film between two similar electrodes and the effect of dielectric constant, (b) an
insulating film between two dissimilar electrodes. [After Simmons 1969.]

(i) *With the electrodes reverse biased.* For this case the potential of the barrier is

$$\phi(x) = \phi_1 + (\Delta\phi - qV)(x/S) - \frac{0.795q^2S}{8\pi\varepsilon x(S-x)} \tag{2.165}$$

and the mean barrier height

$$\overline{\phi} = \left| \frac{1}{S_2 - S_1} \right| \int_{S_1}^{S_2} \phi(x)\,dx$$

$$= \phi_1 + \left(\frac{S_1 + S_2}{2S}\right)(\Delta\phi - qV) - \left[\frac{0.795q^2}{8\pi\varepsilon(S_2 - S_1)}\right]\ln\left[\frac{S_2(S - S_1)}{S_1(S - S_2)}\right] \tag{2.166}$$

The limits S_1 and S_2 are determined by setting $\phi(x) = 0$ in eq. (2.165) and solving for the
roots which, to a good approximation, are

$$\left.\begin{array}{l} S_1 = \left[\frac{9.2q^2\ln 2}{8\pi\varepsilon}\right]\bigg/\left[3\phi_1 + \frac{q^2\ln 2}{2\pi\varepsilon S} - (qV - \Delta\phi)\right] \\[2em] \quad - \frac{1.2q^2\ln 2}{8\pi\varepsilon(\phi_2 - qV)} \\[2em] S_2 = S - \frac{1.2q^2\ln 2}{8\pi\varepsilon(\phi_2 - qV)} \end{array}\right\} \quad 0 < qV \le \Delta\phi \tag{2.167}$$

$$\left.\begin{array}{l} S_1 = \frac{1.2q^2\ln 2}{8\pi\varepsilon\phi_1} \\[2em] S_2 = \left[\frac{9.2q^2\ln 2}{8\pi\varepsilon}\right]\bigg/\left[3\phi_1 + \frac{q^2\ln 2}{2\pi\varepsilon S} - 2(qV - \Delta\phi)\right] \\[2em] \quad + \frac{1.2q^2\ln 2}{8\pi\varepsilon\phi_1} \end{array}\right\} \quad \Delta\phi < qV \le \phi_2 \tag{2.168}$$

$$S_1 = \frac{1.2q^2 \ln 2}{8\pi\varepsilon\phi_1}$$

$$S_2 = \left(\phi_1 - \frac{5.6q^2 \ln 2}{8\pi\varepsilon S}\right)\left(\frac{S}{qV - \Delta\phi}\right) \qquad\qquad qV > \phi_2 \qquad (2.169)$$

(ii) *With the electrodes forward biased.* For this case the potential of the barrier is

$$\phi(x) = \phi_2 - (\Delta\phi + qV)(x/S) - \frac{0.795q^2 S}{8\pi\varepsilon x(S-x)} \qquad (2.170)$$

and the mean barrier height

$$\begin{aligned}
\overline{\phi} &= \frac{1}{|S_2 - S_1|}\int_{S_1}^{S_2}\phi(x)\,dx \\
&= \phi_2 - \left(\frac{S_1 + S_2}{2S}\right)(\Delta\phi + qV) - \left[\frac{0.795q^2}{8\pi\varepsilon(S_2 - S_1)}\right]\ln\left[\frac{S_2(S - S_1)}{S_1(S - S_2)}\right] \qquad (2.171)
\end{aligned}$$

where $S_1 = \dfrac{1.2q^2 \ln 2}{8\pi\varepsilon\phi_2}$

$$S_2 = S - \left[\frac{9.2q^2 \ln 2}{8\pi\varepsilon}\right]\Bigg/\left[3\phi_2 + \frac{q^2 \ln 2}{2\pi\varepsilon S} - 2(qV + \Delta\phi)\right] \left.\begin{array}{c}\\\\\\\\\end{array}\right\} \quad qV \le \phi_1 \qquad (2.172)$$

$$+\frac{1.2q^2 \ln 2}{8\pi\varepsilon\phi_2}$$

$$S_1 = \frac{1.2q^2 \ln 2}{8\pi\varepsilon\phi_2}$$

$$S_2 = \left(\phi_2 - \frac{5.6q^2 \ln 2}{8\pi\varepsilon S}\right)\left(\frac{S}{qV + \Delta\phi}\right) \qquad\qquad qV > \phi_1 \qquad (2.173)$$

Using eqs. (2.166)–(2.169), (2.171)–(2.173), (2.134), and (2.135), it is easy to deduce the expressions for J–V characteristics taking into account the effect of the image force for the low, intermediate, and high voltage ranges. It can be seen that the smaller the value of dielectric constant ε_r, ($\varepsilon = \varepsilon_r\varepsilon_0$), the smaller are the width and the height of the barrier. In other words, the insulating film with a smaller ε_r has a lower tunnel resistance (V/J), that is, a stronger image effect. Therefore, the tunnelling J–V characteristics are functions of the thermal properties of the insulator (through the temperature dependence of ε) as well as of electrodes.

(B) Between similar electrodes

By setting $\phi_1 = \phi_2 = \phi_0$ and $\Delta\phi = 0$, all equations in Section 2.3.2 (A) are applicable to the same cases for similar electrodes. Therefore, we need only to summarize these equations as follows (Fig. 2.23):

(a) LOW VOLTAGE RANGE: $V \simeq 0$

$$
\left.
\begin{aligned}
&|S_2 - S_1| = S \\
&\overline{\phi} = \phi_0 \\
&J = \frac{(2m\phi_0)^{1/2}}{S} \left(\frac{q}{h}\right)^2 \exp\left[-\left(\frac{4\pi S}{h}\right)(2m\phi_0)^{1/2}\right]
\end{aligned}
\right\} \qquad (2.174)
$$

(b) INTERMEDIATE VOLTAGE RANGE: $0 < V < \phi_0/q$

$$
\left.
\begin{aligned}
&|S_2 - S_1| = S \\
&\overline{\phi} = \phi_0 - \frac{qV}{2} \\
&\beta = 1 - (qV)^2/96[\psi_1 + \phi_0 - E_x - qV/2]^2 \simeq 1 \\
&J = \frac{q}{2\pi h(\beta S)^2}\left\{\left(\phi_0 - \frac{qV}{2}\right)\exp\left[-\frac{4\pi\beta S}{h}(2m)^{1/2}\left(\phi_0 - \frac{qV}{2}\right)^{1/2}\right]\right. \\
&\qquad\qquad \left. - \left(\phi_0 + \frac{qV}{2}\right)\exp\left[-\frac{4\pi\beta S}{h}(2m)^{1/2}\left(\phi_0 + \frac{qV}{2}\right)^{1/2}\right]\right\}
\end{aligned}
\right\} \quad (2.175)
$$

(c) HIGH VOLTAGE RANGE: $V \gg \phi_0/q$

$$
\left.
\begin{aligned}
&|S_2 - S_1| \simeq S\phi_0/qV \\
&\overline{\phi} \simeq \phi_0/2 \\
&\beta = 23/24 \\
&J = \frac{2.2q^3 V^2}{8\pi h S^2 \phi_0}\left\{\exp\left[-\frac{8\pi S}{2.96hqV}(2m)^{1/2}\phi_0^{3/2}\right]\right. \\
&\qquad\qquad \left. - \left(1 - \frac{2qV}{\phi_0}\right)\exp\left[-\frac{8\pi S}{2.96hqV}(2m)^{1/2}\phi_0^{3/2}\left(1 + \frac{2qV}{\phi_0}\right)^{3/2}\right]\right\}
\end{aligned}
\right\} \quad (2.176)
$$

(d) THE EFFECT OF THE IMAGE FORCE

$$
\left.
\begin{aligned}
&\phi(x) = \phi_0 - \frac{qVx}{S} - \frac{0.795q^2 S}{8\pi\varepsilon x(S-x)} \\
&\overline{\phi} = \frac{1}{|S_2 - S_1|}\int_{S_1}^{S_2}\phi(x)\,dx \\
&\quad = \phi_0 - \left(\frac{S_1 + S_2}{2S}\right)qV - \left[\frac{0.795q^2}{8\pi\varepsilon(S_2 - S_1)}\right]\ln\left[\frac{S_2(S - S_1)}{S_1(S - S_2)}\right]
\end{aligned}
\right\} \quad (2.177)
$$

where

$$S_1 = \frac{1.2q^2 \ln 2}{8\pi\varepsilon\phi_0}$$

$$S_2 = S_1 + S\left[1 - \frac{9.2q^2 \ln 2}{8\pi\varepsilon S}\left(3\phi_0 + \frac{q^2 \ln 2}{2\pi\varepsilon S} - 2qV\right)^{-1}\right] \qquad\Biggr\}\quad \text{for } qV < \phi_0 \qquad (2.178)$$

$$S_1 = \frac{1.2q^2 \ln 2}{8\pi\varepsilon\phi_0}$$

$$S_2 = \left(\frac{S}{qV}\right)\left(\phi_0 - \frac{5.6q^2 \ln 2}{8\pi\varepsilon S}\right) \qquad\Biggr\}\quad \text{for } qV > \phi_0 \qquad (2.179)$$

Following the same method described in Section 2.3.2(A) we can deduce straightfor-wardly the expressions for J–V characteristics taking into account the effect of the image force for the low, intermediate, and high voltage ranges.

(C) Some comparisons of the theory with experiments

Numerous results of quantum mechanical tunnelling current through thin insulating films have been published in the literature [Fisher and Giaever 1961, Hartman and Chivian 1964, Hartman 1964, Advani *et al.* 1962, Meyerhofer and Ochs 1963, Pollack and Morris 1964, 1965, Handy 1962, McColl and Mead 1965, Simmons and Unterkofler 1963, Chopra 1969, Duke 1969].

In general, electron tunnelling can be observed only in thin films of thickness less than 50 Å and free of pinholes. The techniques for fabricating such very thin films are: (i) vacuum deposition of a metallic thin film on a flat substrate such as on a glass slide, (ii) thermally or anodically grown surface oxide onto this electrode to produce an oxide-insulating film, say Al_2O_3, and (iii) deposition of a counter electrode onto the free surface of the oxide layer. Usually the reproducibility of the results from one specimen to another is poor, even in the same laboratory, and much worse from one laboratory to another. It can be imagined that although the expressions for the tunnelling J–V characteristics are available, the current flowing through an oxide film, which follows one of those expressions, is still not necessarily due to a tunnelling process. This can be easily realized when we study the J–V characteristics due to other electrical transport mechanisms which will be dealt with in Chapters 3–5. However, there are several experimental techniques which can be used to identify the tunnelling process. A more direct and reliable method is to use at least one superconductor electrode to form a normal metal–insulator–superconductor system. In such a system, there will be no tunnelling current if $|qV|$ is smaller than $\frac{1}{2}E_g$, where V is the applied voltage and E_g is the energy gap of the superconductor [Schrieffer 1964, Giaever 1960, McMillan and Rowell 1969], and tunnelling begins when $|qV| > \frac{1}{2}E_g$. The theoretical treatment of J–V characteristics for such a system is referred to the books of Duke [1969], Schrieffer [1964], and McMillan and Rowell [1969]. Other methods, which are less convincing, involve the measurements of the J–V characteristics in a normal metal–insulator–normal metal system as functions of temperature. Pollack and Morris

[1964] have reported that the $J-V$ characteristics for the Al–Al$_2$O$_3$–Al system with the Al$_2$O$_3$ film thermally grown on the aluminium electrode follow closely the analysis of Simmons [1963] given in (A) and (B) of this section as shown in Fig. 2.25. They have also determined the barrier heights at the parent metal–oxide interface and the oxide–counterelectrode interface by fitting the $J-V$ characteristics to theory over 9 decades of current, and obtained $\phi_1 = 1.6$ eV and $\phi_2 = 2.5$ eV, respectively. The difference between ϕ_1 and ϕ_2 for the apparently symmetrical Al–Al$_2$O$_3$–Al system indicates the difference in interface formed by different fabrication processes; the first interface resulting in ϕ_1 is formed by thermally growing an oxide on the aluminium electrode and the other by depositing an aluminium electrode on the free oxide surface. Gundlach and Heldman [1967] have modified Simmons' approach by using Franz's empirical equation [Franz 1956] instead of a simple parabolic equation for the energy–momentum relationship in the band structure of the oxide, and obtained a better fit to their results. They obtained from their results the barrier heights of 1.6 eV and 2.4 eV, which are in excellent agreement with those of Pollack and Morris. The departure of J_1 experimental curve from the theoretical one has been attributed to the existence of a small barrier in the transition region between the parent metal and the oxide layer, which becomes effective at low temperatures.

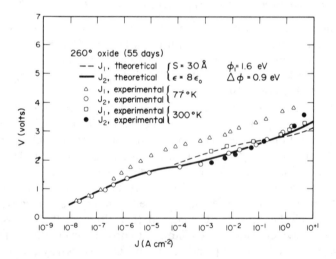

FIG. 2.25. Comparison between the experimental $J-V$ characteristics and the theoretical ones calculated based on an average barrier height of a trapezoidal barrier with image-force corrections, for $T = 0°$K, for an Al–Al$_2$O$_3$–Al system. J_1 and J_2 are the tunnelling currents obtained when the electrode with barrier height ϕ_1 was biased, respectively, at negative and positive voltages. [After Pollack and Morris 1964.]

Using an oxygen glow discharge technique to glow a gaseous anodized film of Al$_2$O$_3$ on the aluminium electrode, Pollack and Morris [1965] have measured ϕ_1 and ϕ_2 by the curve fitting of the experimental and the theoretical $J-V$ characteristics, and obtained $\phi_1 = 1.5$ eV and $\phi_2 = 1.85$ eV for the Al–Al$_2$O$_3$–Al system. These barrier height results are in good agreement with those ($\phi_1 = 1.49$ and $\phi_2 = 1.92$ eV) obtained by Braunstein *et al.* [1965] from Fowler plots of their photoemission results. Pollack

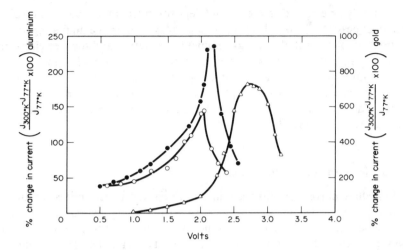

F IG. 2.26. Percentage change in J_1 (●) and J_2 (O) between 77°K and 300°K as a function of voltage for an Al–Al$_2$O$_3$–Al system, and percentage change in J_1 (△) for an Al–Al$_2$O$_3$–Au system. J_1 and J_2 are the tunnelling currents obtained when the electrode with barrier height ϕ_1 was biased, respectively, at negative and positive voltages. [After Pollack and Morris 1965.]

and Morris [1965] have also studied the temperature dependence of J–V characteristics, and obtained the cusps in the percentage change in tunnelling current between two temperatures as a function of applied voltage, as shown in Fig. 2.26. According to Stratton [1962] and Simmons [1964], the current peaks should occur when qV is equal to the barrier heights. From Fig. 2.26 for the Al–Al$_2$O$_3$–Al system, we can see ϕ_1 = 2.0 eV and ϕ_2 = 2.2 eV, which are higher than those (ϕ_1 = 1.5 eV and ϕ_2 = 1.85 eV), obtained from the isothermal results. This discrepancy has been attributed to the electric field penetration of the thin electrodes [Simmons 1965, 1967], since part of the applied voltage is absorbed in the electrodes and, therefore, a higher applied voltage is required to cause the cusp. On the same figure (Fig. 2.26) is also shown the peak for the Al–Al$_2$O$_3$–Au system. The peak is not sharp, possibly because of diffusion of atoms of counterelectrode material into the oxide film [Handy 1962]. However, the difference between the peak of J_1 for the Al–Al$_2$O$_3$–Al system and that for the Al–Al$_2$O$_3$–Au system is about 0.65 V which is close to the difference in work function (about 0.75 V) between aluminium and gold.

There are several factors which should be considered when analysing the tunnelling results. These factors are : (1) the barrier height is dependent not only on applied voltage but also on the thickness of the insulating film [Braunstein *et al.* 1966]; (2) the low and high voltage analyses are lack of self-consistency; (3) the J–V characteristics of a specimen depend on both time and storage conditions—ageing effect, possibly due to the diffusion of the atoms of the counterelectrode metal into the oxide [Handy 1962, Pollack and Morris 1964, Hartman and Chivian 1964. Fisher and Giaever 1961]; (4) both the electrode interface (controlled by the oxidation technique) and counterelectrode interface (controlled by the counterelectrode material and its deposition technique) have great influence on the tunnelling J–V characteristics; (5) because of field penetration in the electrodes [Simmons 1965] the apparent J–V characteristics depend not only on electrode material and deposition technique, but also on electrode area and thickness; (6) the local heating may also play a role in the tunnelling process; (7) the assumption of a parabolic energy–momentum relation and the free electron mass may not be valid; (8) the tunnelling current will be appreciably modified if ionic defects and trapping centres are present in the insulating film; (9) the interaction of the tunnel electrons with the optical phonons of the insulator may yield a highly temperature dependent tunnelling current even at room temperature [Emtage 1967, Savoye and Anderson 1967].

2.3.3. Elastic and inelastic tunnelling

The tunnelling current through a thin insulating film is determined by the supply function, the effective height, and the effective width of the potential barrier which are strongly dependent on the barrier profile. According to energy conservation, the electrons in electrode 1 can undergo the transition at constant energy into empty states in electrode 2 when a voltage V is applied to the system as shown in Fig. 2.27. The tunnelling in which the energies of the electrons remain unchanged during transition is generally referred to as the "elastic" tunnelling. However, the energies of the tunnelling electrons may change at certain applied voltages under the following conditions: (1) if the impurities or traps at a certain energy level in the insulator are ionized by inelastic collisions with the tunnelling electrons—this process will involve changes in both the supply function and the barrier profile; (2) if the tunnelling electrons lose part of their energies to excite the vibrational modes of atomic or molecular species in the insulator through inelastic collisions—this process may be thought of as a perturbation to the barrier profile [Jaklevic and Lambe 1966, Lambe and Jaklevic 1968]; (3) if the electrons tunnel through an MIS (metal–insulator–semiconductor) system, electrons may tunnel to surface states in the semiconductor energy gap with the subsequent recombination of the trapped electrons with holes in the valence band [Shewchun et al. 1967, Gray 1965].

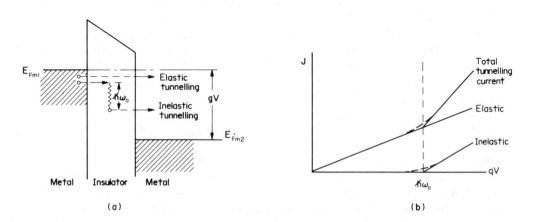

FIG. 2.27. (a) Illustrating elastic tunnelling and inelastic tunnelling via an energy-loss process (such as involving excitation energy $\hbar\omega_0$). (b) Illustrating the J–V characteristics for elastic and inelastic tunnelling. Solid lines are for $T = 0°$ K and dashed lines include the effect of thermal smearing for $T > 0$.

The tunnelling, in which the tunnelling electrons give part of their energy to one of the inelastic transfer processes (1) or (2) or (3) during transition, is referred to as the "inelastic" tunnelling.

When the energy difference ΔE between the energy level of the tunnelling electron in electrode 1 and that of the same electron in electrode 2 is equal to $\hbar\omega_0$, where $\hbar\omega_0$ is the excitation energy of one type of internal mode in the insulator, a new channel for tunnelling opens up and the total tunnelling current will be increased as shown in Fig. 2.27 provided that $qV \geq \Delta E = \hbar\omega_0$ and that empty states in the electrode 2 are available for the tunnelling. If the insulator has more than one type of internal modes that can be excited, more kinks like the one shown in Fig. 2.27 will appear in the J–V curve. It is clear that the tunnelling J–V characteristics may be employed as a "tunnel spectroscopy" to measure the energy levels of adsorbed or absorbed impurities in the metal–insulator–metal or in the MIS systems. Since the magnitude of the tunnelling current resulting from inelastic processes is smaller than that resulting from elastic processes, it is generally necessary to plot the first or second derivatives of the current with respect to voltage versus voltage in order to reveal the inelastic processes.

A theoretical analysis of the inelastic tunnelling current has been given by Scalapino and Marcus [1967] and Lambe and Jaklevic [1968]. The method is mainly based on the independent particle tunnelling model of Harrison [1961]. The expression for the one-dimensional tunnelling current is given by

$$J_x = \frac{2qm}{\hbar^3} \int_0^\infty dE_x \int_0^E |M_{12}|^2 \, g_1 g_2 (f_1 - f_2) dE_\perp \qquad (2.180)$$

where the subscripts 1 and 2 refer, respectively, to electrode 1 and electrode 2; g_1 and g_2 are, respectively, the one-dimensional density of state factors; f_1 and f_2 are the usual Fermi functions; $E_\perp$ is the energy of the electrons with momentum transverse to the barrier; E_x is the energy of the electrons with the component of momentum perpendicular to the barrier; and M_{12} is the matrix element for the transition from electrode 1 to electrode 2.

In the inelastic tunnelling the electrons are expected to tunnel from the states on the left to the empty states on the right with lower energy, and at the same time the impurities in the insulator are excited from their ground state to an excited state

$$\psi_1(\text{el})\phi_{gr}(\text{mol}) \rightarrow \psi_2(\text{el})\phi_{ex}(\text{mol})$$

where $\psi_1(\text{el})$ and $\psi_2(\text{el})$ are the electron wave functions on the left and the right sides of the barrier, and $\phi_{gr}(\text{mol})$ and $\phi_{ex}(\text{mol})$ are the wave functions for the impurities in the ground and excited states, respectively. Employing the approximation for a rectangular barrier and using a WKB approximation in a similar manner for elastic tunnelling, M_{12} for inelastic tunnelling can be expressed as [Scalapino *et al.* 1967]

$$M_{12} \propto \left\langle \phi_{ex} \left| \exp\left(-\int_0^S k_x dx \right) \right| \phi_{gr} \right\rangle \qquad (2.181)$$

where k_x is the absolute value of the imaginary k wave vector in the x-direction inside the barrier. k_x is given by

$$k_x = k_0 \left(1 + \frac{1}{2} \frac{U_{int}}{\phi_0} \right) \qquad (2.182)$$

and
$$k_0 = (2m\phi_0)^2/\hbar \tag{2.183}$$

where U_{int} is the interaction energy. On the basis of this model the excitation of impurities inside the barrier can be interpreted as perturbing the height of the barrier because the height of the barrier depends on the interaction of tunnelling electrons with impurities. The expression for U_{int} depends on the type of internal mode. For example, the electron–dipole interaction energy due to a dipole moment P_x associated with the molecular vibration located at one of the electrodes and taking into account the nearest image of the dipole is

$$U_{int} = 2qxP_x/(x^2 + r_\perp^2)^{3/2} \tag{2.184}$$

while the electron-induced dipole interaction energy due to a dipole induced by the electric field of the electron and taking into account the nearest image of the induced dipole is

$$U_{int} = -4q^2x^2\alpha/(x^2 + r_\perp^2)^3 \tag{2.185}$$

in which the polarizability α usually varies with the vibration of the molecule.

The inelastic portion of the tunnelling current can, therefore, be expressed as

$$J_{in} = \frac{4\pi mq^2}{h^3 c_1} \exp(-2k_0 S) \left\{ \frac{1}{q} \left[\frac{k_0}{2\phi_0} \right]^2 \left| \left\langle \phi_{ex} \middle| \int_0^S U_{int} \, dx \middle| \phi_{gr} \right\rangle \right|^2 \right.$$
$$\left. \times \int_0^\infty f(E)[1 - f(E + qV - \hbar\omega_0)] \, dE \right\} \tag{2.186}$$

for $c_1|qV - \hbar\omega_0| < 1$, where c_1 is defined in eq. (2.142). If $c_1(qV - \hbar\omega_0) \gg 1$, J_{in} is given by [Pollack and Seitchik 1969]

$$J_{in} = \frac{2\pi^2 m}{h^2 \phi_0} \left| \left\langle \phi_{ex} \middle| \int_0^S U_{in} \, dx \middle| \phi_{gr} \right\rangle \right|^2 J_{(elastic\ tunnelling)} \tag{2.187}$$

The inelastic tunnelling due to the excitation of impurities in the barrier has been observed by Jaklevic and Lambe [1966, 1968]. They observed the changes in tunnelling conductance of Al–Al$_2$O$_3$–Pb systems at $T = 4.2°K$ at definite voltages. The voltages have been identified to be $V = \hbar\omega_0/q$ directly relating to vibrational frequencies of bending and stretching vibrational mode (C–H or O–H) of hydrocarbons present in the oxide film as impurities. The coupling of the electron to the impurity in this case may be associated with the interaction of the electron with the dipole moment of the C–H and O–H bonds in the adsorbed hydrocarbons as shown in Fig. 2.28. These investigators have plotted d^2J/dV^2 versus V for the Al–Al$_2$O$_3$–Pb system and compared this result with the infrared spectrum of the bulk specimen of the hydrocarbon which was deliberately adsorbed on the Al–Al$_2$O$_3$–Pb system used for obtaining the tunnelling J–V characteristics. These results are also shown in Fig. 2.28 and are in good agreement with each other.

The deviations from the ideal tunnelling J–V characteristics described in Sections 2.3.1 and 2.3.2 are generally referred to as "tunnelling anomalies". In the presence of a magnetic field, the excitation due to spin–flip scattering of electrons by non-interacting localized magnetic impurities in the barrier will change the tunnelling from elastic to inelastic. Hall *et al.* [1960] were the first to observe a dip in conductance at about zero

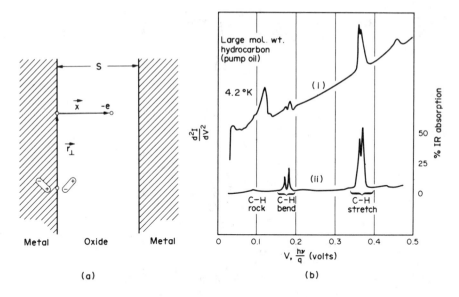

FIG. 2.28. (a) Illustrating the interaction of a tunnelling electron with a dipole in the oxide barrier (insulator) and its image in the metal. (b) (i) The tunnelling d^2I/dV^2 spectrum of an Al–oxide–Pb system with a monolayer of a large molecular weight hydrocarbon adsorbed on the oxide. (ii) The infrared spectrum of the bulk hydrocarbon of which a sample was adsorbed on the oxide used to obtain (i). The origin of the various modes is indicated in the figure. [After Lambe and Jaklevic 1968.]

bias voltage in the conductance–voltage characteristics of III–V compound tunnel diodes. A similar anomaly effect was also noticed by Wyatt [1964] in his studies of the tunnelling from a transition metal electrode through the transition metal oxide to aluminium or silver. Wyatt has reported that the conductance peak has a logarithmic voltage dependence and that the zero bias conductance increases logarithmically with decreasing temperature. Shen and Rowell [1967, 1968] have also investigated the effect of magnetic field on this zero bias anomaly in Ta–insulator–Al and Sn–insulator–Sn systems at $T = 1.4°$ K, and their results are shown in Fig. 2.29. It can be seen that at sufficiently high magnetic fields the zero bias conductance peak splits into two peaks on both sides of the zero bias. Appelbaum [1966, 1967] and Anderson [1966] have explained this phenomenon in terms of a spin–flip scattering process. The tunnelling conductance for the systems used by Shen and Rowell can be regarded as the sum of three parts: (i) the conductance due to electrons which tunnel through the oxide (insulator) without seeing the impurities or are simply scattered without any spin interaction, (ii) the conductance due to electrons which scatter from one metal electrode to the other (for example, from tantalum to aluminium) and so flip the spin of the impurity, and (iii) the conductance due to the interference between the transmitted electrons and the electrons from the transition metal electrode (for example, from tantalum) which are spin–flip scattered back into the transition metal electrode (tantalum electrode). It is this third part which gives rise to the conduction peak. Part (i) is independent of magnetic field, but parts (ii) and (iii) are strongly affected by the magnetic field. The splitting of the zero bias conductance peak into two peak at high

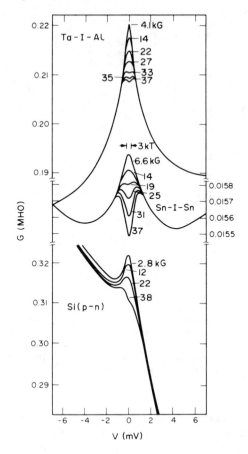

F IG. 2.29. The effect of magnetic field (indicated in the figure) on zero-bias anomaly in the conductance–
voltage characteristics for Ta–insulator–Al, Sn–insulator–Sn and silicon *p–n* tunnel junction
systems at 1.4°K. The voltage corresponding to $3kT$ (about 0.4mV) is also shown in the figure.
[After Shen and Rowell 1967, 1968.]

magnetic fields shown in Fig. 2.29 may be attributed to the spin–flip transition
incorporating an energy loss and to the splitting of the Zeeman levels.

2.3.4. A metal–thin insulating film–semiconductor (MIS) system

Numerous studies of tunnelling phenomena in MIS systems have been reported with
considerable attention to the use of these phenomena as a spectroscopic tool for the
investigation of the electronic band structure of the semiconductor surface [Gray 1962,
1965, Shewchun *et al.* 1967, 1974, 1977, Dahlke and Sze 1967, Freeman and Dahlke
1970, Esaki *et al.* 1965, 1966, 1967, 1968, Card *et al.* 1971, 1972, 1973, 1974, 1975, Green
et al. 1974]. The difference between a metal–semiconductor junction and a MIS system

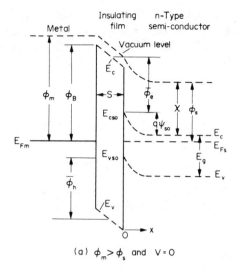

(a) $\phi_m > \phi_s$ and V = 0

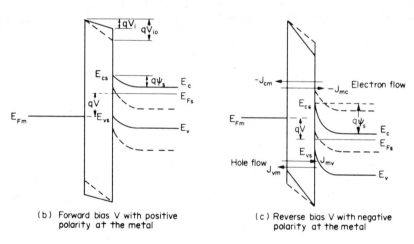

(b) Forward bias V with positive
polarity at the metal

(c) Reverse bias V with negative
polarity at the metal

FIG. 2.30. Energy band diagram for a metal–insulator–semiconductor system. – – – – – at
zero bias, ————— under an applied bias.

can be easily seen from the energy band diagrams shown in Figs. 2.30 and 2.2. To use
Fig. 2.30 to describe the features of the tunnelling through the MIS system, it is
convenient to define some important symbols as follows:

$\overline{\phi}_e, \overline{\phi}_h$ the average barrier heights of the insulator for electron tunnelling
between the metal and the conduction band and for hole tunnelling
between the metal and the valence band of the semiconductor, respectively, which are, in general, functions of applied voltage.

S the barrier width of the insulator which is the thickness of the insulator
(e.g. the oxide layer).

V_i, V_{i0} the voltage drop across the insulator with and without applied voltage,
respectively.

ψ_s, ψ_{s0} the surface potential at the semiconductor surface due to the band bending with and without applied voltage, respectively.

E_{cs}, E_{cs0} the energy level of the conduction band edge at the semiconductor surface with and without applied voltage, respectively.

E_{vs}, E_{vs0} the energy level of the valence band edge at the semiconductor surface with and without applied voltage, respectively.

From Fig. 2.30 the applied voltage V will partly appear across the insulator and partly across the semiconductor. Thus

$$V = V_i + \psi_s + (\phi_m - \phi_s)/q \qquad (2.188)$$

For simplicity, we assume that there are no charges present inside the insulator. Therefore, the electric field in it is uniform and is given by

$$F_i = \frac{V_i}{S} \qquad (2.189)$$

The electric field at the semiconductor surface F_s is determined by the charge in the interface states Q_{ss}, and the charge in the space charge region in the semiconductor Q_{sc}. Since

$$Q_{sc} \simeq -\varepsilon_s F_s \qquad (2.190)$$

according to Gauss' law we can write

$$Q_{ss} - \varepsilon_s F_s = -\varepsilon_i F_i \qquad (2.191)$$

From eqs. (2.189)–(2.191) we obtain

$$V_i = \frac{S}{\varepsilon_i}(\varepsilon_s F_s - Q_{ss}) \qquad (2.192)$$

Similarly, the voltage drop across the insulator at zero bias can be written as

$$V_{i0} = \frac{S}{\varepsilon_i}(\varepsilon_s F_{s0} - Q_{ss0}) \qquad (2.193)$$

in which the subscript 0 denotes the cases for zero bias. It is obvious that to determine V_i and V_{i0}, we have to know F_s and F_{s0} or ψ_s and ψ_{s0}. By solving Poisson's equation, the electric field at the semiconductor is given by [Shewchun *et al.* 1967]

$$F_s = \pm \left(\frac{2kT}{\varepsilon_s}\right)^{1/2}\left\{n_B\left[\exp\left(\frac{q\psi_s}{kT}\right) - 1 - \frac{q\psi_s}{kT}\right] + p_B\left[\exp\left(-\frac{q\psi_s}{kT}\right) - 1 + \frac{q\psi_s}{kT}\right]\right\}^{1/2}$$

$$(2.194)$$

where n_B and p_B are, respectively, the equilibrium concentrations of free electrons and holes in the semiconductor bulk. Since the charge in the interface states is also a function of the surface potential ψ_s, the surface potential is caused by the band bending. Thus the surface potential depends upon the density of interface states, the dopant density in the semiconductor, the thickness of the insulator, the ionic impurities in the insulator (which may result in charge inside the insulator), and the difference in work function between the metal and the semiconductor. All these factors make ψ_s to be a function of applied voltage V.

When $Q_{ss} = 0$ and $\phi_m = \phi_s$, then $\psi_{s0} = V_{i0} = 0$ at zero bias ($V = 0$). This condition is generally referred to as the "flat band" condition. From eqs. (2.188) and (2.193) it can be seen that the difference $\phi_m - \phi_s$ as well as ϕ_{ss} can significantly affect the values of ψ_{s0} and ψ_s, and hence V_i and V_{i0}. Figure 2.31 shows the energy band diagrams for two common electrodes (aluminium and gold) on n-type and p-type silicon with a doping density $N_D = 10^{16}$ cm^{-3} or $N_A = 10^{16}$ cm^{-3} and an oxide (SiO$_2$) thickness of 500 Å, the charge in the interface states Q_{ss} being assumed to be zero. It can be seen that by appropriate choice of the electrode metal, the n-type semiconductor surface can be

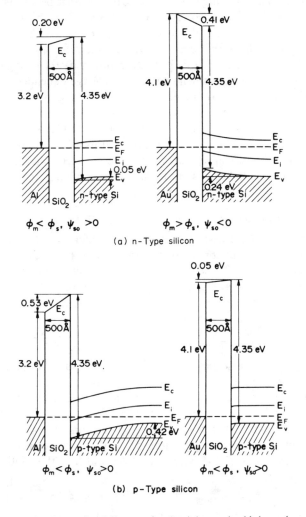

FIG. 2.31. Energy band diagrams for aluminium and gold electrodes on oxidized n-type and p-type silicon specimens at zero bias. [After Deal, Snow, and Mead 1966.]

varied from accumulation to depletion, and the *p*-type semiconductor surface can be varied from flat band to inversion.

By comparing Fig. 2.2 with Fig. 2.30 we can easily find the principal difference between the Schottky barrier without an insulating film (metal–semiconductor intimate contact) and that with a thin insulating film interposed. For the case shown in Fig. 2.2, the distance between the metal Fermi level E_{Fm} and the semiconductor conduction band edge E_c at the interface (Schottky barrier height $= \phi_m - \chi$) is independent of applied voltage. Thus, for a metal–*n*-type semiconductor contact, the applied voltage can regulate mainly the electron flow from the conduction band of the semiconductor to the metal, but does not affect much the electron injection from the metal to the conduction band and the hole injection from the metal to the valence band of the semiconductor because $\phi_m - \chi$ is constant. However, for the case shown in Fig. 2.30 the presence of the insulating film will make the distance between the metal Fermi level E_{Fm} and the semiconductor conduction band edge E_{cs} at the interface dependent on the applied voltage because the voltage drop V_i across the insulating film is dependent on applied voltage and so is E_{cs} or E_{vs}. This causes the voltage dependence of the tunnelling current through the thin insulating film by (i) the voltage dependence of the tunnelling transmission coefficient due to the voltage dependence of the average barrier heights $\overline{\phi}_e$ and $\overline{\phi}_h$ for tunnelling, (ii) the voltage dependence of the supply function due to the voltage dependence of $|E_{Fm} - E_{cs}|$ or $|E_{Fm} - E_{vs}|$.

For the MIS (*n*-type) system as shown in Fig. 2.30, when a forward bias brings E_{Fm} in alignment with E_{vs}, the tunnelling hole current J_{mv} (this means electrons in the valence band see a large number of empty states in the metal—electrons tunnelling from the valence band to the metal is equivalent to the holes tunnelling from the metal to the valence band) will begin to increase. This condition is

$$qV_{w\text{(forward)}} = E_{Fs} - E_{vs} = E_{Fs} - [E_v + q\psi_{sw\text{(forward)}}] \qquad (2.195)$$

With an increase in the forward bias, J_{cm} also increases because the electron concentration n in the conduction band increases; but both J_{mc} (electron flow from the metal to the conduction band) and J_{vm} (hole flow equivalent to electron flow from the metal to the valence band) decreases. When a reverse bias brings E_{Fm} in alignment with E_{cs}, the tunnelling electron current J_{mc} (electron flow from the metal to the conduction band) will begin to increase. This condition is

$$qV_{w\text{(reverse)}} = E_{cs} - E_{Fs} = [E_c + q\psi_{sw\text{(reverse)}}] - E_{Fs} \qquad (2.196)$$

With an increase in the reverse bias, J_{vm} also increases because the band bending results in an increase of the number of empty states in the valence band; but both J_{cm} and J_{mv} become negligible under the reverse bias condition. Therefore, by plotting $J-V$ or $G-V$ characteristics, where G is the dynamic conductance dJ/dV, there is a wide region of low conductance and beyond the ends of this region the conductance rises rapidly. This region of low conductance is generally referred to as the "conductance well" [Shewchun *et al.* 1967], and from eqs. (2.195) and (2.196) the width of the conductance well is

$$V_w = \frac{E_g}{q} + [\psi_{sw\text{(reverse)}} - \psi_{sw\text{(forward)}}] = V_{w\text{(reverse)}} + V_{w\text{(forward)}} \qquad (2.197)$$

The presence of interface states located in the energy band gap at the insulator–semiconductor interface affect V_w through $\psi_{sw\text{(reverse)}}$ and $\psi_{sw\text{(forward)}}$. The charge in the interface states tends to shield the semiconductor from the metal and makes the communication between the metal and the semiconductor more difficult. This leads to a smaller voltage dependence of semiconductor band bending and hence a narrower conductance well. The interface states may contribute to the a.c. conductance (alternating field) due to the time lag in trapping and recombination of carriers in the semiconductor to the metal, but does not affect much the electron injection from the electron tunnelling via interface states [Shewchun *et al.* 1967]. Figure 2.32(a) shows that the low frequency (low enough to be equivalent to d.c.) conductance due to tunnelling through a *p*-type silicon MIS system is insensitive to temperature and the conductance width is larger than E_g/q, and Fig. 2.32(b) shows that the a.c. conductance is frequency dependent with two predominant peaks in the conductance well. One of these two prominent peaks whose location is frequency dependent is caused by the

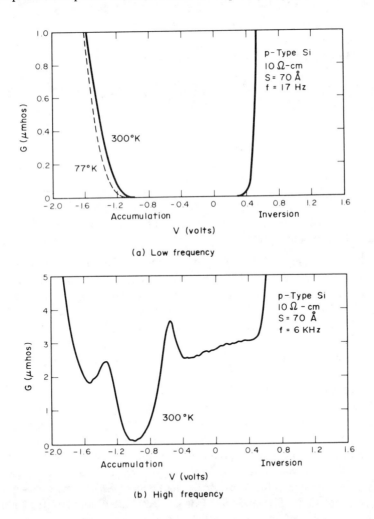

FIG. 2.32. Low and high frequency conductance–voltage characteristics for a *p*-type silicon MIS system. [After Waxman, Shewchun, and Warfield 1967.]

tunnelling between the metal and the interface states, and the charge exchange between the interface states and the majority carriers in the semiconductor; while the other peak is due to the coupling between the internal and external inversion layers [Waxman *et al.* 1967, Nicollian *et al.* 1965]. The internal inversion layer refers to the inversion layer directly under the metal electrode, while the external inversion layer refers to the inversion layer extending out over the surface away from the metal electrode.

For the MIS (*n*-type) system shown in Fig. 2.30, the expressions for the majority carrier (electron) and the minority carrier (hole) tunnelling currents are readily obtained from the one-dimensional theory formulated by Harrison [1961] or by Stratton [1962], and these have been derived by Card and Rhoderick [1971]. The expression for the majority (electron) tunnelling current is

$$J_n = J_{cm} - J_{mc} = A_n T^2 \exp(-\overline{\phi}_e^{1/2} S) \exp[-(q\psi_{so} + E_c - E_{Fs})/kT]$$
$$\times [\exp(qV/n_e kT) - 1] \tag{2.198}$$

and that for the minority carrier (hole) tunnelling current is

$$J_p = J_{mv} - J_{vm} = A_p T^2 \exp(-\overline{\phi}_h^{1/2} S) \left\{ \exp\left[\frac{E_{vo} - E_{Fp0}}{kT}\right] - \exp\left[\frac{E_{vo} - E_{Fm}}{kT}\right] \right\} \tag{2.199}$$

where $A_n = \dfrac{4\pi q m_{te} k^2}{h^3}$ $A_p = \dfrac{4\pi q m_{th} k^2}{h^3}$

m_{te} = the effective mass for electrons with momentum transverse to the barrier
m_{th} = the effective mass for holes with momentum transverse to the barrier
n_e = the ideality factor for electrons which is equal to $V/(\psi_{so} - \psi_s)$
E_{Fp0} = the hole quasi-Fermi level at the semiconductor surface ($x = 0$)

It is obvious that in eq. (2.198) V is positive for the forward bias and negative for the reverse bias, and that in eq. (2.199) the bias dependence of J_p is obtained through the bias dependence of E_{vo} and E_{Fp0}. It should be noted that $(-\overline{\phi}_e^{1/2} S)$ in $\exp(-\overline{\phi}_e^{1/2} S)$ and $(-\overline{\phi}_h^{1/2} S)$ in $\exp(-\overline{\phi}_h^{1/2} S)$ are dimensionless and come from the following expressions [Simmons 1963, Card and Rhoderick 1971]:

$$\exp\left[-\frac{4\pi S}{h}(2m\overline{\phi})^{1/2}\right] \simeq \exp[-1.01\overline{\phi}^{1/2} S] \simeq \exp(-\overline{\phi}^{1/2} S) \tag{2.200}$$

if $\overline{\phi}$ is expressed in eV and S in Å.

In the MIS systems we can adjust the applied voltage to increase the minority carrier injection ratio (defined as the minority carrier current to the total current). This can be easily realized from Fig. 2.30 and eqs. (2.198) and (2.199). On the basis of this principle, Card and Smith (1971) observed green luminescence in an Au–SiO₂–GaP (*n*-type) system under forward bias. Figure 2.33 shows the green light output power of the Au–SiO₂–GaP system as a function of the thickness of SiO₂ films for two values of tunnelling current under forward bias. For a fixed value of current, the light output power (or light intensity) increases with S, reaches a peak, and then decreases with increasing S. This indicates the minority carrier injection ratio $[J_p/(J_n + J_p) \simeq J_p/J_n]$ initially increases with S (when $E_{Fm} \simeq E_{vs}$), but after reaching the peak, it decreases with

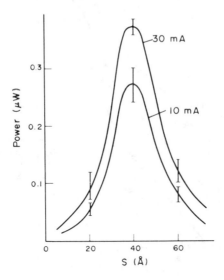

Fɪɢ. 2.33. Green light output power as a function of SiO$_2$ film thickness for Au–SiO$_2$–GaP system for two values of tunnelling current under forward bias. [After Card and Smith 1971.]

S because the barrier width increases, which causes a rapid decrease in the transmission coefficient for tunnelling.

Esaki *et al.* [1966, 1968] fabricated Al–Al$_2$O$_3$–SnTe and Al–Al$_2$O$_3$–GeTe systems by evaporating SnTe and GeTe layer of 4000 Å (highly doped *p*-type semiconductors with typical carrier concentrations of about 8×10^{20} and 2×10^{20} cm^{-3}, respectively) onto oxidized aluminium stripes, and observed a negative differential resistance region in both systems under reverse bias. The typical results for the Al–Al$_2$O$_3$–SnTe systems for three different oxide thicknesses are shown in Fig. 2.34. The negative differential resistance can be explained on the basis of the simple energy band diagram shown in Fig. 2.34. When a negative bias is applied to the metal aluminium, the distance between E_{Fm} and E_v begins to decrease and hence the tunnelling current begins to increase. The tunnelling current will continue to increase with increasing bias until the bias reaches such a value that E_{Fm} is in alignment with E_v [that is until $V = (E_v - E_{Fs})/q$]. After this point a further increase in the bias voltage does not increase the number of electrons for tunnelling from the metal to the valence band of the semiconductor, whereas the barrier height $\bar{\phi}$ for tunnelling is raised, resulting in a decrease in the tunnelling probability. Thus there is a negative differential resistance region in the *J–V* characteristics. When $V > (E_c - E_{Fs})/q$ a new tunnelling current from the metal to the conduction band of the semiconductor begins to flow and, therefore, the current increases again with increasing bias voltage. In Fig. 2.34, Chang *et al.* [1967] have also shown their calculated current–voltage characteristics based on the WKB approximation using the following values for the physical parameters: $E_c - E_v = E_g = 0.3$ eV, $E_v - E_{Fs} = 0.6$ eV, $\phi_B = 1.9$ eV, $qV_{i0} = 1.2$ eV. Their theoretical results are in good agreement with their experimental ones.

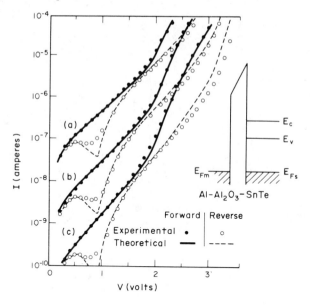

Fig. 2.34. Current–voltage characteristics for Al–Al$_2$O$_3$–SnTe systems at 4.2°K for the values of $(8m_B S^2/h^3)^{1/2}$ equal to (a) 16.3 eV$^{-1/2}$, (b) 18.1 eV$^{-1/2}$, and (c) 20.0 eV$^{-1/2}$. [After Chang, Stiles, and Esaki 1967.]

2.3.5. Effects of space charge and traps, and impurity conduction

In Section 2.3.2 we have discussed the effect of the image force on the profile of metal–insulator–metal potential barriers. In general, the insulating films contain traps; thus the space charge may also seriously affect the profile of the potential barrier and hence the J–V characteristics if the trap density is large. Frank and Simmons [1967] have discussed the effects of space charge and traps on Schottky emission-limited current flow in insulators. Geppert [1962] and Pittelli [1963] have shown theoretically that the space charge and particularly the trapped space charge can severely limit the tunnelling current through a thin insulating film. If there is no space charge, the applied voltage reduces both the effective height and width of the potential barrier for electron tunnelling as shown in Fig. 2.35(b). The method for calculating the tunnelling current for this case has been discussed in Section 2.3.2. However, by assuming the net space charge density to be zero in the region $0 < x < S_2$ and to be constant in the region $S_2 < x < S$, where $|S_2|$ is the barrier width for tunnelling, the space charge is to alter the profile of the potential barrier in a manner of increasing the barrier width so as to limit the electron tunnelling as shown in Fig. 2.35(c). Thus, the space charge effects can be accounted for by extending the straight-line portion of the actual barrier from $x = S_2'$ to $x = S$ and using an effective bias V' in place of the actual bias V to calculate the tunnelling distance $|S_2 - S_1|$ and the average barrier height $\bar{\phi}$.

Pittelli [1963] has calculated $|S_2 - S_1|$ and $\bar{\phi}$ for the thin insulating film containing traps by solving the Poisson equation, and then calculated the J–V characteristics using

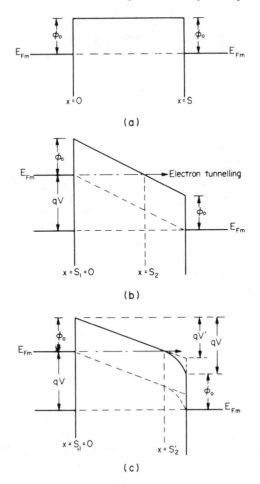

FIG. 2.35. Energy band diagrams for a thin insulating film between two similar metal electrodes with and without space charge effects: (a) rectangular barrier without bias, (b) with bias qV neglecting space charge effects, (c) with bias qV including space charge effects.

the method described in Section 2.3.2 and the values for the following parameters: ϕ_0 = 1 eV, $S = 100$ Å, $\varepsilon = 10\varepsilon_0$, θ [the ratio of the free electron density to the total electron density (free and trapped), see Chapter 3] $=10^{-7}$. Pittelli has concluded on the basis of his results shown in Fig. 2.36 that free carrier space charge without traps is ineffective in lowering the tunnelling current even for extremely low free carrier mobilities, that a high trap density can severely limit the tunnelling current, and that low free carrier mobilities enhance the space charge effects.

Insulating or semiconducting films always contain impurities which may be unavoidably present or may be deliberately doped in the film specimens. These impurities form impurity states in the forbidden energy gap. When the localized electronic wave functions of the impurity states overlap, an electron bound to one impurity state can tunnel to an unoccupied impurity state without involving activation

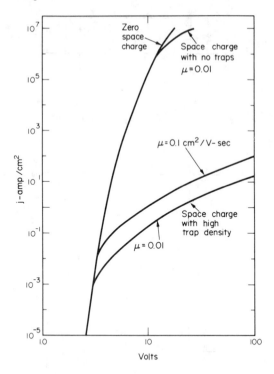

F<small>IG</small>. 2.36. The effects of space charge and traps on the tunnel current–voltage characteristics. [After Pittelli 1963.]

into the conduction band. This tunnelling process between impurity sites is referred to as impurity conduction [Mott and Twose 1961]. As the mobility of an electron moving in the impurity states is very small since it depends on the interaction between widely spaced impurities, this conduction mechanism usually becomes predominant at low temperatures due to low concentration of carriers in the conduction and valence bands. But this depends on impurity concentration and the energy levels of the impurity states, which control the probability of tunnelling from inpurity site to impurity site and the number of electrons taking part in this tunnelling process.

In semiconductors the impurity conduction process is possible only if the material is compensated (that is if the material contains both donor and acceptor impurities). This condition for impurity conduction was put forward by Mott [1956] and Conwell [1956], and confirmed experimentally by Fritzsche *et al.* [1958, 1959, 1960]. For example, if the donor concentration N_D is larger than the acceptor concentration N_A in a compensated *n*-type semiconductor, all the acceptors will be occupied and become negatively charged, and only N_D-N_A donors remain occupied and neutral at low temperatures as shown in Fig. 2.37. If the impurity state A and impurity state B are at the same energy level, the overlap of the wave functions between these two sites will enable the movement of an electron from an occupied to an empty donor site without involving activation into the conduction band. If the impurity state A is located at a lower energy level than the impurity state B, then thermal energy (phonon) supplied by

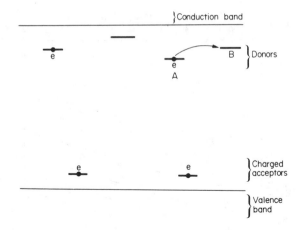

F IG. 2.37. Illustrating an electron from an occupied impurity state *A*
hopping to an empty impurity state *B* due to thermally
assisted tunnelling in a compensated *n*-type semiconductor
with $N_D > N_A$. [After Mott and Davis 1971].

lattice vibrations of the material is required to assist the electron tunnelling from *A* to *B*.
It should be noted that the field created by the charged acceptors and donors will split
the energy levels of donor states, and therefore an electron can tunnel from one
impurity state to another only by exchanging energy with phonons, and that the applied
voltage will alter the energy level difference between sites, thereby making the
tunnelling probability higher in one direction than in the other. A similar process can
readily be realized in compensated *p*-type semiconductors ($N_A > N_D$), but in this case
electrons tunnel through acceptor impurity sites.

The most significant features due to impurity conduction observed in the
resistivity–temperature characteristics are: (1) the resistivity is strongly dependent on
impurity concentration, (2) the plot of ln ρ versus $1/T$ exhibits a finite slope indicating
that a thermal activation energy is required for the electron tunnelling between sites
when the impurity concentration is small, and (3) the activation energy decreases with
increasing impurity concentration and becomes zero when the impurity concentration
reaches a certain critical value or higher, indicating that with the impurity concentrations higher than such a critical value the carriers move freely without involving
thermal activation. Some typical experimental results obtained by Fritzsche and
Cuevas [1960] in compensated *p*-type germanium semiconductors are shown in
Fig. 2.38. A similar impurity conduction phenomenon has also been observed in
tantalum oxide thin films [Mead 1962], in silicon monoxide films [Simmons and
Verderber 1967], in nickel oxide [Bosman and Crevecoeur 1966, Springthorpe *et al.*
1965], in vanadium phosphate glasses [Schmid 1968], and in many other materials
[Mott and Twose 1961, Mott and Davis 1971].

Simmons *et al.* [1967] have reported that an Al–SiO–Au system doped with gold
monovalent ions exhibits a negative differential resistance region in the *J–V*
characteristics as shown in Fig. 2.39, and that if a voltage corresponding to the lowest
point of the negative differential resistance region (8 V in Fig. 2.40) is applied to the
specimen and then reduced to zero in about 0.1 msec (that is an 8 V pulse with a tailing

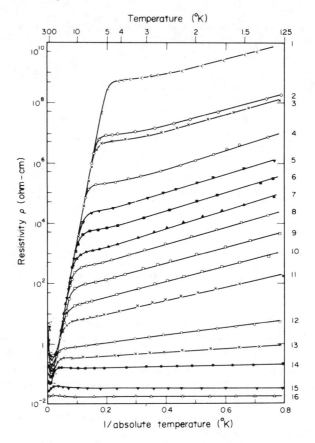

FIG. 2.38. Resistivity ρ in ohm-cm as a function of temperature for compensated p-type germanium semiconductors with compensation $= N_D/N_A = 0.4$ for the acceptor concentrations in cm^{-3}: (1) 7.5×10^{14}; (2) 1.4×10^{15}; (3) 1.5×10^{15}; (4) 2.7×10^{15}; (5) 3.6×10^{15}; (6) 4.9×10^{15}; (7) 7.2×10^{15}; (8) 9.0×10^{15}; (9) 1.4×10^{16}; (10) 2.4×10^{16}; (11) 3.5×10^{16}; (12) 7.3×10^{16}; (13) 1.0×10^{17}; (14) 1.5×10^{17}; (15) 5.3×10^{17}; (16) 1.35×10^{18}. [After Fritzsche and Cuevas 1960.]

edge faster than 0.1 msec), and then a voltage of magnitudes smaller than V_{TH} (the threshold voltage) is reapplied to the specimen, the I–V characteristic follows OE' corresponding to a high impedance state rather than OA corresponding to a low impedance state. The latter phenomenon is a characteristic of memory as shown in Fig. 2.40. The memory state following OE' is obtained by the application of an 8 V pulse with a tailing edge faster than 0.1 msec. Memory states following OC' or OB' can also be obtained by applying, respectively, 6 V and 4.4 V pulses with a tailing edge faster than 0.1 msec. These memory states can be stored without application of electric power for indefinite periods of time (non-volatile memory states), and can be erased by applying a voltage greater than V_{TH} to revert the characteristic to its original low impedance state corresponding to curve OA. Simmons *et al.* [1967] have proposed a model to explain all these phenomena. Here we summarize their explanations using Fig. 2.41 as follows:

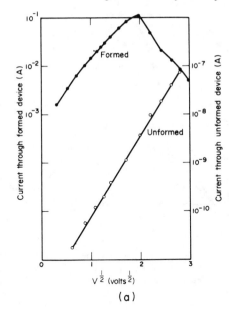

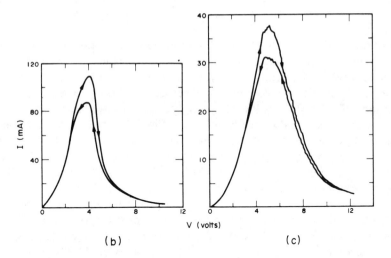

FIG. 2.39. The current–voltage characteristics of Al–SiO–Au systems. Unformed specimens without Au ions and formed specimens with doped Au ions in SiO films: (a) unformed and formed SiO 400 Å thick; (b) formed SiO, 400 Å thick; (c) formed SiO, 820 Å thick. [After Simmons and Verderber 1967.]

(i) The gold ions introduced into the SiO insulating film result in the formation of a broad impurity band of localized states within the forbidden energy gap as shown in Fig. 2.41(a) (between two dashed lines) because of the amorphous nature of the insulating film. Since the insulating film (of the order 200–2000 Å) is too thick for direct tunnelling from one electrode to the other, the very weak temperature dependence of

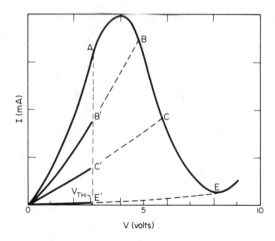

F IG. 2.40. Illustrating the memory phenomenon, low impedance and high impedance states, and the threshold voltage. [After Simmons and Verderber 1967.]

the conduction current suggests that the electric conduction is due to tunnelling of electrons between adjacent impurity sites formed by gold ions (impurity conduction).

(ii) For the case $qV < \phi_i$ the electrons are injected, between the levels E_{FM1} and E_{FM2}, from electrode 1 to the adjacent impurity sites by tunnelling to begin the impurity conduction. The contribution to ·the total current is mainly due to the electrons tunnelling at the levels very close to E_{FM1}, because the effective barrier height for the electron tunnelling at $x = 0$ is lowest at E_{FM1}. As the effective barrier height at $x = 0$ increases with increasing distance below E_{FM1} (Fig. 2.41(d)), the contribution to the total current by electrons at levels below E_{FM1} falls off rapidly with distance below E_{FM1}. This can be easily realized from the tunnelling theory given in Section 2.3.2. Therefore, for $V < \phi_i/q$ the conduction current increases with increasing applied voltage V, and reaches the peak value when $V = \phi_i/q$.

(iii) For the case $qV > \phi_i$, the electrons injected from E_{Fml} can tunnel through impurity sites to the insulating film only up to the point A as shown in Fig. 2.41(c) and hence do not contribute to the conduction current unless these electrons at A can receive sufficient thermal energy to jump to the conduction band at C or to tunnel to the conduction band at B. The probability for the latter process to occur depends on the energy difference AC and the tunnelling distance AB. Thus only energy levels lower than $V - \phi_i$ below E_{Fml} can contribute electrons to the conduction current. This explains why for $V > \phi_i/q$ the conduction current decreases with increasing V and there is a voltage-controlled negative differential resistance region in the I–V characteristics as shown in Fig. 2.39. At higher voltages I rises again. This may be due to the fact that the electron flow via the path OAB and OAC shown in Fig. 2.41(c) becomes predominant.

(iv) The memory phenomenon is attributed to charge storage (electrons trapped) within the insulating thin film when $V > \phi_i$. The clockwise hysteresis loop always

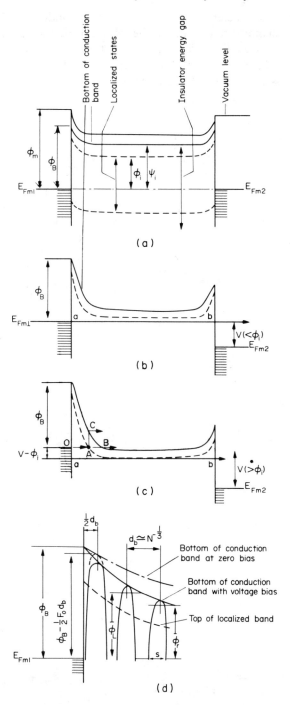

FIG. 2.41. Energy band diagrams for a formed metal–insulator–metal system: (a) applied voltage $V = 0$, (b) $V < \phi_i$, (c) $V > \phi_i$, and (d) localized band and barrier between impurity sites. [After Simmons and Verderber 1967.]

appears in the $I-V$ characteristics as shown in Fig. 2.39 indicating this memory action. The erase of a memory state is due to the release of trapped electrons.

(v) Similar negative differential resistance and memory phenomena have also been observed in Al_2O_3, Ta_2O_5, CaF_2, and MgF_2 thin films using various electrode materials such as Cu, Ag, and Pt [Simmons 1968].

CHAPTER 3

Space Charge Electric Conduction—
One-carrier Current Injection

3.1. SOME CONCEPTS RELEVANT TO SPACE CHARGE ELECTRIC CONDUCTION

Space charge is generally referred to as the space filled with a net positive or negative charge, and it appears in a great variety of situations associated with semiconductors and insulators. In this chapter we are mainly concerned with the limit by such a space charge to the current or the number of charge carriers per second passing across from one electrode to the other. For example, if the cathode emits more electrons per second than the space can accept, the remainder will form a negative space charge which will create a field to reduce the rate of electron emission from the cathode. Thus the current is controlled not by the electron-injecting electrode, but rather by the bulk of the semiconductor or the insulator, or, in other words, by the carrier mobility in the space inside the material. In general, the emitted electrons have a distribution of energies, the material has traps of various distributions, and there exist various high field effects. Actually, the situation is quite complicated and, in order to find a solution that is not too burdensome for the current–voltage (J–V) characteristics, we must resort to simplifying assumptions.

The space-charge-limited (SCL) dark conduction (not through light absorption or heat agitation) occurs when the contacting electrodes are capable of injecting either electrons into the conduction band or holes into the valence band of a semiconductor or an insulator, and when the initial rate of such charge carrier injection is higher than the rate of recombination, so that the injected carriers will form a space charge to limit the current flow. Therefore, the SCL current is bulk limited. In Chapter 2 we have discussed in some detail about the carrier injection from a neutral or a blocking contact. In this section it is desirable to review some possible methods for producing carrier-injecting contacts, which are required for the SCL dark conduction. These are discussed as follows.

(a) The electron injection may take place from the metal into the heavily doped n^+ region through quantum mechanical tunnelling, and then from the n^+ region into the conduction band of an n-type semiconductor through thermal excitation as shown in Fig. 3.1(a). It can be seen that the actual barrier height for the electrons from n^+ region to the conduction band is much smaller than that for the electron from the metal direct to the conduction band.

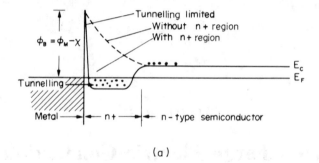

(a)

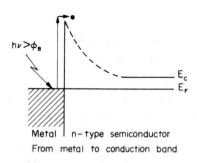

From metal to conduction band

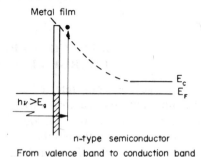

From valence band to conduction band

(b)

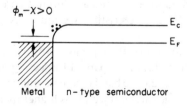

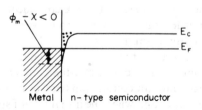

(c)

FIG. 3.1. Energy level diagrams of some contacts for electron injection: (a) degenerate n^+–non-degenerate n-contact, (b) photo-excited carrier-injecting contact, (c) ohmic contacts with $\phi_m < \phi_s$.

(b) The electron injection may take place from the metal into the conduction band of a material through thermionic emission, or field emission, or thermionic field emission as have been discussed in Chapter 2. It should be noted that a metallic point with a sufficiently sharp point in contact with a material may produce an intense electric field near the contact to reduce the blocking barrier width, thus promoting electron tunnelling at relatively small applied voltages.

(c) The electron injection may take place from a metal into a material or from the valence band to the conduction band of the material at the surface in contact with the metal through photoexcitation with the light frequency v such that $hv > \phi_B$ or $hv > E_g$ as shown in Fig. 3.1(b). For illuminating the material surface through the electrode, the electrode must be a thin conducting film so that the light can penetrate through.

(d) The electron injection may take place from the metal to the conduction band of a material through an ohmic contact with $\phi_m < \phi_s$ for n-type materials or with $\phi_m > \phi_s$ for p-type materials. Figure 3.1(c) shows two examples—one for $\phi_m - \chi > 0$ and the other for $\phi_m - \chi < 0$. The physics of such contacts has been discussed in some detail in Chapter 2.

(e) The electron injection may take place from an electrode to the conduction band of a material through a charge transfer process. An electrolyte–insulator contact injects carriers through such a process. Hole injection into anthracene has been observed when an electrolytic electrode containing iodine is used [Pope *et al.* 1962], and electron injection into anthracene when a tetrahydrofuran–anthracene solution containing sodium is used [Williams and Schadt 1970]. Furthermore, dissociation of singlet excitons at the interface between a metal and anthracene results in a charge transfer reaction following the scheme [Singh and Baessler 1974]:

$$M^* + \text{metal} \rightarrow M^+ + e^- \text{ (metal); (oxidative dissociation)}$$
$$M^* + \text{metal} \rightarrow M^- + e^+ \text{ (metal); (reductive dissociation)}$$

where M^* denotes the excited molecule at the interface. This reaction implies that an excited molecule (a Frenkel exciton) can either transfer an electron or a hole to the metal electrode, thus injecting a hole or an electron into the crystal. Such a charge transfer process for carrier injection is commonly used in organic semiconductors.

It should be mentioned that surface states at the interface between a metal and a crystal may assist carrier injection [Takai *et al.* 1975] and a thin insulating film between a metal and a crystal may alter the rate of majority carrier and minority carrier injection due to the potential across the film [Card and Smith 1971].

Once the carrier-injecting contact can provide a reservoir of carriers, the behaviour of the injected carriers and hence the current is controlled by the properties of the material in which the carriers are flowing. In molecular organic crystals, the bandwidth is narrow, the forbidden energy gap is wide, and hence the carrier mobility is low so that the intrinsic resistivity of these materials is high. Thus, the SCL current is easily observed though the carrier-injecting contact is not perfect because the intrinsic resistance of the material is usually much larger than the contact resistance. As there are no perfect crystals existing in this world, traps created by all types of imperfections are always present in the crystals and interact with injected carriers from ohmic contacts, thus controlling the carrier flow and determining the J–V characteristics. In organic crystals two types of carrier trap distributions have been reported [Gutmann and Lyons 1967, Helfrich 1967] and they are (i) the traps confined in discrete energy levels in the forbidden energy gap [Schadt and Williams 1969, Hoesterey and Letson 1963], and (ii) the traps with a quasi-continuous distribution of energy levels (normally following an exponential form or a Gaussian form) having a maximum trap density near the band edges [Helfrich 1967, Sussman 1967]. Both types of traps have been extensively investigated in anthracene crystals [Benderskii and Lavrushko 1971, Helfrich 1967, Mark and Helfrich 1962, Pott and Williams 1969, Reucroft and Mullins

1973, Schadt and Williams 1969, Schmidt and Wedel 1969, Schwob *et al.* 1971, Pope *et al.* 1971, Suna 1970, Sworakowski 1970, Thomas *et al.* 1968, Sussman 1967]. Simultaneous presence of both types of carrier trap distributions have also been observed [Schadt *et al.* 1969, Thomas *et al.* 1968]. Theoretically, several methods can be used to determine experimentally the energetic and kinetic parameters (energy levels and distributions) of carrier traps, such as the SCL current method [Lampert *et al.* 1970], the thermally stimulated current (TSC) method [Thomas *et al.* 1968], or the photoemission method [Caywood 1970]; but they do not provide any information as to the possible physical nature of traps. However, some general considerations have been suggested to relate the discrete trap levels with chemical impurities introduced into the lattice (chemical traps) [Hoesterey and Letson 1963, Lampert *et al.* 1970], and to relate the quasi-continuous trap distribution with the imperfection of the crystal structure (structural traps) [Helfrich 1967, Sworakowski 1970, Reucroft and Mullins 1973].

It should be noted, however, that the surroundings of a given type of trapping are not uniquely defined. It can be imagined that there always exist differences in configuration between nearest neighbours and in character between trapping centres, so that a discrete trap level can be considered "smeared out". Furthermore, the surroundings of an impurity entity are generally inhomogeneous. Several investigators [Lanyon 1963, Unger 1962, Sussman 1967, Simmons *et al.* 1973] have proposed that some types of traps had better be described by a Gaussian distribution function, such as the quasi-continuous trap distribution associated with the statistical dispersion of the charge carrier polarization energy caused by fluctuational structural irregularities of the lattice [Silinsh 1970, Owen *et al.* 1974]. The *J–V* characteristics have been analysed for solids with traps Gaussianly distributed in energy but uniformly distributed in space [Nespurek and Semejtek 1972, Bonham 1973, Grenet *et al.* 1973].

It should also be noted that the spatial distribution of traps can never be homogeneous because there always exist discontinuities between the material and the electrodes [Nicolet 1966, Sworakowski 1970]. The thinner the material specimen used for experimental studies, the more is the influence of the form of spatial distribution of traps on the *J–V* characteristics. The effect of non-uniform spatial trap distribution is important for thin films, and, in fact, this effect has been observed in thin films [Hwang and Kao 1972, Nicolet 1966, Nicolet *et al.* 1968, Covington and Ray 1974], possibly due to surface topography, grain boundaries, non-uniform doping, micro-crystalline defects, etc.

The probability for a trap to capture an electron follows the Fermi–Dirac statistics

$$f_n(E) = 1/\{1 + g_n^{-1} \exp\left[(E - E_{Fn})/kT\right]\} \tag{3.1}$$

and that for a trap to capture a hole follows

$$f_p(E) = 1/\{1 + g_p \exp\left[(E_{Fp} - E)/kT\right]\} \tag{3.2}$$

On the basis of the energy levels of the traps, the traps can be classified into shallow traps and deep traps. The so-called shallow traps refer to the traps whose energy levels $E = E_{tn}$ are located above the quasi-Fermi level E_{Fn} for electron traps, and refer to the traps whose energy levels $E = E_{tp}$ are located below the quasi-Fermi level E_{Fp} for hole traps. It can be seen from eqs. (3.1) and (3.2) that $f_n(E) \ll 1$ or $f_p(E) \ll 1$ if $(E_{tn} - E_{Fn})$ or $(E_{Fp} - E_{tp})$ is much greater than kT. This means that most of the traps may be empty.

Conversely, if E_{tn} is below E_{Fn} or E_{tp} above E_{Fp}, the traps are called "deep traps" in which $f_n(E) \to 1$ or $f_p(E) \to 1$ if $(E_{Fn} - E_{tn})$ or $(E_{tp} - E_{Fp})$ is much greater than kT. This implies that most of the traps are filled with trapped carriers (trapped electrons or trapped holes). Figure 3.2 shows schematically these two cases.

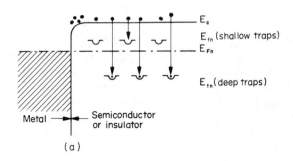

(a)

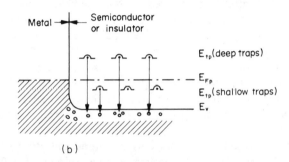

(b)

FIG. 3.2. Schematic energy level digrams for (a) electrons injecting from an electron ohmic contact to a semiconductor (or an insulator) with shallow and deep electron traps, and (b) holes injecting from a hole ohmic contact to a semiconductor (or an insulator) with shallow and deep hole traps.

Carrier injection into a solid is generally classified into (a) single injection, and (b) double injection. Single injection means that the current flow is mainly due to one type of carrier (electrons or holes) injected from a contacting electrode into the solid. These injected carriers would gradually establish a space charge leading to the well-known single-carrier SCL current. Double injection means that the current flow involves two types of carriers—electrons injected from the cathode and holes from the anode. In double injection, recombination kinetics control all the electrical properties. The recombination process may either be bimolecular (i.e. band-to-band electron–hole recombination) or may occur through one or more sets of localized recombination centres. The J–V characteristics are strongly dependent on the concentration and the distribution function of traps inside the specimen and other boundary conditions. In the following we shall begin with the steady state single injection and then double injection and other subjects of space charge electric conduction, with particular emphasis on the work which has not been included in recent review articles.

3.2. ONE-CARRIER (SINGLE) PLANAR INJECTION
IN SOLIDS

3.2.1. Theory

In single crystals the trap energy levels, if there are any, are generally discrete, while in amorphous and polycrystalline materials they are distributed in accordance with certain distribution functions [Sussman 1967]. The latter has been attributed to the intrinsic disorder of the lattice, which is possibly due to the variation of the nearest-neighbour distances. Material specimens in film form produced either by vacuum deposition or by other means are likely to be polycrystalline, and therefore, traps created by defects are generally distributed and their density is rather high even if the material itself is very pure chemically. Furthermore, material specimens always have boundaries such as their surfaces with metallic contacts. The trap distribution near such boundaries would be different from that in the bulk. In the theoretical analysis we confine our discussion to steady state d.c. one-dimensional planar current flow and make the following assumptions, but the treatment is general and therefore can be applied to thick or thin specimens in crystal or in film form of any materials.

(i) The energy band model can be used to treat the behaviour of injected carriers.

(ii) Only injected hole carriers are considered and the ohmic contact to inject them is perfect. (A similar treatment can be easily extended to the case for only injected electron carriers.) This implies that there is no electrode limitation to the current.

(iii) The mobility of the free holes (or free electrons) is independent of electric field and not affected by the presence of traps.

(iv) The free hole (or electron) density follows the Maxwell–Boltzmann statistics, while the trapped hole (or trapped electron) density follows the Fermi–Dirac statistics.

(v) The electric field is so large that the current components due to diffusion and due to carriers thermally generated in the specimen can be neglected. The former is justified if the applied voltage is larger than several kT/q so that the drift term becomes predominant and the diffusion term may be neglected without causing a serious error [Helfrich 1967, Lindmayer *et al.* 1963], while the latter is justified if the contact resistance is much less than the intrinsic resistance of the specimen such as in organic semiconductors.

(vi) The high field effects, such as Poole–Frenkel effect, Onsager model, impact ionization or field dependent mobility, are ignored.

(vii) The treatment is one-dimensional with the plane at $x = 0$ as the hole injecting contact (anode) and that at $x = d$ as the collecting contact, the specimen thickness being d. This implies that $F(x = 0) = 0$, and that the distance W_a between the actual electrode surface and the virtual anode $(-dV/dx = F = 0)$ is so small that we can assume $F(x = W_a \to 0) = 0$ for simplicity.

The distribution function for the trap density as a function of energy level E above the edge of the valence band and distance x from the injecting contact for hole carriers can be written as

$$h(E, x) = N_t(E)S(x) \tag{3.3}$$

where $N_t(E)$ and $S(x)$ represent, respectively, the energy and spatial distribution functions of traps. If the traps capture only holes, the electric field $F(x)$ inside the specimen follows the Poisson's equation

$$\frac{dF(x)}{dx} = \frac{q[p(x) + p_t(x)]}{\varepsilon} = \frac{\rho}{\varepsilon} \tag{3.4}$$

and the current density may be written as

$$J = q\mu_p p(x) F(x) \tag{3.5}$$

where $p(x)$ and $p_t(x)$ are, respectively, the densities of injected free and trapped holes, and they are given by

$$p_t(x) = \int_{E_l}^{E_u} h(E, x) f_p(E) dE \tag{3.6}$$

and

$$p(x) = N_v \exp(-E_{Fp}/kT) \tag{3.7}$$

and $f_p(E)$ is the Fermi–Dirac distribution function which is given by eq. (3.2). In the following we shall consider five general cases.

(A) Without traps (trap free solids—ideal case)

For this case $p_t(x) = 0$. Multiplying both sides of eq. (3.4) with $2F(x)$ and substituting eq. (3.5) into it, we obtain

$$2F(x)\frac{dF(x)}{dx} = \frac{d[F(x)]^2}{dx} = \frac{2J}{\varepsilon\mu_p} \tag{3.8}$$

Integration of eq. (3.8) and use of the boundary condition

$$V = \int_0^d F(x) dx$$

yields

$$J = \frac{9}{8}\varepsilon\mu_p \frac{V^2}{d^3} \tag{3.9}$$

This is the well-known Mott and Gurney equation [Mott and Gurney 1940] and is sometimes referred to as the square law for trap-free SCL currents.

We have ignored the effect of the thermally generated carriers. At low applied voltages the *J–V* characteristics may follow Ohm's law if the density of thermally generated free carriers p_0 inside the specimen (we consider holes only) is predominant such that

$$q p_0 \mu_p \frac{V}{d} \gg \frac{9}{8}\varepsilon\mu_p \frac{V^2}{d^3}$$

The onset of the departure from Ohm's law or the onset of the SCL conduction takes

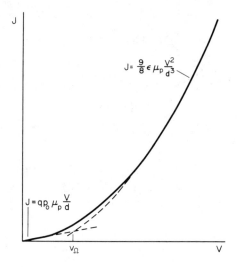

FIG. 3.3. Schematic diagram showing the transition from the ohmic to
the space charge conduction for one carrier (hole) injection in
a trap-free solid.

place when the inequality becomes equal. Thus the applied voltage for this condition to occur is

$$V_\Omega = \frac{8}{9} \frac{qp_0 d^2}{\varepsilon} \tag{3.10}$$

as shown schematically in Fig. 3.3. By rearranging eq. (3.10) in the form

$$\frac{d^2}{\mu_p V_\Omega} = \frac{9}{8} \frac{\varepsilon}{qp_0 \mu_p}$$

we have

$$t_t \simeq \tau_d \tag{3.11}$$

This means that when the transition from the ohmic to the SCL regime takes place, the carrier transit time $t_t = d^2/\mu_p V_\Omega$ at V_Ω (the minimum voltage required for the transition) is approximately equal to the dielectric (or ohmic) relaxation time $\tau_d = \varepsilon/qp_0\mu_p$. If the applied voltage V is less than V_Ω, then $t_t > \tau_d$ implying that the injected carrier density p is small in comparison with p_0 and that the injected carriers will redistribute themselves with a tendency to maintain electric charge neutrality internally in a time comparable to τ_d and they have no chance to travel across the specimen. The redistribution of the charge is known as dielectric relaxation. This means that the injected carriers under this condition do not alter the density of carriers p_0 because the injection of p would be accompanied by an unbalanced space charge which, according to Gauss's law, would give rise to an electric field, thus exerting a force on adjacent electrons so that they move in to neutralize the space charge. The net result is that holes in all parts of the specimen start to drift in such a way that the injected holes flow into the specimen from the injecting contact to replace the holes flowing out at the collecting electrode, so that no appreciable change in hole density occurs anywhere within the specimen. This can be easily understood by solving the following continuity equation

$$q\frac{\partial(p+p_0)}{\partial t} = -\nabla \cdot J \tag{3.12}$$

where current J is given by

$$J = q(p+p_0)\mu_p F - qD_p \nabla(p+p_0) \tag{3.13}$$

in which F produced by injected p follows the Poisson equation. Substituting eq. (3.4) into eq. (3.13) and then into eq. (3.12) and neglecting second-order terms involving pF, we obtain

$$\frac{\partial p}{\partial t} = -\left(\frac{qp\,\mu_p}{\varepsilon}\right)p + D_p \nabla^2 p \tag{3.14}$$

If $p < p_0$ and p spreads uniformly over the specimen in a time comparable to the dielectric relaxation time τ_d, the second term on the right-hand side of eq. (3.14) can be ignored. Thus, the solution of eq. (3.14) yields

$$p(t) = p(t = 0)\exp\left(-\frac{t}{\tau_d}\right) \tag{3.15}$$

τ_d is a measure of the time required for the injected carrier to re-establish equilibrium. Since $t_t > \tau_d$ negligible space charge would appear in the bulk. In most inorganic semiconductors (e.g. τ_d in Ge is about 10^{-22} sec) τ_d is small and the dielectric relaxation is not easy to observe. However, in organic semiconductors such as anthracene whose τ_d is about 10^{-5} sec, the relaxation becomes important.

   Equation (3.15) is no longer valid if $t_t \simeq \tau_d$ because $D_p \nabla^2 p$ in eq. (3.14) can no longer be neglected. When $V \not> V_\Omega$ and $t_t \simeq \tau_d$ (for $V < V_\Omega$, t_t increases with decreasing V but τ_d remains practically constant; while for $V > V_\Omega$, t_t decreases with increasing V, and τ_d also decreases with increasing V since the increase in V causes an increase in free carrier density in the bulk) or $t_t < \tau_d$, the injected excess carriers dominate the thermally generated carriers since the injected carrier transit time is too short for their charge to be relaxed by the thermally generated carriers [Rose 1963, Lampert and Mark 1970]. By rearranging eq. (3.10), we get

$$qp_0 d = \frac{9}{8}V_\Omega\frac{\varepsilon}{d}$$

The physical meaning of the above equation is that the total charge of free carriers is approximately equal to the condenser charge (the product of capacitance and voltage). This is equivalent to saying that the SCL current at $t_t = \tau_d$ is due to the condenser charge driven from one plate through the bulk of the specimen and then into the opposite plate by the applied voltage.

   From a qualitative reasoning we can imagine that there are two parallel current paths in the specimen, one due to the ohmic process and the other the SCL process. Figure 3.3 shows the qualitative physical picture of these two processes and the total current. At

$V = V_\Omega$ the total current can be qualitatively expressed as

$$J = qp_0\mu_p\frac{V_\Omega}{d} + \frac{9}{8}\varepsilon\mu_p\frac{V_\Omega^2}{d^3}$$

$$= qp_0\mu_p\frac{V_\Omega}{d} + qp_0\mu_p\frac{V_\Omega}{d}$$

$$= q(2p_0)\mu_p\frac{V_\Omega}{d} = qp\,\mu_p\frac{V_\Omega}{d} \qquad (3.16)$$

This implies that the onset of $t_t = \tau_d$ is also the condition for doubling the thermally generated free carrier density in the bulk of the specimen as shown in eq. (3.16). It should be noted that eq. (3.16) serves only for explaining the physical picture. It is not logic to consider the two processes existing simultaneously in the same space because the potential distributions are different for these two types of current. It is much better to say that when $t_t > \tau_d$ the ohmic process is predominant and the effect of injected space charge is suppressed, while when $t_t < \tau_d$ the SCL conduction is predominant and the ohmic process is suppressed; and that the potential distribution will adjust itself to suit the dominant process. However, the transition from the ohmic to the SCL conduction is not an abrupt change, but rather a gradual change.

(B) The traps confined in single or multiple discrete energy levels

For this case eq. (3.3)

$$h(E, x) = H_a\delta(E - E_t)S(x) \qquad (3.17)$$

where H_a is the density of traps, E_t is the trap energy level above the edge of the valence band, and $\delta(E - E_t)$ is the Dirac delta function. From eqs. (3.6) and (3.8) we obtain

$$P_t(x) = \int_{E_r}^{E_u} \frac{H_a\delta(E - E_t)S(x)dE}{1 + g_p\exp[(E_{F_p} - E)/kT]}$$

$$\simeq \frac{H_aS(x)}{1 + [H_a\theta_a/p(x)]} \qquad (3.18)$$

in which

$$\theta_a = \frac{g_pN_v}{H_a}\exp(-E_t/kT) \qquad (3.19)$$

Substitution of eq. (3.19) into eq. (3.4) gives

$$\frac{dF(x)}{dx} = \frac{q}{\varepsilon}\left\{p(x) + \frac{H_aS(x)}{1 + [H_a\theta_a/p(x)]}\right\} \qquad (3.20)$$

An analytical solution of eq. (3.20) for J as a function of applied voltage is not possible, although a numerical solution can be obtained for all possible cases separately. For simplicity, we assume that E_t is a shallow trap level located below E_{Fp}. This implies that

$H_a\theta_a > p(x)$. On the basis of this assumption and by multiplying both sides of eq. (3.20) with $2F(x)$ and substituting eq. (3.5) into it, we obtain

$$2F(x)\frac{dF(x)}{dx} = \frac{d[F(x)]^2}{dx} = \frac{2J}{\varepsilon\mu\theta_a}[\theta_a + S(x)] \tag{3.21}$$

Integration of eq. (3.21) and use of the boundary condition

$$V = \int_0^d F(x)dx$$

give

$$J = \frac{9}{8}\varepsilon\mu_p\theta_a\frac{V^2}{d_{\text{eff}}^3} \tag{3.22}$$

in which V is the applied voltage and

$$d_{\text{eff}} = \left\{\frac{3}{2}\int_0^d\left(\int_0^t[\theta_a + S(x)]dx\right)^{1/2}dt\right\}^{2/3} \tag{3.23}$$

Equation (3.22) is similar in form to that derived by Lampert [1956] except that d has been replaced with d_{eff} which can be considered as "effective thickness". The difference between d_{eff} and d can be attributed to the inhomogeneous spatial distribution of free and trapped carriers.

Up to here we shall discuss some important parameters related to the effects of traps. In the following we ignore the effect of non-uniform spatial distribution of traps for simplicity. This means that we use d instead of d_{eff}.

(1) θ_a is, in fact, the ratio of free carrier density to total carrier (free and trapped) density

$$\theta_a = \frac{p}{p + p_t} \tag{3.24}$$

Thus for the trap free case, $p_t = 0$, therefore $\theta_a = 1$. With traps, θ_a is always less than unity and could be as small as 10^{-7}.

(2) When the density of thermally generated free carriers p_0 inside the specimen (we consider holes only) is larger than the density of injected carriers p, the ohmic conduction is predominant. The onset of the transition from the ohmic to the SCL conduction following the same principle used in (A) occurs when the applied voltage reaches

$$V_\Omega = \frac{8}{9}\frac{qp_0d^2}{\theta_a\varepsilon} \tag{3.25}$$

This equation indicates that

(a) The voltage for the transition V_Ω increases with increasing density of thermally generated carriers in the specimen p_0.
(b) The higher the concentration of traps (this means the smaller the value of θ_a), the higher is the value of V_Ω for the transition.
(c) When the free carrier density is changed by injection from p_0 to a new value p,

then in the steady state (when the trapping and detrapping reach a quasi-thermal equilibrium) the density of trapped carriers is p_t, and thus the total density of injected carriers becomes $p_T = p + p_t$. Since μ_p is the mobility of free carriers, we define the effective mobility as

$$\mu_{p\,\mathrm{eff}} = \left(\frac{p}{p+p_t}\right)\mu_p = \theta_a\mu_p \tag{3.26}$$

with the understanding that the effective carrier density for electric conduction is p_T rather than p. On the basis of this definition the effective carrier transit time can be expressed in terms of free carrier transit time t_t at V_Ω as

$$t_{t\,\mathrm{eff}} = \frac{t_t}{\theta_a} = \frac{d^2}{\theta_a\mu_p V_\Omega}$$

$$= \frac{d^2}{\mu_{p\,\mathrm{eff}} V_\Omega} \tag{3.27}$$

The transition from the ohmic to the SCL conduction occurs when $t_{t\,\mathrm{eff}}$ is approximately equal to τ_d.

(3) Even with $V < V_\Omega$, in which the ohmic conduction is predominant in the steady state, there is a transient supply of injected carriers when a voltage is applied across the specimen. The ohmic behaviour can be observed only after these space charge carriers become trapped. This phenomenon has been observed in CdS crystal [Rose 1955, Smith and Rose 1955].

(4) The increase of applied voltage may increase the density of free carriers resulting from injection to such a value that the quasi-Fermi level E_{Fp} moves down below the shallow hole-trapping level E_t, and then most traps are filled up (for hole traps a filled trap means that it has given up an electron to the valence band, while for electron traps a filled trap means that it has captured an electron or it is occupied, so that when most electron traps are filled up, the quasi-Fermi level E_{Fn} moves up above the electron trapping level E_t). The traps-filled limit (TFL) is the condition for the transition from the trapped J–V characteristics to the trap-free J–V characteristics. It can be imagined that after all traps are filled up, the subsequently injected carriers will be free to move in the specimen, so that at the threshold voltage V_{TFL} to set on this transition, the current will rapidly jump from its low trap-limited value to a high trap-free SCL current. V_{TFL} is defined as the voltage required to fill the traps or, in other words, as the voltage at which E_{Fp} passes through E_t. In thermal equilibrium (i.e. in the absence of external perturbation, and for present consideration, in the absence of applied voltage) the density of trapped holes is

$$p_{t0} = \int_{E_u}^{E_l} \frac{H_a\delta(E - E_t)dE}{1 + g_p\exp[(E_{Fp0} - E)/kT]}$$

$$= \frac{H_a}{1 + g_p\exp[(E_{Fp0} - E_t)/kT]} \tag{3.28}$$

and the density of unfilled traps is

$$H_a - p_{t0} = \frac{H_a}{1 + g_p^{-1} \exp[(E_t - E_{Fp0})/kT]} \tag{3.29}$$

where E_{Fp0} is the quasi-Fermi level in thermal equilibrium (i.e. in the absence of applied voltage).

(I) SHALLOW TRAPS

For this case $E_t < E_{Fp0}$, eq. (3.29) can be approximated to

$$H_a - p_{t0} \simeq H_a \tag{3.30}$$

V_{TFL} can be interpreted in such a way that when the unfilled traps are completely filled, the applied voltage reaches the value of V_{TFL}. On the assumption that $H_a \gg p$, then at V_{TFL} we have

$$\frac{dF_{\text{TFL}}}{dx} = \frac{qH_a}{\varepsilon} \tag{3.31}$$

Integration of eq. (3.31) gives

$$V_{\text{TFL}} = \int_0^d F_{\text{TFL}} dx = \frac{qH_a d^2}{2\varepsilon} \tag{3.32}$$

For the cases in which θ_a is not too small so that $H_a/p_0 > \frac{8}{9}\left(\frac{2}{\theta_a}\right)$ we have

$$V_{\text{TFL}} > V_\Omega \tag{3.33}$$

Figure 3.4 shows schematically the variation of V_Ω and V_{TFL} with θ_a and H_a.

(II) DEEP TRAPS

For this case $E_t > E_{Fp0}$, eq. (3.29) can be written as

$$H_a - p_{t0} = g_p H_a \exp[(E_{Fp0} - E_t)/kT] \tag{3.34}$$

Following the same method used in (I) and assuming that $H_a - p_{t0} \gg p$, than at V_{TFL} we have

$$V_{\text{TFL}} = \frac{q(H_a - p_{t0})d^2}{2\varepsilon} \tag{3.35}$$

Since $E_t > E_{Fp0}$, all injected carriers will be used first to fill the traps, and at V_{TFL} all traps will be filled up so that V_{TFL} can also be considered as the voltage to set on the transition from the ohmic to the SCL conduction, that is V_Ω. But for this deep trap case the transition is from the ohmic to the trap-free SCL current because all traps have been filled up. Recalling that [see eq. (3.16)]

$$P(|J|_{V=V_{\text{TFL}}}) \simeq 2p_0 \tag{3.36}$$

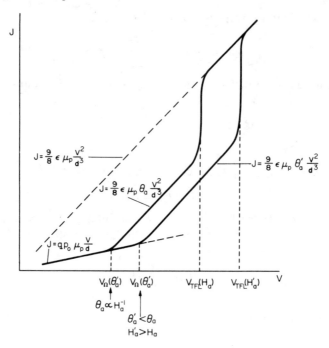

FIG. 3.4. Schematic log–log plot of the J–V characteristics showing the variation of V_Ω and V_{TFL} with θ_a or H_a for a solid with shallow traps confined in a trapping level E_t ($E_t < E_{Fp0}$).

it is now interesting to see the change of p when V is increased to $2V_{TFL}$. Based on the total charge in the specimen, $Q = q(H_a - p_{t0})d$ at $V = V_{TFL}$ and $Q = 2q(H_a - p_{t0})d$ at $V = 2V_{TFL}$ if the capacitance of the specimen is assumed to be unchanged. At $V = V_{TFL}$ the total injected carriers $H_a - p_{t0}$ are trapped carriers, but at $V = 2V_{TFL}$ one half of the injected carriers are trapped and the other half are free carriers so that

$$P(|J|_{V=2V_{TFL}}) = H_a - p_{t0} \tag{3.37}$$

From eqs. (3.36) and (3.37), we have

$$A = \frac{J(2V_{TFL})}{J(V_{TFL})} \simeq \frac{H_a - p_{t0}}{p_0} \tag{3.38}$$

Figure 3.5 shows the schematic log–log plot of the J–V characteristics. The triangle of this plot is sometimes referred to as the Lampert triangle [Lampert 1956, Henderson *et al.* 1972]. A more analytic treatment of this problem taking into account the thermally generated carriers in the specimen has been reported by Lampert and Mark [1970].

(5) If the traps are not confined in only one single discrete energy level, but in several discrete energy levels, such as a solid containing more than two different kinds of impurities, eqs. (3.22) and (3.23) are still applicable provided that θ_a is given by

$$\theta_a^{-1} = \sum_i \theta_i^{-1} \tag{3.39}$$

where

$$\theta_i = \frac{g_{pi}N_v}{H_{ai}} \exp(-E_{ti}/kT) \tag{3.40}$$

in which g_{pi}, H_{ai}, and E_{ti} refer to g_p, H_a, and E_t in the ith single discrete energy level.

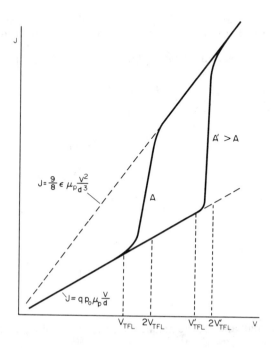

FIG. 3.5   Schematic log–log plot of the J–V characteristics showing the variation of V_{TFL} with the ratio of $A = (H_a - p_{t0})/p_0$ for a solid with deep traps confined in a trapping level E_t ($E_t > E_{Fp0}$).

(C) The traps distributed exponentially within the forbidden energy gap

For this case eq. (3.3) becomes

$$h(E, x) = \frac{H_b}{kT_c} \exp\left(-\frac{E}{kT_c}\right) S(x) \tag{3.41}$$

where H_b is the density of traps and T_c is a characteristic constant of the distribution. If $T_c > T$ we can assume that [Mark and Helfrich 1962] $f_p(E) = 1$ for $E_{Fp} < E < \infty$ and $f_p(E) = 0$ for $E < E_{Fp}$ as if we take $T = 0$. This is a good approximation particularly when T_c is much larger than T. With this assumption we obtain

$$p_t(x) = \int_{E_{Fp}}^{\infty} \frac{H_b}{kT_c} \exp\left(-\frac{E}{kT_c}\right) S(x) dE$$

$$= H_b \exp\left(-\frac{E_{Fp}}{kT_c}\right) S(x)$$

$$= H_b \left(\frac{p}{N_v}\right)^{T/T_c} S(x) \tag{3.42}$$

The upper limit of the integral has been extended to infinity. This is permissible if $E_{Fp}(x)$ is far removed from the Fermi level of the neutral region. By substituting eq. (3.42) into eq. (3.4), letting $T_c/T = l$, and multiplying both sides with $\left(\frac{l+1}{l}\right)[F(x)]^{1/l}$, we obtain

$$\left(\frac{l+1}{l}\right)[F(x)]^{1/l}\frac{dF(x)}{dx} = \frac{d[F(x)]^{(l+1)/l}}{dx}$$

$$= \left(\frac{l+1}{l}\right)\frac{q}{\varepsilon}[pF(x)]^{1/l}$$

$$\times [p^{(1-l)/l} + H_b N_v^{-1/l} S(x)]$$

$$= \left(\frac{l+1}{l}\right)\frac{qH_b}{\varepsilon}\left(\frac{J}{q\mu_p N_v}\right)^{1/l}[\theta_b + S(x)] \tag{3.43}$$

in which

$$\theta_b = \frac{N_v}{H_b}\exp\left[-\frac{E_{Fp}}{kT}\left(\frac{l-1}{l}\right)\right] \tag{3.44}$$

If $p_t \gg p$, θ_b is very small and can be neglected. Integration of eq. (3.43) and use of the boundary condition

$$V = \int_0^d F(x)dx$$

gives

$$J = q^{1-l}\mu_p N_v \left(\frac{2l+1}{l+1}\right)^{l+1}\left(\frac{l}{l+1}\frac{\varepsilon}{H_b}\right)^l\frac{V^{l+1}}{d_{\text{eff}}^{2l+1}} \tag{3.45}$$

in which

$$d_{\text{eff}} = \left\{\frac{2l+1}{l+1}\int_0^d\left(\int_0^t S(x)dx\right)^{1/(l+1)}dt\right\}^{(l+1)/(2l+1)} \tag{3.46}$$

for $p_t > p$.

Equation (3.45) is similar in form to that derived by Mark and Helfrich [1962] except that d has been replaced with d_{eff}. Again, the difference between d_{eff} and d is caused by the inhomogeneous spatial distribution of free and trapped carriers.

Several investigators [Williams and Schadt 1970, Thomas et al. 1968, Reucroft et al. 1974, Hwang and Kao 1974] have reported that some experimental results reveal the exponentially distributed traps scanned at levels above a certain discrete level E_{te}. For this case eq. (3.45) becomes

$$J = q^{1-l} \mu_p N_v \left(\frac{2l+1}{l+1}\right)^{l+1} \left(\frac{l}{l+1} \frac{\varepsilon}{H_b'}\right) \frac{V^{l+1}}{(d_{eff})^{2l+1}} \tag{3.47}$$

The only difference is that H_b has been replaced with H_b' which is given by

$$H_b' = H_b \exp(E_{te}/kT_c) \tag{3.48}$$

This implies that the distribution function for the trap density is written as

$$h(E, x) = \frac{H_b}{kT_c} \exp[-(E - E_{te})/kT_c]S(x) \tag{3.49}$$

It is obvious that eq. (3.47) reduces back to eq. (3.45) if $E_{te} = 0$, that is, the highest trap concentration is at the edge of the valence band.

Following the same procedure given in Section 3.2.1 (B), and using d instead of d_{eff} for simplicity, we obtain V_Ω and V_{TFL} for the present case.

(1) When the density of thermally generated free carriers p_0 inside the specimen (we consider holes only) is larger than that of the injected carriers p_T, the ohmic conduction is predominant. By setting $J = q p_0 \mu_p \frac{V}{d}$ equal to eq. (3.47) we obtain the applied voltage required for the onset of the transition from the ohmic to the SCL conduction, which is given by

$$V_\Omega = \frac{q d^2 H_b'}{\varepsilon} \left(\frac{p_0}{N_v}\right)^{1/l} \left(\frac{l+1}{l}\right) \left(\frac{l+1}{2l+1}\right)^{l+1/l} \tag{3.50}$$

(2) When all traps are filled up, a transition from the trapped SCL current to a trap-free SCL current will take place. By setting eq. (3.9) equal to eq. (3.47) we obtain the TFL threshold voltage [Mark and Helfrich 1962]

$$V_{TFL} = \frac{q d^2}{\varepsilon} \left[\frac{9}{8} \frac{H_b'^l}{N_v} \left(\frac{l+1}{l}\right)^l \left(\frac{l+1}{2l+1}\right)^{l+1}\right]^{1/(l-1)} \tag{3.51}$$

(D) The traps distributed Gaussianly within the forbidden energy gap

For this case eq. (3.3) becomes

$$h(E, x) = \frac{H_d}{(2\pi)^{1/2} \sigma_1} \exp\left[-\frac{(E - E_{tm})^2}{2\sigma_t^2}\right] S(x) \tag{3.52}$$

where $(2\pi)^{1/2}$ is the normalizing factor, E_{tm} is the hole-trapping energy level with a maximum trap density, and σ_t is the standard deviation of the Gaussian function.

(I) SHALLOW TRAPS

The hole traps are considered to be shallow if $E_{tm} < E_{Fp}$. For this case eq. (3.6) becomes [Hwang and Kao 1976]

$$p_t = \int_{E_l}^{E_u} h(E, x) g_p^{-1} \exp[(E - E_{Fp})/kT] dE$$

$$= H_d g_p^{-1} \exp(-E_{Fp}/kT) \exp\left[E_{tm}/kT - \frac{1}{2}(\sigma_t/kT)^2\right] S(x)$$

$$= p\, S(x)/\theta_d \tag{3.53}$$

in which

$$\theta_d = \frac{g_p N_v}{H_d} \exp\left[-\frac{E_{tm}}{kT} + \frac{1}{2}\left(\frac{\sigma_t}{kT}\right)^2\right] \tag{3.54}$$

Substitution of eqs. (3.5) and (3.53) into eq. (3.4) gives

$$\frac{dF}{dx} = \left(\frac{J}{\varepsilon \mu_p F}\right) \frac{1}{\theta_d} [\theta_d + S(x)] \tag{3.55}$$

Integration of eq. (3.55) and use of the boundary condition

$$V = \int_0^d F(x) dx \tag{3.56}$$

give

$$J = \frac{9}{8} \varepsilon \mu_p \theta_d \frac{V^2}{d_{\text{eff}}^3} \tag{3.57}$$

in which

$$d_{\text{eff}} = \left(\frac{3}{2} \int_0^d \left\{\int_0^t [\theta_d + S(x)] dx\right\}^{1/2} dt\right)^{2/3} \tag{3.58}$$

Equation (3.57) is similar in form to that for traps confined in a single discrete energy level [eq. (3.22)] except that θ_a has been replaced with θ_d. It is interesting to note that as $\sigma_t \to 0$ the case for Gaussian trap distribution approaches that for traps confined in a single discrete energy level, and that for the former case the plot of ln J as a function $1/T$ may not be linear.

(II) DEEP TRAPS

The hole traps are considered to be deep if $E_{tm} > E_{Fp}$. By letting $Z = E - E_{tm}$ and using appropriate approximations, eq. (3.6) becomes [Hwang and Kao 1976]

$$p_t \simeq \frac{H_d S(x)}{(2\pi)^{1/2} \sigma_t} \int_0^\infty \frac{\exp(-Z^2/2\sigma_t^2) dZ}{1 + g_p \exp[(E_{Fp} - E_{tm} - Z)/kT]}$$

$$= H_d'(p/N_v)^{1/m} S(x) \tag{3.59}$$

in which

$$H_d' = (H_d/2) g_p^{-1} \exp(E_{tm}/mkT) \tag{3.60}$$

and
$$m = (1 + 2\pi\sigma_t^2/16k^2T^2)^{1/2} \tag{3.61}$$

Substituting eqs. (3.5) and (3.59) into eq. (3.4), and multiplying both sides by $(m+1)$ $[F(x)]^{1/m}/m$, we obtain

$$\left(\frac{m+1}{m}\right)[F(x)]^{1/m}\frac{dF(x)}{dx} = \frac{d[F(x)]^{(m+1)/m}}{dx}$$

$$= \left(\frac{m+1}{m}\right)\frac{qH_d'}{\varepsilon}\left(\frac{J}{q\mu_pN_v}\right)^{1/m}[\theta_d' + S(x)] \tag{3.62}$$

in which

$$\theta_d' = \frac{g_pN_v}{H_d'}\exp\left[-\frac{E_{Fp}}{kT}\left(\frac{m-1}{m}\right)\right]$$

If $p_t \gg p$, θ_d' is very small and can be neglected. Integration of eq. (3.62) and use of the boundary condition given in eq. (3.56) give

$$J = q^{1-m}\mu_pN_v\left(\frac{2m+1}{m+1}\right)^{m+1}\left(\frac{m}{m+1}\frac{\varepsilon}{H_d'}\right)^m\frac{V^{m+1}}{d_{\text{eff}}^{2m+1}} \tag{3.63}$$

in which

$$d_{\text{eff}} = \left\{\frac{2m+1}{m+1}\int_0^d\left[\int_0^t S(x)dx\right]^{m/(m+1)}dt\right\}^{(m+1)/(2m+1)} \tag{3.64}$$

for $p_t \gg p$.

Equation (3.63) is similar in form to that for traps distributed exponentially within the forbidden gap [eq. (3.45)] except that l have been replaced with m. To distinguish between the traps Gaussianly distributed and those exponentially distributed, the technique of measuring thermally stimulated currents as functions of temperature and applied voltage can be used [Unger 1962, Simmons and Tam 1973].

Following the same procedure given in Section 3.2.1(B), and using d instead of d_{eff} for simplicity, we obtain V_Ω and V_{TFL} for this case as follows.

(1) By setting $J = qp_0\mu_p\dfrac{V}{d}$ equal to eq. (3.57) and equal to eq. (3.63), we obtain

$$V_\Omega = \frac{8}{9}\frac{qp_0d^2}{\theta_d\varepsilon} \quad \text{for shallow traps} \tag{3.65}$$

and

$$V_\Omega = \frac{qd^2H_d'}{\varepsilon}\left(\frac{p_0}{N_v}\right)^{1/m}\left(\frac{m+1}{m}\right)\left(\frac{m+1}{2m+1}\right)^{(m+1)/m} \quad \text{for deep traps} \tag{3.66}$$

(2) V_{TFL} for shallow traps. Following eq. (3.28) the density of trapped holes in thermal equilibrium (in the absence of applied voltage) is

$$p_{t0} = \int_{E_l}^{E_u} h(E)g_p^{-1}\exp[(E - E_{Fp0})/kT]dE$$

$$= H_dg_p^{-1}\exp(-E_{Fp0}/kT)\exp[E_{tm}/kT - \tfrac{1}{2}(\sigma_t/kT)^2] \tag{3.67}$$

The density of unfilled traps is

$$N_{t(\text{unfilled})} = \int_{E_l}^{E_u} h(E)\,dE - p_{t0}$$

$$= \frac{H_d}{2}\left[\operatorname{erf}(E_u) - \operatorname{erf}(E_l)\right] - p_{t0} \tag{3.68}$$

If $N_{t(\text{unfilled})} > p$ at V_{TFL}, the Poisson equation is

$$\frac{dF}{dx} = \frac{qN_{t(\text{unfilled})}}{\varepsilon}$$

Thus

$$V_{\text{TFL}} = \int_0^d F\,dx = \frac{qN_{t(\text{unfilled})}d^2}{2\varepsilon} \quad \text{for shallow traps} \tag{3.69}$$

V_{TFL} for deep traps. By setting eq. (3.9) equal to eq. (3.63), we obtain

$$V_{\text{TFL}} = \frac{qd^2}{\varepsilon}\left[\frac{9}{8}\frac{H_d'^m}{N_v}\left(\frac{m+1}{m}\right)^m\left(\frac{m+1}{2m+1}\right)^{m+1}\right]^{1/(m-1)} \quad \text{for deep traps} \tag{3.70}$$

(E) The traps confined in smeared discrete energy levels

In Section 3.2.1(B) we have derived an expression for J as a function of V for the case with traps confined in a discrete energy level. However, this case could also be considered physically as a case with a Gaussian distribution having a very narrow trap energy deviation, σ_t (i.e. $\sigma_t \ll kT$). Supposing that the trap energy level E_{tm} is located below E_{Fp}, which we considered as shallow traps, the J–V characteristics can be easily derived from eqs. (3.57) and (3.58) for the case $\sigma_t \ll kT$. If the traps are confined not in a single smeared discrete energy level but in multiple smeared discrete energy levels, such as a solid containing more than two different kinds of impurities, eqs. (3.57) and (3.58) are still valid provided that θ_d is given by

$$\theta_d^{-1} = \sum_i \theta_i^{-1} \tag{3.71}$$

where

$$\theta_i = \frac{g_{pi}N_v}{H_{di}}\exp\left[-\frac{E_{ti}}{kT} + \frac{1}{2}\left(\frac{\sigma_{ti}}{kT}\right)^2\right] \tag{3.72}$$

in which g_{pi}, H_{di}, σ_{ti}, and E_{ti} refer to g_p, H_d, σ_t, and E_t in the ith single smeared discrete energy level.

If the trap energy level E_{tm} is located above E_{Fp} for the case of deep traps, it is not possible to obtain an exact solution for the J–V characteristics. However, from eqs. (3.61), (3.63), and (3.64) we can obtain an approximate solution for the case $\sigma_t \ll kT$. This is given by

$$J = \frac{9}{8}\varepsilon\mu_p\left(\frac{N_v}{H_d'}\right)\frac{V^2}{d_{\text{eff}}^3} \tag{3.73}$$

(F) The traps distributed uniformly within the forbidden energy gap

This type of trap distribution was first investigated by Rose [1955]. Although physically the uniform distribution of traps within the forbidden energy gap is unlikely to occur in a solid, it is possible that traps due to impurities may not be confined in a single discrete energy level, but rather within a narrow band from E_l to E_u in the forbidden energy gap. For such a case eq. (3.3) can be written as

$$h(E, x) = H_c U(E - E_l) U(E_u - E) S(X) \tag{3.74}$$

where U is the Heaviside step function,

$$U(E - E_l) = 0 \quad \text{if} \quad E < E_l \quad \text{and} \quad U(E_u - E) = 0 \quad \text{if} \quad E > E_u,$$

and

$$U(E - E_l) = 1 \quad \text{if} \quad E > E_l \quad \text{and} \quad U(E_u - E) = 1 \quad \text{if} \quad E < E_u$$

and H_c is the density of traps per unit energy interval. From eqs. (3.6) and (3.74) we obtain

$$
\begin{aligned}
p_t(x) &= \int_{E_l}^{E_u} \frac{H_c U(E - E_l) U(E_u - E) S(x) \, dE}{1 + g_p \exp[(E_{Fp} - E)/kT]} \\
&= H_c \left\{ E_g + kT \ln \frac{1 + g_p \exp[(E_{Fp} - E_u + E_l)/kT]}{1 + g_p \exp(E_{Fp}/kT)} \right\} S(x) \\
&\simeq H_c kT \left(\frac{E_u - E_l - E_{Fp}}{kT} - \ln g_p \right) S(x)
\end{aligned}
\tag{3.75}
$$

By assuming $p_t \gg p$ and substituting eq. (3.75) into eq. (3.4), we obtain

$$
\begin{aligned}
\frac{dF}{dx} &= \frac{q}{\varepsilon} \left\{ P + H_c kT \left(\frac{E_u - E_l - E_{Fp}}{kT} - \ln g_p \right) S(x) \right\} \\
&= \frac{q H_c kT}{\varepsilon} S(x) \ln \left\{ \frac{q \mu_p N_v g_p \exp[-(E_u - E_l)kT]}{J} F \right\}
\end{aligned}
\tag{3.76}
$$

Integration of eq. (3.76) and use of the boundary condition

$$V = \int_0^d F(x) \, dx$$

give [Hwang and Kao 1972]

$$J = 2q\mu_p N_v g_p \frac{V}{d_{\text{eff}}} \exp\left[-\frac{E_u - E_l}{kT} \right] \exp\left(\frac{2\varepsilon V}{q H_c kT d_{\text{eff}}^2} \right) \tag{3.77}$$

in which

$$d_{\text{eff}} = \left\{ 2 \int_0^d \int_0^t S(x) \, dx \, dt \right\}^{1/2} \tag{3.78}$$

Equation (3.77) is similar in form to that derived by Muller [1963] except that d has been replaced with d_{eff}. Again, the difference between d_{eff} and d is caused by the inhomogeneous spatial distribution of free and trapped carriers. Obviously, if the traps are uniformly distributed from E_v to E_c in the forbidden energy gap, $E_u - E_l = E_g$.

From the analysis of experimental results such a uniform trap distribution has so far not been identified to exist in organic semiconductors. However, the experimental results on amorphous Se films [Touraine *et al.* 1972] show that a uniform trap distribution in the forbidden energy gap does exist, and the dark conduction at high fields depends on this form of trap distribution. Touraine *et al.* [1972] have reported that for amorphous selenium the band width of the distribution is 0.25 eV and E_l lies 0.74 eV above the valence band edge, that the trap density per unit energy is $5 \times 10^{14} \, \text{cm}^{-3} \, \text{eV}^{-1}$, and that these values are not affected by the change of temperature in the range from $233°\text{K}$ to $293°\text{K}$.

Following the same procedure given in Section 3.2.1(B) and using d instead of d_{eff} for simplicity, we obtain V_Ω and V_{TFL} for this case as follows:

(1) By setting $J = qp_0\mu_p\dfrac{V}{d}$ equal to eq. (3.77), we obtain

$$V_\Omega = \frac{qH_ckTd^2}{2\varepsilon}\left\{ \ln\left(\frac{p_0}{2g_pN_v}\right) + \frac{E_u - E_l}{kT} \right\} \tag{3.79}$$

(2) For accurate approach, we can obtain V_{TFL} by setting eq. (3.9) equal to eq. (3.77) and using graphical or computer techniques to calculate V_{TFL}, however, using the method of Muller [1963] which states that the exponential trap distribution case approaches to the uniform trap distribution case when $l \to \infty$. Thus by letting $l \to \infty$ in eq. (3.51) we obtain

$$V_{\text{TFL}} = \lim_{l \to \infty} \frac{qd^2}{\varepsilon}\left[\frac{9}{8}\frac{H_b^{\prime l}}{N_v}\left(\frac{l+1}{l}\right)^l \left(\frac{l+1}{2l+1}\right)^{l+1} \right]^{1/(l-1)}$$

$$\simeq \frac{qH_b'd^2}{2\varepsilon} \simeq \frac{qH_c(E_u - E_l)d^2}{2\varepsilon} \tag{3.80}$$

since H_b' is equivalent to $H_c(E_u - E_l)$ when $l \to \infty$.

Tables 3.1 and 3.2 summarize all expressions for V_Ω and V_{TFL} for all cases.

3.2.2. The scaling rule

It can be seen from all expressions for J–V characteristics given in Section 3.2.1 that the general scaling rule [Lampert and Mark 1970, Roberts 1973, Murgatroyd 1973] for one-carrier SCL conduction in any materials with any trap distributions can be expressed in the form of

$$\frac{J}{d_{\text{eff}}} = f\left(\frac{V}{d_{\text{eff}}^2}\right) \tag{3.81}$$

This equation is universally valid provided that the carrier mobility is field independent and the effect of carrier diffusion is ignored. It is also valid for two-carrier (double injection) space charge conduction if the mobilities of both types of carriers are field independent and the effect of carrier diffusion is ignored. The high field effects and the effect of carrier diffusion will be discussed later. In eq. (3.81) the use of d_{eff} (effective

TABLE 3.1. *The expressions for V_Ω for single injection J–V characteristics with and without traps*

Case	V_Ω
Trap free	$\dfrac{9}{8}\dfrac{qp_0 d^2}{\varepsilon}$
Traps confined in a single discrete energy level	$\dfrac{9}{8}\dfrac{qp_0 d^2}{\theta_a \varepsilon}$
Traps distributed exponentially within the forbidden energy gap	$\dfrac{qd^2 H_b'}{\varepsilon}\left(\dfrac{p_0}{N_v}\right)^{1/l}\left(\dfrac{l+1}{l}\right)\left(\dfrac{l+1}{2l+1}\right)^{(l+1)/l}$
Traps distributed Gaussianly within the forbidden energy gap	$\dfrac{qd^2 H_d'}{\varepsilon}\left(\dfrac{p_0}{N_v}\right)^{1/m}\left(\dfrac{m+1}{m}\right)\left(\dfrac{m+1}{2m+1}\right)^{(m+1)/m}$
Traps distributed uniformly within the forbidden energy gap	$\dfrac{qH_c kTd^2}{2\varepsilon}\left[\ln\left(\dfrac{p_0}{2g_p N_v}\right)+\dfrac{E_u-E_l}{kT}\right]$

TABLE 3.2. *The expression for V_{TFL} for single injection J–V characteristics with traps*

Case	V_{TFL}
Traps confined in a single discrete energy level	$\dfrac{qH_a d^2}{2\varepsilon}$ (shallow traps) $\dfrac{q(H_a-p_0)d^2}{2\varepsilon}$ (deep traps)
Traps distributed exponentially within the forbidden energy gap	$\dfrac{qd^2}{\varepsilon}\left[\dfrac{9}{8}\dfrac{H_b'^l}{N_v}\left(\dfrac{l+1}{l}\right)^l\left(\dfrac{l+1}{2l+1}\right)^{l+1}\right]^{1/(l-1)}$
Traps distributed Gaussianly within the forbidden energy gap	$\dfrac{qd^2}{\varepsilon}\left[\dfrac{9}{8}\dfrac{H_d'^m}{N_v}\left(\dfrac{m+1}{m}\right)^m\left(\dfrac{m+1}{2m+1}\right)^{m+1}\right]^{1/(m-1)}$
Traps distributed uniformly within the forbidden energy gap	$\dfrac{qH_c(E_u-E_l)d^2}{2\varepsilon}$

specimen thickness) instead of d (true specimen thickness) in all expressions for J–V characteristics can be thought of as taking into account the effect of non-uniform spatial distribution of traps.

By writing eq. (3.81) in the following form

$$\frac{J}{d}\left(\frac{d}{d_{\text{eff}}}\right)=f\left[\left(\frac{F_{\text{av}}}{d}\right)\left(\frac{d}{d_{\text{eff}}}\right)^2\right] \tag{3.82}$$

we derive the following features:

(i) The factor d/d_{eff} means that a solid with a non-uniform spatial distribution of traps is equivalent to a solid with a uniform spatial distribution of traps if its true thickness d is replaced with an effective thickness d_{eff}.

(ii) J is directly related to the average field $F_{av} = V/d$ because J can always be written as

$$J = q\bar{p}\mu_p F_{av} \qquad (3.83)$$

where $\bar{p}$ is the average density of free carriers. It is this $\bar{p}$ which produces different forms of J–V characteristics. It is obvious that $\bar{p}$ depends upon the distribution of space charge, which in turn depends not only upon the type of trap distributions, but also significantly upon the interaction between the traps and the field which is volume dependent.

(iii) J/d can be interpreted as the flow of one unit volume of free charge carriers per second, while F_{av}/d can be written as $D/\varepsilon d$ in which D is the average charge per unit area on the electrode and can be interpreted as the average charge density in the specimen having a dielectric constant ε. Thus the flow of charge carriers per unit volume per unit second is directly related to the average charge density, which includes the free and trapped charged carriers in the specimen.

(iv) With $\dfrac{J}{d}\left(\dfrac{d}{d_{eff}}\right)$ expressed in terms of $\left[\left(\dfrac{F_{av}}{d}\right)\left(\dfrac{d}{d_{eff}}\right)^2\right]^n$, the value of n will reflect the type of trap distributions. Table 3.3 summarizes the values of n for all cases, and Table 3.4 summarizes all expressions for J–V characteristics for all cases discussed in this chapter.

TABLE 3.3. *The values of n in the factor* $\left[\left(\dfrac{F_{av}}{d}\right)\left(\dfrac{d}{d_{eff}}\right)^2\right]^n$ *for single injection J–V characteristics*

n	Description	References
$\frac{1}{2}$	Minority carrier electron (or hole) injection into a p-type (or n-type) semiconductor Experimental results—e.g. single-crystal trigonal selenium with indium electrodes	Roberts [1968]
1	Traps distributed uniformly within the forbidden energy gap Experimental results—e.g. amorphous selenium films	Touraine *et al.* [1972]
2	Without traps Experimental results—e.g. CdS	Smith and Rose [1955]
2	Traps confined in single discrete energy levels or in smeared discrete energy levels Experimental results—e.g. single crystal anthracene with silver electrodes, β-phthalocyanine with gold electrodes	Helfrich [1967], Schadt and Williams [1969] Barbe and Westgate [1970], Hwang and Kao [1972]
2	Shallow traps distributed Gaussianly within the forbidden energy gap Experimental results—e.g. amorphous selenium films	Lanyon [1963] Hwang and Kao [1976]
$l+1$	Traps distributed exponentially within the forbidden energy gap Experimental results—e.g. anthracene crystals, tetracene crystals	Helfrich [1967], Mark and Helfrich [1962] Reucroft *et al.* [1974], Baessler *et al.* [1969], Sworakowski *et al.* [1969]
$m+1$	Deep traps distributed Gaussianly within the forbidden energy gap Experimental results—e.g. copper phthalocyanine films with gold electrodes	Sussman [1967] Hwang and Kao [1976]

TABLE 3.4. *The expressions for single injection J–V characteristics with and without traps*

Case	Single injection in solids (for hole injection)	References
Trap-free	$J = \dfrac{9}{8}\varepsilon\mu_p \dfrac{V^2}{d^3}$	Mott and Gurney [1940]
Traps confined in a single discrete energy level	$J = \dfrac{9}{8}\varepsilon\mu_p \theta_a \dfrac{V^2}{d_{\text{eff}}^3}$	Helfrich [1969], Lampert and Mark [1970], Hwang and Kao [1972]
Traps distributed exponentially within the forbidden energy gap	$J = q^{1-l}\mu_p N_v \left(\dfrac{2l+1}{l+1}\right)^{l+1}\left(\dfrac{1}{l+1}\dfrac{\varepsilon}{H_b'}\right)^l \dfrac{V^{l+1}}{d_{\text{eff}}^{2l+1}}$	Mark and Helfrich [1962], Hwang and Kao [1972, 1976], Reucroft and Mullins [1973]
Traps distributed Gaussianly within the forbidden energy gap	$J = \dfrac{9}{8}\varepsilon\mu_p \theta_d \dfrac{V^2}{d_{\text{eff}}^3}$ (for shallow traps) $J = q^{1-m}\mu_p N_v \left(\dfrac{2m+1}{m+1}\right)^{m+1}\left(\dfrac{m}{m+1}\dfrac{\varepsilon}{H_d'}\right)^m \dfrac{V^{m+1}}{d_{\text{eff}}^{2m+1}}$ (for deep traps)	Bonham [1973] Hwang and Kao [1976]
Traps distributed uniformly within the forbidden energy gap	$J = 2q\,\mu_p N_v g_p \dfrac{V}{d_{\text{eff}}}\exp\left[-\dfrac{E_u - E_t}{kT}\right]\exp\left[\dfrac{2\varepsilon V}{qH_c kT d_{\text{eff}}^2}\right]H_d$	Muller [1963], Rose [1955], Hwang and Kao [1972, 1976]

The general expressions for the $J-V$ characteristics in a solid with traps uniformly and non-uniformly distributed in space and in energy have been derived using a unified mathematical approach. The analysis technique discussed in this chapter may, in principle, be used to analyse any distribution of traps with space and energy. However, it should be noted that in the derivation both the permittivity and the carrier mobility have been assumed to be constant. For a more rigorous treatment, these two physical parameters may have to be considered to be altered by the charge exchange in traps [Paritskii and Rozental 1967] and by the high field effects.

3.2.3. Some comparisons of the theory with experiments

This section is devoted to the discussion of some typical experimental results in the light of the theory presented in the previous subsection. The typical results are so chosen to serve as experimental confirmation of SCL currents with traps in organic semiconductors, and it is not intended to review all presently available experimental literature which has already grown to a formidable size. References to the work in this branch of research may be found by consulting the Bibliography. The discussion will be divided into six subsections in accordance with the types of trap distributions.

(A) Without traps (trap-free solids)

As there are no perfect crystals existing in this world, there are always traps present in the solids. The ideal case—trap-free solids (without traps)—never exists. The square law given in eq. (3.9) can only be observed after all traps are filled—that means at voltages higher than V_{TFL}. Using ohmic contacts and low fields to avoid field emission from the electrodes, collision ionization and field-enhanced detrapping in the specimen, and using low currents to avoid heating effects, Smith and Rose [1955] have observed that when the applied voltage is suddenly increased, the current will transiently increase to a very high value, and then gradually decay to a much lower steady state value. The sudden application of a voltage across the specimen will force the electrons (or holes) from the injecting ohmic contact into its conduction band (into its valence band for holes), giving rise to a large current burst. If there are no traps, the space charge created by injected carriers will remain in the conduction band (or the valence band) and the peak value of the transient current will not decay but continue as a steady current. But such an ideal case never happens. In actual specimens there are always traps so that after injection the free carriers will be captured by traps, thus causing the current to decrease gradually to a steady state value. In general, the trap concentration (of the order of 10^{15} cm^{-3}) is higher than the free carrier concentration. Smith and Rose [1955] have used the experimental arrangement shown in Fig. 3.6 and found that the charge carriers injected into the specimen are trapped and remained in the specimen even when both electrodes are grounded before the specimen is released and dropped into the electrometer pan.

However, the trap-free square law can be seen when $V > V_{TFL}$ as shown in Figs. 3.10 and 3.11. It should be noted that in organic semiconductors E_g is large and hence the resistivity in high; the SCL currents should be easily identified in these materials.

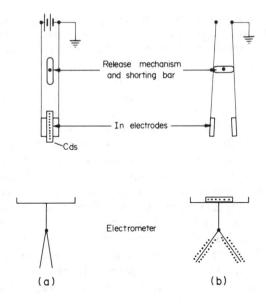

FIG. 3.6. Schematic diagram illustrating the experimental arrangement for measuring the injected
trapped charge in an insulator. [After Smith
and Rose 1955.]

(B) The traps confined in single or multiple discrete energy levels

The general expression for $J-V$ characteristics for single injection given in Section 3.2.2 is similar to that for double injection (Section 4.3.5). The only difference between single injection and double injection is the effective carrier mobility. To distinguish between these two cases, the following experiments may be used.

(i) To measure V_{TFL}. The $J-V$ characteristics follow the trap-free square law for $V > V_{TFL}$ for single injection, while for double injection it may exhibit a current-controlled negative differential resistance phenomenon. However, the V_{TFL} measurement will enable the determination of the trap concentration in the specimen.

(ii) To measure the temperature dependence of the SCL currents. The trapping level can be determined from the plot of $\ln(J)$ as a function of $1/T$, the slope of which is E_t/k. We can do this easily for single injection because θ_a is an exponential function of E_t, while for double injection μ_{eff} is a complex function.

Many investigators [Helfrich 1967, Schadt and Williams 1969] have studied the $J-V$ characteristics for organic semiconductors with traps confined in discrete energy levels. Hoesterey and Letson [1963] have reported that tetracene molecules doped into an anthracene crystal form shallow discrete traps both for electrons and for holes, and that anthraquinone molecules in anthracene do not form effective discrete traps for electrons. Many investigators have reported that organic semiconductors behave like inorganic semiconductors and that the electric conduction in them with carrier-injecting contacts is space charge limited. The $J-V$ characteristics in metal-free phthalocyanine single crystals have been found to follow Ohm's law at low applied voltages and to follow the general square law at higher voltages [Heilmeier and

Warfield 1963, Barbe and Westgate 1970]. The metal-free phthalocyanine is known to exist in several polymorphic forms and the crystal structure plays an important role in electric conduction [Sharp and London 1968]. Using gold electrodes as electron-injecting contacts and guard rings to avoid the effect of surface leakages, Barbe and Westgate [1970] have measured the $J-V$ characteristics in β-phthalocyanine with the applied field in parallel with the c' direction and in various ambients. Their experimental results are shown in Fig. 3.7. It can be seen that except in hydrogen the $J-V$ characteristics of the crystals in vacuum, air, and oxygen follow the square law in the SCL region. The diffusion of gases (air, oxygen, and hydrogen) into the crystals introduces additional trapping levels, so that the measurements of temperature dependence of SCL currents and ohmic currents will enable the determination of trapping levels. Figure 3.8 shows the typical temperature dependence of SCL and ohmic currents of β-phthalocyanine in vacuum. Using this method, Barbe and Westgate have determined the trapping level E_t and the electron trap density H_a for the crystals after being kept in a gas chamber for 24 hours to stabilize the gas diffusion, and their results are summarized in Table 3.5. Since the dominant trapping level for hydrogen ambient located below the middle of the forbidden gap and the trap density is relatively low, the trap filling starts to be effective at a lower voltage and this may explain why in hydrogen

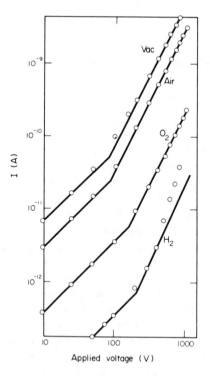

Fɪɢ. 3.7. The current–voltage characteristics of β-phthalocyanine single crystals in vacuum, air, oxygen, and hydrogen ambients at $T = 300°K$. [After Barbe and Westgate 1970.]

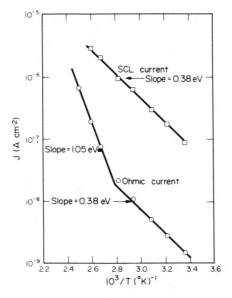

F IG. 3.8. Temperature dependence of SCL currents
and ohmic currents in β-phthalocyanine in
vacuum. [After Barbe and Westgate 1970.]

TABLE 3.5. *Mean values of the dominant trapping levels in*
β-phthalocyanine [after Barbe and Westgate 1970]

Ambient	$E_c - E_t$ (eV)	H_a (cm^{-3})
Vacuum	0.38 ± 0.004	5×10^{19}
Oxygen	0.82 ± 0.005	1×10^{14}
Hydrogen	0.95 ± 0.010	1×10^{13}

The energy gap = 1.68 eV.

ambient the SCL current departs from the square law. Barbe and Westgate [1970] have
also measured the thickness dependence of SCL and ohmic currents in β-
phthalocyanine in vacuum and in air at $T = 300°$K, and their typical results in vacuum
are shown in Fig. 3.9. They found that in vacuum or in air $J \propto d^{-1}$ in the ohmic region
and $J \propto d^{-3}$ in the SCL region, and this is in good agreement with the theory presented
in Section 3.2.1(B).

Campos [1972] has also observed the SCL currents in naphthalene single crystals
using silver paste coated on the specimen surfaces as electron-injecting contacts.
Typical results showing the temperature dependence of V_Ω and V_{TFL}, ohmic current,
trapped SCL current, and trap-free SCL current are given in Fig. 3.10. Based on the
theory for traps confined in a single discrete level for shallow traps, Campos has
calculated some physical parameters. From V_{TFL} results he found H_a to be about 10^{16}
$- 10^{17}$ trapping centres per cm^3, from temperature dependence of SCL currents he

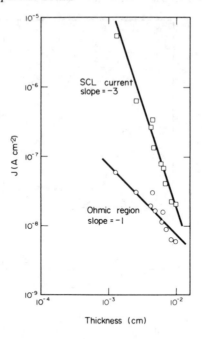

FIG. 3.9. Specimen thickness dependence of SCL currents and ohmic currents in β-phthalocyanine
in vacuum at $T = 300°$K. [After Barbe and
Westgate 1970.]

found $E_c - E_t$ to be about 0.65 eV, and from the following equations

$$S = \text{capture cross-section} \simeq \frac{1}{H_a v \tau} \qquad (3.84)$$

$$v = \text{frequency of escape} \simeq \frac{\theta_a}{\tau} \exp[(E_c - E_t)/kT] \qquad (3.85)$$

and taking τ (trapping time) $= 10^{-5}$ sec and v (thermal velocity) $= 10$ cm sec^{-1}, he
found S and v to be 10^{-17} cm^2 and 10^9 sec^{-1}, respectively. E_g of this material is 1.4 eV.
   For deep traps Henderson *et al.* [1969, 1972] have observed the Lampert triangle in
neutron-irradiated silicon and their results are shown in Fig. 3.11. To further support
the agreement of their results with the simple SCL-model, they have also measured the
V_{TFL} as a function of specimen thickness and the results are shown in Fig. 3.12, which
follows well eq. (3.32). It should be noted that the slope in the TFL region increases in
verticality with increasing ratio of the unfilled equilibrium trap density to equilibrium
free carrier density, and that TFL behaviour is greatly affected by high field effects and
double injection effects. This will be discussed in Section 5.3.

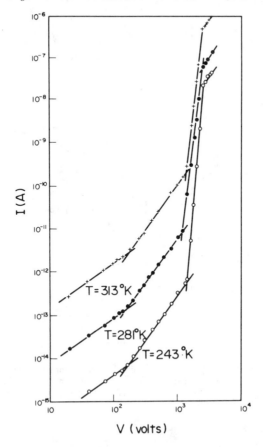

F IG. 3.10. The current–voltage characteristics of naphthalene single crystals as functions of temperature. [After Campos 1972.]

(C) The traps distributed exponentially within the forbidden energy gap

Measurements of the steady state SCL currents in single crystals and polycrystalline specimens of naphthalene [Lohmann and Mehl 1969, Campos 1972], anthracene [Mark and Helfrich 1962, Pott and Williams 1969, Reucroft *et al.* 1973, Schadt and Williams 1969, Sworakowski *et al.* 1969, Thomas *et al.* 1968, Hwang and Kao 1972, 1974], tetracene [Baessler *et al.* 1969], Perylene [Sworakowski *et al.* 1969], *p*-terphenyl [Szymanski 1968], *p*-quaterphenyl [Szymanski 1968], and phthalocyanine [Hamann 1968, Sussman 1967, Tantzscher and Hamann 1975], as well as in inorganic semiconductors [Rose 1955], have shown that the currents are controlled by traps distributed exponentially within the forbidden energy gap following the distribution function given by eq. (3.41). This implies that the traps in those specimens may be due to structural defects or due to perturbed molecules in the lattice causing the charge of polarization energy in the perturbed regions, which may also tend to lower the bottom

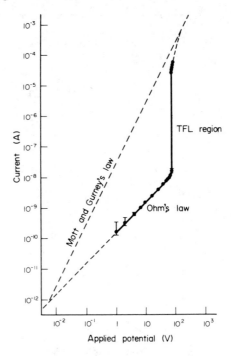

F IG. 3.11. Experimental evidence of the Lampert triangle in silicon irradiated to 1.1×10^{16} neutrons cm^{-2} (> 0.1 MeV) with a steep transition at V_{TFL}. [After Henderson, Ashley, and Shen 1972.]

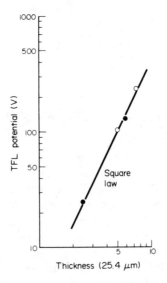

F IG. 3.12. The TFL threshold voltage (V_{TFL}) for neutron-irradiated silicon as a function of specimen thickness. [After Henderson, Ashley, and Shen 1972.]

edge of the conduction band or to raise the upper edge of the valence band [Reucroft and Mullins 1973, 1974]. From the measurements of V_{TFL} as a function of temperature we can determine the trap density H_b' and H_b, and from the log–log plot of the J–V characteristics we can determine l and hence T_c from the following equation

$$l = \frac{T_c}{T} = 1 + \frac{d(\ln J)}{d(\ln V)} \tag{3.86}$$

Since $l > 1$ the dependence of J on V follows a power law and is stronger than that for the cases without traps or with discrete traps in which $J \propto V^2$. The typical J–V characteristics for α-copper phthalocyanine with an exponential trap distribution as functions of temperature are shown in Fig. 3.13. These results were obtained using gold electrodes as electron–ohmic contacts, and after the specimens had been heat-treated in high vacuum at 200° C for about 16 hours [Hamann 1968]. These results show that (i)

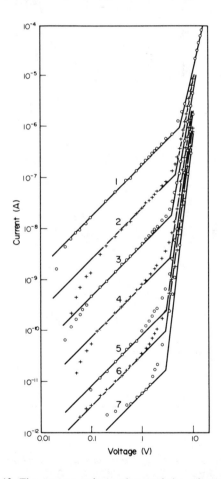

FIG. 3.13. The current–voltage characteristics of α-copper phthalocyanine films of thickness of $4\,\mu$m: (1) 144.6°C, (2) 116.9°C, (3) 96.8°C, (4) 76.5°C, (5) 54.2°C, (6) 33.8°C, (7) 20.9°C, highest heat treatment at 200°C. [After Hamann 1968.]

the slope in the non-ohmic region decreases with increasing temperature, (ii) V_Ω the transition voltage between the ohmic and non-ohmic regions increases with increasing temperature, and (iii) the total trap density is about 10^{15} cm^{-3}.

Using gold and palladium for hole-injecting ohmic contacts, and sodium and barium for electron-injecting ohmic contacts, Baessler *et al.* [1969] have observed the exponential trap distribution in sublimation grown tetracene crystals based on the $J-V$ characteristics shown in Fig. 3.14 in which $j \propto V^5$ implying that $l = (T_c/T) = 4$ from eq. (3.45). Taking $T_c \simeq 1200^\circ \mathrm{K} \mu_p \simeq 0.5$ cm^2 V^{-1} sec^{-1}, $N_v = 4 \times 10^{21}$ cm^{-3} (assuming

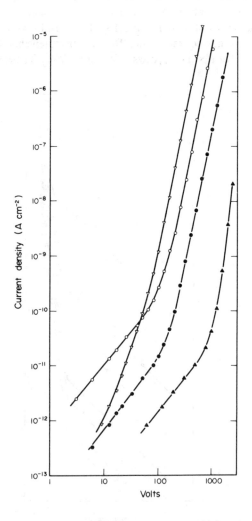

F IG. 3.14. The current–voltage characteristics of tetracene crystals with various contact materials and various crystal thicknesses. –∇–∇–: gold contacts for both electrodes ($d = 80\ \mu$m); –O–O–: KI/I$_2$ (positive) and Mg(OH) (negative) ($d = 110\ \mu$m); –●–●–: Barium contacts for both electrodes ($d = 220\ \mu$m); –▲–▲–: lead contacts for both electrodes ($d = 240\ \mu$m). [After Baessler, Herrmann, Riehl, and Vaubel 1969.]

that μ_n and N_c are, respectively, the same values), $\varepsilon = 3.8 \, \varepsilon_0$ and $S(x) = 1$; Baessler *et al.* have estimated the value of H_b from eq. (3.45) to be about $3 \times 10^{15} \, \text{cm}^{-3}$, and concluded that the same defects are able to trap electrons and holes. For an exponential trap distribution the quasi-Fermi level which depends on the magnitude of the stored charge and hence on the applied voltage V is given by

$$E_F(V) = kT_c \ln \left[\frac{(l+1)^2}{l(2l+1)} \frac{qH_b d^2}{\varepsilon V} \right]$$

(3.87)

which is measured from the edge of the valence band for the hole injection or from the edge of the conduction band for the electron injection. Some results shown in Fig. 3.15 agree well with eq. (3.87). To further confirm the exponential trap distribution, Baessler *et al.* [1969] have also found that for a fixed average field $F_{av} = V/d$, J is proportional to d^{-l} in which $l \simeq 4$.

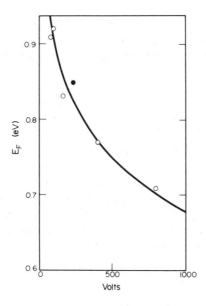

FIG. 3.15. Activation energy of the SCL currents in tetracene versus applied voltage (crystal thickness $d = 250 \, \mu$m). Open circles: hole injection from gold and palladium; full circles: electron injection from barium. The solid line is the theoretical curve based on eq. (3.87) with V normalized to the experimental point at 400 V and the quasi-Fermi level from either the valence or the conduction band edge ($E_g = 3$ eV). [After Baessler, Herrmann, Riehl, and Vaubel 1969.]

Many investigators [e.g. Schadt and Williams 1969, 1970, Reucroft and Mullins 1973, 1974] have suggested that the exponential trap distribution may not extend to the band edge. For this case J–V characteristics follow eq. (3.47) rather than eq. (3.45). Using a solid hole-injecting contact (made of a solution containing saturated nitromethane with aluminium chloride and anthracene, and after slow evaporation of the solvent) stable in the temperature range 400–77°K, Schadt and Williams [1969] have measured the J–V characteristics of melt grown anthracene crystals, and their results

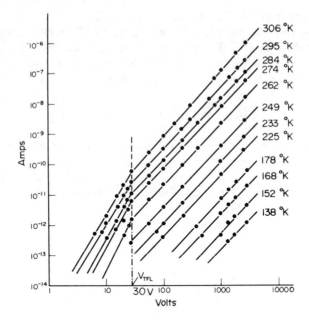

F IG. 3.16. The current–voltage characteristics of anthracene crystals of thickness of 2 mm at various temperatures using a solid hole-injecting
electrode. [After Schadt and Williams 1969.]

are shown in Fig. 3.16. These results show that for $V < 30$ V, $J \propto V^{l+1}$ with $l > 2$; while
for $V > 30$ V, $J \propto V^2$; and $V_{\text{TFL}} = 30$ remain practically constant—independent of
temperature. For $V < 30$ V the values for T_c determined from the results given in Fig.
3.16 and based on eq. (3.86) are given in Table 3.6. They have also reported that
H'_b is about 10^{16} cm^{-3} and the quasi-Fermi level E_{Fp} (measured from the edge of
the valence band) is constant for $V < V_{\text{TFL}}$ and decreases with increasing voltage for
$V > V_{\text{TFL}}$, that this voltage dependence of E_{Fp} is consistent with the filling of traps of
exponential distribution which does not extend to the edge of the valence band, and
that there are two discrete shallow hole-trapping levels with energies 0.53 eV and 0.2 eV
from the edge of the valence band.

Reucroft and Mullins [1973, 1974] have suggested that the trap distribution in
anthracene crystals may arise from anthracene molecules perturbed from their lattice

T ABLE 3.6 *Values for T_c calculated using eq. (3.86) and the
experimental results of Fig. 3.16 for $V < V_{\text{TFL}}$ [after
Schadt and Williams 1969]*

Temperature (°K)	$\dfrac{d(\ln J)}{d(\ln V)}$	T_c
306	3.0	620
295	3.0	590
284	3.2	625
274	3.4	655
262	3.3	605

sites due to distortions caused by impurity molecules "squeezing" into the crystal lattice, and that the trap distribution starts at the discrete trapping level and extends exponentially to deeper levels in the forbidden gap. By analysing the experimental J–V characteristics using eqs. (3.47) and (3.86) to obtain values of H'_b and KT_c, then E_{te} can be determined from the slope of the plot of ln H'_b versus $1/kT_c$ based on eq. (3.48) and H_b can be determined by extending the plot to $1/kT_c = 0$ as shown in Fig. 3.17. It can be easily shown that from Fig. 3.17, $E_{te} = 0.71$ eV (from the edge of the valence band) and $H_b = 10^{17}$ m^{-3} for melt grown and vapour grown crystals, and $E_{te} = 0.69$ eV and $H_b = 10^{19}$ m^{-3} for solution grown crystals. H_b and H'_b can be interpreted as the total trap density above the valence band edge and the total trap density above the discrete trap level E_{te}, respectively. Using four sets of hole trap distribution data (H'_b and kT_c) for four melt grown anthracene crystals to plot four curves of ln(h) as a function of E (trap depth measured from the valence band edge) based on eq. (3.49) for $S(x) = 1$, the four curves all cross at about 0.7 eV and at a mean trap density of 10^{17} m^{-3} ($H_b = hKT_c$ when $E = E_{te}$). These results agree well with the discrete hole-trapping level at 0.6–0.8 eV above the valence band edge revealed from thermally stimulated current measurements, and indicate that a perturbed molecule is introduced for every 10^5 host molecules and the total density of perturbed molecules is constant at 10^{17} m^{-3} for melt grown and vapour grown anthracene crystals [Reucroft and Mullins 1974]. However, there is still no direct evidence whether a trap distribution from the valence band edge to the discrete trapping level exists physically.

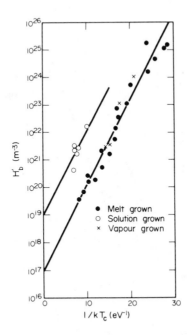

FIG. 3.17. H'_b as a function of $1/kT_c$ for exponential distribution of hole traps in the forbidden gap in melt grown and solution grown anthracene crystals. [After Reucroft and Mullins 1974.]

So far very little attention has been devoted to non-uniform or inhomogeneous spatial distribution of traps. It can be easily imagined that the trap concentration at and near discontinuities (surfaces) is higher than that in the bulk partly due to unsaturated diffusion of impurities from the surroundings through the surfaces. The thinner the material specimen used for experimental studies, the more serious is the influence of the form of inhomogeneous spatial distribution of traps on the results. To illustrate the effect of inhomogeneous spatial trap distribution, Hwang and Kao [1972] have used anthracene films, each of which is sandwiched between two silver plane electrodes of 2 mm in diameter deposited step by step on a glass substrate under a vacuum of 10^{-6} torr at $-60°C$ [Northrop and Simpson 1956], and measured the $J-V$ characteristics as functions of film thickness. Their results are shown in Fig. 3.18. For a given thickness, J is proportional to V^3, indicating that the traps are distributed exponentially within the forbidden energy gap following eq. (3.45) with $l = 2$. This is reasonable because the structure of anthracene films is likely to be polycrystalline, and the traps created by defects due to such a structure generally have an exponential distribution in energy. Since the carriers injected from silver electrodes to anthracene are holes, we can now use eq. (3.46) to calculate d_{eff} and then determine $S(x)$ from the thickness dependence of J for a given applied voltage V. We have computed d_{eff} for three most probable distribution functions of $S(x)$ as follows.

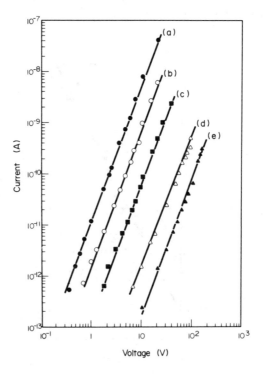

FIG. 3.18. The current voltage characteristics of anthracene film of thickness (a) 1.62 μm, (b) 2.49 μm, (c) 3.90 μm, (d) 8.80 μm, and (e) 11.50 μm. [After Hwang and Kao 1972.]

(I) UNIFORM SPATIAL DISTRIBUTION

The distribution function may be written as

$$S(x) = 1 \tag{3.88}$$

d_{eff} for this distribution function is therefore

$$d_{\text{eff}} = \left[\frac{2l+1}{l+1} \int_0^d \left(\int_0^t dx \right)^{1/(l+1)} dt \right]^{(l+1)/(2l+1)}$$

$$= d \tag{3.89}$$

(II) EXPONENTIAL SPATIAL DISTRIBUTION WITH THE MAXIMUM DENSITY AT THE INJECTING ELECTRODE (AT $x = 0$)

The distribution function may be written as

$$S(x) = 1 + A \exp\left(-\frac{x}{x_0} \right) \tag{3.90}$$

where A and x_0 are constants. Substitution of eq. (3.90) into eq. (3.46) gives

$$d_{\text{eff}} = \left\{ \frac{2l+1}{l+1} \int_0^d \left[\int_0^t \left\{ 1 + A \exp\left(-\frac{x}{x_0} \right) \right\} dx \right]^{\frac{l}{(l+1)}} dt \right\}^{(l+1)/(2l+1)}$$

By setting $p = l/(l+1)$, $w = \exp(-t/x_0)$ and $w_d = \exp(-d/x_0)$, we obtain

$$\frac{d_{\text{eff}}}{d} = \frac{x_0}{d}(1+p)^{p/(1+p)} \left\{ \int_{w_d}^1 \frac{[A(1-w) - \ln w]^p}{w} dw \right\}^{1/(p+1)} \tag{3.91}$$

which can be easily evaluated.

(III) EXPONENTIAL SPATIAL DISTRIBUTION WITH THE MAXIMUM DENSITY AT BOTH ELECTRODES (AT $x = 0$ AND $x = d$)

The distribution function may be written as

$$S(x) = 1 + B \left[\exp\left(-\frac{x}{x_0} \right) + \exp\left(-\frac{d-x}{x_0} \right) \right] \tag{3.92}$$

where B is a constant. Similarly, substituting eq. (3.92) in eq. (3.46) and simplifying it, we obtain

$$\frac{d_{\text{eff}}}{d} = \frac{x_0}{d}(1+p)^{p/(1+p)} \left\{ \int_{w_d}^1 \frac{[B(1-w-w_d+(w_d/w)) - \ln w]^p}{w} dw \right\}^{1/(p+1)} \tag{3.93}$$

From the experimental results given in Fig. 3.18 we have $l = 2$. Using this value for l, we have computed d_{eff}/d as a function of d/x_0, and the results are given in Fig. 3.19 for various $S(x)$. From Fig. 3.18 and eq. (3.45) we would expect J to be proportional to d^{-5} for a given V if the spatial distribution of traps is uniform. But the plot of J as a function

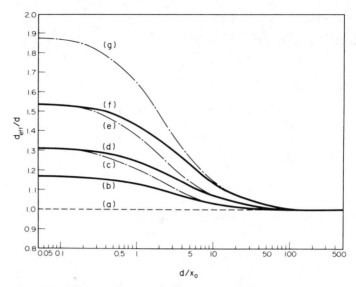

FIG. 3.19. d_{eff}/d as a function of d/x_0 for traps distributed exponentially in
energy with $l = 2$ and distributed exponentially in space with
(a) $S(x) = 1$, (b) $S(x) = 1 + 0.5 \;\; \exp(-x/x_0)$, (c) $S(x) = 1 + 0 \cdot 5$
$\{\exp(-x/x_0) + \exp[-(d-x)/x_0]\}$, (d) $S(x) = 1 + \exp(-x/x_0)$,
(e) $S(x) = 1 + \{\exp(-x/x_0) + \exp[-(d-x)/x_0]\}$, (f) $S(x) = 1 + 2$
$\exp(-x/x_0)$, (g) $S(x) = 1 + 2\{\exp(-x/x_0) + \exp[-(d-x)/x_0]\}$.
[After Hwang and Kao 1972.]

of d for a given V in Fig. 3.20 shows that I is proportional to d^{-n}, in which n is
dependent on d. This indicates that the trap distribution in the anthracene films
under investigation is spatially inhomogeneous. Since I can be simply expressed as
$I = M d^{-n} = M d_{\text{eff}}^{-5}$ for a given V and $l = 2$, we can determine from Fig. 3.20 the values
of n and d_{eff}/d for various values of d. The result of the latter is also plotted in the same
figure.

It is reasonable to assume that if the inhomogeneity of the spatial trap distribution is
caused by the metal–anthracene contacts, then the distribution function should be
independent of film thickness because the same technique was used to fabricate all
specimens. Since the material used for both electrodes is silver, we would expect that the
trap distribution at $x = 0$ should be symmetrical with that at $x = d$. Thus the most
probable spatial distribution function would be exponential with the maximum trap
density at $x = 0$ and $x = d$. From Figs. 3.19 and 3.20 we have estimated that for $S(x)$
independent of d, the value of x_0 is between 0.3 and 0.4 μm, and the value of B is
between 1 and 2. It should be noted that the number of thicknesses used for this
experimental investigation is only 5; the curves drawn in Fig. 3.20 are clearly not unique.
For accurate determination of $S(x)$ more results on thickness dependence of I–V
characteristics and an iterative procedure to find x_0, B, or A are necessary.

(D) The traps distributed Gaussianly within the forbidden
energy gap

Many investigators [Sussman 1967, Lanyon 1963] have suggested that in disordered

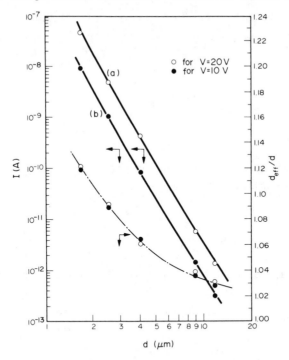

F IG. 3.20. I and d_{eff}/d as functions of film thickness for
anthracene films for (a) $V = 20$ V, and (b) V
$= 10$ V. The theoretical curve is based on the
exponential distribution of traps in energy and
in space with $l = 2$. [After Hwang and Kao
1972.]

solids the trap distribution had better be described by a Gaussian distribution function.
Since eq. (3.57) is very similar to eq. (3.22) for shallow traps and eq. (3.63) is similar to eq.
(3.47) for deep traps the $J-V$ characteristics alone are not sensitive to distinguish
between shallow discrete traps and shallow Gaussian traps, or between exponential
traps and deep Gaussian traps. However, on the assumption that the results of Sussman
[1967] for evaporated films of copper phthalocyanine are due to traps having a
Gaussian distribution in trapping energy levels, and that the traps are distributed non-
uniformly in space and also follow a Gaussian distribution, we have computed d_{eff} using
eq. (3.64) and determined some trap parameters as follows.

(I) GAUSSIAN SPATIAL DISTRIBUTION WITH THE MAXIMUM
DENSITY AT THE INJECTING ELECTRODE (AT $x = 0$)

The distribution function may be written as

$$S(x) = 1 + c_1 \exp\left(-\frac{x^2}{x_0^2}\right) \qquad (3.94)$$

where c_1 and x_0 are constants. Substitution of eq. (3.94) into eq. (3.64) gives

$$\frac{d_{\text{eff}}}{d} = \left[\frac{2m+1}{m+1}\right]^{(m+1)/(2m+1)} \left\{\int_0^1 \left[t + c_1 \frac{\sqrt{\pi}}{2} \frac{x_0}{d} \operatorname{erf}\left(\frac{d}{x_0}t\right)\right]^{m/(m+1)} dt\right\}^{(m+1)/(2m+1)}$$

(3.95)

which can be easily evaluated.

(II) GAUSSIAN SPATIAL DISTRIBUTION WITH THE MAXIMUM
DENSITY AT BOTH ELECTRODES

The distribution function may be written as:

$$S(x) = 1 + c_1 \exp\left(-\frac{x^2}{x_0^2}\right) + c_2 \exp\left[-\left(\frac{d-x}{x_0}\right)^2\right]$$

(3.96)

where c_1 and c_2 are constants. Substituting eq. (3.96) into eq. (3.64) and simplifying it, we obtain

$$\frac{d_{\text{eff}}}{d} = \left(\frac{2m+1}{m+1}\right)^{(m+1)/(2m+1)} \left\{\int_0^1 \left[t + c_1 \frac{\sqrt{\pi}}{2} \frac{x_0}{d} \operatorname{erf}\left(\frac{d}{x_0}t\right)\right.\right.$$
$$\left.\left. + c_2 \frac{\sqrt{\pi}}{2} \frac{x_0}{d} \operatorname{erf}\left(\frac{d}{x}(1-t)\right)\right]^{m/(m+1)} dt\right\}^{(m+1)/(2m+1)}$$

(3.97)

For Gaussian spatial distribution, m as a function of σ_t/kT based on eq. (3.61) has been computed and the result is shown in Fig. 3.21. Since d_{eff}/d is a function of both m and d/x_0, we have computed d_{eff}/d as a function of d/x_0 for a special case based on eq. (3.97) for $c_1 = c_2 = 1$ and $c_1 = c_2 = 2$ and for various values of m in order to see the effect of m. These results are also given in Fig. 3.21.

From the experimental results of Sussman [1967] for evaporated films of copper phthalocyanine, we have found that for his results $m = 3$. Using this value for m and the Gaussian–Quadratic method, we have computed d_{eff}/d as a function of d/x_0 for various $S(x)$, and the results are given in Fig. 3.22. From eq. (3.63), we would expect J to be proportional to d^{-7} for a given V if the spatial distribution of traps is uniform. But the plot of J as a function of d for a given V in Fig. 3.23 shows that J is proportional to d^{-n}, in which n is dependent on d. This indicates that the trap distribution in the copper phthalocyanine films is spatially inhomogeneous. Since J can be simply expressed as $J = Md^{-n} = Md_{\text{eff}}^{-7}$ for a given V and $m = 3$, we can determine from Fig. 3.23 the values of n and d_{eff}/d for various values of d. The results of the latter are also plotted in the same figure.

It is reasonable to assume that if the inhomogeneity of the spatial trap distribution is caused by the metal–copper phthalocyanine contact, the distribution function should be independent of film thickness because the same technique was used to fabricate all specimens. Since the material used for both electrodes was gold [Sussman 1967], we would expect that the trap distribution at $x = 0$ should be symmetrical with that at $x = d$. Thus the most probable spatial distribution would be Gaussian with the maximum trap density at $x = 0$ and $x = d$. From Figs. 3.22 and 3.23 we have estimated

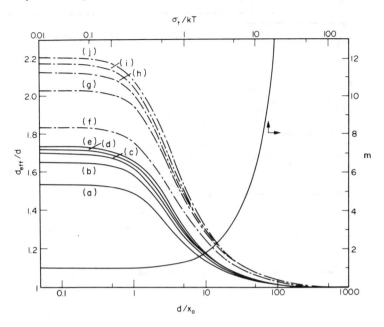

FIG. 3.21. d_{eff}/d as a function of d/x_0 and m as a function of σ_t/kT for traps distributed
Gaussianly in energy and in space based on $S(x) = 1 + \exp\left[-\left(\dfrac{x}{x_0}\right)^2\right] + \exp$
$\left[-\left(\dfrac{d-x}{x_0}\right)^2\right]$ with (a) $m = 1$, (b) $m = 2$, (c) $m = 3$, (d) $m = 4$, and (e) $m = 5$,
and for $S(x) = 1 + 2\left\{\exp\left[-\left(\dfrac{x}{x_0}\right)^2\right] + \exp\left[-\left(\dfrac{d-x}{x_0}\right)^2\right]\right\}$ with (f) $m = 1$,
(g) $m = 2$, (h) $m = 3$, (i) $m = 4$, and (j) $m = 5$. [After Hwang and Kao 1976.]

that for $S(x)$ independent of d, the value of x_0 is between 0.03 and 0.05 μm, and $c_1 = c_2$
and its value is between 1.0 and 2.0. We have also examined the experimental results for
Lanyon [1963] for vitreous selenium films and found that the trap distribution in both
energy and space in these films follows also a Gaussian function. However, it should be
noted that the number of thicknesses used for the experimental copper phthalocyanine
results [Sussman 1967] is only 4; the curves drawn in Fig. 3.23 are clearly not unique.
For accurate determination of $S(x)$ more results on thickness dependence of $J-V$
characteristics and the use of an iterative procedure to find x_0, c_1, and c_2 are necessary.

(E) The traps confined in smeared discrete energy levels

Again, eq. (3.73) is very similar to eq. (3.22) and the $J-V$ characteristics alone are not
sensitive to distinguish between a single discrete energy level and a smeared discrete
energy level. However, it is logical to think that a given type of traps originated either
from chemical or structural defects does not have a unique defined environment
because the disturbance due to such defects is different at least in degree from one
trapping site to another or in general from domain to domain, and that as a result of this

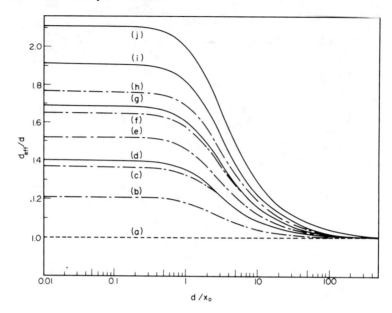

FIG. 3.22. d_{eff}/d for $m = 3$ as a function of d/x_0 for traps distributed Gaussianly in energy and in space with:
(a) $S(x) = 1$, (b) $S(x) = 1 + 0.5 \exp\left[-\left(\frac{x}{x_0}\right)^2\right]$, (c) $S(x) = 1 + \exp\left[-\left(\frac{x}{x_0}\right)^2\right]$, (d) $S(x) = 1$
$+ 0.5 \left\{ \exp\left[-\left(\frac{x}{x_0}\right)^2\right] + \exp\left[-\left(\frac{d-x}{x_0}\right)^2\right] \right\}$, (e) $S(x) = 1 + 1.5 \exp\left[-\left(\frac{x}{x_0}\right)^2\right]$, (f) $S(x) = 1$
$+ 2 \exp\left[-\left(\frac{x}{x_0}\right)^2\right]$, (g) $S(x) = 1 + \left\{ \exp\left[-\left(\frac{x}{x_0}\right)^2\right] + \exp\left[-\left(\frac{d-x}{x_0}\right)^2\right] \right\}$, (h) $S(x) = 1$
$+ 2.5 \exp\left[-\left(\frac{x}{x_0}\right)^2\right]$, (i) $S(x) = 1 + 1.5 \left\{ \exp\left[-\left(\frac{x}{x_0}\right)^2\right] + \exp\left[-\left(\frac{d-x}{x_0}\right)^2\right] \right\}$, (j) $S(x) = 1$
$+ 2 \left\{ \exp\left[\left(-\frac{x}{x_0}\right)^2\right] + \exp\left[-\left(\frac{d-x}{x_0}\right)^2\right] \right\}$. [After Hwang and Kao 1976.]

effect a dominant trapping level will be smeared out with the trap distribution close to a Gaussian distribution [Lampert and Mark 1970]. Therefore, it is most likely that the experimental results discussed in Section 3.2.3(B) are due to smeared discrete traps rather than single discrete traps.

(F) The traps distributed uniformly within the forbidden energy gap

As has been mentioned in Section 3.2.1(F), Touraine *et al.* [1972] have observed this case. A typical set of their results is shown in Fig. 3.24. It can be seen that J/V is an exponential function of V following eq. (3.77).

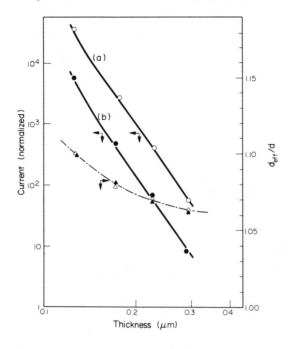

F IG. 3.23. Current and d_{eff}/d as functions of film thickness for copper
phthalocyanine films for (a) $V = 2$ V, and (b) $V = 1$ V. The
experimental results are from the work of Sussman [1967]
and the theoretical curve is based on the Gaussian distribution of traps in energy and in space with $m = 3$. [After
Hwang and Kao 1976.]

3.3. ONE-CARRIER (SINGLE) INJECTION FROM A NON-PLANAR CONTACTING ELECTRODE

In Section 3.2 we have discussed in some detail one-carrier injection from a planar
contacting electrode. In practice, non-planar injecting contacts such as a point contact
and a cylindrical contact are commonly used. Therefore, this section is devoted to the
analysis of two most important geometries—spherical and cylindrical. We use the same
assumptions given in Section 3.2.1 except that the injecting contact is non-planar and
that the injected carriers are electrons rather than holes. A similar treatment can be
easily extended to the case for which the injected carriers are holes.

3.3.1. Spherical injecting contacts without considering
thermally generated free carriers

Figure 3.25 shows the concentric sphere configurations. When $r_b - r_a \ll r_a$, the
concentric sphere configuration degenerates to a planar configuration and when r_a is
very small and r_b very large the concentric sphere configuration degenerates to a point
planar configuration. In the following analysis the current components due to diffusion
and due to carriers thermally generated in the specimen are ignored.

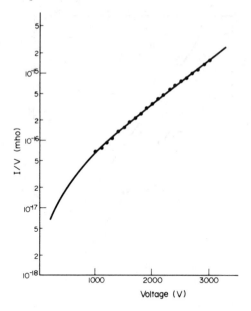

FIG. 3.24. I/V as a function of V for amorphous selenium films of thickness of about 4000 Å with gold–electrode separation of 150 μm. The good agreement between the experimental results and the theoretical solid curve indicates a uniform trap distribution of band width of 0.24 eV located at 0.7 eV above the valence band edge. [After Touraine, Vautier, and Carles 1972.]

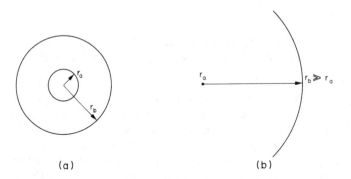

FIG. 3.25. Two concentric spherical electrodes: (a) normal concentric sphere configuration with $r_b > r_a$, and (b) concentric sphere configuration degenerates to a point planar configuration when r_a is extremely small compared with r_b.

The behaviour of single injection in a solid with a concentric sphere configuration is governed by the current flow equation

$$I = \text{total current}$$
$$= q\mu_n n F (4\pi r^2) \tag{3.98}$$

and the Poisson's equation

$$\frac{1}{r^2}\frac{d}{dr}(r^2 F) = \frac{q(n+n_t)}{\varepsilon} \tag{3.99}$$

where n and n_t are, respectively, the densities of injected free and trapped electrons, n and n_t are given by

$$n = N_c \exp\left[(E_{Fn}-E_c)/kT\right] \tag{3.100}$$

and

$$n_t = \int_{E_l}^{E_u} h(E, r)f(E)\,dE \tag{3.101}$$

in which

$$h(E, r) = N_t(E)S(r) \tag{3.102}$$

and

$$f(E) = \frac{1}{1+g_n^{-1}[(E-E_{Fn})/kT]} \tag{3.103}$$

(A) Without traps

For this case $n_t = 0$. By eliminating n from eqs. (3.98) and (3.99), we obtain

$$\frac{d}{dr}(r^2 F)^2 = \frac{I r^2}{2\pi\varepsilon\mu_n} \tag{3.104}$$

With the inner sphere as the ohmic injecting contact, we can use the boundary

$$F(r_a) = 0 \tag{3.105}$$

Thus, integration of eq. (3.104) yields

$$F(r) = \left[\frac{I(r^3 - r_a^3)}{6\pi\varepsilon\mu_n r^4}\right]^{1/2} \tag{3.106}$$

The maximum field occurs at $r_m = 4^{1/3}r_a$ where $dF/dr = 0$, and this field is given by

$$F_m(r_m) = \left[\frac{3}{6\pi\varepsilon\mu_n(4^{4/3}r_a)}\right]^{1/2} \tag{3.107}$$

Using the boundary condition

$$V = \int_{r_a}^{r_b} F(r)\,dr \tag{3.108}$$

we obtain for $r_b \gg r_a$

$$I = \frac{3\pi}{2}\varepsilon\mu_n\frac{V^2}{r_b} \tag{3.109}$$

For the case with the outer sphere as the ohmic injecting contact, we use the following boundary condition

$$F_r(r_b) = 0 \qquad (3.110)$$

Then $F_r(r)$ and I_r, the field and the current in reverse direction for $r_b \gg r_a$, are, respectively, given by [Lampert and Mark 1970]

$$F_r(r) = \left[\frac{I_r(r_b^3 - r^3)}{6\pi\varepsilon\mu_n r^4} \right]^{1/2} \qquad (3.111)$$

$$I_r = I(2r_a/r_b)^2$$

$$= \frac{3\pi}{2}\varepsilon\mu_n \frac{V^2}{r_b} \left(\frac{2r_a}{r_b} \right)^2 \qquad (3.112)$$

The rectification ratio which is of geometric origin only is

$$\frac{I}{I_r} = \left(\frac{r_b}{2r_a} \right)^2 \qquad (3.113)$$

which can be very large if r_b is made much larger than r_a.

(B) With shallow traps confined in single discrete energy levels

For this case eq. (3.102) becomes

$$h(E) = H_e\delta(E - E_t) \qquad (3.114)$$

for uniform spatial trap density distribution. From eqs. (3.101), (3.103), and (3.114) we obtain

$$n_t = \int_{E_l}^{E_u} \frac{H_e\delta(E - E_t)dE}{1 + g_n^{-1} \exp[(E - E_{Fn})/kT]}$$

$$= \frac{H_e}{1 + \dfrac{H_e\theta_e}{n}} \qquad (3.115)$$

in which

$$\theta_e = \frac{N_c}{g_n H_e} \exp[-(E_c - E_t)/kT] \qquad (3.116)$$

For shallow traps, $H_e\theta_e \gg n$, substitution of eq. (3.115) into eq. (3.99), elimination of n from eq. (3.98) and the use of the boundary condition given in eq. (3.105) yield

$$F(r) = \left[\frac{I(r^3 - r_a^3)}{6\pi\varepsilon\mu_n\theta_e r^4} \right]^{1/2} \qquad (3.117)$$

Using the boundary condition given in eq. (3.108), we obtain for the case with the inner sphere as the injecting contact and $r_b \gg r_a$

$$I = \frac{3\pi}{2} \varepsilon \mu_n \theta_e \frac{V^2}{r_b} \tag{3.118}$$

Similarly, for the case with the outer sphere as the injecting contact, it can be easily shown that the reverse current is given by

$$I_r = I \left(\frac{2r_a}{r_b} \right)^2$$

$$= \frac{3\pi}{2} \varepsilon \mu_n \theta_e \frac{V^2}{r_b^2} \left(\frac{2r_a}{r_b} \right)^2 \tag{3.119}$$

(C) With traps distributed exponentially within the forbidden energy gap

For this case eq. (3.102) becomes

$$h(E) = \frac{H'_f}{kT_c} \exp\left[-\left(\frac{E_c - E}{kT_c} \right) \right] \tag{3.120}$$

in which

$$H'_f = H_f \exp\left(\frac{E_c - E_{te}}{kT_c} \right) \tag{3.121}$$

This means that the exponentially distributed traps are scanned at $E_c - E_{te}$. Using the same method adopted in Section 3.2.1(C), it can be easily shown that for the case with the inner sphere as the electron-injecting contact $[F(r_a) = 0]$ and $r_b \gg r_a$ the SCL current is given by

$$I = q^{1-l}(4\pi\mu_n N_c)\left(\frac{2l-1}{l+1} \right)^{l+1} \left(\frac{l}{l+1} \frac{3\varepsilon}{H'_f} \right)^l \left(\frac{V^{l+1}}{r_b^{2l-1}} \right) \tag{3.122}$$

where $l = T_c/T > 1$.

Similarly, for the case with the outer sphere as the electron-injecting contact $[F(r_b) = 0]$ and $r_b \gg r_a$ it can be easily shown that the reverse current is given by

$$I_r = I \left[\left(\frac{l+1}{2l-1} \right) \frac{r_a}{r_b} \right]^{l+1}$$

$$= q^{1-l}(4\pi\mu_n N_c)\left(\frac{r_a}{r_b} \right)^{l+1} \left(\frac{l}{l+1} \frac{3\varepsilon}{H'_f} \right)^l \left(\frac{V^{l+1}}{r_b^{2l-1}} \right) \tag{3.123}$$

3.3.2. Cylindrical injecting contacts without considering thermally generated free carriers

In the following analysis the current components due to diffusion and due to carriers thermally generated in the specimen are ignored. The behaviour of single injection in a solid with a concentric cylinder configuration is governed by the current flow equation

$$I = \text{total current}$$

$$= q\mu_n n F (2\pi r L) \tag{3.124}$$

where L is the length of the cylinder, and the Poisson's equation

$$\frac{dF}{dr} + \frac{1}{r} F = \frac{q(n + n_t)}{\varepsilon} \tag{3.125}$$

(A) Without traps

For this case, $n_t = 0$. By eliminating n from eqs. (3.124) and (3.125), we obtain

$$\frac{d}{dr}(r^2 F^2) = \frac{Ir}{\pi \varepsilon \mu_n L} \tag{3.126}$$

With the inner cylinder of radius r_a as the ohmic injecting contact we can use the boundary condition

$$F (r_a) = 0$$

Then integration of eq. (3.126) yields

$$F (r) = \left[\frac{I(r^2 - r_a^2)}{2\pi \varepsilon \mu_n L r^2} \right]^{1/2} \tag{3.127}$$

using the boundary condition

$$V = \int_{r_a}^{r_b} F (r)\, dr$$

and $r_b \gg r_a$, we obtain

$$I = 2\pi \varepsilon \mu_n L \frac{V^2}{r_b^2} \tag{3.128}$$

For the case with the outer cylinder as the ohmic injecting contact we use the following boundary condition

$$F (r_b) = 0$$

It can be easily shown that the reverse current is given by

$$I_r = I \left[\ln\left(\frac{r_a}{r_b}\right) \right]^2$$

$$= 2\pi \varepsilon \mu_n L \frac{V^2}{r_b^2} \left[\ln\left(\frac{r_a}{r_b}\right) \right]^2 \tag{3.129}$$

(B) With shallow traps confined in single discrete energy levels

Using the same method used in Section 3.3.1(B), I and I_r for this case are obtained simply by multiplying eqs. (3.128) and (3.129) by θ_e given in eq. (3.116). Thus, for the case with the inner cylinder of radius r_a as the electron-injecting contact $[F(r_a) = 0]$ and $r_b \gg r_a$, the SCL current is given by

$$I = 2\pi\varepsilon\mu_n\theta_e L \frac{V^2}{r_b^2} \tag{3.130}$$

and for the case with outer cylinder of radius r_b as the electron-injecting contact $[F(r_b) = 0]$ and $r_b \gg r_a$. The reverse current is given by

$$I_r = 2\pi\varepsilon\mu_n\theta_e L \frac{V^2}{r_b^2} \left[\ln\left(\frac{r_a}{r_b}\right) \right]^2 \tag{3.131}$$

(C) With traps distributed exponentially within the forbidden energy gap

Similar to the approach given in Section 3.3.1(C) it can be easily shown that for the case with the inner cylinder as the electron-injecting contact $[F(r_a) = 0]$ and $r_b \gg r_a$ the SCL current is given by

$$I = q^{1-l}(2\pi\mu_n N_c L)\left(\frac{2l}{l+1}\right)^{l+1}\left(\frac{l}{l+1}\frac{2\varepsilon}{H'_f}\right)^l\left(\frac{V^{l+1}}{r_b^{2l}}\right) \tag{3.132}$$

and that for the case with the outer cylinder as the electron-injecting contact $[F(r_b) = 0]$ and $r_b \gg r_a$ the reverse current is given by

$$I_r = I\left[\frac{l+1}{2l}\ln\left(\frac{r_a}{r_b}\right)\right]^{l+1}$$

$$= q^{1-l}(2\pi\mu_n N_c L)\left[\ln\left(\frac{r_a}{r_b}\right)\right]^{l+1}\left(\frac{l}{l+1}\frac{2\varepsilon}{H'_f}\right)^l\left(\frac{V^{l+1}}{r_b^{2l}}\right) \tag{3.133}$$

In Table 3.7 are summarized all the cases for SCL currents without considering thermally generated free carriers. In most organic semiconductors the forbidden energy gap is large and the concentration of thermally generated free carriers is usually very small and can be ignored. Therefore all expressions given in Table 3.7 are generally applicable to insulating materials. It is interesting to note that the purely injected SCL current–voltage characteristics are not contingent on the configuration of the injecting contact, but are governed by the trap distribution in the forbidden energy gap. However, the thermally generated free carriers do affect the I–V characteristics. The effect of thermally generated free carriers is important and should be taken into account for those semiconductors with a small forbidden energy gap (or with low resistance). In the following we shall show this effect for the concentric sphere configuration.

TABLE 3.7. *The expressions for I–V characteristics for single injection (electron carriers) from a non-planar contacting electrode into a solid without thermally generated free carriers*

Case	Spherical	Cylindrical
Trap-free	$I = \dfrac{3\pi}{2}\varepsilon\mu_n \dfrac{V^2}{r_b}$ $I_r = \dfrac{3\pi}{2}\varepsilon\mu_n \dfrac{V^2}{r_b}\left(\dfrac{2r_a}{r_b}\right)^2$	$I = 2\pi\varepsilon\mu_n L \dfrac{V^2}{r_b^2}$ $I_r = 2\pi\varepsilon\mu_n L \dfrac{V^2}{r_b^2}\left[\ln\left(\dfrac{r_a}{r_b}\right)\right]^2$
Traps confined in a single discrete energy level within the forbidden energy gap	$I = \dfrac{3\pi}{2}\varepsilon\mu_n \theta_e \dfrac{V^2}{r_b}$ $I_r = \dfrac{3\pi}{2}\varepsilon\mu_n \theta_e \dfrac{V^2}{r_b}\left(\dfrac{2r_a}{r_b}\right)^2$	$I = 2\pi\varepsilon\mu_n \theta_e L \dfrac{V^2}{r_b^2}$ $I_r = 2\pi\varepsilon\mu_n \theta_e L \dfrac{V^2}{r_b^2}\left[\ln\left(\dfrac{r_a}{r_b}\right)\right]^2$
Traps distributed exponentially within the forbidden energy gap	$I = q^{1-l}(4\pi\mu_n N_c)\left(\dfrac{2l-1}{l+1}\right)^{l+1}\left(\dfrac{1}{l+1}\dfrac{3\varepsilon}{H_f'}\right)^l\left(\dfrac{V^{l+1}}{r_b^{2l-1}}\right)$ $I_r = q^{1-l}(4\pi\mu_n N_c)\left(\dfrac{1}{l+1}\dfrac{3\varepsilon}{H_f'}\right)^l\left(\dfrac{r_a}{r_b}\right)\left(\dfrac{V^{l+1}}{r_b^{2l-1}}\right)$	$I = q^{1-l}(2\pi\mu_n N_c L)\left(\dfrac{2l}{l+1}\right)^{l+1}\left(\dfrac{1}{l+1}\dfrac{2\varepsilon}{H_f'}\right)^l\left(\dfrac{V^{l+1}}{r_b^{2l}}\right)$ $I_r = q^{1-l}(2\pi\mu_n N_c L)\left[\ln\left(\dfrac{r_a}{r_b}\right)\right]^{l+1}\left(\dfrac{1}{l+1}\dfrac{2\varepsilon}{H_f'}\right)^l\left(\dfrac{V^{l+1}}{r_b^{2l}}\right)$

3.3.3. Spherical injecting contacts with thermally generated free carriers

For this case the behaviour of single injection in a solid with a concentric sphere configuration is governed by the following current flow equation:

$$I = \text{total current}$$

$$= q\mu_n(n+n_0)F(4\pi r^2) \tag{3.134}$$

and the Poisson equation

$$\frac{1}{r^2}\frac{d}{dr}(r^2 F) = \frac{q}{\varepsilon}[n+n_t] \tag{3.135}$$

and also

$$n+n_0 = N_c \exp[-(E_c - E_{Fn})/kT] \tag{3.136}$$

$$n_t + n_{t0} = \int_{E_l}^{E_u} \frac{h(E)\,dE}{1+g_n^{-1}[(E-E_{Fn})/kT]} \tag{3.137}$$

where n and n_t are, respectively, the densities of injected free and trapped electrons; n_0 and n_{t0} are, respectively, the densities of thermally generated free and trapped electrons which are independent of r. Supposing that the inner sphere is the electron injecting contact, the injected carrier density may be greater than n_0 for $r < r_{cr}$ and less than n_0 for $r > r_{cr}$ because of the distribution of traps and the quantity of injected carriers, r_{cr} being the critical point to separate these two regions. To solve such problems it is convenient to use the "regional approximation method" [Lampert *et al.* 1964, Lampert and Schilling 1970, Lampert and Mark 1970]. We shall analyse the following cases using this method.

(A) Without traps

Since for this case $n_t = 0$, there are two regions as follows.

Region I. Space charge region $r_a \le r \le r_{cr}$ (Fig. 3.26)

In this region $n > n_0$ and n_0 can be neglected. Thus we have

$$I \simeq q\mu_n n F(4\pi r^2) \tag{3.138}$$

$$\frac{1}{r^2}\frac{d}{dr}(r^2 F) = \frac{qn}{\varepsilon} \tag{3.139}$$

From eqs. (3.138) and (3.139) we can obtain $F(r)$ which is the same as that given by eq. (3.106). Substituting eq. (3.106) into eq. (3.138) we obtain

$$n = \left(\frac{3\varepsilon I}{8\pi q^2 \mu_n}\right)^{1/2}\frac{1}{(r^3 - r_a^3)^{1/2}} \tag{3.140}$$

Using the boundary condition

$$n(r_{cr}) = n_0 \tag{3.141}$$

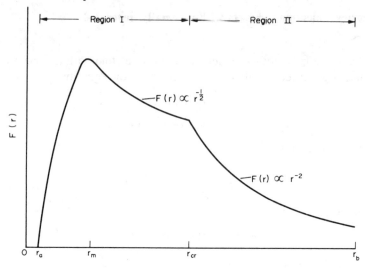

FIG. 3.26. The electric field as a function of radial distance between two concentric spherical electrodes with the inner sphere as the carrier-injecting contact for a trap-free solid with thermally generated free carriers.

we obtain from eq. (3.140)

$$r_{cr} = \left(\frac{3\varepsilon I}{8\pi q^2 \mu_n n_0^2} + r_a^3\right)^{1/3}$$

$$\simeq \left(\frac{3\varepsilon I}{8\pi q^2 \mu_n n_0^2}\right)^{1/3} \tag{3.142}$$

if I is large or r_a is very small. From eqs. (3.106), (3.108), and (3.142) and for $r_{cr} \gg r_a$ the voltage across Region I is given by

$$V_{\text{Region I}} = \int_{r_a}^{r_{cr}} F(r)dr \simeq \left[\frac{2Ir_{cr}}{3\pi\varepsilon\mu_n}\right]^{1/2}$$

$$= \left[\frac{I^2}{3\pi^2 q\varepsilon\mu_n^2 n_0}\right]^{1/3} \tag{3.143}$$

Region II. Ohmic region $r_{cr} < r < r_b$ (Fig. 3.26). In this region $n < n_0$ and n can be neglected. Thus we have

$$I \simeq q\mu_n n_0 F(4\pi r^2) \tag{3.144}$$

$$\frac{1}{r^2}\frac{d}{dr}(r^2 F) \simeq 0 \tag{3.145}$$

From eqs. (3.144) and (3.145) we obtain

$$F = \frac{I}{4\pi q\mu_n n_0 r^2} \tag{3.146}$$

Assuming $r_{cr} \ll r_b$, the voltage across Region II is given by

$$V_{\text{Region II}} = \int_{r_{cr}}^{r_b} F(r)dr \simeq \frac{I}{4\pi q \mu_n n_0 r_{cr}}$$

$$= \left(\frac{I^2}{24\pi^2 q \varepsilon \mu_n^2 n_0}\right)^{1/3} = \frac{1}{2} V_{\text{Region I}} \tag{3.147}$$

Therefore the total voltage between the inner and outer spherical electrodes is

$$V = V_{\text{Region I}} + V_{\text{Region II}} \simeq \frac{3}{2} V_{\text{Region I}}$$

$$\simeq \frac{3}{2}\left(\frac{I^2}{3\pi^2 q \varepsilon \mu_n^2 n_0}\right)^{1/3}$$

or

$$I \simeq \frac{2\pi \sqrt{2}}{3} q^{1/2} \varepsilon^{1/2} \mu_n n_0^{1/2} V^{3/2} \tag{3.148}$$

This is known as the 3/2 power law which has the following features:

(i) This 3/2 power law is valid only over the following range of currents. I must be larger than a certain critical value below which Ohm's law holds. From eq. (3.142) if $r_{cr} \rightarrow r_a$ then Region II extends to r_a and the current from eq. (3.147) becomes

$$I = 4\pi q \mu_n n_0 r_a V \tag{3.149}$$

The critical value of I above which the 3/2 power law is valid is the intersection of eqs. (3.148) and (3.149) which gives

$$I_{cr_a} = \frac{72\pi q^2 \mu_n n_0^2 r_a^3}{\varepsilon} \tag{3.150}$$

Also, I must be smaller than a certain critical value above which purely SCL square law holds. From eq. (3.143), if $r_{cr} \rightarrow r_b$ then Region I extends to r_b, and the current from eq. (3.143) becomes

$$I = \frac{3\pi}{2} \varepsilon \mu_n \frac{V^2}{r_b} \tag{3.151}$$

Thus, the critical value of I below which the 3/2 power law is valid is the intersection of eqs. (3.148) and (3.151) which gives

$$I_{cr_b} = \frac{2^9 \pi q^2 \mu_n n_0^2 r_b^3}{3^7 \varepsilon} \tag{3.152}$$

So the 3/2 power law is valid in the range of current $I_{cr_a} < I < I_{cr_b}$. The larger the ratio of r_b/r_a the larger is the range of I for the validity of this 3/2 power law.

(ii) This 3/2 power law is universally true for all cases with and without traps involving a concentric sphere configuration with thermally generated free carriers in which there are two regimes—SCL regime and ohmic regime. We can write [Lampert *et al.* 1964, 1970]

$$I = KV^{3/2} \tag{3.153}$$

where K is a constant which depends on the distribution of traps and is independent of r_a and r_b.

(B) With traps

Lampert *et al.* [1964, 1970] have derived a general expression using the regional approximation method for a solid with traps between two concentric spherical electrodes with the inner one as the electron-injecting contact based on eqs. (3.134)–(3.137). It is given by

$$I = 4\pi q \mu_n n_0 \left[\frac{3\varepsilon}{(1+c_1)^3 c_2 q(n_0 + n_{t0})} \right]^{1/2} V^{3/2}$$

$$= KV^{3/2} \tag{3.154}$$

The 3/2 power law is valid over the range of current between I_{cr_a} and I_{cr_b} ($I_{cr_a} < I < I_{cr_b}$) which are given by

$$I_{cr_a} = \frac{4\pi(1+c_1)^3 c_2 q^2 \mu_n n_0 (n_0 + n_{t0}) r_a^3}{3\varepsilon} \tag{3.155}$$

and

$$I_{cr_b} = \frac{(1+c_1)^3 c_2 q(n_0 + n_{t0}) r_a^2}{3\varepsilon} \tag{3.156}$$

The transition radius r_{cr} for the SCL region ($r < r_{cr}$) to the ohmic region ($r > r_{cr}$) is given by

$$r_{cr} = \left[\frac{3\varepsilon V}{(1+c_1)c_2 q(n_0 + n_{t0})} \right]^{1/2} \tag{3.157}$$

The capacitance C due to the total injected charge Q is

$$C = \frac{Q}{V} = \frac{1}{V} \int_{r_a}^{r_{cr}} q[n + n_t] 4\pi r^2 dr$$

$$= \frac{4\pi \varepsilon r_{cr}}{1+c_1} = \left[\frac{48\pi^2 \varepsilon^3 V}{(1+c_1)^3 c_2 q(n_0 + n_{t0})} \right]^{1/2} \tag{3.158}$$

where c_1 and c_2 are constants depending on the distribution of traps. Generally they are within the following ranges:

$$\tfrac{1}{2} \le c_1 \le 2 \quad \text{and} \quad 1 \le c_2 \le 2 \tag{3.159}$$

The lower limits and the upper limits of c_1 and c_2 hold, respectively, for deep traps and shallow traps (or without traps).

The expressions for I–V characteristics for various types of trap distributions are

TABLE 3.8. *The expressions for I–V characteristics for single injection (electron carriers) from the inner sphere in a concentric sphere configuration to a solid with thermally generated free carriers*

Case	$y \to$	c_1	$m \to$	c_2	I–V characteristics $I = KV^{3/2}$
Trap-free	$-\dfrac{1}{2}$	2	$-\dfrac{3}{2}$	2	$I \simeq \dfrac{2\pi\sqrt{2}}{3} q^{1/2}\varepsilon^{1/2}\mu_n n_0^{1/2} V^{3/2}$
Shallow traps at energy level E_t with $E_t - E_{Fn0} \gg kT$	$-\dfrac{1}{2}$	2	$-\dfrac{3}{2}$	2	$I \simeq \dfrac{2\pi\sqrt{2}}{3} q^{1/2}\varepsilon^{1/2}\mu_n \theta_e^{1/2} n_0^{1/2} V^{3/2}$
					$\theta_e = \dfrac{n}{n + n_t} \simeq \dfrac{n_0}{n_0 + n_{t0}}$
Deep traps at energy level E_t with $E_{Fn0} - E_t \gg kT$	1	$\dfrac{1}{2}$	0	1	$I \simeq \dfrac{8\pi\sqrt{2}}{3} q^{1/2}\varepsilon^{1/2}\mu_n \left(\dfrac{n_0^2}{H_e - n_{t0}}\right)^{1/2} V^{3/2}$
					taking $n_{t0} \simeq H_e - n_{t0} =$ unoccupied trap density in thermal equilibrium with the quasi-Fermi level E_{Fn0}, and $n_{t0} \gg n_0$
Traps distributed exponentially within the forbidden energy gap	$\dfrac{l-2}{l+1}$	$\dfrac{l+1}{2l-1}$	$-\dfrac{3}{l+1}$	$\dfrac{l+1}{l}$	$I \simeq 4\pi q^{1/2}\mu_n n_0 \left(\dfrac{2l-1}{3l}\right)^{3/2} \left(\dfrac{1}{l+1}\dfrac{3\varepsilon}{1+H_b}\right)^{1/2} V^{3/2}$

$F \propto r^y,\; y > -1,\; c_1 = \dfrac{1}{y+1}$

$n + n_t \propto r^m,\; 0 \geq m \geq -3,\; c_2 = \dfrac{3}{m+3}$

E_{Fn0} is the quasi-Fermi level in thermal equilibrium (in the absence of applied voltage).

summarized in Table 3.8 and some typical experimental results following $3/2$ power law are given in Fig. 3.27. The details of the derivation of these expressions are given in publications of Lampert *et al.* [1964, 1970]. In determining $I-V$ characteristics only Region I with $r_a \leq r \leq r_{cr}$ needs to be considered. Again, for all expressions it is assumed that $r_{cr} \gg r_a$.

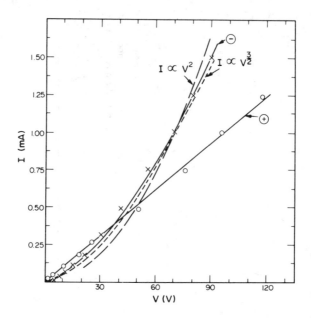

F IG. 3.27. The current–voltage characteristics of CdS crystals (resistivity = 20 ohm-cm) with a point-injecting contact (point radius = 1 μm) measured at room temperature. The crosses represent data taken with the point biased negatively, while the open circles represent data taken with the point biased positively. The dashed curve is a $3/2$ power-law curve fitted to the data at one point $I = 1$ mA. The straight line labelled $\oplus$ is an Ohm's law fitted to the data. [After Lampert, Many, and Mark 1964.]

3.4. THE ELECTROLYTE–INSULATOR SYSTEMS

When an insulator or a semiconductor is in contact with an electrolyte, the redistribution of charge at the interface causes the formation of a double layer in which the total net ionic charge in the electrolyte is equal to the net space charge of opposite polarity in the insulator (or the semiconductor). In insulators the concentration of thermally generated carriers in the bulk is negligibly small compared with the concentration of injected carriers. Many investigators [Klier 1962, Skinner 1955, Wright 1961, Dousmanis and Duncan 1958, Frankl 1965, Kingston and Neustadter 1955, Seiwatz and Green 1958, MacDonald 1964] have studied the effect of the redistribution of charge on the spatial distribution of space charge and the associated electric fields. The equilibrium between an electrolyte and an insulator is dynamic with continuous exchange of charge carriers between the ions of the electrolyte and the surface of the insulator. Thus, at equilibrium, the forward and backward rates of the reaction [Mehl and Hale 1967]

$$D \rightleftarrows A \pm C \tag{3.160}$$

are equal. If C represents the electrons in the conduction band of the insulator, $(+)$ sign is adopted. Then D represents in the electrolyte the donor (or "reduced") species which transfer electrons from the electrolyte to the conduction band of the insulator, and A represents the acceptor (or "oxidized") species. Since in this case the injected carriers are electrons, the current is electron current and the contact is a rectifying contact and the bias is forward when the applied voltage is negative at the electrolyte electrode with respect to the collecting electrode. Similarly, if C represents the holes in the valence band of the insulator, $(-)$ sign is adopted. Then A represents in the electrolyte the acceptor (or "oxidized") species which accepts electrons from the valence band of the insulator, thus creating holes (or injecting holes) into the valence band of the insulator, and D represents the donor (or "reduced") species. Since in this case the injected carriers are holes, the current is hole current, and the bias is forward when the applied voltage is positive at the electrolytic electrode with respect to the collecting electrode.

The rate of the charge flow in each direction through the interface at equilibrium is called the "exchange current". The application of a forward bias voltage is to reduce the rate of the charge flow in the reverse direction, thus resulting in a net charge flow in the forward direction. The current flow through an insulator under a forward bias voltage is limited by space charge (bulk limited) or by the rate of carrier injection through the interface from charged species in the electrolyte (electrode limited). The onset of the electrode limited condition is that the rate of the charge flow in the reverse direction is reduced to zero, and the forward current is saturated at the value equal to the exchange current.

3.4.1. The exchange currents

Gerischer [1961] has treated the redox reactions between an electrolyte and a semiconductor based on the assumption that the electrons are tunnelling between the electrolyte and the semiconductor surface at levels of equal energy so that there is no radiation nor exchange of energy involved in such an electron exchange process. Equation (3.160) represents a redox reaction which can also be written as

$$\text{red} \rightleftarrows \text{ox} + e$$

or

$$\text{red} \rightleftarrows \text{ox} - h \tag{3.161}$$

where e and h denote, respectively, the electron and the hole. By analogy the red species which donate electrons correspond to electron-occupied states in a valence band whereas the ox species which accept electrons correspond to empty states in a conduction band. The energy required for an electron to transfer from the red species to the ox species is analogous to the energy, equal to an energy band gap, required for an electron to be excited from the valence band to the conduction band in an intrinsic semiconductor. During the actual electron transfer between two orbitals, which follows the Franck–Condon principle, there is no movement of atoms in the system. Because of the thermal fluctuation, the energy distributions of the red and ox states in a redox electrolyte are

$$g_{red}(E) = \frac{g_{redox}(E)}{\exp\left[\dfrac{E - E_{F(redox)}}{kT}\right] + 1} = g_{redox} f[E - E_{F(redox)}] \qquad (3.162)$$

$$g_{ox}(E) = \frac{g_{redox}(E)}{\exp\left[\dfrac{E_{F(redox)} - E}{kT}\right] + 1} = g_{redox} f[E_{F(redox)} - E] \qquad (3.163)$$

and

$$g_{redox} = g_{red} + g_{ox} \qquad (3.164)$$

where g_{redox} is the density of electron energy states in the redox electrolyte, g_{red} is the density of occupied electron energy states (reduced species) and g_{ox} is the density of unoccupied electron energy states (oxidized species), f is the Fermi–Dirac distribution function, and $E_{F(redox)}$ is the Fermi energy level which is given by

$$E_{F(redox)} = E_{F0(redox)} + kT \ln(C_{red}/C_{ox}) \qquad (3.165)$$

where C_{red} and C_{ox} are, respectively, the concentrations of the reduced and oxidized species.

In a cathodic process the electrons will transfer from the valence band of the semiconductor to the redox electrolyte, so the rate of such an electron transfer or the exchange current is proportional to the number of occupied states in the semiconductor and the number of empty states in the redox electrolyte. Thus the current transfer is

$$J^- = q \int_{-\infty}^{\infty} v^-(E) g_s(E) f(E - E_{Fs}) g_{redox}(E) f[E_{F(redox)} - E] dE \qquad (3.166)$$

Similarly, in an anodic process the current transfer due to electron transfer from reduced species to the semiconductor is given by

$$J^+ = \int_{-\infty}^{\infty} v^+(E) g_s(E) f(E_{Fs} - E) g_{redox}(E) f[E - E_{F(redox)}] dE \qquad (3.167)$$

At thermal equilibrium without applied voltage, $E_{Fs} = E_{F(redox)}$, then

$$J^- = J^+ = J_0 \qquad (3.168)$$

where J_0 is the exchange current density, $v^-(E)$ and $v^+(E)$ represent the probability functions for electron transfer, and the subscript s refers to the semiconductor. J_0 is not a net reaction rate but is an equilibrium rate of forward electron transfer equal to that of reverse electron transfer. The electron exchange on the semiconductor surface takes place in the overlaps between appropriate functions of densities of states [Gerischer 1961, Vijh 1973] as shown in Fig. 3.28. The exchange current in an insulator is either proportional to the concentration of the reduced species for conduction band reactions or proportional to the concentration of the oxidized species for valence band reactions. It should be noted that for the electrolyte–insulator systems simultaneous exchange

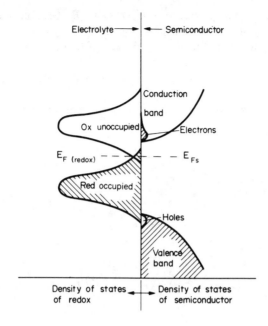

FIG. 3.28. The overlaps between the appropriate functions of the densities of states in an electrolyte–semiconductor system.

with both bands at an equal rate is not likely to happen because the energy band gap is so large that the appreciable injection is possible only when the band bending at the interface is bringing one of the bands close to the Fermi level and moving the other band away from the Fermi level so as to decrease the probability of carrier exchange with that band. In general, the redox couples tend to inject electrons to the conduction band of an insulator when the redox potentials are relatively negative, and tend to inject holes to the valence band of an insulator when the redox potentials are relatively positive.

3.4.2. Equilibrium properties of the electrolyte–insulator interface

(A) Theory

In the present analysis we make the following assumptions, but the treatment is general and therefore can be applied to thick or thin specimens of any materials.

(i) The energy band model can be used to treat the behaviour of injected carriers.

(ii) Only injected hole carriers are considered and the ohmic contact to inject them is perfect (a similar treatment can be easily extended to the case for only injected electron carriers).

(iii) For simplicity, the treatment is one-dimensional with the plane at $x = 0$ as the injecting contact and that at $x = d$ as the blocking contact.

(iv) The effect of image forces is ignored.

(v) The free hole density follows the Maxwell–Boltzmann statistics, while the trapped hole density follows the Fermi–Dirac statistics.

(vi) The mobility and the diffusion coefficient of the holes are not affected by the presence of traps.

(vii) The edge of the valence band E_v is chosen as the reference level, and is made equal to zero.

The steady state behaviour of single injection in a solid is governed by the current flow equation

$$J = q\mu_p p F - q D_p \frac{dp}{dx}$$ (3.169)

and the Poisson equation

$$\frac{dF}{dx} = \frac{q}{\varepsilon}(p + p_t)$$ (3.170)

in which

$$p = N_v \exp(-E_{Fp}/kT)$$ (3.171)

and

$$p_t = \int_{E_l}^{E_u} \frac{h(E, x)dE}{1 + g_p \exp[(E_{Fp} - E)/kT]}$$ (3.172)

where $h(E, x)$ is the distribution function for the trap density as a function of energy level E above the edge of the valence band and the distance x from the hole-injecting contact, which can be written as [Hwang and Kao 1972]

$$h(E, x) = N_t(E)S(x)$$ (3.173)

In the absence of applied voltages and at thermal equilibrium, $J = 0$; thus, eq. (3.169) becomes

$$\frac{dp}{dx} = \frac{\mu_p}{D_p} p F$$ (3.174)

Using the boundary conditions

$$F = 0 \quad \text{at} \quad x = d$$ (3.175)

and

$$V_0 = \int_0^d F dx$$ (3.176)

we can solve eqs. (3.170) and (3.174).

To treat this problem we can divide the region $0 \le x \le d$ into M equal intervals as shown in Fig. 3.29. The average value of $S(x)$ over this region can be approximated by

$$\langle S(x) \rangle_j \to b_j = \frac{M}{d} \int_{x_{j-1}}^{x_j} S(x)dx$$ (3.177)

To calculate the voltage across the interval between x_{j-1} and x_j, we solve eq. (3.172)

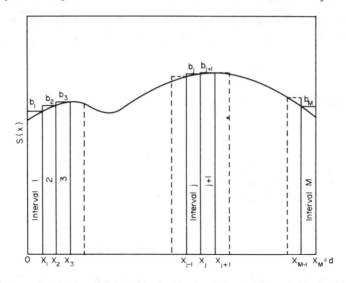

F IG. 3.29. Illustrating the discrete regional approximation for an arbitrary
spatial distribution function of traps.

first by taking

$$h_j(E, x) = b_j N_t(E) \qquad (3.178)$$

and then solve eqs. (3.170) and (3.174) as if the traps are uniformly distributed. By
matching the solutions at the respective boundaries and applying the boundary
conditions given in eqs. (3.175) and (3.176), we can obtain a complete solution
for the whole region $0 \le x \le d$. After determining the electric field F_j due to
the trap distribution h_j [eq. (3.178)], the voltage across the interval between x_{j-1}
and x_j is

$$V_j = \int_{x_{j-1}}^{x_j} F_j dx \qquad (3.179)$$

Using the same procedure, we can find the voltages across other intervals. The total
voltage across the specimen is then

$$V_0 = \sum_{j=1}^{M} V_j \qquad (3.180)$$

In the following we shall consider two general cases.

(I) THE TRAPS CONFINED IN A SINGLE DISCRETE ENERGY LEVEL

For this case $h(E, x)$ can be written as

$$h(E, x) = H_a \delta(E - E_t) S(x) \qquad (3.181)$$

For shallow traps, $E_t < E_F$. Then substitution of eq. (3.181) into eq. (3.172) and integration gives

$$p_t = \frac{H_a}{g_p} \exp[(E_t - E_{Fp})/kT]S(x) \tag{3.182}$$

Thus, the total charge carrier density is

$$p = p + p_t$$

$$= \left[1 + \frac{H_a}{g_p N_v} \exp(E_t/kT)S(x) \right] p$$

$$= K_{ax} p \tag{3.183}$$

where

$$K_{ax} = 1 + \frac{H_a}{g N_v} \exp(E_t/kT)S(x) \tag{3.184}$$

Using the method described in eqs. (3.177)–(3.180), we can write K_{ax} for the interval between x_{j-1} and x_j as

$$K_{aj} = 1 + \frac{H_a}{g_p N_v} b_j \exp(E_t/kT) \tag{3.185}$$

Equations (3.170) and (3.174) due to K_{aj} become

$$\frac{dF_j}{dx} = \frac{q}{\varepsilon} K_{aj} P_j \tag{3.186}$$

and

$$\frac{dp_j}{dx} = p_j F_j / V_T \tag{3.187}$$

where

$$V_T = D_p/\mu_p = kT/q \tag{3.188}$$

Up to here we can easily deduce the following cases.

(a) Without traps

For this case we set

$$\left. \begin{array}{ll} F_j = F, & K_{aj} = K_{ax} = 1 \\ p_j = p, & S(x) = 1 \end{array} \right\} \tag{3.189}$$

Solving eqs. (3.186) and (3.187) with the boundary conditions given in eqs. (3.175) and (3.176), we obtain

$$F(x) = \frac{2V_T}{d} [\cos^{-1} \exp(-V_0/2V_T)] \tan[(1 - x/d) \cos^{-1} \exp(-V_0/2V_T)] \tag{3.190}$$

and

$$p(x) = \frac{2\varepsilon V_T}{qd^2}[\cos^{-1}\exp(-V_0/2V_T)]^2 \sec^2[(1-x/d)\cos^{-1}\exp(-V_0/2V_T)]$$

$$(3.191)$$

At the surface, $x = 0$. Thus, we have

$$F(0) = \frac{2V_T}{d}[\cos^{-1}\exp(-V_0/2V_T)][\exp(V_0/V_T)-1]^{1/2} \qquad (3.192)$$

and

$$p(0) = \frac{2\varepsilon V_T}{qd^2}[\cos^{-1}\exp(-V_0/2V_T)]^2 \exp(V_0/V_T) \qquad (3.193)$$

Equations (3.192) and (3.193) are similar to those derived by other investigators [Kingston and Neustadter 1955, Seiwatz and Green 1958, MacDonald 1958].

(b) With uniform spatial distribution of traps

For this case we set

$$\left.\begin{array}{ll} F_j = F, & K_{aj} = K_{ax} = K_a \\ p_j = p, & S(x) = 1 \end{array}\right\} \qquad (3.194)$$

Solving eqs. (3.186) and (3.187) with the boundary conditions given in eqs. (3.175) and (3.176), we obtain

$$F(x) = \frac{2V_T}{d}[\cos^{-1}\exp(-V_0/2V_T)]\tan[(1-x/d)\cos^{-1}\exp(-V_0/2V_T)]$$

$$(3.195)$$

and

$$p(x) = \frac{2\varepsilon V_T}{K_a qd^2}[\cos^{-1}\exp(-V_0/2V_T)]^2 \sec^2[(1-x/d)\cos^{-1}\exp(-V_0/2V_T)]$$

$$(3.196)$$

At the surface $x = 0$. Thus we have

$$F(0) = \frac{2V_T}{d}[\cos^{-1}\exp(-V_0/2V_T)][\exp(V_0/V_T)-1]^{1/2} \qquad (3.197)$$

$$p(0) = \frac{2\varepsilon V_T}{K_a qd^2}[\cos^{-1}\exp(-V_0/2V_T)]^2 \exp(V_0/V_T) \qquad (3.198)$$

and

$$p_T(0) = K_a p(0) \qquad (3.199)$$

(c) With non-uniform spatial distribution of traps

For this case the solution of eqs. (3.186) and (3.187) yields

$$F_j = A_j^{1/2} \left(\frac{K_{aj} q V_T}{\varepsilon} \right)^{1/2} \tan \left[A_j^{1/2} \left(\frac{K_{aj} q}{\varepsilon V_T} \right)^{1/2} x + B_j \right] \qquad (3.200)$$

where A_j and B_j are the integration constants. Applying the boundary conditions given by eqs. (3.175) and (3.176) in the interval $x_{M-1} \le x \le x_M$ with $x_M = d$, then we obtain

$$B_M = -A_M^{1/2} \left(\frac{K_{aM} q}{\varepsilon V_T} \right) d \qquad (3.201)$$

Since F and $p = \dfrac{\varepsilon}{K_a q} \dfrac{dF}{dx}$ are continuous at the boundary between any two intervals, we can write

$$F_j(x = x_j) = F_{j+1}(x = x_j) \qquad (3.202)$$

and

$$\left[\frac{1}{K_{aj}} \frac{dF_j}{dx} \right]_{x = x_j} = \left[\frac{1}{K_{aj+1}} \frac{dF_{j+1}}{dx} \right]_{x = x_j} \qquad (3.203)$$

This implies that

$$A_j = A_{j+1} \left\{ 1 + \left(1 - \frac{K_{aj+1}}{K_{aj}} \right) \tan^2 \left[A_{j+1}^{1/2} \left(\frac{K_{aj+1} q}{\varepsilon V_T} \right)^{1/2} \frac{dj}{M} + B_{j+1} \right] \right\} \qquad (3.204)$$

and

$$B_j = \sec^{-1} \left\{ \left(\frac{A_{j+1}}{A_j} \right)^{1/2} \sec \left[A_{j+1} \left(\frac{K_{aj+1} q}{\varepsilon V_T} \right)^{1/2} \frac{dj}{M} + B_{j+1} \right] \right\} - A_j \left(\frac{K_{aj} q}{\varepsilon V_T} \right)^{1/2} \frac{dj}{M} \qquad (3.205)$$

Equations (3.204) and (3.205) are the recurrence formulae for determining A_j's and B_j's for successive intervals. It can be seen that all these constants are functions of A_M. To determine A_M, we use the boundary condition given in eq. (3.176). Integration of eq. (3.200) based on eq. (3.179) gives the voltage across the interval between x_{j-1} and x_j.

$$V_j = \int_{x_{j-1}}^{x_j} F_j dx = V_T \ln \left\{ \frac{\cos \left[A_j^{1/2} \left(\frac{K_{aj} q}{\varepsilon V_T} \right)^{1/2} \frac{d(j-1)}{M} + B_j \right]}{\cos \left[A_j^{1/2} \left(\frac{K_{aj} q}{\varepsilon V_T} \right)^{1/2} \frac{dj}{M} + B_j \right]} \right\} \qquad (3.206)$$

Using eq. (3.180) to obtain the total voltage across the entire specimen, we obtain

$$\exp(-V_0/V_T) = \prod_{j=1}^{M} \frac{\cos \left[A_j^{1/2} \left(\frac{K_{aj} q}{\varepsilon V_T} \right)^{1/2} \frac{d(j-1)}{M} + B_j \right]}{\cos \left[A_j^{1/2} \left(\frac{K_{aj} q}{\varepsilon V_T} \right)^{1/2} \frac{dj}{M} + B_j \right]} \qquad (3.207)$$

As has been mentioned, eq. (3.207) is a function of A_M only, so the solution of it by means of the error minimization techniques will give the value of A_M. Once A_M is found, all the A_j's and B_j's can be easily determined, and hence $F(x)$, $p(x)$, and $p_t(x)$ can be obtained. At the surface, $x = 0$; then we have

$$F(0) = A_1^{1/2}\left(\frac{K_{a1}qV_T}{\varepsilon}\right)^{1/2} \tan B_1 \tag{3.208}$$

$$p(0) = A_1 \sec^2 B_1 \tag{3.209}$$

$$p_T(0) = A_1 K_{a1} \sec^2 B_1 \tag{3.210}$$

(II) THE TRAPS DISTRIBUTED EXPONENTIALLY WITHIN THE FORBIDDEN ENERGY GAP

For this case $h(E, x)$ can be written as

$$h(E, x) = \frac{H_b}{kT_c}\exp(-E/kT_c)S(x) \tag{3.211}$$

If $T_c > T$ we can assume [Hwang and Kao 1972] that $f(E) = 1$ for $E_F < E < \infty$ and $f(E) = 0$ for $E < E_F$ as if we take $T = 0$. This is a good approximation particularly when T_c is much larger than T. With this assumption, integration of eq. (3.172) gives

$$p_t = \int_{E_F}^{\infty}\frac{H_b}{kT_c}\exp(-E/kT_c)S(x)dE$$

$$= H_b\left(\frac{p}{N_v}\right)^{1/l}S(x) \tag{3.212}$$

where

$$l = T_c/T \tag{3.213}$$

For thin films the trap density is generally high and therefore we can assume $p_t \gg p$. Thus, the total charge carrier density can be approximated to [Rose 1955]

$$p_T = p + p_t \simeq p_t = K_b p^{1/l}S(x) = K_{bx}p^{1/l} \tag{3.214}$$

where

$$K_{bx} = \frac{H_b S(x)}{N_v^{1/l}} \tag{3.215}$$

For the interval between x_{j-1} and x_j, eq. (3.215) becomes

$$K_{bj} = \frac{H_b b_j}{N_v^{1/l}} \tag{3.216}$$

Equations (3.170) and (3.174) due to K_{bj} become

$$\frac{dF_j}{dx} = \frac{q}{\varepsilon}K_{bj}p_j^{1/l} \tag{3.217}$$

and

$$\frac{dp_j}{dx} = p_j F_j / V_T \tag{3.218}$$

Up to here we can easily deduce the following cases.

(a) With uniform spatial distribution of traps

For this case we set

$$\left.\begin{array}{ll} F_j = F, & K_{bj} = K_{bx} = K_b \\ p_j = p, & S(x) = 1 \end{array}\right\} \tag{3.219}$$

Solving eqs. (3.217) and (3.218) with the boundary conditions given in eqs. (3.175) and (3.176), we obtain

$$F(x) = \frac{lV_T}{d}\left[\cos^{-1}\exp(-V_0/lV_T)\right]\tan\left[(1-x/d)\cos^{-1}\exp(-V_0/lV_T)\right] \tag{3.220}$$

and

$$p(x) = \left\{\frac{l\varepsilon V_T}{K_b q d^2}\left[\cos^{-1}\exp(-V_0/lV_T)\right]^2 \sec^2\left[(1-x/d)\cos^{-1}\exp(-V_0/lV_T)\right]\right\}^l \tag{3.221}$$

At the surface, $x = 0$. Thus, we have

$$F(0) = \frac{lV_T}{d}\left[\cos^{-1}\exp(-V_0/lV_T)\right]\left[\exp(2V_0/lV_T)-1\right]^{1/2} \tag{3.222}$$

$$p(0) = \left(\frac{l\varepsilon T}{K_b q d^2}\right)^l\left[\cos^{-1}\exp(-V_0/lV_T)\right]^{2l}\exp(2V_0/V_T) \tag{3.223}$$

and

$$p_T(0) = K_b p^{1/l} \tag{3.224}$$

(b) With non-uniform spatial distribution of traps

For this case the solution of eqs. (3.217) and (3.218) yields

$$F_j = C_j^{1/2}(lV_T)^{1/2}\tan\left[C_j^{1/2}(lV_T)^{-1/2}x + D_j\right] \tag{3.225}$$

In the interval $x_{M-1} \le x \le x_M$ with $x_M = d$ and based on eq. (3.175) we have

$$D_M = C_M^{1/2}(lV_T)^{-1/2}d \tag{3.226}$$

The recurrence formulae for C_j's and D_j's are

$$C_j = C_{j+1}\tan^2\left[C_{j+1}^{1/2}(lV_T)^{-1/2}\frac{dj}{M} + D_{j+1}\right]$$

$$\times\left\{\frac{K_{bj}}{K_{b\,j+1}}\sec^2\left[C_{j+1}^{1/2}(lV_T)^{-1/2}\frac{dj}{M} + D_{j+1}\right]-1\right\}^{-1} \tag{3.227}$$

and

$$D_j = \sec^{-1}\left\{\left(\frac{K_{bj}}{K_{b\,j+1}}\right)^{1/2}\sec\left[C_{j+1}^{1/2}(lV_T)^{-1/2}\frac{dj}{M}+D_{j+1}\right]\right\}$$
$$-C_j^{1/2}(lV_T)^{-1/2}\frac{dj}{M} \tag{3.228}$$

All C_j's and D_j's are functions of C_M. From eqs. (3.179), (3.180), and (3.225), we obtain

$$\exp(-V_0/lV_T) = \prod_{j=1}^{M}\left\{\frac{\cos\left[C_j^{1/2}(lV_T)^{-1/2}\frac{d(j-1)}{M}+D_j\right]}{\cos\left[C_j^{1/2}(lV_T)^{-1/2}\frac{dj}{M}+D_j\right]}\right\} \tag{3.229}$$

Equation (3.229) is a function of C_M only. So we can determine C_M and hence C_j's and D_j's, and obtain $F(x)$, $p(x)$, and $p_t(x)$. At the surface, $x = 0$. Thus, we have

$$F(0) = C_1^{1/2}(lV_T)^{1/2}\tan D_1 \tag{3.230}$$

$$p(0) = \left[\frac{C_1\varepsilon}{k_{b1}q}\sec^2 D_1\right]^l$$

and

$$\left.\begin{array}{l}\\ \\ \\ \\ \\ \end{array}\right\} \tag{3.231}$$

$$p_T(0) = \frac{C_1\varepsilon}{d}\sec^2 D_1$$

(B) Computed results and discussion

There are no experimental results available to compare with the theory presented in Section 3.4.2(A). This may be due to the fact that experiments in the absence of applied voltages and at zero external current are not easy to perform. However, the theoretical analysis will be used in conjunction with Section 3.4.3 dealing with the effects of traps on the current–voltage characteristics of insulator–electrolyte systems. In this section we have computed $F(0)$, $p(0)$, and $p_T(0)$ for some simple cases with the aim of showing the effect of the non-uniformity of spatial distribution of traps. In thin films the traps are generally distributed exponentially in energy [Hwang and Kao 1972], so we consider only the traps with an exponential energy distribution and with the following forms of spatial distribution.

(a) Uniform spatial distribution

For this case the distribution function may be written as

$$S(x) = 1 \tag{3.232}$$

So we can use eqs. (3.222)–(3.224) for computing $F(0)$, $p(0)$, and $p_T(0)$.

(b) Linear spatial distribution with the maximum density at the injecting electrode (at $x = 0$)

For this case the distribution function may be written as

$$S(x) = 1 + A - x/x_0 \tag{3.233}$$

where A and x_0 are constants. Substitution of eq. (3.233) into eq. (3.177) gives

$$b_j = 1 + A - \frac{d(j + \frac{1}{2})}{Mx_0} \tag{3.234}$$

(c) Exponential spatial distribution with the maximum density at the injecting electrode (at $x = 0$)

For this case the distribution function may be written as

$$S(x) = 1 + B \exp(-x/x_0) \tag{3.235}$$

where B is a constant. Similarly, substitution of eq. (3.235) into eq. (3.177) gives

$$b_j = 1 + \frac{BMx_0}{d} \exp\left(-\frac{dj}{Mx_0}\right)[\exp(-d/Mx_0) - 1] \tag{3.236}$$

(d) Exponential spatial distribution with the maximum densities at both electrodes (at $x = 0$ and $x = d$)

For this case the distribution function may be written as

$$S(x) = 1 + C \exp(-x/x_0) + D \exp[-(d - x)/x_0] \tag{3.237}$$

where C and D are constants. Similarly, substitution of eq. (3.237) into eq. (3.177) gives

$$b_j = 1 - \frac{Mx_0}{d} \Big\{ C \exp(-dj/Mx_0)[\exp(-d/Mx_0) - 1]$$

$$- D \exp\left[-\frac{d(M-j)}{Mx_0}\right][\exp(d/Mx_0) - 1] \Big\} \tag{3.238}$$

In order to show the effect of traps on the space charge created by an injecting contact, we choose anthracene films as an example. $F(0)$, $p_T(0)$, and $p(0)$ have been computed using 20 intervals ($M = 20$) for the discrete regional approximation and the following physical parameters for anthracene films at 300°K [Gutmann and Lyons 1967, Mehl 1965]:

$$N_v = 2.4 \times 10^{25} \text{ m}^{-3} \qquad H_b = 10^{25} \text{ m}^{-3}$$
$$\varepsilon = 3.2 \times 10^{-11} \text{ F m}^{-1} \qquad T_c = 600°\text{K}$$
$$\mu_p = 8 \times 10^{-5} \text{ m}^2 \text{ V}^{-1} \text{ sec}^{-1} \qquad V_0 = 0.125 \text{ V}$$

In all the figures, the subscript A denotes the case in absence of traps, the subscript U denotes the case with traps exponentially distributed in energy and uniformly distributed in space, and the subscript NU denotes the cases with traps exponentially distributed in energy and non-uniformly distributed in space. Figure 3.30 shows

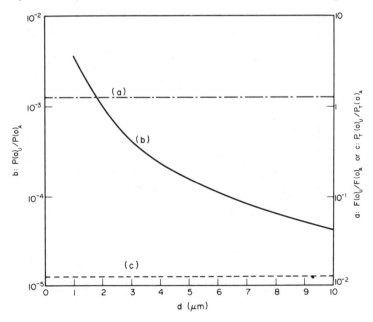

FIG. 3.30. $F(0)_U/F(0)_A$, $p_T(0)_U/p_T(0)_A$, and $p(0)_U/p(0)_A$ as functions of d for the exponential energy distribution and uniform spatial distribution of traps. [After Elsharkawi and Kao 1976.]

(a) $F(0)_U/F(0)_A$, (b) $p(0)_U/p(0)_A$, and (c) $p_T(0)_U/p_T(0)_A$ as functions of specimen thickness. It can be seen that the presence of traps affects greatly the values of $F(0)$, $p_T(0)$, and $p(0)$. Figures 3.31 shows (a) $F(0)_{NU}/F(0)_U$, (b) $p_T(0)_{NU}/p_T(0)_U$, and (c) $p(0)_{NU}/p(p)_U$ as functions of specimen thickness for the exponential spatial distribution of traps based on eq. (3.237). These results indicate that the thinner the specimen the more significant is the effect of the non-uniformity of spatial distribution, and that for very thick specimens this effect may become negligible. The critical value of d for this effect to become negligible depends on the distribution function $S(x)$. However, for thin films this effect should not be ignored.

3.4.3. The current–voltage characteristics of an
electrolyte–insulator system with traps of various distributions
in energy and in space inside the insulator

To form ohmic injecting contacts for the measurements of SCL currents, electrolytes have been used as electrodes in both organic [Mark and Helfrich 1962, Kallmann and Pope 1962, Heilmeier and Warfield 1963] and inorganic [Schmidt 1968, Vijh 1973, Beckmann and Memming 1969] insulating solids. Using such electrodes, the current flow through an insulator is limited by space charge when carriers in excess of those thermally generated in the bulk can be injected through the insulator–electrolyte interface, as well as by the rate of carrier supply from the charged species in the electrolytic electrode. Furthermore, the current–voltage $(I–V)$ characteristics are

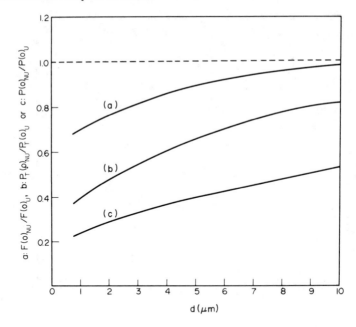

F$_{IG}$. 3.31. $F(0)_{NU}/F(0)_U$, $p_T(0)_{NU}/p_T(0)_U$, and $p(0)_{NU}/p(0)_U$ as functions of d for the exponential distribution of traps in energy and in space based on eq. (3.237) with $x_0 = 0.5$ μm, $C = 0.9$, and $D = 0.5$. [After Elsharkawi and Kao 1976.]

strongly affected by the form of trap distribution in energy and in space [Hwang and Kao 1972, Nicolet 1966, Sworakowski 1970]. In this section we shall present a unified analysis of the steady state I–V characteristics of an insulator–electrolyte system with traps distributed uniformly or non-uniformly in energy and in space, and to show that the computed results based on the expressions derived are in good agreement with the experimental results for some organic insulating film–electrolyte systems.

(A) Theory

In the theoretical analysis we use the same assumptions listed in Section 3.4.2(A) and, in addition, we assume that the carriers injected are controlled by the rate of carrier supply from the charge species in the electrolytic electrode, and that the electric field is so large that the current components due to diffusion and due to carriers thermally generated in the specimen can be neglected. The steady state behaviour of single injection in a solid is governed by the current flow equation

$$J = q\mu_p p^* F \tag{3.239}$$

and the Poisson equation

$$\frac{dF}{dx} = \frac{q}{\varepsilon}(p^* + p_t) \tag{3.240}$$

in which

$$p^* = N_v \exp(-E_{Fp}/kT) \tag{3.241}$$

and

$$p_t = \int_{E_l}^{E_u} \frac{h(E, x)dE}{1 + g_p \exp[(E_{Fp} - E)/kT]} \tag{3.242}$$

The boundary conditions used to solve eqs. (3.239) and (3.240) are:

$$p^*(0) = p(0)[1 - J/J_0] \tag{3.243}$$

and

$$V + V_0 = \int_0^d F dx \tag{3.244}$$

where $p^*(0)$ and $p(0)$ are, respectively, the free hole densities at $x = 0$ with and without applied fields. This implies that at high fields when J approaches J_0, the concentration of carriers at the surface approaches zero; and that the saturation current J_0 is due mainly to the exchange of carriers between the surface of the insulator and the ions of the electrolyte [Hale and Mehl 1966, Mehl and Hale 1967]. Equation (3.244) implies that the total potential across the insulator specimen from $x = W_a$, where $F = 0$ to $x = d$ (W_a is very small and can be assumed to be zero), is the sum of the applied voltage V and the equilibrium potential difference created inside the specimen by carrier injection from the electrolyte V_0.

In Section 3.4.2(A) we have derived the expressions for $p(0)$ in the absence of applied voltages for two general cases. In the following we shall consider also the same two general cases.

(I) THE TRAPS CONFINED IN A SINGLE DISCRETE ENERGY LEVEL

For this case $h(E, x)$ can be written as

$$h(E, x) = H_a \delta(E - E_t)S(x) \tag{3.245}$$

For simplicity, we assume that E_t is a shallow trap level located below E_{Fp}. Then we can write

$$p^* + p_t = \Phi_a p^* \tag{3.246}$$

where

$$\Phi_a = 1 + \frac{H_a}{g_p N_v} \exp(E_t/kT)S(x)$$

$$= K_a\{1 + G[S(x) - 1]\} \tag{3.247}$$

in which

$$G = \frac{H_a}{K_a g_p N_v} \exp(E_t/kT) \tag{3.248}$$

Substituting eq. (3.246) into eq. (3.240), and then solving eqs. (3.239) and (3.240) with the boundary conditions given in eqs. (3.243) and (3.244), we obtain

$$F = F_a(0)\left[1+\frac{\gamma_a(x)}{x_{0a}}\right]^{1/2}$$

(3.249)

and

$$V + V_0 = F_a(0)\int_0^d \left[1+\gamma_a(x)/x_{0a}\right]^{1/2}dx$$

(3.250)

where

$$F_a(0) = \left[q\mu_p p(0)\left(\frac{1}{J}-\frac{1}{J_0}\right)\right]^{-1}$$

(3.251)

$$x_{0a} = \frac{\varepsilon}{2K_a q^2 \mu_p p^2(0)}\frac{J}{(1-J/J_0)^2}$$

(3.252)

$$\gamma_a(x) = \int_0^x \{1+G[S(t)-1]\}dt$$

(3.253)

and

$$K_a = 1 + \frac{H_a}{g_p N_v}\exp(E_t/kT)$$

(3.254)

Up to here we can easily deduce the following cases.

(a) Without traps

For this case we set

$$\left.\begin{array}{ll} G = 0, & \gamma_a(x) = x \\ \phi_a = K_a = 1, & S(x) = 0 \end{array}\right\}$$

(3.255)

Then eq. (3.250) becomes

$$V + V_0 = \frac{2}{3}F_a(0)x_{0a}[1-(1+d/x_{0a})^{3/2}]$$

(3.256)

(i) *Low injection (or low current) case.* In this case, we can assume that $x_{0a} \ll d$ and $1 - J/J_0 \simeq 1$. After simplification, the result is

$$J = \frac{9}{8}\varepsilon\mu_p\frac{(V+V_0)^2}{d^3}$$

(3.257)

This equation has been derived by other investigators [Mehl and Hale 1967]. It should be noted that with the forward bias voltage V to be positive, V_0 is negative and that eq. (3.257) is valid only for $|V| > |V_0|$ since V must be sufficiently large $V > (kT/q)\ln[p^*(0)/p^*(d)]$ to justify the neglect of the diffusion current.

(ii) *High injection (or high current) case.* In this case we can assume that $x_{0a} \gg d$. From eqs. (3.251), (3.252), and (3.256) we obtain

$$1/J = 1/J_0 - d/q\mu_p p(0)[V+V_0]$$

(3.258)

where $p(0)$ is given by eq. (3.193) which is rewritten as follows:

$$p(0) = \frac{2\varepsilon V_T}{qd^2}[\cos^{-1}\exp(-V_0/2V_T)]^2\exp(V_0/V_T) \tag{3.259}$$

in which

$$V_T = kT/q \tag{3.260}$$

Equation (3.258) has been derived by other investigators [Mehl and Hale 1967]. It indicates that J approaches J_0 as a saturation current when V increases to such a value that the second term on the right-hand side of eq. (3.258) becomes negligibly small. This implies that under such a condition the concentration of donor species in the electrolyte becomes independent of the overpotential, or in other words, the charge carrier supply is exhausted.

(b) With uniform spatial distribution of traps

For this case we set

$$\left.\begin{array}{ll} G = 1 - 1/K_a, & \gamma_a(x) = x \\ \phi_a = K_a, & S(x) = 1 \end{array}\right\} \tag{3.261}$$

Then eq. (3.250) becomes

$$V + V_0 = \frac{2}{3}F_a(0)x_{0a}[1 - (1 + d/x_{0a})^{3/2}] \tag{3.262}$$

For the case of low injection, we can assume that $x_{0a} \ll d$ and $1 - J/J_0 \simeq 1$. After simplification, eq. (3.262) becomes

$$J = \frac{9\varepsilon\mu_p}{8 K_a}\frac{(V + V_0)^2}{d^3} \tag{3.263}$$

Again, eq. (3.263) is valid only for $|V| > |V_0|$.

For the case of high injection, we can assume that $x_{0a} \gg d$, and then simplification of eq. (3.262) yields

$$1/J = 1/J_0 - d/q\mu_p p(0)[V + V_0] \tag{3.264}$$

where $p(0)$ is given by eq. (3.198) which is rewritten as follows:

$$p(0) = \frac{2\varepsilon V_T}{K_a qd^2}[\cos^{-1}\exp(-V_0/V_T)]^2\exp(V_0/V_T) \tag{3.265}$$

Equations (3.263) and (3.264) are, respectively, similar in form to eqs. (3.257) and (3.258) except that the former involve K_a, which is equal to 1 for the insulator without traps, and larger than 1 for the insulator with traps.

(c) With non-uniform spatial distribution of traps

For the case of low injection in which $x_{0a} \ll d$, eq. (3.250) reduces to

$$J = \frac{9}{8} \frac{\varepsilon \mu_p}{K_a} \frac{(V + V_0)^2}{d_{\text{eff}}^3} \tag{3.266}$$

where

$$d_{\text{eff}} = \left\{ \frac{3}{2} \int_0^d \left(\int_0^x [1 + G(S(x) - 1)] \, dt \right)^{1/2} dx \right\}^{2/3} \tag{3.267}$$

Equation (3.266) is similar in form to eq. (3.263) except that d has been replaced with d_{eff} which can be considered as "effective thickness" to take into account the effect of the inhomogeneous spatial distribution of free and trapped carriers.

For the case of high injection in which $x_{0a} \gg d$, eq. (3.264) is applicable except that for this case $p(0)$ is given by eq. (3.209) which is rewritten as follows:

$$p(0) = A_1 \sec^2 B_1 \tag{3.268}$$

where A_1 and B_1 are constants, the values of which depend on the form of the spatial trap distribution function.

(II) THE TRAPS DISTRIBUTED EXPONENTIALLY WITHIN THE FORBIDDEN ENERGY GAP

For this case $h(E, x)$ can be written

$$h(E, x) = \frac{H_b}{k T_c} \exp(-E/k T_c) S(x) \tag{3.269}$$

If $T_c > T$, we can assume that $f(E) = 1$ for $E > E_F$ and $f(E) = 0$ for $E < E_F$, as if we take $T = 0$. This is good approximation particularly when T_c is much larger than T. With this assumption we obtain

$$p_t = H_b \left(\frac{p^*}{N_v} \right)^{1/l} S(x) \tag{3.270}$$

where $l = T_c/T$. For thin films we can assume $p_t \gg p^*$. By introducing the following parameter:

$$\Phi_b = K_b S(x) \tag{3.271}$$

eq. (3.240) becomes

$$\frac{dF}{dx} = \frac{q}{\varepsilon} \Phi_b p^{*1/l} \tag{3.272}$$

where

$$K_b = \frac{H_b}{N_v^{1/l}} \tag{3.273}$$

Solution of eqs. (3.239) and (3.272) with the aid of eqs. (3.243) and (3.244) yields

$$F = F_b(0) \left[1 + \frac{\gamma_b(x)}{x_{0b}} \right]^{l/(l+1)} \tag{3.274}$$

and

$$V + V_0 = F_b(0) \int_0^d [1 + \gamma_b(x)/x_{Ob}]^{1/(l+1)} dx \tag{3.275}$$

where

$$F_b(0) = \left[q\mu_p p(0) \left(\frac{1}{J} - \frac{1}{J_0} \right) \right]^{-1} \tag{3.276}$$

$$x_{Ob} = \left(\frac{l}{l+1} \right) \frac{\varepsilon \mu_p^{1/l} J}{K_b q^{(l-1)/l}} [q\mu_p p(0)(1 - J/J_0)]^{-(l+1)/l} \tag{3.277}$$

and

$$\gamma_b(x) = \int_0^x S(t)dt \tag{3.278}$$

Up to here we can easily deduce the following cases.

(a) With uniform spatial distribution of traps

For this case we set

$$\phi_b = K_b, \qquad \gamma_b(x) = x, \qquad S(x) = 1 \tag{3.279}$$

Then eq. (3.275) becomes

$$V + V_0 = \frac{l+1}{2l+1} F_b(0) x_{Ob} [1 - (1 + d/x_{Ob})^{[(2l+1)/(l+1)]}] \tag{3.280}$$

For the case of low injection, we can assume that $x_{Ob} \ll d$ and $1 - J/J_0 \simeq 1$, and the simplification of eq. (3.280) yields

$$J = q^{1-l} \mu_p N_v \left(\frac{2l+1}{l+1} \right)^{l+1} \left(\frac{l}{l+1} \frac{\varepsilon}{H_b} \right)^l \frac{(V + V_0)^{l+1}}{d^{2l+1}} \tag{3.281}$$

This equation is similar in form to that derived by Hwang and Kao [1972] for metallic ohmic contacts, except that V has been replaced with $V + V_0$, which takes into account the equilibrium potential difference created due to the presence of the electrolytic electrode. Equation (3.281) is valid only for $|V| > |V_0|$.

For the case of high injection, we can assume that $x_{Ob} \gg d$, and then simplification of eq. (3.275) yields

$$\frac{1}{J} = \frac{1}{J_0} - \frac{d}{q\mu_p p(0)[V + V_0]} \tag{3.282}$$

where $p(0)$ is given by eq. (3.223) which is rewritten as follows

$$p(0) = \left(\frac{l\varepsilon V_T}{K_p q d^2} \right)^l [\cos^{-1} \exp(-V_0/lV_T)]^{2l} \exp(2V_0/V_T) \tag{3.283}$$

Equation (3.282) is similar in form to eq. (3.258) for the case without traps and to eq. (3.264) for the case with traps confined in a single discrete energy level, except that the expressions for $p(0)$ are different for these three cases.

(b) With non-uniform spatial distribution of traps

For the case of low injection in which $x_{ob} \ll d$, eq. (3.275) reduces to

$$J = q^{1-l} \mu_p N_v \left(\frac{2l+1}{l+1} \right)^{l+1} \left(\frac{l}{l+1} \frac{\varepsilon}{H_b} \right)^l \frac{(V+V_0)^{l+1}}{d_{\text{eff}}^{2l+1}} \qquad (3.284)$$

where

$$d_{\text{eff}} = \left\{ \frac{2l+1}{l+1} \int_0^d \left[\int_0^x S(t)dt \right]^{1/(l+1)} dx \right\}^{(l+1)/(2l+1)} \qquad (3.285)$$

Equation (3.284) is similar in form to eq. (3.283) except that d has been replaced with d_{eff}. Again, the difference between d_{eff} and d is caused by the inhomogeneous spatial distribution of free and trapped carriers.

For the case of high injection, in which $x_{ob} \gg d$, eq. (3.282) is applicable except that for this case $p(0)$ is given by eq. (3.231) which is rewritten as follows

$$p(0) = \left[\frac{C_1 \varepsilon}{K_{b1} q} \sec^2 D_1 \right]^l \qquad (3.286)$$

where C_1, D_1, and K_{b1} are constants, the values of which depend on the form of the spatial trap distribution function.

If the exponentially distributed traps scanned at levels above a certain discrete level E_{te}, all expressions given above are valid if H_b is replaced with H_b' given by eq. (3.48).

(B) Some typical results and discussion

In thin films the traps are generally exponentially distributed in energy [Hwang and Kao 1972, Helfrich et al. 1965]. We shall consider the same forms of spatial trap distribution given by eqs. (3.232), (3.233), (3.235), and (3.237) in Section 3.4.2(B).

In order to show the effect of traps on the steady state current–voltage characteristics, we choose anthracene films as an example and use the same physical parameters for anthracene films at $300°K$ for the computed results given in Figs. 3.30 and 3.31, and in addition we assume also $J_0 = 2 \times 10^{-2} \text{ A m}^{-2}$. The current as a function of d at a fixed applied average field (V/d) of 0.6 MV m^{-1} for the electrode, $0.1 \text{ M Ce(SO}_4)_2$ in $0.5 \text{ M H}_2\text{SO}_4$, and $l = 2$ using eqs. (3.275)–(3.278) have been computed. This particular average field and this particular electrode were chosen because these computed results were used to correlate some experimental results of Mehl [1965]. In all figures, the subscript A denotes the case of absence of traps, the subscript U denotes the case with traps exponentially distributed in energy and uniformly distributed in space, and the subscript NU denotes the cases with traps exponentially distributed in energy and non-uniformly distributed in space. Figure 3.32 shows the ratio of J_U/J_A at a constant applied average field of 0.6 MV m^{-1} as a function of specimen thickness d. It can be seen that the presence of traps affects greatly the value of J. Figure 3.33 shows J_{NU}/J_U as a function of specimen thickness for four different types of non-uniform spatial

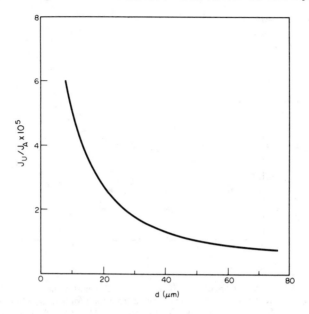

FIG. 3.32. J_U/J_A as a function of d for a fixed value of $V/d = 0.6$ MV m^{-1} for the exponential energy distribution and uniform spatial distribution of traps. [After Elsharkawi and Kao 1976.]

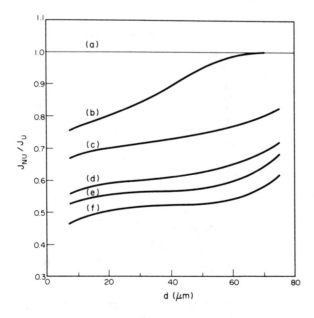

FIG. 3.33. J_{NU}/J_U as a function of d for a fixed value of $V/d = 0.6$ MV m^{-1} for the exponential energy distribution of traps and for various forms of spatial distribution of traps. (a) $S(x) = 1$; (b) $S(x) = 1 + 0.9 - x/x_0$, $x_0 = 70$ μm; (c) $S(x) = 1 + 0.6$ exp$(-x/x_0)$, $x_0 = 60$ μm; (d) $S(x) = 1 + \exp(-x/x_0) + 0.6$ exp$[-(d-x)/x_0]$, $x_0 = 60$ μm; (e) $S(x) = 1 + \exp(-x/x_0) + \exp[-(d-x)/x_0]$, $x_0 = 60$ μm; and (f) $S(x) = 1 + 2$ exp$(-x/x_0) + 1.5$ exp$[-(d-x)/x_0]$, $x_0 = 60$ μm. [After Elsharkawi and Kao 1976.]

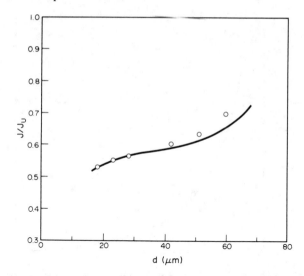

FIG. 3.34. Comparison of J_{NU}/J_U with J/J_U. J denotes the experimental currents in anthracene extracted from the results of Mehl [1965]. Solid line: theoretical, O: experimental. [After Elsharkawi and Kao 1976.]

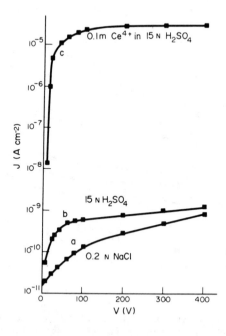

FIG. 3.35. The current–voltage characteristics of anthracene and the exchange current (saturation-electrode limited) in various electrolytes (hole-injecting contacts). The blocking contact is provided by a 0.2 N NaCl solution. [After Mehl 1965.]

distribution of traps. These results indicate that the thinner the specimen the more significant is the effect of non-uniformity of spatial distribution, and that for very thick specimens this effect may be negligible. The critical value of d for this effect to become negligible depends on the distribution function $S(x)$. However, for thin specimens this effect should not be ignored.

By plotting the ratio J/J_U as a function of d with the values of J extracted from the experimental results obtained by Mehl [1965] and comparing the ratio J_{NU}/J_U as a function of d, in Fig. 3.34, it can be seen that the most probable spatial distribution function of the specimens used by Mehl may be exponential with the maximum trap densities at $x = 0$ and $x = d$. Based on this comparison and for $S(x)$ independent of d, the value of x_0 is about 50 ± 5 μm, C about 0.9, and D about 0.3. For a more rigorous treatment, the curve fitting techniques of the experimental data to find the values of the unknown parameters C, D, and x_0 of the spatial trap distribution function can be used.

Mehl and his co-workers in a series of their publications [1965–71] have reported in detail the behaviour of electrolyte–insulator systems. Since most insulators have a large energy band gap such as anthracene, thermally generated carriers at room temperature can be neglected. Figure 3.35 shows that the current at low voltages is SCL and increases roughly with the third power of the applied voltage indicating an exponential

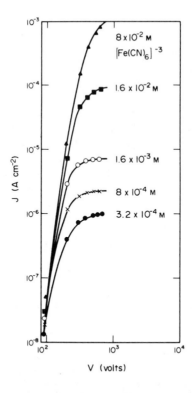

FIG. 3.36. The current–voltage characteristics of pery- lene crystals of thickness of 15 μm as func- tions of the rate of hole injection by $[Fe(CN)_6]^{3-}$ ions. [After Gerischer 1970.]

distribution of traps in energy. At higher voltages the current saturates at the value of exchange current and its value depends on the type of electrolyte. Mehl [1965] has also shown that the current in anthracene crystals with electrolyte of 0.1 M $Ce(SO_4)_2$ in 0.5 M H_2SO_4 as a hole injecting electrode depends on the specimen thickness while the saturation current (exchange) current is independent of the crystal thickness. The saturation current is strongly dependent on the rate of hole injection as shown in Fig. 3.36. However, it should be noted that in certain cases the current due to injected carriers could be larger than the exchange current [Lohmann and Mehl 1968], and this effect will be discussed in Section 5.3—High field effects.

CHAPTER 4

Space Charge Electric Conduction—Two-carrier Current Injection

4.1. DEFINITIONS AND PHYSICAL CONCEPTS OF TERMS RELATED TO TRAPPING AND RECOMBINATION PROCESSES

Charge carriers injected into an insulator or a semiconductor through electron and hole emission from the contacting electrodes (double injection) or through absorption of light will contribute to the increase in electric conductivity. Under such a condition $np > n_i^2$, implying that there are excess carriers in the material, where $n_i^2 = n_0 p_0$ and n_0 and p_0 are the thermally generated electron and hole densities inside the material. Contrarily, it is also possible that $np < n_i^2$, but under this condition the carriers are extracted from the material. These two conditions are generally referred to as non-equilibrium conditions. However, when the carrier density distributions are disturbed from their thermal equilibrium values, they will tend to return to equilibrium through a recombination process in the case of carrier injection or through a generation process in the case of carrier extraction.

The injected carriers in the material will either be temporarily captured at trapping centres or lost permanently through recombination centres. As there are no perfect crystals, completely free of defects (or imperfections), in existence, materials always contain localized states which may be confined in smeared discrete levels or distributed in the forbidden energy gap. These localized states form so-called trapping and recombination centres. Therefore, trapping and recombination processes play one of the most important roles in double injection, photoconduction, and luminescence in solids (insulators and semiconductors), and in all solid state electron devices. This section is devoted to clarifying some physical concepts related to these processes.

For single injection presented in Section 3.2, we are concerned mainly with traps acting as trapping centres, but for double injection the so-called "traps" in the case of single injection may act as trapping (or simply traps) or as recombination centres. To avoid confusion, we start with a clear definition of these two centres. A trapping centre (or simply a trap) is a centre which captures a free carrier, and after a while this captured (or trapped) carrier has a greater probability of being thermally re-excited to the nearest allowed band to become a free carrier again than of recombining with a carrier of opposite sign at the centre. The trapping centres which capture electrons only are called electron traps, and those capturing holes only are hole traps. The occupancy of such centres is determined by the thermal equilibrium interchange of the carriers of one particular sign between the centres and the nearest allowed band. A recombination centre is a centre which also captures a free carrier, but the captured carrier has a greater probability of recombining with a carrier of opposite sign, resulting in the annihilation of both, than of being thermally re-excited to the nearest allowed band. The recombination centres in which the localized states are normally empty capture electrons first and then recombine with holes, while those in which the localized states are normally filled capture holes (or, in other words, give up electrons to the valence band) first and then recombine with electrons. The occupancy of such centres is governed by the kinetic recombination processes. A localized state may act as a trapping or a recombination centre depending on its location in the forbidden energy gap (governed by the nature of impurities and temperature), the concentrations of free electrons and holes, and the capture cross-sections for capturing electrons and holes. Thus, the distinction between a trapping and a recombination centre is a quantitative rather than a qualitative one.

In the steady state the rate of generation of electrons and holes must be equal to the rate of recombination, and the rate of trapping must be equal to the rate of detrapping (or re-excitation). To analyse carrier transport problems involving these processes, it is important to set a quantitative criterion to separate trapping centres and recombination centres. Rose [1955, 1963] has used the "demarcation levels" to separate them. The demarcation level for electron traps E_{Dn} is defined as the level at which a captured electron has equal probabilities of being excited into the conduction band and of recombining with a hole from the valence band. Similarly, the demarcation level for hole traps E_{Dp} is defined as the level at which a captured hole has equal probabilities of being excited into the valence band and of recombining with an electron from the conduction band. The localized states located between E_C and E_{Dn} will act predominantly as electron traps, those located between E_V and E_{Dp} will act predominantly as hole traps, and those located between E_{Dp} and E_{Dn} will act predominantly as recombination centres.

In the following we shall discuss the terms which are frequently used in describing the trapping and recombination processes. Figures 4.1 and 4.2 illustrate the physical concepts of these terms.

4.1.1. Capture rates and capture cross-sections

The electron capture rate is defined as the rate at which electrons are captured from the conduction band by traps following the equation

$$\frac{dn}{dt} = -C_n n N_n \tag{4.1}$$

where n is the free (or conduction) electron density in the conduction band, N_n is the density of empty electron traps, and C_n is the electron capture rate constant (or simply the electron capture coefficient). The capture cross-section of an electron trapping centre σ_n is defined as a cross-section, through which a moving electron has to come to the centre to be captured. Assuming that all the electrons have the same energy E and the same velocity v, then within a time Δt the volume of space which electrons pass through and will be captured is $\sigma_n v \Delta t \, n$, and the number of electrons per unit volume to be captured within a time Δt is

$$\Delta n = v \, \Delta t \, \sigma_n n N_n \left.\begin{matrix} \\ \\ \\ \\ \\ \end{matrix}\right\}$$

and

$$v = (2E/m_e^*)^{1/2}$$

(4.2)

From eqs. (4.1) and (4.2), we obtain

$$C_n = v \sigma_n \tag{4.3}$$

Experimentally, we do not measure directly σ_n, but rather measure C_n. The measured value of C_n is an average value of $C_n(E)$ taking into account the actual energy distribution of electrons. Thus, for a thermal equilibrium distribution we may write

$$C_n = \langle v \rangle \sigma_n = \langle v \sigma_n \rangle \tag{4.4}$$

where $\langle v \rangle$ is the average velocity of the carriers which is given by

$$\langle v \rangle = \frac{\displaystyle\int_0^\infty \left(\frac{2E}{m_e^*}\right)^{1/2} f(E)\, g(E)\, dE}{\displaystyle\int_0^\infty f(E)\, g(E)\, dE}$$

$$= \left(\frac{2}{m_e^*}\right)^{1/2} \frac{\displaystyle\int_0^\infty E \exp(-E/kT)\, dE}{\displaystyle\int_0^\infty E^{1/2} \exp(-E/kT)\, dE}$$

$$= (4kT/\pi m_e^*)^{1/2} \tag{4.5}$$

and

$$\sigma_n = \frac{\langle v \sigma_n(E) \rangle}{\langle v \rangle}$$

$$= \frac{\displaystyle\int_0^\infty v \sigma_n(E) f(E) g(E)\, dE \Big/ \int_0^\infty f(E) g(E)\, dE}{\displaystyle\int_0^\infty v f(E) g(E)\, dE \Big/ \int_0^\infty f(E) g(E)\, dE}$$

$$= \frac{\displaystyle\int_0^\infty E \sigma_n(E) \exp(-E/kT)\, dE}{\displaystyle\int_0^\infty E \exp(-E/kT)\, dE} \tag{4.6}$$

in which the electron energy E is measured from E_c (the edge of the conduction band). Since the thermal velocity of the carriers is given by

$$v_{\text{th}} = (3kT/m_e^*)^{1/2} \tag{4.7}$$

the electron capture cross-section most frequently quoted in the literature is the root-mean-square cross-section [Lax 1960]

$$\sigma_{n(\text{r.m.s.})} = C_n/(3kT/m_e^*)^{1/2} \tag{4.8}$$

Similarly, the hole capture rate constant (or simply the hole capture coefficient) C_p and the hole capture cross-section σ_p can be expressed as

$$C_p = \langle v \rangle \sigma_p = \langle v\sigma_p \rangle \tag{4.9}$$

$$\sigma_p = \frac{\langle v\sigma_p(E) \rangle}{\langle v \rangle} \tag{4.10}$$

where

$$\langle v \rangle = (4kT/\pi m_h^*)^{1/2} \tag{4.11}$$

4.1.2. Recombination rates and recombination cross-sections

Recombination occurs by (i) direct band-to-band recombination of free electrons and free holes without involving recombination centres, and by (ii) indirect recombination through recombination centres as a stepping-stone—free carriers of one type being captured first at the centres and then recombined with free carriers of opposite sign. Theoretically, both recombination mechanisms exist simultaneously, but in most cases the indirect recombination is predominant; the direct band-to-band recombination becomes important only when both electron and hole densities are high.

The direct band-to-band recombination rate R can be defined by the following equation:

$$\frac{dn}{dt} = \frac{dp}{dt} = -R = -C_r np \tag{4.12}$$

Similar to the expressions for C_n and C_p, we can write the direct band-to-band recombination rate constant C_r as

$$C_r = \langle v\sigma_R \rangle \tag{4.13}$$

where v in this case is the microscopic relative velocity of an electron and a hole, σ_R is their recombination cross-section. Thus the measured value of C_r is the average value of $v\sigma_R$ over the two velocity distributions.

For indirect recombination through a set of recombination centres of acceptor type, the centres will capture electrons first and then recombine with holes; and the rate of capturing electrons at the centres must be equal to the rate of capturing holes at the centres for recombination there. Thus the recombination rate is

$$R_a = \langle v\sigma_n \rangle n(N_{ra} - n_{ra})$$

$$= \langle v\sigma_p \rangle p n_{ra} \tag{4.14}$$

where N_{ra} and n_{ra} are, respectively, the densities of total acceptor-type recombination centres [including occupied (or filled) and unoccupied (or empty) localized states] and captured electrons (filled localized states). If the recombination centres are of donor-type, then the centres will capture holes first and then recombine with electrons there.

For this case, the recombination rate is

$$R_d = \langle v\sigma_n \rangle n n_{rd}$$
$$= \langle v\sigma_p \rangle p(N_{rd} - n_{rd}) \qquad (4.15)$$

where N_{rd} and n_{rd} are, respectively, the densities of total donor-type recombination centres [including occupied (or filled) and unoccupied (or empty) localized states] and captured holes (empty localized states).

It should be noted that in eq. (4.12) we have ignored the effect of thermally generated carriers $n_0 p_0 = n_i^2$ and that in eqs. (4.14) and (4.15) we have ignored the probability of thermal re-excitation of the captured carriers in the recombination centres to the nearest allowed band (instead of recombination with carriers of opposite sign). However, for large energy gap materials such as organic crystals, n_0 and p_0 are small and can be ignored without causing a great error for most cases. If the location of the recombination centres is far away from E_{Fn} for the acceptor-type or far away from E_{Fp} for the donor-type, so that the thermal re-excitation can be ignored, then eqs. (4.14) and (4.15) are valid. We shall come back to this topic in Section 4.2—Kinetics of recombination processes.

4.1.3. Demarcation levels

We first consider acceptor-type centres of density N_n located at an energy level E_t in the forbidden energy gap. In thermal equilibrium, if there are no other sinks to take the free electrons away, the rate of capturing free electrons by the empty centres is equal to the rate of thermally re-exciting the captured electrons from the occupied centres to the conduction band. Thus we can write

$$n \langle v\sigma_n \rangle (N_n - n_t) = n_t v_n \exp\left[-(E_c - E_t)/kT \right] \qquad (4.16)$$

where n_t is the trapped electron density and v_n is the attempt-to-escape frequency which, in a classical physical concept, represents a number of times per second a captured electron attempts to absorb sufficient energy from the lattice vibration and to surmount the potential barrier of the trap. By expressing n, n_t, and $(N_n - n_t)$ as

$$n = N_c \exp\left[-(E_c - E_{Fn})/kT \right] \qquad (4.17)$$
$$n_t = N_n \{ 1 + g_n^{-1} \exp\left[(E_t - E_{Fn})/kT \right] \}^{-1} \qquad (4.18)$$
$$N_n - n_t = N_n \{ 1 + g_n \exp\left[(E_{Fn} - E_t)/kT \right] \}^{-1} \qquad (4.19)$$

substitution of eqs. (4.17)–(4.19) into eq. (4.16) yields

$$v_n = g_n^{-1} N_c \langle v\sigma_n \rangle \qquad (4.20)$$

The rate of thermal re-excitation of captured electrons is

$$g_n^{-1} n_t N_c \langle v\sigma_n \rangle \exp\left[-(E_c - E_t)/kT \right]$$

To derive an expression for the electron demarcation level E_{Dn} we assume $g_n = 1$ for simplicity and replace E_t with E_{Dn} for $E_t = E_{Dn}$, and then set the rate of thermal reexcitation of captured electrons to the conduction band equal to the rate of recombination of these captured electrons with free holes from the valence band. Thus we can write

$$n_t N_c \langle v\sigma_n \rangle \exp[-(E_c - E_{Dn})/kT] = n_t p \langle v\sigma_p \rangle \tag{4.21}$$

From eqs. (4.17) and (4.21) we obtain

$$E_{Dn} = E_{Fn} + kT \ln\left(\frac{p\sigma_p}{n\sigma_n}\right) \tag{4.22}$$

Using the same argument, if the same localized states are located below the middle of the forbidden energy gap, then these states may act as hole traps, and we can set the rate of thermal re-excitation of captured holes to the valence band equal to the rate of recombination of these captured holes $(N_n - n_t)$ with free electrons from the conduction band.

$$(N_n - n_t)N_v \langle v\sigma_p \rangle \exp[-(E_{Dp} - E_v)/kT] = (N_n - n_t)n \langle v\sigma_n \rangle \tag{4.23}$$

Thus we obtain

$$E_{Dp} = E_{Fp} + kT \ln\left(\frac{p\sigma_p}{n\sigma_n}\right) \tag{4.24}$$

Derivation of eqs. (4.22) and (4.24) is based on the assumption that $\langle v\sigma_n \rangle = v\sigma_n$ $\langle v\sigma_p \rangle = v\sigma_p$.

There is one set of demarcation levels (E_{Dn} and E_{Dp}) for a particular type of imperfection (or for one set of localized states) characterized by a particular set of capture cross-sections (σ_n and σ_p) as shown in Fig. 4.1. From eqs. (4.22) and (4.24) we have

$$E_{Dn} - E_{Fn} = E_{Dp} - E_{Fp} \tag{4.25}$$

Supposing that electron traps distributed exponentially between E_{te1} and E_{te2} within the forbidden energy gap following the equation

$$h(E) = \frac{H_b}{kT_c} \exp[-(E_{te1} - E)/kT_c] \tag{4.26}$$

as shown in Fig. 4.1(a); these traps will act as electron traps for single injection, but their behaviour will be quite different for double injection. In the following we shall show their behaviour under various conditions based on eqs. (4.22), (4.24), and (4.25).

(a) When the injected carrier densities (injected by either optical or electrical means) exceed the thermally generated carrier densities such as in most insulators ($n > n_0$ and $p > p_0$), the carrier densities n and p are generally expressed in terms of quasi-Fermi levels E_{Fn} and E_{Fp} under such non-equilibrium conditions. If the injected carrier densities are of such values that $n\sigma_n = p\sigma_p$, then the electron demarcation level E_{Dn} coincides with E_{Fn}, and the hole demarcation level E_{Dp} coincides with E_{Fp} as shown in Fig. 4.1(b). If the injected carrier densities are of such values that $n\sigma_n < p\sigma_p$, then E_{Dn} separates from E_{Fn} and is located above it by an amount of $kT \ln(p\sigma_p/n\sigma_n)$; and from eq. (4.25) E_{Dp} is located above E_{Fp} by the same amounts as shown in Fig. 4.1(c).

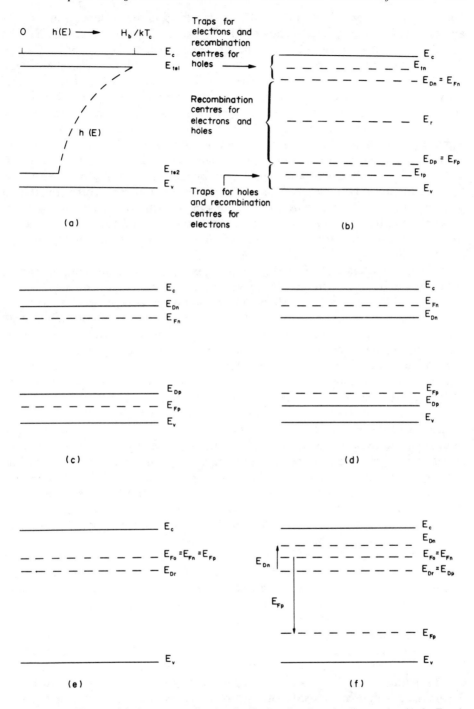

F<small>IG</small>. 4.1. (a) Exponential electron trap density distribution function; (b) demarcation levels, Fermi levels, energy levels for trapping and recombination centres for $n\sigma_n = p\sigma_p$; (c) for $n\sigma_n < p\sigma_p$; (d) for $n\sigma_n > p\sigma_p$; (e) for an n-type semiconductor with $n_0 > \Delta n$, $p_0 > \Delta p$, and $n_0 > p_0$; and (f) for an n-type semiconductor with $n_0 > \Delta n$, $p_0 < \Delta p$, and $n_0 > p_0$.

Similarly, if $n\sigma_n > p\sigma_p$, E_{Dn} and E_{Dp} will be, respectively, located below E_{Fn} and E_{Fp} by an amount of $kT\ln(n\sigma_n/p\sigma_p)$ as shown in Fig. 4.1(d). In the interval between E_{Dn} and E_{Dp} the occupancy of the localized states [the so-called "traps" in Fig. 4.1(a)] is determined by the kinetic recombination processes and, therefore, the localized states within this interval act as recombination centres. The occupancy of the localized states above E_{Dn} is determined by E_{Fn}, and they act as electron-trapping centres; while the occupancy of the localized states below E_{Dp} is determined by E_{Fp} and they act as hole traps. It should be emphasized that each set of recombination centres has its own set of demarcation levels, and that when there are two sets of recombination centres due to two different types of imperfections, each characterized by its own pair of capture cross-sections, σ_n and σ_p, the demarcation levels are displaced independently for each set from the common quasi-Fermi levels E_{Fn} and E_{Fp} [Rose 1963]. It can also be seen that in single injection the electron traps (e.g. acceptor-type) distributed as shown in Fig. 4.1(a) act as electron traps, but in double injection only the traps located above E_{Dn} can be considered as electron traps; the rest act as recombination centres between E_{Dn} and E_{Dp}, and act as hole traps between E_{Dp} and E_v.

(b) With the increase of carrier injection either by increasing the light intensity in optical excitation or by increasing the applied field in double electric contact injection, both E_{Fn} and E_{Fp}, E_{Dn} and E_{Dp}, will be shifted towards the band edges and, therefore, some localized states acting as traps will be transformed into recombination centres. Such a process in producing recombination centres is sometimes referred to as "electronic doping" because it involves electronic excitation or injection.

(c) An increase in temperature will shift the quasi-Fermi levels and hence the demarcation levels away from the band edges, and so some localized states acting as recombination centres will be transformed into trapping centres.

(d) When the injected carrier densities are smaller than the thermally generated carrier densities, then n_0 and p_0 are predominant in the conduction and valence bands. In this case the quasi-Fermi levels coincide with each other $E_{Fn} = E_{Fp} = E_{F0}$, and the demarcation levels also coincide with each other $E_{Dn} = E_{Dp} = E_{Dr}$ as shown in Fig. 4.1(e) for the n-type semiconductors. By considering E_{Dr} as E_{Dp}, all electron-occupied states located between E_{Dr} and E_{F0} act mainly as recombination centres for holes and those above E_{F0} as electron traps and below E_{Dr} as hole traps. But by considering E_{Dr} as E_{Dn}, then the unoccupied localized states located above E_{Dr} act as electron traps and those below E_{Dr} as recombination centres for electrons.

(e) If the injected electron density is still smaller than the thermally generated electron density n_0 but the injected hole density is larger than the thermally generated hole density p_0, then E_{Fn} remains practically at E_{F0} but E_{Fp} separates from E_{F0} and moves towards E_v. At the same time, E_{Dp} remains practically at E_{Dr} but E_{Dn} moves towards E_c at the same rate as E_{Fp} moves towards E_v as shown in Fig. 4.1(f).

(f) If the injected electron density is much larger than n_0 and p_0, and there is no hole injection such as single injection (electron injection) into an insulator, then the term $kT\ln(n\sigma_n/p\sigma_p)$ becomes very large and much larger than E_g and E_{Dn}; E_{Dp} and E_{Fp} will disappear in the forbidden energy gap. In this case the traps located above E_{Fn} are shallow traps and those below E_{Fn} are deep traps. There are no recombination centres.

(g) It should be noted that E_{Fn}, E_{Fp}, E_{Dn}, and E_{Dp} are functions of distance from the injecting contacts, which are not shown in Fig. 4.1 for clarity.

4.1.4. Coulombic traps

Trapping can be considered as a process of energy storage by localizing spatially electrons and holes at certain sites so as to hinder them from moving freely (this means to stop them contributing to electric conduction). These captured electrons and holes may be released to be free again by absorbing sufficient thermal or optical energy, or be lost through recombination by giving up their stored energy. Any centres formed by localized states capable of capturing carriers are called "traps" or "trapping centres". After the capture of a carrier the subsequent action determines whether the trap acts as a trapping centre or as a recombination centre, and this has been discussed in Section 4.1.3.

A trap can be considered as an entity with a certain charge, positive, neutral, or negative, when empty or unoccupied. With M to represent one trap, e to represent an electron, h to represent a hole, and the superscript to denote the charge, $(-)$ negative, (0) neutral, and $(+)$ positive, Table 4.1 shows that the behaviour of the traps can be grouped into three types, namely: (a) Coulombic attractive centres, (b) Coulombic neutral centres, and (c) Coulombic repulsive centres. The variation of potential energy of these three types of Coulombic traps is shown schematically in Fig. 4.2. In general, the trap densities in solids range from about 10^{12} cm^{-3} in the highly pure single crystals to about 10^{19} cm^{-3} in the imperfect large band gap insulators; the capture cross-sections range from about 10^{-11} cm^2 for Coulombic attractive centres to about or less than 10^{-21} cm^2 for Coulombic repulsive centres; and the carrier lifetimes based on a carrier thermal velocity of 10^7 cm sec^{-1} could range from 10^2 sec to 10^{-14} sec. The behaviour of these three types of Coulombic traps for inorganic semiconductors has been discussed in detail in Milnes's book [1973].

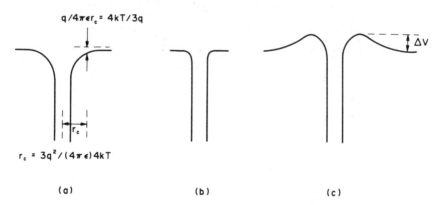

$q/4\pi\epsilon r_c = 4kT/3q$

$r_c = 3q^2/(4\pi\epsilon)4kT$

(a)   (b)   (c)

FIG. 4.2. Schematic diagrams illustrating the variation of potential energy of (a) Coulombic attractive centre, (b) Coulombic neutral centre, and (c) Coulombic repulsive centre.

The capture cross-section of a centre is determined by the variation of potential energy in the vicinity of the centre. For attractive centres we can assume that a free carrier will be captured when it approaches to the centre at a distance r from the centre having a drift velocity

$$v_d = F\mu = \frac{q\mu}{4\pi\varepsilon r^2} \qquad (4.27)$$

TABLE 4.1. *The trapping and detrapping processes for three types of traps*

Coulombic trapping centre	Type of carriers to be trapped	Trapping and detrapping processes		Remarks
		Charge before trapping (trap unoccupied)	Charge after trapping (trap filled)	
Attractive	Electron (e)	$M^+ + e \;\xrightleftharpoons[\text{Detrapping}]{\text{Trapping}}\; M^0$		Deep donors with compensation for electron traps. The detrapping process is known as the Poole–Frenkel effect
Attractive	Hole (h)	$M^- + h \;\xrightleftharpoons[\text{Detrapping}]{\text{Trapping}}\; M^0$		Deep acceptors with compensation for hole traps. The detrapping process is known as the Poole–Frenkel effect
Neutral	Electron (e)	$M^0 + e \;\xrightleftharpoons[\text{Detrapping}]{\text{Trapping}}\; M^-$		The field dependence of the detrapping process for electrons or for holes is small
Neutral	Hole (h)	$M^0 + h \;\xrightleftharpoons[\text{Detrapping}]{\text{Trapping}}\; M^+$		
Repulsive	Electron (e)	$M^- + e \;\xrightleftharpoons[\text{Detrapping}]{\text{Trapping}}\; M^{--}$		Double donors or double acceptors with one level compensated
Repulsive	Hole (h)	$M^+ + h \;\xrightleftharpoons[\text{Detrapping}]{\text{Trapping}}\; M^{++}$		

The average diffusion velocity of a particle executing Brownian motion at a distance r from the centre is given by

$$\bar{v} = \frac{2\lambda}{3r}v \tag{4.28}$$

By equating v_d and $\bar{v}$, we obtain the radius of the attractive centre [Rose 1963]

$$r_c = \frac{3q\mu}{(4\pi\varepsilon)2\lambda v} \tag{4.29}$$

Since

$$v = \lambda/\tau \qquad \mu = q\tau/m^* \qquad kT = m^*v^2/2 \tag{4.30}$$

the capture cross-section can be written as

$$(\sigma)_{\text{attractive}} = \pi r_c^{\,2}$$

$$= \frac{9\pi q^4}{(4\pi\varepsilon)^2\, 16(kT)^2} \tag{4.31}$$

where λ is the carrier mean free path between collisions for energy loss, τ is the carrier lifetime, and v is the carrier thermal velocity. If $\lambda < r_c$, many collisions can occur within the critical radius, and the interaction between the centre and the carrier is diffusion limited [Lax 1960]. Equations (4.29) and (4.31) are derived on the basis of this assumption. For this condition $(\sigma)_{\text{attractive}} \propto (\varepsilon)^{-2}(T)^{-2}$ and in most materials with a large band gap and a low carrier mobility $\lambda < r_c$, such as in organic semiconductors. However, for silicon and most inorganic semiconductors $\lambda > r_c$, then eq. (4.31) is not applicable but can be used as a guide by multiplying it by the ratio of $2r_c/\lambda$ [Rose 1963]. Thus we have

$$\underset{(\lambda > r_c)}{(\sigma)_{\text{attractive}}} \simeq \frac{2r_c}{\lambda}(\pi r_c^2) \tag{4.32}$$

This implies that the capture cross-section is roughly proportional to r_c^3 and T^{-2} if λ can be assumed to be proportional to μ which is proportional to $T^{-3/2}$. For $\lambda > r_c$ it can be imagined that a free carrier may pass through the centre a number of times before being captured. This implies that the effective capture cross-section is reduced. If such an attractive centre is not deformed by the field, the radius of the centre is approximately proportional to $F^{-1/2}$ so that its capture cross-section should decrease with field as $F^{-3/2}$ [Dussel and Bube 1966].

It should be noted that the rapid increase in capture cross-section with decreasing temperature may be associated with a large capture rate into highly excited states followed by a cascade process in which a certain fraction of the captured carriers reaches the ground state [Lax 1960].

For neutral centres the polarizability of the centre provides a quasi-long-range interaction with the carrier. A charge q at a distance r from the centre with polarizability α will produce a dipole moment of $\alpha q/4\pi\varepsilon r^2$. This dipole will in turn produce an attractive force of $2(\alpha q/4\pi\varepsilon r^2)(q/4\pi\varepsilon r^3)$ on the charge so that an attractive potential is

$$V(r) = -A/r^4 \tag{4.33}$$

where A is a constant equal to $\alpha q^2/2(4\pi\varepsilon)^2$. Different centres (different impurities) in the same host lattice or the same impurities in different host lattices will have different values of α. Since the radius r at which the potential energy is kT varies with $T^{-1/4}$, the capture cross-section will be less sensitive to temperature, and since the capture cross-section is much smaller than that of the attractive centre, it is more sensitive to the potential profile at short ranges so that it is more sensitive to the chemical nature of the centre [Lax 1960].

For repulsive centres, either $M^- + e \to M^{--}$ or $M^+ + h \to M^{++}$, the capturing action is repulsive. An electron approaching to a centre which has already captured an electron will see a repulsive potential barrier that it must surmount thermally or tunnel through before being captured. In Fig. 4.2(c) it can be seen that when the free electron reaches the potential top (ΔV) it sees an attractive potential field. Such a trapping process indicates that the capture cross-section of the repulsive centre would be very small and very sensitive to temperature. The increase in electric field will cause an increase in number of high energy electrons to overcome the repulsive potential barrier and to be trapped. This phenomenon is generally referred to as "field-enhanced trapping" which produces a negative differential resistance region and current oscillations in n-type GaAs [Ridley and Watkins 1961, Law and Kao 1970, Elsharkawi and Kao 1973].

In general, $\sigma_{\text{attractive}} > \sigma_{\text{neutral}} > \sigma_{\text{repulsive}}$. When an electron is captured it has to lose its energy which is carried away by (a) a photon-radiative capture or recombination, or (b) phonons (optical and acoustic phonons)—non-radiative capture or recombination, or (c) another electron or hole—Auger recombination. We shall discuss these loss processes in Section 4.2.

4.1.5. Free electron–free hole recombination

It is not necessary to separate direct and indirect band-to-band transition to determine the recombination constant, since this depends only on the total absorption coefficient due to band-to-band transitions. The simple van Roosbroeck–Shockley relation [van Roosbroeck and Shockley 1954] states that in thermal equilibrium the rate of optical generation of electron–hole pairs is equal to their rate of recombination. Thus for the frequency interval dv we can write

$$R(v)dv = P(v)\rho(v)dv \tag{4.34}$$

where $P(v)$ is the probability per unit time for absorbing a photon of energy hv and $\rho(v)dv$ is the density of photons of frequency v in the interval dv. The latter, according to Planck's black-body radiation law, is given by

$$\rho(v)dv = \frac{8\pi v^2 n_c^3}{c^3} \frac{1}{\exp(hv/kT)-1} dv \tag{4.35}$$

and the former is given by

$$P(v) = \frac{1}{\tau(v)} = \alpha(v)v = \alpha(v)c/n_c \tag{4.36}$$

where c is the light velocity in vacuum, $\tau(v)$ is the mean lifetime of the photon in the solid, n_c is the index of refraction which is assumed to be independent of v, and $\alpha(v)$ is the absorption coefficient $[1/\alpha(v)$ is the mean free path of a photon in the solid] which is given by

$$\alpha(v) = 4\pi v k_c(v)/c \tag{4.37}$$

in which $k_c(v)$ is the extinction coefficient (it is the imagined part of the complex index of refraction $n_c^* = n_c - jk_c$). From eqs. (4.35) and (4.36), eq. (4.34) can be written as

$$R(v)\,dv = \frac{32\pi^2 n_c^2 k_c(v) v^3}{c^3 [\exp(hv/kT - 1)]}\,dv \tag{4.38}$$

This equation gives the fundamental relation between the absorption spectrum and the emission spectrum. To obtain the total recombination rate, we introduce the parameter

$$\mu = hv/kT$$

$$= \frac{h}{kT}\left[\frac{c\alpha(v)}{4\pi k_c(v)}\right] \tag{4.39}$$

Thus, integration of eqn. (4.38) over all photon frequencies gives

$$R = \frac{8\pi n_c^2 (kT)^3}{h^3 c^2}\int_0^\infty \frac{\alpha(v)\mu^2}{e^\mu - 1}\,d\mu \tag{4.40}$$

Roosbroeck and Shockley [1954] and Varshni [1967] have used eqs. (4.38) and (4.40) to calculate the recombination rate of germanium.

In thermal equilibrium, $np = n_0 p_0 = n_i^2$, and from eq. (4.12), we obtain the recombination rate constant

$$\langle v\sigma_R \rangle = \frac{8\pi n_c^2 (kT)^3}{h^3 c^2 n_i^2}\int_0^\infty \frac{\alpha(v)\mu^2}{e^\mu - 1}\,d\mu \tag{4.41}$$

Using a similar approach and assuming $v \simeq 10^7$ cm sec^{-1} Rose [1963] and Bube [1974] have estimated the recombination cross-section σ_R to be about 10^{-19} cm^2. It should be noted that R is proportional to np. Supposing that $n = n_0 + \Delta n$ and $p = p_0 + \Delta p$ then the recombination rate

$$R + \Delta R = \frac{(n_0 + \Delta n)(p_0 + \Delta p)}{n_0 p_0}R \tag{4.42}$$

For small signals, $n_0 > \Delta n$, $p_0 > \Delta p$, and $\Delta n = \Delta p$, we have

$$\frac{\Delta R}{R} = \frac{\Delta n}{n_0} + \frac{\Delta p}{p_0} = \frac{\Delta n(n_0 + p_0)}{n_0 p_0} \tag{4.43}$$

Thus the lifetime of the excess carriers is

$$\tau = \frac{\Delta n}{\Delta R} = \frac{1}{R}\frac{n_0 p_0}{n_0 + p_0} = \frac{1}{\langle v\sigma_R \rangle(n_0 + p_0)} \tag{4.44}$$

It can be seen that τ varies from $[\langle v\sigma_R \rangle(2n_i)]^{-1}$ to about 10^{-6} sec when $n_0 \gg p_0$ or $p_0 \gg n_0$ (for example, $n_0 = 10^6 p_0$ or $p_0 = 10^6 n_0$ but $n_0 p_0 = n_i^2$). For very large signals

$\Delta n \gg n_0$, $\Delta p \gg p_0$, and $\Delta n = \Delta p$. We have

$$\tau = \frac{1}{\langle v\sigma_R \rangle (n_0 + p_0 + \Delta n)}$$

$$\simeq \frac{1}{\langle v\sigma_R \rangle \Delta n} \tag{4.45}$$

for this case, if $\Delta n = \Delta p = 10^{20}$ cm^{-3} and τ may reach 10^{-8} sec. Therefore, for high carrier concentrations in solids free electron–free hole recombination should be able to compete with the indirect recombination through localized recombination centres.

4.1.6. Characteristic times

Rose [1963] has discussed the following four characteristic times:

τ carrier lifetime
τ_d dielectric relaxation time
t_t carrier transit time
τ_r response time

and pointed out their important role in the trapping and recombination processes. These characteristic times have been used to describe a number of various quantities, and it is therefore necessary to have a clear definition and physical concept for these terms.

The carrier lifetime (or simply the lifetime) is generally referred to as the time during which a charge carrier is free to move and so to contribute to electric conduction. We can define τ_n as the time that an excited electron spends in the conduction band and τ_p as the time that an excited hole spends in the valence band. Supposing that a uniform excitation generates G electron–hole pairs per second per unit volume in a solid, then the generated electron and hole densities in the conduction band and valence band are, respectively,

$$\Delta n = G\tau_n \tag{4.46}$$

$$\Delta p = G\tau_p \tag{4.47}$$

If these carriers are trapped and then thermally re-excited, the time spent in the traps is not included in τ_n and τ_p. In the steady state the rate of generation is equal to the rate of trapping. Thus we have

$$\tau_n = \frac{1}{\langle v\sigma_n \rangle (N_r - n_r)} \tag{4.48}$$

$$\tau_p = \frac{1}{\langle v\sigma_p \rangle n_r} \tag{4.49}$$

where N_r and n_r are respectively, the total (occupied and unoccupied) and the occupied trapping or recombination centres. These equations are valid provided that the carrier mean free path is larger than the diameter of the capture cross-section $\left(2\sqrt{\sigma_n/\pi}\right.$ or $2\sqrt{\sigma_p/\pi}\left.\right)$. For small mean free paths, τ_n or τ_p will be increased by a factor of the order of the ratio of half the spacing between capturing centres to the mean free path [Rose 1963].

τ_n or τ_p is constant if $\Delta n \propto G$ or $\Delta p \propto G$. This implies that n and p are smaller than the density of trapping or recombination centres. This is true for most insulators in which the density of localized states is usually greater than 10^{15} cm^{-3} and n or p less than 10^{15} cm^{-3}. For this case the photoconductivity is linear with light intensity. However, in some materials and under certain conditions Δn or Δp may vary with G^a. Then the photoconductivity is said to be superlinear if $a > 1$ and sublinear if $a < 1$. For such cases τ_n and τ_p are no longer constant but depend also on G.

There are several "lifetime" terms used frequently in the literature as follows [Bube 1960, 1974]:

Free lifetime—the lifetime of a free carrier excluding any time spent by the carrier in the traps.

Excited lifetime—the total lifetime of an excited carrier including both the free lifetime and the time spent in the traps (trapping time or capturing time), or in other words, the total time between the action of excitation and the action of recombination.

Minority carrier lifetime—the free lifetime of a minority carrier, electron or hole, present in lower density.

Majority carrier lifetime—the free lifetime of a majority carrier, electron, or hole, present in higher density. If free carrier densities n and p are much larger than the density of recombination centres, the majority carrier lifetime is equal to the minority carrier lifetime. If n and p are smaller than the density of recombination centres such as in most insulators, the majority carrier lifetime is much large than the minority carrier lifetime.

Electron–hole pair lifetime—the free lifetime of the carrier (usually the minority carrier) first captured in traps.

Diffusion-length lifetime (or recombination lifetime), τ_0— based on the relation $\tau_0 = L_0^2/D_0$, where D_0 is the ambipolar diffusion coefficient and L_0 is the diffusion length.

In Section 3.2.1(A) we have discussed the physical meaning of τ_d. The dielectric relaxation time ($\tau_d = \varepsilon/q\mu_n n$ for electrons and $\tau_d = \varepsilon/q\mu_p p$ for holes) is defined as the time necessary for the re-establishment of quasi-neutrality after an injection of carriers into the solid. The transit time t_t is defined as the time required for a carrier to travel across a specimen, which includes the total time spent as a free carrier and the total time spent as a trapped carrier in the traps during the transit. It is obvious that only for $\tau_d < t_t$ and $\tau_d < \tau_0$ can the condition of local space charge neutrality be used as a good approximation for electric transport calculations.

The response time is defined as the time required for the photocurrent to reach a steady state value (or an appropriate fraction of the steady state value such as $1 - 1/e$) after the light excitation is switched on. This is also the time required for the photocurrent to decay to the same fraction ($1/e$) of its steady state value after the light excitation is switched off. The response times are given by

$$\tau_{rn} = \left(1 + \frac{n_t}{n}\right)\tau_n \qquad (4.50)$$

$$\tau_{rp} = \left(1 + \frac{p_t}{p}\right)\tau_p \qquad (4.51)$$

It can be seen that (a) if there are no trapping or recombination centres, or the free carrier densities n and p are much larger than the density of trapping or recombination centres the response time is equal to the carrier lifetime; (b) if n and p are comparable or less than the density of trapping or recombination centres, $\tau_{rn} > \tau_n$ and $\tau_{rp} > \tau_p$. Therefore from the expressions for n_t and p_t [eqs. (4.129) and (4.130)] a temperature rise or an increase of light intensity may reduce the difference between the response time and the carrier lifetime. $\tau_r > \tau$ because the carrier injection has to supply not only carriers into the bands for electric conduction, but also to pour carriers to trapping centres; and when the light excitation is switched off, the time is required not only for the free carriers to be recombined in the recombination centres, but also for the trapped carriers to be detrapped and then recombined in the recombination centres via thermal excitation and subsequent capture and recombination processes.

4.2. KINETICS OF THE RECOMBINATION PROCESSES

The recombination processes by which free carriers recombine can be classified in several ways. The most convenient way is to divide them into two major divisions, namely, (a) radiative transition—the released energy is emitted as radiation which can lead to luminescence, and (b) non-radiative transition—the released energy is dissipated ultimately as heat by various mechanisms. Each division can then be subdivided into several groups according to whether the recombination is direct band-to-band or through an imperfection, as follows.

(a) RADIATIVE TRANSITION

(i) Band-to-band recombination by collision between a free electron and a free hole. There are two possibilities: (1) Direct band-gap material—the energy band extremes (the minimum of the conduction band and the maximum of the valence band) occur at the same value of k (wave vector) in the $E-k$ space. Since the momentum of the electron does not change between the initial and the final state of the electron, the transition probability is high and the mean free lifetime of injected carriers is very short. (2) Indirect band-gap materials—the energy band extremes occur at different values of k. For this case the principle of conservation of momentum requires phonon participation in the transition process as shown in Fig. 4.3.
(ii) Recombination through a recombination centre.
(iii) Recombination between two localized centres.
(iv) Annihilation of excitons.

(b) NON-RADIATIVE TRANSITION

(v) The released energy is dissipated as a shower of phonons without the involvement of localized centres (free electron–free hole recombination). This is usually considered to be a highly unlikely event.

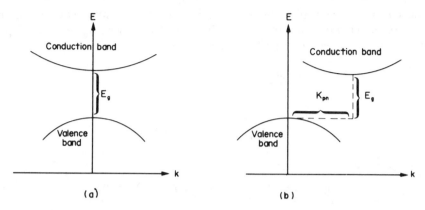

F<small>IG</small>. 4.3. Radiative transition via (a) direct, and (b) indirect transition. In (a) the radiative frequency is $\omega = E_g/\hbar$, but in (b) it is $E_g/\hbar - \omega(K_{pn})$, where K_{pn} is the wave vector of the phonon. The frequency of the emitted light or the photon energy is less than that corresponding to E_g, the difference $\hbar\omega(K_{pn})$ being used to create the phonon to take up the difference in crystal momentum.

(vi) The released energy is dissipated as phonons with the involvement of localized centres.

(vii) Auger or impact recombination which can be intrinsic involving electron–electron collision or hole–hole collision, or extrinsic involving electron impurity (or other imperfection) collision.

(viii) Recombination through surface states.

The recombination processes may involve one body, two bodies, and three bodies. The so-called "bodies" may mean the "quasi-particles" which obey the Fermi–Dirac statistics, such as electrons and holes. The recombination which involves one free carrier at a time, such as the indirect recombination through a recombination centre (for example, an electron captured by a recombination centre and then recombined with a hole, each process involving only one carrier), is generally referred to as the monomolecular recombination. The recombination which involves two free carriers simultaneously, such as direct band-to-band recombination, is generally referred to as the bimolecular recombination. The recombination which involves three free carriers simultaneously such as a three-body collision in the Auger intrinsic recombination in which one electron in the conduction band recombines with a hole in the valence band and the energy released is taken up by a third particle electron is generally referred to as the trimolecular recombination or three body recombination (or simply as Auger or impact recombination).

In the following we shall discuss the kinetics of these recombination processes.

4.2.1. Without recombination centres and without traps
(band-to-band recombination)

An intrinsic semiconductor or insulator may have excess carriers due to thermal excitation or external stimulation (e.g. optical or electrical injections). Such excess carriers will disappear through either recombination or carrier flow to the collecting contacts. If there are no collecting contacts, the only way to limit the excess carriers is through recombination. Thus the rates of the change of carrier densities due to optical excitation with a generation rate G can be written as

$$\frac{dn}{dt} = \frac{dp}{dt} = G - R$$

$$= G - \langle v\sigma_R \rangle np \qquad (4.52)$$

$$= G - C_r np$$

where

$$n = n_0 + \Delta n, \qquad p = p_0 + \Delta p, \qquad \Delta n = \Delta p \qquad (4.53)$$

Δn and Δp are the excess carrier densities generated by optical excitation. We shall consider the following three cases.

Case 1. If $\Delta n \gg n_0$ and $\Delta p \gg p_0$, then the thermally generated carriers can be neglected and $n = p$. Thus we have:
In equilibrium:

$$\frac{dn}{dt} = 0 \qquad (4.54)$$

$$(\Delta n)_0 = (G/C_r)^{1/2} \qquad (4.55)$$

Rate of growth:

$$\frac{dn}{dt} = G - C_r n^2 \qquad (4.56)$$

$$\Delta n = (\Delta n)_0 \tanh[(GC_r)^{1/2} t] \qquad (4.57)$$

Rate of decay:

$$\frac{dn}{dt} = -C_r n^2 \qquad (4.58)$$

$$\Delta n = (\Delta n)_0 [1 + (\Delta n)_0 C_r t]^{-1} \qquad (4.59)$$

This case is a typical example of bimolecular recombination. If n_0 and p_0 are taken into account, the bimolecular recombination rate is determined not by eqs. (4.55), (4.57), and (4.59), but by equations given in cases 2 and 3 below.

Case 2. If $\Delta n \ll n_0$ and $\Delta p \ll p_0$, then the thermally generated carriers are predominant and $n \simeq n_0$ and $p \simeq p_0 \simeq n_i^2/n_0$. For this case we have:
In equilibrium:

$$\frac{dn}{dt} = G - C_r(n_0 + \Delta n)(p_0 + \Delta p) = 0$$

$$(\Delta n)_0 = \frac{G - C_r n_i^2}{C_r(n_0 + p_0)} \qquad (4.60)$$

Rate of growth:

$$\frac{dn}{dt} = G - C_r(n_0 + \Delta n)(p_0 + \Delta p)$$

or
$$\frac{d(\Delta n)}{dt} = G - C_r[n_i^2 + \Delta n(n_0 + p_0)]$$

$$\Delta n = (\Delta n)_0[1 - \exp(-t/\tau)] \tag{4.61}$$

and
$$\tau = [C_r(n_0 + p_0)]^{-1}$$

Rate of decay:

$$\frac{d(\Delta n)}{dt} = -C_r \Delta n(n_0 + p_0)$$

$$\Delta n = (\Delta n)_0 \exp(-t/\tau) \tag{4.62}$$

Case 3. Both Δn and n_0, and Δp and p_0 play equally important roles in the recombination processes. For this case we have:

In equilibrium:

$$\frac{dn}{dt} = G - C_r(n_0 + \Delta n)(p_0 + \Delta p)$$

$$= G - C_r[n_0 p_0 + \Delta n(n_0 + p_0) + (\Delta n)^2]$$

$$= 0$$

$$\left.\begin{aligned}
(\Delta n)_0 &= \frac{2(G - C_r n_i^2)}{C_r(n_0 + p_0) + A} \\[2mm]
(\Delta n)_0 &= \frac{G - C_r n_i^2}{C_r(n_0 + p_0)}
\end{aligned}\right\} \tag{4.63}$$

if $C_r^2(n_0 + p_0)^2 \gg 4C_r(G - C_r n_i^2)$

$$(\Delta n)_0 = \left(\frac{G - C_r n_i^2}{C_r}\right)^{1/2}$$

if $C_r^2(n_0 + p_0)^2 \ll 4C_r(G - C_r n_i^2)$

Rate of growth:

$$\frac{d(\Delta n)}{dt} = (G - C_r n_i^2) - C_r(n_0 + p_0)\Delta n - C_r(\Delta n)^2$$

$$(\Delta n) = (\Delta n)_0 \left\{ \frac{C_r(n_0 + p_0) + A}{C_r(n_0 + p_0) + A \coth(At/2)} \right\} \tag{4.64}$$

Rate of decay:

$$\frac{d(\Delta n)}{dt} = -C_r[n_i^2 + (n_0 + p_0)\Delta n + (\Delta n)^2]$$

$$(\Delta n) = (\Delta n)_0 \left\{ \frac{(n_0 + p_0)\exp(-t/\tau)}{n_0 + p_0 + (\Delta n)_0[1 - \exp(-t/\tau)]} \right\} \tag{4.65}$$

where

$$A = [C_r^2(n_0 + p_0)^2 + 4C_r(G - C_r n_i^2)]^{1/2}$$

$$\tau = [C_r(n_0 + p_0)]^{-1}$$

From eq. (4.65) it can be seen that when $(\Delta n)_0 < n_0 + p_0$ the decay is virtually exponential throughout its course. Even if $(\Delta n)_0 > n_0 + p_0$, only the initial decay is hyperbolic when $t < \tau$, and it eventually becomes exponential for $t > \tau$. It is of interest to note that no matter how large $(\Delta n)_0$ may be, the hyperbolic decay will rapidly bring (Δn) down to a value less than $(n_0 + p_0)$ within the time interval τ.

4.2.2. With a single set of recombination centres but without traps

Shockley and Read [1952] were the first to analyse in detail the recombination kinetics for semiconductors with one type of recombination centres but without traps based on the basic processes shown in Fig. 4.4. The centre is assumed to be neutral when empty and negatively charged when occupied by an electron. The four processes are: (a) the rate of electron capture by neutral centres of density N_r with capture coefficient C_n is $C_n n(N_r - n_r)$, (b) the rate of thermal re-excitation of the captured electrons from the centres to the conduction band is $C_n n_r N_c \exp[-(E_c - E_r)/kT]$ based on eqs. (4.16) and (4.20), (c) the rate of hole capture by the electron-occupied centres with capture coefficient C_p leading to recombination is $C_p p n_r$, and (d) the rate of thermal excitation of holes from the neutral centres to the valence band (thermal excitation of electrons from the valence band to the unoccupied centres) is $C_p(N_r - n_r)N_v \exp[-(E_r - \dot{E}_v/kT]$. Thus the net rate of electron capture by the recombination centre is

$$\frac{dn_r}{dt} = C_n\{n(N_r - n_r) - n_r N_c \exp[-(E_c - E_r)/kT.]\} \tag{4.66}$$

Similarly, the net rate of hole capture by the recombination centres is

$$\frac{d(N_r - n_r)}{dt} = C_p\{p n_r - (N_r - n_r)N_v \exp[-(E_r - E_v)/kT]\} \tag{4.67}$$

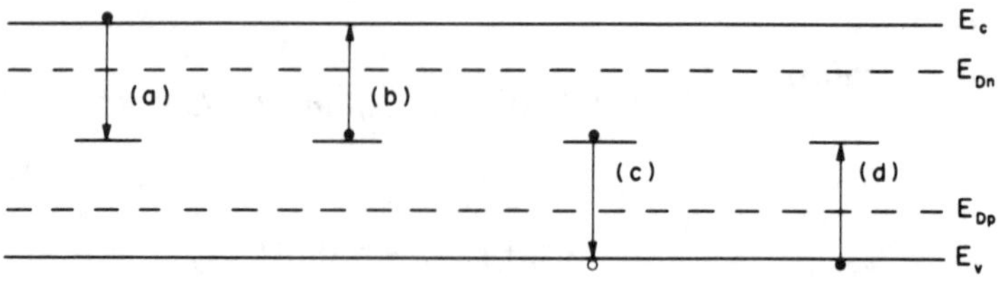

Fig. 4.4. The Shockley–Read recombination and generation model. The recombination centres are neutral when empty and negatively charged when filled with a captured electron: (a) capture of an electron from the conduction band by a neutral centre, (b) thermal excitation of the captured electron from the centre to the conduction band, (c) capture of a hole from the valence band by a filled centre (recombination), and (d) thermal excitation of a hole to the valence band from a neutral centre (capture of an electron from the valence band by a neutral centre—generation).

where n_r is the density of electron-occupied centres and E_r is the energy level of the centres. In the steady state, eqs. (4.66) and (4.67) must be equal to each other. By denoting f_r as the probability for a recombination centre to be occupied, we can write

$$n_r = N_r f_r \tag{4.68}$$

$$(N_r - n_r) = N_r (1 - f_r) \tag{4.69}$$

$$n_1 = N_c \exp[-(E_c - E_r)/kT] \tag{4.70}$$

$$p_1 = N_v \exp[-(E_r - E_v)/kT] \tag{4.71}$$

Substitution of eqs. (4.68)–(4.71) into eqs. (4.66) and (4.67), and solution of them after being set equal to each other, yield

$$f_r = \frac{C_n n + C_p p_1}{C_n (n + n_1) + C_p (p + p_1)} \tag{4.72}$$

Thus the recombination rate for these acceptor-type recombination centres is

$$R_a = \frac{C_n C_p N_r (np - n_1 p_1)}{C_n (n + n_1) + C_p (p + p_1)} \tag{4.73}$$

By introducing

$$\tau_{n0} = \frac{1}{C_n N_r} \tag{4.74}$$

$$\tau_{p0} = \frac{1}{C_p N_r} \tag{4.75}$$

and since

$$n_1 p_1 = N_c N_v \exp[-(E_c - E_v)/kT]$$

$$= n_0 p_0 = n_i^2 \tag{4.76}$$

eq. (4.73) can be expressed as

$$R_a = [\tau_{p0} (n + n_1) + \tau_{n0} (p + p_1)]^{-1} (np - n_i^2) \tag{4.77}$$

This is the Shockley–Read equation. It can be seen that the driving force for recombination is $np - n_i^2$, which is in fact the deviation from the equilibrium condition.

To obtain exact equations for the carrier lifetimes τ_n and τ_p for arbitrary values of excess carrier densities Δn and Δp and recombination centre density N_r is difficult, but equations that are useful for most cases with a very good approximation have been derived by Blakemore [1962] and they are

$$\tau_n = \frac{\tau_{p0} (n_0 + n_1 + \Delta n) + \tau_{n0} (p_0 + p_1 + \Delta p) + \tau_{n0} N_r \left[\dfrac{p_1 (n_0 + n_1 + \Delta n) + 2p_0 \Delta n}{(p_0 + p_1)(n_0 + n_1 + \Delta n)} \right]}{(n_0 + p_0 + \Delta n) + N_r \left[\dfrac{p_0 (n_0 + \Delta n)}{(p_0 + p_1)(n_0 + n_1 + \Delta n)} \right]} \tag{4.78}$$

$$\tau_p = \frac{\tau_{p0}(n_0 + n_1 + \Delta n) + \tau_{n0}(p_0 + p_1 + \Delta p) + \tau_{p0} N_r \left[\dfrac{p_0(p_0 + p_1 + \Delta p) + 2p_1 \Delta p}{(p_0 + p_1)(p_0 + p_1 + \Delta p)} \right]}{(n_0 + p_0 + \Delta p) + N_r \left[\dfrac{p_1(p_0 + \Delta p)}{(p_0 + p_1)(p_0 + p_1 + \Delta p)} \right]}$$

(4.79)

Case I. Δn and Δp are small as compared with n_0 and p_0 $(n = n_0 + \Delta n, p = p_0 + \Delta p)$ but $\tau_n \neq \tau_p$. For this case we can obtain τ_n and τ_p by setting $\Delta n \to 0$ and $\Delta p \to 0$ in eqs. (4.78) and (4.79). Thus we have

$$\tau_n = \frac{\tau_{p0}(n_0 + n_1) + \tau_{n0}[p_0 + p_1 + N_r p_1/(p_0 + p_1)]}{n_0 + p_0 + N_r[n_0 p_0/(n_0 + n_1)(p_0 + p_1)]}$$

(4.80)

$$\tau_p = \frac{\tau_{n0}(p_0 + p_1) + \tau_{p0}[n_0 + n_1 + N_r p_0/(p_0 + p_1)]}{n_0 + p_0 + N_r[p_0 p_1/(p_0 + p_1)^2]}$$

(4.81)

and

$$R_a = \frac{n_0 \Delta p + p_0 \Delta n}{\tau_{p0}(n_0 + n_1) + \tau_{n0}(p_0 + p_1)}$$

(4.82)

Case II. Δn and Δp are small as compared with n_0 and p_0, but $\Delta n = \Delta p$ and $\tau_n = \tau_p$. For this case we can obtain $\tau_n = \tau_p$ by setting the terms in N_r to zero in eqs. (4.80) or (4.81). Thus we have

$$\tau_n = \tau_p = \tau_{p0}(n_0 + n_1)/(n_0 + p_0) + \tau_{n0}(p_0 + p_1)/(n_0 + p_0)$$

(4.83)

and

$$R_a = \frac{\Delta n(n_0 + p_0) \cdot}{\tau_{p0}(n_0 + n_1) + \tau_{n0}(p_0 + p_1)}$$

(4.84)

Case III. Δn and Δp are much larger than n_0 or p_0, but still $\Delta n = \Delta p$ and $\tau_n = \tau_p$. For this case, we can obtain $\tau_n = \tau_p$ by setting the terms in N_r to zero in eqs. (4.78) or (4.79). Thus we have

$$\tau_n = \tau_p = \tau_{p0}(n_0 + n_1 + \Delta n)/(n_0 + p_0 + \Delta n) + \tau_{n0}(p_0 + p_1 + \Delta n)/(n_0 + p_0 + \Delta n)$$

(4.85)

and

$$R_a = \frac{\Delta n(n_0 + p_0 + \Delta n)}{\tau_{p0}(n_0 + n_1 + \Delta n) + \tau_{n0}(p_0 + p_1 + \Delta n)}$$

(4.86)

Letting τ' denote the lifetime for vanishingly small values of Δn as given by eq. (4.83), then eq. (4.85) can be written in the form

$$\tau = \tau' \frac{1 + a \, \Delta n}{1 + b \, \Delta n}$$

(4.87)

where

$$a = \frac{\tau_{p0} + \tau_{n0}}{\tau_{p0}(n_0 + n_1) + \tau_{p0}(p_0 + p_1)}$$

(4.88)

$$b = (n_0 + p_0)^{-1}$$

(4.89)

If $a > b$, then τ increases monotonically with increasing Δn, giving rise to superlinear photoconductivity; if $a < b$, then τ decreases monotonically with increasing Δn, giving rise to sublinear photoconductivity. The limiting value for τ as Δn approaches infinity is [Shockley and Read 1952]

$$\tau_\infty = \tau_{p0} + \tau_{n0} \tag{4.90}$$

In fact, Case III is close to most cases for insulators or organic semiconductors in which the energy band gap is large and the mobility is small. Supposing that $\Delta n = \Delta p$ and $\Delta n > n_0$ and $\Delta n > p_0$, then we can write

$$\frac{d(\Delta n)}{dt} = G - R$$

$$= G - \frac{\Delta n}{\tau} \tag{4.91}$$

In the steady state,

$$G = \frac{(\Delta n)_0}{\tau'} \left[\frac{1 + b(\Delta n)_0}{1 + a(\Delta n)_0} \right] \tag{4.92}$$

If now the optical excitation, after Δn has reached its steady state value $(\Delta n)_0$, is switched off, Δn will decay following the expression

$$\Delta n = (\Delta n)_0 \exp(-t/\tau) \tag{4.93}$$

where τ is given by eq. (4.87). For this case the response time is approximately equal to the lifetime.

By comparing eq. (4.93) with eq. (4.59) it is clear that for $\Delta n = \Delta p \gg n_0$ and $\gg p_0$ the excess carrier density or photocurrent decays exponentially with time if the recombination is monomolecular (indirect recombination through recombination centres), and decays hyperbolically with t for $t \ll \tau$ if the recombination is bimolecular (direct band-to-band recombination). This is generally used as a criterion to distinguish experimentally these two types of recombination processes.

4.2.3. With a single set of recombination centres and with traps

The model is shown in Fig. 4.5, which includes a level of electron trapping centres E_{tn} above E_{Dn} and a level of hole-trapping centres E_{tp} below E_{Dp}. Under an optical excitation and if the densities of traps in the solid are high, then $\Delta n \neq \Delta p$ and the transient decay will not correspond to an exponential function with time [see eq. (4.93)], when the optical excitation is switched off after the steady state has been established. The response time is no longer equal to the lifetime, but is greater than the lifetime (see Section 4.1.6).

Since the shallow traps are in thermal equilibrium with free carriers, the ratios of trapped to free carrier densities can easily be shown to be

$$\frac{n_t}{n} = \left(\frac{N_{tn}}{N_c} \right) \exp[-(E_{tn} - E_c)/kT] \tag{4.94}$$

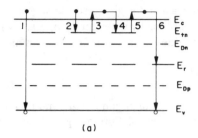 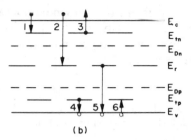

(a)   (b)

FIG. 4.5. Electron behaviour in a solid with a single set of recombination centres and with traps.
(a) With electron traps only: 1, direct recombination of an electron with a hole; 2, electron
trapping; 3–6, electron trapping and detrapping from traps at E_{tn} and final recombination
with a hole via a recombination centre at E_r. (b) With both electron and hole traps: 1 and 3;
electron trapping and detrapping from traps at E_{tn}; 4 and 6, hole trapping and detrapping
from traps at E_{tp}; 2 and 5, electron and hole recombined at the recombination centre at E_r.

and

$$\frac{p_t}{p} = \left(\frac{N_{tp}}{N_v}\right) \exp[-(E_v - E_{tp})/kT] \tag{4.95}$$

where N_{tn}, n_t, N_{tp}, and p_t are, respectively, the densities of total and electron-occupied
electron traps (trapped electrons), and total and empty hole traps (trapped holes). It can
be seen that these ratios decrease with increasing temperature and with increasing
intensity of photoexcitation. Thus at sufficiently high intensity of photoexcitation the
shallow trapping centres may become part of the recombination centres (see Section
4.1.3) and the model reverts then to Case III in Section 4.2.2. However, the general and
special cases for solids with a single set of recombination centres and with traps have
been analysed by several investigators [Rose 1951, 1955, 1963, Curtis 1968, Curtis and
Srour 1974, Sah and Shockley 1958, Bube 1974].

4.2.4. With two levels of recombination centres but without traps

This model is shown in Fig. 4.6 in which we call the two levels of centres as Class 1 and
Class 2 centres. The behaviour of these centres depends on the carrier injection level.
The redistribution of electrons and holes among these centres is free to take place, and it
is this redistribution that causes sensitization, infrared quenching, and superlinearity in
photoconduction processes. A theoretical analysis of this model is not a simple problem
because it involves four capture cross-sections and four states in both Class 1 and Class
2 centres. In this section we limit ourselves to a qualitative discussion to show how the
recombination competition between these two classes of centres with different capture
coefficients would lead to various phenomena.

In Fig. 4.6 we assume that Class 1 centres with relatively large capture coefficients for
both electrons and holes are initially present in the material, and Class 2 centres are
formed by adding a special type of impurities to the material. We shall consider the
following phenomena in terms of these doping impurities.

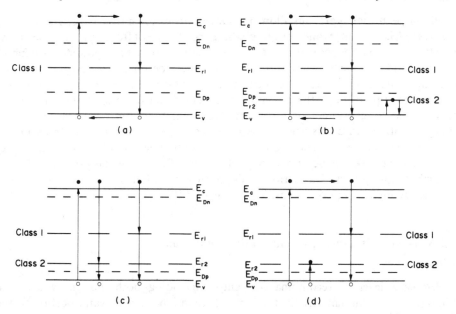

F<small>IG</small>. 4.6. The recombination processes for the case with two levels of recombination centres but without traps: (a) fast recombination through recombination centres and short electron lifetime, (b) the introduction of Class 2 centres does not affect condition (a) if these centres are located below the hole demarcation level E_{Dp}, (c) Class 2 centres act as sensitizing centres to cause an increase in electron lifetime and hence in photoconductivity if these centres are located above E_{Dp}, and (d) a secondary excitation that creates holes due to the presence of class 2 centres results in a decrease in electron lifetime and in optical quenching of the photoconductivity.

(a) *Sensitization.* If the Class 2 centres are absent, the Class 1 centres alone will make the recombination through the recombination centres fast, the lifetime of free carriers short, and the material insensitive to excitation intensity. The introduction of Class 2 centres near the valence band will not affect the lifetime of electrons, and the situation is practically unchanged from that without Class 2 centres if they are located below E_{Dp} (demarcation level) at low excitation levels. But when the E_{Dp} moves below E_{r2} due to the increase of the excitation levels or to the decrease of temperature, then Class 2 centres have a much larger cross-section for capturing free holes than they have subsequently for capturing electrons, thus increasing the lifetime of free electrons so as to increase the photosensitivity of the material. Such Class 2 centres are generally referred to as activators or activating centres [Rose 1955, 1963].

(b) *Infrared quenching.* When E_{Dp} is below E_{r2}, Class 2 centres act like activators. But if a second light source of infrared region can excite electrons from the valence band to hole-captured Class 2 centres, releasing these holes to be captured by Class 1 centres, the effect of the second infrared light source will reduce the photoconductivity or convert Class 2 centres from their function as activator to that as "quenching" or "poisoning" centres. This effect is sometimes called "optical quenching" or "thermal quenching" because an increase in temperature above a certain value may also produce this quenching effect [Bube 1960, 1974].

(c) *Superlinearity.* As the excitation intensity is further increased, the E_{Dn} and E_{Dp}

levels move further apart thus allowing more electrons to be transferred from Class 2 centres to Class 1 centres and hence increasing the lifetime of free carriers. Under such a condition the photoconductivity increases rapidly with a high power of excitation intensity and this phenomenon is referred to as the superlinearity. However, when the excitation intensity is increased to such a value for which the free carrier concentrations are large compared with the recombination centre concentrations, then $n \simeq p$, and there is no longer interaction between recombination centres. Each centre acts independently to adjust its behaviour to the free carrier concentrations. In particular, each centre will contain an electron $\tau_n/(\tau_n + \tau_p)$ of the time and a hole of $\tau_p/(\tau_n + \tau_p)$ of the time. These occupancy fractions will not be affected by the presence of other recombination centres. This action tends to cause the photosensitivity to saturate at sufficiently high excitation intensities [Rose 1955].

4.2.5. With multilevels of recombination centres but without traps

For each level of recombination centres including both monomolecular and bimolecular recombinations, the total rate of recombinations is [van Roosbroeck and Casey 1972]

$$R = \{C_r + [\tau_{p0}(n + n_1) + \tau_{n0}(p + p_1)]^{-1}\}(np - n_i^2) \tag{4.96}$$

For recombination centres of more than one type, R still has $(np - n_i^2)$ as the driving force for recombination, but the coefficient of $(np - n_i^2)$ is the sum of similar terms for each type (or each level).

For each level of recombination centres (acceptor-type) the rates of change of carrier densities are given by

$$\frac{dn}{dt} = G - C_r(np - n_i^2) - C_n[n(N_r - n_r) - n_1 n_r] \tag{4.97}$$

$$\frac{dp}{dt} = G - C_r(np - n_i^2) - C_p[pn_r - p_1(N_r - n_r)] \tag{4.98}$$

If such recombination centres are distributed following a distribution function $h(E)$ per unit volume per unit energy throughout the forbidden energy gap, then the terms in eqs. (4.97) and (4.98) have to be changed to

$$pn_r = p \int_{E_v}^{E_{Dn}} h(E) f_n(E) dE \tag{4.99}$$

$$n(N_r - n_r) = n \int_{E_{Dp}}^{E_c} h(E)[1 - f_n(E)] dE \tag{4.100}$$

$$n_1 n_r = N_c \int_{E_v}^{E_{Dn}} h(E) f_n(E) \exp[-(E_c - E)/kT] dE \tag{4.101}$$

$$p_1(N_r - n_r) = N_v \int_{E_{Dp}}^{E_c} h(E)[1 - f_n(E)] \exp[-(E - E_v)/kT] dE \tag{4.102}$$

where

$$f_n(E) = \frac{1}{1 + g_n^{-1} \exp[(E - E_{Fn})/kT]} \tag{4.103}$$

By setting eq. (4.97) equal to eq. (4.98), we can also express the probability of occupation of a recombination centre at any energy level E as

$$f_n(E) = \frac{C_n n + c_p p_1(E)}{C_n[n + n_1(E)] + C_p[p + p_1(E)]} \tag{4.104}$$

which is the same as eq. (4.72) [Simmons and Taylor 1971].

For distributed recombination centres, R in eq. (4.96) has to be expressed in terms of eqs. (4.99)–(4.102). This gives rise only to the change of the coefficient of the factor $(np - n_i^2)$.

4.2.6. The Auger effect

In the Auger effect the energy released from a recombining electron is immediately imparted by way of collision to another carrier which then dissipates this energy by emitting phonons. A number of Auger recombination processes may take place depending on the free carrier concentration and the nature of the recombination. A process, in which the second carrier after receiving the energy released from the recombining electron dissipates its energy radiatively is normally not considered as an Auger effect. Such a process is generally referred to as a "resonant absorption". In the following we shall discuss briefly the intrinsic and extrinsic Auger recombination processes.

(i) *Intrinsic Auger processes.* The three-body collision involves only free carriers: two electrons and one hole, or two holes and one electron, are converted into one hot carrier. The model which explains the Auger processes is shown in Fig. 4.7. These processes are: (a) electron–electron (ee) collisions result in the fall of electron 1 from the conduction band to the valence band to recombine with a hole, and in the rise of electron 2 to a higher energy level in the conduction band, the rate at which the electron–hole pairs recombine due to electron–electron collisions is denoted by R_{ee}; (b) the rate of electron–hole pair generation resulting from impact ionization by hot electrons [the inverse of process (a)] is denoted by G_{ee}; (c) hole–hole (hh) collisions results in the lowering of hole 2 to a lower energy level in the valence band in a manner similar to process (a), the rate at which the electron–hole recombine due to hole–hole collision is denoted by R_{hh}; (d) the rate of electron–hole pair generation due to the inverse of process (c) is denoted by G_{hh}.

At thermal equilibrium the recombination rates for the Auger processes can be expressed as

$$(G_{ee})_0 = (R_{ee})_0 = \beta_e n_0^2 p_0 \tag{4.105}$$

$$(G_{hh})_0 = (R_{hh})_0 = \beta_h n_0 p_0^2 \tag{4.106}$$

For non-degenerate semiconductors or insulators, the recombination rates, when the

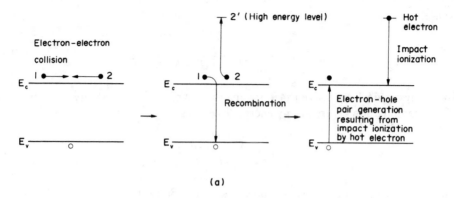

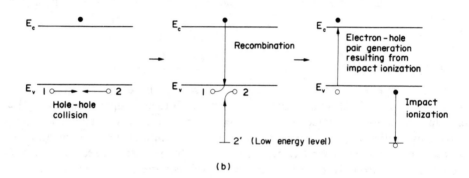

F<small>IG</small>. 4.7. Illustrating the intrinsic Auger recombination processes (a) by electron–electron collision, and (b) by hole–hole collision.

material contains excess carrier pairs of concentration $\Delta n = \Delta p = n - n_0 = p - p_0$, are given by

$$R_{ee} = \beta_e n^2 p = (G_{ee})_0 \left(\frac{np}{n_i^2}\right)\left(\frac{n}{n_0}\right) \tag{4.107}$$

and

$$R_{hh} = \beta_h np^2 = (G_{hh})_0 \left(\frac{np}{n_i^2}\right)\left(\frac{p}{p_0}\right) \tag{4.108}$$

Similarly, the generation processes due to impact ionization under the same condition can be expressed as

$$G_{ee} = (G_{ee})_0 \left(\frac{n}{n_0}\right) \tag{4.109}$$

$$G_{hh} = (G_{hh})_0 \left(\frac{p}{p_0}\right) \tag{4.110}$$

The net recombination rate under photoexcitation due to both *ee* and *hh* processes is

$$R = (R_{ee} - G_{ee}) + (R_{hh} - G_{hh})$$

$$= \frac{(np - n_i^2)(G_{ee}np_0 + G_{hh}n_0p)}{n_i^4} \tag{4.111}$$

The Auger lifetime can then be defined as [Blakemore 1962]

$$\tau_A = \frac{\Delta n}{R} = \frac{\Delta p}{R}$$

$$= \frac{n_i^4}{(n_0 + p_0 + \Delta n)(G_{ee}np_0 + G_{hh}n_0p)} \tag{4.112}$$

The expressions for $(G_{ee})_0$ and $(G_{hh})_0$ have been derived by Beattie and Landsberg [1959]. For non-degenerate semiconductors or insulators they are given by

$$(G_{ee})_0 = \frac{2(2\pi)^{1/2}q^4 m_e |F_1 F_2|^2 n_0 (kT/E_g)^{3/2}}{h^3 \varepsilon^2 (1+\mu)^{1/2}(1+2\mu)} \exp\left[-\left(\frac{1+2\mu}{1+\mu}\right)\frac{E_g}{kT}\right] \tag{4.113}$$

$$(G_{hh})_0 = \frac{2(2\pi)^{1/2}q^4 m_h |F_1 F_2|^2 p_0 (kT/E_g)^{3/2}}{h^3 \varepsilon^2 (1+1/\mu)^{1/2}(1+2/\mu)} \exp\left[-\left(\frac{2+\mu}{1+\mu}\right)\frac{E_q}{kT}\right] \tag{4.114}$$

where m_e and m_h are the effective electron and hole masses, respectively; $\mu = m_e/m_h$; E_g is the energy band gap; and F_1 and F_2 are the overlap integrals of the periodic parts of Bloch functions which are given by [Blakemore 1962, Beattie and Landsberg 1959, 1960].

$$F_1 = \int U_c^* (k_1, r) U_v (k_1' r) dr \tag{4.115}$$

$$F_2 = \int U_c^* (k_2, r) U_v (k_2' r) dr \tag{4.116}$$

(ii) *Extrinsic Auger processes.* These processes involve the capture of a free carrier by a localized state and the excitation of another free carrier to a higher energy level; the net result is the transfer of a free carrier into a localized state and the dissipation of the released energy as heat. The model which explains these processes is shown in Fig. 4.8. The Auger lifetime can be expressed as

$$\frac{1}{\tau_A} = Anp + Bn^2 \tag{4.117}$$

for extrinsic *n*-type semiconductors, and as

$$\frac{1}{\tau_A} = Anp + Bp^2 \tag{4.118}$$

for extrinsic *p*-type semiconductors [Bube 1960, 1974, Pankove 1971]. The first term on the right-hand side of these equations expresses Auger excitation of minority carriers, while the second term expresses Auger excitation of majority carriers. In most extrinsic materials the second term dominates and the constant *B* represents the Auger recombination coefficient. By equating the Auger recombination rate and impact

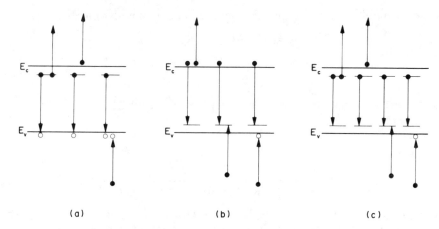

(a)   (b)   (c)

Fɪɢ. 4.8. Illustrating the possible extrinsic Auger recombination processes (a) involving one
level of localized states near the conduction band, (b) involving one level of localized
states near the valence band, and (c) involving two levels of localized states. ● and ○
represent electrons and holes respectively.

ionization rate for the localized states formed by impurities and acting like hydrogen
atoms, the expression for B has been derived and is given by [Burstein *et al.*
1956]

$$B = \frac{2.44 \times 10^{-19}}{E_I T^2} \left(\frac{m_0}{m^*}\right)^2 \left(\frac{1 + 0.522 \log \varepsilon}{\varepsilon}\right) \qquad (4.119)$$

where E_I is the ionization energy of the impurity centres in eV, m^*/m_0 is the effective
mass ratio for the relevant carrier, and ε is the relative permittivity (dielectric constant).
Using these units, B is in cm^6 sec^{-1}. If $A \gg B$ the first term dominates, then $1/\tau_A \simeq Anp$
$= An_i^2$ and should be independent of impurity doping [Bube 1960].

It should be noted that relatively little is known about the Auger processes. In
general, Auger recombination is important only in narrow band gap semiconductors.
Hynes and Hornbeck [1955] have suggested that the dependence of minority carrier
lifetime on the square of the majority carrier density in silicon is an indication of Auger
processes. Galkin *et al.* [1971] have observed band-to-band Auger recombination
processes in InAs. So far, there is no experimental evidence about the Auger effect in
large band gap materials such as in organic semiconductors. However, a variety of more
complicated Auger processes can be conceived. For example, an electron may
recombine with a centre and the energy may be used up to create a hole by promoting an
electron to a neighbouring centre. It is theoretically possible for many Auger processes
involving traps or recombination centres to take place, but their importance in
experiments performed so far is uncertain [Landsberg 1970].

Other recombination processes such as those involving excitations, surface states and
multiphonons will be discussed later in Chapters 6 and 7. Regarding the techniques for
measuring experimentally the lifetimes and capture cross-sections of trapping or
recombination centres, which are beyond the scope of this book, the reader is referred
to some excellent review articles [Bullis 1968, Milnes 1973].

4.3. TWO-CARRIER (DOUBLE) PLANAR INJECTION IN SOLIDS

The theory of one-carrier current injection in a solid having an arbitrary distribution of traps in energy and in space has been discussed in detail in Chapter 3. In double injection, charge carriers of both types (electrons and holes) are present and the problem becomes much more complicated because in this case the recombination which controls the current–voltage (J–V) characteristics may either be bimolecular—direct (band-to-band) recombination—or occur through one or more sets of localized traps—indirect recombination. Double-injection and electron–hole recombination in the intrinsic region of a p–i–n structure (or p–π–n structure) have been discussed by Hall [1952] and studied by Herlet and Spenke [1955] and Kleinman [1956]. Since then, the theory has been further developed by Lampert and Rose [Lampert 1959, 1962, Lampert and Rose 1961, Rose 1964] for the cases with monomolecular recombination, and by Parmenter and Ruppel [1959] for the cases with bimolecular recombination. Their work later led to a more detailed investigation in this field [Lampert and Mark 1970, Baron and Mayer 1970] for partially or completely filled deep traps (recombination centres). Lampert and Schilling [1970] have derived the J–V characteristics using the regional approximation method. With this method the specimen is divided into regions which either satisfy the quasi-neutrality approximation or are dominated by trapped or free space charge. Double injection may result in two important phenomena, namely, negative differential resistance [Holonyak *et al.* 1962, 1963, Ashley and Milnes 1964, Brown and Jordan 1966] and electroluminescence [Hwang and Kao 1973, 1974]. These phenomena will be discussed in later sections.

For double injection, the traps behave as traps rather than recombination centres if they are located above E_{Dn} for electron traps or below E_{Dp} for hole traps, and the traps located between E_{Dn} and E_{Dp} can be considered as recombination centres. E_{Dn} and E_{Dp} are demarcation levels for electron traps and hole traps, respectively; and they have been discussed in detail in Section 4.1.3. In this section we shall present the theory of double carrier current injection in accordance with the locations and distributions of traps. In the theoretical analysis we shall confine ourselves to steady state d.c. one-dimensional planar current flow and make the following assumptions, but the treatment is general and therefore can be applied to thick or thin specimens in crystal or in film form of any materials.

(i) The energy band model can be used to treat the behaviour of injected carriers.

(ii) Both the anode for injecting holes and the cathode for injecting electrons are perfect ohmic-injecting contacts located at $z = 0$ and $z = d$, respectively, the specimen thickness being d.

(iii) The electric field is so large that the current components due to diffusion and due to carriers thermally generated in the specimen can be neglected.

(iv) The free hole and electron densities and trapped hole and electron densities in shallow traps follow the Maxwell–Boltzmann statistics, while the trapped hole and electron densities in deep traps follow the Fermi–Dirac statistics.

(v) The mobilities of the free holes and electrons are independent of field and are not affected by the presence of traps and recombination centres.

(vi) The planes perpendicular to the z-axis, at which the field is zero, are located at

$z = w_a$ and $z = w_c$ which are, respectively, very close to the injecting contacts at $z = 0$ and $z = d$; so that

$$F(z = w_a \simeq 0) = F(z = w_c \simeq d) = 0 \qquad (4.120)$$

The behaviour of double injection in a solid is governed by the current flow equations

$$J_n = q\mu_n nF \qquad (4.121)$$

$$J_p = q\mu_p pF \qquad (4.122)$$

$$J = J_n + J_p \qquad (4.123)$$

the continuity equations

$$\frac{1}{q}\frac{dJ_n}{dz} = R \qquad (4.124)$$

$$-\frac{1}{q}\frac{dJ_p}{dz} = R \qquad (4.125)$$

and Poisson's equation

$$\frac{dF}{dz} = \frac{q}{\varepsilon}[p(z) + p_t - n(z) - n_t] = \frac{\rho}{\varepsilon} \qquad (4.126)$$

where n and p are given by

$$n = N_c \exp[-(E_c - E_{Fn})/kT] \qquad (4.127)$$

$$p = N_v \exp[-(E_{Fp} - E_v)/kT] \qquad (4.128)$$

and n_t and p_t are given by

$$n_t = \int_{E_l}^{E_u} h_n(E, z)\{1 + g_n^{-1}\exp[(E - E_{Fn})/kT]\}^{-1}dE \qquad (4.129)$$

$$p_t = \int_{E_l}^{E_u} h_p(E, z)\{1 + g_p\exp[(E_{Fp} - E)/kT]\}^{-1}dE \qquad (4.130)$$

in which $h_n(E, z)$ and $h_p(E, z)$ are the distribution function for electron and hole trap densities, respectively.

In the following we shall consider four major cases and state any additional assumptions which have to be made when dealing with individual cases.

4.3.1. Without recombination centres and without traps (trap-free solids—ideal case)

For this case $n_t = p_t = 0$, $R = np\langle v\sigma_R\rangle$. To simplify the mathematical treatment, we introduce the following parameters:

$$\left.\begin{array}{ll} S_0 = q\mu_n nF/J, & \alpha = 2q/\varepsilon \\ T_0 = q\mu_p pF/J, & \beta_0 = \langle v\sigma_R\rangle \\ U_0 = qF^2/J \end{array}\right\} \qquad (4.131)$$

Using these parameters, eqs. (4.123)–(4.126) can be written as

$$S_0 + T_0 = 1 \tag{4.132}$$

$$\frac{dS_0}{dz} = \beta_0 S_0 T_0 / \mu_n \mu_p U_0 \tag{4.133}$$

$$\frac{dT_0}{dz} = -\beta_0 S_0 T_0 / \mu_n \mu_p U_0 \tag{4.134}$$

$$\frac{dU_0}{dz} = \alpha[T_0/\mu_p - S_0/\mu_p] \tag{4.135}$$

The solution of the above equations gives

$$U_0 = C_0 S_0^{\alpha \mu_n/\beta_0}(1 - S_0)^{\alpha \mu_p/\beta_0} \tag{4.136}$$

where C_0 is the integration constant. Since the entire current at the anode is carried by holes, thus $S_0 = 0$; and the entire current at the cathode is carried by electrons, thus $S_0 = 1$. So substituting eq. (4.136) into eq. (4.133) and then integrating it, we obtain

$$C_0 = \beta d \left[\mu_n \mu_p \int_0^1 S_0^{\alpha \mu_n/\beta_0 - 1}(1 - S_0)^{\alpha \mu_p/\beta_0 - 1} dS_0 \right]^{-1} \tag{4.137}$$

Using the boundary condition

$$V = \int_0^d F \, dz \tag{4.138}$$

and from eqs. (4.131) and (4.137), it can easily be shown that the relation between J and V is

$$J = \frac{9}{8} \varepsilon \mu_{\text{eff}} \frac{V^2}{d^3} \tag{4.139}$$

where

$$
\mu_{\text{eff}} = \frac{8q}{9\varepsilon} \frac{\mu_n \mu_p}{\langle v\sigma_R \rangle} \frac{\left[\int_0^1 S_0^{\alpha \mu_n/\beta_0 - 1}(1 - S_0)^{\alpha \mu_p/\beta_0 - 1} dS_0 \right]^3}{\left[\int_0^1 S_0^{3\alpha \mu_n/2\beta_0 - 1}(1 - S_0)^{3\alpha \mu_p/2\beta_0 - 1} dS_0 \right]^2}
$$

$$
= \frac{8q}{9\varepsilon} \frac{\mu_n \mu_p}{\langle v\sigma_R \rangle} \frac{\left[B\left(\dfrac{\alpha \mu_n}{\beta_0}, \dfrac{\alpha \mu_p}{\beta_0} \right) \right]^3}{\left[B\left(\dfrac{3\alpha \mu_n}{2\beta_0}, \dfrac{3\alpha \mu_p}{2\beta_0} \right) \right]^2}
$$

$$
= \frac{8q}{9\varepsilon} \frac{\mu_n \mu_p}{\langle v\sigma_R \rangle} \frac{\left[\left(\dfrac{\alpha \mu_n}{\beta_0} - 1 \right)! \left(\dfrac{\alpha \mu_p}{\beta_0} - 1 \right)! \right]^3}{\left[\dfrac{\alpha}{\beta_0}(\mu_n + \mu_p) - 1 \right]!}
$$

$$\times \left[\frac{\left[\dfrac{3\alpha}{2\beta_0}(\mu_n + \mu_p) - 1 \right]!}{\left(\dfrac{3\alpha\mu_n}{2\beta_0} - 1 \right)! \left(\dfrac{3\alpha\mu_p}{2\beta_0} - 1 \right)!} \right]^2 \tag{4.140}$$

in which $B(m, n)$ is the β-function. This is the general equation for solids without recombination centres and without traps. This equation for μ_{eff} can be simplified for some simple cases as follows.

(a) INJECTED PLASMA

For this case $\langle v\sigma_R \rangle$ is small and so $(\alpha\mu_n/\beta_0) \gg 1$ and $(\alpha\mu_p/\beta_0) \gg 1$. Thus, with the aid of Stirling's formula,

$$(m - 1)! \simeq m! \simeq (m/e)^m (2\pi m)^{1/2} \tag{4.141}$$

eq. (4.140) can be approximated to

$$\mu_{\text{eff}} = \frac{2}{3} \left[\frac{4\pi q \mu_n \mu_p (\mu_n + \mu_p)}{\varepsilon \langle v\sigma_R \rangle} \right]^{1/2} \tag{4.142}$$

Equations (4.139) and (4.142) are exactly the same expressions originally derived by Parmenter and Ruppel [1959].

If $\langle v\sigma_R \rangle$ is small, the recombination cross-section σ_R is small. This implies that recombination is not a barrier to hinder both electrons and holes from completing their penetration of the material specimen, and that the charge neutrality condition prevails throughout the bulk, and this can be seen in Fig. 4.9(a) in which $n \simeq p$ throughout the bulk.

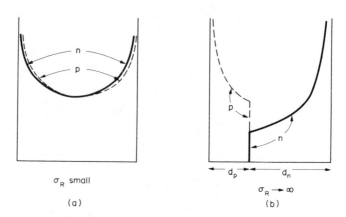

FIG. 4.9. Schematic spatial distribution of the injected electron and hole densities for double injection into a trap-free insulator for two limiting cases: (a) the small σ_R limit corresponding to an injected plasma, and (b) the $\sigma_R \to \infty$ limit corresponding to back-to-back single carrier SCL currents; the particular case shown corresponding to the choice of $\mu_n/\mu_p = 2$. [After Rosenberg and Lampert 1970.]

(b) SPACE-CHARGE-LIMITED CURRENTS

For this case $\langle v\sigma_R \rangle$ is very large so $\alpha\mu_n/\beta_0 \ll 1$ and $\alpha\mu_p/\beta_0 \ll 1$. Using the relation

$$(m-1)! \simeq 1/m \quad \text{for} \quad 0 < m \ll 1 \tag{4.143}$$

eq. (4.140) can be approximated to

$$\mu_{\text{eff}} = \mu_n + \mu_p \tag{4.144}$$

This equation and eq. (4.139) indicate that the total current is simply the sum of the two separate one-carrier space-charge-limited (SCL) currents.

When $\langle v\sigma_R \rangle$ is very large and approaches infinity, σ_R also approaches infinity. This implies that there is no region in which electrons and holes can overlap since there would be an infinite recombination current if such an overlap region occurs. Under such a condition the electron current will exist only on the cathode side and the hole current on the anode side, and they meet and annihilate at a certain plane dividing these two regions as shown in Fig. 4.9(b). This case can be applied to perfect organic crystals.

(c) ONE-CARRIER SPACE-CHARGE-LIMITED CURRENTS

For this case, $\alpha\mu_n/\beta_0 \gg 1$ but $\alpha\mu_p/\beta_0 \ll 1$. Using the same technique for simplifying eq. (4.140), we obtain

$$\mu_{\text{eff}} \simeq \mu_n \tag{4.145}$$

Thus the total current will be mainly electron SCL current, the hole current being negligible because the holes are so sluggish that most of them will be annihilated by recombination at the anode.

(d) GENERAL CASE

For intermediate values of $\langle v\sigma_R \rangle$ between two extreme limits, the regional approximation method developed by Rosenberg and Lampert [1970] can be used to estimate the current flow. The schematic diagram illustrating this method is shown in Fig. 4.10. Supposing that $\mu_n/\mu_p > 2$, the whole specimen is divided into four regions and they are as follows:

(i) *Region I* $(0 \leq z \leq z_1, p > \mu_n n/\mu_p)$. In this space charge region the charge carriers are dominated by holes. Thus

$$J \simeq J_p = q\mu_p p E = \text{constant} \tag{4.146}$$

and

$$\frac{dF}{dz} = \frac{qp}{\varepsilon} \tag{4.147}$$

It is obvious that using the boundary condition $F = 0$ at $z = 0$ the solution is simply a hole SCL current given by eq. (3.9). In this region eq. (4.123) has been replaced by eq. (4.146), so p and F as functions of z can be determined from the solution of eqs. (4.146) and (4.147), but n as a function of z has to be determined from eq. (4.124).

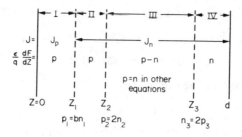

FIG. 4.10. Schematic regional approximation diagram for the problem
of double injection into a perfect, trap-free insulator for the
case $b = \mu_n/\mu_p > 2$. At $z = 0$: $p/n \to \infty$; at z_1: $p = \mu_n n/\mu_p$; at
z_2: $p = 2n$; at z_3: $p = n/2$; at $z = d$: $p/n \to 0$. [After Rosenberg
and Lampert 1970.]

(ii) *Region II* $(z_1 \leq z \leq z_2, \mu_n n/\mu_p \geq p \geq 2n)$. In this hybrid region we have

$$J \simeq J_n = q\mu_n nF = \text{constant} \tag{4.148}$$

$$\frac{dF}{dz} = \frac{\varepsilon p}{q} \tag{4.149}$$

$$-\mu_p \frac{d(pF)}{dz} = np \langle v\sigma_R \rangle \tag{4.150}$$

This implies that the current is determined solely by the electrons and the space charge
solely by the holes. Using eqs. (4.148) and (4.149) to eliminate n and p in eq. (4.150), we
obtain

$$-F^2 \frac{d^2(F^2)}{dz^2} = \left(\frac{J \langle v\sigma_R \rangle}{q\mu_n \mu_p} \right) \frac{d(F^2)}{dz} \tag{4.151}$$

This differential equation can be solved in terms of the exponential integral function
[Rosenberg and Lampert 1970]. n and p as functions of z can be determined from eqs.
(4.148) and (4.149).

(iii) *Region III* $(z_2 \leq z \leq z_3, 2n > p > n/2)$. In this injected plasma region we have

$$J \simeq J_n = q\mu_n nF = \text{constant} \tag{4.152}$$

$$\frac{dF}{dz} = q(p-n)/\varepsilon \tag{4.153}$$

$$\mu_n \frac{d(nF)}{dz} = n^2 \langle v\sigma_R \rangle \tag{4.154}$$

$$-\mu_p \frac{d(pF)}{dz} = n^2 \langle v\sigma_R \rangle \tag{4.155}$$

Eliminating n and p from eqs. (4.152)–(4.155), we obtain

$$-F^2 \frac{d^2(F^2)}{dz^2} = \frac{2(\mu_n + \mu_p)J^2 \langle v\sigma_R \rangle}{q\varepsilon \mu_p \mu_n^3} \tag{4.156}$$

Similarly, from these equations F, n, and p as functions of z in this region can be readily determined [Rosenberg and Lampert 1970].

(iv) *Region IV* $(z_3 \leq z \leq d, n/2 > p)$. In this space charge region the charge carriers are dominated by electrons. Thus

$$J \simeq J_n = q\mu_n nF = \text{constant} \tag{4.157}$$

and

$$\frac{dF}{dz} = -\frac{qn}{\varepsilon} \tag{4.158}$$

Using the similar method, F, n, and p as functions of z can be readily determined.

The solution for F in these four regions must satisfy the following boundary condition:

$$\left. \begin{aligned} F_I(z_1^-) &= F_{II}(z_1^+) \\ F_{II}(z_2^-) &= F_{III}(z_2^+) \\ F_{III}(z_3^-) &= F_{IV}(z_3^+) \end{aligned} \right\} \tag{4.159}$$

Rosenberg and Lampert [1970] have calculated F for various cases. Figure 4.11 shows the normalized F as a function of z/d for $\alpha\mu_n/\beta_0 = 1/2$ and $\alpha\mu_p/\beta_0 = 1/16$, and Fig. 4.12 shows the factor μ_{eff}/μ_n as a function of $\alpha\mu_p/\beta_0$ for $\mu_n/\mu_p = 8$. It should be noted that for $\mu_n/\mu_p < 1/2$ we still need four regions, but for this case the role of electrons and holes are interchanged. For the value of μ_n/μ_p between $1/2$ and 2, Region II can be ignored and the problem can be solved with three regions (Regions I, III, and IV).

However, if $\langle v\sigma_R \rangle$ is finite, a certain charge overlap will occur. Supposing that there

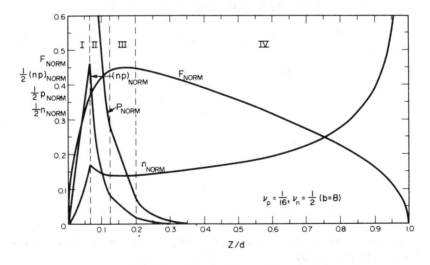

FIG. 4.11. Normalized electric field F_{NORM}, injected carrier densities n_{NORM} and p_{NORM}, and the product of the densities $(np)_{\text{NORM}}$ as functions of z/d for double injection in a perfect insulator with $b = \mu_n/\mu_p = 8$, $v_p = \alpha\mu_p/\beta_0 = 1/16$, and $v_n = \alpha\mu_n/\beta_0 = 1/2$. The vertical dashed lines demarcate the separate regions. [After Rosenberg and Lampert 1970.]

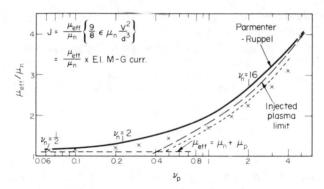

Fɪɢ. 4.12. μ_{eff}/μ_n as a function of v_p ($v_p = \alpha\mu_p/\beta_0$, $v_n = \alpha\mu_n/\beta_0$) for double injection into a perfect insulator with $b = \mu_n/\mu_p = 8$. Solid curve is based on eq. (4.140), upper dashed curve is the injected-plasma limit based on eq. (4.142), and the crosses are points obtained from the regional approximation method (the crosses are asymptotes to the lower dashed curve which is $b/(b+1)$ times the upper dashed curve). El. M–G curr. denotes the Mott–Gurney SCL electron current. The horizontal dashed line corresponds to the $\sigma_R \to \infty$ limit. [After Rosenberg and Lampert 1970.]

is a recombination zone in which the electric field F can be considered to be constant because the width of the zone is assumed to be small compared with the specimen thickness d; then in this zone the hole current density decreases with increasing z as

$$\frac{dJ_p}{dz} = q\mu_p F \frac{dp}{dz} = -qnp\langle v\sigma_R \rangle \qquad (4.160)$$

and the electron current density increases with increasing z as

$$\frac{dJ_n}{dz} = q\mu_n F \frac{dn}{dz} = qnp\langle v\sigma_R \rangle \qquad (4.161)$$

From eqs. (4.160), (4.161), and (4.123) we obtain

$$\frac{dn}{dz} = \frac{n\langle v\sigma_R \rangle}{\mu_n F}\left(\frac{J}{q\mu_p F} - \frac{\mu_n n}{\mu_p}\right) \qquad (4.162)$$

From Fig. 4.9, for $\langle v\sigma_R \rangle \to \infty$ we have

$$d_p + d_n = d$$
$$V_p + V_n = V \qquad (4.163)$$

For a finite value of $\langle v\sigma_R \rangle$ there will be a recombination zone. With n_∞ being the value of n far from the recombination zone (but not so far that the charge density inhomogeneity of one-carrier SCL current flow becomes important) the solution of eq. (4.162) gives [Helfrich 1967]

$$n(z) = \frac{n_\infty}{1 + \exp[(J\langle v\sigma_R \rangle/q\mu_n\mu_p F^2)(d_p - z)]} \qquad (4.164)$$

In a similar way we can obtain $p(z)$. These expressions have the form of Fermi–Dirac distribution function. So the width of the recombination zone w_r can be defined as twice the reciprocal of the coefficient of $(d_p - z)$. Thus we can write [Helfrich 1967]

$$w = \frac{2q\mu_n\mu_p F^2}{J\langle v\sigma_R \rangle} \tag{4.165}$$

By expressing $F = 3/2(V/d)$ based on eqs. (4.139) and (4.144), we obtain

$$w_r = \frac{4q\mu_n\mu_p d}{\varepsilon\langle v\sigma_R \rangle (\mu_n + \mu_p)} \tag{4.166}$$

It is interesting to note that the ratio w_r/d is independent of applied voltage and specimen thickness. Although eq. (4.166) is valid only if $w_r \ll d_n, d_p$, this equation indicates qualitatively that if $\mu_p \gg \mu_n$ or $\mu_n \gg \mu_p$ the recombination zone is limited to some particular position of the specimen, and this effect may be more apparent at low temperatures owing to different temperature dependence of mobility and trapping effects. Also, the smaller the value of $\langle v\sigma_R \rangle$ or the larger the value of w_r/d, the wider is the injected plasma region as expected.

4.3.2. Without recombination centres but with traps

From eqs. (4.22) and (4.24) it is not possible to have both deep electron traps and deep hole traps simultaneously. If $\sigma_n = \sigma_p$ and $n = p$ there will be no deep traps at all. But in most cases for double injection in organic semiconductors the concentration of one type of carrier (say, holes) is much larger than that of the other type (say, electrons), then it is possible that one type of trap (say, hole traps) may be located between E_{Fp} and E_{Dp} which may be considered deep traps (the depth depends on the ratio of $\sigma_p p/\sigma_n n$ and T) and the other type (say, electron traps) must be shallow traps located above E_{Fn} and above E_{Dn}. In general, traps which act as traps in double injection are shallow traps because deep traps existing under single injection conditions will become recombination centres under double injection conditions. Therefore, for double injection there may be shallow traps of both types, or shallow traps of one type and deep traps of the other type. In the following we shall consider only shallow traps.

The distribution functions for electron trap density $h_n(E, z)$ and for hole trap density $h_p(E, z)$ as functions of energy level E and the distance z from the hole-injecting contact are given by

$$h_n(E, z) = N_{tn}(E)S_n(z) \tag{4.167}$$

$$h_p(E, z) = N_{tp}(E)S_p(z) \tag{4.168}$$

where $N_{tn}(E)$, $N_{tp}(E)$, and $S_n(z)$, $S_p(z)$ represent, respectively, the energy and spatial distribution functions of electron and hole traps. If $S_n(z)$ and $S_p(z)$ are continuous and non-singular functions in the region between $z = 0$ and $z = d$, we can employ the regional approximation method to solve this problem by dividing the whole region $0 \le z \le d$ into M equal intervals as shown in Fig. 3.29. (In Fig. 3.29 the electrode axis is along the x-direction but in here it is along the z-direction.) The average values of $S_n(z)$ or $S_p(z)$ over a region between z_{i-1} and z_i can be approximated by the following constants:

$$b_{ni} = \frac{1}{z_i - z_{i-1}} \int_{z_{i-1}}^{z_i} S_n(z)dz = \frac{M}{d} \int_{z_{i-1}}^{z_i} S_n(z)dz \qquad (4.169)$$

and

$$b_{pi} = \frac{1}{z_i - z_{i-1}} \int_{z_{i-1}}^{z_i} S_p(z)dz = \frac{M}{d} \int_{z_{i-1}}^{z_i} S_p(z)dz \qquad (4.170)$$

To determine the $J-V$ characteristics for the region between z_{i-1} and z_i, we use

$$h_{ni}(E, z) = b_{ni} N_{tn}(E) \qquad (4.171)$$

and

$$h_{pi}(E, z) = b_{pi} N_{tp}(E) \qquad (4.172)$$

instead of eqs. (4.167) and (4.168), as if the traps are uniformly distributed in space. The voltage across the ith interval between z_{i-1} and z_i is given by

$$V_i = \int_{z_{i-1}}^{z_i} F_i dz \qquad (4.173)$$

where F_i is the electric field within the ith interval. Thus we can obtain a complete solution for the whole system by matching the solutions of individual intervals at the respective boundaries based on the continuity of individual fields F_i. The total voltage across the whole specimen is the applied voltage V, and thus

$$V = \sum_{i=1}^{M} V_i \qquad (4.174)$$

As there are no recombination centres for this case, the direct band-to-band (bimolecular) recombination is predominant. Thus the continuity equations (4.124) and (4.125) can be written as

$$\frac{1}{q}\frac{dJ_n}{dz} = R = np\langle v\sigma_R \rangle \qquad (4.175)$$

$$-\frac{1}{q}\frac{dJ_p}{dz} = R = np\langle v\sigma_R \rangle \qquad (4.176)$$

(A) The traps confined in single or multiple discrete energy levels

For a solid specimen with shallow traps confined in two discrete energy levels, the distribution functions for the electron trap and the hole trap densities may be written as

$$h_n(E, z) = H_{an}\delta(E - E_{tn})S_n(z) \qquad (4.177)$$

and

$$h_p(E, z) = H_{ap}\delta(E - E_{tp})S_p(z) \qquad (4.178)$$

where H_{an} and H_{ap} are the densities of electron and hole traps, respectively; E_{tn} and E_{tp}

are the electron and hole-trapping energy levels, respectively; $\delta(E - E_{tn})$ and $\delta(E - E_{tp})$ are the Dirac delta functions. Substitution of eqs. (4.177) and (4.178) into eqs. (4.129) and (4.130) gives

$$n_t = H_{an} g_n \exp[(E_{Fn} - E_{tn})/kT] S_n(z) \tag{4.179}$$

and

$$p_t = H_{ap} g_p^{-1} \exp[(E_{tp} - E_{Fp})/kT] S_p(z) \tag{4.180}$$

To simplify the mathematical treatment we introduce the following parameters:

$$\theta_{an}^{-1} = 1 + g_n H_{an} N_c^{-1} \exp[(E_c - E_{tn})/kT] S_n(z) \tag{4.181}$$

$$\theta_{ap}^{-1} = 1 + g_p^{-1} H_{ap} N_v^{-1} \exp[(E_{tp} - E_v)/kT] S_p(z) \tag{4.182}$$

$$v_{an} = (\mu_n \theta_{an})^{-1} \tag{4.183}$$

$$v_{ap} = (\mu_p \theta_{ap})^{-1} \tag{4.184}$$

$$S_a = q\mu_n n F/J \tag{4.185}$$

$$T_a = q\mu_p p F/J \tag{4.186}$$

$$U_a = qF^2/J \tag{4.187}$$

$$\alpha = 2q/\varepsilon \tag{4.188}$$

$$\beta = \langle v\sigma_R \rangle \theta_{an} \theta_{ap} \tag{4.189}$$

Using these parameters, eqs. (4.123)–(4.126) can be written as

$$S_a + T_a = 1 \tag{4.190}$$

$$\frac{dS_a}{dz} = \beta S_a T_a v_{an} v_{ap}/U_a \tag{4.191}$$

$$\frac{dT_a}{dz} = -\beta S_a T_a v_{an} v_{ap}/U_a \tag{4.192}$$

$$\frac{dU_a}{dz} = \alpha[T_a v_{ap} - S_a v_{an}] \tag{4.193}$$

Since v_{an} and v_{ap} are functions of the independent parameters z through the spatial distribution functions of traps $S_n(z)$ and $S_p(z)$, we have to employ the regional approximation method described earlier. Using this method, the problem reduces to the solution of eqs. (4.190)–(4.193) in the intervals $z_{i-1} \leq z \leq z_i$, within each of which v_{ani} and v_{api} are replacing v_{an} and v_{ap} and are constants. Thus, in the ith interval the solution is given by

$$U_{ai} = C_{ai} S_{ai}^{\alpha/v_{ani}\beta_i} (1 - S_{ai})^{\alpha/v_{ani}\beta_i} \tag{4.194}$$

where C_{ai} is the integration constant, and v_{ani} and v_{api} are, respectively, given by

$$v_{ani} = (\mu_n \theta_{ani})^{-1} = \mu_n^{-1} \{1 + g_n H_{an} N_c^{-1} \exp[(E_c - E_{tn})/kT] b_{ni}\}^{-1} \tag{4.195}$$

$$v_{api} = (\mu_p \theta_{api})^{-1} = \mu_p^{-1} \{1 + g_p^{-1} H_{ap} N_v^{-1} \exp[(E_{tp} - E_v)/kT] b_{pi}\}^{-1} \tag{4.196}$$

Since the electric field is continuous at each boundary between intervals, it can be easily shown that for

$$(F_i)_{z=z_i} = (F_{i+1})_{z=z_i} \tag{4.197}$$

the recurrence formula for C_{ai} is given by

$$C_{a(i+1)} = C_{ai} \frac{S_{ai}^{\alpha/v_{ani}\beta_i}[1 - S_{ai}]^{\alpha/v_{api}\beta_i}}{S_{a(i+1)}^{\alpha/v_{an(i+1)}\beta_{(i+1)}}[1 - S_{a(i+1)}]^{\alpha/v_{ap(i+1)}\beta_{(i+1)}}} \tag{4.198}$$

and for each interval, substitution of eq. (4.194) into eq. (4.191) gives

$$C_{ai} = \beta \, dM^{-1} v_{ani} v_{api} \left[\int_{S_{a(i-1)}}^{S_{ai}} \left\{ S_a^{\alpha/v_{ani}\beta_i - 1}(1 - S_a)^{\alpha/v_{api}\beta_i - 1} \right\} dS_a \right]^{-1} \tag{4.199}$$

We have M unknown C_{ai}'s and $M+1$ unknown S_{ai}'s. However, we can assume that the entire current at the anode is carried by holes and the entire current at the cathode is carried by electrons, therefore $S_{a0} = 0$ and $S_{aM} = 1$. Hence the number of unknowns reduces to $2M-1$. From eqs. (4.198) and (4.199) we have $2M-1$ simultaneous equations, so that these unknowns can be solved by an iterative technique.

From eqs. (4.173), (4.174), (4.187), and (4.194) we obtain the current–voltage relation [Kao and Elsharkawi 1976]

$$J = \frac{9}{8} \varepsilon \mu_{eff} \frac{V^2}{d_{eff}^3} \tag{4.200}$$

in which

$$\frac{\mu_{eff}}{d_{eff}^3} = \frac{8}{9} \frac{q\mu_n\mu_p}{\varepsilon \langle v\sigma_R \rangle} \sum_{i=1}^{M} \frac{\left[\int_{S_{a(i-1)}}^{S_{ai}} S_a^{\alpha/v_{ani}\beta_i - 1}(1 - S_a)^{\alpha/v_{api}\beta_i - 1} dS_a \right]^3 (M/d)^3}{\left[\int_{S_{a(i-1)}}^{S_{ai}} S_a^{3\alpha/2v_{ani}\beta_i - 1}(1 - S_a)^{3\alpha/2v_{api}\beta_i - 1} dS_a \right]^2} \tag{4.201}$$

and μ_{eff} and d_{eff} can be considered as the effective mobility of carriers and the effective thickness of the specimen, respectively. The ratio of μ_{eff}/d_{eff}^3 has taken into account the effect of non-uniform spatial distribution of traps. It can be easily shown that eqs. (4.200) and (4.201) reduce to those for a solid with uniform spatial distribution of traps by setting $b_{ni} = b_{pi} = 1$, and to those for a solid without traps by setting $b_{ni} = b_{pi} = 0$. It should also be noted that eqs. (4.200) and (4.201) can be considered to be the general equations for double injection with shallow traps of any type of trap distributions but without recombination centres. Different trap distribution functions have different θ_{an} and θ_{ap}, but the general J–V characteristics remain the same.

For uniform spatial distribution of traps, $S_n(z) = S_p(z) = 1$. This is equivalent to setting $b_{ni} = b_{pi} = 1$ in eqs. (4.200) and (4.201). By doing so, we obtain

$$J = \frac{9}{8} \varepsilon \mu_{eff} \frac{V^2}{d^3} \tag{4.202}$$

where

$$\mu_{eff} = \frac{8}{9} \frac{q\mu_n\mu_p}{\varepsilon \langle v\sigma_R \rangle} \frac{[B(\alpha/\beta v_{an}, \alpha/\beta v_{ap})]^3}{[B(3\alpha/2\beta v_{an}, 3\alpha/2\beta v_{ap})]^2} \tag{4.203}$$

where $B(m, n)$ is the β-function. Equation (4.203) is similar in form to eq. (4.140) except that β_0, μ_n, and μ_p have been replaced with β, v_{an}^{-1}, and v_{ap}^{-1}.

If the traps are confined in multiple discrete energy levels, such as a solid containing more than two different kinds of impurities, eqs. (4.202) and (4.203) are still valid provided that θ_{an} and θ_{ap} are expressed as

$$\theta_{an}^{-1} = \sum_j \theta_{anj}^{-1} \tag{4.204}$$

$$\theta_{ap}^{-1} = \sum_j \theta_{apj}^{-1} \tag{4.205}$$

where

$$\theta_{anj}^{-1} = 1 + g_{nj} H_{anj} N_c^{-1} \exp[(E_c - E_{tnj})/kT] \tag{4.206}$$
$$\theta_{apj}^{-1} = 1 + g_{pj}^{-1} H_{apj} N_v^{-1} \exp[(E_{tpj} - E_v)/kT] \tag{4.207}$$

in which g_{nj}, g_{pj}, H_{anj}, H_{apj}, E_{tnj}, and E_{tpj} refer to g_n, g_p, H_{an}, H_{ap}, E_{tn}, and E_{tp} in the jth single discrete energy level.

(B) The traps distributed Gaussianly in energy and uniformly in space

For this case the distribution functions for electron and hole trap densities are, respectively, given by

$$h_n(E) = \frac{H_{dn}}{(2\pi)^{1/2}\sigma_{tn}} \exp\left[-\frac{(E - E_{tn})^2}{2\sigma_{tn}^2}\right] \tag{4.208}$$

and

$$h_p(E) = \frac{H_{dp}}{(2\pi)^{1/2}\sigma_{tp}} \exp\left[-\frac{(E - E_{tp})^2}{2\sigma_{tp}^2}\right] \tag{4.209}$$

where $(2\pi)^{1/2}$ is the normalizing factor. For shallow traps we can assume that $E_{tn} > F_{Fn}$, $E_{tn} > E_{Dn}$, $E_{tp} < E_{Fp}$, and $E_{tp} < E_{Dp}$; that $n_t \gg n$ and $p_t \gg p$; and that there are no recombination centres so $R = np \langle v\sigma_R \rangle$. Thus we can write

$$n + n_t \simeq n_t = \theta_{dn}^{-1} n \tag{4.210}$$

$$p + p_t \simeq p_t = \theta_{dp}^{-1} p \tag{4.211}$$

where

$$\theta_{dn} = \frac{N_c}{g_n H_{dn}} \exp\left[(E_{tn} - E_c)/kT - \frac{1}{2}\left(\frac{\sigma_{tn}}{kT}\right)^2\right] \tag{4.212}$$

$$\theta_{dp} = \frac{g_p N_v}{H_{dp}} \exp\left[(E_v - E_{tp})/kT + \frac{1}{2}\left(\frac{\sigma_{tp}}{kT}\right)^2\right] \tag{4.213}$$

Using the same method employed in (A), we can obtain the equations for J–V characteristics in exactly the same form as eqs. (4.202) and (4.203) except that θ_{an} and θ_{ap}

in β, v_{an}, and v_{ap} have to be replaced with θ_{dn} and θ_{dp} given by eqs. (4.212) and (4.213).

If $\langle v\sigma_R \rangle$ is large such as the cases for most organic semiconductors, eq. (4.203) reduces to $\mu_{\text{eff}} = \theta_{dn}\mu_n + \theta_{dp}\mu_p$ and J in eq. (4.202) becomes simply the sum of electron and hole SCL currents.

(C) The traps confined in smeared discrete energy levels

Equations (4.202) and (4.203) are for the case with electron and hole traps confined, respectively, in discrete energy levels E_{tn} and E_{tp}. However, this case could also be considered as a case with a Gaussian distribution having very narrow trap energy deviations, σ_{tn} and σ_{tp} (i.e. σ_{tn}, $\sigma_{tp} \ll kT$). For such smeared shallow traps, eqs. (4.202) and (4.203) are applicable except that θ_{an} and θ_{ap} in β, v_{an}, and v_{ap} have to be replaced with θ'_{dn} and θ'_{dp} given below:

$$\theta'_{dn} = \frac{N_c}{g_n H_{dn}} \exp[(E_{tn} - E_c)/kT] \qquad (4.214)$$

$$\theta'_{dp} = \frac{g_p N_v}{H_{dp}} \exp[(E_v - E_{tp})/kT] \qquad (4.215)$$

Finally, we would like to mention that an alternative approach to this double injection problem can also be obtained by employing free carrier densities instead of local coordinates as the independent variables. Based on this alternative approach, we can easily deduce from eqs. (4.121)–(4.126) the following equation [Schwob and Zschokke-Gränacher 1972]:

$$\frac{d(\ln n)}{d(\ln p)} = \frac{\rho - \varepsilon R/\mu_n n}{\rho + \varepsilon R/\mu_p p} \qquad (4.216)$$

If the recombination rate constant, trap densities, and their distribution function are known, n can be calculated in terms of p. Thus we can write

$$n = G(p) \qquad (4.217)$$

and the crystal specimen thickness and the applied voltage in the form

$$d = \int_0^d dz = \frac{J\varepsilon}{q} \int_{p(z=0)}^{p(z=d)} \frac{\mu_p dp}{(\mu_n n + \mu_p p)(\mu_p p\rho + \varepsilon R)} \qquad (4.218)$$

$$V = \int_0^d F dz = \frac{J^2\varepsilon}{q^2} \int_{p(z=0)}^{p(z=d)} \frac{\mu_p dp}{(\mu_n n + \mu_p p)^2(\mu_p p\rho + \varepsilon R)} \qquad (4.219)$$

From eqs. (4.216)–(4.219) we can easily deduce expressions of J–V characteristics for all cases described above.

4.3.3. With recombination centres but without traps

Lampert [1962], Lampert and Mark [1970], Ashley and Milnes [1964], Migliorato *et al.* [1976] and also other investigators [Lampert and Schilling 1970, Baron and Mayer 1970] have used a model shown in Fig. 4.13 to analyse double injection in a

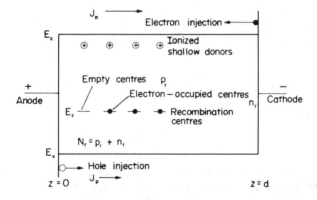

FIG. 4.13. Illustrating schematically a single set of recombination centres in a perfectly compensated semiconductor. N_r denotes the density of total recombination centres, n_r and p_r denote, respectively, the densities of electron-occupied and empty recombination centres.

perfectly compensated semiconductor. In this model the deep acceptor-like traps act as recombination centres, and shallow donors provide compensating electrons for these deep acceptors so that local electrical neutrality is preserved. A recombination centre is negatively charged when it is occupied by a compensating electron, but this negative charge is balanced out by the positive charge of the ionized shallow donor, maintaining local neutrality in thermal equilibrium. Similarly, the existence of a neutral unoccupied recombination centre implies the existence of an unionized shallow donor in thermal equilibrium. Double injection will disturb this thermal equilibrium local neutrality condition. The shallow donors can be assumed to provide only compensating electrons in thermal equilibrium and to play no role in the recombination and conduction processes. The recombination centres play the major role in the double injection. Under such a condition the capture cross-section for capturing holes is greater than the capture cross-section for capturing electrons in the recombination centres. This implies that the electron lifetime is greater than the hole lifetime. In this section we shall use this model to describe and to explain some important features of double injection.

In the steady state, the behaviour of double injection in a solid is governed by the current flow equations

$$J_n = q\mu_n nF + qD_n \frac{dn}{dz} \tag{4.220}$$

$$J_p = q\mu_p pF - qD_p \frac{dp}{dz} \tag{4.221}$$

$$J = J_n + J_p \tag{4.222}$$

the continuity equations

$$\frac{1}{q}\frac{dJ_n}{dz} = R \tag{4.223}$$

$$-\frac{1}{q}\frac{dJ_p}{dz} = R \tag{4.224}$$

the Poisson equation

$$\frac{dF}{dz} = \frac{q}{\varepsilon}\left[(p-p_0)-(n-n_0)+(p_r-p_{r0})\right] \tag{4.225}$$

and

$$V = \int_0^d F\,dz \tag{4.226}$$

The above equations include the current components due to diffusion and due to carriers thermally generated in the specimen. In thermal equilibrium $n = n_0, p = p_0, p_r = p_{r0}$, and the whole specimen is electrically neutral (there is no net space charge). The double injection causes the change to these thermal equilibrium quantities by an amount of

$$\Delta n = n - n_0 \qquad \Delta p = p - p_0 \qquad \Delta p_r = p_r - p_{r0} \tag{4.227}$$

Thus Δn and Δp are densities of injected free electron and hole carriers, respectively, and Δp_r is the density of injected trapped holes. If n_0, p_0 and p_{r0}, which are thermally generated, are ignored, then automatically $\Delta n = n$, $\Delta p = p$, and $\Delta p_r = p_r$.

Equations (4.220)–(4.226) are completely general and apply to all electrical transport problems in semiconductors and insulators. If the monomolecular recombination is predominant, and the bimolecular recombination can be ignored, then the rate of recombination may be written based on eqs. (4.66) or (4.67) as

$$\begin{aligned}
R &= C_n\left[(n_0 + \Delta n)(p_{r0} + \Delta p_r) - (n_{r0} - \Delta p_r)n_1\right] \\
&= C_n\left[\Delta n p_{r0} + \Delta p_r(\Delta n + n_0 + n_1)\right]
\end{aligned}$$

or

$$\begin{aligned}
&= C_p\left[(p_0 + \Delta p)(n_{r0} - \Delta p_r) - (p_{r0} + \Delta p_r)p_1\right] \\
&= C_p\left[\Delta p n_{r0} - \Delta p_r(\Delta p + p_0 + p_1)\right]
\end{aligned} \tag{4.228}$$

Analytical solutions of eqs. (4.220)–(4.226) are not possible without involving restrictive assumptions. In the following discussion we shall make such assumptions when necessary in order to simplify the problem so as to provide an insight into the physical nature of the behaviour of double injection. The carrier lifetimes τ_n and τ_p depend on the injection level and hence on Δp_r, and it is the changes of these lifetimes that are responsible for various features of the J–V characteristics.

(A) With small density of recombination centres (small N_r)

Because of small N_r we assume that Δp_r can be neglected in the Poisson equation. Thus by combining eqs. (4.223) and (4.224), and from eqs. (4.220)–(4.222), we obtain [Baron and Mayer 1970]

$$\underbrace{\left(\frac{b+1}{\mu_n}\right)R = (n_0 - p_0)\frac{dF}{dz}}_{A} \underbrace{-\frac{\varepsilon}{q}\frac{d}{dz}\left[F\left(\frac{dF}{dz}\right)\right]}_{B} + \underbrace{V_T\frac{d^2}{dz^2}(n+p)}_{C} \tag{4.229}$$

and

$$J = \frac{q\mu_n}{b}\left[(bn_0 + p_0)F + (b\Delta n + \Delta p)F + V_T \frac{d}{dz}(bn - p) \right] \qquad (4.230)$$

where $\quad b = \mu_n/\mu_p \quad$ and $\quad V_T = kT/q = D_n/\mu_n = D_p/\mu_p \qquad (4.231)$

Figure 4.14 shows the possible regimes that may appear in double injection J–V characteristics. The mechanisms for each regime is described as follows:

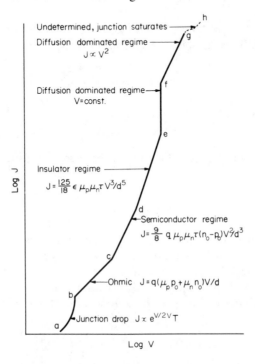

FIG. 4.14. Schematic illustration of possible regimes in the J–V characteristics due to double injection in a solid with small density of recombination centres. Depending on material parameters, some of these regimes may not appear. [After Baron and Mayer 1970.]

Regime a–b (junction-drop regime): the current is junction limited and proportional to $\exp(qV/2kT)$.

Regime b–c (ohmic regime): the thermally generated carriers n_0 and p_0 are predominant; and the current is simply given by

$$J = q(\mu_n n_0 + \mu_p p_0)V/d \qquad (4.232)$$

Regime c–d (semiconductor regime): of three terms on the right of eq. (4.229), only term A is dominant for semiconductors. Thus eq. (4.229) reduces to

$$\left(\frac{b+1}{\mu_n}\right)R = (n_0 - p_0)\frac{dF}{dz} \qquad (4.233)$$

If we assume that for applied voltages higher than the applied voltage corresponding to point c in Fig. 4.14, Δn and Δp are much greater than any of the quantities n_0, p_0, n_1, p_1, N_r, and the space charge $\Delta p - \Delta n$; then the recombination rate R can be expressed as

$$R = \frac{\Delta n}{\tau_n} \simeq \frac{n}{\tau_n}$$

$$= \frac{\Delta p}{\tau_p} \simeq \frac{p}{\tau_p} \tag{4.234}$$

Since it is assumed that $\Delta n, \Delta p \gg |\Delta p - \Delta n|$, we can assume $\Delta n \simeq \Delta p$ and hence $\tau_n \simeq \tau_p = \tau$, which leads to [see eqs. (4.74), (4.75), and (4.90)]

$$\tau = \tau_{p0} + \tau_{n0} = \frac{C_p + C_n}{C_p C_n N_r} \tag{4.235}$$

And eq. (4.230) reduces to

$$J = q\mu_p(b+1)nF \tag{4.236}$$

Thus, from eqs. (4.226), (4.233), (4.236), and the boundary condition given by eq. (4.120), we obtain

$$J = \frac{9}{8}q\mu_n\mu_p(n_0 - p_0)\tau\frac{V^2}{d^3} \tag{4.237}$$

It should be noted that the assumption that the space charge is small compared with the injected charge is valid as long as $\varepsilon(b+1)/q\mu_n(n_0 - p_0) \ll \tau$ based on eq. (4.233), and that an increase in applied voltage will cause an increase in injection level and hence n, which will in turn lead to breakdown of this condition.

Regime d–e (insulator regime): for this regime, term B is dominant. Thus eq. (4.229) reduces to

$$\frac{(b+1)}{\mu_n}\frac{n}{\tau} = -\frac{\varepsilon}{q}\frac{d}{dz}\left[F\left(\frac{dF}{dz}\right)\right] \tag{4.238}$$

This regime can occur in insulators in which $n_0 - p_0$ is very small or approaches to zero or for semiconductors at sufficiently high injection levels. Equations (4.234)–(4.236) are still applicable for this regime. Thus, solution of eqs. (4.236) and (4.238) with the aid of eqs. (4.120) and (4.226) yields [Baron and Mayer 1970]

$$J = \frac{125}{18}\varepsilon\mu_n\mu_p\tau\frac{V^3}{d^5} \tag{4.239}$$

In this regime the space charge located near the carrier-injecting contacts controls the current and the recombination kinetics.

Regime e–f (diffusion-dominated regime): if $n = p$, and n and p become sufficiently large at very high injection levels, the diffusion-dominated region near the contacts becomes predominant in the conduction processes. When this regime is set on, eqs. (4.229) and (4.230) reduce to

$$\frac{b+1}{\mu_n}\frac{n}{\tau} = 2V_T\frac{d^2n}{dz^2}$$

or

$$\frac{d^2n}{dz^2} = \frac{n}{L_a^2} \tag{4.240}$$

and

$$J = q\mu_p \left[(b+1)nF + V_T(b-1)\frac{dn}{dz} \right] \tag{4.241}$$

where L_a is the ambipolar diffusion length which is given by

$$L_a = \left[\frac{2V_T\mu_n\tau}{b+1} \right]^{1/2} \tag{4.242}$$

Several investigators [Herlet *et al.* 1955, Kleinman 1956, Lampert and Rose 1961, Baron 1965, Baron and Mayer 1970] have attempted to solve eqs. (4.240) and (4.241) under various boundary conditions. The term "diffusion dominated" means that the densities of injected carriers are determined primarily by diffusion processes rather than by the applied electric fields. The solution of eq. (4.240) is

$$n = \frac{n(0)\sinh[(d-z)/L_a] + n(d)\sinh(z/L_a)}{\sinh(d/L_a)} \tag{4.243}$$

where $n(0)$ and $n(d)$ are, respectively, the electron densities at $z = 0$ and $z = d$. The effective field (the diffusion field) in the first diffusion length L_a is $kT/qL_a = V_T/L_a$. To preserve a constant total current, the field along the specimen would be

$$F(z) \simeq (V_T/L_a)\exp(z/L_a), \quad 0 \le z \le d/2 \tag{4.244}$$

and so the voltage across one half the specimen would be

$$V' \simeq \int_0^{d/2} F\,dz = (kT/q)[\exp(d/2L_a) - 1]$$

$$\simeq (kT/q)\exp(d/2L_a),\ d \gg L_a$$

$$\simeq (kT/q)(d/2L_a),\ d \ll L_a \tag{4.245}$$

The significant feature about the diffusion-dominated double injection is that once the voltage across the specimen reaches twice the value given by eq. (4.245) the voltage across the specimen remains constant as the current increases rapidly. The additional applied voltage is consumed at the contacts to increase the injected carrier densities at the contacts and gives rise to an exponential increase in current as shown in regime *e–f* of Fig. 4.14. The analysis of Herlet *et al.* [1955] has indicated that at voltages smaller than that given by eq. (4.245) the *J–V* characteristic is approximately ohmic, and that when the applied voltage approaches the threshold needed for the onset of the diffusion-dominated solution, namely

$$V = 2(kT/q)[\exp(d/2L_a) + d/2L_a] \tag{4.246}$$

the current increases exponentially as

$$J = (\mu n_0 kT/L_a)\exp(q\Delta V/2kT) \tag{4.247}$$

where ΔV is the applied voltage in excess of that given by eq. (4.246). In eq. (4.246) the first term on the right is the voltage across the section required for the diffusion solution, and the second term on the right is the voltage required across each junction to give the enhanced density [Herlet *et al.* 1955, Rose 1964].

Regime *f–g* (diffusion-dominated regime with high injection): following Kleinman's analysis [1956] for high injection levels, the *J–V* characteristics follow the relatively simple relation [Rose 1964, Baron and Mayer 1970]

$$J = AV^2 \qquad (4.248)$$

The square-law dependence occurs when the voltage drop across the bulk exceeds $(2kT/q)\exp(d/2L_a)$ and is large compared with the voltage drop across the junction contacts.

Regime *g–h* (undetermined, junction saturation regime): for applied voltages higher than that corresponding to point *g* in Fig. 4.14 the current becomes junction limited and controlled by the carrier injection process. It tends to reach the junction saturation current.

For a detailed discussion of all these regimes, the reader is referred to the excellent articles of Rose [1964] and Baron and Mayer [1970]. It should be noted that some of these regimes may not appear depending on the structure of the material specimens and the carrier-injecting contacts.

(B) With large density of recombination centres (Large N_r)

Because of large N_r we cannot neglect Δp_r in the Poisson equation. As Δp_r depends on its own energy level and the injection level, the inclusion of the term Δp_r complicates the problem. For partially or completely filled recombination centres, Lampert and Schilling [1970] have analysed this problem using the regional approximation method. The specimen is divided into regions which either satisfy the quasi-neutrality approximation or are dominated by trapped or free space charge. For partially filled recombination centres with the effect of thermally generated carriers ignored, Ashley and Milnes [1964] have derived the *J–V* characteristics which is generally referred to as the Ashley–Milnes space charge regime.

(i) *Ashley–Milnes space charge regime.* Ashley and Milnes [1964] have analysed the double injection problem using the same model shown in Fig. 4.13 and the following assumptions:

(a) The location of the recombination centres is sufficiently deep so that the densities of thermally generated carriers (n_0 and p_0) are much less than N_r.

(b) The injected free electron density $\Delta n > n_0$ and $\Delta n > p_0$ but the injected hole density $\Delta p \simeq p_0$ because of the recombination barrier to hole injection. Thus $n = \Delta n + n_0 \simeq \Delta n$, $p = \Delta p + p_0 \simeq p_0$.

(c) The lifetime for electrons is constant (for the present case the major carriers are electrons because the deep acceptor-like recombination centres act as a barrier to hole injection) $\tau_n \simeq \tau_{n0} = (\langle v\sigma_n \rangle p_{r0})^{-1}$.

(d) Since Δp is small, $d(p_0 F)/dz \gg d(\Delta pF)/dz$.

(e) Because of the above assumptions $\sigma_p \gg \sigma_n$ and hence $\tau_p \ll \tau_n$.
(f) The current component due to diffusion is neglected.

The behaviour of double injection in a solid is governed by the current flow equation

$$J = J_n + J_p = q(\mu_n n + \mu_p p_0)F \tag{4.249}$$

the continuity equations

$$-\frac{1}{q}\frac{dJ_p}{dz} = R = \frac{n}{\tau_n} = np_r \langle v\sigma_n \rangle \tag{4.250}$$

and the Poisson equation

$$\frac{dF}{dz} = \frac{q}{\varepsilon}[p - n + p_r - p_{r0}] \simeq \frac{q}{\varepsilon}(\Delta p_r - n) \tag{4.251}$$

From eq. (4.249) we obtain

$$\frac{dJ_p}{dz} = q\mu_p p_0 \frac{dF}{dz} \tag{4.252}$$

From eqs. (4.250)–(4.252) it can easily be shown that

$$\Delta p_r = \left(1 - \frac{\tau_d}{\tau_n}\right)n \tag{4.253}$$

where τ_d is the dielectric relaxation time which is given by

$$\tau_d = \varepsilon/q\mu_p p_0 \tag{4.254}$$

Equation (4.253) means that the trapped space charge density is proportional to the free electron density. By neglecting the current contribution of holes in eq. (4.249) and using the boundary condition $V = \int_0^d F\,dz$, and from eqs. (4.249), (4.251), and (4.253), we obtain

$$J = \frac{9}{8}\frac{\tau_n}{\tau_d}\varepsilon\mu_n\frac{V^2}{d^3} \tag{4.255}$$

From eqs. (4.253) and (4.255) the following conclusions can be drawn:
(1) If $\tau_d/\tau_n \gg 1$ the space charge is large and mainly due to trapped charge in the centres. This may affect the validity of assumption (c). This also means that $\Delta p_r < 0$ so that τ_p tends to decrease and τ_n tends to increase with increasing injection level. When τ_n approaches infinity and τ_p approaches zero, the transition from the Ashley–Milnes regime to trap free single carrier SCL current ensues. However, this trap-free SCL current regime will, in turn, have a transition to a higher level double injection regime when more injected holes traverse the bulk resulting in a decrease of electron lifetime and an increase of hole lifetime as the injection level increases. Whenever $\tau_d/\tau_n \gg 1$, J is smaller than the trap-free single carrier SCL current. However, J would be larger than the trap-free single carrier SCL current if the centres are shallow traps only.
(2) If $\tau_d/\tau_n \ll 1$, the space charge is small and the quasi-neutrality assumption is valid. This implies that $\Delta p_r > 0$, so that τ_p tends to increase and τ_n to decrease with increasmg injection level, resulting in the transition to a higher level double injection regime.

(3) When $n < p_0$, Ohm's law predominates and

$$J = q\mu_p p_0 V/d \qquad (4.256)$$

The two regimes, Ohm's law and square law, intersect at the transition voltage

$$V_\Omega \simeq \frac{d^2}{\mu_n \tau_n} \qquad (4.257)$$

This occurs when the electron lifetime is approximately equal to the electron transit time $[t_t \simeq d/(V_\Omega \mu_n/d)]$.

Figure 4.15(a) shows the expected $J–V$ characteristics and Fig. 4.15(b) shows the typical $J–V$ characteristics of the indium-doped silicon p^+–π–n^+ structure as functions of temperature. These results indicate a very good agreement between the experiment and the theory of Ashley and Milnes.

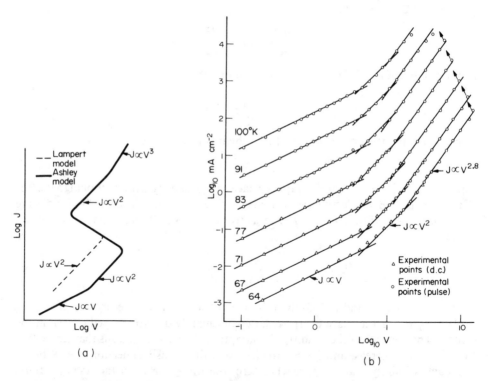

FIG. 4.15. (a) Expected $J–V$ characteristics based on the Ashley–Milnes model, and (b) the experimental $J–V$ characteristics for double injection in 15.9 mil-thick indium-doped silicon of p^+–π–n^+ structure as a function of temperature. Triangles are experimental points taken with direct current and circles taken with pulse techniques. [After Wagener and Milnes 1965.]

(ii) *Negative differential resistance region.* In the square law region prior to the negative differential resistance (NDR) region, the important difference between the model of Ashley and Milnes and that of Lampert [1962] is the fact that in the former model the space charge is predominantly that of trapped injected electrons in the recombination centres rather than the free electrons themselves, while in the Lampert's model the current is a trap-free SCL electron current which recombines with injected holes within a diffusion length and is proportional to the square of voltage, the space charge being mainly that of free electrons. However, at a certain critical voltage the recombination barrier to the flow of holes decreases and hence the hole lifetime increases with increasing injection level. The more the holes are injected, the easier it is for them to traverse the solid. The tendency is to allow more holes to flow at lower voltages, and subsequently to set on the current-controlled NDR region. This phenomenon was first predicted by Stafeev [1959] and Lampert and Rose [1961] and it has been confirmed experimentally to occur in Si, Ge, GaAs, InSb, and CdS by many investigators [Stafeev 1962, Holonyak and Bevacque 1963, Holonyak 1962, Weiser and Levitt 1964, Litton and Reynolds 1964, Barnett and Milnes 1966].

On the basis of the following assumptions:

(a) the current is volume controlled and no constraints on the current are imposed by the carrier-injecting contacts;
(b) the structure is a p^+–i–n^+ long structure with the i-region several ambipolar diffusion lengths long so that the diffusion current component can be neglected;
(c) the carrier mobilities are field independent; and
(d) the reombination centres are acceptor-like deep traps,

the behaviour of the double injection current can be described by the following equations:

$$J = q(\mu_n n + \mu_p p)F \qquad (4.258)$$

$$\mu_n \frac{d(nF)}{dz} = R \qquad (4.259)$$

$$\frac{dF}{dz} = \frac{q}{\varepsilon}[(p - p_0) - (n - n_0) + (p_r - p_{r0})] \qquad (4.260)$$

where

$$R = \frac{N_r C_n C_p (pn - n_i^2)}{C_n(n + n_1) + C_p(p + p_1)} \qquad (4.261)$$

$$N_r - p_r = \frac{N_r(C_n n + C_p p_1)}{C_n(n + n_1) + C_p(p + p_1)} \qquad (4.262)$$

An analytical and exact solution of eqs. (4.258)–(4.262) is not possible. However, Deuling [1970] has obtained an exact numerical solution using a Rung–Kutta method with a computer for gold-doped silicon. The comparison of this theoretical exact solution with Lampert's quasi-neutrality approximation (dF/dz is set equal to zero) [Lampert 1962] is shown in Fig. 4.16. Deuling has used the model similar to that shown in Fig. 4.13 and the following physical constants for gold-doped silicon for his calculations:

N_r (gold impurities in silicon) $= 10^{16}$ cm^{-3}
$E_F - E_c = 0.485$ eV (fixed Fermi level)

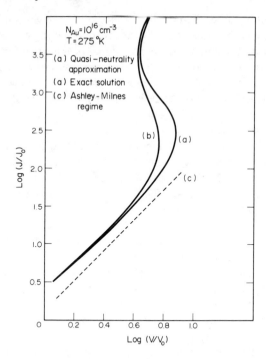

F IG. 4.16. The calculated *J–V* characteristics of gold-doped silicon based on (a) the quasi-neutrality approximation, (b) the exact numerical solution, and (c) the square law for the Ashley–Milnes regime. [After Deuling 1970.]

$E_r - E_v = 0.620$ eV (acceptor-like recombination centre energy level)
$C_n = 1.65 \times 10^{-9}$ cm^3 sec^{-1} (electron capture coefficient)
$C_p = 1.15 \times 10^{-7}$ cm^3 sec^{-1} (hole capture coefficient)

The voltage V is normalized to $V_0 = d^2(C_n N_r/\mu_n)$ and the current J is normalized to $J_0 = \varepsilon \mu_n d(C_n N_r/\mu_n)^2$. It can be seen that in Fig. 4.16 at low voltages the curves agree rather well and exhibit approximately a square law. In the NDR region there is a large discrepancy between the curves. The dotted curve for the Ashley–Milnes regime is also included for comparison purposes. However, the quasi-neutrality approximation gives a general shape of *J–V* characteristics although it overemphasizes the NDR region. Deuling [1970] has also calculated the *J–V* characteristics as functions of the temperature, and his results are shown in Fig. 4.17. The lower the temperature, the longer is the NDR region, and this agrees with the experimental results of Bykovskii *et al.* [1968].

It should be noted that when the injection level becomes so high in the NDR region, a transition will occur to convert the NDR back to the positive differential resistance but at higher current levels. After such a transition, $J \propto V^2$ in the semiconductor regime and $J \propto V^3$ in the insulator regime follow a similar way to that shown in Fig. 4.14. If at such high injection levels, $n, p > N_r$, the *J–V* curves may be very close to those of high current levels in Fig. 4.14.

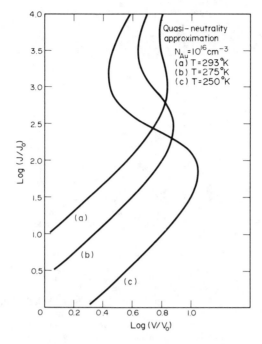

FIG. 4.17. The negative differential resistance regions of the J–V characteristics of gold-doped silicon as functions of temperature calculated in the quasi-neutrality approximation. [After Deuling 1970.]

For details about the regional approximation method based on the quasi-neutrality approximation, the reader is referred to the review of Lampert and Schilling [1970] and Lampert and Mark [1970]. For details about other approximation methods to deal with double injection problems the reader is referred to the review of Baron and Mayer [1970].

Some typical experimental results showing the transition from the square law region to the NDR region are given in Fig. 4.18 for gold-doped *n*-GaP. In organic semiconductor anthracene crystals with oleum as hole-injecting contact and lithium in $(CH_2)_2(NH_2)_2$ as electron injecting contact for double injection, Mehl and Büchner [1965] have observed a cubic law dependence in the J–V characteristics [eq. (4.239)] as shown in Fig. 4.19.

So far we have considered only the solids containing only one set of recombination centres. For double injection in solids with multivalent recombination centres, Zwicker *et al.* [1970] have calculated the J–V characteristics of p^+–i–n^+ structures in which the *i*-region contains multivalent recombination centres based on a numerical solution of the general equations of conduction and recombination in a manner similar to that adopted by Deuling [1970] for treating the case with recombination centres of a single discrete energy level.

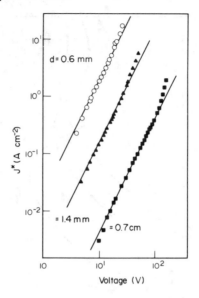

Fɪɢ. 4.18. The *J–V* characteristics of gold-doped *n*-GaP crystals with various specimen thicknesses. The data are taken from the results of Migliorato *et al.* [1973] and the straight lines with slope of 2 are drawn to indicate the square law. $J^* = (J - q\mu_n n_0 V/d)^3/J^2$. [After Cerrina, Margaritondo, Migliorato, Perfetti, and Salusti 1975.]

(C) The effect of magnetic field on negative differential resistance

The effect of magnetic field on double injection NDR has been studied and observed experimentally by several investigators [Melngailis and Rediker 1962, Nordman and Kvinlaug 1968, Merinsky *et al.* 1968, Karakushan and Stafeev 1961]. Recently, Otani *et al.* [1970] have presented a more quantitative theory on this effect. They considered a p^+–π–n^+ structure in which the π-region consists of a semiconductor (or a semi-insulator) of large dimension relative to the diffusion length of injected minority carriers, and in which there are donor-like recombination centres which are empty in thermal equilibrium so that $\tau_n > \tau_p$ (this type of recombination centre is different from what we have discussed so far in this section). The equations governing the conduction and recombination processes are:

$$\mathbf{J}_n = q\mu_n n\mathbf{F} + qD_n\nabla n - \mu_n\mathbf{J}_n \times \mathbf{B} \tag{4.263}$$

$$\mathbf{J}_p = q\mu_p p\mathbf{F} - qD_p\nabla p + \mu_p\mathbf{J}_p \times \mathbf{B} \tag{4.264}$$

$$\frac{\partial n}{\partial t} = (1/q)\nabla \cdot \mathbf{J}_n - R \tag{4.265}$$

$$\frac{\partial p}{\partial t} = (-1/q)\nabla \cdot \mathbf{J}_p - R \tag{4.266}$$

$$\nabla \cdot \mathbf{F} = (q/\varepsilon)\left[(p - p_0) - (n - n_0) - n_r\right] \tag{4.267}$$

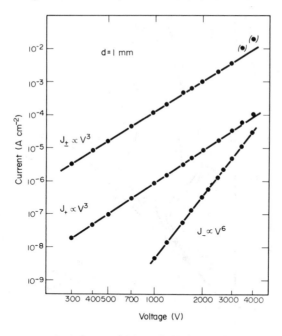

FIG. 4.19. Current–voltage characteristics for double hole $(J_{\pm})$, single hole (J_{+}), and single electron (J_{-}) current injections in anthracene crystals with oleum as hole-injecting contact and lithium in $(CH_2)_2(NH_2)_2$ as electron-injecting contact. [After Mehl and Büchner 1965.]

where **B** is the magnetic field, n_r is the density of electrons captured in the recombination centres, ∇ is the del operator, $\rightarrow$ indicates the quantities expressed in vector form, and the other symbols have the usual meanings. By assuming that the diffusion terms in eqs. (4.263) and (4.264) are small and can be neglected, Otani *et al.* [1970] have solved eqs. (4.263)–(4.267) by means of numerical computations. Some of their typical results are shown in Fig. 4.20. These results were computed using the following parameters

$$b = \mu_n/\mu_p = 50 \quad (\mu_n = 5 \times 10^5, \; \mu_p = 10^4 \text{ cm}^2/\text{V-cm})$$

$$d = 0.2 \text{ cm}$$

$$p_0 \gg n_0$$

$$p_0 = 10^{13} \text{ cm}^{-3}$$

$$N_r = 10^{14} \text{ cm}^{-3}$$

$$\beta = \langle v\sigma_n \rangle / \langle v\sigma_p \rangle = 10^3 \quad (\langle v\sigma_p \rangle = 10^{-8} \text{ cm}^3 \text{ sec}^{-1})$$

The voltage V is normalized to $V_1 = d^2 \langle v\sigma_p \rangle N_r/\mu_n$ and J is normalized to $J_1 = q\langle v\sigma_p \rangle p_0 N_r d$. For InSb, $J_1 = 0.32 \text{ A cm}^{-2}$, and $V_1 = 0.08 \ V$ in Fig. 4.20. These theoretical results are in reasonable agreement with the experimental results of Melngailis and Rediker [1962] in InSb.

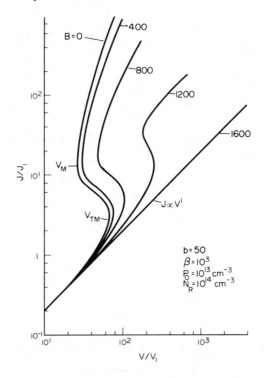

FIG. 4.20. Computed results showing the effect of magnetic field on the J–V characteristics of a p^+–π–n^+ structure. The results are in reasonably close agreement with the experimental results of Melngailis and Rediker [1962] in InSb. [After Otani, Matsubara, and Nishida 1970.]

The major conclusions from the work of Otani *et al.* [1970] are:

(a) The threshold voltage for the onset of the NDR region is strongly dependent on the ratio of recombination centre density N_r to the majority carrier density p_0 and the magnetic field.

(b) The effect of magnetic field in the double injection regime mainly depends on the recombination process for the injected minority carriers which are electrons for the present case. This differs from the effect in the ohmic regime where majority carriers contribute to the current.

(c) The current J in the double injection regime beyond a minimum voltage V_M in Fig. 4.20 follows a $J \propto V^{2+s}$ law where s is a function of the mobility ratio μ_n/μ_p and is approximately equal to $2[(\mu_n/\mu_p)^{\log 2} - 1]$.

The effect of magnetic field on single injection SCL currents has also been investigated by several investigators [Mortensen *et al.* 1971, Korn 1969].

4.3.4. With recombination centres and with traps

In the previous sections we have discussed the occurrence of the following regimes in sequence when the injection current level is increased: (1) linear Ohm's law; (2) lower level square law; (3) vertical transition; (4) breakover or NDR; (5) upper level square law; and (6) cube law. The double injection current is limited by recombination alone or by recombination in conjunction with space charge. The smaller the carrier lifetime, the greater is the carrier annihilation through electron–hole recombination and hence the greater will be the departure from the plasma condition (i.e. $n \simeq p$). Conversely, the smaller the carrier transit time t_t the more are the carriers to contribute to the formation of a plasma. Thus, the quasi-neutrality approximation or the injected plasma condition can be satisfied only when $t_{tn}/\tau_n \ll 1$ and $t_{tp}/\tau_p \ll 1$. Obviously, the presence of traps located above E_{Dn} or below E_{Dp} will affect t_t and τ, and thus affect not only the space charge distribution but also the recombination and conduction processes. Keating [1963, 1964] and Ashley *et al.* [1973] have considered the double injection involving both recombination centres and shallow traps. By expressing all the basic equations governing the behaviour of double injection in terms of a very useful variable, the ratio

$$u = n/p \qquad (4.268)$$

Keating [1964] has analysed the J–V characteristics by means of numerical calculations for a solid containing a set of deep recombination centres, and shallow hole and electron traps. His major results are: (a) the existence of a NDR region is a consequence of unequal lifetimes of both types of carriers in the absence of trapping. (b) The threshold voltage for the onset of double injection current flow with space charge neutrality is infinite if the recombination centres are not completely filled in the absence of injected carriers. (c) The J–V characteristics in the low level region for initially completely filled centres are very different from those for initially partly filled centres. (d) The relative effectiveness of hole and electron traps in storing carriers is an important factor in controlling the current because of the space charge which can be stored in traps and which has to be neutralized by carriers of the other type.

Schwob and Zschokke-Gränacher [1971] have also reported that the traps play an important role in the double injection J–V characteristics, and that free-trapped carrier recombination is the dominant recombination mechanism. By dividing the crystal specimen into three regions: (a) hole injection region ($p \gg n$ and n can be neglected), (b) space charge free region (traps are emptied by recombination and the thickness of this region is proportional to applied voltage), and (c) electron injection region ($n \gg p$ and the thickness of this region is much smaller than the specimen thickness), and employing the regional approximation method, they have derived a simple expression for double injection J–V characteristics and it is given by

$$J = A \left(\frac{V_A}{V_A - V} \right)^{m-1} \frac{V^m}{d^{2m-1}} \qquad (4.269)$$

where A is a function of trap parameters and physical parameters of the crystal, V_A is

$$V_A = d^2/\mu_n \tau_n (\Delta \ln n) \qquad (4.270)$$

$\Delta \ln n$ is the difference in value of $\ln n$ between the two boundaries of the space charge free region. Equation (4.269) agrees qualitatively with the experimental results for

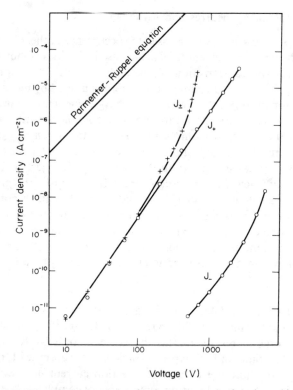

FIG. 4.21. The *J–V* characteristics of anthracene crystals of specimen thickness
$d = 0.62 \pm 0.06$ mm. J_- denotes single electron injection, J_+ single
hole injection, and $J_\pm$ double injection. For double injection +
represents the experimental points and the solid line represents the
theoretical curve based on eq. (4.269). [After Schwob and Zschokke-
Grànacher 1971.]

anthracene crystals with the value of *m* between 3 and 4 for low injection levels and with
$m > 6$ for higher injection levels as shown in Fig. 4.21.

4.3.5. The scaling rule

We can now realize that the double injection problem is so complex that a unified
approach taking into account all the effects of trapping and recombination centres,
even with the effects of surface states and contact potential barriers at the interfaces
between the material and the carrier-injecting contacts ignored, is simply not available.
However, several approximation methods to handle this problem are available and
have been briefly discussed in Sections 4.3.1–4.3.4. In Section 3.2.2 we have discussed
the scaling law for the theory of single injection SCL currents. For the general,
simplified theory of double injection currents, Lampert and Mark [1970] have also
derived the scaling law which is given by

$$\frac{J}{d} = f\left(\frac{V}{d^2}\right) \tag{4.271}$$

This equation is similar to eq. (3.81). For details about the scaling law the reader is referred to the book of Lampert and Mark [1970] and Section 3.2.2. It can be imagined, though there is no direct proof, that if the traps are distributed non-uniformly in space, the scaling law for double injection would be in exactly the same form as the scaling law for single injection

$$\frac{J}{d_{\text{eff}}} = f\left(\frac{V}{d_{\text{eff}}^2}\right) \tag{4.272}$$

with d replaced with d_{eff}. Equation (4.200) may be considered as an example following this scaling law.

CHAPTER 5

Electrical Transport under Special Conditions

5.1. FILAMENTARY CHARGE CARRIER INJECTION IN SOLIDS

It is well known that for either single or double injections, different materials used for electrodes or different techniques employed for preparation of the surfaces of semiconductors or insulators result in different values of critical voltages for transition from regime to regime. This indicates that there are no perfect ohmic contacts, and that different electrode materials in contact with a crystal surface will form different potential barriers for carrier injection. It can also be imagined that the interface between an electrode and a crystal surface which is not microscopically identical from domain to domain in asperity and surface conditions is never homogeneous and uniform. Thus there may be one or more microregions at which the potential barrier has a profile more favourable to carrier injection than at other regions of the interface. Furthermore, the crystal itself is never microscopically homogeneous and uniform. For all these unavoidable imperfections the current passing through a crystal specimen is filamentary at least from a microscopic point of view and particularly under high fields. For an electric field applied to the specimen longitudinally, the field will not be uniformly longitudinally due to the effect of space charge, and the current density will not be uniform radially due to the formation of filamentary paths. The current filaments formed in Si, GaAs, ZnTe, $GaAs_xP_{1-x}$, and polycrystalline silicon have been observed by Barnett et al. [1966, 1968, 1969, 1970] in anthracene and in chalcogenide glasses observed by Kao and his co-workers [1973, 1974, 1975, 1976]. In the following we present a theoretical model for the filamentary injection and to show that the model can be used to explain some experimental aspects of single and double injection.

5.1.1. Filamentary one-carrier (single) injection

In the theoretical analysis we consider only electron injection and make the following assumptions:

    (i) At the voltage of or higher than the threshold voltage for the onset of carrier injection there may be one filament or more than one formed between electrodes. But for mathematical simplicity we use cylindrical coordinates and consider only one filament formed along the z-axis, which coincides with the central line joining the two circular plane electrodes. The filament radius is r_d. The whole system is symmetrical about the z-axis.

(ii) In the filament the longitudinal component of the diffusion current can be ignored because of the large longitudinal component of the electric field, and the radial component of the drift current can be ignored because of the small radial component of the electric field.

(iii) The free electron density follows the Maxwell–Boltzmann statistics, while the trapped electron density follows the Fermi–Dirac statistics.

(iv) The mobility of the free electrons μ_n is not affected by the presence of traps nor by the high electric field.

(v) The treatment is two-dimensional with the plane at $z = 0$ as the electron-injecting contact and that at $z = d$ as the electron-collecting contact, the specimen thickness being d.

(vi) The thermally generated electron concentration is negligible.

The behaviour of single injection in the steady state is governed by the current flow equation

$$\mathbf{J}(r) = \mathbf{J}_z + \mathbf{J}_r$$

$$= q\mu_n nF\,\mathbf{i}_z - qD_n\left(\frac{\partial n}{\partial r}\right)\mathbf{i}_r \tag{5.1}$$

the continuity equation

$$\mu_n\left(\frac{\partial}{\partial z}\right)(nF) \simeq 0 \tag{5.2}$$

$$\mu_n\frac{\partial(nF)}{\partial r} - \frac{D_n}{r}\frac{\partial}{\partial r}\left(r\frac{\partial n}{\partial r}\right) = 0 \tag{5.3}$$

and the Poisson equation

$$\nabla F = \frac{q}{\varepsilon}(n + n_t) \tag{5.4}$$

where $\mathbf{i}_z$ and $\mathbf{i}_r$ are, respectively, the unit vectors in the z- and r-directions, and the other symbols have the usual meanings.

We shall solve first eq. (5.3) for the radial variation of n. Since F in the radial direction is assumed to be negligibly small but dF/dr is not, it should follow the Poisson equation. Thus, eq. (5.3) becomes

$$\frac{d^2n}{dr^2} + \frac{1}{r}\frac{dn}{dr} = \frac{q\mu_n}{\varepsilon D_n}n(n + n_t) \tag{5.5}$$

For trap free solids $n_t = 0$ and for solids with shallow traps we can write

$$\theta_n = \frac{n}{n + n_t} \tag{5.6}$$

An examination of eq. (5.5) shows that when r approaches to zero, n would approach to infinity. Physical reality requires n to be finite for all values of r, and this demands that $dn/dr \to 0$ when $r \to 0$. However, eq. (5.5) cannot be rigorously solved without the aid of numerical computation. To emphasize the physical picture of the problem, a simple analytical solution is far better than a computer solution. Therefore, we have to make an approximation. It is postulated [Barnett 1970] that the solution of eq. (5.5) by neglecting the term $\dfrac{1}{r}\dfrac{dn}{dr}$ is a good approximation to the exact solution of eq. (5.5). On the basis of this approximation and the assumption that the traps are shallow traps, eq. (5.5) reduces to

$$\frac{d^2 n}{dr^2} = \frac{q\mu_n}{\varepsilon D_n}\theta_n^{-1}n^2 \tag{5.7}$$

The solution of eq. (5.7) is

$$\frac{dn}{dr} = -\left(\frac{2}{3}\frac{q\mu_n n^3}{D_n \varepsilon \theta_n}+C_1\right)^{1/2} \tag{5.8}$$

where C_1 is the constant of integration. Using the boundary conditions

$$r \to 0, \; n \to n_0,$$

$$r \to \infty, \quad n \to 0, \quad \text{and} \quad \frac{dn}{dr} \to 0 \tag{5.9}$$

we can solve for C_1. Then integrating again, we obtain

$$n = n_0 \bigg/ \left(1+\frac{n_0}{6}\frac{q\mu_n}{D_n \varepsilon \theta_n}r\right)^2 \tag{5.10}$$

where n_0 is the electron density at the centre of the filament, which can be determined by solving eq. (5.2) and relating it to the current density J_{z0} at $r = 0$. Thus we have

$$n_0 = J_{z0}/q\mu_n F \tag{5.11}$$

Using the boundary condition

$$F \to F_c \quad \text{when} \quad z \to 0 \tag{5.12}$$

and

$$V = \int_0^d F\,dz = F_{av}d \tag{5.13}$$

the solution of eqs. (5.4) and (5.11) at $r = 0$ for $n_t \gg n$ yields [Kao 1976]

$$J_{z0} = \frac{9}{8}\varepsilon\mu_n\theta_n\frac{F_{av}^2}{d}M = \frac{9}{8}\varepsilon\mu_n\theta_n\frac{V^2}{d^3}M \tag{5.14}$$

where

$$M = -\frac{4\,Y^2-3}{6}+\left[\left(-\frac{4\,Y^2-3}{6}\right)^2-\frac{16\,Y^3(\,Y-1)}{27}\right]^{1/2} \tag{5.15}$$

$$Y = F_c/F_{av} \tag{5.16}$$

V is the applied voltage, F_{av} is the average value of F along the z-direction, and F_c is the electric field at the cathode ($z = 0$).

Equation (5.11) can also be expressed by approximation as

$$J_{z0} = q\mu_n \bar{n}_0 F_{av} \qquad (5.17)$$

where $\bar{n}_0$ is the average carrier density at the centre of the filament (at $r = 0$). Comparing eq. (5.17) with eq. (5.14) we obtain

$$\bar{n}_0 = \frac{9}{8} \frac{\varepsilon \theta_n F_{av}}{qd} M \qquad (5.18)$$

Using $\bar{n}_0$ for n in eq. (5.10), we obtain the average electron density as a function of r as

$$\bar{n} = \frac{9}{8} \frac{\varepsilon \theta_n F_{av}}{qd} M \left[1 + \left(\frac{3M\mu_n F_{av}}{16 D_n d} \right)^{1/2} r \right]^{-2} \qquad (5.19)$$

Thus the average current density is

$$J_z(r) = q\mu_n \bar{n} F_{av}$$

$$= J_{z0} \left[1 + \left(\frac{3M}{16} \frac{\mu_n V}{D_n} \right)^{1/2} \frac{r}{d} \right]^{-2}$$

$$= J_{z0} \left[1 + \left(\frac{J_{z0}}{6} \frac{1}{\theta_n D_n \varepsilon} \frac{d}{V} \right)^{1/2} r \right]^{-2} \qquad (5.20)$$

and the total current by

$$I = \int_0^{2\pi} \int_0^{r_d} J_z(r) r \, dr \, d\theta$$

$$= 2\pi J_{z0} \int_0^{r_d} \left\{ r \, dr \Big/ \left[1 + \left(\frac{J_{z0}}{6} \frac{d}{\theta_n D_n \varepsilon V} \right)^{1/2} r \right]^2 \right\}$$

$$= 12\pi \theta_n D_n \varepsilon \frac{V}{d} \left\{ \ln \left[1 + \left(\frac{J_{z0}}{6} \frac{d}{\theta_n D_n \varepsilon V} \right)^{1/2} r_d \right] \right.$$

$$\left. + \left[1 + \left(\frac{J_{z0}}{6} \frac{d}{\theta_n D_n \varepsilon V} \right)^{1/2} r_d \right]^{-1} - 1 \right\} \qquad (5.21)$$

It should be noted that multiple current filaments may simultaneously exist between two parallel electrodes. For such a case the total current between the plane electrodes may be expressed as

$$I_T = I_{\text{domain 1}} + I_{\text{domain 2}} + \ldots = \sum_n I_n \simeq HI \qquad (5.22)$$

This means that the total current can be represented by the current in one filament I multiplied by a parameter H which may be field dependent.

With $\theta_n = 1$ Kao [1976] has computed M as a function of Y based on eq. (5.15) and $J_z(r)/J_{z0}$ as a function of r based on eq. (5.20), and his results are shown in Fig. 5.1. The

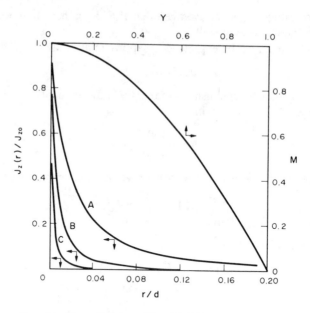

F IG. 5.1. M as a function of Y and $J_z(r)/J_{z0}$ as a function of r/d. A, $M = 16/300$, $d = 0.02$ cm, and $E_{av} = 10^5$ V·cm^{-1}; B, $M = 16/300$, $d = 0.02$ cm, and $E_{av} = 10^6$ V·cm^{-1}, or $M = 16/30$, $d = 0.02$ cm, and $E_{av} = 10^5$ V·cm^{-1}; C, $M = 16/30$, $d = 0.02$ cm, and $E_{av} = 10^6$ V·cm^{-1}. [After Kao 1976.]

larger the value of Y the smaller are the values of M and J_{z0}. It can also be seen that $J_z(r)$ depends strongly on the values of M and F_{av}, and that $J_z(r)$ decreases very rapidly with increasing r. For a perfect ohmic contact we may assume $F(z = 0) = F_c \simeq 0$. Then $Y \rightarrow 0$ and $M \rightarrow 1$, and eq. (5.14) reduces to the equation for single injection space-charge-limited (SCL) current with shallow traps. However, for general cases, to evaluate J_{z0} and hence I, we have to determine first the value of Y and then the value of M for a given condition. By assuming that the electron injection is mainly due to tunnelling through a potential barrier at the cathode following eq. (2.106) [field emission],

$$J_c = a F_c^2 \exp(-b/F_c) \tag{5.23}$$

and making $J_c = J_{z0}$ since the current in the z-direction is continuous, then from eqs. (5.14)–(5.16) and (5.23) we obtain

$$Y = \frac{8}{27[Qe^{-s} + 2/3]^2 + 4} + \left\{ \left(\frac{8}{27[Qe^{-s} + 2/3]^2 + 4} \right)^2 + \frac{27Qe^{-s}}{27[Qe^{-s} + 2/3]^2 + 4} \right\}^{1/2} \tag{5.24}$$

where

$$Q = 8ad/q\varepsilon\mu_n \quad \text{and} \quad S = b/YF_{av} \tag{5.25}$$

a and b are constants depending on the interfacial condition between the cathode and the material specimen, and J_c and F_c are, respectively, the current density and the field

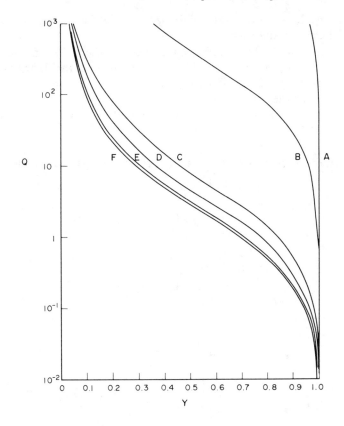

FIG. 5.2. Y as a function of Q. A, $S = 10$; B, $S = 5$; C, $S = 1$; D, $S = 0.5$; E, $S = 0.1$; and F, $S = 0$. [After Kao 1976.]

strength at the cathode. The plot of Y as a function of Q for various values of S is shown in Fig. 5.2. It can be seen that for a given material specimen and a given carrier-injecting contact (cathode electrode) Y decreases (or M increases) with increasing d and F_{av}, implying that the ratio F_c/F_{av} or the space charge effect depends on both the specimen thickness and the applied voltage. Kao [1976] has used this model to explain semiquantitatively the high field electric conduction and breakdown phenomena in dielectric liquids. Of course, this model may also be used for high field electric conduction in solids involving a field emission contact.

5.1.2. Filamentary two-carrier (double) injection

For the theoretical analysis of filamentary double injection, we make, in addition to the assumptions given in Section 4.3, the following assumptions:

(i) We use cylindrical coordinates and consider only one filament formed along the z-direction which coincides with the central line joining the two circular plane electrodes, the effective radius of the filament being r_d. The whole system is symmetrically about the z-axis with the hole-injecting contact located at $z = 0$ and the electron-injecting contact at $z = d$.

(ii) In the filament the longitudinal component of the diffusion current can be ignored because of the large longitudinal component of the electric field, and the radial component of the drift current can be ignored because of the small radial component of the electric field.

(iii) The recombination rate R consists of a longitudinal component R_z and a radial component R_r.

The behaviour of double injection in a solid is governed by the current flow equations

$$J_n = J_{nz} + J_{nr}$$

$$= q\mu_n n F \mathbf{i}_z + q D_n \frac{\partial n}{\partial r} \mathbf{i}_r \tag{5.26}$$

$$J_p = J_{pz} + J_{pr}$$

$$= q\mu_p p F \mathbf{i}_z - q D_p \frac{\partial p}{\partial r} \mathbf{i}_r \tag{5.27}$$

the continuity equations

$$\mu_n \frac{\partial}{\partial z}(nF) = -\mu_p \frac{\partial}{\partial z}(pF) = np \langle v\sigma_R \rangle_z = R_z \tag{5.28}$$

$$D_n \left[\frac{1}{r} \frac{\partial}{\partial r} \left(r \frac{\partial n}{\partial r} \right) \right] = D_p \left[\frac{1}{r} \frac{\partial}{\partial r} \left(r \frac{\partial p}{\partial r} \right) \right] = np \langle v\sigma_R \rangle_r = R_r \tag{5.29}$$

and the Poisson equation

$$\nabla F = \frac{q}{\varepsilon} [p + p_t - n - n_t] \tag{5.30}$$

For simplicity, we introduce the following parameters

$$A = q^{-1} \left[\frac{\Omega_n \Omega_p (\mu_n D_p + \mu_p D_n)}{D_n D_p (\mu_n \Omega_n + \mu_p \Omega_p)^2} \right] \langle v\sigma_R \rangle \tag{5.31}$$

$$S_z = q\mu_n n F_z / J_{z0} \tag{5.32}$$

$$T_z = q\mu_p p F_z / J_{z0} \tag{5.33}$$

$$W = J_z / J_{z0} = S_z + T_z \tag{5.34}$$

$$\Omega_n = 1 + n_t/n, \quad \Omega_p = 1 + p_t/p \tag{5.35}$$

where J_{z0} is the filamentary current density at the centre of the filament $r = 0$. It is likely that holes are injected from one asperity on the anode and electrons from the other on the cathode, both asperities having a very small injection area. If this is true, it is reasonable to assume that J_{z0} follows the normal expression for SCL currents from two planar-injecting electrodes because the density of carriers diffusing away from $r = 0$ would be very small as compared with that at $r = 0$. Thus J_{z0} is the same as J in eq. (4.200).

It is assumed that $\langle v\sigma_R \rangle_z = \langle v\sigma_R \rangle_r = \langle v\sigma_R \rangle$, that the traps are shallow traps, and that the radial variation of F_z is negligibly small and may be considered to be equal to zero. Then, eq. (5.30) becomes

$$\frac{dF}{dr} = \frac{q}{\varepsilon}\left(\Omega_p p - \Omega_n n \right) = 0 \tag{5.36}$$

Thus we have

$$T_z = \left(\frac{\mu_p \Omega_n}{\mu_n \Omega_p} \right) S_z \tag{5.37}$$

and

$$W = \left(1 + \frac{\mu_p \Omega_n}{\mu_n \Omega_p} \right) S_z = \left(1 + \frac{\mu_n \Omega_p}{\mu_p \Omega_n} \right) T_z \tag{5.38}$$

From eqs. (5.31)–(5.35) and (5.37)–(5.38), eq. (5.29) can be rewritten, in terms of the above parameters, as

$$\frac{1}{r}\frac{\partial}{\partial r}\left(r\frac{\partial W}{\partial r} \right) = (AJ_{z0}/F_z)W^2 \tag{5.39}$$

This equation implies that the radial distribution of W is a function of z. To obtain an equation for an average profile, we average the parameters over the specimen thickness d. Thus eq. (5.39) becomes

$$\frac{\partial^2 \overline{W}}{\partial r^2} + \frac{1}{r}\frac{\partial \overline{W}}{\partial r} = (AJ_{z0}/F_{av})\overline{W}^2 \tag{5.40}$$

where $\overline{W}$ is the average value of W in the z-direction, and F_{av} is the average value of F_z in the z-direction which is given by

$$F_{av} = \frac{1}{d}\int_0^d F_z\,dz = \frac{V}{d} \tag{5.41}$$

An examination of eq. (5.40) shows that when r approaches to zero, $\overline{W}$ would approach to infinity. Physical reality requires $\overline{W}$ to be finite for all values of r, and this demands that $d\overline{W}/dr \to 0$ when $r \to 0$. However, eq. (5.40) cannot be rigorously solved without the aid of numerical computation. Using the same approximation adopted in Section 5.1.1, we neglect the term $(1/r)\,(dW/dV)$. Thus eq. (5.40) reduces to

$$\frac{d^2\overline{W}}{dr^2} = (AdJ_{z0}/V)\overline{W}^2 \tag{5.42}$$

Using the boundary condition

$$r \to 0, \quad \overline{W} \to 1$$

$$r \to \infty, \quad \overline{W} \to 0, \quad \text{and} \quad \frac{d\overline{W}}{dr} \to 0$$

and assuming Ωn and Ωp to remain unchanged with r for $0 < r < r_d$, the solution of eq. (5.42) gives

$$\overline{W} = \left[1 + \left(\frac{1}{6}\frac{AdJ_{z0}}{V} \right)^{1/2} r \right]^{-2} \tag{5.43}$$

From eqs. (5.34) and (5.43) the average current density over the specimen thickness d is given by

$$\bar{J}_z(r) = J_{z0}\left[1 + \left(\frac{1}{6}\frac{Ad\,J_{z0}}{V}\right)^{1/2}r\right]^{-2} \qquad (5.44)$$

and the total current

$$I = \int_0^{2\pi}\int_0^{r_d}\bar{J}_z(r)r\,dr\,d\theta$$

$$= \frac{12\pi}{A}\frac{V}{d}\left\{\ln\left[1 + \left(\frac{1}{6}\frac{Ad\,J_{z0}}{V}\right)^{1/2}r_d\right] + \left[1 + \left(\frac{1}{6}\frac{Ad\,J_{z0}}{V}\right)^{1/2}r_d\right]^{-1} - 1\right\} \quad (5.45)$$

This equation is similar to eq. (5.21) for the case of single injection.

In most cases we can assume that the term $\left(\dfrac{1}{6}\dfrac{Ad\,J_{z0}}{V}\right)^{1/2}r_d$ is much smaller than unity. On the basis of this assumption, eq. (5.45) can be approximated to

$$I = C_1 V^2 + C_2 V^{5/2} \qquad (5.46)$$

In most cases $C_1 \gg C_2$ and therefore I is approximately proportional to V^2.

It should be noted that the filamentary current density $J_z(r)$ and current I are dependent on the form of spatial distribution of traps through J_{z0}, which is the same as J given by eq. (4.200), the effect of non-uniform spatial distribution of traps being included in eq. (4.201).

Similar to filamentary single injection, the total current for the case with multiple current filaments is HI as given by eq. (5.22). Equations (5.44) and (5.45) are applicable to all cases involving only shallow traps regardless of the form of their distributions (i.e. following any distribution function in energy and in space), provided that there are no deep traps (i.e. no recombination centres).

To demonstrate the existence of filamentary conduction, Barnett et al. [1966, 1968, 1969, 1970] have observed the development and growth of filaments in Si, GaAs, ZnTe, GaAs$_x$P$_{1-x}$ and polycrystalline silicon by means of recombination radiation photography. Usually a stable filament is formed after the transition from a low current regime to a high current level regime, and sometimes after the occurrence of a current-controlled negative differential resistance region. Saji and Kao [1975, 1976] and Ballard and Christy [1975] have reported that prior to switching the current in the OFF state is dependent on the electrode area but after switching to the ON state the current is practically independent of the electrode area, indicating that electric conduction takes place along a filamentary path in the "switching on" state. This is true for all materials with a switching phenomenon. Typical J–V characteristics showing a filamentary breakdown for a p^+–π–n^+ indium-doped silicon structure at $71°$ K is shown in Fig. 5.3. It should be noted that in most materials, if there is a negative differential resistance region or a switching transition, it appears to be associated with the development of a current filament between electrical contacts. Strictly speaking, theoretical treatment of double injection problems for high current levels as have been discussed in Section 4.3 should take into account the processes of the development and growth of filaments. It should also be noted that most switching phenomena involve mixed electronic and thermal processes [Owen and Robertson 1973, Kroll 1974, Vezzoli et al. 1975,

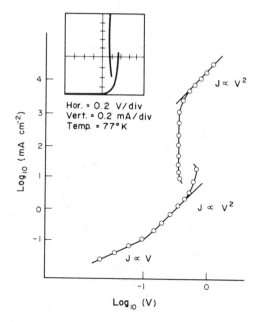

FIG. 5.3. Typical *J–V* characteristics of a $p^+–\pi–n^+$ indium-doped silicon structure at 71°K showing the filamentary breakdown current. The circles are experimental points for which the current density was obtained by dividing the current by the total electrode area. Inset is an oscillograph showing the breakdown and hysteresis in which the maximum pre-breakdown current exceeds the minimum post-breakdown current. [After Wagener and Milnes 1964.]

Nakashima and Kao 1979], particularly in amorphous semiconductors. Therefore, to analyse filamentary conduction processes, the electronic–thermal mutual feedback mechanism should not be ignored. There are a great deal of experimental results on the switching filaments in chalcogenide glasses [Fritzsche 1974]. Recently, Ballard and Christy [1975] have observed the memory switching in electron-beam-deposited tetrabutyltin polymer films, and their results are shown in Fig. 5.4. Switching phenomena have also been observed in several organic thin films [Kevorkien *et al.* 1971, Szymanski *et al.* 1969, Carchano *et al.* 1971, Stafeev *et al.* 1968, Garrett *et al.* 1974] and in biological materials [Hanai *et al.* 1965, Hodgkin 1964]. Typical results showing the switching and memory phenomena in anthracene thin films observed by Elsharkawi and Kao [1977] are given in Fig. 5.5. At V_{TH} the specimen is rapidly switched from its "low conductivity" or OFF state to a "high conductivity" or ON state corresponding to point *a*. The ON state will be retained as long as the current is not reduced below the holding current corresponding to point *b*. But when the current is reduced below point *b* the specimen returns to its OFF state. After one or several switching cycles some specimens are changed from their "switching-on" state to a "memory" state, as shown by the curve extended beyond point *a* and point *b* in Fig. 5.5. This "memory" state could persist for quite a long time and could be erased by passing a current pulse of either polarity through the specimen, but, generally, a pulse with the polarity opposite to that

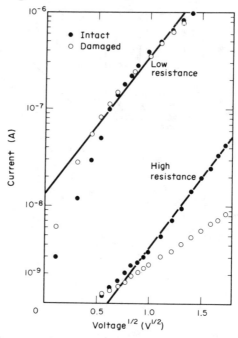

F IG. 5 4. Current versus square root of voltage characteristics of a tetrabutyltin polymer film in the low resistance and the high resistance states. Electrodes were silver. Lines are drawn with slope $3.4V^{-\frac{1}{2}}$. [After Ballard and Christy 1975.]

of the applied voltage initiating the "memory" state is more effective. Although the true mechanisms responsible for such switching and memory phenomena are not yet fully understood, these phenomena are generally believed to be associated with the formation of filaments and the electrothermal processes [Saji and Kao 1975, 1976].

5.2. THE EFFECT OF CARRIER DIFFUSION

In Sections 3.2–3.4, 4.3, and 5.1 we have assumed that the diffusion current component can be neglected for mathematical simplicity in the theoretical analysis. In this section we shall examine the validity of this assumption and the conditions under which such an assumption can be applied without causing a serious error.

5.2.1. One-carrier (single) injection

The total current density for hole injection from the injecting contact (anode) at $x = 0$ is the sum of drift and diffusion current densities and is given by

$$J = J_{dr} + J_{diff}$$

$$= q\mu_p\left(pF + V_T\frac{dp}{dx}\right) \qquad (5.47)$$

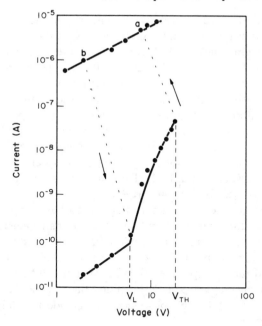

FIG. 5.5. Typical d.c. J–V characteristics showing the switching and memory actions in anthracene thin films (6000 Å in thickness) with silver electrodes at 20°C. [After Elsharkawi and Kao 1977.]

where

$$V_T = \frac{D_p}{\mu_p} = \frac{kT}{q} \qquad (5.48)$$

based on Einstein's relation. From eq. (5.47) it is obvious that J_{diff} can be neglected if and only if $J_{dr} \gg J_{\text{diff}}$. This means

$$|pF| \gg \left| V_T \frac{dp}{dx} \right| \qquad (5.49)$$

Supposing that the drift current component has to be at least 10 times the diffusion current component before we can justify neglecting the diffusion current component, then we have

$$\left| \int_0^d F \, dx \right| = V = F_{\text{av}} d$$

$$\geq 10 V_T \left| \int_0^d \frac{dp}{p} \right| \geq 10 V_T \ln \frac{p(x=0)}{p(x=d)} \qquad (5.50)$$

This equation implies that only when $p(x = 0) = p(x = d)$ can we completely neglect the diffusion contribution, and that when $p(x = 0) > p(x = d)$ the applied voltage V must be larger than $10 V_T \ln [p(x = 0)/p(x = d)]$ before we can neglect the diffusion term. For example, if $p(x = 0)/p(x = d) = 10^4$, the applied voltage must be larger than about

$100 V_T$, that is larger than 2.5 volts at $300\,°K$. The magnitude of applied voltage alone does not give a clear criterion—it must go together with the specimen thickness. If we want to apply an average field of $3\,kV\,cm^{-1}$ across the specimen, then the specimen must be larger than $2.5/3 \times 10^3 \simeq 10^{-3}$ cm, or $10\,\mu m$. This simple example indicates that for a given applied average field F_{av} the larger the specimen thickness the less important is the diffusion contribution. This is why the diffusion current component can sometimes be ignored for long specimens but not for thin specimens. For thin films the diffusion current may become large enough to produce various effects. For example, the diffusion tends to move the "virtual anode" (the plane of which $F = 0$) away from the hole-injecting contact, thus shortening the effective specimen thickness and hence enhancing the field towards the cathode. The same argument given above is applicable to the case for only injected electron carriers.

However, for organic semiconductors the energy band gap is large; the carrier mobility, the carrier diffusion coefficient, and the thermally generated carrier density are all very small. In practice, it is not difficult to satisfy the condition given by eq. (5.50) for the diffusion term to be neglected for organic semiconductors and for most insulators provided that the specimen thickness is not too thin.

For inorganic semiconductors and thin films the diffusion current may play an important role in the conduction processes. To include the diffusion term and to express p_t in the form

$$p_t = Ap^B \qquad (5.51)$$

and Poisson equation as

$$\frac{dF}{dx} = \frac{q}{\varepsilon}(p + p_t) \simeq \frac{q}{\varepsilon} Ap^B \qquad (5.52)$$

the elimination of the hole density p from eqs. (5.47) and (5.52) yields

$$(V_T/B)\left(\frac{dF}{dx}\right)^{1/B-1}\frac{d^2 F}{dx^2} + F\left(\frac{dF}{dx}\right)^{1/B} = J/q\mu_p(qA/\varepsilon)^{1/B} \qquad (5.53)$$

where A and B are constants depending on the trap parameters of the material under study. The following cases may be encountered in most semiconductors:

    (i) Without traps: $A = 1$, $B = 1$.
    (ii) With shallow traps: $A = \theta_p^{-1}$, $B = 1$.
    (iii) With exponentially distributed traps: $A = H_b^{-1} N_v^l$, $B = l^{-1}$, and $l = T_c/T$.

It can be seen that eq. (5.53) cannot be solved analytically. Several investigators [Wright 1961, Lindmayer *et al.* 1963, Sinharay and Meltzer 1964, Lampert and Edelman 1964, Rosental and Sapar 1974] have attempted to solve eq. (5.53) for cases (i) and (ii). Wright has obtained a general solution in terms of Bessel functions, the solution obtained by Lindmayer *et al.* involves a power series, and that obtained by Sinharay *et al.* involves Airy functions. For all these solutions, use of a computer is necessary. Lampert and Edelman [1964] have reported that based on their computer solutions the inclusion of the diffusion term tends to soften the steep rise of current with voltage as predicted on the basis of Lampert's simplified theory [Lampert 1956] with the diffusive current flow neglected. A similar finding about the effect of the diffusion term has also been reported by Roberts and Tredgold [1964]. Recently, Rosental and Sapar [1974] have

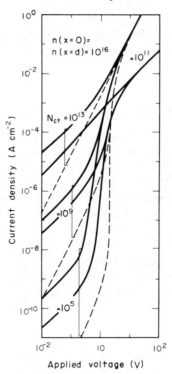

FIG. 5.6. Current–voltage characteristics for insulators containing electron traps of concentration $N_t = 10^{12} \text{cm}^{-3}$ but of different depths characterized by $N_{ct} = N_c \exp[(E_t - E_c)/kT]$ cm $^{-3}$. The solid curves are computed, including diffusion, for two values of free electron boundary concentrations $n(x = 0) = n(x = d) = 10^{16}$ and 10^{11} cm $^{-3}$. The dashed curves are computed on the basis of Lampert's simplified theory without diffusion contribution. [After Rosental and Sapar 1974.]

complemented the results of Lampert and Edelman by considering the corresponding layers of non-metallic solids provided with two equal metal contacts of reservoir type, and their computed results show that the overlap of the reservoir contacts in symmetrical structures causes the further softening of the steepness of the TFL (trap-filled limit) curves as shown in Fig. 5.6. They employed an iterative method in conjunction with a computer to compute the J–V characteristics using the following parameters:

N_t = electron trap concentration = 10^{12} cm $^{-3}$

d = specimen thickness = 5×10^{-3} cm

μ_n = electron mobility = 200 cm^2/V-sec

$\varepsilon = \varepsilon_r \varepsilon_0 = 11\varepsilon_0$

$n(x = 0) = n(x = d) = 10^{16}$ cm $^{-3}$ or 10^{11} cm $^{-3}$

$N_{ct} = N_c \exp[-(E_c - E_t)/kT]$

E_t = electron-trapping energy level

They stressed the importance of the overlap of the reservoir contacts in the insulation (the overlapping accumulation region in which the space charge extends throughout the insulator) [Simmons 1971] and the importance of the diffusion corrections for short structures (thin specimens).

5.2.2. Two-carrier (double) injection

If $d \gg L_a$ in long structure insulators or semiconductors such as in long p^+-i-n^+ structures, the diffusion-dominated regions near the junctions (or contacts) may be neglected so that the boundary condition for $F = 0$ can be considered to be at the junctions. In Section 4.3.3(A) the use of the junction boundary conditions is based on the assumption that the carrier concentration (or the injection level) is very high and the field is very low near the junction, so that the portions of the i (insulator) region close to the junctions are dominated by the diffusion term and the carrier concentration decreases exponentially away from the junctions. In the centre of the i-region the diffusion term becomes negligible, and the insulator regime based on the simple double injection theory is dominant in this region. Typical patterns for carrier concentration and F for $d/L_a \simeq 28$ are shown in Fig. 5.7 in which $\eta = n/n^*$ and $\xi = F/F^*$ where $n^* = [\varepsilon J^2/2kTq(\mu_n+\mu_p)^2]^{1/3}$ and $F^* = [2kTJ/q\varepsilon(\mu_n+\mu_p)]^{1/3}$ and $n = p$. It can be seen that nearly all of the voltage drop is across the central portions and that the effect of the diffusion wings is to reduce the geometric specimen thickness by $2QL_a$, where QL_a is the length of one of the diffusion wings and Q is about 2–4. Thus the effective specimen thickness becomes $d - 2QL_a$. Since $J \propto d^{-5}$ in eq. (4.239) the replacement of d with $d - 2QL_a$ in eq. (4.239) implies that the correction factor for the J–V

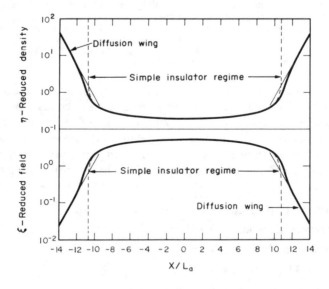

FIG. 5.7. Typical carrier density and field distribution for a long p^+-i-n^+ structure (insulator regime) showing the diffusion wings extending from the junctions into the specimen and the central high field region. [After Baron and Mayer 1970.]

characteristics due to the diffusion term can be very significant [Baron and Mayer 1970].

For a detailed discussion of the effects of carrier diffusion and thermally generated carriers in insulators and semiconductors, the reader is referred to the work of Lampert and Schilling [1970], Lampert and Mark [1970], Baron [1965, 1968], Baron and Mayer [1970], and Tredgold [1966].

5.3. HIGH FIELD EFFECTS, POOLE–FRENKEL, AND ONSAGER MODELS

The effects of high fields on the $J-V$ characteristics are mainly caused either by the change of the distribution function and mobility of carriers, or by the change of the rate of carrier generation or injection, or by both. The high field effects have been extensively studied both theoretically and experimentally by many investigators since 1950 [Conwell 1967, Simmons 1971, Mott 1971, Nag 1972, O'Dwyer 1973, Jonscher 1973, Marshall and Miller 1973]. For high resistivity solids (or low mobility solids) such as organic semiconductors and most insulators, the high field effects can be divided into three categories:

(A) *Electrode effects.* The high field effects are caused by the field dependent rate of carrier injection or emission from electrodes through either Schottky-type thermionic emission or tunnelling through a potential barrier near the injecting contact or across a thin insulating film.

(B) *Bulk effects.* The high field effects are caused by the field dependent carrier mobility due to various scatterings; by the field dependent carrier density due to field dependent trapping probability, detrapping process (e.g. the Poole–Frenkel effect), and tunnelling of trapped carriers; and by the field dependent thermal effect arising from the Joule heating of the specimen.

(C) *Combined electrode and bulk effects.* The high field effects may be changed from the bulk-limited to the electrode-limited or from the electrode-limited to the bulk-limited processes when the applied field is increased.

Some of the high field effects mentioned in (A) and (C) have been discussed in Chapter 2, and some will be discussed in Chapter 6. In the present section we shall confine ourself to the high field effects related to space charge electric conduction [mainly category (B)].

5.3.1. Transition from the bulk-limited to the electrode-limited conduction processes

The current injection from a normal metallic contact into an insulator follows the general thermionic emission equation given by eq. (2.33) which is

$$J = A^* T^2 \exp(-\phi_B/kT) \tag{5.54}$$

The SCL current in an insulator containing shallow traps is given by eq. (3.22) which, for traps distributed uniformly in space and for electron current only, is

$$J = \frac{9}{8} \varepsilon \mu_n \theta_a V^2/d^3 \tag{5.55}$$

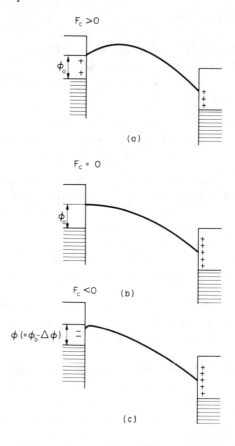

FIG. 5.8. Energy diagrams for an insulator with ohmic contacts
under various voltage biases: (a) $V < V_c$, (b) $V = V_c$,
and (c) $V > V_c$. [After Simmons 1971.]

Figure 5.8 shows the energy diagrams for an insulator with ohmic contacts under various voltage bias conditions. At low voltages the injecting contact can supply more electrons than required to replenish those collected by the opposite electrode and thus there are a negative space charge in the vicinity of the cathode and a positive charge on the cathode surface indicating that the electric conduction is bulk limited (SCL) as shown in Fig. 5.8(a). When the applied voltage is increased to such a value that the number of electrons that the cathode can inject into the insulator per second are exactly equal to what the anode can collect per second, then there is no space charge in the insulator, nor any charge on the cathode surface, and the electric conduction process begins to change from the bulk-limited to the electrode-limited process. The critical voltage to set on this transition can be obtained by setting eq. (5.54) equal to eq. (5.55), and it is [Simmons 1971]

$$V_c = T\left(\frac{8 A^* d^3}{9\mu_n \varepsilon \theta_a}\right)^{1/2} \exp\left(-\frac{\phi_B}{2kT}\right) \tag{5.56}$$

If Φ_B is independent of applied field, then the electrode-limited current would saturate to a value given by eq. (5.54). If ϕ_B is lowered due to the combination of the applied electric field and the image force [the Schottky effect—eq. (2.25)], then the electrode-limited current for V larger than V_c is given by

$$J = A^* T^2 \exp(-\phi_B/kT) \exp(\beta F_c^{1/2}/kT) \tag{5.57}$$

and there exists a negative charge on the cathode surface. The different polarity of charge on the cathode surface is generally used as a major distinction between the bulk-limited (SCL) and the electrode-limited (the injection-limited) processes. Since F_c is the field at the cathode and is not linear dependent on V, the calculation of J–V characteristics has to be carried out by means of numerical methods. For further details about this topic the reader is referred to the literature [Frank and Simmons 1967, Simmons 1971].

The transition from the bulk-SCL to the electrode-limited conduction has been observed experimentally in inorganic insulators such as in Mylar, SiO and Ta_2O_5 films by Schug et al. [1970], and in organic semiconductors such as in anthracene crystals with silver electrodes by Johnston and Lyons [1970], in anthracene and naphthalene crystals with electrolytic electrodes by Mehl et al. [Mehl and Hale 1967, Lohmann and Mehl 1969] and by Bonham and Lyons [1973]. Figure 5.9 shows the transition from

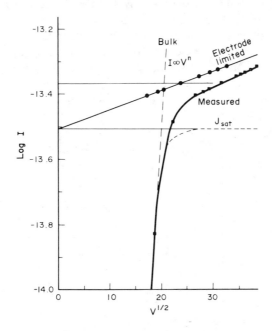

FIG. 5.9. Plot of log I versus $V^{1/2}$ showing the total measured current (■) and the calculated electrode-limited current (●), and indicating the transition from SCL current to electrode-limited current. The theoretical saturation current I_{sat} is due to Richardson thermionic emission in the absence of the Schottky effect. $T = 314.8°K$; crystal: H480 anthracene zone refined with 115 passes; area of injecting electrode: 14 mm²; specimen thickness: 2.5 mm. [After Johnston and Lyons 1970.]

space-charge-limited currents at low fields to electrode-limited currents at high fields where the Schottky effect controls the current–voltage characteristics for anthracene crystals with silver paste electrodes. The bulk $I \propto V^n$ curve with $n > 2$ indicates that the traps in the anthracene specimen may be distributed exponentially in energy following eq. (3.45). The theoretical electrode barrier J–V line is calculated by subtracting the extrapolated "bulk" voltage from the total voltage for each measured current, and I_{sat} represents the saturation current due to thermionic emission in the absence of the Schottky effect. Their results are in good agreement with the theoretical model of Frank and Simmons [1967].

In Section 3.4.3 we have mentioned that with electrolytic-injecting electrodes the current is limited by the build-up of space charge within the crystal at low applied fields until the current becomes equal to the forward rate of the injection, at which the current saturates as shown in Figs. 3.35 and 3.36. The magnitude of the current is determined by the oxidation potential of the electrolyte relative to the valence band of the crystal for hole injection and by the energy needed to rearrange the solvation sheath around the oxidizing species when it is reduced. Figure 5.10 shows the schematic J–V characteristics for a molecular crystal with electrolytic electrodes. Curve A is predicted by Mehl and Hale theory [1967], but this type of curve is rarely observed. Most available experimental J–V curves either follow curve B, which is very nearly constant close to the saturation point, but begins to rise gradually and then rapidly at higher voltages, or follow curve C, which is not flat over any range of voltages and rises quite steeply with increasing applied voltage. Typical experimental results following curve B in Fig. 5.10 are shown in Fig. 5.11 which were obtained using $Ce(SO_4)_2$ in 1 M H_2SO_4 as a hole-injecting electrolyte in contact with an anthracene crystal. Typical experimental results following curve C in Fig. 5.10 are shown in Fig. 5.12, which were obtained using 10^{-3} M KI_3 as a hole-injecting electrolyte in contact with an anthracene crystal. Table

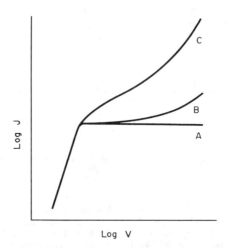

FIG. 5.10. Schematic J–V curves for a molecular crystal (such as anthracene) with electrolytic electrodes. A, predicted by Mehl and Hale theory; B and C, experimental. [After Bonham and Lyons 1973.]

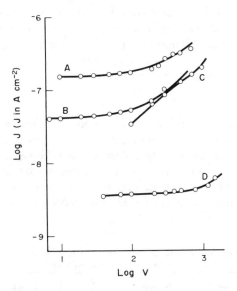

FIG. 5.11. Saturation currents for hole injection from $Ce(SO_4)_2$ into anthracene specimen with the following concentration of $Ce(SO_4)_2$ in $1\,M$ of H_2SO_4 and specimen thicknesses; *A*, $10^{-1}\,M$ $Ce(SO_4)_2$, $18\,\mu m$ thick; *B*, $10^{-2}\,M$ $Ce(SO_4)_2$, $18\,\mu m$ thick; *C*, $10^{-2}\,M$ $Ce(SO_4)_2$, $76\,\mu m$ thick; *D*, $10^{-3}\,M$ $Ce(SO_4)_2$, $76\,\mu m$ thick. The SCL current part of the curves has not been plotted. [After Bonham and Lyons 1973.]

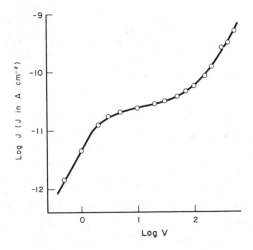

FIG. 5.12. *J–V* curve for hole injection from $10^{-3}\,M\,KI_3$ into anthracene specimen of thickness of $27\,\mu m$. [After Bonham and Lyons 1973.]

5.1 gives a list of electrolyte–insulator systems in which the saturation current continues to rise with increasing voltage in the saturation region. Bonham and Lyons [1973] have proposed that two mechanisms may operate in the saturation region: one following the theory of Mehl and Hale [1967] with a flat saturation curve and one following the Schottky equation. The current may be expressed as

$$J = J_S + J_0 \exp[b(V/d)^{1/2}] \tag{5.58}$$

Schottky emission is usually applied to carrier injection from a metal into an insulator. For an electrolyte–insulator system Bonham and Lyons [1973] have proposed that the initial charge transfer occurs from the electrolyte to a deep energy state on the crystal surface, which may be a physical defect on the surface or an oxidized anthracene molecule. Such trapping centres lying between $x = 0$ and $x = (qd/16\pi\varepsilon V)^{1/2}$ [see eq. (2.23)] would affect strongly the carrier injection processes because within this region the carrier's image force is significant.

TABLE 5.1. *Some electrolyte–insulator systems in which the saturation current is a function of voltage [after Bonham and Lyons 1973]*

Electrolyte	Insulator	References
10^{-3} M KI_3	Anthracene	Bonham and Lyons [1973]
10^{-1} M $SnCl_4$	Anthracene	Bonham and Lyons [1973]
Conc. H_2SO_4	Anthracene	Mehl and Hale [1967]
0.2 M NaCl	Anthracene	Mehl and Hale [1967]
$Ce(SO_4)_2$	Anthracene	Mehl and Hale [1967]
$CH_3NO_2 + AlCl_3$	Anthracene	Mehl and Hale [1967]
Ethanol	Anthracene	Yamagishi and Soma [1970]
Ethanol + phthalonitrile	Anthracene	Yamagishi and Soma [1970]
$Ce(SO_4)_2$	Anthracene	Yamagishi and Soma [1970]
$Ce(SO_4)_2$	Tetracene	Yamagishi and Soma [1970]
$K_3Fe(CN)_6$	Perylene	Michel-Beyerle *et al.* [1969]
$Ce(SO_4)_2$	Phenanthrane	Yamagishi and Soma [1970]
$Ce(SO_4)_2$	p-Terphenyl	Yamagishi and Soma [1970]
Sat. I_2/1 M NaI	p-Terphenyl	Helfrich and Mark [1962]
Sat. I_2/1 M NaI	p-Quaterphenyl	Helfrich and Mark [1962]

5.3.2. Transition from the electrode-limited to the bulk-limited conduction processes

When an insulator has shallow donors as well as deep traps and a blocking contact with a very narrow depletion region as shown in Fig. 5.13, all electrons from shallow donors will be captured by deep traps. Therefore, at low voltages the contact resistance is much higher than the resistance of the bulk and the J–V characteristics due mainly to field emission follow eq. (2.106) and will be practically independent of specimen thickness. Under such conditions electric conduction is electrode limited. If the applied voltage is increased, the bulk resistance tends to decrease and the contact resistance also tends to decrease but the rate of the decrease for the latter is much smaller than that for the former. At a certain critical voltage V_c, the bulk resistance becomes higher than the contact resistance and it is at this critical voltage the transition from the electrode-limited to the bulk-limited conduction occurs. At $V > V_c$ the J–V characteristics may become controlled by the space charge effect and the Poole–Frenkel effect, and then the

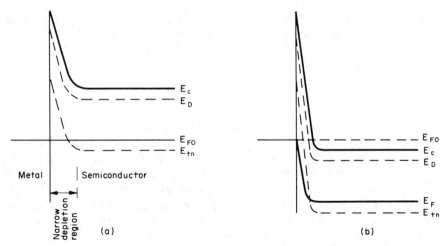

F IG. 5.13. Schematic diagrams showing the energy levels for electron tunnelling–field emission from a blocking contact into an insulator with deep traps of trapping level E_{tn} (a) without bias in thermal equilibrium, and (b) under a bias for tunnelling injection. This illustrates the possible transition from the electrode-limited to the bulk-limited conduction processes by increasing applied fields.

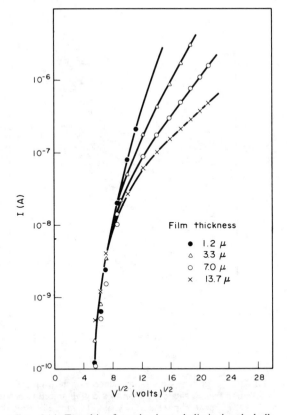

F IG. 5.14. Transition from the electrode-limited to the bulk-limited electric conduction in Al–SiO Al films. Electrode area: 0.1 cm². [After Stuart 1967.]

current becomes dependent on specimen thickness [Simmons 1971]. Typical experimental results demonstrating such a transition are shown in Fig. 5.14.

5.3.3. Transition from a single injection to a double injection process

If an insulator has a pair of identical electrodes or a pair of different electrodes, it is possible that one of these electrodes is definitely electron injecting and the other is unable to inject holes efficiently, or vice versa, at low applied fields. But such a single injection would become double injection at a critical applied field. This high field effect has been reported by Kao and his co-worker [Hwang and Kao 1973, Kunkel and Kao 1976] and Sworakowski *et al.* [1974]. Typical experimental results for anthracene with silver electrodes are shown in Fig. 5.15. For voltages below the threshold voltage V_{TH} for the onset of electroluminescence, J is proportional to V^2 following the relation for the single injection into a solid containing hole traps in a discrete energy level. The $\ln J - 1/T$ plot for $V = 2.0\,\mathrm{kV}$ gives an activation energy E_{act} of 0.56 eV, which can be

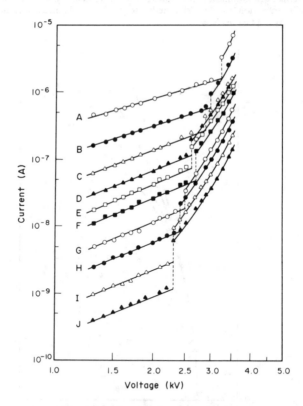

FIG. 5.15. The d.c. current–voltage characteristics showing the transition from a single injection to a double injection process at high fields. Specimen thickness, 0.9 mm; electrode diameter, 1.6 mm; temperature: *A*, 87°C; *B*, 63°C; *C*, 51°C; *D*, 41°C; *E*, 34.5°C; *F*, 20°C; *G*, 10°C; *H*, 0°C; *I*, −10°C; *J*, −20°C. [After Hwang and Kao 1973.]

interpreted as this discrete energy level above the valence band. For voltages above V_{th}, J is proportional to V^n with $n > 6$, implying that the large current may be associated with the field-enhanced electron injection and the release of trapped carriers due to the reabsorption of electroluminescence. E_{act} for $V > V_{th}$ tends to increase with increasing V, and such a change in E_{act} may be attributed to the effect of the reabsorption of electroluminescence. The transition occurs when sufficient space charge builds up to modify the potential barrier at the cathode to allow electron tunnelling from it.

5.3.4. Field dependent carrier mobilities

The field dependence of the mean carrier mobility in semiconductors has been extensively reviewed by Conwell [1967], and applied particularly to germanium, silicon, and some III–V compounds. For these semiconductors there exists a great deal of information about their energy band structures and lattice vibration modes. Such information is simply not available for organic semiconductors and insulators, and therefore practically no theoretical and experimental results on field dependent carrier mobilities are available for these materials. The calculation of the field dependent carrier mobility requires a knowledge of the energy distribution function of the carriers, and the mathematical solution for this is usually lengthy and complex even for simple semiconductors. For simplest models of electric conduction, O'Dwyer [1973] has briefly reviewed the methods for calculating the field dependent carrier mobilities for polar and non-polar crystals. For further details about such calculations the reader is referred to the articles of Heller [1951], Franz [1952, 1956], Fröhlich and Paranjape [1956], Stratton [1957, 1958, 1961], O'Dwyer [1954, 1957, 1973], Paranjape [1961], Shumka et al. [1964], Canali et al. [1971], Nag [1972], Ottaviani et al. [1973] and Tandon et al. [1975]. In this section we shall discuss briefly the physical concept of field dependent carrier mobilities.

Under an applied field, an electron tends to gain energy from the field due to acceleration until it encounters a scattering either with lattice vibration waves or with other particles. If there is a net gain in energy, the energy of the electron increases. For an electron in an insulator or in a non-degenerate semiconductor, an increase in energy by an amount kT is usually considered a large change in the mean energy. By assuming the electron energy distribution to be Maxwellian, the mean velocity of the electron may be written as

$$v = (2kT_e/m^*)^{1/2} \tag{5.59}$$

where T_e is defined as the electron temperature based on the expression of the electron energy in terms of kT_e. It can be seen from this simple equation that the electron mobility would be independent of field only if $T_e = T$, and that if $T_e > T$ then the electron becomes hot and the electron mobility becomes field dependent. For organic molecular crystals the mean carrier mobility and the carrier mean free path are generally very small, the carrier will not be able to gain much energy from the field unless the applied field is extremely high in which other high field effects may be predominant. It is reasonable to assume that the carrier mobilities in organic molecular crystals are practically independent of field. However, for other materials, particularly for inorganic semiconductors, the field dependent mobility effect cannot be ignored.

Lampert [1958] and Lampert and Mark [1970] have used the following formula to calculate the field dependent electron mobility (μ–F relation):

$$\mu(F) = \tfrac{1}{2}\mu_l(F_1/F)\{[1+(4F/F_1)]^{1/2} - 1\} \tag{5.60}$$

and the field dependent electron velocity ($v - F$ relation)

$$F = v/\mu_l - F_0 - [F_0 v_s/(v - v_s)] \tag{5.61}$$

where μ_l is the low field electron mobility which is independent of field, F_0, F_1, and v_s are parameters. Equation (5.60) is a useful mathematical representation of the mobility for constant mean free path problems, while eq. (5.61) is a useful mathematical representation of the v–F relation in problems where a transition is made directly from the ohmic low field relation to velocity saturation [Lampert and Mark 1970]. Wagener and Milnes [1965] have used three ranges of field dependent mobility: (a) $\mu = \mu_l$ at low fields, (b) $\mu \propto F^{-1/2}$ at moderate fields, and (c) $\mu \propto F^{-1}$ at high fields in a manner similar to eq. (5.60) to analyse the Ashley–Milnes space charge regime including high field effects.

From simple current flow equation and Poisson equation [eqs. (3.4) and (3.5)], a solution in parametric form for the single injection J–V characteristics of a solid containing shallow traps and having a field dependent carrier mobility can be written as [Tandon *et al.* 1975]

$$J = \frac{\varepsilon}{\Omega_p d} \int_{F(x=0)}^{F(x=d)} \mu_p(F)\,F\,dF \tag{5.62}$$

and

$$V = \frac{\varepsilon}{\Omega_p J} \int_{F(x=0)}^{F(x=d)} \mu_p(F)\,F^2\,dF = \frac{d\displaystyle\int_{F(x=0)}^{F(x=d)} \mu_p(F)\,F^2\,dF}{\displaystyle\int_{F(x=0)}^{F(x=d)} \mu_p(F)\,F\,dF} \tag{5.63}$$

where $\Omega_p = 1 + p_t/p$ for the hole injection case. If the field dependence of μ_p is known, the J–V characteristics will be readily derived from eq. (5.63).

By assuming that electron motion occurs through thermally activated jumps from one trap to another with an average trap depth (or an average barrier height) E_t separated by an average distance λ, then the effective carrier mobility measured as the average carrier velocity divided by the field F at any point in a specimen is given by [Bagley 1970]

$$\mu = \frac{2\lambda v}{F} \exp\left(-\frac{E_t}{kT}\right) \sinh\left(\frac{q\lambda F}{2kT}\right) \tag{5.64}$$

where v is the attempt to escape frequency. At low fields, the mobility μ_l, from eq. (5.64), may reduce to

$$\mu_l = \frac{\lambda v}{F} \exp\left(-\frac{E_t}{kT}\right) \tag{5.65}.$$

which is independent of applied field. At very high fields, eq. (5.64) may be written as

$$\mu(F) = \frac{\lambda v}{F} \exp\left[-\left(\frac{E_t}{kT} - \frac{q\lambda F}{2kT}\right)\right] \qquad (5.66)$$

Using eq. (5.66) for the field dependent carrier mobility to replace the assumption of constant carrier mobility, eqs. (5.14) and (5.21) for J_{z0} and I for the filamentary one-carrier injection case can be written in the following form [Kao and Rashwan 1978]:

$$J_{z0} = \frac{2kTv\varepsilon}{qd}\left\{\exp\left[1 - \frac{E_t - q\lambda F_{av}/2}{kT}\right] - \exp\left[\frac{E_t - q\lambda YF_{av}/2}{kT}\right]\right\} \qquad (5.67)$$

and

$$I = J_{z0}(2\pi d^2)\left\{\frac{2qd}{9\lambda}\left[1 + \frac{q\lambda F_{av}}{4kT}(1 - Y)\right]\right\}^{-1}$$

$$\times \left\{\ln\left[1 + \left(\frac{2qd}{9\lambda}\left[1 + \frac{q\lambda F_{av}}{4kT}(1 - Y)\right]\right)^{1/2}\frac{r_d}{d}\right]\right.$$

$$+ \left[1 + \left(\frac{2qd}{9\lambda}\left[1 + \frac{q\lambda F_{av}}{4kT}(1 - Y)\right]\right)^{1/2}\frac{r_d}{d}\right]^{-1} - 1\right\} \qquad (5.68)$$

The basic difference between the approach given in Section 5.1.1 and eqs. (5.67) and (5.68) is as follows: in Section 5.1.1 we have assumed that at any moment there are n free electrons travelling across the specimen with the mobility μ_n and trapped electrons n_t always stationary in the traps. This assumption is reasonable for the cases in which the carrier transit time is smaller than the time in which the trapped electron remains in the trap so that during the carrier transit time we can see only n electrons contributing to electric conduction. But in eqs. (5.67) and (5.68) the assumption implies that each of all injected carriers (say electrons) will spend a total time τ_n as a free carrier and a total time τ_{tn} in the trap so that the transit time is

$$t_t = \tau_n + \tau_{tn} \qquad (5.69)$$

and the effective carrier mobility is

$$\mu_{eff} = \frac{\tau_n\mu}{\sigma_n + \sigma_{tn}} \qquad (5.70)$$

If τ_{tn} is field dependent, τ_n and hence μ_{eff} will also be field dependent even if the mobility of free carriers μ is independent of field. In physical concept μ_{eff} in eq. (5.70) is the same as μ in eq. (5.64), and both are associated with the field-enhanced detrapping processes by lowering the potential barrier height and hence reducing τ_{tn}. Equation (5.70) also implies that during the transit time we can see all injected electrons contributing to electric conduction, and there is no separation between free and trapped carriers. This assumption is reasonable for the cases in which τ_n or $\tau_{tn} < t_t$. Kao and Rashwan [1978] have reported that eqs. (5.67) and (5.68) explain well their experimental results on high field I-V characteristics and field dependent activation energy for electric conduction for hydrocarbon liquids.

5.3.5. The Poole–Frenkel detrapping model

The Poole–Frenkel effect is sometimes called the "internal Schottky effect" since the mechanism of this effect is associated with the field-enhanced thermal excitation (or detrapping) of trapped electrons or trapped holes from traps, which is very similar to the Schottky effect in the thermionic emission. The effect of applied field in lowering the potential barrier for a trapped electron to escape in a one-dimensional model is shown in Fig. 5.16, which is very similar to Fig. 2.12. Both effects are due to the Coulombic interaction between the escaping electron and a positive charge, but they differ in that the positive charge is fixed for the Poole–Frenkel trapping barrier, while the positive charge is a mobile image for the Schottky barrier. This results in the barrier lowering due to the Poole–Frenkel effect twice that due to the Schottky effect. The amount of the barrier lowering due to the Poole–Frankel effect is

$$\Delta E_{pF} = \left(\frac{q^3 F}{\pi \varepsilon} \right)^{1/2} = \beta_{pF} F^{1/2} \tag{5.71}$$

where β_{pF} is called the Poole–Frenkel constant

$$\beta_{pF} = \left(\frac{q^3}{\pi \varepsilon} \right)^{1/2} \tag{5.72}$$

Equation (5.71) differs from $\Delta\phi_B$ given by eq. (2.24) by a factor of 2 because the Coulombic attractive force to the electron is $q^2/4\pi\varepsilon(r_{pF})^2$ for the Poole–Frenkel effect, and is $q^2/4\pi\varepsilon(2x_m)^2$ for the Schottky effect. Thus the Poole–Frenkel effect is effective only for the traps which are neutral when filled and positively charged when empty. Traps which are neutral when empty and charged when filled will not manifest this effect for lack of Coulombic interaction, the degree of field dependent detrapping for various types of traps being summarized in Table 4.1.

Many investigators have used the Poole–Frenkel model to interpret many high field

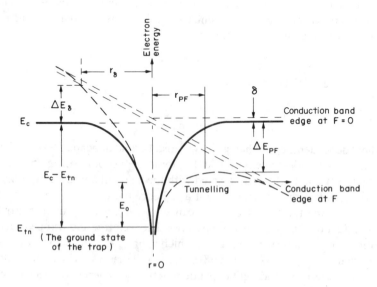

F ıɢ. 5.16. Schematic diagram illustrating the Poole–Frenkel effect.

transport phenomena in insulators and semiconductors [Frenkel 1938, Mead 1962, Simmons 1967, Jonscher 1967, Yeargan and Taylor 1968, Seki 1970, Ieda *et al.* 1971, Murgatroyd 1970, 1973, Hill 1971, Antula 1972, Connell *et al.* 1972, Vollmann 1974, Adamec and Calderwood 1975]. It is obvious that the Poole–Frenkel effect is observed when the electric conduction is bulk limited, and the Schottky effect is observed when the electric conduction is electrode limited. For both cases $\ln \sigma \propto F^{1/2}$. Thus to distinguish one case from the other, we have to rely on the comparison between the experimental and the theoretical values of β_{Sc} and β_{pF}, which can be calculated quite accurately if the value of ε for high frequencies, which should be approximately equal to η^2 (the refractive index of the material), is known. However, many investigators have reported that the value of β_{pF} determined from the slope of the $\ln \sigma - F^{1/2}$ plots does not agree with the theoretical one for the whole range of the experimental *J–V* characteristics. This implies that there may be several slopes in the $\ln \sigma - F^{1/2}$ plots and that even from the best fit curve the value of ε determined from it is usually not close to what would be expected. A trapped electron can be thermally released not only in the forward direction of the applied field in which the potential barrier is lowered by the field, but it can also be thermally released in other directions though the probability is smaller. The relative probabilities for a trapped electron to be thermally released in the forward direction and in the reverse direction of the applied field are field dependent. The simple expression for the barrier lowering given in eq. (5.71) derived by Frenkel [1938] is based on a one-dimensional (planar) model. Hartke [1968] and Ieda *et al.* [1971] have shown that a three-dimensional treatment gives the field and temperature dependent conductivity different from the following original Poole–Frenkel relation

$$\sigma = \sigma_0 \exp(\beta_{pF} F^{1/2}/2kT) \tag{5.73}$$

where σ_0 is the low field conductivity. From Fig. 5.16 it can be seen that the potential energy of an electron which is attracted to a positively charged trap located at $r = 0$ under the influence of a uniform field **F** may be written as

$$\phi = -q^2/4\pi\varepsilon r - q(\mathbf{F} \cdot \mathbf{r}) \tag{5.74}$$

By setting $d\phi/dr = 0$, we obtain the barrier lowering

$$\Delta\phi_{pF} = \beta_{pF}(F \cos \theta)^{1/2} \tag{5.75}$$

which occurs at

$$r_{pF} = \left(\frac{q}{4\pi\varepsilon F \cos \theta}\right)^{1/2} \tag{5.76}$$

where θ is an angle between **F** and **r**. The potential barrier is lowered only in the forward direction for $0 \le \theta \le \pi/2$. In the reverse direction, Ieda *et al.* [1971] have assumed that there is a state denoted by δ in which, by the interaction with phonons, the transition probability of an electron to a distance r_δ to become a free carrier is much larger than that to the ground state. They have derived an expression for r_δ as

$$r_\delta = \frac{q^2}{4\pi\varepsilon \delta} \tag{5.77}$$

and that for the increase of the potential barrier in the reverse direction as

$$\Delta E_\delta = \beta_{pF}^2 F \cos \delta / 4\delta \qquad (5.78)$$

Using the effective barrier lowering in the forward direction for $r_{pF} \le r_\delta$

$$\Delta E_{pF} = \beta_{pF} (F \cos \theta)^{1/2} - \delta \qquad (5.79)$$

and that for $r_{pF} \ge r_\delta$

$$\Delta F_{pF} = \beta_{pF}^2 F \cos \theta / 4\delta \qquad (5.80)$$

Ieda *et al.* [1971] have modified eq. (5.73) based on a one-dimensional treatment to the following equation based on a three-dimensional treatment

$$\sigma = \sigma_0 \left(\frac{4\gamma}{\alpha^2 F} \right) \sinh(\alpha^2 F / 4\gamma) \quad \text{for} \quad \alpha F^{1/2} \le 2\gamma \qquad (5.81)$$

and

$$\sigma = \sigma_0 \left(\frac{1}{\alpha^2 F} \right) [(\alpha F^{1/2} - 1) \exp (\alpha F^{1/2} - \gamma) - 2\gamma \exp(-\alpha^2 F / 4\gamma)$$
$$+ \exp(\gamma)] \quad \text{for} \quad \alpha F^{1/2} \ge 2\gamma \qquad (5.82)$$

where

$$\alpha = \beta_{pF} / 2kT \qquad (5.83)$$

$$\gamma = \delta / 2kT \qquad (5.84)$$

It can be seen that at low fields, $\alpha F^2 / 4\gamma \ll 1$, the conductivity follows a simple Ohm's law and that at high fields eq. (5.82) approaches asymptotically to eq. (5.73). In the intermediate fields the field dependent σ follows eqs. (5.81) and (5.82), which are quite different from eq. (5.73). Ieda *et al.* [1971] have reported that eqs. (5.81) and (5.82) are very consistent with the *J*–*F* characteristics as functions of temperature in polyacrylonitrile films measured by Hirai and Nakada [1968] and in SiO films measured by Hartman *et al.* [1966].

There are many other modifications of the original Poole–Frenkel model. By assuming that the barrier is lowered only for $0 \le \theta \le \frac{1}{2}\pi$ in the forward direction, and detrapping rate is field independent for $\frac{1}{2}\pi \le \theta \le \pi$ in the reverse direction, Hartke [1968] has modified eq. (5.73) to the following form

$$\sigma = \sigma_0 \left\{ \left(\frac{kT}{\beta_{pF} F^{1/2}} \right)^2 \left[1 + \left(\frac{\beta_{pF} F^{1/2}}{kT} - 1 \right) \exp \left(\frac{\beta_{pF} F^{1/2}}{kT} \right) \right] + \frac{1}{2} \right\} \qquad (5.85)$$

This equation is very similar to that derived by Jonscher [1967] based on a similar assumption. However, Hill [1971] has considered that the detrapping rate remaining constant in the reserve direction with the probability of $\exp[-(E_c - E_{tn})/kT]$ is not realistic because the height of the barrier in that direction (Fig. 5.16) has been increased by the applied field. By assuming that it has been increased by the same factor as the barrier in the forward direction has been decreased, $\beta_{pF} F^{1/2}$, Hill has derived expressions for the *J*–*F* characteristics. In a semi-crystalline material, detrapping could take place preferentially along the field direction, and thus *J* is given by

$$J = 2qN_t (kT)^4 \beta_{pF}^{-2} \mu_1 \exp[-(E_c - E_{tn})/kT] \alpha_{pF}^2 \sinh \alpha_{pF} \qquad (5.86)$$

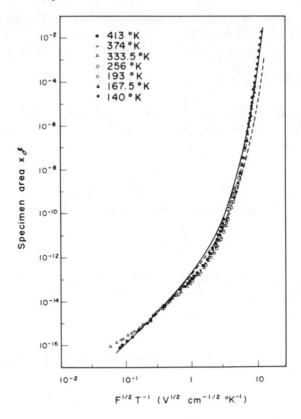

F_{IG}. 5.17. $\mathscr{J}$ versus $F^{1/2}T^{-1}$ for silicon monoxide for high temperatures obtained by Servini and Jonscher [1969]. $\mathscr{J} = JT^{-4}\exp(0.35$ per $kT)$ indicates $E_c - E_{tn} = 0.35$ eV. The solid curve is based on eq. (5.86) and the dashed curve is also based on eq. (5.86) but with $\alpha_{PF}\cosh\alpha_{PF} - \sinh\alpha_{PF}$ to replace $\alpha_{PF}^2\sinh\alpha_{PF}$. [After Hill 1971.]

where

$$A\overline{\phi}^{1/2} = -\frac{4\pi[m(E_c - E_{tn})^3]^{1/2}}{hqF}\lambda\left\{1 - \frac{\gamma F}{2\lambda}\ln\left[\frac{1+\lambda}{1-\lambda}\right]\right\}^{1/2} \tag{5.93}$$

$$B = \frac{4\pi[m(E_c - E_{tn})]^{1/2}}{hqF}\left\{1 - \frac{\gamma F}{2\lambda}\ln\left[\frac{1+\lambda}{1-\lambda}\right]\right\}^{1/2} \tag{5.94}$$

$$\lambda = (1 - \gamma F)^{1/2} \tag{5.95}$$

$$\gamma = [\beta_{pF}/(E_c - E_{tn})]^2 \tag{5.96}$$

and $N_i v_i$ is a product comparable to the product of thermal free carrier density and mobility in free carrier conduction. Hill [1971] has reported that eq. (5.92) fits well the SiO results obtained at high fields by Servini and Jonscher [1969] and Klein and Lisak [1966]. At high temperatures a carrier may possess an energy E_0 due to thermal excitation but this energy is less than the energy of the peak of the Poole–Frenkel barrier. For such a thermally assisted tunnelling, Hill [1971] has suggested that the probability of tunnelling through the reduced barrier at E_0 in Fig. 5.16 can be given approximately by replacing $(E_c - E_{tn})$ with $(E_c - E_{tn} - E_0)$ in the exponential term of eq. (5.92) and in eqs. (5.93)–(5.96).

Using a simplifying calculation involving three mutually perpendicular axes, one of them being in the direction of the applied field, Adamec and Calderwood [1975] have derived the following expression for the electric conductivity σ taking account of six escape directions,

$$\sigma \simeq \sigma_0 \left\{ \frac{1}{3} \left[2 + \cosh \left(\frac{\beta_{pF} F^{1/2}}{2kT} \right) \right] \right\} \tag{5.97}$$

For the case in which there are two trapping sites located so closely that the carrier detrapped from a trap will go to an adjacent trap rather than to the conduction band, several investigators have treated this case [Hill 1971, Dussel and Böer 1970, Vollmann 1974]. For details of this case the reader is referred to their original papers.

It is important to note that so far all expressions for the Poole–Frenkel effect are based on a uniform field. But for most cases, the field F is non-uniform spatially, particularly for the cases involving carrier injection from electrodes and space charge effects. For example, supposing that the electron trap distribution is of Gaussian type and the electron-injecting contact is ohmic as the case treated in Section 3.2.1 (D) except that in the present case the traps are electron traps. For shallow traps and $n_t > n$ we can follow eqs. (3.53) and (3.54), and write

$$n_t \simeq n_t + n = n/\theta_{dn0} \tag{5.98}$$

and

$$\theta_{dn0} = \frac{N_c}{g_n H_{dn}} \exp \left[-\frac{(E_c - E_{tn})}{kT} + \frac{1}{2} \left(\frac{\sigma_t}{kT} \right)^2 \right] \tag{5.99}$$

for zero-applied field. At high fields and with the Poole–Frenkel effect involved in the conduction processes, then θ_{dn0} has to be changed to θ_{dn} which becomes field dependent

$$\theta_{dn}(F) = \frac{N_c}{g_n H_{dn}} \exp \left[-\frac{(E_c - E_{tn})}{kT} + \frac{1}{2} \left(\frac{\sigma_t}{kT} \right)^2 + f_{pF}(F) \right]$$

$$= \theta_{dn0} \exp[f_{pF}(F)] \tag{5.100}$$

where $f_{pF}(F)$ is the effective lowering of potential barrier by the field F. Therefore

$$f_{pF}(F) = \beta_{pF} F^{1/2}/kT \tag{5.101}$$

for the original one-dimensional Poole–Frenkel model,

$$f_{pF}(F) = \ln \left\{ \left(\frac{kT}{\beta_{pF} F^{1/2}} \right)^2 \left[\left(\frac{\beta_{pF} F^{1/2}}{kT} - 1 \right) \exp \left(\frac{\beta_{pF} F^{1/2}}{kT} \right) + 1 \right] + \frac{1}{2} \right\} \tag{5.102}$$

for Hartke's three-dimensional model, and

$$f_{pF}(F) = \ln \left\{ \frac{1}{3} \left[2 + \cosh \left(\frac{\beta_{pF} F^{1/2}}{2kT} \right) \right] \right\} \tag{5.103}$$

for the model of Adamec and Calderwood, and so on. From Poisson equation, we can write

$$dz = \frac{\varepsilon\theta_{dn}(F_\cdot)dF}{qn} = \mu_n\varepsilon\theta_{dn0}\,F\,\exp[f_{pF}(F)]dF/J \tag{5.104}$$

Integration of eq. (5.104) gives

$$d = \int_0^d dz = \frac{\mu_n\varepsilon\theta_{dn0}}{J}\int_{F(z=0)}^{F(z=d)} F\,\exp[f_{pF}(F)]dF \tag{5.105}$$

Using the boundary condition $V = \displaystyle\int_0^d F\,dz$ it can easily be shown that

$$J = \frac{9}{8}\varepsilon\mu_n\theta_{dno}\frac{V^2}{d^3}\left\{\frac{8}{9}\frac{\left[\displaystyle\int_{F(z=0)}^{F(z=d)} F\,\exp[f_{pF}(F)]dF\right]^3}{\left[\displaystyle\int_{F(z=0)}^{F(z=d)} F^2\,\exp[f_{pF}(F)]dF\right]^2}\right\} \tag{5.106}$$

For deep traps and $n_t > n$ we can follow eqs. (3.59)–(3.61), and write

$$n_t = (H_{dn}/2)g_n\left(\frac{n}{N_c}\right)^{1/m}\exp\left(\frac{E_c - E_{tn}}{mkT} - \frac{f_{pF}(F)}{m}\right) \tag{5.107}$$

where

$$m = \left[1 + \frac{\pi}{8}\left(\frac{\sigma_t}{kT}\right)^2\right]^{1/2} \tag{5.108}$$

Following the same mathematical treatment for deriving eq. (3.63), it can easily be shown that

$$J = q^{1-m}\mu_n N_c\left(\frac{2\varepsilon}{g_n H_{dn}}\right)^m\exp\left[-\frac{(E_c - E_{tn})}{kT}\right]\theta'_{pF}(F)\frac{V^{m+1}}{d^{2m+1}} \tag{5.109}$$

where

$$\theta'_{pF}(F) = \frac{\left[\displaystyle\int_{F(z=0)}^{F(z=d)} F^{1/m}\,\exp[f_{pF}(F)/m]dF\right]^{2m+1}}{\left[\displaystyle\int_{F(z=0)}^{F(z=d)} F^{1/m+1}\,\exp[f_{pF}(F)/m]dF\right]^{m+1}} \tag{5.110}$$

The J–V characteristics for the cases involving shallow traps [eq. (5.106)] or for the cases involving deep traps [eqs. (5.109) and (5.110)] can easily be computed using any $f_{pF}(F)$ given in eqs. (5.101)–(5.103) or other forms of $f_{pF}(F)$. However, although $F(z = 0)$ can be assumed to be zero for ohmic contact or to follow a Schottky type thermionic emission or a Nordheim and Fowler type tunnelling injection, $F(z = d)$ is not easy to determine, and this is the basic disadvantage of the above approach to the solution of this problem. The other approach is to integrate eq. (5.104) to obtain z as a function of F, but in this way the functional relation cannot be analytically inverted to give F as a function of z. However, this approach does not need to involve the unknown $F(z = d)$, but numerical computation with the aid of a computer to find $F = F(z, J)$ and then to compute $V = E_{\text{av}}d = \displaystyle\int_0^d F(z, J)\,dz$ to yield the J–V characteristics is necessary.

Antula [1972] has modified the models of Frenkel, Hartke, and Hill by including the hot electron concept introduced by Fröhlich [1947] based on the fact that if the rate of energy exchange between electrons and lattice is much smaller than the rate of energy exchange of trapped and conduction electrons among themselves, the electron temperature T_e will be greater than the lattice temperature T. Applying this hot electron concept to the Poole–Frenkel detrapping process and assuming that the solid contains deep donor-like traps located at $E_c - E_{tn}$ and numerous shallow traps within the energy range ΔE immediately below the bottom edge of the conduction band, Antula has obtained

$$\left(\frac{\sigma}{\sigma_0}\right)_H = \left(\frac{\sigma}{\sigma_0}\right)\exp\{\,[\,(E_c - E_{tn}) - \beta_{pF}F^{1/2}](F/F_0)^2(\Delta E)^{-1}\} \qquad (5.111)$$

where (σ/σ_0) is given in eqs. (5.73), (5.81), (5.82), (5.85), or (5.97) or those derived from eqs. (5.86) or (5.87) and F_0 is defined by

$$\Delta E/kT - \Delta E/kT_e \simeq (F/F_0)^2 \qquad (5.112)$$

He has also reported that eq. (5.111) fits well the experimental results for SiO films obtained by Hartman *et al.* [1966] and Servani and Jonscher [1969], and also the experimental results for Al_2O_3 films. He used eq. (5.85) for σ/σ_0 for his computation of eq. (5.111).

Chan and Jayadevaiah [1972] have considered the screening effects on the Poole–Frenkel conduction in amorphous solids. By considering a donor trap surrounded by a cloud of neutral trapping levels, spread out spatially, they have suggested using the effective Poole–Frenkel constant β_{eff} to replace the usual β_{pF}, which is given by

$$\beta_{\text{eff}} = \beta_{pF}\left(\frac{F}{F + q^3N_t/8\pi\varepsilon^2 kT}\right) \qquad (5.113)$$

and is field and temperature dependent.

As has been mentioned at the beginning of this subsection, a trap which will experience effectively the Poole–Frenkel effect must be neutral when filled and positively charged when empty. However, Jonscher [1967] has pointed out that electrons escaping from neutral traps, which are neutral when empty and negatively charged when filled, would also be affected by the field-enhanced lowering of the trap potential barrier. Recently, Arnett and Klein [1975] have proposed that for such neutral traps in which the potential ϕ can be represented by an inverse power law

$$\phi \propto r^{-n} \qquad (5.114)$$

the field-enhanced lowering of the potential barrier is

$$\Delta E_{pF} = -2(nV_0F^n)^{1/(n+1)} \qquad (5.115)$$

where V_0 characterizes the magnitude of the attractive potential of the neutral trap.

5.3.6. The Onsager model

Unlike the Poole–Frenkel models, which are based on the assumption that the probability for the trapped carriers to be thermally released increases with increasing applied field because of the field-enhanced lowering of the potential barrier of traps, the Onsager model is based on a completely different assumption that the probability $p(r, \theta, F)$ for an electron–hole pair (or electron and donor-like trap) thermalized with an initial separation r and orientation θ relative to the applied field F to escape initial recombination increases with increasing applied field. For the former model the carriers are regenerated after they have been captured in the traps, while for the latter model the carriers are separated before they are trapped. On the basis of the postulate that the theory of geminate (or initial) recombination reduces to the problem of Brownian motion of one particle under the action of the Coulombic attraction and the applied electric field, Onsager [1934, 1938] has approached this problem by solving the equation of Brownian motion

$$\frac{\partial f}{\partial t} = \frac{kT}{q}\mu_n \mathrm{div}\left[\exp\left(-\frac{U}{kT}\right)\mathrm{grad}\,f\exp\left(\frac{U}{kT}\right)\right] \tag{5.116}$$

where f is a probability function, μ_n is the electron mobility, and U is the Coulombic potential modified by the applied field and is given by

$$U = -(q^2/4\pi\varepsilon r) - qFr\cos\theta \tag{5.117}$$

Using the boundary condition of zero initial separation between the electron and the positively charged centre, the probability of ionization under the steady state condition $(\partial f/\partial t = 0)$ in the presence of the applied field is increased by the ratio [Pai 1975, Pai and Enck 1975]

$$\frac{P(F)}{P(0)} = \frac{J_1(j\alpha)}{j\alpha/2}$$

$$= 1 + \frac{1}{2!}\left(\frac{\alpha^2}{4}\right) + \frac{1}{2!3!}\left(\frac{\alpha^2}{4}\right)^2 + \frac{1}{3!4!}\left(\frac{\alpha^2}{4}\right)^3 + \cdots \tag{5.118}$$

where J_1 is the Bessel function of the first order and

$$\alpha = \left(\frac{q^3}{\pi\varepsilon}\right)^{1/2}\frac{F^{1/2}}{kT} \tag{5.119}$$

which is the same as α_{pF} given in eq. (5.88). If an initial separation between the electron and the positively charged centre being r_0 rather than zero is used as the boundary condition, eq. (5.118) is modified to [Pai 1975]

$$\frac{P(F)}{P(0)} = 1 + \frac{1}{2!}\left(\frac{\alpha^2}{4}\right) + \frac{1}{2!3!}\left(\frac{\alpha^2}{4}\right)^2\left(1 - \frac{2r_0}{r_c}\right)$$

$$+ \frac{1}{3!4!}\left(\frac{\alpha^2}{4}\right)^3\left(1 + 3!\frac{r_0^2}{r_c^2} - 3!\frac{r_0}{r_c}\right) + \cdots \tag{5.120}$$

where r_c is the cut-off separation distance to separate the bound and the free carriers and is defined by

$$r_c = \frac{q^2}{4\pi\varepsilon kT} \tag{5.121}$$

Carriers which are within the separation distance $0 \le r \le r_c/2$ are bound charge carriers. When $r_0 = 0$, eq. (5.120) reduces to eq. (5.118). Figure 5.18 shows the difference between eq. (5.120) and eq. (5.118) and also the comparison of the Onsager model with the Poole–Frenkel model. It can be seen that the $P(F)/P(0)$ ratio is smaller for the Onsager model for a given field and that the larger the value of r_0 the smaller is this ratio.

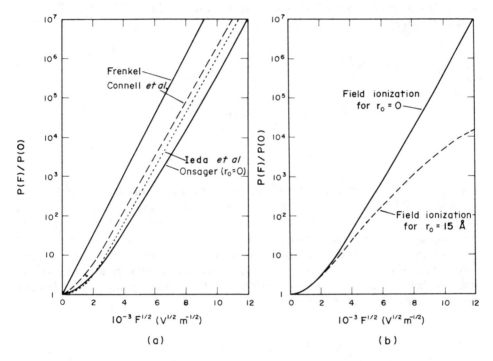

FIG. 5.18. (a) Electric-field-enhanced probability of ionization as a function of the square root of the applied electric field for a material with a relative dielectric constant of 3 at $T = 293°$K calculated from eq. (5.118) based on the Onsager model and compared with those calculated from various expressions [Frenkel 1938, Connell *et al.* 1972, Ieda *et al.* 1971] based on the Poole–Frenkel effect. (b) Electric-field-enhanced probability of ionization as a function of the square root of the applied electric field for a material with a relative dielectric constant of 3 at $T = 293°$K calculated for $r_0 = 0$ from eq. (5.118) and for $r_0 = 15$ Å from eq. (5.120). [After Pai 1975.]

Using the Onsager model Pai [1975] has derived an expression for the field dependent electric conductivity, and it is

$$\sigma(F) = K_1 \left[\frac{J_1(j\alpha)}{j\alpha/2} \right]^{1/2} \exp\left(\frac{E_c - E_{tn}}{2kT} \right) \tag{5.122}$$

and the field dependent current density

$$J(F) = K_1 F \left[\frac{J_1(j\alpha)}{j\alpha/2} \right]^{1/m} \exp\left(-\frac{E_c - E_{tn}}{mkT} \right) \tag{5.123}$$

where K_1 is a constant and the parameter m ranges between 1 and 2 depending on the complexity of the distribution of states in the forbidden band gap (donors, acceptors, traps, and the like). Equations (5.122) and (5.123) are based on a uniform field distribution in a solid specimen. If the conduction involves carrier injection from the electrodes and space charge effects, an approach similar to that described in Section 5.3.5 [see eqs. (5.98)–(5.110)] to modify these two equations is necessary.

Onsager [1934, 1938] has also calculated the probability $p(r, \theta, F)$. The expression of $p(r, \theta, F)$ is given by [see also Batt *et al.* 1968, Geacintov and Pope 1971, Melz 1972, Pai and Enck 1975]

$$p(r, \theta, F) = \exp(-A) \exp(-B) \sum_{m=0}^{\infty} \sum_{n=0}^{\infty} \frac{A^m}{m!} \frac{B^{m+n}}{(m+n)!} \tag{5.124}$$

where

$$A = \frac{q^2}{4\pi \varepsilon k T r}$$

$$B = (qFr/2kT)(1 + \cos\theta) \tag{5.125}$$

Many investigators [Batt *et al.* 1968, Chance and Braun 1973, Braun and Chance 1974, Melz 1972, Pfister and Williams 1974] have used the Onsager model to explain photoconduction and photogeneration processes in organic semiconductors. By defining ϕ_{p0} as the ionization quantum yield [the efficiency of production of thermalized ion pairs per absorbed photon] and $g(r, \theta)$ as the initial spatial distribution of thermalized pair configuration [separation between ions of each ion pair—or between electron and ionized donor], the carrier quantum yield [carrier generation efficiency] is given by

$$\phi_p(r_0, F) = \phi_{p0} \int p(r, \theta, F) g(r, \theta) d^3 r \tag{5.126}$$

By assuming that ϕ_{p0} is independent of the applied electric field and $g(r, \theta)$ is an isotropic δ function and is expressed as

$$g(r, \theta) = (4\pi r_0^2)^{-1} \delta(r - r_0), \tag{5.127}$$

substituting eqs. (5.124) and (5.127) in eq. (5.126) and carrying out the integration, we obtain [e.g. Pai and Enck 1975]

$$\phi_p(r_0, F) = \phi_{p0} \left(\frac{kT}{qFr_0} \right) \exp(-A) \sum_{m=0}^{\infty} \frac{A^m}{m!}$$

$$\times \sum_{n=0}^{\infty} \left[1 - \exp(-qFr_0/kT) \sum_{l=0}^{m+n} \left(\frac{qFr_0}{kT} \right)^l \frac{1}{l!} \right] \tag{5.128}$$

The first few terms of eq. (5.128) can be written as

$$\phi_p(r_0, F) = \phi_{p0} \exp\left(-\frac{r_c}{r_0}\right)\left[1 + \left(\frac{q}{kT}\right)\frac{1}{2!}Fr_c\right.$$

$$+ \left(\frac{q}{kT}\right)^2 \frac{1}{3!} F^2 r_c \left(\frac{1}{2}r_c - r_0\right)$$

$$+ \left(\frac{q}{kT}\right)^3 \frac{1}{4!} F^3 r_c \left(r_0^2 - r_0 r_c + \frac{1}{6}r_c^2\right)$$

$$\left. + \ldots\right] \tag{5.129}$$

These investigators have reported that eq. (5.129) agrees well with their experimental results on electric field and temperature dependence of hole and electron quantum yields in anthracene and tetracene crystals [Geacintov and Pope 1971, Chance and Braun 1973] and in trinitrofluorenone-poly (*N*-vinylcarbazole) and in poly (*N*-vinylcarbazole) [Melz 1972, Pfister and Williams 1974]. Further discussion on the photoconduction and generation processes will be given in Chapter 6.

We would like to point out that the Onsager theory, because of its inherent microscopic diffusive nature, takes into account the integrated ionization probability within the 4π solid angle. The physical soundness of the Onsager theory is to include the reverse ionization (or escape) probability in a most natural way devoid of any approximation [Pai 1975].

5.4. THERMALLY STIMULATED DETRAPPING AND NON-EXTRINSIC ELECTRIC CONDUCTION

An important and convenient method for determining trapping levels, capture cross-sections and other trapping parameters in semiconductors and insulators is by means of thermally stimulated detrapping. This method is generally referred to as the thermal glow (TG) method which includes thermoluminescence (TL) and thermally stimulated current (TSC) measurements. In a broad sense this method also includes the thermally stimulated capacitor depolarization (TSCD) (sometimes called the "thermally stimulated capacitor discharge" or the "thermally stimulated depolarization current", and thus a symbol "TSDC" is sometimes used). Other measurements, such as the photostimulated current and the isothermal current decay (ICD), are also within this category.

Trapping energy levels can sometimes be determined by measuring the activation energies in the ohmic and non-ohmic regions, which can, in turn, be used to distinguish between intrinsic, extrinsic, and non-extrinsic conductions. In the following we shall review briefly the theory and experiments of TSC, TSCD, ICD, and non-extrinsic conduction.

5.4.1. Thermally stimulated currents

When a specimen is mounted in a vacuum cryostat and excited at a sufficiently low temperature so that the generated carriers will be captured by traps, then upon ceasing the excitation the trapped carriers cannot be freed by the thermal energy available at that temperature. If the excitation is caused by light illumination and if the specimen is heated in the dark by raising the temperature at a constant rate after the removal of the excitation, the carriers liberated by such a thermal action may recombine radiatively to produce TL, or contribute under an applied field to an excess current which is called the TSC. The luminescence intensity or the excess current measured as a function of temperature during heating yields a thermally stimulated glow curve (TSG) or a TSC curve. The positions of the peaks in these curves which are associated with the trapping levels and the capture cross-sections depend on the heating rate; and, therefore, by varying the heating rate these trapping parameters can be determined. For the case of TSC the excess current first increases exponentially due to increasing probability of thermal detrapping of the trapped carriers, and then returns to zero due to recombination of the released carriers. The TSG method was first introduced by Urbach [1930], but did not come into extensive use until the first theoretical treatment was reported by Randall and Wilkins [1945]. Since then, many investigators have found this method useful for studying trap parameters in solids [Nicholas and Woods 1964, Dussel and Bube 1967, Chen 1969, 1971, Simmons *et al.* 1973, Helfrich *et al.* 1964, Devaux and Schott 1967, Westgate and Warfield 1967].

Supposing that a solid contains one single set of electron traps located at an energy level ΔE eV below the conduction band ($\Delta E = E_c - E_t$), then the probability P that a trapped electron will escape from the trap to the conduction band at the temperature T is given by

$$P = v \exp(-\Delta E/kT) \qquad (5.130)$$

where v is the attempt-to-escape frequency. By assuming $g_n = 1$ in eq. (4.20), it is given by

$$v = N_c \langle v\sigma_n \rangle \qquad (5.131)$$

To determine the TSC curve, the solid specimen is cooled to a temperature T_0 and excited so that the density of electrons trapped in the traps is n_{t0} which is assumed to be smaller than the total density of traps N_n. If now the specimen is heated at a constant rate

$$\beta = dT/dt \qquad (5.132)$$

then at some time t after the heating has begun, the rate of change of free electron density is

$$\frac{dn}{dt} = -\frac{n}{\tau} - \frac{dn_t}{dt} \qquad (5.133)$$

where τ is the electron lifetime which is determined by recombination processes. The first term on the right represents the recombination rate of the free electrons, and the second term represents the rate of change of trapped electron density in the traps, which is given by

$$\frac{dn_t}{dt} = -n_t v \exp(-\Delta E/kT) + n(N_n - n_t)\langle v\sigma_n \rangle \qquad (5.134)$$

in which the first term on the right represents the rate of thermal release of trapped electrons, and the second term represents the rate of retrapping of free electrons.

Most theories are based on the assumption that τ is short so that $n/\tau > dn/dt$. With this assumption the general solution of eqs. (5.133) and (5.134) yields $n(t)$ and hence the thermally stimulated conductivity

$$\sigma(T) = q\mu_n n = -q\mu_n \tau \frac{dn_t}{dt} = -q\mu_n \tau \, \beta \frac{dn_t}{dT}$$

$$= \frac{q\mu_n \tau N_c \langle v\sigma_n \rangle}{1+\tau N_n \langle v\sigma_n \rangle} n_{t0} \exp \left[-\frac{\Delta E}{kT} - \frac{1}{\beta} \int_{T_0}^{T} \frac{N_c \langle v\sigma_n \rangle \exp(-\Delta E/kT) dT}{1+\tau N_c \langle v\sigma_n \rangle} \right] \quad (5.135)$$

where $T(t) = T_0 + \beta t$. Equation (5.135) is similar to that derived by Saunders and Jewitt [1965]. To find the temperature at which σ is a maximum, we have to know the temperature dependence of $\mu_n(T), \tau(T), N_c(T), v(T)$, and $\sigma_n(T)$, which depend on energy band structure and carrier scattering and recombination processes, and thus vary from material to material. By assuming the following temperature dependence of those parameters

$$\begin{aligned} N_c &= AT^{3/2}, & \mu_n &= DT^{-b} \\ v &= BT^{1/2}, & \tau &= \text{constant} \\ \sigma_n &= CT^{-a} \end{aligned} \quad (5.136)$$

and setting $d\sigma/dT = 0$ for the occurrence of the peak (i.e. $\sigma = \sigma_m$) at $T = T_m$, we obtain

$$\frac{\Delta E}{kT_m} = \ln \left(\frac{T_m^2}{\beta} \right) + \ln \left(\frac{kN_c \langle v\sigma_n \rangle}{\Delta E} \right) - \ln(1+\tau N_n \langle v\sigma_n \rangle) \quad (5.137)$$

for cases $\Delta E > kT_m$ [Milnes 1973]. Three special cases are considered as follows:

(A) *Monomolecular recombination.* For this case there is no retrapping or slow retrapping and so we can assume $\tau N_n \langle v\sigma_n \rangle \ll 1$. Thus eqs. (5.135) and (5.137) reduce to those derived by Randall and Wilkins [1945].

$$\sigma(T) = q\mu_n \tau N_c \langle v\sigma_n \rangle n_{t0} \exp \left[-\frac{\Delta E}{kT} - \frac{1}{\beta} \int_{T_0}^{T} N_c \langle v\sigma_n \rangle \exp(-\Delta E/kt) dT \right]$$

$$(5.138)$$

and

$$\frac{\Delta E}{kT_m} = \ln \left(\frac{kT_m^2 N_c \langle v\sigma_n \rangle}{\beta \Delta E} \right) \quad (5.139)$$

Since T_m depends on the heating rate β, Booth [1954] and Bohun [1954] have proposed to use two heating rates to determine ΔE from eq. (5.139). Thus

$$\Delta E = \frac{kT_{m_1} T_{m_2}}{(T_{m_1} - T_{m_2})} \ln \left(\frac{\beta_1 T_{m_2}}{\beta_2 T_{m_1}} \right) \quad (5.140)$$

Later, Hoogenstraaten [1958] has suggested using a number of heating rates so that $\ln(T_m^2/\beta)$ as a function of $1/T_m$ can be plotted and from this plot ΔE and hence σ_n can be determined. And Keating [1961], following an argument similar to that of Randall and

Wilkins, has derived the following formula for determining ΔE:

$$\frac{kT_m}{\Delta E} = \frac{T''-T'}{T_m}(1.2\gamma - 0.54) + 5.5 \times 10^{-3} - \left(\frac{\gamma - 0.75}{2}\right)^2 \qquad (5.141)$$

where $\gamma = (T''-T_m)/(T_m - T')$, and T' and T'' are temperatures at which $\sigma(T)$ attains the value $\frac{1}{2}\sigma_m(T)$ on either side of T_m. Equation (5.141) is a good approximation when $10 < \Delta E/kT_m < 35$ and $0.75 < \gamma < 0.9$.

(B) *Fast retrapping.* For this case the recombination is mainly bimolecular and the free electrons can be assumed to be in thermal equilibrium with the trapped electrons in the traps and $\tau N_n \langle v\sigma_n \rangle \gg 1$. Thus eq. (5.135) reduces to

$$\sigma(T) = \frac{q\mu_n N_c}{N_n} n_{t0} \exp\left[-\frac{\Delta E}{kT} - \frac{1}{\beta \tau N_n} \int_{T_0}^{T} N_c \exp(-\Delta E/kT)\,dT \right] \qquad (5.142)$$

This equation does not involve σ_n and it is not possible to determine σ_n from the measured glow curve. However, Böer *et al.* [1958] have shown that the magnitude of σ_R can be estimated by the following equation:

$$v\sigma_R = \frac{\Delta E}{kT_m} \frac{\beta}{n_m T_m} \qquad (5.143)$$

where n_m is the density of free electrons at $T = T_m$. By setting $dn_t/dt = 0$ in eq. (5.134) we obtain the condition for the occurrence of the peak in the glow curve and the following equation for determining ΔE:

$$\frac{\Delta E}{kT_m} = \ln\left(\frac{N_c}{n_m}\right) + \ln\left(\frac{n_t}{N_n - n_t}\right) \qquad (5.144)$$

By assuming that the peak occurs when the quasi-Fermi level coincides with the trapping energy level, the ratio $n_t/N_n = 1/2$ and eq. (5.144) becomes [Bube 1955, 1960]

$$\frac{\Delta E}{kT_m} = \ln\left(\frac{N_c}{n_m}\right) \qquad (5.145)$$

Thus, a plot of $\ln(n_m)$ as a function of $1/T$ should yield a straight line of slope ΔE.

(C) *Intermediate retrapping.* Garlick and Gibson [1948] have considered the case in which a free electron has equal probability of recombining or of being retrapped, and the TSC is given by

$$\sigma(T) = \frac{q\mu_n \tau N_c \langle v\sigma_n \rangle n_{t0}^2 \exp(-\Delta E/kT)}{N_n \left[1 + \frac{N_c \langle v\sigma_n \rangle}{N_n \beta} n_{t0} \int_{T_0}^{T} \exp(-\Delta E/kT)\,dT \right]^2} \qquad (5.146)$$

Under this condition the recombination is mainly bimolecular. It is important to note that T_m depends on the ratio of n_{t0}/N_n (the fraction of traps occupied) for bimolecular recombination, but it is independent of n_{t0}/N_n for monomolecular recombination.

Bryant *et al.* [1959], Kokado and Schneider [1964], Weisz *et al.* [1964], Bree and Kydd [1964], Thomas *et al.* [1968], Garofano and Morell [1973] have used the measurements of TSC to determine the trapping levels in anthracene. Typical glow curves for solution-grown anthracene crystals are shown in Fig. 5.19. Thomas *et al.*

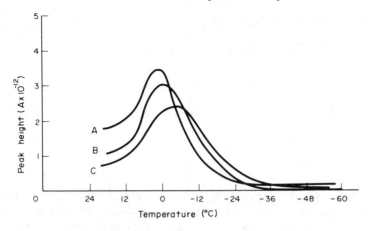

[1968] have reported that the hole traps in solution-grown anthracene are located at a level of 0.7 eV above the valence band edge.

So far, we have considered only one trapping level; the TSC becomes complicated if two or more trapping levels are involved. For more trapping levels we may have to resort to a computer solution [Dussel and Bube 1967, Sacks 1970, Milnes 1973]. However, it should be noted that in spite of many inherent advantages of the TSC and TL methods, there is still little evidence that these methods can provide consistent quantitative information of trapping levels and other trapping parameters even for one single set of traps [Devaux and Schott 1967, Land 1969, Kelly *et al.* 1970, 1971, Braunlich and Kelly 1970]. This may be due to the following reasons:

(a) The traps may not be confined in single discrete energy levels, but rather distributed; and then analytical solution of such a problem becomes intractable.
(b) The kinetic processes of the thermally stimulated recombination cannot be simply described by τ. Even if we can use τ, the dependence of τ on T, n, n_t, and other parameters is not known.
(c) The simple treatment given above is based on one carrier (e.g. electron) kinetics, but the problem is clearly a two-carrier (both electron and hole) problem. It is not the electron carriers alone to contribute to TSC, the role of holes in TSC cannot be ignored.

Several investigators have suggested that the correlation between TL and TSC based on the basic simple model would yield a new technique for the analysis of experimental data [Kelly *et al.* 1971]. Recently, Fields and Moran [1974] have derived a correlation expression for TL and TSC. In the following we summarize their analysis.

Equation (5.134) can be written as

$$\frac{dn_t}{dt} = -n_t v \exp(-\Delta E/kT) + n(N_n - n_t)\langle v\sigma_n \rangle$$

$$= -Pn_t + C_n n(N_n - n_t) \qquad (5.147)$$

and the rate of diminution of both free and trapped carriers associated with the particular glow peak is

$$\frac{d(n + n_t)}{dt} = -C_R n(n + n_t + M) \qquad (5.148)$$

where C_n and C_R are, respectively, temperature and time independent capture coefficient of a trapping centre and that of a recombination centre for a free charge carrier; N_n is the density of traps associated with the peak of interest; and M is the density of deep traps associated with thermally disconnected peaks having higher activation energies; $n + n_t$ on the right of eq. (5.148) is the number of states emptied due to the creation of n and n_t. The TSC (σ) and TL (B) may be expressed as

$$\sigma(T) = q\mu_n n \qquad (5.149)$$

$$B(T) = -\eta \frac{d(n + n_t)}{dt} \qquad (5.150)$$

where η is the probability that the decay of a charge carrier results in the emission of a photon. Their analysis from the above basic equations leads to the following theoretical predictions:

(i) $\dfrac{dB}{dt} = -BC_R n$ for $T = T_{m(TSC)}$ $\qquad (5.151)$

This implies that the TL is decreasing at the TSC peak, or in other words, the TSC peak occurs at a higher temperature than the TL peak.

(ii) The correlation equation relating the integrated TL at any measurement time t from the initial time 0 just prior to the start of heating to the ratio of TL/TSC at the same time t is given by

$$\int_0^t B(T)dt = \eta(n_0 + n_{t0} + M) - \left(\frac{q\varepsilon}{q'}\right)\left[\frac{B(T)}{\sigma(T)}\right] \qquad (5.152)$$

where n_0 and n_{t0} are the values of n and n_t at $t = 0$, respectively, and q' is the effective charge of the recombination centre.

However, they have reported that experimental results in LiF are in striking disagreement with the theoretical correlation equation which is based on the simple trapped carrier-free carrier recombination model. This demonstrates that the measurements of TL and TSC do not provide sufficient information on all the relevant parameters entering the kinetic equations to determine the properties of TL and TSC curves. They have also suggested that the actual system probably possesses substantial spatial correlation between trapped carriers and TL active recombination centres, and that this may account for the discrepancy between the experiment and the theory.

We can say that the TL and TSC methods have not yet been fully explored. The trapping parameters deduced from the TL and TSC peaks in conjunction with the simple model may be doubtful. Further studies should take into account the spatial correlation between trapped carriers and recombination centres, two-carrier conduction processes, the temperature and time dependent recombination kinetics, and the form of trap distributions.

5.4.2. Thermally stimulated depolarization currents

Several investigators [Kokado and Schneider 1964, Devaux and Schott 1967, Agarwal and Fritzsche 1974, Agarwal 1974] have suggested that the TSC method is not suitable for use in materials in which the dark current is of the same order of or larger than the TSC excess current, or in which the steady state photocurrent is an exponentially increasing function of temperature under a constant illumination and applied voltage, so that the time required for a sufficient filling of traps at a low temperature (e.g. at 77° K) is large. Furthermore, surface photoeffects, together with trapping and recombination processes at the surface, may affect the filling of trapping centres in the bulk by means of photoexcitation at a low temperature, thus making the TSC measurements difficult or inaccurate. For all of these reasons, the TSDC method was introduced. Driver and Wright [1963] and Kirov and Zhelev [1965] were among the first investigators to use the TSDC method to study trapping centres in CdS crystals. Since then, this method has been widely used for investigating trapping centres in inorganic and organic crystalline and amorphous materials [Simmons et al. 1972, 1973, Devaux and Schott 1967, Andriesh et al. 1975]. With this method the traps are first filled at some high temperature by the application of an electric field across the solid specimen, which is provided with one blocking and one ohmic contact, and the specimen is then cooled down to a temperature T_0 with the field applied so that the carriers injected into the specimen through the ohmic contact will be trapped. Upon heating and applying a field of reverse sign (or by short-circuiting the electrodes on heating), the trapped carriers are released, giving rise to the so-called thermally stimulated depolarization current. This method is similar to the TSC method; the only difference is the way of filling the traps so that the techniques similar to those described in Section 5.4.1 for the TSC can be used to analyse the TSDC curves.

Devaux and Schott [1967] have studied the trapping levels in copper phthalocyanine using the TSDC method. The specimen with a mylar to separate the metallic electrode as blocking contact and a metallic electrode as ohmic contact was polarized with a voltage of 3000 V at a temperature range from 100° C to 150° C, then cooled down to room temperature, and heated under a reverse lower voltage up to about 150° C. Typical results show J_m and T_m (the peak current J_m at $T = T_m$) as functions of β in Fig. 5.20(a), J_m and T_m as functions of applied voltage during heating in Fig. 5.20(b), and J_m and T_m as functions of polarization voltage in Fig. 5.20(c). These results follow closely the following equations [Devaux and Schott 1967]:

$$\frac{\Delta E}{kT_m} \simeq \ln\frac{AkT_m^2}{\beta\Delta E} \tag{5.153}$$

and

$$J_m \simeq \frac{q\mu_n vF_0}{2}\exp\left(-\frac{\Delta E}{kT_m}\right) \tag{5.154}$$

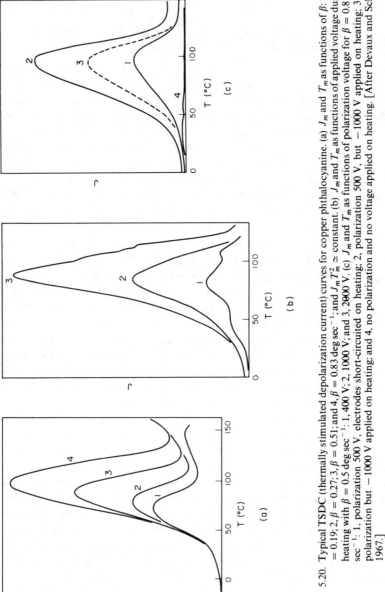

Fig. 5.20. Typical TSDC (thermally stimulated depolarization current) curves for copper phthalocyanine. (a) J_m and T_m as functions of β: 1, $\beta = 0.19$; 2, $\beta = 0.27$; 3, $\beta = 0.51$; and 4, $\beta = 0.83$ deg sec^{-1}; and $J_m T_m^2 \simeq$ constant. (b) J_m and T_m as functions of applied voltage during heating with $\beta = 0.5$ deg sec^{-1}: 1, 400 V; 2, 1000 V; and 3, 2000 V. (c) J_m and T_m as functions of polarization voltage for $\beta = 0.8$ deg sec^{-1}: 1, polarization 500 V, electrodes short-circuited on heating; 2, polarization 500 V, but -1000 V applied on heating; 3, no polarization but -1000 V applied on heating; and 4, no polarization and no voltage applied on heating. [After Devaux and Schott 1967.]

where A is a constant and F_0 is the initial field resulting from a polarization. Equation (5.153) is similar to eq. (5.139), and eq. (5.154) is similar to eq. (5.138) when $\sigma(T_m) = \sigma_m$. To decompose a broad peak by fractional heating into two peaks, Devaux and Schott found two trapping levels, 0.41 and 0.66 eV, located below the conduction band edge.

The depolarization current observed in the TSDC experiment consists of two parts: (i) the dielectric relaxation current, and (ii) the trap-limited current due to carriers thermally released from the traps. Only part (ii) would yield information about trap parameters. If part (i) is dominant, the TSDC current is mainly the dielectric relaxation current and becomes virtually useless for studying trap parameters. Recently Agarwal [1974] has demonstrated that the position, shape, and number of the depolarization peaks caused by the dielectric relaxation current depend on the geometry of the specimen and the electrodes, and on the conductivity of the material. He has pointed out that the behaviour of the dielectric relaxation current peaks is quite similar to the trap-limited current peaks, and that the analysis of the TSDC curves may lead to wrong conclusions if part (i) is not or cannot be subtracted from the total TSDC to extract part (ii)—the pure trap-limited current. He has suggested that the trap-limited TSDC may not be observed in those materials which have a low drift mobility, a short screening length, and a large density of localized states near the Fermi level, such as chalcogenide glass semiconductors. However, to use TSDC for studying trap parameters, the effect of dielectric relaxation should not be ignored.

5.4.3. Isothermal current decay

The trap parameters may also be determined by the so-called "isothermal current decay" (ICD) method. The method is similar to the TSC method except that after the traps are filled and the photoexcitation has ceased, the temperature is kept constant and the isothermal current decay is measured as a function of time. Using eqs. (5.130) and (5.134) and defining $\tau_{dt} = 1/P$ as the mean detrapping time, and assuming that the carrier lifetime is longer than the transit time so that the retrapping and recombination processes may be ignored, the current due to charge carriers liberated from the traps within dx at the distance x from the collecting electrode inside the specimen is given by

$$d\,J(t) = \frac{q n_t}{\tau_{dt}} \frac{x\,dx}{d} \tag{5.155}$$

By solving these equations, Garofano and Corazzari [1975] have derived the following expression for the decay J–t characteristics

$$J(t) = \frac{1}{2} q n_{t0} dv \exp\left[-\frac{\Delta E}{kT}\right] \exp\left\{-\left[v \exp\left(-\frac{\Delta E}{kT}\right)\right]t\right\}$$

$$= \frac{q n_{t0} d}{2\tau_{dt}} \exp\left[-t/\tau_{dt}\right] \tag{5.156}$$

They have also used this method to determine the trapping levels in anthracene crystals. Their specimens were between 1 and 3 mm in thickness, provided with silver electrodes. An average field of 2000–6000 V cm^{-1} was applied and the specimen was excited for a fairly short time (about 30–60 sec), and then the excitation was removed and the current

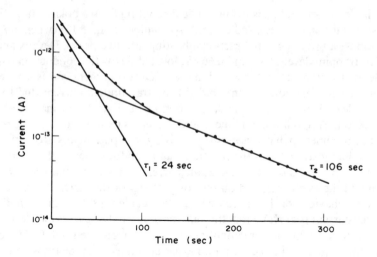

FIG. 5.21. Typical isothermal current decay curve for melt-grown anthracene crystals. Thickness: 2.12 mm, photoexcitation time: 60 sec, applied field: 2750 V cm^{-1} (illuminated electrode positive); temperature: 250°K; circuit overall time constant: 2.5 sec. [After Garofano and Corazzari 1975.]

decay was recorded. A typical ICD curve is shown in Fig. 5.21. The current decay is exponential and is characterized by two time constants τ_1 and τ_2. τ_1 is a function of T, and according to eq. (5.156) the exponential dependence of τ_1 on T gives directly ΔE and v. For the melt-grown anthracene crystals, they found $\Delta E = 0.7$ eV and $v = 6.5 \times 10^{14}$ sec^{-1}, which are in good agreement with those obtained using other methods [Aprilest *et al.* 1971, Garofano 1974]. They attributed the origin of τ_2 either to the existence of trap distributions or to electrode injection phenomena.

5.4.4. Non-extrinsic electric conduction

For the one-carrier (electron) injection in solids with shallow traps, the current in the ohmic region ($V < V_\Omega$, cf. Section 3.2.1 B) is given by

$$J_\Omega = qn_0\mu_n\frac{V}{d} \tag{5.157}$$

and that in the SCL region by

$$J_{sc} = \frac{9}{8}\varepsilon\mu_n\theta_n\frac{V^2}{d^3} \tag{5.158}$$

where θ_n is the fraction of total electrons which are free $[= n/(n+n_t)]$. The transition voltage at which $J_\Omega = J_{sc}$ is [cf. eq. (3.25)]

$$V_\Omega = \frac{8}{9}\frac{qn_0d^2}{\theta_n\varepsilon} \tag{5.159}$$

It is clear that both J_Ω and J_{sc} are thermally activated and their corresponding activation energies are contained in n_0 and θ_m, and V_Ω has a thermal activation energy equal to the difference between the activation energies in the ohmic and SCL regions. Schmidlin and Roberts [1968, 1974] and Roberts and Schmidlin [1969] have shown that measurements of thermal activation energies for both ohmic and SCL conduction over a wide temperature range enable the determination of localized levels in semi-insulators, and measurements of the activation energies for ohmic conduction alone are rarely definitive for wide band gap materials. In general, the activation energy in the ohmic region ΔE_Ω is equal to that in the space charge region ΔE_{sc} and V_Ω is independent of temperature only when the material is extrinsic, i.e. when the material has either a dominant donor level or a dominant acceptor level. For wide band gap materials, particularly when compensated, the dominant levels which contribute overwhelmingly to partition functions of the electrons and holes may be some localized levels within the energy band gap. The locations and concentrations of such localized states, together with the concentration of free electrons contributed by excess donors, or with the concentration of free holes contributed by excess acceptors, determine the location of the Fermi level. If the excess donor concentration or the excess acceptor concentration is small as compared with the concentration of electrons excited from the dominant hole level to the dominant electron level, as shown in Fig. 5.22, then the Fermi level is located between the two dominant levels in such a way as to make the concentration of electrons in the dominant electron level equal to the concentration of holes in the dominant hole level. This condition is called the non-extrinsic condition [Roberts and Schmidlin 1969]. It is similar to the familiar intrinsic condition when the dominant levels are the transport bands (conduction and valence bands). However, for narrow band gap semiconductors with $E_g < 40\,kT$, the dominant levels tend to be the transport bands, and in such narrow band gap materials the non-extrinsic condition may not occur.

From Fig. 5.22 we can define the total donor concentration above the Fermi level E_F as

$$N_d = N_c \exp[-(E_c - E_F)/kT] + \sum_i N_i \exp[-(E_i - E_F)/kT]$$

$$= Q_n X \tag{5.160}$$

$$\begin{array}{ll}
\rule{4cm}{0.4pt} & E_c \\[1em]
\text{- - - - - - - -} & E_i \\[0.3em]
\uparrow \Delta E_i & \\[0.3em]
\text{- - - - - - - -} & \phi \\[0.3em]
\uparrow \Delta E_j & \\[0.3em]
\text{- - - - - - - -} & E_j \\[1em]
\rule{4cm}{0.4pt} & E_v
\end{array}$$

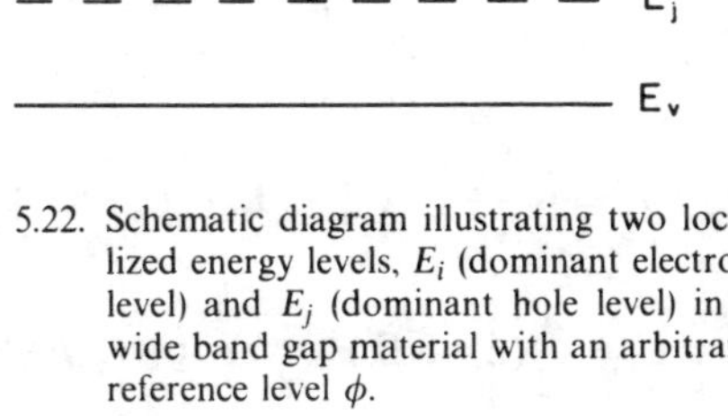

FIG. 5.22. Schematic diagram illustrating two localized energy levels, E_i (dominant electron level) and E_j (dominant hole level) in a wide band gap material with an arbitrary reference level ϕ.

and the total acceptor concentration below the Fermi level E_F as

$$N_a = N_v \exp[-(E_F - E_v)/kT] + \sum_j N_j \exp[-(E_F - E_j)/kT]$$
$$= Q_p X^{-1} \tag{5.161}$$

where

$$X = \exp[(E_F - \phi)/kT] \tag{5.162}$$

$Q_n =$ the partition function for electrons

$$= N_c \exp[-(E_c - \phi)/kT] + \sum_i N_i \exp[-(E_i - \phi)/kT]$$

$$\simeq N_m \exp[-\Delta E_m/kT] \tag{5.163}$$

$Q_p =$ the partition function for holes

$$= N_v \exp[-(\phi - E_v)/kT] + \sum_j N_j \exp[-(\phi - E_j)/kT]$$

$$\simeq N_q \exp[-\Delta E_q/kT] \tag{5.164}$$

ΔE_m and ΔE_q are measured from the reference level ϕ, $\Delta E_m = E_m - \phi$, and $\Delta E_q = \phi - E_q$. The criterion to distinguish between the extrinsic and non-extrinsic conduction for an *n*-type material is as follows [Schmidlin and Roberts 1974]:

(i) Extrinsic conduction:

$$(N_d - N_a)^2 \gg 4Q_n Q_p \tag{5.165}$$

$$n_0 = \frac{(N_d - N_a)N_c}{N_m} \exp[-(E_c - E_m)/kT] \tag{5.166}$$

$$\theta_n = \left(\frac{N_c}{N_m}\right) \exp[-(E_c - E_m)/kT] \tag{5.167}$$

$$E_F = E_m + kT \ln[(N_d - N_a)/N_m] \tag{5.168}$$

This indicates that the activation energies $\Delta E_\Omega = \Delta E_{sc}$. It is easy to extend this case for a *p*-type material.

(ii) Non-extrinsic conduction:

$$(N_d - N_a)^2 \ll 4Q_n Q_p \tag{5.169}$$

$$n_0 = (N_q/N_m)^{1/2} N_c \exp\left\{-\frac{\left[(E_c - E_m) + \frac{1}{2}(E_m - E_q)\right]}{kT}\right\} \tag{5.170}$$

$$\theta_n = \left(\frac{N_c}{N_m}\right) \exp[-(E_c - E_m)/kT] \tag{5.171}$$

$$E_F = \frac{1}{2}(E_m + E_q) + \frac{1}{2}kT \ln(N_q/N_m) \tag{5.172}$$

This indicates that the activation energies $\Delta E_\Omega \neq \Delta E_{sc}$. It is easy to extend this condition for a *p*-type material.

The extrinsic condition generally means the condition for which the Fermi level depends completely on the excess donor concentration $N_d - N_a$ (or on the excess acceptor concentration $N_a - N_d$); while the non-extrinsic condition can be considered to include "intrinsic" condition (both transport bands are dominant levels, mainly for narrow band gap materials), "half-intrinsic" condition (one transport band is one of dominant levels), and "compensated" condition (neither transport band is dominant, i.e. the dominant levels are within the band gap). It should be noted that the carrier mobility μ_n (or μ_p) is generally temperature dependent. In order to achieve well-defined activation energies in different temperature regions from the plot of log J as a function of $1/T$, we may have to remove any weak temperature dependence of carrier mobility or effective density of states by multiplying J by T to an appropriate power.

Figure 5.23 shows the typical J–V characteristics of HgS crystals at various temperatures and the activation energies in the ohmic and SCL regions and that of V_Ω. That the activation energy for V_Ω is approximately equal to $\Delta E_\Omega - \Delta E_{sc}$ indicates that the material is non-extrinsic. From these results and from the voltage required to cause the trap-filled limit V_{TFL} [cf. eq. (3.32)], Roberts and Schmidlin [1969] have estimated the following parameters for HgS: $E_c - E_m = 0.62$ eV for the dominant electron level, $E_c - E_q = 1.08$ eV for the dominant hole level, $N_m = 10^{13}$ cm^{-3}, $N_q = 10^{16}$ cm^{-3}, and $\mu_n = 10$ cm^2/V-sec, and the material is *n*-type.

It should be noted that in the present section we do not intend to review the techniques for the measurements of trap parameters which are beyond the scope of this book. We include TSC, TSDC, IC, and non-extrinsic conduction because they are part of conduction phenomena. For measurement techniques the reader is referred to some review articles [Bullis 1968, Milnes 1973].

5.5. CURRENT TRANSIENT PHENOMENA

So far we have considered only steady state injection currents. Prior to the attainment of the steady state, a variety of current transient phenomena may occur immediately after the application of an electric field. Such phenomena may provide useful information about transport, trapping, recombination, and photogeneration processes in solids. Many investigators [Spear 1957, 1960, 1969; Wertheim 1958; Kepler 1960, 1962; Many *et al.* 1961, 1962; Helfrich and Mark 1962; Helfrich and Schneider 1966; Lampert and Mark 1970] have studied these transient phenomena. On the basis of the total charge Q injected into the solid, the current transient phenomena can be classified into: (a) the space-charge-free (SCF) transient for which Q is much smaller than that required to influence significantly the electric field near the injecting electrode, $Q \ll 2\varepsilon F$; (b) the SCL transient for which Q is large enough to create a charge reservoir near the injecting electrode, $Q \gg 2\varepsilon F$; and (c) the space-charge-perturbed (SCP) transient for which the values of Q are intermediate between (a) and (b). The current transients can be induced by the application of a voltage pulse, electron beam bombardment, corona discharge, or light pulse, etc. In this section we shall review briefly the theory and applications of the current transient phenomena.

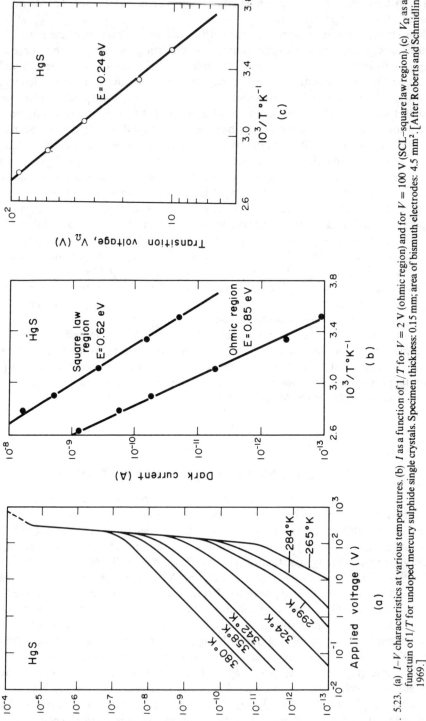

FIG. 5.23. (a) *I–V* characteristics at various temperatures. (b) *I* as a function of $1/T$ for $V = 2$ V (ohmic region) and for $V = 100$ V (SCL–square law region). (c) V_Ω as a functuin of $1/T$ for undoped mercury sulphide single crystals. Specimen thickness: 0.15 mm; area of bismuth electrodes: 4.5 mm². [After Roberts and Schmidlin 1969.]

5.5.1. One-carrier (single) planar injection transient currents

Supposing that a step function voltage of the form

$$V(t) = 0 \quad \text{when} \quad t < 0$$

$$V(t) = V \quad \text{when} \quad t \geq 0$$

is applied to a solid specimen provided with an electron-injecting contact at $x = 0$ as cathode, then the current transient begins at $t = 0$ and its behaviour is governed by the current flow equation

$$J(x, t) = J_c(x, t) + \varepsilon \frac{\partial F(x, t)}{\partial t}$$

$$= q\mu_n n(x, t) F(x, t) - q D_n \frac{\partial n(x, t)}{\partial x} + \varepsilon \frac{\partial F(x, t)}{\partial t} \qquad (5.173)$$

the Poisson equation

$$\frac{\partial F(x, t)}{\partial x} = \frac{q}{\varepsilon} [n(x, t) - n_0 + n_t(x, t) - n_{t0}] \qquad (5.174)$$

the rate equation [cf. eq. (4.66)]

$$\frac{\partial n_t(x, t)}{\partial t} = C_n \{ n(x, t) [N_n - n_t(x, t)] - N_c \exp[-(E_c - E_t)] \cdot n_t(x, t) \}$$

$$= C_n \{ n(x, t) [N_n - n_t(x, t)] - n_1 n_t(x, t) \}$$

$$= \tau_n^{-1} [n(x, t) - \theta_0 n_t(x, t)] \qquad (5.175)$$

and the continuity equation

$$\frac{\partial J(x, t)}{\partial x} = \frac{\partial J_c(x, t)}{\partial x} + \varepsilon \frac{\partial}{\partial x} \left[\frac{\partial F(x, t)}{\partial t} \right] = 0$$

or

$$-\frac{1}{q} \frac{\partial J_c(x, t)}{\partial x} = \frac{\partial n(x, t)}{\partial t} + \frac{\partial n_t(x, t)}{\partial t} \qquad (5.176)$$

where n_0 and n_{t0} are, respectively, the free electron and trapped electron densities under thermal equilibrium and they are assumed to be uniformly distributed throughout the specimen, C_n is the electron capture coefficient, and N_n is the density of electron traps in a discrete set of localized states located at E_t in the band gap. τ_n is the trapping time defined as the mean free time of a mobile free carrier before it is trapped, which is

$$\tau_n = [C_n(N_n - n_{t0})]^{-1} \qquad (5.177)$$

and n_1 and θ_0 are defined by

$$n_1 = N_c \exp[-(E_c - E_t)/kT] \qquad (5.178)$$

$$\theta_0 = n_0/n_{t0} = n_1/(N_n - n_{t0}) \qquad (5.179)$$

and $(N_n - n_{t0}) \gg n_t(x, t) - n_{t0}$. Equation (5.175) may be extended for any forms of trap distribution provided that the total trapping rate is larger than the total detrapping rate.

It is obvious that an analytical solution of eqs. (5.173)–(5.176) cannot be obtained without making some simplifying assumptions. By assuming $n \gg n_0$ and $n_t \gg n_{t0}$, integration of eq. (5.173) yields

$$J(t) = \frac{\varepsilon\mu_n}{2d}[F_a^2(t) - F_c^2(t)] - \frac{q\mu_n}{d}\int_0^d n_t(x, t)F(x, t)dx$$

$$-\frac{\varepsilon D_n}{d}\left\{\left[\frac{\partial F(x, t)}{\partial x}\right]_{x=d} - \left[\frac{\partial F(x, t)}{\partial x}\right]_{x=0}\right\}$$

since

$$+\frac{q D_n}{d}\left\{[n_t(x, t)]_{x=d} - [n_t(x, t)]_{x=0}\right\} \tag{5.180}$$

$$(\varepsilon/d)\frac{\partial}{\partial t}\int_0^d F(x, t)dx = (\varepsilon/d)\frac{\partial V}{\partial t} = 0$$

by assumption, where $F_a(t) = F(d, t)$ and $F_c(t) = F(0, t)$ are, respectively, the fields at the anode and at the cathode. Integration of eq. (5.174) yields the relation between $F_c(t)$ and $F_a(t)$ in terms of the total charge $Q(t)$ per unit area in the specimen at time t.

$$F_a(t) = F_c(t) + Q(t)/\varepsilon \tag{5.181}$$

In the following we shall consider three major cases:

(A) The space-charge-free (SCF) transient

CASE I: IN THE ABSENCE OF TRAPS AND DIFFUSION

For this case eq. (5.180) becomes

$$J(t) = \frac{\varepsilon\mu_n}{2d}[F_a^2(t) - F_c^2(t)]$$

$$= \frac{\mu_n Q(t)}{2d}[2F_a(t) - Q(t)/\varepsilon]$$

$$= \frac{\mu_n Q(t)}{2d}[2F_c(t) + Q(t)/\varepsilon] \tag{5.182}$$

and $Q \ll 2\varepsilon F_c$ by assumption. This implies that Q is smaller than the maximum charge that the specimen can store (the maximum charge $= V\varepsilon/d$), and that this case corresponds to the case of a blocking contact cathode and a weak light pulse to produce the injected charge. From eq. (5.181), we have

$$F_a \simeq F_c \simeq F_{av} = \frac{V}{d} \tag{5.183}$$

Thus, eq. (5.182) becomes

$$J = \frac{\mu_n Q F_c}{d} = \frac{\mu_n Q V}{d^2} = \text{constant}$$

$$= \frac{Q}{t_{t0}} \tag{5.184}$$

where t_{t0} is the SCF electron transit time.

The simplest example for this case is the measurement of the time-to-flight of carriers injected at the cathode for the determination of carrier mobilities. The specimen is initially subject to a constant applied voltage V between two non-injecting electrodes in the dark; and at $t = 0$ a light pulse is applied at the cathode to produce a sheet of charge Q which drifts from $x = 0$ at $t = 0$ to $x = d$ at $t = t_{t0}$ so that the external current rises suddenly at $t = 0$ and reaches a constant value following eq. (5.184) at $t = t_{t0}$. It should be noted that the so-called "sheet of charge" has a finite thickness due to the finite duration of the light pulse. The basic experimental arrangement using a light pulse is shown in Fig. 5.24. Kepler [1960, 1962] has used this technique to measure the charge carrier mobilities in anthracene crystals. The electrodes in this arrangement do not act like normal electrodes but act like the plates of a parallel plate capacitor, no physical contact to the crystal being necessary. With the absence of a third electrode around the middle of the crystal, a typical SCF current transient produced by a light pulse of 1 to 2 μsec duration, incident on the crystal specimen of 2 mm in thickness through one of the transparent electrodes, is shown in Fig. 5.25(a). Kepler used the wavelengths of the light shorter than the absorption edge so that it would not penetrate the specimen more than 10^{-3} mm which is negligibly small compared with the specimen thickness. In Fig. 5.25(b) the slow rise of the transient is caused by the time constant of the measuring circuit, while the long tail is due to the spreading of the charge carrier sheet by diffusion. The times $t = 0$, t_1, t_2, t_3 correspond to the positions of the charge sheet along the x-axis shown in Fig. 5.25(c). The third electrode around the centre of the specimen shown in Fig. 5.24 is used to eliminate any possible surface leakage current, or to check whether the charge carriers under observation are moving on the surface of the specimen or whether the charge carriers are moving in the right direction. Using this technique, Kepler has determined the drift mobilities of electrons and holes in anthracene crystals to be of the order from 0.3 to 3.0 cm^2/V-sec, at room temperature depending on the crystal orientation. The carrier mobilities increase with decreasing temperature.

A similar technique has been reported earlier by Brown [1955], Le Blanc [1960], and Spear [1957, 1969]. In general, the concentration of imperfection centres is higher near

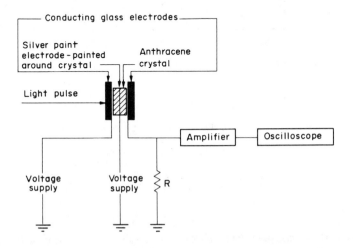

Fig. 5.24. Basic experimental arrangement for the time-to-flight measurements using light excitation. [After Kepler 1960.]

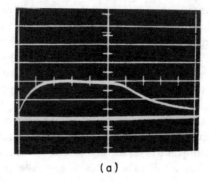

(a)

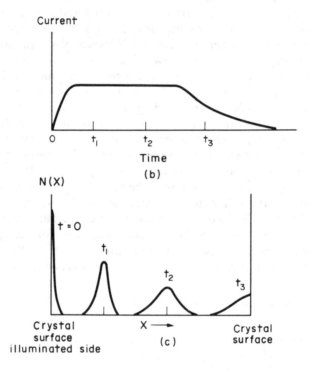

F IG. 5.25. Typical SCF current transient in anthracene crystals:
(a) Oscillogram, each division on the horizontal axis
represents 50 μsec; (b) interpretation of the observed
current transient; (c) the positions of the charge sheet
along the x-axis in the crystal specimen corresponding to $t = 0, t_1, t_2$, and t_3. [After Kepler 1962.]

a surface, and these, coupled with any other possible surface states, will considerably
shorten the carrier lifetime in the generation region. It is not easy to find a light source
which can produce an intense flash of duration much shorter than the carrier transit

time and yet generate sufficient electron–hole pairs within less than 1 μm from the illuminated surface. This problem becomes more difficult for wide band gap materials. Spear [1957, 1969] has suggested the use of electron beam excitation instead of optical excitation for the following advantages: (i) the intensity of the beam is high and its duration can be made sufficiently short for the generation of sufficient carriers, (ii) the depth of the generation region below the top surface can be varied to any desired value simply by adjusting the accelerating potential, (iii) it does not involve the absorption in the electrode, and (iv) for wide band gap materials for which the optical excitation is not feasible, electron beam or α-particle excitation is the only means for generating electron–hole pairs. Figure 5.26 shows the basic experimental arrangement used in drift mobility measurements under electron beam excitation. This system can provide a single short pulse or a series of pulses at repetition rates of 50 or 100 pulses per sec. The electron beam is focused by the magnetic lens.

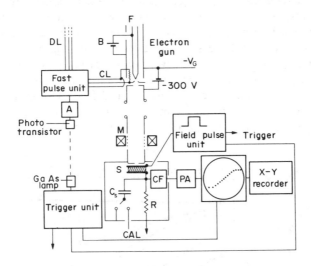

Fig. 5.26. Basic experimental arrangement for the drift mobility measurements using electron beam excitation. *F*, Hairpin filament; *DL*, delay line; *CL*, connecting line; *M*, magnetic lens; *S*, specimen; *CF*, cathode follower; *PA*, pre-amplifier. [After Spear 1969.]

CASE II: IN THE ABSENCE OF DIFFUSION BUT WITH TRAPS

The presence of traps may limit the applicability of the transient method for drift mobility measurements. Supposing that μ_n is the mobility of the free electron carriers whose concentration is n. But the trapped carrier concentration is n_t so that the measured drift mobility μ' is given by

$$n\mu_n = (n + n_t)\mu_n' \tag{5.185}$$

If the traps are shallow electron traps located at a single discrete level E_t, then in thermal equilibrium we have

$$\frac{n_t}{n} = \frac{N_n}{N_c}\exp[(E_c - E_t)/kT] \tag{5.186}$$

From eqs. (5.185) and (5.186) we obtain

$$\mu'_n = \mu_n\left\{1 + \frac{N_n}{N_c}\exp[(E_c - E_t)/kT]\right\}^{-1} \tag{5.187}$$

At very high temperatures $\mu'_n \simeq \mu_n$ and at low temperatures

$$\mu'_n \simeq \frac{N_c\mu_n}{N_n}\exp[-(E_c - E_t)/kT] \tag{5.188}$$

Similar expressions are applicable to the case for hole carriers except that n, n_t, μ_n, μ'_n and $(E_c - E_t)$ are replaced with p, p_t, μ_p, μ'_p, and $(E_t - E_v)$, respectively. Typical results showing the measured hole mobility as a function of temperature and concentration of doped naphthacene impurities in anthracene crystals (trap concentration) are given in Fig. 5.27. These results are in reasonable agreement with eq. (5.187).

For shallow traps the average time for a trapped carrier to stay in the trap before thermal release (the trap release time) $\tau_{tr} \ll t_t$ (transit-time), and the lifetime $\tau \ll t_t$ (depending on the specimen thickness and applied field). However, for deep traps $\tau_{tr} \gg t_t$; thus, the effect of traps on the measurements of drift mobilities depends on whether $\tau < t_t$ or $\tau > t_t$. It is obvious that the presence of traps has little effect if $\tau > t_t$,

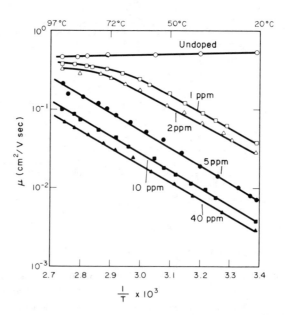

FIG. 5.27. The drift hole mobility in undoped and naphthacene-doped anthracene crystals as a function of temperature. The results show $E_t - E_v = 0.43$ eV and that the mobilities vary approximately as the reciprocal of the impurity content (trap concentration). [After Hoesterey and Letson 1963.]

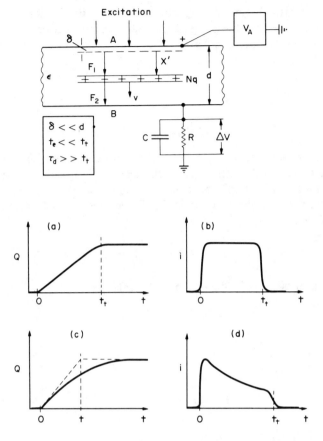

F<small>IG</small>. 5.28. The top diagram illustrates the principle of drift mobility measurements on a high resistivity material. A and B are metallic electrodes, V_A is the applied voltage with the positive polarity at A with respect to ground, t_e is the duration of the excitation pulse, t_t is the transit time, τ_d is the dielectric relaxation time, Nq is the narrow charge sheet due to N carriers generated at the surface, F_1 and F_2 are, respectively, the electric field behind and in front of the charge sheet. Typical pluse shapes in these time-to-flight measurements with the carriers generated close to the top electrode, $\delta \ll d$, are: (a) integrated signal, $CR \gg t_t$; and (b) current pulse, $CR \ll t_t$. In both (a) and (b) deep traps are absent. (c) and (d) show the corresponding pulse shapes for the cases with deep traps. [After Spear 1969.]

and makes the measurements impossible if $\tau < t_t$ for deep traps. With a short lifetime $\tau \simeq t_t$, for deep traps, the concentration of free carriers will gradually decrease during the transit from one electrode to the other, and typical transient responses for this case are shown in Fig. 5.28(c) and (d).

(B) The space-charge-limited (SCL) transient

CASE I: IN THE ABSENCE OF TRAPS AND DIFFUSION

For this case the injected charge Q is large enough to create an electron charge carrier reservoir at the cathode (or at the anode for holes) due either to intense optical excitation at the cathode or to strong carrier injection from the ohmic contact cathode. Since $Q \gg 2\varepsilon F_c$ for this case, we have

$$F_c(t) = 0 \tag{5.189}$$

and eq. (5.180) becomes

$$J(t) = \frac{\varepsilon \mu_n}{2d} F_a^2(t) \tag{5.190}$$

Supposing that t_1 is the time required for the leading front of the injected charge to arrive at the anode (the transit time of the leading front), then during the period $0 \le t \le t_1$ the total current is simply the displacement current at the anode because the conduction current J_c which depends on the charge collected at the anode is practically equal to zero. Thus

$$J(t) = \varepsilon \frac{\partial F_a(t)}{\partial t} \tag{5.191}$$

The combination of eqs. (5.190) and (5.191) yields

$$\frac{dF_a}{dt} = \frac{\mu_n}{2d} F_a^2 \tag{5.192}$$

Using the boundary condition, $F(x, 0) = V/d$ since there is no injected charge at $t = 0$, and the following relations

$$t_{t0} = d^2/\mu_n V, \qquad Q_c = \frac{\varepsilon V}{d} \tag{5.193}$$

the solution of eq. (5.192) yields

$$\varepsilon F_a(t) = Q(t) = \frac{Q_c}{1 - (t/2t_{t0})} \tag{5.194}$$

Hence we obtain

$$J(t) = J_0/[1 - (t/2t_{t0})]^2 \tag{5.195}$$

where

$$J_0 = J(0) = \frac{\varepsilon \mu_n V^2}{2d^3} \tag{5.196}$$

Since the field at the leading front of the charge at time $t < t_1$ can be considered to be the same as the field at the anode at time t, the value of t_1 can be determined by the following relation

$$d = \int_0^{t_1} \mu_n F_a(t)\, dt = (2t_{t0}\mu_n V/d) \ln\left[1 - (t_1/2t_{t0})\right]^{-1} \tag{5.197}$$

with the aid of eq. (5.194). From eq. (5.193) this gives

$$t_1 = 2(1 - e^{-1/2})t_{t0} \simeq 0.786\, t_{t0} \tag{5.198}$$

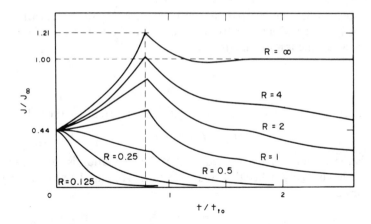

FIG. 5.29   The theoretical SCL current transients plotted in current density as a function
of time for various trapping times τ_n for insulating crystals in which θ_0
$= n_0/n_{t0} = 0$. $R = \tau_n/t_{t0}$, t_{t0} is the space charge free transit time, J_∞ is the
steady state current without trapping, $R = \infty$ corresponds to the absence of
trapping. [After Many and Rakavy 1962.]

Letting $J_1 = J(t_1)$ and $J_\infty = J(\infty) = 9\varepsilon\mu_n V^2/8d^3 =$ the trap-free steady state SCL
current, eqs. (5.195) and (5.198), yield

$$J_1/J_0 = e = 2.72, \qquad J_1/J_\infty = 4e/9 = 1.21 \qquad (5.199)$$

The curve for $R = \tau_n/t_{t0} = \infty$ shown in Fig. 5.29 corresponds to the case of the absence
of trapping. It can be seen that $J_0/J_\infty = 0.44$ occurs at $t/t_{t0} = 0$ and $J_1/J_\infty = 1.21$
occurs at $t/t_{t0} = 0.786$. The time rate of change of the current at $t = t_1^-$ just before it
reaches the peak is

$$\left.\frac{dJ}{dt}\right|_{t=t_1^-} = \frac{e^{1/2}J_1}{t_{t0}} = 1.65 J_1/t_{t0} \qquad (5.200)$$

and that at $t = t_1^+$ just after it passes the peak is

$$\left.\frac{dJ}{dt}\right|_{t=t_1^+} = \frac{1-2e^{-1/2}}{1-e^{-1/2}}\left(\left.\frac{dJ}{dt}\right|_{t=t_1^-}\right) = -0.90 J_1/t_{t0} \qquad (5.201)$$

At $t = t_1$, when the leading front reaches the anode ($x = d$), the specimen contains the
maximum amount of space charge and $J(t)$ attains its peak value. Thereafter $J(t)$ decays
towards J_∞ because for $t > t_1$ more space charge leaves the specimen than that injected
into it, and the space charge distribution at $t = t_1$ relaxes towards its steady state
configuration with the relaxation time inversely proportional to V and of the order of
t_1. Since at $t = t_1$ the specimen contains more space charge than it can hold under
steady state conditions, an undershot occurs in the decay due to the inhibition of charge
injection in the vicinity of t_1 because of the overshot of charge already present at the
time. This may cause the oscillations of the decaying current [Many and Rakavy 1962,
Baron *et al.* 1966, Schilling and Schachter 1967].

   Supposing that a light flush of duration much shorter than the transit time produces a very thin sheet (much thinner than the specimen thickness) of charge carriers at $x = 0$ and at $t = 0$, then the field is V/d in front of the sheet and zero behind it. Thus, the width of the sheet as a function of time is

$$S(t) = \frac{V\mu_n t}{d} \tag{5.202}$$

and the current density of such sheet current is given [Helfrich 1967] by

$$J(t) = J_0 \exp(t/t_{t0})/2 \tag{5.203}$$

where J_0 is the same as J_0 given by eq. (5.196). This equation is valid as long as the leading front has not yet reached the anode. The time t_2 required for the leading front to reach the anode is

$$t_2 = 0.792 t_{t0} \tag{5.204}$$

The current transient has a cusp at $t = t_2$. The time rate of change of the current at $t = t_2^-$ just before it reaches the peak is

$$\left. \frac{dJ}{dt} \right|_{t = t_2^-} = \frac{J(t_2)}{t_{t0}} \tag{5.205}$$

and that at $t = t_2^+$ just after it passes the peak is

$$\left. \frac{dJ}{dt} \right|_{t = t_2^+} = -2.3 J(t_2)/t_{t0} \tag{5.206}$$

Theoretically, another transit time is required for all the space charge to leave the specimen. However, it is interesting to note that the initial charge distribution has a negligible effect on the general shape of the transient provided that it is confined in a sheet very thin compared with the specimen thickness, and that the width of the charge sheet increases with time and becomes almost equal to the specimen thickness when the leading front arrives at the collective electrode which is anode for this case, and therefore the transients of SCL current and SCL sheet current are so similar [Helfrich 1967].

For details about the derivation of the above equations, the reader is referred to the work of Many et al. [1961, 1962], Helfrich and Mark [1962], Schwartz and Hornig [1965], and Helfrich [1967].

CASE II: IN THE ABSENCE OF DIFFUSION BUT WITH TRAPS

The SCL transient in a solid with traps has been studied by several investigators [Many and Rakavy 1962, Blakney and Grunwald 1967, Batra and Seki 1970]. Many and Rakavy have computed numerically $J(t)$ as a function of time for a solid containing traps, and their results are shown also in Fig. 5.29. The smaller the value of $R = \sigma_n/t_{t0}$, the faster is the current decay and the higher is the value of t_1. This implies that the concentration of trapped carriers near $x = 0$ increases with decreasing value of R, giving rise to a correspondingly lower field under which the leading front moves. It is important to note that there always exists a current cusp in the presence of trapping provided that the trapping is not too fast ($R = \sigma_n/t_{t0} > 0.5$), that the peak of the current occurs at about the same time t_1, and that the initial current $J_0 = J(t = 0)$ is independent of R, i.e. independent of the trapping time.

In the following we shall follow the analysis of Batra and Seki [1970] by considering the traps with a uniform concentration N_n and located at a single discrete energy level E_t and assuming that: (i) prior to optical excitation the specimen is electrically neutral, (ii) the equilibrium concentration of free n_0 and trapped n_{t0} charge carriers is negligibly small, (iii) the injection is accomplished by illuminating the specimen through the transparent cathode at $x = 0$, (iv) the sheet of charge carriers produced at $x = 0$ drifts towards the collecting anode at $x = d$ under an applied field, and (v) $N_n > n_t$. For this case the mean free time of a mobile cirrier (the trapping time) is

$$\tau_n = 1/C_n N_n \qquad (5.207)$$

and the rate at which the concentration of trapped carriers changes with time is given by

$$\frac{\partial n_t(x, t)}{\partial t} = \frac{n(x, t)}{\tau_n} - \frac{n_t(x, t)}{\tau_{tr}}$$

$$= \tau_n^{-1}[n(x, t) - \theta n_t(x, t)] \qquad (5.208)$$

where

$$\theta = \frac{\tau_n}{\tau_{tr}}$$

$$= \frac{n}{n_t} \text{ (in thermal equilibrium)} \qquad (5.209)$$

τ_{tr} is the mean time of dwell for a carrier in the trap (the trap release time or $1/\tau_{tr}$ is the probability per unit time for thermal release of a trapped carrier into the conduction band). The parameter θ measures the extent of detrapping.

By introducing the following parameters to reduce the quantities to dimensionless ones for simplicity and convenience

$$\begin{matrix} x' = x/d, & \rho = qnd^2/\varepsilon V \\ t' = t/t_{t0}, & F' = F/F_{av} = Fd/V \\ R = \tau_n/t_{t0}, & j = J dt_{t0}/\varepsilon V \end{matrix} \qquad (5.210)$$

eqs. (5.173), (5.174), (5.176), and (5.208) reduce, respectively, to

$$j(x', t') = j_c(x', t') + \frac{\partial F'(x', t')}{\partial t'} \qquad (5.211)$$

$$\frac{\partial F'(x', t')}{\partial x'} = \rho(x', t') + \rho_t(x', t') \qquad (5.212)$$

$$\frac{\partial j_c(x', t')}{\partial x'} = -\left(\frac{\partial}{\partial t'}\right)[\rho(x', t') + \rho_t(x', t')] \qquad (5.213)$$

and

$$\frac{\partial \rho_t}{\partial t'} = R^{-1}[\rho(x', t') - \theta \rho_t(x', t')] \qquad (5.214)$$

Ignoring the contribution by diffusion, j_c can be written as

$$j_c(x', t') = \rho(x', t')F'(x', t') \tag{5.215}$$

Using the following boundary conditions:

$$\int_0^1 [\rho(x', t') + \rho_t(x', t')]dx' = Q_0, \quad t'_1 \geq t' \geq 0$$

$$\left.\begin{array}{ll} \int_0^1 F'(x', t')dx' = 1 & x' > 0 \\ F'(x', 0) \quad = 1 & x' \geq 0 \\ \rho(x', 0) \quad = 0 & x' > 0 \\ \rho_t(x', 0) \quad = 0 & x' \geq 0 \end{array}\right\} \tag{5.216}$$

and making the weak trapping approximation ($\rho \gg \rho_t$), Batra and Seki [1970] have solved eqs. (5.211)–(5.215) for $j(t')$ which is given by

$$j(t') = \frac{Q_0(1 - Q_0/2)}{Q_0 + (1 + \theta)/R}$$

$$\times \{(Q_0 + \theta/R)\exp(Q_0 t') + (1/R)\exp[-(1 + \theta)t'/R]\} \tag{5.217}$$

which is valid for $0 < t' < t'_1$, where $t'_1 = t_1/t_{t0}$ and t_1 is the transit time required for the leading front to arrive at the collecting electrode. The value of t'_1 can be determined by the following equation

$$\exp(Q_0 t'_1) + [Q_0^2/(2 - Q_0)]t'_1 = (2 + Q_0)/(2 - Q_0) \tag{5.218}$$

To solve both eqs. (5.217) and (5.218) we need to know Q_0 which is $0 \leq Q_0 \leq 1$. In the following we shall consider two limiting cases:

(a) Small signal mode

For this case, $Q_0 \ll 1$ and eq. (5.217) becomes

$$j(t') = Q_0(1 + \theta)^{-1}\exp[-(1 + \theta)t'/R] + Q_0\theta/(1 + \theta) \tag{5.219}$$

In the absence of trapping and detrapping, $\theta \to 0$ and $R \to \infty$, then $j(t') = Q_0$. In dimensioned units it is

$$J = q\mu_n V/d \tag{5.220}$$

which is independent of time and behaves according to Ohm's law as expected. For the calculation of the transit time, we use eq. (5.218) which yields $t'_1 = 1$ for this case. In dimensioned units it is

$$t_1 = t_{t0} = d^2/\mu_n V \tag{5.221}$$

(b) SCL mode

For this case, we set $Q_0 = 1$. Then eq. (5.217) becomes

$$j(t') = [2(R+\theta+1)]^{-1}\{(R+\theta)\exp(t') + \exp[-\{(1+\theta)/R\}t']\} \quad (5.222)$$

In the absence of trapping, $R \to \infty$, eq. (5.222) reduces to

$$j(t') = \frac{1}{2}\exp(t') \quad (5.223)$$

In the presence of deep traps, $\theta = 0$, eq. (5.222) reduces to

$$j(t') = [1/2(R+1)][R\exp(t') + \exp(-t'/R)] \quad (5.224)$$

The transit time can easily be determined by setting $Q_0 = 1$ in eq. (5.218). Figure 5.30(a) shows t_1' as a function of Q_0 based on eq. (5.218), (b) shows $j(t')$ as a function of t' for various R and fixed $\theta = 0$, and (c) shows $j(t')$ as a function of t' for various θ and fixed $R = 2$.

A typical oscillogram of SCL current transients in anthracene crystals with pronounced trapping and with an aqueous solution of KI and I_2 as ohmic contact for hole injection is shown in Fig. 5.31. A similar transient behaviour in iodine crystals has also been reported by Many *et al.* [1962].

CASE III: INCLUDING DIFFUSION BUT WITHOUT TRAPS

It is obvious that the two simplifying assumptions: (1) the diffusion term is neglected, and (2) the electric field at the injecting contact is zero, may not be justified in the analysis of current transients, particularly for insulators provided with a good ohmic injecting contact because there always exists a reverse electric field in the insulator due to carrier injection from the contact even when no voltage is applied. Several investigators have considered the effect of diffusion [Rosen 1966, 1967; Schilling and Schachter 1967; Boudry 1968; Rosental and Lember 1970; Silver 1974]. The equations governing the transient behaviour with the diffusion but without trapping effects are repeated here for convenience:

$$J(x, t) = q\mu_n n(x, t)F(x, t) - qD_n\frac{\partial n(x, t)}{\partial x} \quad (5.225)$$

$$\frac{\partial F(x, t)}{\partial x} = \frac{qn(x, t)}{\varepsilon} \quad (5.226)$$

$$-\frac{1}{q}\frac{\partial J(x, t)}{\partial x} = \frac{\partial n(x, t)}{\partial t} \quad (5.227)$$

By assuming that the thickness of the specimen is much larger than the Debye length $L_D = (\varepsilon kT/q^2 n_c)^{1/2}$, that the density of injected carriers at the injecting contact (cathode) n_c is constant for all times $n(0, t) = n_c$, and that the density of carriers at the collecting electrode (anode) is zero for all times up to the transit time of the leading front of the injected carriers $n(d, t) = 0$, integration of eq. (5.227) twice gives

$$J(t) = \frac{\varepsilon\mu_n}{2d}[F_a^2(t) - F_c^2(t)] + \mu_n kTn_c/d$$

$$= \frac{\mu_n}{2d}\{Q(t)[F_a(t) + F_c(t)] + Q_0 F_\infty\} \quad (5.228)$$

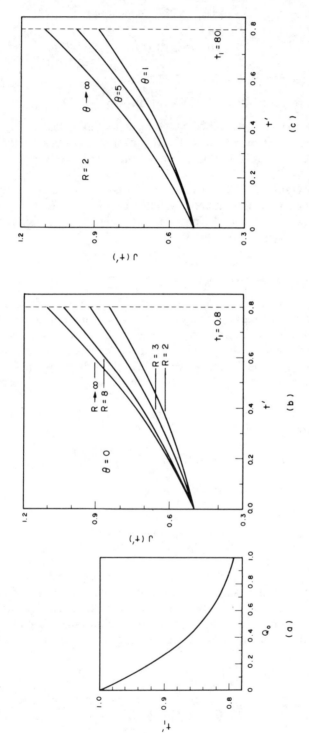

FIG. 5.30. (a) t_1' as a function of injection charge Q_0. (b) Photocurrent $j(t')$ as a function of t' in the SCL mode ($Q_0 = 1$) for various lifetimes in terms of R and fixed θ. (c) Photocurrent $j(t')$ as a function of t' in the SCL mode ($Q_0 = 1$) for various θ and fixed R. [After Batra and Seki 1970.]

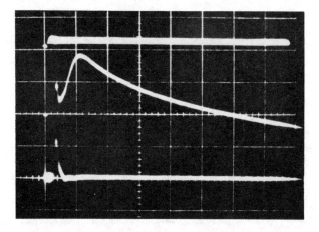

F IG. 5.31. A typical oscillogram of the SCL hole current transients showing the pronounced trapping effect in anthracene crystals. Specimen thickness: $d = 0.01$ cm, applied voltage $V = 20$ V, electrode area: 0.2 cm^2, vertical scale: 5 μA/division, horizontal scale: 5 μsec/division. Upper trace is the applied voltage, middle trace is the SCL current, and lower trace is the zero current obtained when the polarity of the applied voltage was reversed. The irregularities of the two current transients at the very beginning were caused by the measuring circuit. [After Helfrich and Mark 1962.]

where E_∞ is the reverse field at the injecting contact due to charge Q_0 injected to the specimen prior to the application of the voltage, and $Q(t)$ is the total charge in the specimen. This means that $F(0, t) = F_\infty$ prior to the application of the voltage, and that the applied voltage causes an increase in $F(0, t) = E_c$, thus increasing the charge Q so long as $F(0, t) > E_\infty$. Therefore, Q can be expressed as

$$Q(t) = Q_0 + (\mu_n Q_0/L_D) \int_0^t \left[F_a(t) - F_\infty - Q(t)/\varepsilon \right] dt \qquad (5.229)$$

Equations (5.228) and (5.229) are valid for $0 < t < t_1$, where t_1 is the transit time for the leading front to arrive at the collecting electrode. Using the parameters $F_a' = F_a/(V/d)$, $t' = t/t_{t0}$, $Q'(t') = Q(t)/\varepsilon(V/d)$, $Q_0' = Q_0/\varepsilon(V/d)$, and $d' = d/L_D$, Silver [1974] has obtained the following expression based on eqs. (5.228) and (5.229) and the first order approximation:

$$\frac{\partial F_a'}{\partial t'} = Q_0' F_a' \exp(-Q_a' d' t') + \frac{F_a'}{2}[1 - 2 \exp(-2Q_0' d' t')] \qquad (5.230)$$

Silver [1974] has also integrated eq. (5.230) numerically for $d' = d/L_D = 100$ and Q_0' varying from 10^{-2} to 10^2, and some of his computed results are shown in Fig. 5.32. It is interesting to note that from the very beginning to the time when the first minimum of the SCL current appears, the current is diffusion dominated. The current decreases with time during this period because the gradient of the carrier concentration decreases in the short distance where the cloud of injected carriers has advanced and the drift current component is increased. For the times beyond this period the current becomes

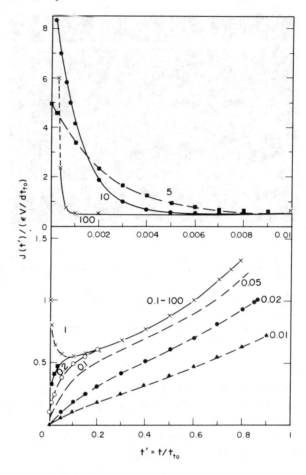

F_{IG}. 5.32. Normalized SCL current transients (including diffusion but without traps) as functions of normalized time for various values of Q'_0 (from 0.01 to 100). [After Silver 1974.]

drift dominated. The main role of the diffusion is to cause the initial dispersion of the leading front of the injected space charge, i.e. to cause a "smearing out" of the sharp leading front and a rounding of the cusp as shown in Fig. 5.33.

CASE IV: INCLUDING DIFFUSION AND WITH TRAPS

Recently, Rosental [1973] has studied the SCL current transients taking into account both the effects of diffusion and trapping. The effect of the trapping is to reduce $Q'_0 [= Q_0/\varepsilon(V/d)]$. Rosental has computed $J(t)$ as a function of t for hole injection SCL current transients in solids containing traps confined in a single discrete energy level, and the results are shown in Fig. 5.33. The physical constants used for the computation are:

$$\mu_p = 2 \times 10^{-2} \, m^2 \, V^{-1} \, s^{-1}$$
$$\varepsilon = 10^{-10} \, F \, m^{-1}$$

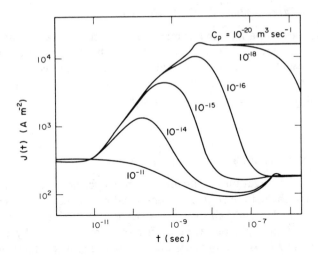

$C_p = 10^{-20}$ m^3 sec^{-1}

10^{18}

10^{-16}

10^{-15}

10^{-14}

10^{-11}

FIG. 5.33. SCL current transients in insulators including diffusion characterized by various trapping coefficients C_p. [After Rosental 1973.]

$$\phi_B(0, 0) = \phi_B(d, 0) = 0.25 \text{ V}$$
$$N_p = 10^{24} \text{ m}^{-3}$$
$$p_1 = N_v \exp(-E_t/kT) = 10^{22} \text{ m}^{-3} \text{ at } 293° \text{ K}$$
$$T = 293° \text{ K}$$
$$d = 10^{-4} \text{ m}$$
$$V = 90 \text{ V}$$
$$C_p = \langle v\sigma_p \rangle \simeq 1/\tau_p N_p \text{ [cf. eq. (5.207)]}$$

In Fig. 5.33, C_p is the hole capture coefficient or the hole-trapping coefficient. For example, the value of C_p equal to 10^{-15} m^3 sec^{-1} corresponds to the value of $R = 0.2$ in Fig. 5.29. The physical constants are so chosen that the steady state SCL current would obey the shallow trap square law.

The peaks on the curves for $C_p = 10^{-20}, 10^{-18}$, and 10^{-11} m^3 sec^{-1} are substantially identical with those reported by Many and Rakavy [1962] (cf. Fig. 5.29), except the rounding of the peaks when diffusion is taken into account. In other respects, however, there are some conspicuous features not predicted by the simplified theory, and they are [Rosental 1973]: (i) the shape of the space charge distributions along the specimen, (ii) the SCL current in the very beginning, (iii) the minimum before the increase of the SCL current, (iv) the position of the maximum (the peak) moving on the time axis with C_p, and (v) the occurrence of the minimum at high values of C_p.

Finally, it should be noted that for a rigorous treatment of the current transients the effect of the image force [Silver 1974] and the effect of the charge exchange in traps on the permittivity and the carrier mobility [Paritskii and Rozental 1967] should be considered.

(C) *The space-charge-perturbed (SCP) transient*

In most practical situations the ohmic contact is not so perfect and the carrier reservoir is not large enough to maintain zero field at the injecting electrode, either part of the time or throughout the duration of the current. In this subsection we shall discuss the current transient which is partly space-charge-controlled and partly electrode limited. Two such cases are considered as follows.

(I) CARRIER GENERATION BY A STRONG LIGHT PULSE

The intensity of the light pulse is assumed to be sufficiently great such that initially zero field is established at the electrode illuminated by the light pulse and the current initially follows a SCL form [Schwartz and Hornig 1965, Weisz et al. 1968]. But after the termination of the light pulse the carrier reservoir is gradually diminished by recombination and $F(0, t) = 0$ for times up to t_A, where t_A is the duration of the light pulse and $t_A \leq t_1$.

By assuming that the carrier lifetime $\tau < t_A$ so that the carrier reservoir at the cathode collapses completely at $t = t_A$, and that $F(0, t) = 0$ for $t \leq t_A$ and $F(0, t) \neq 0$ for $t > t_A$, then for the time interval $t < t_A$ the current as a function of t is given by eqs. (5.195) and (5.196), and for the time interval $t > t_A$ Weisz et al. [1968] have derived the expression for $J(t)$ for the case with shallow traps but excluding diffusion, and it is

$$J(t) = \frac{\varepsilon \mu_n V^2}{2d^3}[1 - (t_A/2t_{t0})]^2 \exp[(t - t_A)/(t_{t0} - t_1/2)] \quad \text{for} \quad t_A \leq t \leq t_1$$

(5.231)

and

$$J = -(M/t)\{\beta(t) + \beta^2(t)[\beta(t)/y^2 - 1/y^2]\} \quad \text{for} \quad t_1 < t < t_2 \qquad (5.232)$$

where

$$\left.\begin{aligned} \beta(t) &= y[J_1(y) + cH_1^{(1)}(y)]/[J_0(y) + cH_0^{(1)}(y)] \\ y &= \frac{2\mu_n Mt}{\varepsilon d} \\ M &= -\frac{\varepsilon V}{d} \quad \text{for} \quad t_A = 0 \end{aligned}\right\} \qquad (5.233)$$

and $J_0(y)$ and $J_1(y)$ are the zero and the first order Bessel functions, respectively; $H_0^{(1)}(y)$ and $H_1^{(1)}(y)$ are the zero and first order Hankel functions of the first kind, respectively; and c is the constant of integration; t_1 and t_2 are the transit times for the leading front and the tail of the injected charge to arrive at the anode, respectively. For $t_A \geq 0$, M and c can be determined from the following equations:

$$t\left(\frac{dQ}{dt}\right) + Q = \frac{\mu_n}{2\varepsilon dQ^2} + M$$

$$Q(t_A) = \frac{\varepsilon V}{d}[1 - (t_A/2t_{t0})]^{-1}$$

$$t\frac{dQ}{dt} = -\varepsilon F(d, t)$$

$$F(d, t) = (V/2d)[1 - (t_A/2t_{t0})]^{-1}\{1 + \exp[(t - t_A)/(t_{t0} - t_A/2)]\}$$

$$tQ = \frac{\varepsilon dy}{\mu_n}\{[J_1(y) + cH_1^{(1)}(y)]/[J_0(y) + cH_0^{(1)}(y)]\}$$

$$\left.\rule{0pt}{80pt}\right\} \quad (5.234)$$

Weisz *et al.* [1968] has computed $J(t)$ as a function of t for $t_A = 0, 0.5t_1, t_1$, and ∞, and the results are shown in Fig. 5.34. In the top four curves the current is initially SCL, but for $t > t_A$ the current changes more rapidly with time than that predicted by Many and Rakavy [1962] (cf. Fig. 5.29). This explains well the experimental results on anthracene [Helfrich and Mark 1962] (cf. Fig. 5.31) and iodine [Many *et al.* 1962].

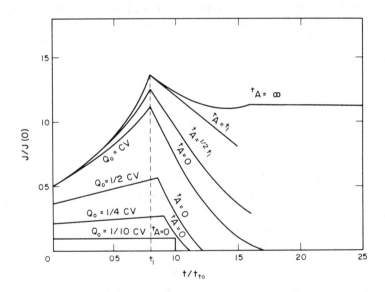

FIG. 5.34. Current as a function of time due to injected charge Q_0 in insulators. The top four curves are considered due to strong light excitation and the bottom three due to weak light excitation. The physical constants for computing these curves are: $\mu_n = 1$ cm^2/V-sec, absorption coefficient: (light penetration length)$^{-1} = 10^3$ cm^{-1}, $\varepsilon/\varepsilon_0 = 2$, and $d = 0.1$ cm. [After Weisz, Cobas, Trester, and Many 1968.]

(II) CARRIER GENERATION BY A WEAK LIGHT PULSE

The light intensity is assumed to be so weak or the pulse duration so short compared to the transit time that the charge Q_0 injected by the light pulse is smaller than

$CV = \varepsilon V/d$. In this case $F(0, t) \neq 0$ for $t < t_A$, but $F(0, t) = 0$ for $t > t_A$. Weisz et al. [1968] have analysed this case and their analysis yields

$$J(t) = (\mu_n Q_0/d)(V/d - Q_0/2\varepsilon) \exp(\mu_n Q_0 t/\varepsilon d) \quad \text{for} \quad t \leq t_1 \qquad (5.235)$$

and if $Q_0 \ll CV$ then eq. (5.235) becomes

$$J(t) = \mu_n V Q_0/d^2 \quad \text{for} \quad t \leq t_1 \qquad (5.236)$$

For $t_1 < t < t_2$ the value of $J(t)$ can be calculated from eq. (5.232). Weisz et al. have also computed $J(t)$ for $Q_0 < CV$ and their results are shown in Fig. 5.34 (the bottom three curves). From eq. (5.235) the initial current at $t = 0$ for $Q_0 < CV$ is

$$J(0) = (\mu_n Q_0 V/d^2) - (\mu_n Q_0^2/2\varepsilon d) \qquad (5.237)$$

Theoretically $J(0)$ is directly proportional to V, but experimentally it is proportional to V^2 over a considerable range of V indicating that Q_0 depends on V even if the intensity and duration of the light pulse are constant. Of course, the higher the applied voltage, the more will the carriers of the appropriate type be swept into the bulk of the specimen. In the illuminated electrode region carriers of both types are present and they encounter carrier recombination as well as carrier trapping, whereas in the bulk only carrier trapping is possible. Thus the lifetime of a carrier in the electrode region can be much smaller than that in the bulk. This implies that Q_0 increases with increasing V. Weisz et al. [1968] have estimated the voltage dependence of Q_0, and Q_0 can be expressed as

$$Q_0(V) \simeq [2K/(1 + K)]CV \qquad (5.238)$$

and $J(0)$ becomes

$$J(0) \simeq [4K/(1 + K)^2](\varepsilon \mu_n V^2/2d^3) \qquad (5.239)$$

where

$$K = \rho(0)\mu_n \tau_n \exp(-2)/2\varepsilon \qquad (5.240)$$

$\rho(0)$ is the charge density produced by a light flash at $t = 0$ and at $x = 0$. Equations (5.238) and (5.239) are valid only for $K \leq 1$. The intensity of light at which the initial current undergoes a transition from the electrode-limited to the SCL regimes is obtained by the condition $K = 1$, which is

$$\rho(0) \exp(-2) = 2\varepsilon/\mu_n \tau_n \qquad (5.241)$$

For details of SCP transients the reader is referred to the papers of Weisz et al. [1968], Schwartz and Hornig [1965], Many et al. [1961, 1962], Helfrich and Mark [1962], Silver et al. [1963], Bogus [1965], Papadakis [1967], Tabak and Scharfe [1970].

Tabak and Scharfe [1970] have observed a transition from emission- (or electrode-) limited to SCL current transient in amorphous selenium films, and their results are shown in Fig. 5.35. If the light is absorbed close to the surface, only the carriers with the same polarity as the illuminated surface will transit across the specimen. If $Q_0 < CV$, a fraction of the total charge during carrier transport will be trapped, and this fraction is determined by the ratio of the average distance travelled before trapping to the specimen thickness. For a constant applied voltage the accumulation of space charge will cause the electric field at the illuminated electrode to approach zero and the current

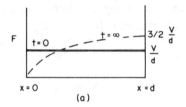

(a)

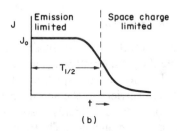

(b)

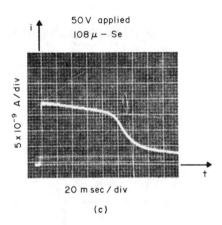

(c)

FIG. 5.35. Illustrating a transition from the emission-limited to the SCL current transients: (a) theoretical electric field distribution at $t = 0$ and at $t = \infty$, (b) theoretical J–t curve, and (c) experimental J–t curve for an amorphous selenium film with $d = 108\ \mu$m and the illuminated (4500 Å light) electrode positive so that holes are the injected carriers, and $V = 50$ V. [After Tabak and Scharfe 1970.]

to become SCL, and eventually at $t = \infty$, F varies with $x^{1/2}$. The dividing line between the emission-limited and the SCL regimes is labelled $T_{1/2}$ in Fig. 5.35, which is defined as the time required for the current to decay to half of its initial value, and is also the time taken to trap one CV of charge [Tabak and Scharfe 1970, Rose 1955].

5.5.2. Two-carrier (double) planar injection transient currents

Many investigators [Baron *et al.* 1966, Dean 1968, 1969, van der Ziel 1970, Baron and Mayer 1970, Gill and Batra 1971, Weber *et al.* 1971] have studied double injection current transients. It can be imagined that the treatment including diffusion and trapping under double injection conditions would be considerably more difficult. In this subsection we shall consider the optical double injection current transients, i.e. transients produced by simultaneous carrier injection at both electrodes by illumination with highly absorbed light under an applied bias voltage. For simplicity the diffusion term is neglected for the analysis. The effect of diffusion may not be significant provided that the specimens are sufficiently thick (long structures) and the applied average field is sufficiently large to justify this assumption. We shall consider two general cases as follows.

(A) In the absence of traps and diffusion

For this case we shall follow closely the work of Gill and Batra [1971]. Figure 5.36 is the schematic diagram showing the geometric arrangement used for the analysis. The electrodes are blocking electrodes and semi-transparent to the light used to illuminate the electrodes for generating free carriers in the specimen near the electrodes. The wavelength of the light is so chosen that it is highly absorbed so that the penetration depth is much smaller than the specimen thickness. The transient behaviour is governed by the current flow equation

$$J(x, t) = q\mu_n n(x, t) F(x, t) + q\mu_p p(x, t) F(x, t)$$

$$+ \varepsilon \frac{\partial F(x, t)}{\partial t} = J_{cp} + J_{cn} + \varepsilon \frac{\partial F(x, t)}{\partial t} \tag{5.242}$$

the Poisson equation

$$\frac{\partial F(x, t)}{\partial x} = \frac{q}{\varepsilon}[p(x, t) - n(x, t)] \tag{5.243}$$

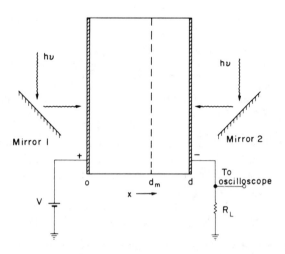

FIG. 5.36. The geometrical arrangement used for the analysis of optical double injection current transients. [After Gill and Batra 1971.]

and the continuity equation

$$\frac{\partial n(x, t)}{\partial t} = \frac{1}{q}\left(\frac{\partial J_{cn}}{\partial x}\right) - \langle v\sigma_R \rangle n(x, t)p(x, t) \tag{5.244}$$

$$\frac{\partial p(x, t)}{\partial t} = -\frac{1}{q}\left(\frac{\partial J_{cp}}{\partial x}\right) - \langle v\sigma_R \rangle n(x, t)p(x, t) \tag{5.245}$$

CASE I: PULSE INJECTION

Supposing that electrons and holes injected into the specimen by a highly absorbed pulsed light initially extend over a region very small compared to the specimen thickness, the total number of injected electrons and holes per unit area, n_i and p_i, are given by

$$n_i = \int_0^d n(x, t)dx \tag{5.246}$$

$$p_i = \int_0^d p(x, t)dx \tag{5.247}$$

where $0 \leq qn_i$, $qp_i \leq CV$, and $C = \varepsilon/d$. From eqs. (5.242)–(5.247) and using the following boundary conditions,

$$\int_0^d F(x, t)dx = V$$

$$p(x, 0) = n(x, 0) = 0$$

$$F(x, 0) = F_{av} = \frac{V}{d}$$

Gill and Batra [1971] have derived the following expression for $J(t)$:

$$J(t) = \left(\frac{\varepsilon V}{d^2}\right)\left[F(d_m, t)(\mu_n \rho_{ni} + \mu_p \rho_{pi}) - \left(\frac{V}{2d}\right)(\mu_n \rho_{ni}^2 + \mu_p \rho_{pi}^2)\right] \tag{5.248}$$

in which

$$F(d_m, t) = F_{av}\left[1 - \frac{\mu_n \rho_{ni}^2 + \mu_p \rho_{pi}^2}{\mu_n \rho_{ni} + \mu_p \rho_{pi}}\right]\exp\left[\left(\frac{\mu_n \rho_{ni} + \mu_p \rho_{pi}}{\mu_n + \mu_p}\right)\left(\frac{t}{t_{tb}}\right)\right]$$

$$+ \frac{\mu_n \rho_{ni}^2 + \mu_p \rho_{pi}^2}{2(\mu_n \rho_{ni} + \mu_p \rho_{pi})} \tag{5.249}$$

$$\rho_{ni} = qn_i/CV \tag{5.250}$$

$$\rho_{pi} = qp_i/CV \tag{5.251}$$

$$t_{tb} = \frac{d^2}{(\mu_n + \mu_p)V} \tag{5.252}$$

and $x = d_m$ is the plane at which the leading fronts of injected electrons and holes meet at time t_m, and it is

$$d_m = \mu_p \int_0^{t_m} F(d_m, t)dt = d - \mu_n \int_0^{t_m} F(d_m, t)dt \qquad (5.253)$$

Equation (5.253) leads to the following simple relation:

$$d_m = [\mu_p/(\mu_n + \mu_p)]d \qquad (5.254)$$

This implies that the location of the plane at which both types of injected carriers meet is independent of the charge carrier injection level (the magnitude of n_i and p_i) because the leading fronts of both types of carriers experience the same field $F(d_m, t)$ for $t \leq t_m$. Substitution of eq. (5.249) into eq. (5.253) and the integration of eq. (5.253) yield a transcendental equation which can be solved numerically to yield t_m.

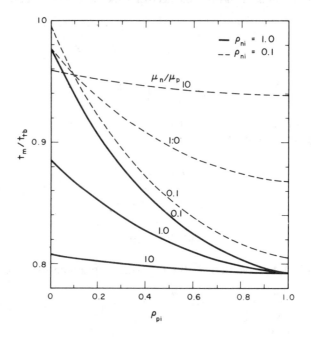

FIG. 5.37. The normalized meeting time t_m/t_{tb} as a function of the hole injection level ρ_{pi} for various ratios μ_n/μ_p and $\rho_{ni} = 1.0$ and 0.1. [After Gill and Batra 1971.]

The above analysis is valid for times up to t_m. Figure 5.37 shows t_m as a function of the carrier injection level and the mobility ratio. It is interesting to note that if $\rho_{ni} = \rho_{pi} = \rho_i$, integration of eq. (5.253) yields

$$\exp(\rho_i t_m/t_{tb}) + [\rho_i^2/(2-\rho_i)](t_m/t_{tb}) = (2+\rho_i)/(2-\rho_i) \qquad (5.255)$$

which is identical to eq. (5.218) for one-carrier injection current transients. But it should be noted that the solutions do not reduce to the one-carrier solutions by setting either ρ_{ni} or ρ_{pi} equal to zero because this would violate the original assumption for finite double injection. However, one-carrier solutions can be deduced by setting either μ_n or

μ_p equal to zero. The case of equal electron and hole injection can be analysed by treating a pair of one-carrier injection systems placed back to back. Gill and Batra [1971] have also obtained the expressions for the spatial extents of the injected electron pulse and the injected hole pulse, which are, respectively, given by

$$x_2(t) - x_1(t) = \mu_p \rho_{pi} F_{av} t \tag{5.256}$$

$$x_3(t) - x_4(t) = \mu_n \rho_{ni} F_{av} t \tag{5.257}$$

where $x_2(t)$ and $x_3(t)$ are the locations of the leading fronts, and $x_1(t)$ and $x_4(t)$ are the locations of the trailing edges of the injected hole and electron pulses, respectively. Equations (5.256) and (5.257) indicate that due to Coulombic interactions the carrier pulse widths grow linearly with time. The carrier distribution within the pulse can be determined using eqs. (5.242) and (5.243) and the concept of flow lines [Many and Rakavy 1962, Helfrich and Mark 1962, Gill and Batra 1971] which is defined by

$$\frac{dx(t)}{dt} = \mu F[x(t), t] \tag{5.258}$$

Up to here, it should be pointed out that for $t > t_m$ the injected electron and hole space charges overlap, and the problem is analytically intractable because this will involve complicated recombination processes.

Figure 5.38 shows some transient results for anthracene crystals obtained with a highly absorbed light pulse of duration of about 100 nsec (half-width with a rise time of 40 nsec), and the wavelengths of the range 2500–5000 Å. The specimen thickness is 270 μm. For the specimens used for obtaining these results, the hole quantum yield is much greater than the electron quantum yield due to the dominant extrinsic generation process in these particular crystals. Figure 5.38(a) shows the transient due to hole transport for $\rho_{pi} = 1$ which is very similar to Fig. 5.31. The drift mobility of holes determined from the transit time is about 0.97 cm^2/V-sec. Figure 5.38(b) shows the transient due to both electron and hole transport for $\rho_{pi} = \rho_{ni} = 1$. The cusp represents the meeting of the leading fronts of the electron and hole pulses, and from this the value of $\mu_n + \mu_p$ has been determined to be 1.36 cm^2/V-sec. This implies that the drift mobility of electrons is about 0.39 cm^2/V-sec, in good agreement with measured values in similar crystals [Kepler 1960, Le Blanc 1960]. Figure 5.38(c) shows the transient due to electron transport for $\rho_{ni} = 1$. In this case, hole injection corresponding to $\rho_{pi} < 1$ is also present in the specimen due to the presence of weakly absorbed light in the light pulse and the high hole generation efficiency in the specimen. A cusp at $t = 600$ nsec represents the meeting of the electron and hole pulse, and a break at $t = 2$ μsec corresponds to the arrival of the leading front of the remaining electrons (electrons encounter a rapid recombination with holes after they meet at $x = d_m$) at the positive electrode. The electron mobility determined from the break is about 0.42 cm^2/V-sec, again in close agreement with that determined from Fig. 5.38(b).

CASE II: STEP FUNCTION INJECTION

Gill and Batra [1971] have obtained an exact analytic solution of the transient response for step function illumination under the conditions which would lead finally to SCL electric conduction. Using eqs. (5.242) and (5.243) and the boundary conditions

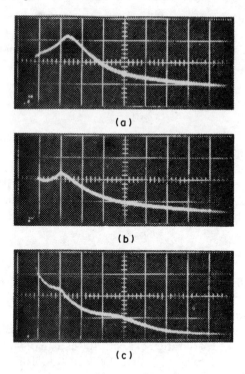

(a)

(b)

(c)

FIG. 5.38. Optical double injection current transients in anthracene
obtained with intense, highly absorbed pulse light. Trace (a)
is for illumination of the positive electrode, trace (b) is for
simultaneous illumination of both electrodes, and trace (c) is
for illumination of the negative electrode. Applied voltage V
$= 700$ V, horizontal scale: 500 nsec/division for all traces,
vertical scales: 0.1, 0.2, and 0.04 mA/div for traces (a), (b), and
(c), respectively. [After Gill and Batra 1971.]

$F(0, t) = F(d, t) = 0$ for $t > 0$, they have solved for $J(t)$ which is given by

$$J(t) = \frac{\varepsilon(\mu_n + \mu_p)V^2}{2d^3}\left[1 - \left(\frac{t}{2t_{tb}}\right)\right]^{-2}, \quad 0 \le t \le t_m \tag{5.259}$$

This equation is equivalent to eq. (5.195) for one-carrier SCL transients. For this case
the field at the meeting plane $x = d_m$ is

$$F(d_m, t) = (V/d)[1 - (t/2t_{tb})]^{-1} \tag{5.260}$$

(B) In the absence of diffusion but with traps

The effect of deep traps on the transient behaviour in semiconductors has been
pointed out by Baron *et al.* [1966], Ancker-Johnson [1967], and Abduvakhidov *et al.*
[1967], and investigated by Dean [1969] and Weber *et al.* [1971]. In this subsection we
shall discuss the small signal transient response in a specimen biased in the
semiconductor regime in which the steady state current–voltage relation follows

$$J = \frac{9}{8}\varepsilon\mu_n\mu_p(n\tau_p - p\tau_n)(V^2/d^3) \tag{5.261}$$

This equation is, in fact, the same as eq. (4.237). The small signal transient behaviour is governed by the current flow equation

$$j = j_n + j_p = q(\mu_n n + \mu_p p)F_t + q(\mu_n \Delta n + \mu_p \Delta p)F \tag{5.262}$$

the continuity equations

$$\frac{1}{q}\frac{\partial j_n}{\partial x} = \frac{\partial(\Delta n)}{\partial t} + \frac{\Delta n}{\tau_n} \tag{5.263}$$

$$-\frac{1}{q}\frac{\partial j_p}{\partial x} = \frac{\partial(\Delta p)}{\partial t} + \frac{\Delta p}{\tau_p} \tag{5.264}$$

and the equations of the recombination rates for electrons and holes

$$\frac{\partial(\Delta n_t)}{\partial t} = \frac{\Delta n}{\tau_n} \tag{5.265}$$

$$\frac{\partial(\Delta p_t)}{\partial t} = \frac{\Delta p}{\tau_p} \tag{5.266}$$

where n and p are, respectively, the electron and hole densities in the steady state; J and F are, respectively, the current density and the electric field in the steady state, the latter being given by

$$F = (3V/2d)(1 - x/d)^{1/2} \tag{5.267}$$

Δn, Δp, Δn_t, and Δp_t are, respectively, the small signal time dependent components of the free electron, free hole, trapped electron, and trapped hole densities; j, j_n, and j_p are, respectively, the time varying total, electron and hole current densities; and F_t is the time dependent field which is caused by the application of a small voltage pulse ΔV to superimpose to the voltage V which is already applied to yield a steady state current in the high level semiconductor regime. On the basis of the following assumptions: (i) the time-varying quantities are small compared with their steady state values, (ii) the changes in the carrier recombination rates due to the changes in Δn_t and Δp_t are neglected because of large concentration of deep traps by assumption, (iii) the carrier lifetimes τ_n and τ_p are independent of position, (iv) the quasi-neutrality approximation is used instead of Poisson's equation, thus $\Delta p - \Delta n = \Delta n_t - \Delta p_t$, and this implies that the solutions are valid only for the carrier transit time larger than the dielectric relaxation time (or for frequencies lower than the dielectric relaxation rate which is $q(\mu_n n + \mu_p p)/\varepsilon$), and (v) the displacement current $\varepsilon\partial F_t/\partial t$ is neglected; Weber et al. [1971] have solved eqs. (5.262)–(5.266) and obtained the following expression for $j(t)$:

$$j(t) = J\{1 + A_n[1 - \exp(-t/\tau_n)] + A_p[1 - \exp(-t/\tau_p)]\} \tag{5.268}$$

where

$$A_n = \mu_n\tau_n/(\mu_n\tau_n + \mu_p\tau_p) \tag{5.269}$$

$$A_p = \mu_p\tau_p/(\mu_n\tau_n + \mu_p\tau_p) \tag{5.270}$$

An equation similar to eq. (5.268) for the trap free case has been derived earlier by Baron et al. [1966]. The so-called "small signal" case means that the specimen is initially biased in one of the high level regimes (e.g. in the semiconductor regime) and the final state after the application of a small voltage pulse is still in the same high level regime. The solutions based on small signal approximation remain valid over the range of magnitudes of voltage pulses or steps for which the definition of the small signal case is not violated. The investigation of the transient response is a powerful tool for determining the carrier lifetime and for distinguishing between double injection and other modes of electrical transport processes. Using this small signal transient response technique, the carrier lifetimes for a series of silicon $p–\pi–n$ and $p–i–n$ structures have been measured by Mayer et al. [1965, 1968], and the carrier lifetimes for gold-doped silicon $p–i–n$ structures by Weber et al. [1971].

Dean [1969] has extended the theory of the "small signal" double injection transient to the "large signal" case. In the "large signal" case the specimen is initially unbiased or biased in a low level regime, and the application of a large signal (e.g. a large voltage step) will lead the final state to a different regime (e.g. to the high level semiconductor regime). Dean has found that there exists a cusp in the time derivative of the current, which occurs at the time equal to the arrival time of the leading front of the injected carriers at the far end, which is $t'_1 = (5/6)t_a$, where $t_a = d^2/\mu_a V$ in the ambipolar transit time, and $\mu_a = \mu_n\mu_p(n+p)/(\mu_n n + \mu_p p)$ is the ambipolar mobility. The measurement of the time for the occurrence of the cusp can be used to determine the minority carrier mobility. For example, the electron mobility in p-type germanium can be determined by applying a step function voltage to a long specimen between two injecting contacts and measuring the time at which the cusp occurs. After the occurrence of the cusp the current varies exponentially with time, with the time constant equal to the carrier lifetimes. The presence of deep traps can significantly affect the transient behaviour. If the minority carrier transit time is much shorter than all capture times (corresponding to a high applied step voltage), the leading front of the injected carriers will arrive at the far end of the specimen long before the traps are filled up, so that the transient response for $t < t'_1$ is similar to that for the trap free semiconductors. The presence of traps will modify the trap free transient for $t > t'_1$ by stretching out the transient and increasing the steady state current. If the minority carrier transit time is longer than the initially minority carrier capture time but shorter than all other capture times (corresponding to an intermediate applied step voltage), the traps will be filled up as soon as the leading front of the injected carriers reaches them so that the trapping effects play a role in the transient behaviour immediately. They alter the early transient field profile, reduce the velocity of the propagating injected charge, and cause the current to exhibit a long delay followed by a sharp rise. If the minority carrier transit time is much longer than all capture times (corresponding to low applied step voltage), the velocity of the propagating injected charge drops sharply towards zero because of recombination through the deep traps, and the injected charge barely reaches the far end of the specimen. Dean [1969] has verified his theoretical prediction by experiments on p-type InSb at liquid nitrogen temperature. For details of the large signal case the reader is referred to the original papers of Dean [1969].

5.6. SPACE CHARGE ELECTRIC CONDUCTION
UNDER TIME-VARYING FIELDS

5.6.1. One-carrier (single) planar injection

From the analysis given in Section 5.5 it can be seen that the high frequency admittance of a solid specimen provided with a single-injecting electrode depends on the carrier transit time t_t between the injecting and collecting electrodes. Several investigators [Shao and Wright 1961, Brojdo 1963, van der Ziel 1968, Wright 1966] have studied this problem. In Section 3.2 we have mentioned that if the current due to injected carriers is SCL, the carrier transit time should be much less than the dielectric relaxation time. In the following we shall discuss the frequency dependence of the current on an applied small a.c. voltage

$$e = V_1 \sin \omega t \tag{5.271}$$

superimposed on a steady bias voltage V_s. With the following assumptions: (i) the concentration of thermally generated carriers is negligibly small compared with that of injected carriers, (ii) there is no trapping effect, (iii) the magnitude of V used is such that the neglect of the diffusion term is justified, and (iv) the electron mobility μ_n is constant implying that the mean time between collisions of the space charge electrons with the crystal lattice is much shorter than the period of the time-varying applied voltage, the behaviour of the singly injected electrons from the cathode under a time-varying electric field is governed by the following three basic equations:

$$J = qnv + \varepsilon \frac{\partial F}{\partial t} \tag{5.272}$$

$$\frac{\partial F}{\partial x} = \frac{qn}{\varepsilon} \tag{5.273}$$

and
$$v = \mu_n F \tag{5.274}$$

The injected electrons experience a time-varying field, the time rate of change of which is given by

$$\frac{dF}{dt} = \frac{\partial F}{\partial x}\frac{dx}{dt} + \frac{\partial F}{\partial t}$$

$$= \frac{qnv}{\varepsilon} + \frac{\partial F}{\partial t} \tag{5.275}$$

Substitution of eq. (5.275) into eq. (5.272) gives

$$J = \varepsilon \frac{dF}{dt} = \frac{\varepsilon}{\mu_n}\frac{dv}{dt} \tag{5.276}$$

This equation implies that the total current J can be expressed in terms of the time rate of change of the electric field or of the carrier velocity. Following the method of Benham [1928] and Llewellyn [1935], expressing the time-varying total current as

$$J = J_s + J_1 \exp(j\omega t) \tag{5.277}$$

and the time-varying time interval $t - t_c$ (the average carrier transit time from $x = 0$ to

$x = x$ is $T_s = t - t_c$) as

$$t - t_c = T_s + T_1 \exp(j\omega t) \tag{5.278}$$

and using Taylor's expansion for $\exp\{j\omega[(t - T_s) - T_1 \exp(j\omega t)]\}$ and retaining only first order terms, Shao and Wright [1961] have obtained the following equation:

$$v = \frac{\mu_n J_s T_s}{\varepsilon} + \frac{\mu_n}{\varepsilon T_s}\left[\frac{J_1 T_s}{j\omega} - \frac{J_1}{(j\omega)^2} + \frac{J_1 \exp(-j\omega T_s)}{(j\omega)^2}\right]\exp(j\omega t)$$

$$= v_s + v_1 \exp(j\omega t) \tag{5.279}$$

where

$$v_s = \mu_n J_s T_s/\varepsilon \tag{5.280}$$

$$v_1 = \frac{\mu_n}{\varepsilon T_s}\left[\frac{J_1 T_s}{j\omega} - \frac{J_1}{(j\omega)^2} + \frac{J_1 \exp(-j\omega T_s)}{(j\omega)^2}\right] \tag{5.281}$$

t_c is the time at which the electrons left the cathode with $v = 0$; J_1, T_1, and v_1 are, respectively, the amplitudes of the first order current, time and velocity variation due to the application of the small a.c. voltage, and the corresponding parameters J_s, T_s, and v_s are due to the steady bias voltage V_s. Fron eq. (5.280) the mean distance x to which the electrons have travelled in the time interval $t - t_c$ is

$$x = \frac{\mu_n}{\varepsilon}(J_s T_s^2/2) \tag{5.282}$$

Since $dx = \mu_n J_s T_s dT_s/\varepsilon$, the total applied voltage can be written as

$$V = V_s + V_1 \exp(j\omega t)$$

$$= \int_0^d F dx = \frac{1}{\mu_n}\int_0^d v dx$$

$$= \frac{1}{\varepsilon}\int_0^{t_t}[v_s + v_1 \exp(j\omega t)]J_s T_s dT_s \tag{5.283}$$

$$v_s = \frac{1}{\varepsilon}\int_0^{t_t} v_s J_s T_s\, dT_s \tag{5.284}$$

$$V_1 = \frac{1}{\varepsilon}\int_0^{t_t} v_1 J_s T_s dT_s \tag{5.285}$$

where $t_t = t_d - t_c$ is the transit time from the virtual cathode to the anode. Using eqs. (5.281) and (5.285), the incremental admittance is given by [Shao and Wright 1961, Wright 1966]

$$Y_1 = J_1/V_1 = G_1 + jB_1$$

$$G_1 = \frac{g\theta^2}{6}\frac{\theta - \sin\theta}{(\theta - \sin\theta)^2 + (\theta^2/2 + \cos\theta - 1)^2} \tag{5.286}$$

$$B_1 = \frac{g\theta^3}{6}\frac{\theta^2/2 + \cos\theta - 1}{(\theta - \sin\theta)^2 + (\theta^2/2 + \cos\theta - 1)^2} \tag{5.287}$$

where g is the steady current incremental conductance defined by

$$g = \frac{\partial J_s}{\partial V_s} = \frac{3C}{t_t} \qquad (5.288)$$

and θ is the transit angle defined by

$$\theta = \omega t_t \qquad (5.289)$$

and C is the geometric capacitance which is equal to ε/d per unit area.

Equation (5.287) may be separated into two components corresponding to the geometrical or "cold" capacitance and to the "electronic" capacitance due to transit time effects. The susceptance due to the geometrical capacitance is

$$B_{1c} = \omega C = \theta C/t_t = g\theta/3 \qquad (5.290)$$

and the susceptance due to transit time effect is

$$B_{1e} = B_1 - B_{1c} \qquad (5.291)$$

Figure 5.39 shows G_1/g, B_{1c}/g, and B_{1e}/g as functions of θ. It is interesting to note that in the frequency range shown in Fig. 5.39 B_{1e} is negative and acts as an inductive susceptance since the current induced by the carriers lags behind the signal voltage because of the finite times of transit [Wright 1966]. Figure 5.40 shows the comparison of the computed results based on the above analysis and the experimental results of the incremental conductance and incremental capacitance for cadmium sulphide. The experiments are in good agreement with the theory.

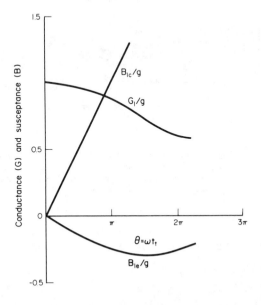

FIG. 5.39. Frequency dependence of conductance G_1, geometric capacitive susceptance B_{1c}, and electronic capacitive susceptance B_{1e} showing the transit time effects in the SCL silicon diodes. [After Wright 1966.]

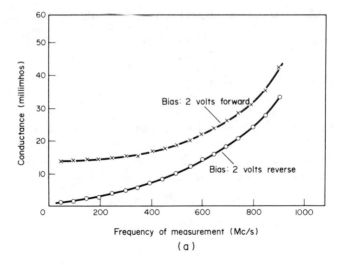

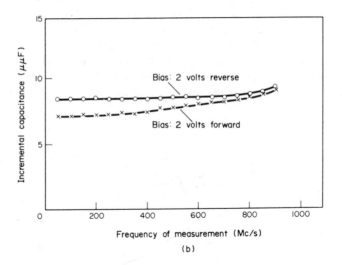

F IG. 5.40. The variation with frequency of incremental conductance and of incremental input capacitance for cadmium sulphide SLC dielectric diodes under forward and reverse bias conditions. [After Shao and Wright 1961.]

Wright [1966] has also extended this theory to obtain a second order solution and reported that the rectified current is much less frequency dependent than the first order current J_1.

5.6.2. Two-carrier (double) planar injection

Several investigators [Baron and Mayer 1967, van der Ziel 1969, 1970, Bilger *et al.* 1968, Nordman and Greiner 1963] have studied the SCL electric conduction at high frequencies based on the small signal theory. Again, the diffusion contribution is ignored for simplicity. This is acceptable for long structures and at fields such that $qFd \gg kT$. In the following we shall consider the high level injection in the semiconductor regime with small concentration of recombination centres. Therefore, in the steady state, eqs. (4.229) and (4.230) are applicable to the present analysis. Because of high level injection we can assume $\tau_n = \tau_p = \tau$ and that the steady state concentrations of electrons and holes can be expressed as

$$\left.\begin{array}{l} n_s = n_0 + \Delta n \simeq \Delta n \\[6pt] p_s = p_0 + \Delta p \simeq \Delta p \\[6pt] \Delta n = \Delta p \simeq n_s \simeq p_s \end{array}\right\} \tag{5.292}$$

Here the subscript s denotes the parameters under the steady state conditions, and later we use the subscript 1 to denote the small signal parameters. Using the same assumptions as for Section 4.3.3(A), the equations describing the time dependent current and time dependent carrier concentration for the high level semiconductor regime are

$$J = q\mu_n \left(\frac{b+1}{b}\right) nF + \varepsilon \frac{\partial F}{\partial t} \tag{5.293}$$

and

$$\left(\frac{b+1}{\mu_n}\right)\frac{\partial n}{\partial t} = -\left(\frac{b+1}{\mu_n}\right)\frac{n}{\tau} + (n_0 + p_0)\frac{\partial F}{\partial x} \tag{5.294}$$

Following the first order small signal approximation, the total current and carrier concentration consisting of the small signal a.c. response can be written in the form

$$J = J_s + J_1 \exp(j\omega t) \tag{5.295}$$

$$n = n_s + n_1 \exp(j\omega t) \tag{5.296}$$

Substituting eqs. (5.295) and (5.296) into eqs. (5.293) and (5.294), collecting the terms proportional to $\exp(j\omega t)$ for the first order approximation and neglecting all higher order terms so as to linearize the equations, Baron and Mayor [1970] have obtained F_1 and J_1. Using the boundary condition $V_1 = \int_0^d F_1 dx$, their analysis leads to the following expression for the small signal admittance Y_1:

$$(Y_1)^{-1} = \frac{V_1}{J_1}$$

$$= 3R_s(1+j\omega\tau)\int_0^1 dz' \int_0^{z'} dz$$

$$\times \left\{ z'\left(\frac{z}{z'}\right)^{(1+j\omega\tau)} \exp[j\omega t_r(1+j\omega\tau)(z-z')] \right\} \tag{5.297}$$

where

$$R_s = V_s/J_s = 1/G_s, \qquad t_r = 3CR_s/2, \qquad z = (x/d)^{1/2} \tag{5.298}$$

and t_r is of the order of the average high level dielectric relaxation time. Two cases are considered as follows:

CASE I: $\omega t_r \ll 1$

For this case eq. (5.297) can be integrated by expanding the exponential function in powers of ωt_r and ignoring high power terms. Thus, integration of eq. (5.297) gives

$$Y_1 \simeq \frac{1}{R_s[1+3t_r/4\tau]} + \frac{1}{R_s+j\omega\tau R_s} + \frac{j9\omega C}{8} + \frac{1}{4\tau R_s/3t_r + j4\omega\tau^2 R_s/9t_r} \quad (5.299)$$

The equivalent circuit for this case is shown in Fig. 5.41(a). If ωt_r is small enough to be neglected, then eq. (5.299) reduces to

$$Y_1 \simeq \frac{1}{R_s} + \frac{1}{R_s+j\omega\tau R_s} \quad (5.300)$$

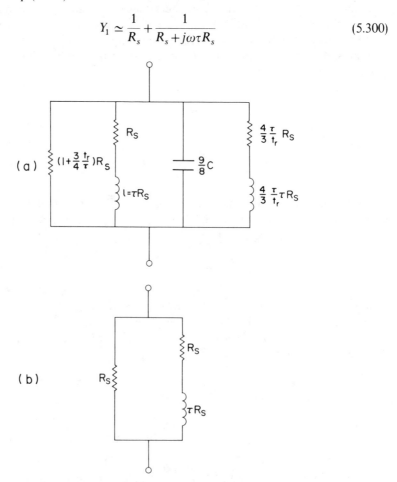

FIG. 5.41. (a) Equivalent circuit for a double injection device biased in the semiconductor regime based on eq. (5.299). (b) Simplified equivalent circuit based on eq. (5.300) for the cases in which ωt_r is negligibly small and terms of order ωt_r in the impedance are neglected. [After Baron and Mayer 1970.]

The equivalent circuit for this case is shown in Fig. 5.41(b). It is interesting to note that the double injection devices exhibit an inductive behaviour. On the basis of the equivalent circuit shown in Fig. 5.41(b), Baron and Mayer [1970] have also calculated Y_1 as a function of $\omega\tau$ and the results are shown in Fig. 5.42. Since $Y_1 = G_1 + jB_1$ the quality factor Q is given by

$$Q = \frac{|B_1|}{G_1} \tag{5.301}$$

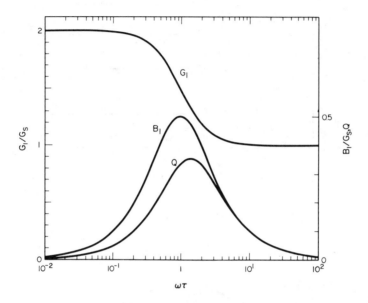

FIG. 5.42. The small signal conductance G_1, susceptance B_1, and $Q = B_1/G_1$ of a double injection device biased in the high level semiconductor regime as functions of frequency. [Baron and Mayer 1970.]

It can be seen that B_1 tends to become zero at very low and at very high frequencies. This is because the carrier movement follows the varying field at low frequencies and cannot follow the varying field at high frequencies. At frequencies around $\omega\tau = 1$, B_1 is large because the carrier movement lags behind the field variation.

Bilger *et al.* [1968] have measured the small signal admittance of a silicon p–π–n structure as a function of frequency, and their results are shown in Fig. 5.43 in which the solid lines are least squares fits to the experimental data based on the fitting function derived from the equivalent circuits shown in Fig. 5.41. Their experimental results are in reasonable agreement with the theoretical prediction shown in Fig. 5.42.

CASE II: $\omega t_r \gg 1$

For this case, eq. (5.297) can be integrated by expanding the exponential function in powers of $1/\omega t_r$ and ignoring high power terms. Thus, integration of eq. (5.297) gives

$$Y_1 \simeq 4/3R_s + j\omega C \tag{5.302}$$

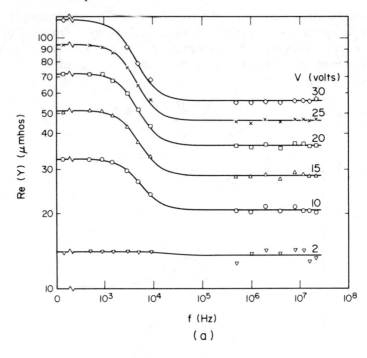

f (Hz)

(a)

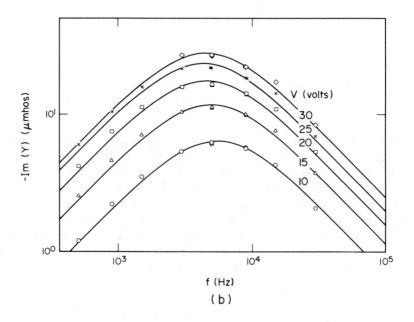

f (Hz)

(b)

FIG. 5.43. (a) The conductance (real part of Y_1), and (b) the susceptance (imaginary part of Y_1) as functions of frequency for a silicon p–π–n structure. The solid curves are least-squares fits to the experimental data. [After Bilger, Lee, Nicolet, and McCarter 1968.]

At such high frequencies, the space charge distribution does not have time to change and thus the susceptance is mainly due to the geometric capacitance. The conductance is higher than G_s because the time-varying field does not have time to relax to a $x^{1/2}$ dependence and becomes spatially independent [Baron and Mayer 1970].

For the small signal admittance in the ohmic regime, insulator regime, and diffusion-dominated regime, the reader is referred to the work of Baron and Mayer [1970], Nordman and Greiner [1963], van der Ziel [1969, 1970], Driedonks and Zijlstra [1968].

5.7. CHARGE CARRIER TRANSPORT IN RELAXATION SEMICONDUCTORS

In Section 1.10 we have introduced the concept of relaxation electric conduction. The relaxation semiconductors are generally referred to the materials in which the dielectric relaxation time τ_d is larger than carrier recombination time or diffusion length lifetime τ_0 [van Roosbroeck 1960, van Roosbroeck and Casey 1972]. This inequality $\tau_d > \tau_0$ implies that when minority carriers are injected in such semiconductors there exist (i) a majority carrier depletion region, (ii) a recombination front, (iii) a low conductivity region in which the resistivity is larger than that expected from the bulk resistivity, and (iv) a sublinear current–voltage characteristic because of the persistence of space charge. Kiess and Rose [1973] have shown that the van Roosbroeck model for the depletion of majority carriers by the injection of minority carriers may lead to an inconsistency in the requirement of the current continuity, and predicted that the inclusion of diffusion does not alter their argument. Both van Roosbroeck, and Kiess and Rose, did not take into account the effect of diffusion in their analysis. However, Döhler and Heyszenau [1973] have reported that the neglect of the diffusion term in the differential equation governing the relaxation regime may lead to false conclusions even in the high field limit. Recently, Popescu and Henisch [1975, 1976] have analysed the complete transport equations (including the diffusion term) by numerical procedures and reported that the majority carrier depletion predicted by van Roosbroeck can in fact occur without violating the current continuity condition. They have avoided the definitions which link the character of a relaxation semiconductor to $np = n_i^2$ (the mass action law). The following discussion is mainly based on their analyses.

5.7.1. Minority carrier injection in relaxation semiconductors

(A) Without traps

For this case the carrier recombination is bimolecular. The electrical transport is governed by the electron and hole current flow equations

$$\mathbf{J}_n = q\mu_n n\mathbf{F} + qD_n \nabla n \tag{5.303}$$

$$\mathbf{J}_p = q\mu_p p\mathbf{F} - qD_p \nabla p \tag{5.304}$$

the continuity equations

$$\frac{\partial n}{\partial t} = q^{-1} \nabla \cdot \mathbf{J}_n + G - R \tag{5.305}$$

$$\frac{\partial p}{\partial t} = -q^{-1} \nabla \cdot \mathbf{J}_p + G - R \tag{5.306}$$

and the Poisson equation

$$\nabla \cdot \mathbf{F} = \frac{q}{\varepsilon}(p + \hat{p} - n - \hat{n})$$

$$= \frac{q}{\varepsilon}(\Delta p + \Delta\hat{p} - \Delta n - \Delta\hat{n}) \tag{5.307}$$

where $\hat{p}$ and $\hat{n}$ are the concentrations of fixed positive and negative charges, such as ionized donors and acceptors. $(\Delta\hat{p} - \Delta\hat{n})$ is zero only for intrinsic semiconductors and $\Delta\hat{p} - \Delta\hat{n} \neq 0$ for extrinsic semiconductors, and departs from its thermal equilibrium value if traps are present. The recombination terms R can be written as

$$R = (np - n_0 p_0)/\tau_0(n_0 + p_0) \tag{5.308}$$

where n_0 and p_0 are the thermal equilibrium concentrations of electrons and holes, respectively.

For this discussion we assume that holes are minority carriers injected into an n-type relaxation semiconductor. We define the Debye length for equilibrium electrons as

$$L_{Dn} = (kT\varepsilon/q^2 n_0)^{1/2} \tag{5.309}$$

the majority carrier (electron for the present case) dielectric relaxation time for equilibrium electrons as

$$\tau_{dn} = \varepsilon/\sigma_n = \varepsilon/q\mu_n n_0, \tag{5.310}$$

the dielectric relaxation time for the n-type material in equilibrium as

$$\tau_d = \varepsilon/q(\mu_n n_0 + \mu_p p_0) \simeq \tau_{dn}, \tag{5.311}$$

and a parameter A as

$$A = \tau_{dn}/\tau_0(1 + p_0/n_0) \tag{5.312}$$

Using the following dimensionless quantities

$$
\left.
\begin{array}{ll}
N = n/n_0 & S = J_n/(qD_p p_0/L_p) \\
p = p/n_0 & T = J_p/(qD_p p_0/L_p) \\
P_0 = p_0/n_0 & \Delta P = P - P_0, \Delta N = N - 1 \\
N_0 = n_0/n_0 = 1 & X = x/L_{Dn} \\
U = F/(kT/qL_{Dn}) & t' = t/\tau_{Dn}
\end{array}
\right\} \tag{5.313}
$$

where L_p is the diffusion length of holes, eqs. (5.303)–(5.307) can be rewritten as

$$S = \frac{\mu_n}{\mu_p} \frac{L_p}{L_{Dn}} \frac{1}{P_0}\left(NU + \frac{dN}{dX}\right) \tag{5.314}$$

$$T = \frac{L_p}{L_{Dn}} \frac{1}{P_0} \left(PU - \frac{dP}{dX} \right) \tag{5.315}$$

$$\frac{\partial N}{\partial t'} = \frac{\partial (NU)}{\partial X} + \frac{\partial^2 N}{\partial X^2} - A(NP - P_0) \tag{5.316}$$

$$\frac{\partial P}{\partial t'} = -\frac{\mu_p}{\mu_n} \left[\frac{\partial (PU)}{\partial X} - \frac{\partial^2 P}{\partial X^2} \right] - A(NP - P_0) \tag{5.317}$$

$$\frac{\partial U}{\partial X} = (P - P_0) - (N - 1) \tag{5.318}$$

for the one-dimensional treatment with $G = 0$. It is obvious that the parameter $A \simeq \tau_{dn}/\tau_0$ for n-type materials governs the balance between recombination, on the one hand, and neutrality restoration, on the other. Supposing that $\Delta p = \Delta p_0$ and $\Delta N = \Delta N_0 = 0$ at $t = 0$, we can have the following cases.

Case (i). $A \ll 1$ corresponding to the conventional lifetime regime. For this case the restoration of neutrality takes place first, followed by recombination of electrons and holes in equal number. Thus in the initial stage, the electron concentration increases $(dN/dt' > 0)$, while the hole concentration decreases $(dP/dt' < 0)$ as shown in Fig. 5.44(a) and (b). After ΔN approaches ΔP (that is, after neutrality is restored), the process is governed by recombination in which $dN/dt' = dP/dt' < 0$ and the contour is IBO. If $A = 0$, the contour becomes $IB'O$ in Fig. 5.44(a).

Case (ii). $A \gg 1$ corresponding to the relaxation regime. For this case the recombination takes place first following the contour IC with its rate diminishing with time as shown in Fig. 5.44(a) and (c). The point at C (actually at C') corresponds to the zero recombination point. After this point is reached, neutrality is restored along the contour CO. Majority carrier depletion is associated with all stages of this process. When $A \to \infty$, the point C approaches C'.

Case (iii). $A = 1$ corresponding to the boundary between cases (i) and (ii). For this case the rate at which electrons are brought from the outside in order to restore neutrality is equal to the rate at which they recombine with the injected excess holes, thus the electron concentration remains unchanged. The process of restoring neutrality follows the line IO in Fig. 5.44(a).

Using the boundary conditions given below,

$$S = 0 \quad \text{and} \quad U = 0 \quad \text{at} \quad X = 0$$

$$\left. \begin{array}{l} \dfrac{dU}{dX} = 0 \quad \text{and} \quad \dfrac{dN}{dX} = \dfrac{dP}{dX} = 0 \\[2mm] P = P_0 \quad \text{and} \quad N = N_0 = 1 \end{array} \right\} \quad \text{at} \quad X = \infty$$

for a semi-infinite trap free n-type material with the minority carriers (holes) injected at $X = 0$, Popescu and Henisch [1975] have computed the steady state carrier concentration and field contours from eqs. (5.314)–(5.318) by setting $\dfrac{\partial N}{\partial t'} = \dfrac{\partial P}{\partial t'} = 0$.

Figure 5.45(a) and (b) show the difference in such contours between a moderate lifetime case and a relaxation case. By comparing these two cases, it can be seen that there is an electron depletion region in Fig. 5.45(b). In this region (a) because of the

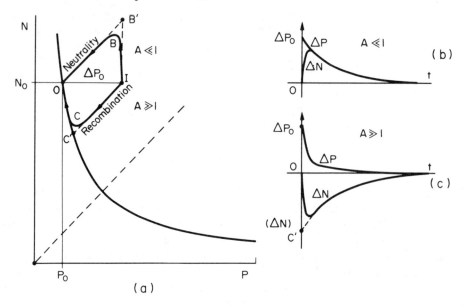

FIG. 5.44. Restoration of equilibrium in the homogeneous material: (a) restoration process in the *N–P* plane, (b) restoration process as a function of time for the lifetime case, and (c) restoration process as a function of time for the relaxation case. [After Popescu and Henisch 1975.]

reverse electron concentration gradient the electron diffusion current adds to the electron drift current, which is completely different from those in Fig. 5.45(a), (b) the electron depletion increases the positive space charge which builds up a field, and (c) the recombination rate is reduced through electron depletion. Popescu and Henisch have concluded on the basis of their analysis for the trap free cases that the electron depletion can become complete by enhancing the recombination rate, either by increasing the value of *A* or by increasing the injected current carriers. Under such conditions there is a recombination front. Between the injecting contact and the recombination front the current is minority carrier SCL; beyond the recombination front it is majority carrier dominated. The computed *J–V* characteristics show an extended linear region, resulting from the opposed tendencies of majority carrier depletion (sublinear) and minority carrier injection (superlinear), but do not exhibit the feature predicted by van Roosbroeck that the overall resistance in the relaxation regime is higher than that calculated on the basis of the unperturbed bulk resistivity at least under the conditions investigated by them. However, they have reported that such a feature may occur in semiconductors with traps, and this is considered below.

(B) With traps

For this case the carrier recombination is predominantly monomolecular. The traps in a semiconductor play two important roles in the electric conduction processes—one as recombination centres and the other as charge storage localities. Therefore the boundary conditions for the present case would be different from those for the trap free case. In the following we shall consider two cases.

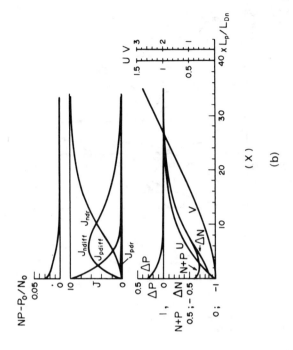

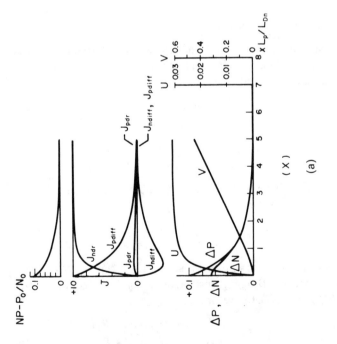

Fig. 5.45. For legend, see page 380.

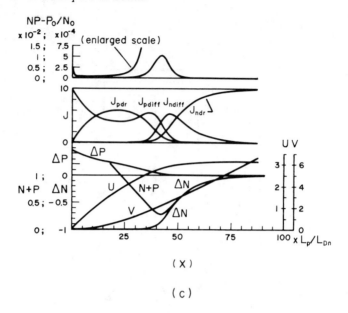

Fig. 5.45. Spatial contours of carrier concentrations, field, voltage drop, current components, and recombination for an *n*-type semiconductor with $J = S + T = 10$, $\mu_n = \mu_p$, and $p_0/n_0 = 10^{-2}$. All magnitudes are normalized on the basis of eq. (5.313) and $NP - N_0 P_0$ is proportional to the recombination rate: (a) for a moderate lifetime case in which $\tau_{dn}/\tau_0 = 0.1$ ($A = 0.099$), (b) for a relaxation case in which $\tau_{dn}/\tau_0 = 10^2$ ($A = 99$), and (c) for a relaxation case in which $\tau_{dn}/\tau_0 = 10^3$ ($A = 990$). [After Popescu and Henisch 1975.]

(I) WITH SMALL DENSITY OF TRAPS

When the density of traps is so low that their space charge can be neglected, the recombination rate through such traps as recombination centres follows the Shockley–Read equation [cf. eq. (4.77)]

$$R = \frac{np - n_0 p_0}{\tau_0(n_0 + p_0)}$$

$$= \frac{np - n_0 p_0}{\tau_{p0}(n + n_1) + \tau_{n0}(p + p_1)} \qquad (5.319)$$

where τ_{n0}, τ_{p0}, n_1, and p_1 have been defined in Section 4.2.2. For small departures from equilibrium, eq. (5.319) can be written as

$$R_0 = \frac{np - n_0 p_0}{\tau_{p0}(n_0 + n_1) + \tau_{n0}(p_0 + p_1)} \qquad (5.320)$$

Using the normalization factors given in eq. (5.313) except that J_n and J_p are now normalized with respect to $qD_n n_0/L_{Dn}$ instead of $qD_p p_0/L_p$ in Section 5.7.1(A), and from eqs. (5.311), (5.319), and (5.320), we obtain

$$A = A_0 \frac{1 + N_1 + (\tau_{n0}/\tau_{p0})(P_0 + P_1)}{N + N_1 + (\tau_{n0}/\tau_{p0})(P + P_1)} \qquad (5.321)$$

where

$$A_0 = \frac{\tau_{dn}}{\tau_{po}(1+N_1)+\tau_{no}(P_0+P_1)}$$

$$N_1 = n_1/n_0 \tag{5.322}$$

$$P_1 = p_1/n_0$$

Equation (5.321) shows that A decreases as N and P increase. Using $\tau_{dn}/\tau_0 = 10^3$ corresponding to $A \simeq 990$, $J = S + T = \sqrt{10}$, $\mu_n = \mu_p$, $p_0/n_0 = 10^{-2}$, $\tau_{no}/\tau_{po} = 1$, and assuming that the recombination centres are located in the middle of the forbidden gap making $N_1 = P_1 = N_i(N_i = n_i/n_0)$, Popescu and Henisch [1976] have computed the recombination contours for two cases: (A) the carrier lifetime is independent of injection level $A = A_0$, and (B) the carrier lifetime is dependent on injection level. Their results are shown in Fig. 5.46. It can be seen that the recombination front is sharpened for case (B).

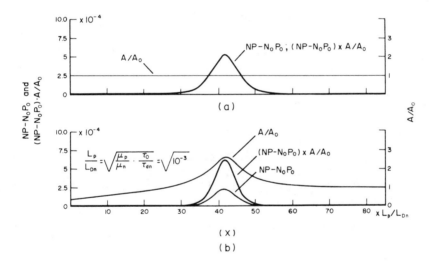

FIG. 5.46. Recombination contours for the relaxation regime with $\tau_{dn}/\tau_0 = 10^3$ corresponding to $A = 990$ for an n-type semiconductor with $J = S + T = (10)^{1/2}$, $\mu_n = \mu_p$, and $p_0/n_0 = 10^{-2}$. All magnitudes are normalized and $(NP - N_0 P_0)(A/A_0)$ is proportional to the recombination rate. The abscissae here are given in multiples of the diffusion length L_p. (a) The carrier lifetime is independent of injection level, $A = A_0$, and (b) the carrier lifetime is dependent on injection level according to the Shockley–Read model [After Popescu and Henisch 1976.]

(II) WITH LARGE DENSITY OF TRAPS

When the density of traps is large, the space charge in the traps cannot be neglected. Supposing that the traps of density N_n are located at a single discrete energy level E_t, the electrical transport is governed by the following equations:

$$\frac{\partial N}{\partial t'} = \frac{\partial(NU)}{\partial X} + \frac{\partial^2 N}{\partial X^2} - A_n \left[N \left(1 - \frac{M_t}{M} \right) - \left(\frac{M_t}{M} \right) N_1 \right]$$

$$= \frac{\partial S}{\partial X} - A_n \left[N \left(1 - \frac{M_t}{M} \right) - \frac{M_t}{M} N_1 \right] \tag{5.323}$$

$$\frac{\partial P}{\partial t'} = -\frac{\mu_p}{\mu_n} \left[\frac{\partial(PU)}{\partial X} - \frac{\partial^2 P}{\partial X^2} \right] - A_p \left[P \left(\frac{M_t}{M} \right) - \left(1 - \frac{M_t}{M} \right) P_1 \right]$$

$$= -\frac{\partial T}{\partial X} - A_p \left[P \left(\frac{M_t}{M} \right) - \left(1 - \frac{M_t}{M} \right) P_1 \right] \tag{5.324}$$

$$\frac{\partial M_t}{\partial t'} = A_n \left[N \left(1 - \frac{M_t}{M} \right) - \left(\frac{M_t}{M} \right) N_1 \right]$$

$$- A_p \left[P \left(\frac{M_t}{M} \right) - \left(1 - \frac{M_t}{M} \right) P_1 \right] \tag{5.325}$$

$$\frac{\partial U}{\partial X} = (P - P_0) - (N - 1) - (M_t - M_0) \tag{5.326}$$

$$\Delta Q_t = -(M_t - M_0) = M_0 \frac{\Delta P(\tau_{n0}/\tau_{p0}) - \Delta N[N_1 + (\tau_{n0}/\tau_{p0})P_0]/[1 + (\tau_{n0}/\tau_{p0})P_1]}{1 + N_1 + (\tau_{n0}/\tau_{p0})(P_0 + P_1) + \Delta N + (\tau_{n0}/\tau_{p0})\Delta P} \tag{5.327}$$

where

$$A_n = \tau_{dn}/\tau_{n0} \tag{5.328}$$

$$A_p = \tau_{dn}/\tau_{p0} \tag{5.329}$$

$$M_t = \frac{n_t}{n_0} = M\{[N + (\tau_{n0}/\tau_{p0})P_1]/[N + N_1 + (\tau_{n0}/\tau_{p0})(P + P_1)]\} \tag{5.330}$$

$$M_0 = \frac{n_{t0}}{n_0} = M\{[1 + (\tau_{n0}/\tau_{p0})P_1]/[1 + N_1 + (\tau_{n0}/\tau_{p0})(P_0 + P_1)]\} \tag{5.331}$$

$$M = \frac{N_n}{n_0} \tag{5.332}$$

in which n_t is the concentration of traps filled with electrons (trapped electrons), and n_{t0} is the equilibrium value of n_t. The above equations are normalized using the normalization procedures similar to those given above. According to the computed results for τ_{n0}/τ_{p0} and $N_1 = P_1 = N_i$ obtained by Popescu and Henisch [1976], the boundary between the lifetime and the relaxation regimes should be defined by

$$A_0/(M + 1) \simeq 1 \tag{5.333}$$

for materials with significant charge trapping. This implies that $A_0/(M + 1) > 1$ corresponds to the relaxation regime and $A_0/(M + 1) < 1$ to the lifetime regime. In other words, the presence of traps reduces greatly majority carrier depletion. Thus the condition $A_0 > 1$ for trap free materials implies majority carrier depletion, but the presence of traps can make $A_0/(M+1) < 1$ and thereby lead to majority carrier augmentation. It should be noted that since $\tau_{n0} = 1/C_n N_n = 1/C_n M n_0$, the condition for majority carrier augmentation is $\tau_{dn}/\tau_{n0} < M$ which requires

$$C_n < 1/\tau_{dn} n_0 = q\mu_n/\varepsilon \qquad (5.334)$$

This means that the electron capture coefficient should be lower than the Langevin recombination coefficient in order to produce an excess of majority carriers when minority carriers are injected.

The computations carried out by Popescu and Henisch [1976] have shown that (i) the presence of traps changes the boundary between the majority carrier depletion $(n - n_0 < 0)$ and the majority carrier augmentation $(n - n_0 > 0)$, moving the system more towards the latter; (ii) trapping of injected minority carriers leads to an increase of majority carrier diffusion in the direction opposite to the normal current direction, thus causing the occurrence of a higher than normal resistance; (iii) if a higher than normal resistance occurs anywhere, it will not occur in the majority carrier depletion region, but can be at some location where the free majority carriers have a maximum decreasing slope as shown in Fig. 5.47; (iv) since the main requirement for the occurrence of a higher than normal resistance is the negative majority carrier diffusion gradient, such a situation can happen not only in relaxation semiconductors with traps but also in lifetime semiconductors with traps; (v) a trap free relaxation semiconductor can be brought into the lifetime regime by introduction of traps into it and by minority injection to create the condition for the occurrence of a higher than normal resistance.

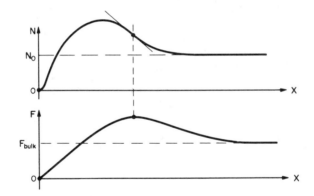

F IG. 5.47. Possible conditions for the occurrence of a higher-than-normal resistance in the presence of majority carrier depletion. [After Popescu and Henisch 1976.]

It should be emphasized that more work, both theoretical and experimental, is required in order to understand fully the behaviour of the relaxation regime and the role it would play in the electrical transport.

5.7.2. Some comparisons of the theory with experiments

There are some experimental results obtained in relaxation semiconductors, which may be related to the theory given in Section 5.7.1. They are now discussed briefly as follows.

(A) Current–voltage characteristics

Roberts [1968] has reported that in the dark current–voltage characteristics of trigonal selenium with indium electrodes there exists a sublinear region following the ohmic region. In this sublinear region $J \propto V^{1/2}$ indicating that there is minority electron carrier injection from the indicum electrode into the p-type selenium which contains traps. Earlier, Ludwig and Zeise [1964] have observed such a sublinear region in zinc sulphide, and Edwards *et al.* [1965] have observed it in aluminium nitride. A typical forward current–voltage characteristic of a single-crystal n-type GaAs (oxygen-doped and compensated crystal) diode with such a sublinear region is shown in Fig. 5.48. This p–n junction diode was formed by diffusing zinc to create a p-layer and indium–gold were evaporated and alloyed to form ohmic contacts on two opposite

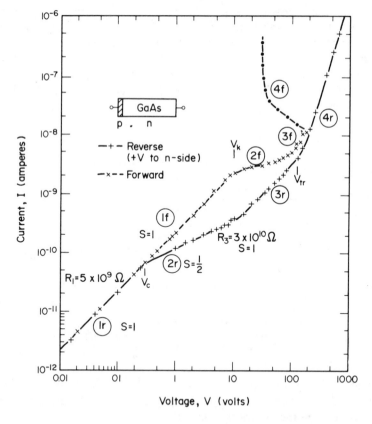

F IG. 5.48. Current–voltage characteristics for a GaAs p–n junction diode in the relaxation regime measured in darkness at temperature 22°C. The solid line is for reverse bias with positive voltage to the n-region. The dashed line is for forward bias. Regions "1r" to "4f" have different resistances indicated by the slopes $S = d \ln I/d \ln V$. V_c, V_k, and V_{tr} denote the critical voltages for the transition from one region to another. [After Queisser, Casey, and van Roosbroeck 1971.]

surfaces [Queisser *et al.* 1971]. In Fig. 5.48, for the sublinear region "2r", $I \propto V^{0.5}$, while for the sublinear region "2f", $I \propto V^{0.4}$. The occurrence of such regions can be attributed to the formation of a majority carrier depletion region, followed with a negative majority carrier diffusion gradient as shown in Fig. 5.47.

It should be noted, however, that an alternative explanation for the occurrence of the sublinear region is also possible. Considering a material specimen provided with two blocking contacts whose effect is masked at low voltages by the series resistance of the specimen, we would expect that at low voltages the *I–V* characteristic curve is linear because most of the voltage is across the bulk series resistance. As the voltage is increased, the depletion layer near the contacts will extend to the bulk and eventually convert the current from being bulk limited to electrode limited. When this occurs, the ideal reverse-biased diode characteristic with $I \propto V^{1/2}$ is followed [Roberts 1973].

(B) Other properties

On the basis of their theoretical analysis for minority carrier injection into a trap free relaxation semiconductor under uniform illumination, Popescu and Henisch [1976] have reported that photoconductive decay, which is generally used to measure the lifetime τ_0 for lifetime semiconductors, does not measure τ_0 for relaxation semiconductors because, besides recombination, the movement of space charge is also involved in the restoration of neutrality.

The Hall effect in relaxation materials is different from that in lifetime materials [Queisser 1971] because in the former the Hall effect may be seriously altered by space charge effects. If a majority carrier depletion occurs, it is likely that a negative Hall coefficient results regardless of the initial doping. This situation may have the bearing on the anomalous Hall effect and anomalous Seebeck effect always observed in amorphous films [Kolomiets and Nazarova 1960, Edmond 1966]. Queisser [1972] has also observed a negative magneto-resistance in relaxation semiconductor GaAs. This phenomenon may also be caused by this majority carrier depletion.

van Roosbroeck [1972, 1973] has explained the switching phenomena in amorphous semiconductors in terms of recombinative injection of minority carriers in the relaxation regime and considered it as a unifying principle for various electronic switching and charge storage effects. However, for switching phenomena, other mechanisms such as the formation of a switching filament and the thermoelectric processes should not be ignored [Saji and Kao 1975, 1976, 1977, Nakashima and Kao 1979].

It should be noted that the above are just some examples of many semiconductor properties which may be associated with the relaxation regime when $\tau_d > \tau_0$. The properties of relaxation materials are much more dominated by recombination statistics and other collective effects, such as the dielectric relaxation time, space charge configuration, etc. Simmons and Taylor [1972], Simmons and Nadkarni [1972], and Nadkarni and Simmons [1972, 1976] have analysed the dielectric relaxation effects on the thermal and isothermal electrical properties in insulators with traps. Further analysis along this direction and further development of experimental techniques to investigate the behaviour of the relaxation regime may lead to new applications of relaxation materials (high-resistivity semiconductors or insulators) and better understanding of the electrical transport in low mobility materials.

CHAPTER 6

Photoelectronic Processes

6.1. BASIC CONCEPTS

The processes which are responsible for a variety of phenomena created due to the absorption or the emission of light in a solid are generally referred to as the photoelectronic processes. Obviously, the limited size of the present chapter cannot treat all the subjects related to these processes in great detail. Therefore, we shall confine ourselves to deal with some subjects in three major topics—photoconduction, photoemission, and photovoltaic effects—which are directly related to electrical transport processes and important to organic semiconductors.

The carrier generation by photoexcitation is usually caused by (i) the transition of electrons from the valence band to the conduction band, (ii) the transition of electrons from localized states at certain energy levels in the forbidden energy gap to the conduction band or from the valence band to unoccupied localized states, and (iii) the production of excitons which then contribute to carriers when the excitons are dissociated or ionized. The photoelectronic processes depend on the quantum yield which is defined as the number of free electron–hole pairs, or the number of free electrons or holes produced per absorbed light quantum. The quantum yield depends on the structure and the imperfections (chemical and structural defects) of the crystal, the temperature, the wavelength of excitation light, the electrical contacts, and the applied field. In organic semiconductors, the probability of the direct transition from band to band is so small that it may be obscured by the molecular absorption processes. For most organic semiconductors the quantum yield is within the range of 10^{-7}–10^{-3}. Such low quantum yield values indicate that the carrier generation must be derived from excitons via some secondary processes, such as exciton dissociation processes which will be discussed later in more detail.

To study the photoelectronic processes it is important to understand the definitions and the basic concepts of some important terms generally used in the literature. These are now given in this section.

6.1.1. Electronic states of molecules

Molecular orbitals are generally polycentric and those of interest and most concern to us are π-orbitals which are formed from the combination of the appropriate p-atomic orbitals. The one-electron molecular orbital wave function ψ_n corresponding to a particular molecular orbital can be expressed as a linear combination of atomic orbital wave functions ϕ_i in the following form:

$$\psi_n = C_1\phi_1 + C_2\phi_2 + \ldots = \sum_i C_i\phi_i \tag{6.1}$$

To determine ψ_n and its corresponding energy level, we have to calculate the coefficients $C_1, C_2, C_3, \ldots$. This can be done by means of the variational method. Since the Schrödinger equation is

$$H\psi_n = E_n\psi_n = E_n\sum_i C_i\phi_i \tag{6.2}$$

the energy E_n can be calculated by

$$E_n = \frac{\int \overline{\psi}_n H\psi_n \, d\tau}{\int \psi_n^2 \, d\tau} \tag{6.3}$$

Using the variational method, E_n is minimized with respect to all the coefficients

$$\left.\begin{array}{c} \dfrac{\partial E_n}{\partial C_1} = 0 \\[2ex] \dfrac{\partial E_n}{\partial C_2} = 0 \\[2ex] \vdots \\[1ex] \dfrac{\partial E_n}{\partial C_i} = 0 \end{array}\right\} \tag{6.4}$$

Thus, substitution of eq. (6.2) into eq. (6.3) introduces the following four important integrals.

(i) Coulomb integral

$$\alpha_i = \int \phi_i H\phi_i \, d\tau \tag{6.5}$$

It represents the energy of a localized electron in an atom and depends on the electronic structure of the atom and its surroundings. The energy may be regarded as the energy required to remove an electron from the ith atom.

(ii) Resonance or exchange integral

$$\beta_{ij} = \int \phi_i H\phi_j \, d\tau \tag{6.6}$$

It represents the energy of the overlap portion of orbitals ϕ_i and ϕ_j, and depends on the amount of overlapping and hence on the interatomic distance. It is called the exchange integral because the electron is changing from the ith to the jth atom and then back to the ith atom continuously around both nuclei involving an interchange of the coordinates of the electrons. This integral may be considered as the energy of coupling for quantum mechanical resonance between two atomic orbitals, which tends to stabilize the bonding of molecular orbital and to destabilize the antibonding of molecular orbital. β_{ij} is a negative quantity.

(iii) Overlap or non-orthogonality integral

$$S_{ij} = \int \phi_i \phi_j \, d\tau \tag{6.7}$$

It measures the extent to which ϕ_i and ϕ_j overlap. $S = 0$ means that the two atoms are far apart or the positive and negative overlap regions cancel out resulting in a zero overlap integral. $S > 0$ means that ϕ_i and ϕ_j overlap considerably leading to a strong bonding, and $S < 0$ means that the combination of ϕ_i and ϕ_j tends to increase the system energy leading to an antibonding.

(iv) Normalization integral

$$S_{ii} = \int \phi_i \phi_i \, d\tau \tag{6.8}$$

When $S_{ii} = 1$ the atomic orbitals are said to be normalized.

It should be noted that electrons can be exchanged not only between atomic orbitals in the same molecule, but also between molecular orbitals (between neighbouring molecules). The molecular exchange integral depends on the amount of overlapping of the molecular orbitals and hence on the molecular shape and the intermolecular distance. In molecular organic crystals only π-electrons which are essentially delocalized can be exchanged between molecules.

There are several approximation methods to calculate the energy levels of various electronic states of molecules. The simplest ones are based on the valence bond theory, molecular orbital theory, and free electron theory. The relatively straightforward and simple method is based on the Hückel molecular orbital approximation. This approximation is based on the following assumptions: (i) all $S_{ii} = 1$, (ii) all $S_{ij} = 0$ if $i \neq j$, (iii) α_i is the same for each atom so $\alpha_i = \alpha$, (iv) $\beta_{ij} = \beta$ if the atom i is bonded directly to the neighbouring atom j, and $\beta_{ij} = 0$ if the atom i is not bonded directly to the atom j. Following this method, it is quite straightforward to calculate ψ_n and their corresponding E_n from eqs. (6.1)–(6.4). Taking butadiene molecule $(CH_2 = CH-CH = CH_2)$ as an example, it can be easily shown that

$$\psi_1 = 0.37\phi_1 + 0.60\phi_2 + 0.60\phi_3 + 0.37\phi_4, \quad E_1 = \alpha + 1.62\beta$$

$$\psi_2 = 0.60\phi_1 + 0.37\phi_2 - 0.37\phi_3 - 0.60\phi_4, \quad E_2 = \alpha + 0.62\beta$$

$$\psi_3 = 0.60\phi_1 - 0.37\phi_2 - 0.37\phi_3 + 0.60\phi_4, \quad E_3 = \alpha - 0.62\beta$$

$$\psi_4 = 0.37\phi_1 - 0.60\phi_2 + 0.60\phi_3 - 0.37\phi_4, \quad E_4 = \alpha - 1.62\beta$$

Since the value of β is negative, the lowest energy level is E_1 and the next energy level is E_2, and each of these two levels can hold two electrons of opposite spins. A butadiene molecule has 4 π-electrons. Thus ψ_1 and ψ_2 are called the bonding orbitals which are occupied in the ground state, while the other two, ψ_3 and ψ_4, with much higher energy levels, E_3 and E_4, are called the antibonding orbitals which are normally empty and are regarded as excited states. The total energy of the 4 π-electrons in the ground state is $E = 2(E_1 + E_2) = 4\alpha + 4.48\beta$. If an electron is localized in the region of one bond between atom 1 and atom 2, its molecular wave function would be $\psi = \phi_1 + \phi_2$ corresponding to energy $\alpha + \beta$. Four electrons of this kind would therefore have total energy $4\alpha + 4\beta$. It can be seen that the difference between this total energy and the total energy of the 4 π-electrons in the ground state for the butadiene molecule is 0.48β. This energy 0.48β is called the delocalization energy resulting from the delocalization of the π-electrons, and it is this energy which stabilizes the structure of the butadiene molecule.

The major weakness of the Hückel's approximation method is that it does not take into account the electron interaction and that the potential of one electron is assumed to be independent of the positions of other electrons. Many approximation methods have been put forward either to modify or to replace the Hückel's method [cf. books of Murrell *et al.* 1965, Salem 1966]. We use this simple Hückel's method in order to show clearly the basic concept of calculating the energy level E_n and the relative π-electron distribution in a molecule. (The density of π-electrons is proportional to $|\psi_n|^2$.)

On the basis of the molecular orbital theory, the number of molecular orbitals is equal to the number of π-electrons. The electronic states of a molecule depends on the coupling of the total orbital angular momentum L with the total spin quantum number of the π-electrons S, which results in a resultant angular momentum J. The possible values of S are $Ns, Ns - 1, Ns - 2, \ldots$, to 0 for N to be even, and to 1/2 for N to be odd, where N is the number of electrons and $s = 1/2$. The possible values of L are 0, 1, 2, 3, ... The quantity $2S + 1$ is referred to as the multiplicity of a state because it gives the quantum mechanically allowed number of component levels (degeneracy) due to spin. The possible values of J are $L + S, L + S - 1, L + S - 2, \ldots$, to $L - S$. Taking the butadiene molecule as an example, in the ground state $S = 0$, and the multiplicity $2S + 1 = 1$, so there·is no degeneracy and this is why this state is called the "ground singlet state, S_0". In the excited singlet state, again $S = 0$ because the transition of an electron from S_0 to S_1 does not involve the change in spin. For this case $2S + 1 = 1$, $L = 1$ and hence $J = 1$, and this excited state is generally called the singlet state or singlet denoted by S_1 or $^1(\pi, \pi^*)$ state. If $S = 1$, then $2S + 1 = 3$, implying that there are threefold degeneracies due to spin. For $L = 1$, three orientations are allowed ($J = 2, 1, 0$), giving rise to three energy levels slightly separated, which are generally referred as the triplet states or triplets denoted by T_1 or $^3(\pi, \pi^*)$. The relative energy levels of the ground and excited states for the butadiene molecule as an example are shown in Fig. 6.1. The triplet energy levels are lower than the corresponding singlet energy level because the former involve a change in spin caused by the spin–orbit coupling. Excited singlet and triplet states coexist in organic molecules. The probability for an electron to be excited from S_0 to S_1 or T_1 depends on the atomic number of the atoms in the molecule. Generally, the higher the atomic numbers (or the heavier the atoms), the higher is the probability for the $S_0 \to T_1$ transition.

Since the magnetic properties depend on the electron spin, a molecule is said to be

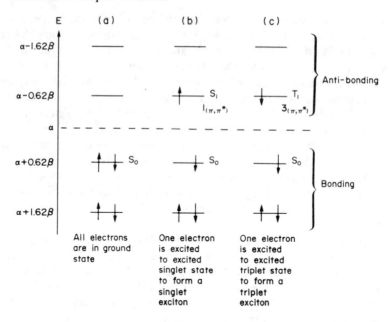

FIG. 6.1. Electron energy levels and states of the butadiene molecule: (a) all electrons in ground state—total spin = 0, multiplicity $2S + 1 = 1$, diamagnetic; (b) one-singlet exciton—total spin = 0, $2S + 1 = 1$, diamagnetic; (c) one-triplet exciton whose energy is lower than the singlet exciton—total spin = 1, $2S + 1 = 3$, paramagnetic.

diamagnetic when its total electron spin $S = 0$, while it is paramagnetic when $S > 0$. Most stable molecules contain an even number of electrons which are paired with opposite spin, $S = 0$, in the ground state; thus they are diamagnetic in the ground state. If they are excited to the triplet state, $S = 1$, then they become paramagnetic.

On the basis of the Hückel's approximation method, eqs. (6.3) and (6.4) will lead to a secular determinant equation. It can be easily shown that the solution of the secular equation will yield

$$E_n = \alpha + 2\beta \cos \frac{n\pi}{A + 1} \tag{6.9}$$

for linear chain hydrocarbons, where A is the number of carbon atoms and $n = 1, 2, 3, \ldots, A$ representing the energy levels. Similarly, it can be shown that for even alternant cyclic (aromatic) hydrocarbons, E_n is

$$E_n = \alpha + 2\beta \cos \frac{2n\pi}{A} \tag{6.10}$$

where $n = 0, \pm 1, \pm 2, \pm 3, \ldots, A - 1$. From eqs. (6.9) and (6.10)·it can be seen that the higher the value of A (the higher the number of π-electrons), the smaller is the coefficient of β. This implies that ΔE between the highest filled state (ground state) and the lowest empty state (first excited state) decreases with increasing A. This is also consistent with the tendency of the first ionization energy, which is that the energy required to remove an electron from the highest filled state (ground state) to infinity decreases with an increase in the number of atoms in the molecule. For example, the first ionization energies are 9.24 eV for benzene (C_6H_6), 8.20 eV for naphthalene ($C_{10}H_8$), 7.60 eV for anthracene ($C_{14}H_{10}$), and 7.15 eV for tetracene (or naphthacene) ($C_{18}H_{12}$). Of course, the structure of the molecule also affects the ionization energy. For example, phenanthrene has the same number of carbon atoms as tetracene but with a bent structure; its ionization energy is 8.04 eV, higher than that of tetracene.

Apart from the most common excited states S_1 and T_1, there are some other electronic states which play an important role in the photoelectronic processes. If a molecule contains atoms with non-bonding electrons, called n-electrons (or lone pairs of electrons) which do not contribute to the bonding of atoms, these n-electrons will add to the π-electron clouds and thus less energy is required to excite an n-electron to the antibonding π^* orbitals to form (n, π^*) states. Hydrocarbons having a substituent N, O, or S, such as acridine $\left(\vcenter{\hbox{\includegraphics{acridine}}}\right)$ and phenazine $\left(\vcenter{\hbox{\includegraphics{phenazine}}}\right)$ with the same molecular skeleton as that of anthracene $\left(\vcenter{\hbox{\includegraphics{anthracene}}}\right)$ but also with one N or two N in it, always have singlet $^1(n, \pi^*)$ states and triplet $^3(n, \pi^*)$ states as shown in Fig. 6.2.

If donor and acceptor groups are present for the intramolecular charge transfer, the charge transfer (CT) states will be formed. The relative energy levels of the states $^1(\pi, \pi^*)$, $^3(\pi, \pi^*)$, $^1(n, \pi^*)$, $^3(n, \pi^*)$, $^1(CT)$ and $^3(CT)$ are shown in Fig. 6.2. More details about the behaviour of excited electrons in those states are given in Figs. 7.1 and 7.3 of Chapter 7.

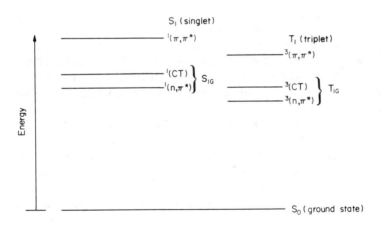

FIG. 6.2. Relative energy levels of $^1(\pi, \pi^*)$, $^3(\pi, \pi^*)$, $^1(CT)$, $^3(CT)$, $^1(n, \pi^*)$, and $^3(n, \pi^*)$ states.

6.1.2. Dissociation of excitons

Since in organic crystals molecules are held together by weak van der Waals forces, the properties of individual molecules are preserved despite that they are bound into a lattice. This distinct feature is generally used to distinguish the organic crystals from the inorganic crystals. In fact, unlike inorganic crystals, the photogeneration of charge carriers in organic crystals is mainly due to the dissociation of excitons, such as in anthracene and tetracene. When an organic crystal specimen is excited by light in the absorption region, Frenkel-type singlet and triplet excitons will be generated. They will move and collide with electrode–crystal interfaces, impurities, or structural defects, causing dissociation of excitons. This process normally creates one type of carriers (say, holes) and capture the other type of carriers (say, electrons) at electrode surfaces or at impurity or structural defect sites. Such an exciton dissociation process is a very efficient channel for non-radiative decay of excitons. Exciton–exciton annihilation or other multiphoton interactions will also lead to the creation of carriers. In the following we shall discuss briefly various exciton dissociation processes.

(A) Exciton dissociation at electrode–semiconductor interfaces

Many investigators have reported that exciton dissociation at electrode surfaces is a common process for carrier generation in organic crystals [Lyons 1955, Kommandeur 1961, Pope and Kallmann 1971, Kallmann *et al.* 1971, Killesreiter and Baessler 1971, 1972, Vaubel *et al.* 1971, Baessler *et al.* 1972]. Excitons produced in the crystal due to the light absorption will diffuse towards the electrodes. If the electrodes act as deep electron traps, then excitons approaching the electrode surface will be dissociated, resulting in subsequent trapping of the electrons and the freeing of the holes. It has been reported [Braun 1968, Johnston and Lyons 1970] that if oxygen is rigorously removed under ultra-high vacuum from a purified anthracene crystal surface, then the free hole density produced by the dissociation of singlet excitons at the crystal surface is markedly reduced, and that reintroduction of oxygen results in a large increase in hole density. This indicates clearly that oxygen adsorbed at the interface plays a dominant role in exciton dissociation to generate holes. It is obvious that the closer to the interface the excitons originate, the better the chance they have of reaching the interface and the more the charge carriers are created by exciton dissociation. This is equivalent to saying that the higher the absorption coefficient of the incident light, the thinner is the layer in which most of the light is absorbed and the closer is this layer (with excitons) to the interface.

Killesreiter *et al.* [1971], Baessler *et al.* [1971], and Vaubel *et al.* [1971] have reported that at the aluminium–anthracene interface a singlet exciton ejects its electron to an empty state in the metal through a tunnelling process, yielding a hole in the anthracene (oxidative dissociation), and that at the aluminium–p–chloranil interface a $^3(n, \pi^*)$ triplet exciton dissociates with one electron injected from the metal to the conduction band of the crystal (reductive dissociation). For electrolyte electrodes, Mehl and Hale [1968], Mulder [1968], and Hale and Mehl [1971] have treated this subject in some detail. In general, the charge transfer reaction between the organic semiconductor and the electrolyte solution, which proceeds slowly in the dark, can be accelerated significantly by illumination of the semiconductor with light of suitable wavelength. Rosseinsky *et al.* [1974] have studied the competition of electron and hole photoinjection from redox electrolytes to anthracene crystals.

The exciton–surface dissociation process can be summarized by the following reaction equations:

$$S_1 + \text{surface} \rightarrow e\,(\text{or } h) \tag{6.11}$$

$$T_1 + \text{surface} \rightarrow e\,(\text{or } h) \tag{6.12}$$

where e and h denote, respectively, the free electron and free hole; surface could be either metal or electrolyte electrode surface, or dye layer surface.

It should be noted that the non-radiative decay of a Frenkel exciton striking the interface not only results in exciton dissociation to produce a free carrier as discussed above—charge transfer process, but also results in non-radiative energy transfer to cause a hot electron or hole injecting from the metal in analogy to the photoemission process—energy transfer process. In fact, these two processes compete with each other. The potential barrier at the interface and the surface conditions determine which process is dominant. It should also be noted that although some investigators [Vaubel *et al.* 1969, 1970, Weisz *et al.* 1973] have suggested the involvement of triplet exciton dissociation at electrode surfaces in the photogeneration of carriers, the inconsistency between the triplet surface impurity interaction and the spectral response of the photocurrent and the related fluorescence quantum yield results does not favour this suggestion.

(B) Exciton dissociation due to interaction of excitons and trapped carriers

A trapped carrier may directly absorb a photon and thereby undergo a transition from the trapping level to one of the bands (electron to the conduction or the hole to the valence band). There is an indirect detrapping process in which an exciton is first produced by the absorption of a photon and it then transfers its energy to a trapped carrier. The efficiency of these detrapping processes for photogeneration of carriers is dependent on the concentration and the energy distribution of traps in the semiconductor, and the intensity and wavelength of the exciting light. Direct photon detrapping should be independent of the wavelength of the exciting light. The excellent correspondence about the spectral response between the triplet absorption and the photogeneration of carriers supports the hypothesis that the excess photocarriers are generated by the interaction of triplet excitons and trapped carriers following the process

$$T_1 + e_t(\text{or } h_t) \xrightarrow{K_{Tct}} S_0 + e(\text{or } h) \tag{6.13}$$

where e_t and h_t are, respectively, the trapped electron and trapped hole, and K_{Tct} is the rate constant for the triplet exciton detrapping. Weisz *et al.* [1969] have determined K_{Tct} for anthracene to be 3.2×10^{-11} cm^3 sec^{-1}, which is of the same order as the triplet–triplet interaction rate constant (5×10^{-11} cm^3 sec^{-1}) indicating that the cross-sections for these two different processes are of the same order. They find little or no evidence for singlet detrapping. A similar conclusion has also been reported by Frankevich and Balabanov [1965] and Geacintov *et al.* [1970] based on their studies of magnetic field effects on the photoconductivity. It is reasonable to say that the carrier generation at the surface is mainly due to singlet exciton dissociation, while the carrier generation in the bulk is due to detrapping by triplet excitons.

The above discussion does not rule out the possibility that singlet excitons interacting with trapped carriers can also cause detrapping. Several investigators [Kalinowski and Godlewski 1974, Schott and Berrehar 1973, Coret and Fort 1974] have studied the involvement of singlet excitons in the detrapping process. Similarly to eq. (6.13), we can write the singlet-exciton detrapping process as

$$S_1 + e_t \,(\text{or } h_t) \xrightarrow{K_{Sct}} S_0 + e \,(\text{or } h) \tag{6.14}$$

where K_{Sct} is the rate constant for the singlet exciton detrapping. If singlet exciton–trapped hole pair formation is diffusion limited, then $K_{Sct} = 4\pi D_S R$, where R is the distance between the centres of two molecules which is about 6×10^{-8} cm for anthracene. Thus $K_{Sct} = 4 \times 10^{-9}$ cm^3 sec^{-1} for $S_1 + h_t \to S_0 + h$ [Kalinowski and Godlewski 1974]. This value is close to the rate constant for the energy transfer of singlet to a guest activator G (e.g. to tetracene, $\alpha_{HG} = 2 \times 10^{-9}$ cm^3 sec^{-1}). This indicates that the cross-section for the singlet exciton–trapped hole interaction does not differ much from that for the singlet exciton–neutral trap interaction, and that the probability for the $S_1 + h_t$ (or e_t) interaction is very close to that for the $S_1 +$ trap interaction.

The detrapping effect in the bulk can also occur due to the interaction of (n, π^*) states with trapped carriers [Oyama and Nakada 1968, Itoh 1975]. It is likely that this process involves $^3(n, \pi^*)$ triplets rather than $^1(n, \pi^*)$ singlets because almost complete intersystem crossing takes place from the lowest $^1(n, \pi^*)$ singlet state to the $^3(n, \pi^*)$ triplet state. The photogeneration of carriers in acridine $\left(\vcenter{\hbox{\includegraphics{acridine}}}\right)$ and phenazine $\left(\vcenter{\hbox{\includegraphics{phenazine}}}\right)$ which contain non-bonding electrons may be due to this detrapping effect [Itoh 1975].

(C) Auto-ionization (AI) processes

The AI processes are generally referred to as the processes in which excitons are produced by the absorption of two or more photons, which then either spontaneously dissociate into free carriers or decay to low-lying non-ionizing states [Pope and Kallmann 1971]. Several mechanisms are responsible for electron–hole pair generation (or intrinsic charge carrier generation) through light absorption in organic crystals, and they are described as follows.

(i) *Exciton auto-ionization.* For direct excitation to a metastable state A, which is above the direct threshold to the conduction band, the decay of the metastable state A will yield free charge carriers. This process may involve one photon or two photons [Bergman and Jortner 1974]

$$S_0 + h\nu \,(\text{or } 2h\nu) \to A \to e + h \tag{6.15}$$

This process competes with the internal conversion back to S_1

$$S_0 + h\nu \,(\text{or } 2h\nu) \to A \to S_1 + \text{phonons} \tag{6.16}$$

With the involvement of excited states located below the direct threshold to the conduction band (about 4 eV for anthracene) the following processes are important for photogeneration of free electron–hole pairs.

(ii) *Exciton photoionization.* The excited states located below E_c can be excited to a metastable state A which is located above E_c. The decay of the metastable state A will yield charge carriers via AI. For the singlet exciton photoionization, the process is [Bergman and Jortner 1974]

$$S_0 + hv \text{ (or } 2hv) \rightarrow S_1 \tag{6.17}$$

$$S_1 + hv \rightarrow A \rightarrow e + h$$

This process is competing with the internal conversion back to S_1

$$S_1 + hv \rightarrow A \rightarrow S_1 + \text{phonons} \tag{6.18}$$

It should be noted that the process given in eq. (6.17) is not limited to the involvement of singlet excitons only. It can be a three-photon process involving also triplet excitons such as

$$\left.\begin{cases} S_0 + 2hv \rightarrow T_1 + T_1 \\ T_1 + T_1 \rightarrow S_1 \\ S_1 + hv \rightarrow A \rightarrow e + h \end{cases}\right\} \tag{6.19}$$

or

$$S_0 + 3hv \rightarrow e + h$$

For the triplet exciton photoionization, the process is [Holzman *et al.* 1967]

$$T_1 + hv \rightarrow A \rightarrow e + h \tag{6.20}$$

Of course, this process is competing with the internal conversion back to T_1

$$T_1 + hv \rightarrow A \rightarrow T_1 + \text{phonons} \tag{6.21}$$

(iii) *Exciton collision ionization.* Two vibrationally relaxed singlet excitons collide and then yield a metastable state A, the decay of which produces charge carriers via AI following the process [Bergman and Jortner 1974]

$$\begin{aligned} S_0 + hv \text{ (or } 2hv) &\rightarrow S_1 \\ 2S_1 &\rightarrow A \rightarrow e + h \end{aligned} \tag{6.22}$$

This process is competing with the internal conversion process

$$2S_1 \rightarrow A \rightarrow S_1 + \text{phonons} \tag{6.23}$$

or

$$2S_1 \rightarrow A \rightarrow T_1^* + T_1^* + \text{phonons}$$

It should be noted that the process given in eq. (6.22) can also be as follows:

$$\left.\begin{cases} S_0 + 2hv \rightarrow T_1 + T_1 \\ T_1 + T_1 \rightarrow S_1 \\ S_1 + S_1 \rightarrow A \rightarrow e + h \end{cases}\right\} \tag{6.24}$$

or

$$S_0 + 4hv \rightarrow e + h$$

Fourny *et al.* [1968] have reported that singlet–triplet collision can also lead to AI following the process

$$S_1 + T_1 \rightarrow A \rightarrow e + h \tag{6.25}$$

Energetically, this is a feasible process since the sum of energies of one singlet and one triplet is about 5 eV in anthracene, which is above the direct threshold to the conduction band. Of course, this process is competing with the internal conversion process

$$S_1 + T_1 \rightarrow T_2^* + S_0 \tag{6.26}$$

The photocarrier generation processes are shown in Fig. 6.3.

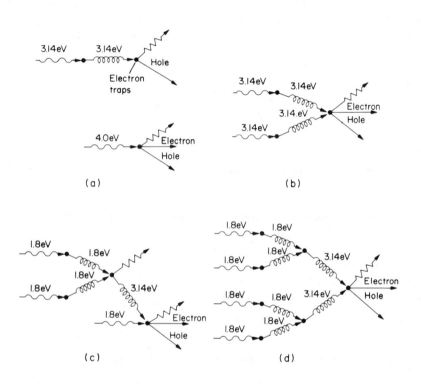

FIG. 6.3. Photocarrier generation processes: (a) one-quantum process, (b) two-quantum process, (c) three-quantum process, and (d) four-quantum process. ──▶ electron or hole; ─ⱮⱮ▶ phonon; ∿∿▶ photon; ∞∞▶ exciton.

6.1.3. Quantum yield and quantum efficiency for photoconduction

Quantum yield η is generally referred to as the number of free electron–hole pairs generated per light quantum absorbed for intrinsic photoconduction, and as the number of free electrons or holes per light quantum absorbed for extrinsic photoconduction.

Quantum efficiency g_{ph} is defined as the ratio of the number of charge carriers generated by photoexcitation passing through a solid between two electrodes to the number of light quanta absorbed by this solid during the same period of time [Rose

1963]. Supposing that in the steady state, free electrons of density Δn and free holes of density Δp are generated during the continuous photoexcitation, then we can write

$$\Delta n = G_n \tau_n \tag{6.27}$$

$$\Delta p = G_p \tau_p \tag{6.28}$$

where G_n and G_p are, respectively, the generation rates of electrons and holes per unit volume and τ_n and τ_p are, respectively, the lifetimes of electrons and holes. For perfect crystals, both the excitation and the recombination are of a direct band-to-band process, and in this case $\Delta n = \Delta p$, $G_n = G_p$ and hence $\tau_n = \tau_p$ (intrinsic photoconduction). For real crystals in which there are traps and recombination centres introduced by imperfections (chemical or structural), then $\Delta n \neq \Delta p$, $G_n \neq G_p$, and hence $\tau_n \neq \tau_p$ (extrinsic photoconduction).

For the general case the photocurrent density can be written as

$$J_{\text{ph}} = q\left(\frac{G_n \tau_n d}{t_{tn}} + \frac{G_p \tau_p d}{t_{tp}}\right) \tag{6.29}$$

or

$$J_{\text{ph}} = q(G_n X_n + G_p X_p) \tag{6.30}$$

where t_{tn} and t_{tp} are, respectively, the transit time for an electron and a hole to travel across the specimen of thickness d; X_n and X_p are the distances which an electron and a hole have travelled in the direction of the electric field F during their lifetime before they are trapped. Since

$$X_n = \tau_n \mu_n F, \qquad t_{tn} = d/\mu_n F \tag{6.31}$$

$$X_p = \tau_p \mu_p F, \qquad t_{tp} = d/\mu_p F$$

it can be seen that at low fields J_{ph} is proportional to F because in this case X_n, $X_p < d$, and τ_n, $\tau_p < t_{tn}$, t_{tp}. At high fields such that $\tau_n \simeq t_{tn}$ and $\tau_p \simeq t_{tp}$, the photocurrent becomes saturated at a constant value (independent of applied field) if both the cathode and the anode electrodes are of blocking contacts (this means a contact collecting holes cannot inject electrons and vice versa; cf. Chapter 2).

The quantum efficiency g_{ph} (called the photoconductive gain) can be expressed as

$$
\begin{aligned}
g_{\text{ph}} &= \frac{J_{\text{ph}}/q}{(G_n/\eta_n + G_p/\eta_p)d} \\
&= \frac{G_n X_n + G_p X_p}{[G_n/\eta_n + G_p/\eta_p]d} \\
&= \frac{G_n(\tau_n/t_{tn}) + G_p(\tau_p/t_{tp})}{G_n/\eta_n + G_p/\eta_p}
\end{aligned} \tag{6.32}
$$

where η_n and η_p are, respectively, the quantum yields for electrons and holes. It is obvious that with blocking contacts the maximum value of $X_n + X_p$ is d. If it is assumed that $G_n = G_p$ and $\eta_n = \eta_p = \eta$, the maximum gain is $g_{\text{ph}} = \eta$.

If both electrodes are of ohmic contacts (this implies that the cathode will inject electrons and the anode holes to the semiconductor), then the situation is quite different. In this case the gain can be higher than unity (if η is assumed to be unity). The charge carriers injected into the semiconductor from the electrodes are continuously neutralized by dielectric relaxation if the carrier transit time is larger than the dielectric relaxation time τ_d, and under this condition there is negligible space charge near the electrodes and the photocurrent is linearly proportional to the applied voltage. If the applied voltage is increased to such values that the carrier transit time becomes equal to τ_d, and both decrease together with increasing applied voltage, then the photocurrent becomes space-charge-limited (SCL) and behaves in a manner similar to that described in Chapters 3 and 4. If the semiconductor is a perfect crystal free of traps, the condition for the onset of the SCL photocurrent is the carrier transit time equal to τ_d. If the semiconductor is a real crystal containing traps, the threshold voltage for the onset of the SCL current increases because the presence of traps reduces the average carrier mobility and makes the experimentally observed decay time of the photocurrent after the removal of excitation longer than the carrier lifetime (in trap free perfect crystals the observed decay time is equal to the carrier lifetime). This is due to two reasons: (a) the thermally released carriers from traps prolong the observed decay time if n_t and p_t (trapped electron and hole concentrations) are larger than the corresponding Δn and Δp, and (b) the traps reduce the effective drift mobility and carrier lifetime.

Supposing that $\Delta p \gg \Delta n$ and the minority carriers (electrons) are trapped immediately by traps after excitation, then eq. (6.32) reduces to

$$g_{ph} = \eta_p(\tau_p/t_{tp}) \tag{6.33}$$

Assuming $\eta_p = 1$, $g_{ph} = \tau_p/t_{tp}$. This indicates the importance of the carrier lifetime for the photosensitivity of a solid. It is obvious that the value of g_{ph} is field dependent, temperature dependent, specimen thickness dependent, and purity and structure dependent even for a given material. In Chapter 4 we have discussed the carrier lifetimes for perfect crystals free of traps, $\tau_p = (\langle v\sigma_p \rangle \Delta n)^{-1}$, and for real crystals with $\tau_p = (\langle v\sigma_p \rangle n_r)^{-1}$, where n_r is the electron-occupied trap concentration ready to capture holes (occupied recombination centre concentration). If $n_r > \Delta n$, then τ_p (with recombination centres) is about $\Delta n/n_r$ times τ_p (without recombination centres). If the semiconductor contains also shallow hole traps, then the effective hole mobility is decreased and its transit time is increased. For this case, the transit time has to be reduced to about $t_{tp} = (\Delta p/p_t)\tau_d$ or the applied voltage has to raise to $p_t/\Delta p$ times that of the case without traps for the onset of the SCL photocurrent [Bube 1960, Rose 1963]. For semiconductors with ohmic contacts the maximum gain for photoconduction is set when the SCL current starts to be comparable to the photocurrent. Thus

$$(g_{ph})_{max} = \frac{\tau_p}{t_{tp}} = \frac{\tau_p}{(\Delta p/p_t)\tau_d}$$

$$= \frac{\tau_{rp}}{\tau_d} \tag{6.34}$$

where τ_{rp} is the observed decay time or the response time which is

$$\tau_{rp} = (p_t/\Delta p)\tau_p \tag{6.35}$$

For most cases the observed decay time τ_{rp} does not exceed τ_d, so $(g_{ph})_{max}$ is limited to unity. But $(g_{ph})_{max} > 1$ can be expected if the energy levels of traps are such that the effect of p_t on t_{tp} is greater than the effect of p_t on τ_{rp}. Thus we can rewrite eq. (6.34) as [Rose 1963]

$$(g_{ph})_{max} = \frac{\tau_{rp}}{\tau_d} \frac{(p_t)_{\text{SCL injection}}}{(p_t)_{\text{light excitation}}} \tag{6.36}$$

$$= \frac{\tau_{rp}}{\tau_d} M$$

where $(p_t)_{\text{SCL injection}}$ is the filled trap density due to SCL injection and tends to increase the transit time, and $(p_t)_{\text{light excitation}}$ is the filled trap density due to light excitation and tends to increase the observed decay time. It is obvious that M can be made larger than unity.

6.1.4. Non-equilibrium steady state statistics

To analyse theoretically the photoelectronic processes, it is necessary to know the non-equilibirium steady state statistics which govern all significant excitation processes as well as all radiative and non-radiative recombination processes. The absorption region of most real crystals always extends beyond the absorption edge towards the long wavelengths as a tail of the absorption. Such a tail region indicates that in the case of organic semiconductors and crystalline insulators the energy levels of trapping and recombination centres are distributed throughout the forbidden energy band gap. In Section 4.2 we have discussed the kinetics of recombination processes based on the assumption that the trapping or the recombination centres are confined in dominant discrete energy levels. This assumption is a rough approximation and valid only for some cases. However, it helps to reveal clearly the physical picture of these processes without involving complicated mathematics.

Any imperfection site can act as a trap or as a recombination centre depending on its location in the energy band gap with respect to the demarcation levels [cf. Section 4.1.3]. Rose [1963] has used the quasi-Fermi levels E_{Fn} and E_{Fp} for the determination of free electron and hole concentrations, and the demarcation levels E_{Dn} and E_{Dp} for the distinction between traps and recombination centres. Referring Section 4.2.2 and Fig. 4.4 for semiconductors with a single set of recombination centres located between E_{Dn} and E_{Dp}, the rate of electron capture is

$$\gamma_a = C_n n N_r (1 - f_r) \tag{6.37}$$

the rate of thermal re-excitation of the captured electrons is

$$\gamma_b = e_n N_r f_r \tag{6.38}$$

the rate of hole capture is

$$\gamma_c = C_p p N_r f_r \tag{6.39}$$

and the rate of thermal re-excitation of the captured holes is

$$\gamma_d = e_p N_r (1 - f_r) \tag{6.40}$$

where

$$C_n = \langle v\sigma_n \rangle \tag{6.41}$$

$$C_p = \langle v\sigma_p \rangle \tag{6.42}$$

$$e_n = C_n N_c \exp[-(E_c - E_r)/kT] = C_n n_1 \tag{6.43}$$

$$e_p = C_p N_v \exp[-(E_r - E_v)/kT] = C_p p_1 \tag{6.44}$$

and f_r is the probability for a recombination centre to be occupied. In the thermal equilibrium steady state

$$\gamma_a = \gamma_b \qquad \gamma_c = \gamma_d \tag{6.45}$$

Thus we have

$$f_{r0} = \{1 + \exp[(E_r - E_{F0})/kT]\}^{-1} \tag{6.46}$$

where E_{F0} is the equilibrium Fermi level.

In the non-equilibrium steady state, the condition becomes

$$\gamma_a - \gamma_b = \gamma_c - \gamma_d \tag{6.47}$$

and f_r becomes

$$f_r = \frac{C_n n + C_p p_1}{C_n n + C_n n_1 + C_p p + C_p p_1} \tag{6.48}$$

The physical concept of $C_n n$ (or $C_p p$) is that it represents the rate of electron capture (or hole capture) per unfilled (or filled) recombination centre. The physical concept of $C_n n_1$ (or $C_p p_1$) is that it represents the rate of re-excitation of the captured electrons (or the captured holes) into the conduction (or the valence) band per filled (or unfilled) recombination centre. In other words, $C_n n$ is the probability for an electron to be captured by an unfilled centre per unit time, while $C_n n_1$ is the probability for re-excitation of a captured electron into the conduction band per unit time. A similar description can be easily applied to $C_p p$ and $C_p p_1$.

Supposing that the localized states are not confined in a single discrete level E_r but are distributed with the concentration per unit volume per unit energy $h(E)$, then for those at a particular energy level $E_j > E_{F0}$, it is obvious that in this case we can neglect $C_p p_1$ in eq. (6.48) since

$$C_p p_1 < C_n n_1$$

$$C_p p_1 < C_n n, \ C_p p \tag{6.49}$$

Thus eq. (6.48) can be written as

$$f_r(E_j) = \frac{C_n n}{C_n n_1 + C_n n + C_p p}$$

$$= \frac{C_n n}{C_n n + C_p p} \left[1 \bigg/ \left(1 + \frac{C_n n_1}{C_n n + C_p p} \right) \right]$$

$$= \frac{C_n n}{C_n n + C_p p} \left\{ \frac{1}{1 + \exp[(E_j - E'_{Fn})/kT]} \right\} \tag{6.50}$$

Simmons and Taylor [1971] were the first to derive eq. (6.50). They called E_{Fn}^t the quasi-Fermi level for trapped electrons, $C_n n(C_n n + C_p p)^{-1}$ the modulation factor, and $\{1 + \exp[(E_j - E_{Fn}^t)/kT]\}^{-1}$ the Fermi–Dirac function about an energy E_{Fn}^t. This means that at E_{Fn}^t we have

$$C_n n_1 = C_n n + C_p p \tag{6.51}$$

By assuming that the thermal velocity for electrons is equal to that for holes, it can be easily shown that

$$E_{Fn}^t = E_{F0} + kT \ln[(n\sigma_n + p\sigma_p)/n_0 \sigma_n]$$
$$= E_{Fn} + kT \ln[(n\sigma_n + p\sigma_p)/n\sigma_n] \tag{6.52}$$

where n_0 is the thermal equilibrium electron concentration which is

$$n_0 = N_c \exp[-(E_c - E_{F0})/kT] \tag{6.53}$$

and n and p are, respectively, the non-equilibrium electron and hole concentrations, and they are

$$n = N_c \exp[-(E_c - E_{Fn})/kT] \tag{6.54}$$
$$p = N_v \exp[-(E_{Fp} - E_v)/kT] \tag{6.55}$$

It can be seen that E_{Fn}^t is always located above the demarcation level E_{Dn} which is determined on the basis of $C_n n_1 = C_p p$ and also above E_{Fn}. The physical significance of eq. (6.51) is that E_{Fn}^t is located at the level at which the net probability for thermal re-excitation of a trapped electron to the conduction band $[(C_n n_1 - C_n n) = $ re-excitation probability minus trapping probability for electrons] is equal to the probability for recombination of a trapped electron with a hole from the valence band. According to Simmons and Taylor [Simmons and Taylor 1971, 1975, Taylor and Simmons 1972, 1975] the quasi-Fermi level for trapped electrons provides a better way of distinguishing between shallow traps and recombination centres. This implies that for traps at energy levels above E_{Fn}^t, eq. (6.50) can be approximated to

$$f_r(E_j) = \frac{C_n n}{C_n n + C_p p} \exp[-(E_j - E_{Fn}^t)/kT] \tag{6.56}$$

This means that the occupancy of traps follows a simple Boltzmann's distribution and they are essentially in thermal equilibrium with the conduction band. For traps at energy levels below E_{Fn}, eq. (6.50) can be approximated to

$$f_r(E_j) = \frac{C_n n}{C_n n + C_p p} \tag{6.57}$$

This means that the traps lose their thermal communication with the conduction band and the occupancy becomes independent of the trapping energy level E_j. Under this condition the trapped electrons will remain in the traps until they recombine with holes from the valence band so that the traps located below E_{Fn}^t act essentially as recombination centres.

Using similar procedures we can easily treat the case with arbitrary trapping energy

E_i below E_{F0}. For this case $C_n n_1$ in eq. (6.48) can be neglected since

$$C_n n_1 < C_p p_1$$

$$C_n n_1 < C_n n, C_p p \tag{6.58}$$

Then from eq. (6.48) the probability for hole occupancy (emptying a filled trap) is

$$1 - f_r = \frac{C_n n_1 + C_p p}{C_n n + C_n n_1 + C_p p + C_p p_1}$$

$$\simeq \frac{C_p p}{C_n n + C_p p + C_p p_1}$$

$$= \frac{C_p p}{C_n n + C_p p} \left[1 \bigg/ \left(1 + \frac{C_p p_1}{C_n n + C_p p} \right) \right]$$

$$= \frac{C_p p}{C_n n + C_p p} \{ 1 + \exp[(E'_{Fp} - E_i)/kT] \}^{-1} \tag{6.59}$$

where E'_{Fp} is the quasi-Fermi level for trapped holes which is given by

$$E'_{Fp} = E_{F0} - kT \ln[(n\sigma_n + p\sigma_p)/p_0 \sigma_p]$$

$$= E_{Fp} - kT \ln[(n\sigma_n + p\sigma_p)/p\sigma_p] \tag{6.60}$$

where p_0 is the thermal equilibrium hole concentration given by

$$p_0 = N_v \exp[-(E_{F0} - E_v)/kT] \tag{6.61}$$

The remarks for f_r and E'_{Fn} for $E_j > E_{F0}$ pertain to $1 - f_r$ and E'_{Fp} for $E_i < E_{F0}$, and they are (a) $E'_{Fp} < E_{Fp} < E_{Dp}$, and (b) traps located at energy levels below E'_{Fp} act as shallow hole traps and those above E'_{Fp} act essentially as recombination centres.

If the trapping levels are distributed or there are many trapping levels within the energy band gap as shown in Fig. 6.4, and if the traps become negative when they are occupied by electrons, and become positive when they capture holes, then under non-equilibrium conditions the total number of trapped electrons is

$$n_t = \int_{E_{F0}}^{E'_{Fn}} h(E) f_r dE + \int_{E'_{Fn}}^{E_c} h(E) f_r dE$$

$$= \mathrm{I} + \mathrm{II}$$

$$= \text{trapped electrons in recombination centres (I)}$$
$$+ \text{trapped electrons in shallow traps (II)} \tag{6.62}$$

and the total number of trapped holes is

$$p_t = \int_{E'_{Fp}}^{E_{F0}} h(E)[1 - f_r] dE + \int_{E_v}^{E'_{Fp}} h(E)[1 - f_r] dE$$

$$= \mathrm{III} + \mathrm{IV}$$

$$= \text{trapped holes in recombination centres (III)}$$
$$+ \text{trapped holes in shallow traps (IV)} \tag{6.63}$$

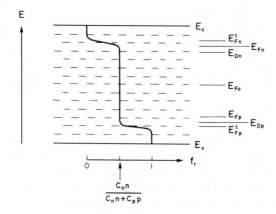

FIG. 6.4. The relative energy levels of E_{Fn}^t, E_{Fp}^t, E_{Fn}, E_{Fp}, E_{Dn}, and E_{Dp}, and the values of f_r for an arbitrary distribution of traps.

Under thermal equilibrium conditions, eqs. (6.62) and (6.63) become

$$n_{t0} = \int_{E_{F0}}^{E_{Fn}^t} h(E) f_{r0} \, dE + \int_{E_{Fn}^t}^{E_c} h(E) f_{r0} \, dE$$

$$= \mathrm{I}_0 + \mathrm{II}_0 \tag{6.64}$$

and

$$p_{t0} = \int_{E_{Fp}^t}^{E_{F0}} h(E) [1 - f_{r0}] \, dE + \int_{E_v}^{E_{Fp}^t} h(E) [1 - f_{r0}] \, dE$$

$$= \mathrm{III}_0 + \mathrm{IV}_0 \tag{6.65}$$

The non-equilibrium conditions are generally created by carrier injection into the semiconductor from an external source, such as photoexcitation, single- or double-carrier injection from electrical contacts (electrodes). The densities of excess electrons and holes generated by photoexcitation are

$$\Delta n = n - n_0 \tag{6.66}$$

$$\Delta p = p - p_0 \tag{6.67}$$

The portion of photogenerated trapped electrons $\mathrm{I} - \mathrm{I}_0$ [cf. eqs. (6.62) and (6.64)] is ready to recombine with holes, and the portion of photogenerated holes $\mathrm{III} - \mathrm{III}_0$ [cf. eqs. (6.63) and (6.65)] is ready to recombine with electrons. If only the traps located between E_{Fp}^t and E_{Fn}^t are considered to be recombination centres, the recombination rate can be expressed as

$$R = \frac{C_n n C_p p}{C_n n + C_p p} \int_{E_{Fp}^t}^{E_{Fn}^t} h(E) \, dE \tag{6.68}$$

For this case the charge neutrality condition imposes the relation

$$\left.\begin{array}{c} \Delta n + n_t - n_{t0} = \Delta p + p_t - p_{t0} \\[2mm] n_0 + n_{t0} = p_0 + p_{t0} \end{array}\right\} \tag{6.69}$$

It should be noted that the capture cross-sections for electrons and holes alone are sufficient to deal with non-equilibrium steady state statistics, but they are not sufficient for setting the charge neutrality condition and for calculating the field distribution due to the formation of space charge caused either by carrier injection from electrodes, or by non-uniform photoexcitation or high field effects. For example, eqs. (6.62)–(6.65) and (6.69) are valid for the cases with traps to be neutral before capturing either electrons or holes. If the traps are of acceptor-type, these traps become negatively charged when they have captured electrons. These traps will normally not capture holes when they are empty because the hole capture cross-section is extremely small. However, when such traps are filled with trapped electrons, they will caputre holes with a much larger hole capture cross-section, but under this condition the traps become neutral again after capturing holes. In this case eqs. (6.62) and (6.64) are still valid, but eqs. (6.63) and (6.65) have to be interpreted as the number of neutral traps (or the number of filled traps being empty). In electric charge, they are neutral. Therefore the charge neutrality condition becomes

$$\Delta n + n_t - (n_{t0} + N_{A0}^-) + \left[\int_{E_r}^{E_{F0}} h(E)dE - p_t \right] = \Delta p$$

(6.70)

$$n_0 + (n_{t0} + N_{A0}^-) = p_0$$

where

$$N_{A0}^- = \int_{E_r}^{E_{F0}} h(E)[1 + g_n^{-1} \exp(E - E_{F0})/kT]^{-1} dE$$

(6.71)

In this case though the expression for p_t in eq. (6.70) is the same as eq. (6.63) it is neutral in charge. It can be said that $\int_{E_r}^{E_{F0}} h(E)dE$ is the electron-filled trap density below E_{F0} and negatively charged. Under non-equilibrium conditions this portion of the electron-filled trap density is reduced by a quantity of p_t. Generally, eq. (6.68) for the recombination rate is valid also for this type of trap. Similar remarks pertain to donor-type traps.

It should also be noted that in Chapter 4 we have discussed the demarcation levels for the distinction between traps and recombination centres for solving double injection problems. It is possible that the use of the quasi-Fermi levels for trapped electrons and holes instead of the demarcation levels may be better for treating double injection problems.

6.2. PHOTOCONDUCTION

In perfect crystals completely free of traps, it is easily shown that the steady state carrier concentration is

$$n = p = (G/\gamma)^{1/2}$$

(6.72)

where γ is the biomolecular (band-to-band) recombination rate constant (or recombination coefficient), and G is the photogeneration rate of electron–hole pairs. The recombination rate is given by

$$R = \gamma np = \gamma n^2 = \gamma p^2 \qquad (6.73)$$

The steady state photoconductivity is given by

$$\sigma_{ph} = q(G/\gamma)^{1/2}(\mu_n + \mu_p) \qquad (6.74)$$

and the steady state photocurrent by

$$J_{ph} = q(G/\gamma)^{1/2}(\mu_n + \mu_p) F_{av} \qquad (6.75)$$

where F_{av} is the average applied electric field which is equal to V/d at low fields.

In real crystals containing various traps, the recombination through the traps becomes dominant and the calculation of n or p becomes quite mathematically involved. Rose [1951, 1955, 1961, 1963] was the first to give insight into the understanding of photoconduction in real materials. For an arbitrary distribution of traps as shown in Fig. 6.4 we can write the following rate equations:

$$\frac{\partial n}{\partial t} = G + \int_{E_v}^{E_c} e_n h(E) f_r(E) \, dE - \int_{E_v}^{E_c} C_n n \, h(E) [1 - f_r(E)] dE - \nabla J_n/q \qquad (6.76)$$

$$\frac{\partial p}{\partial t} = G + \int_{E_v}^{E_c} e_p h(E) [1 - f_r(E)] dE - \int_{E_v}^{E_c} C_p p \, h(E) f_r(E) dE - \nabla J_p/q \qquad (6.77)$$

$$J_n = q\mu_n n F + qD_n \nabla n \qquad (6.78)$$

$$J_p = q\mu_p p F - qD_p \nabla p \qquad (6.79)$$

$$J_{ph} = J_n + J_p \qquad (6.80)$$

$$\nabla F = q(p + p_t - n - n_t) \qquad (6.81)$$

where n_t and p_t are given by eqs. (6.62) and (6.63). In these equations we have ignored the bimolecular (band-to-band) recombination term because this term is suppressed to a negligibly small magnitude by the traps. The first term on the right-hand side of eq. (6.76) or eq. (6.77) represents the density of carriers generated per unit time by the light excitation, the second term represents the rate of re-excitation of trapped carriers, and the third term represents the carrier loss due to trapping in the shallow traps and in the recombination centres.

The solution of the above equations [eqs. (6.76)–(6.81)] for a set of specific boundary conditions is generally formidable even with the aid of numerical techniques. Therefore, it is necessary to make certain assumptions in order to reduce the mathematical complexity. By neglecting the diffusion effect for the case of uniform excitation at low applied fields, the steady state recombination rate is equal to the photocarrier generation rate

$$R = G = \int_{E_v}^{E_c} \{C_n n \, h(E) [1 - f_r(E)] - e_n h(E) f_r(E)\} dE \qquad (6.82)$$

Using charge neutrality conditions given in eqs. (6.69) and (6.82), n and p can be determined and the photocurrent

$$J_{ph} = q(\mu_n n + \mu_p p) F \qquad (6.83)$$

can also be determined.

It should be noted that eqs. (6.75) and (6.83) include the background current due to thermally generated n_0 and p_0, and that the above is only the outline of the general procedures for treating photoconduction problems. However, it is clear that if the carrier mobilities μ_n and μ_p are assumed to be unaffected by light excitation, the photocurrent depends only on how the free carriers are produced. The photocurrent is always dependent on the wavelength and intensity of the exciting light, temperature, applied fields (electric and magnetic), the surface conditions, and the ambient atmosphere, because n and p are dependent on these parameters. Thus the study of photoconduction under various experimental conditions would reveal the mechanisms for photocarrier generation and the information about the distribution and concentration of traps. In general, the photocarrier generation results from one or more of the following processes:

I. The processes involving the dissociation of excitons:
    (a) exciton dissociation at electrode–semiconductor interface;
    (b) exciton dissociation due to the interaction of excitons and trapped carriers;
    (c) auto-ionization processes.
II. The processes without involving excitons:
    (d) direct excitation of trapped carriers into the conduction or valence bands;
    (e) direct band-to-band transition.

Processes (a), (b), and (c) have been discussed in Section 6.1.2. Processes (d) and (e) are self-explanatory and are the main sources for photocarriers in inorganic semiconductors.

Photoconduction can be divided into (i) extrinsic photoconduction and (ii) intrinsic photoconduction depending on what are the dominant photocarrier generation processes. The former is mainly due to processes (a), (b), and (d), and usually occurs at lower excitation energies (longer wavelengths), while the latter is mainly due to processes (c) and (e) with a very low quantum efficiency and strong field dependence. In the following we shall discuss these two types of photoconduction.

6.2.1. Extrinsic photoconduction

Extrinsic photoconduction is generally unipolar, that is, involves mainly one type of carrier (either electrons or holes are dominant). The photogeneration processes for extrinsic photoconduction are discussed as follows:

(A) Processes involving trapping and recombination centres

(a) RISE TIME AND DECAY TIME OF PHOTOCURRENTS

In a semiconductor the time required for the photoconductivity (or the photocurrent) to change with changes in exciting light intensity is determined by the free carrier lifetime. The occurrence of the rise time and decay time much greater than the carrier lifetime is caused by the trapping of free carriers. A rise time is required for the traps to capture photogenerated free carriers and for the steady state to be established between the new density of free carriers and the new occupancy of traps after an exciting light is switched on, and a decay time is required for the trapped carriers to be thermally released after the excitation is terminated.

The observed decay time (or response time) can be related to the carrier lifetime by the following approximate equation [Rose 1963]:

$$\tau_{rn} = \tau_n(1 + n_t/n) \tag{6.84}$$

$$\tau_{rp} = \tau_p(1 + p_t/p) \tag{6.85}$$

depending on whether electrons or holes are the dominant carriers. Only for perfect crystals free of traps is the decay time equal to the lifetime. In general, the lifetimes τ_n and τ_p are insensitive to light intensity. If $p_t \gg p$ for the hole dominant photoconductors, eq. (6.85) reduces to

$$\tau_{rp} = \tau_p(p_t/p) \tag{6.86}$$

From eq. (6.28), we can obtain

$$p_t = G_p\tau_{rp} \tag{6.87}$$

assuming $\Delta p = p - p_0 \simeq p$. Thus we can estimate the concentration of trapped carriers. Since p_t is very sensitive to both temperature and light intensity, τ_{rp} is expected to decrease with increasing temperature and increasing light intensity. The measurements of the photocurrent decay as a function of time at various temperatures, wavelengths, and intensities of the exciting light will allow the determination of trap parameters. For details about the response time the reader is referred to Rose's work [1951, 1963].

(b) PHOTOCURRENT-VOLTAGE CHARACTERISTICS

(i) With carrier-injecting ohmic contacts

At low fields the photocurrent increases linearly with increasing field, following closely Ohm's law. At higher fields the carriers injected from electrodes will become appreciable and add to the carriers generated by light excitation. For organic semiconductors, the energy band gap is usually quite large and therefore the concentration of thermally generated carriers (n_0 or p_0) is very small and in most cases can be neglected. If the photogenerated carrier density is higher than the injected carrier density from the electrodes, the J–V characteristics still obey Ohm's law. This implies that the critical voltage for the transition from the linear ohmic region to the superlinear SCL region V_Ω shifts towards a higher voltage as the exciting light intensity is increased. Such a shift has been observed, for example, in triphenylene [Almeleh and Harrison 1966]. Studies of photoenhanced SCL currents in anthracene and tetracene crystals have clearly shown that the long wavelength photoconductivity in the wavelength region beyond the S–T (singlet–triplet) absorption is determined by direct detrapping (or photoionization of trapped carriers) [Jansen *et al.* 1964, Helfrich 1967, Weisz *et al.* 1964, 1971, Benderskii and Lavrushko 1971, Belkind *et al.* 1972].

For the traps confined in a single discrete energy level the SCL current in the dark is given by [cf. eq. (3.22)]

$$J = \frac{9}{8} \varepsilon \mu_p \theta_a \frac{V^2}{d^3}$$ (6.88)

If the exciting light intensity is low, the thermal detrapping dominates the detrapping processes and the SCL current will be the same as in the dark. When the light intensity is increased to such a level that the optical detrapping becomes predominant, then the current depends on the light intensity and θ_a is no longer a function of temperature but rather a function of light intensity. To obtain expressions for SCL photocurrent, it is necessary to know the amount of charge carriers remaining trapped after photo-excitation. This can be calculated by assuming that there is a demarcation level E_{DOp} below which the trapped carriers (trapped holes) are released mainly by the thermal process, and above which the trapped holes are released mainly by the optical detrapping [Helfrich 1964]. Obviously, at E_{DOp} the rate of thermal excitation is equal to the rate of optical excitation from the traps. Considering the hole traps to be distributed

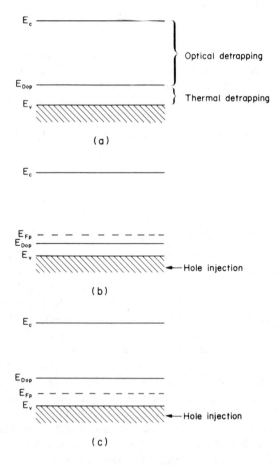

FIG. 6.5. Illustrating the three general cases: (a) thermal equilibrium, (b) low injection $E_{Dop} < E_{Fp}$, and (c) high injection $E_{Dop} > E_{Fp}$.

exponentially [cf. eq. (3.41) of Chapter 3] the trapped carriers located above E_{DOp} with $E_{DOp} < E_{Fp}$ for the low injection case can be released by light as shown in Fig. 6.5(b), resulting in an increase in current under light excitation. If the injection is high such that $E_{DOp} > E_{Fp}$ as shown in Fig. 6.5(c), the concentration of trapped holes above E_{DOp} is small and the amount of free carriers released by light is small so that the effect of light is negligible.

The demarcation level E_{DOp} is defined as the level at which the rate of thermal re-excitation of the trapped holes γ_{th} equals the rate of optical detrapping of the trapped holes γ_{0p}.

$$\gamma_{th} = \gamma_{0p}$$

or

$$p_t v_p \exp\left[-(E_t - E_v)/kT\right] = p_t \sigma_{0p} i(h\nu) \qquad (6.89)$$

which leads to

$$E_{DOp} = E_t^* = E_v + kT \ln\left[v_p/\sigma_{0p} i(h\nu)\right] \qquad (6.90)$$

where v_p is the attempt-to-escape frequency [cf. Section 4.1.3], σ_{0p} and i are, respectively, the total photoionization cross-section and the exciting light intensity (photon flux density) at a given photon energy $h\nu$. Since the total detrapping rate is $\gamma_{th} + \gamma_{0p}$, the SCL current can be increased by the light excitation only when

$$i(h\nu) > (v_p/\sigma_{0p}) \exp\left[-(E_t - E_v)/kT\right] \qquad (6.91)$$

If eq. (6.91) is satisfied, eq. (6.88) is valid for SCL photocurrent provided that θ_a is changed to θ_a', which is given by [Lampert and Mark 1970]

$$\theta_a' = \sigma_{0p} i / N_t^* \langle v\sigma_p \rangle \qquad (6.92)$$

where N_t^* is the effective trap concentration.

On the basis of the same principle, by replacing θ_a with a new θ_a' of eq. (6.88), it can easily be shown that for the case with the traps distributed exponentially, the effective trap concentration can be calculated from eq. (3.41) as follows:

$$N_t^* = \int_{E_v}^{E_{F0}} h(E)\, dE$$

$$= \int_{E_v}^{E_{F0}} \frac{H_b}{kT_c} \exp\left(-\frac{E}{kT_c}\right) dE$$

$$\simeq H_b \exp\left[(E_v - E_{F0})/kT_c\right] \qquad (6.93)$$

From eqs. (6.90), (6.92), and (6.93), and then (6.88), the SCL photocurrent is given by [Helfrich 1964, Lampert and Mark 1970]

$$J = \frac{9\varepsilon\mu_p N_v^{1/l}}{8} \frac{}{H_b} \left(\frac{\sigma_{0p} i}{\langle v\sigma_p \rangle}\right)^{(l-1)/l} \frac{V^2}{d^3} \qquad (6.94)$$

where $l = T_c/T$. It can be seen that $J \propto V^2/d^3$ under light excitation is quite different from $J \propto V^{l+1}/d^{2l+1}$ in the dark [cf. eq. (3.45)]. Furthermore, the SCL photocurrent is proportional to $(i)^{(l+1)/l}$. However, the SCL photocurrent–voltage curve will merge the dark SCL current–voltage curve at the current level J at which $E_{Fp} = E_{DOp}$. For $E_{Fp} < E_{DOp}$ the light does not change the J–V characteristics. Mark and Helfrich [1962] and Jansen *et al.* [1964] have reported that the experimental SCL photocurrent in anthracene follow V^2 and $(i)^{(l-1)/l}$ dependence as predicted by eq. (6.94). A similar phenomenon has also been observed in tetracene containing hole traps of exponential distribution in energy [Belkind *et al.* 1972].

Helfrich [1964] has also derived an expression for SCL photocurrent for the case with the traps distributed uniformly throughout the energy band gap using a similar analysis. Again his expression shows a V^2 dependence of the SCL photocurrent, which is quite different from the exponential dependence on V for the dark SCL current [cf. eq. (3.77)]. It should be noted that if the light excitation energy is of the order of energy band gap, then the light absorbed in the fundamental absorption band produces not only majority carriers but also minority carriers. The analysis of such cases is much more complicated and eqs. (6.76)–(6.82) may have to be used.

With carrier-injecting contacts, the photocurrent can become saturated if the concentration of the majority carriers injected from the ohmic contact is suppressed at high fields by minority carrier injection from the opposite blocking contact through a recombination process [Kiess and Rose 1973, Queisser 1972, Popescu and Henisch 1975, 1976], or if the bulk-limited condition is changed to an electrode-limited condition at high fields [Pope *et al.* 1962, Simmons 1971].

(ii) With non-injecting blocking contacts

It is obvious that with non-injecting blocking contacts the photocurrent would become saturated when all photogenerated carriers are extracted from the semiconductor. The photoconductive gain is maximum when the field reaches a value at which the photocurrent is saturated. In the following we shall discuss the photocurrent–voltage characteristics for three general cases.

Supposing that the contacts are non-injecting for holes and electrons, and electron–hole pairs are photogenerated uniformly throughout the specimen, then all photogenerated carriers are extracted if the mean electron drift length X_n $(= \tau_n \mu_n F)$ and the mean hole drift length X_p $(= \tau_p \mu_p F)$ are equal or larger than the specimen thickness d. If either $X_n < d$ or $X_p < d$ or both are smaller than d, then space charge will form due to accumulated trapped carriers, which makes the applied field non-uniform. For showing the space charge effects, we consider a specific case in which $\mu_p \tau_p > \mu_n \tau_n$ and $X_n < d$. This implies that trapped electrons tend to increase the field near the anode and to decrease the field near the cathode as shown in Fig. 6.6. A high field zone is formed near the anode, and its width can be assumed to be equal to the electron drift length in the zone, which is given by

$$X_1 = \tau_n \mu_n F_1 \tag{6.95}$$

There is also a low field zone; its width is

$$X_2 = \tau_n \mu_n F_2 \tag{6.96}$$

where F_1 and F_2 are, respectively, the fields in zone 1 and zone 2. At low applied voltages X_1 is small and F_1 is not much different from V/d. Then J_{ph} follows Ohm's law:

$$J_{ph} = qG_p \mu_p \tau_p V/d \tag{6.97}$$

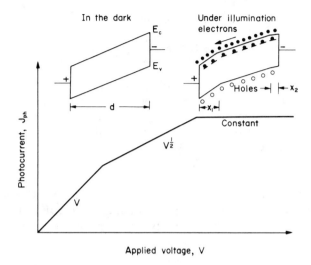

FIG. 6.6. Illustrating the behaviour of blocking contacts and the general photocurrent–voltage characteristics.

The contribution of electrons is neglected here because $\mu_p\tau_p$ is assumed to be much larger than $\mu_n\tau_n$. At higher voltage, X_1 increases. By assuming the major portion of the applied voltage is across X_1, then

$$F_1 = V/X_1 \tag{6.98}$$

From eq. (6.30) with $G_pX_p \gg G_nX_n$, it can be easily shown that

$$J_{ph} = qG_p(\mu_p\tau_p)^{1/2} V^{1/2} \tag{6.99}$$

At this particular voltage range J_{ph} is proportional to $V^{1/2}$. At still higher voltage, X_1 will extend to the cathode and $X_1 \simeq d$, then the limiting photocurrent becomes

$$J_{ph} = qG_pd \tag{6.100}$$

The J_{ph}–V characteristics for these three voltage ranges are shown schematically in Fig. 6.6. A similar and more detailed analysis for this case has been reported by Goodman and Rose [1971].

If the contacts are not perfectly blocking, but do inject electrons and holes and if the hole injection is dominant, then a SCL current will flow following the equation

$$J = \frac{9}{8}\varepsilon\mu_p\theta\frac{V_1^2}{X_1^3} \tag{6.101}$$

This current should also be consistent with the relation

$$J_{ph} = qG_pX_1 \tag{6.102}$$

Combination of eqs. (6.101) and (6.102) yields

$$X_1 = \left(\frac{9\varepsilon\mu_p\theta}{8qG_p}\right)^{1/4} V_1^{1/2} \tag{6.103}$$

Substitution of eq. (6.103) into eq. (6.102) gives

$$J_{ph} = q \left(\frac{9 \varepsilon \mu_p \theta}{8} \frac{1}{q} \right)^{1/4} (G_p)^{3/4} V^{1/2} \tag{6.104}$$

for $V_1 \to V$. Again, $J_{ph} \propto V^{1/2}$ in this voltage range. It should be noted that for the completely non-injecting cases J_{ph} is proportional to G_p or the light intensity, while for the SCL cases J_{ph} is proportional to $(G_p)^{3/4}$ or to the 3/4 power of light intensity.

If the contacts are of the Schottky barrier contacts, the width of the Schottky barrier can be approximately expressed as [cf. eq. (2.26)]

$$W = A + B V^{1/2} \tag{6.105}$$

where A and B are constants depending on the work functions of the semiconductor and the metallic contacts. If the light penetration depth $L \gg W$ and photogenerated carriers do not affect the space charge density in the depletion region of the barrier, then

$$J_{ph} = q G_p W \tag{6.106}$$

$$\propto V^{1/2}$$

In this case J_{ph} is also proportional to $V^{1/2}$. At higher voltages at which $L \ll W$, then J_{ph} becomes independent of applied voltage because photocarriers are generated mainly within the depletion region and are completely extracted.

(c) LIGHT INTENSITY DEPENDENCE

The photocurrent J_{ph} and the light intensity i are related by

$$J_{ph} \propto i^w \tag{6.107}$$

In the steady state the carrier generation rate is always equal to the sum of the recombination rates through various recombination channels. Generally, for a given range of i, one recombination channel dominates and it is this particular channel which determines the value of w in eq. (6.107). To explain this, we consider the recombination centres to be of acceptor type; they are neutral when empty, and are negatively charged when occupied by electrons (when they capture electrons). Referring to Fig. 4.4 and eqs. (6.37)–(6.44), and assuming the recombination centres are the active ones, the generation rate G in the steady state can be written as

$$G = \gamma_a - \gamma_b = \gamma_c - \gamma_d$$
$$= C_n n N_{rp} - C_n n_1 N_{rn} = C_p p N_{rn} - C_p p_1 N_{rp} \tag{6.108}$$

where N_{rn} is the concentration of occupied recombination centres (or capture electrons), $N_{rn} = N_r f_r$ which are negatively charged, N_{rp} is the concentration of unoccupied recombination centres, and $N_{rp} = N_r (1 - f_r)$, which are neutral. Obviously $N_r = N_{rn} + N_{rp}$ is the total concentration of active recombination centres. It can be easily shown then

$$N_{rn} = \frac{(C_n n + C_p p_1) N_r}{C_n (n + n_1) + C_p (p + p_1)} \tag{6.109}$$

$$N_{rp} = \frac{(C_p p + C_n n_1) N_r}{C_n (n + n_1) + C_p (p + p_1)} \tag{6.110}$$

and

$$G = \frac{C_n C_p (np - n_1 p_1) N_r}{C_n (n + n_1) + C_p (p + p_1)} \tag{6.111}$$

Equation (6.111) leads to the following conclusions [Stockmann 1969, 1970]:

(1) At low light intensities, $pn - p_1 n_1 \ll n_i^2$, where $n_1 p_1 = n_i^2$. Letting $n = n_{\mathrm{ph}} + n_0$, $p = p_{\mathrm{ph}} + p_0$, eq. (6.111) can be linearized, giving n_{ph} which is linearly proportional to G or i.

(2) At high light intensities, $pn - p_1 n_1 \gg n_i^2$, $p \simeq p_{\mathrm{ph}}$ and $n \simeq n_{\mathrm{ph}}$. If $C_p p$ is larger than the three other terms in the denominator of eq. (6.111), then n is proportional to G or i. This case implies that $N_{rp} \simeq N_r$ and they are located above E_{Fn}^t.

(3) If $pn - p_1 n_1 \gg n_i^2$ and $C_n n$ is larger than the three other terms in the denominator of eq. (6.111), then p is proportional to G or i with $w = 1$. This case implies that $N_{rn} = N_r$ and they are located below E_{Fp}^t.

(4) If $pn - p_1 n_1 \gg n_i^2$, and $C_n n + C_p p$ are dominant in the denominator of eq. (6.111), then np is proportional to G or i.

Stockmann [1969, 1970] has used

$$n \propto G^v \tag{6.112}$$

$$n \propto p^{v*} \tag{6.113}$$

$$v = \frac{G}{n} \frac{dn}{dG} \tag{6.114}$$

$$v* = \frac{p}{n} \frac{dn}{dp} = -dE_{Fn}/dE_{Fp}$$
$$= h_p(E_{Fp})/h_n(E_{Fn}) \tag{6.115}$$

where $h_p(E_{Fp})$ and $h_n(E_{Fn})$ are, respectively, the concentrations per energy level of hole traps at the quasi-Fermi level E_{Fp} and of electron traps at the quasi-Fermi level E_{Fn}, and he has determined the value of v and $v*$ for many cases with various combinations of the following traps and recombination centres:

(1) electron traps if $\gamma_a \simeq \gamma_b \gg \gamma_c$, γ_d and N_r becomes N_{tn} (acceptor-type traps)
(2) hole traps if $\gamma_c \simeq \gamma_d \gg \gamma_a$, γ_b and N_r becomes N_{tp} (donor-type traps)
(3) recombination centres if $\gamma_a \simeq \gamma_c \gg \gamma_b$, γ_d and N_r can be of acceptor type or donor type.

If one type of free carriers is dominant, for example, $n \gg p$, then $J_{\mathrm{ph}} \propto n \propto G^v$. Since $G \propto i$ so in this case $w = v$ [cf. eqs. (6.107) and (6.112)].

The superlinear photoconductivity is caused by a superlinear dependence of the majority carrier concentration on G making $v > 1$. According to Stockmann [1969] the necessary and sufficient conditions for this to occur in an n-type material are as follows: (1) There is a set of electron traps of a sufficiently large concentration, which are essentially unoccupied in the dark (located above F_{Fn}), but become fully occupied by electrons with increasing G (located below E_{Fn})—trapping action is changed to recombination action. This is equivalent to saying that if all traps are occupied, the free carrier concentration and hence the carrier lifetime increase. (2) There must not simultaneously be a set of hole traps fully occupied by holes. (3) The rate-determining process in the recombination channels is the capture of holes by the recombination centres. This set of electron traps acts to sensitize the photoconductor, that is, to increase the electron lifetime. Rose [1963] has called this action the sensitization by "electronic doping". While these traps are being converted into recombination centres the electron lifetime is continuously increasing and the photocurrent increases superlinearly with increasing light intensity. After the conversion is completed, the photocurrent again increases linearly with light intensity [Rose 1963].

Regarding superlinear photoconductivity and infrared quenching, the reader is referred to Section 4.2.4, and the work of Bube [1960], Rose [1963], and Stockmann [1969, 1970]. It should be noted that the measurements of J_{ph} as a function of light intensity at various temperatures and light wavelengths would provide useful information about the nature, the energy distribution, and the concentration of imperfections (traps and recombination centres).

(d) LIGHT WAVELENGTH DEPENDENCE

By measuring the spectral response of the steady state photocurrent we can detect the major trapping levels. Using hole- or electron-injecting contacts to provide trapped carriers and then using light of suitable wavelengths to release them, direct optical detrapping in anthracene crystals has been observed [Adolph et al. 1964, Many et al. 1969]. The major hole-trapping level is about 1.1 eV from the valence band and the major electron-trapping level is about 0.9 eV from the conduction band. For anthracene crystals with non-injecting contacts, optical means have been used to cause carrier injection from the contact surfaces to provide trapped carriers in the specimen [Itoh 1973, Itoh and Anzai 1974]. However, this method is inaccurate or invalid for determining the depths of shallow traps which correspond to energies of the infrared light region because this is the absorption region for molecular vibrations.

To avoid the difficulty of separating the bulk carrier generation due to optical detrapping from those due to photoemission from carrier-injecting contacts, Itoh and Takeishi [1974] have used SnO_2 non-injecting electrodes to measure the enhanced photocurrent due to optical detrapping in anthraquinone-doped anthracene as a function of exciting light wavelength. The anthraquinone-doped anthracene crystals have many deep traps. The charging of the deep traps was carried out by illuminating the positive electrode for the creation of trapped holes, or by illuminating the negative electrode for the creation of trapped electrons in the deep traps at the light wavelength of 395 nm under a high field for about an hour. The trapped carriers can be removed by short-circuiting both electrodes and then illuminating the specimen with a tungsten lamp for about 30 min. The field dependence of the spectral response of the hole and electron photocurrents is shown in Fig. 6.7. At low fields the peaks occur at wavelengths between 420 nm and 440 nm, while at high fields the peak appears at about 395 nm. The peak at wavelengths between 420 nm and 440 nm also occurs at high fields if the deep traps have been charged prior to the photocurrent measurement. These results show that at low fields most carriers are generated by the optical detrapping and at high fields the carrier generation by the exciton surface interaction becomes dominant. The spectral response of the hole and electron photocurrent for various charging treatments prior to the photocurrent measurement is shown in Fig. 6.8. The direct optical

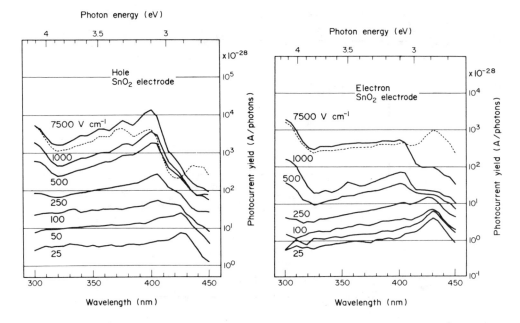

F IG. 6.7. Electric field dependence of spectral response curves of hole and electron photocurrents in anthraquinone-doped anthracene crystals. The dotted curves are the spectral response curves for the photocurrent measured at $7500\,\mathrm{V\,cm^{-1}}$ after charging treatment. [After Itoh and Takeishi 1974.]

detrapping takes the form of shoulders in the spectrum of the enhanced photocurrent [Many *et al.* 1969, Itoh and Takeishi 1974]. Thus, from the rising points of the shoulders in Fig. 6.8(b) it can be seen that there are three electron-trapping levels, which are 1 eV, 1.6 eV, and 2.3 eV from the conduction band. For detrapping of holes [Fig. 6.8(a)] the spectrum below 1.5 eV is due to direct optical detrapping from a trapping level at about 1 eV from the valence band, and the spectrum between 1.5 eV and 2.5 eV is due to triplet exciton dissocation and that at about 2.8 eV due to singlet exciton dissociation.

It is important to mention that in the measurements of photocurrents the effect of space-charge-type polarization, which persists after the removal of both the light and the applied field, should not be ignored, particularly when measuring transient photocurrents using light pulses [Kallmann and Rosenberg 1955, Freeman *et al.* 1961, Itoh 1973]. This effect can be suppressed by using high light intensities for the photocurrent measurements, or by cleaning out the trapped carriers before commencing the photocurrent measurements.

(e) TEMPERATURE DEPENDENCE

Although J_{ph} commonly increases exponentially with T, the activation energy ΔE_{ph} is usually small as compared with the dark electric conduction activation energy ΔE_{σ}. For example, $\Delta E_{\mathrm{ph}} \simeq 0.17$ eV for anthracene [Compton *et al.* 1957] as compared with $\Delta E_{\sigma} \simeq 2.6$ eV for the same material [Kommandeur 1965]; $\Delta E_{\mathrm{ph}} \simeq 0.44$ eV and ΔE_{σ}

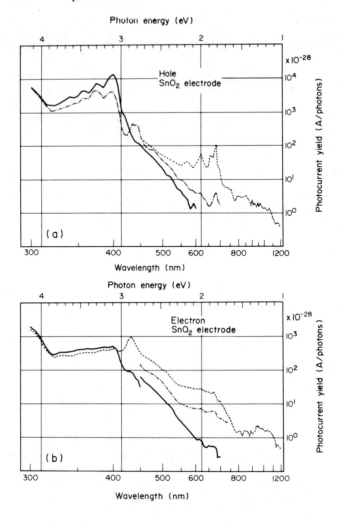

F IG. 6.8. Spectral response of (a) hole photocurrent, and (b) electron photocurrent enhanced due to optical detrapping in anthraquinone-doped anthracene crystals measured at 7500 V cm^{-1}. The dotted curves represent those with charging treatment before measurements, the chain curves represent those for which part of trapped carriers have been cleaned out before measurements, and the solid curves are those with most trapped carriers cleaned out before measurements. [After Itoh and Takeishi 1974.]

$\simeq 1.5$ eV for orthochrome T dye [Meier 1974]. The temperature dependence of J_{ph} is associated with $d(\mu_n + \mu_p)/dT$ and $d(n + p)/dT$, which are dependent on the density distribution of the traps, on the temperature, on the exciting quantal energy, on the applied field, and on the ambient atmosphere. Northrop and Simpson [1958] have reported that J_{ph} is independent of T in anthracene for temperatures between 0°C and 80°C. Nakada [1965] has reported that ΔE_{ph} depends on the type of carriers and on the ambient atmosphere in which it is measured. In anthracene $\Delta E_{ph} \simeq 0.11$ eV in nitrogen and $\Delta E_{ph} \simeq 0.13$ eV in vacuum for photogenerated electrons, and $\Delta E_{ph} \simeq 0.14$ eV in nitrogen for photogenerated holes when the specimen is excited at quantal energy of 3.15–3.5 eV. In general, ΔE_{ph} increases with decreasing quantal energy. Nakada and Ishihara [1964] have reported that the photocarrier generation in anthracene increases with increasing temperature at low fields, but at high fields it is independent of temperature. It can be imagined that there is no simple way to interpret ΔE_{ph}. The possible factors contributing to ΔE_{ph} are, however, listed below [Meier 1974].

(i) The energy required for carrier generation by excitons.
(ii) The activation energy of the quantum yield.
(iii) The mean free path for carriers to travel without encountering a collision.
(iv) The thermalization distance to escape the Coulombic field by diffusion according to the Onsager model.
(v) The temperature dependent shift of the quasi-stationary equilibrium between the trapped carriers n_t, p_t and the free carriers n, p, where the temperature dependence of J_{ph} is determined by the density distribution of the traps.

(f) EFFECTS OF SURFACE CONDITIONS AND AMBIENT ATMOSPHERE

Surface states existing at the interface between the electrode and the specimen play a very important role in the carrier recombination processes. If the surface recombination rate R_S is small as compared with the bulk recombination rate R_B, the carriers produced in a thin surface layer by strongly absorbed light will recombine slowly through the surface states, and under the action of the applied field of sufficient magnitude excess photocarriers will be drawn to the bulk to contribute to the photocurrent. But if the surface recombination rate is large as compared with the bulk recombination rate, the carriers produced in a thin surface layer by strongly absorbed light will recombine through the surface states so rapidly that they cannot contribute much to the photocurrent. In this case, only the carriers produced by weakly absorbed light (large penetration depth) and produced at a region far from the surface can contribute to the photocurrent. Thus for $R_S \ll R_B$ the photocurrent spectrum has a good correlation with the absorption spectrum, while for $R_S \gg R_B$ such a correlation disappears.

It is well known that the presence of oxygen or air or other gases will influence the photoconductivity and its spectral response in organic crystals [Chynoweth 1954, Kawasaki *et al.* 1966, Johnston and Lyons 1970, Pope and Kallmann 1971, Meier 1974]. The adsorbed gases on the surface give rise to the formation of deep traps, thus increasing the surface recombination rate. Oxygen and iodine, for example, would form acceptor-type electron traps, while hydrogen and NH_3 would form donor-type hole traps. On the basis of this tendency we can conclude that [cf. Meier 1974]:

(i) With the illuminated surface at the positive polarity of the applied voltage, the photocurrent (mainly hole current) increases if it is measured in the oxygen atmosphere and decreases if it is in the hydrogen atmosphere.
(ii) With the illuminated surface at the negative polarity of the applied voltage, the photocurrent (mainly electron current) increases if it is measured in the hydrogen atmosphere and decreases if it is in the oxygen atmosphere.

This rule can be used to explain most experimental results involving the effects of ambient atmosphere, such as these effects on photocurrent in anthracene crystals [Chynoweth 1954] and in phenazine crystals [Itoh 1975].

(B) Processes involving excitons

(a) EXCITON DISSOCIATION AT ELECTRODE–SEMICONDUCTOR INTERFACES

The efficiency of photogeneration of holes or electrons by exciton dissociation at the electrode–semiconductor interfaces depends on the polarity and the surface conditions of the illuminated electrode [Steketee and De Jonge 1962, Mulder 1968, Mehl and Hale 1968, Vaubel *et al.* 1971]. Such kind of dissociation requires the presence of traps at the surface, to which the excitons (the excited molecules) can transfer their electrons or holes [Pope and Kallmann 1971, Baessler *et al.* 1971]. The discontinuity of the dielectric constant across the interface between a metal and a semiconductor enhances the molecular polarization energy and thus makes the surface molecules act as traps for excitons. However, a metal alone can act as a charge and energy acceptor because it has empty states above and occupied states below the Fermi level. The non-radiative decay of excitons at the interface can be caused by (i) energy transfer to metal electrons by dipole–dipole coupling—exchange interaction, (ii) intersystem crossing to a state of different multiplicity induced by the presence of unpaired metal electrons, and (iii) exciton dissociation according to

$$M^* + \text{metal} \rightarrow M^+ + e^- \text{(metal) (oxidative dissociation)}$$
$$M^* + \text{metal} \rightarrow M^- + e^+ \text{(metal) (reductive dissociation)} \qquad (6.116)$$

where M^* denotes an excited molecule; this process may lead to conversion of optical to electrochemical energy [Killesreiter and Baessler 1972]. Processes (i) to (iii) can also occur at the interface between an electrolytic solution and an organic semiconductor.

If the observed photocurrent transient is mainly associated with exciton dissociation at the electrode surface following eqs. (6.11) and (6.12), the fast component is due to the singlet–surface interaction and the slow component due to the triplet–surface interaction. We shall use the one-dimensional diffusion model to analyse the dissociation processes. We consider first the singlet exciton dissociation. The rate of creation, decay, and diffusion of singlet excitons can be described by the following kinetic equation:

$$\frac{\partial [S]}{\partial t} = \varepsilon_S i_S \exp\left(-\varepsilon_S x\right) - \frac{[S]}{\tau_S} + D_S \frac{\partial^2 [S]}{\partial x^2} \qquad (6.117)$$

where $[S]$ is the concentration of singlet excitons, D_S is the singlet diffusion constant, and τ_S is the singlet lifetime, ε_S is the absorption coefficient for the light intensity i_S for singlet exciton generation. Using the boundary conditions for perfect quenching of singlet at the surface

$$[S]_{x=0} = 0 \qquad (6.118)$$

and complete absorption within the semiconductor specimen

$$[S]_{x\to\infty} = 0 \qquad (6.119)$$

the steady state solution of eq. (6.117) with the reabsorption effect excluded is [Chance and Prock 1973]

$$[S] = \frac{\varepsilon_S i_S \tau_S}{1 - \varepsilon_S^2 D_S \tau_S} [\exp(-\varepsilon_S x) - \exp(-x/\sqrt{D_S \tau_S})] \qquad (6.120)$$

The fraction of singlet excitons which will diffuse to the illuminated electrode at $x = 0$ is

$$F_S = L_S (L_a + L_s)^{-1} \qquad (6.121)$$

where L_S is the singlet diffusion length $[L_S = (D_S \tau_S)^{1/2}]$ and L_a is the penetration length of the absorbed light $(L_a = \varepsilon_S^{-1})$.

The singlet exciton distribution can be significantly affected by the reabsorption of fluorescent light because it causes an increase in the effective singlet exciton lifetime and the generation of excitons deep inside the specimen. By including the reabsorption effect, eq. (6.117) under steady state conditions becomes

$$D_S \frac{d^2[S]}{dx^2} - \frac{[S]}{\tau_S} + \varepsilon_S i_S \exp(-\varepsilon_S x) + R(x) = 0 \qquad (6.122)$$

where $R(x)$ is a reabsorption term which is given by [Mulder 1968]

$$R(x) = \frac{1}{2\pi\tau_S} \sum_{i=1}^{n} q_i \varepsilon_i \int_{x_1 = -\infty}^{+\infty} [S]_{x = x_1}$$

$$\int_{y = -\infty}^{\infty} \frac{\exp\{-\varepsilon_i[(x-x_1)^2 + y^2]^{1/2}\}}{[(x-x_1)^2 + y^2]^{1/2}} dy \, dx_1 \qquad (6.123)$$

in which ε_i and q_i are, respectively, the average absorption coefficient and the weighting factor proportional to the "re-absorption free" fluorescence intensity in the ith small wavelength interval in the fluorescence spectrum; x and x_1 are measured along the c' crystal direction (e.g. anthracene) and y perpendicular to x corresponding to a crystal direction. Using the same boundary conditions given by eqs. (6.118) and (6.119), the solution of eq. (6.122) is [Chance and Prock 1973]

$$[S]_{\substack{\text{with} \\ \text{reabsorption}}} = [S]_{\substack{\text{without} \\ \text{reabsorption}}} + \frac{\tau_S}{2L_S}$$

$$\times \int_0^\infty \{\exp[-|u-x|/L_S] - \exp[-(\mu+x)/L_S]\} R(u) \, du \qquad (6.124)$$

Equation (6.124) can be solved by an iterative numerical technique. To measure the reabsorption effect, a parameter R_S is used, which is defined by

$$R_S = \frac{d[S]_{\text{with reabsorption}}/dx|_{x=0}}{d[S]_{\text{without reabsorption}}/dx|_{x=0}} \qquad (6.125)$$

Thus with the reabsorption effect included, eq. (6.121) is changed to

$$F_{S \text{(with reabsorption)}} = R_S L_S (L_a + L_s)^{-1} \qquad (6.126)$$

The quantum yield for the carrier generation due to singlet exciton dissociation at the illuminated electrode is [Chance and Prock 1973]

$$\eta_{Sd} = \sum_{i=1}^{n} \chi K_{SCq} R_S^i F_S^i \qquad (6.127)$$

where χ is the charge separation probability which is assumed to be independent of the path of charge pair formation but dependent on the external field, and K_{SCq} is the fraction of singlet excitons quenched via electron transfer at the surface.

The most important decay channels for singlet excitons near metal surfaces are electron transfer and energy transfer. By including the rates of electron transfer and energy transfer, the steady state rate equation becomes

$$D_S \frac{d^2[S]'}{dx^2} - \frac{[S]'}{\tau_S} - K_{CT}(x)[S]' - K_q(x)[S]' + \varepsilon_S i_S \exp(-\varepsilon_S x) = 0 \qquad (6.128)$$

where K_{CT} is the rate constant for electron transfer, which depends exponentially on x from the metal surface based on a tunnelling mechanism; and K_q is the rate constant for energy transfer, which is proportional to x^{-3} according to Kuhn's theory [Kuhn 1970]. By solving (6.128) for $d[S]'/dx$, the singlet-electron transfer efficiency can be estimated approximately by

$$K_{SCq} \simeq \frac{d[S]'/dx|_{x=0}}{d[S]/dx|_{x=0}} \qquad (6.129)$$

Phenomenologically, the photocurrent resulting from the carriers generated by singlet exciton dissociation at the electrode surface with strongly absorbed light can be expressed as [Killesreiter and Baessler 1971, 1972]

$$J_{ph} = qi_s \left[\frac{L_S}{L_a + L_S} \right] \left[1 + \frac{L_S}{c\tau_\infty \Sigma K} \right] \left[\frac{K_{CT}}{\Sigma K} \right] \chi \qquad (6.130)$$

where c is the difference between neighbouring ab lattice planes, τ_∞ is the intrinsic exciton lifetime, and $\Sigma K = K_{CT} + K_q + K_{OX} + K_{dd}$ in which K_{CT} and K_q have been defined above, K_{OX} and K_{dd} are, respectively, the rate constants for exciton quenching by oxygen molecules and for energy transfer to metal electrons by dipole–dipole coupling.

Using a fatty acid monolayer inserted between an anthracene crystal specimen and an aluminium contact to increase the energy transfer distance, several investigators [Vaubel *et al.* 1971, Killesreiter and Baessler 1971] have studied the singlet-exciton reaction at the anthracene–metal interface. The energy transfer rate constant K_q and the exciton-quenching velocity ($V_q = K_q c$) as functions of the thickness d of the fatty acid monolayer are shown in Fig. 6.9. The photocurrent and the exciton dissociation constant χK_{CT} as functions of d are shown in Fig. 6.10. Their experimental results are in good agreement with the theory.

Chance and Prock [1973] have also treated the triplet exciton dissociation at the electrode surface. The quantum yield for the carrier generation due to triplet exciton dissociation at the illuminated electrode, similar to eq. (6.127), can be written as

$$\eta_{Td} = \sum_{i=1}^{n} \chi K_{TCq} R_T^i F_T^i \qquad (6.131)$$

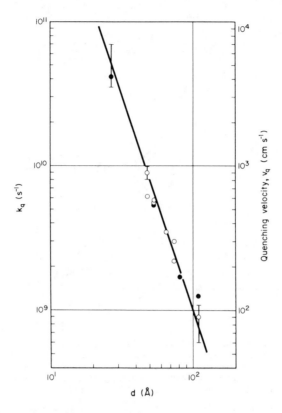

FIG. 6.9. The energy transfer rate constant K_q and the quenching velocity v_q for singlet excitons as functions of the thickness of the fatty acid monolayer inserted between anthracene and aluminium. $v_q = K_q c$, where c is the distance between neighbouring *ab*-lattice planes. The full circles are obtained by normalizing the fluorescence output to the incident photon flux. $K_q = K_{q(\text{metal})} + K_{q(\text{impurities})}$ with $K_{q(\text{metal})} \gg K_{q(\text{impurities})}$. [After Vaubel, Baessler, and Mobius 1971.]

where [Chance and Prock 1973]

$$F_T = \frac{K_{ST}\tau_S}{i_T} \int_0^\infty [S]P(x)\,dx \tag{6.132}$$

$$P(x) = \frac{1}{2\sqrt{\pi}} \int_0^\infty \sigma^{-1/2} \exp(-x^2/D_T\tau_T\sigma - \sigma/4)\,d\sigma \tag{6.133}$$

$$R_T = \frac{\displaystyle\int_0^\infty [S]_{\text{with reabsorption}}\, P(x)\,dx}{\displaystyle\int_0^\infty [S]_{\text{without reabsorption}}\, P(x)\,dx} \tag{6.134}$$

$$K_{TCq} = AK_{SCq} \tag{6.135}$$

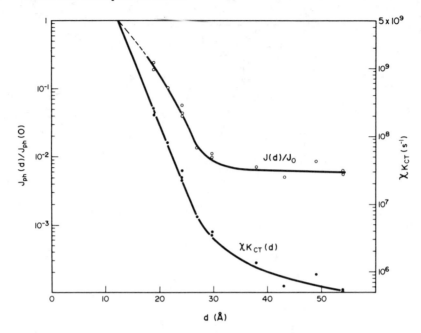

FIG. 6.10. The normalized photocurrent $J_{ph}(d)/J_{ph}(0)$ and the singlet exciton dissociation constant χK_{CT} as functions of the thickness d of the fatty acid layer inserted between anthracene and aluminium. $J_{ph}(0)$ is the photocurrent in the absence of a fatty acid layer and data for $d < 30$ Å are taken with single monolayer spacers. [After Killesreiter and Baessler 1971.]

and A is a constant. Chance and Prock [1973] and Weisz *et al.* [1973] have reported the involvement of triplet exciton dissociation in the carrier generation at the anthracene–metal interface. However, the triplet–surface impurity interaction cannot explain the spectral response of the photocurrent and the related fluorescence quantum yield results. Further experiments are obviously required to examine their results.

(b) EXCITON DISSOCIATION DUE TO INTERACTION OF TRIPLET EXCITONS AND TRAPPED CARRIERS

In Section 6.1.2(b) we have discussed the basic concept of the carrier generation due to the interaction of triplet excitons and trapped carriers. On the assumption that the traps are confined in a single discrete energy level and that the thermal detrapping is neglected, the free carrier density (say hole density) p generated by the triplet exciton dissociation in the bulk can be described by the following equation:

$$\frac{dp}{dt} = \eta_T K_{TCt} [T]p_t + \eta'_T \varepsilon_{0p} i_T p_t - \frac{p}{\tau_T} \tag{6.136}$$

where τ_T is the triplet lifetime, K_{TCt} is the rate constant for the interaction between triplet excitons and trapped holes, ε_{0p} is the absorption coefficient for optical detrapping, η_T and η'_T are, respectively, the efficiencies for carrier generation by exciton dissociation and by optical detrapping, and $[T]$ is the triplet exciton concentration which is governed by

$$\frac{\partial[\mathrm{T}]}{\partial t} = \varepsilon_T i_T - (\beta_1 + \beta_2)[T] - \gamma_T[T]^2 - K_{TC}[T]p - K_{TC_t}[T]p_t + D_T \frac{\partial^2[T]}{\partial x^2}$$

$$(6.137)$$

in which ε_T, β_1, β_2, γ_T, and K_{CT} have the usual meanings [cf. Table 7.1 of Chapter 7]. For the cases of thick specimens the diffusion term can be ignored. If the light intensity is low, the bimolecular term $\gamma_T[T]^2$ can be neglected. Thus the steady state triplet concentration under these conditions is

$$[T] = \varepsilon_T i_T [(\beta_1 + \beta_2) + K_{TC}p + K_{TC_t}p_t]^{-1} \qquad (6.138)$$

For most cases it can be assumed that $(\beta_1 + \beta_2) \gg K_{TC}p + K_{TC_t}p_t$ and therefore eq. (6.138) reduces to

$$[T] \simeq \varepsilon_T i_T (\beta_1 + \beta_2) \qquad (6.139)$$

Thus substitution of eq. (6.139) into eq. (6.136) yields the steady state concentration of free holes

$$p = \tau_T i_T p_t [\eta_T K_{TC_t} \varepsilon_T (\beta_1 + \beta_2)^{-1} + \eta'_T \varepsilon_{0p}] \qquad (6.140)$$

Since J_{ph} is proportional to p, the spectral response of J_{ph} should be similar to the $S_0 \rightarrow T_1$ absorption spectrum if the first term on the right-hand side of eq. (6.140), which represents the contribution of the triplet-trapped carrier interaction, is dominant.

A similar treatment can be easily extended to the cases in which free electrons (rather than holes) are generated by the triplet exciton-trapped electron interaction. In fact, the triplet lifetime decreases with increasing concentration of traps (or increasing trapped carriers) can be considered as a direct evidence that the interaction between triplet excitons and trapped carriers takes place.

Using a ruby laser, which provides photons of energy corresponding to 14,400 cm^{-1}, Itoh and Izumi [1972, 1973, also Izumi and Itoh 1972] have studied the photocurrent in 9,10-dichloroanthracene (DCA) and 9,10-dibromoanthracene (DBA). Since the photon energy is much smaller than the energy band gap of DCA or DBA, which is about 4 eV (about 32,000 cm^{-1}), but is close to the energies of their triplet states which are about 13,800 cm^{-1} and 14,100 cm^{-1}, respectively, the photocarriers are not generated by the direct band-to-band transition. They have proved using various experiments that the photocarriers are generated by the indirect interaction of trapped carriers with triplet excitons following

$$S_0 + h\nu \text{ (laser photon)} \rightarrow T_1$$

$$T_1 + \text{trapped carriers} \rightarrow \text{free carriers}$$

Figure 6.11 shows that the spectral response of J_{ph} is similar to the absorption spectrum and that the photogenerated carriers increase linearly with light intensity for DCA. Their results agree well with eq. (6.140).

As has been mentioned in Section 6.1.2, the magnetic field dependence of photocurrent is another strong experimental evidence of the carrier generation by the triplet exciton-trapped carrier interaction [Geacintov *et al.* 1970, Ern *et al.* 1971, Rusin *et al.* 1969, Frankevich *et al.* 1970, 1972]. Figure 6.12 shows the J_{ph}–V curves at various light intensities. The magnetic field has no effect on the dark current due to the injection

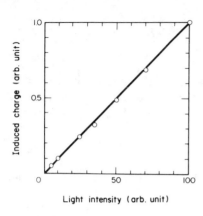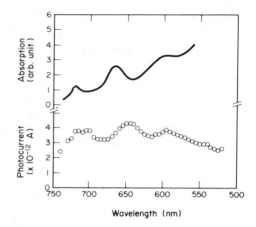

FIG. 6.11. Pulsed photocarrier generation as a function of laser light intensity, and steady state photocurrent and the $S_0 \rightarrow T_1$ absorption as functions of light wavelength in 9,10-dichloro-anthracene (DCA). The absorption spectrum was measured in ethyl iodine solution by McGlynn *et al.* [1964]. [After Itoh and Izumi 1973.]

of holes from iodine to anthracene, but has a strong effect on the photocurrent at high light intensities. At low applied voltages $J_{ph} \propto V^2$, and at higher voltages J_{ph} tends to become saturated. The $J_{ph} \propto V^2$ relationship has also been observed by other investigators [e.g. Mark and Helfrich 1962]. If $p_t \gg p$ and the space charge is homogeneous, $p_t \propto V$ [Helfrich 1967]. Thus, if τ_T does not vary with V, then it can be expected that $J_{ph} \propto V^2$ based on eq. (6.140).

(c) EXCITON DISSOCIATION DUE TO INTERACTION OF SINGLET EXCITONS AND TRAPPED CARRIERS

Although it is generally accepted that the carrier generation by the interaction of trapped carriers with triplet excitons is far more important than that with singlet excitons [Helfrich 1967, Bauser *et al.* 1969, 1972], there is some indirect evidence that the singlet exciton-trapped carrier interaction does contribute to the photoconduction processes [Pope and Kallmann 1971, Kalinowski and Godlewski 1974]. In tetracene the singlet exciton concentration is also magnetic field dependent, and it has been shown that definitely the triplet excitons, and probably the singlet excitons as well, participate in the observed carrier detrapping [Pope *et al.* 1970]. Many investigators have studied the involvement of the singlet exciton-trapped carrier interaction in the carrier detrapping [Weisz *et al.* 1969, Schott and Berrehar 1973, Coret and Fort 1974].

6.2.2. Intrinsic photoconduction

Photoconductivity of organic semiconductors may have several origins. We have discussed some origins responsible for the extrinsic photoconductivity in the previous section. If the density of excitons is sufficiently high, either the exciton photoionization through a multiphoton process or the exciton–exciton collision can lead to intrinsic photoconduction. However, single-photon intrinsic photoconduction can be observed only when the photon energies exceed the optical absorption threshold.

In contrast with the extrinsic photoconduction for which the action spectrum is

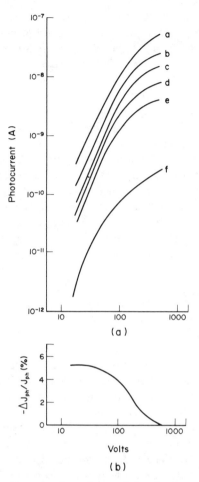

FIG. 6.12. (a) Current–voltage characteristics of anthracene crystals of
0.6 mm in thickness: light wavelength: $\lambda > 420$ nm with an
anthracene crystal of 0.7 mm in thickness as filter. Light
intensities in quanta/cm^2 sec: a, 3.1×10^{17}; b, 1.1×10^{17}; c,
3.8×10^{16}; d, 1.3×10^{16}; e, 4.5×10^{15}; f, dark current due to
injection of holes from the iodine electrode. (b) The relative
photocurrent decrease as a function of applied voltage
corresponding to curve a at a magnetic field of 3000 oersted.
[After Frankevich and Sokolik 1970.]

similar in shape to the optical absorption spectrum, there is no direct relation between
the intrinsic photocurrent and the absorption spectrum. When a crystal is excited
optically to a higher energy state, the excited electron in that state is unstable and its
lifetime is very short (of the order of 10^{-14}–10^{-12} sec). It rapidly loses its energy either
via intramolecular non-radiative transition (internal conversion) to a lower exciton
state or through AI to produce an electron–hole pair. In general, the efficiency of
internal conversion is higher than that of AI, and this is considered as the indication
that the higher energy states ($> E_c$) are localized as the lower lying exciton states ($< E_c$).

Several investigators [Jortner 1968, 1971, Jortner and Bixon 1969, Geacintov and
Pope 1967, 1969, Bergman and Jortner 1974, Bergman *et al.* 1967, 1972] have analysed

the two competitive and simultaneous decay processes (internal conversion and AI) of the metastable exciton states located above the direct threshold to the conduction band based on a modified Fano's theory [Fano 1961]. In organic crystals, the probability for the optical transition from band to band is small as compared with that to a (zero order) bound exciton state located above E_c based on the following experimental facts: (a) In anthracene, for example, the quantum yield for the production of charge carriers with the photon energies over the range of 4–6 eV is about 10^{-4} carriers, per photon, whereas the quantum yield for the production of fluorescent excitons (3.1 eV) is close to unity indicating that the internal conversion is dominant. (b) In the optical one- and two-photon spectra of crystalline anthracene no evidence has ever been obtained for Fano-type anti-resonances in the optical line shapes indicating that the transition moment for direct transition to the conduction band is small [Jortner 1968, Jortner and Baxon 1969]. (c) The theoretically estimated cross-section for interband transition is small [Druger 1972]. However, any way of combinations of excitons and photons that has a total energy greater than the threshold to the conduction band (e.g. 3.9 eV for anthracene) would give rise to carrier generation and hence intrinsic photoconduction.

(A) Band-to-band transition

This process implies that the absorbed photons create directly free electron–hole pairs due to the band-to-band transition. This process generally occurs in inorganic semiconductors, and also occur, to a much less extent, in organic semiconductors. Using a pulse technique, Castro and Hornig [1965] have observed a peak in the spectral response curves of hole and electron quantum yields in anthracene, which occurs at the wavelength of 278 nm. They have attributed this peak to intrinsic charge carrier generation by a single photon process based primarily on the fact that this peak does not correspond to any known optical transition and that its magnitude is insensitive to the crystal surface conditions. A similar phenomenon has also been observed by Chaiken and Kearns [1966]. Their results on the spectral response of d.c. hole and electron photocurrent in anthracene are shown in Fig. 6.13. It can be seen that in the spectral range 310–400 nm, both the hole and electron photocurrent curves resemble the absorption curve which represents the vibrational structure of the $S_0 \rightarrow S_1$ absorption spectrum. The similarity between the spectral response of J_{ph} and the absorption spectrum and the sensitivity of J_{ph} to the crystal surface conditions indicate that in this wavelength region the carrier generation is extrinsic and involves the interaction of singlet excitons with surface or near-surface impurity sites. In the region between 250–310 nm, however, such a similarity between J_{ph} and absorption spectra disappears. The peak at 280 nm corresponds to a maximum intrinsic photocarrier yield at 4.4 eV ($\sim$ 280 nm) and the direct ionization occurs at 4.0 eV ($\sim$ 310 nm) in anthracene. It should be noted that it has been generally believed that for anthracene the energy level 4.0 eV corresponds to the first narrow conduction band and the energy level 4.4 eV corresponds to the second wide conduction band [Cook and Le Comber 1971].

The high field and temperature effects on photocurrents in organic semiconductors have been studied by several investigators based on the Onsager model [Batt *et al.* 1968, Geacintov and Pope 1971, Chance and Braun 1973, Braun and Chance 1974, Melz 1972, Lakatos 1975]. In Section 5.3.6 we have discussed briefly the Onsager model. The probability $p(r, \theta, F)$ that an isolated, thermalized pair with initial separation r and

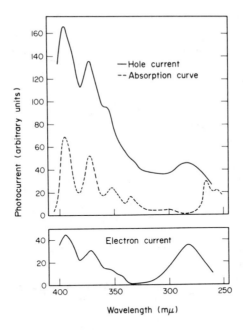

FIG. 6.13. Spectral response of d.c. hole and electron photocurrents in anthracene at temperature of 179°C. Applied voltage: 300 V d.c.; specimen thickness: 0.5 mm; light intensity: 6 × 10¹² photons/sec. The arbitrary photocurrent units are the same for both the hole and electron curves. [After Chaiken and Kearns 1966.]

orientation θ relative to applied field F in an isotropic medium of dielectric constant ε will escape geminate recombination is given by eq. (5.124), and the quantum yield by eq. (5.129). For fields below about 10^4 V cm^{-1}, it is a good approximation to keep terms only to the first order in the electric field. Thus from eq. (5.129) we can write

$$\eta(r_0 F) \simeq \eta_0 \exp[-r_c(T)/r_0]\left[1 + \left(\frac{q}{kT}\right)\frac{1}{2!}r_c F\right] \qquad (6.141)$$

where η_0 is the quantum yield of quasi-free electron–hole pairs per photon and assumed to be independent of applied field. The quantum yield should be linearly dependent on the electric field in this region and should have a slope to intercept ratio of

$$S/I = q^3(8\pi\varepsilon k^2 T^2)^{-1} \qquad (6.142)$$

For anthracene, $\varepsilon = 3.2\varepsilon_0$, $S/I \simeq 3.4 \times 10^{-5}$ cm V^{-1} at $T = 298$°K. Using light pulses to limit the number of absorbed photons to about 2.6×10^{10} per flush for their photocurrent measurements, Chance and Braun [1973] have reported that in anthracene at room temperature $S/I = 3.21 \times 10^{-5}$ cm V^{-1} and $\eta_0 = 8.75 \times 10^{-5}$, for electrons and $S/I = 3.02 \times 10^{-5}$ cm V^{-1} and $\eta_0 = 9.16 \times 10^{-5}$ for holes, and that S/I varies approximately with T^{-2}. Their results agree essentially quantitatively with the prediction of Onsager theory. It should be noted that only for virgin specimens are

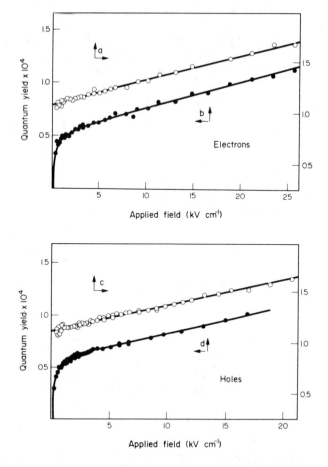

FIG. 6.14. Electron and hole carrier quantum yields at 255 nm as functions of applied
field. Curves *a* and *c* are for previously unirradiated or "virgin" anthracene
crystal specimens. Curves *b* and *d* are for the specimens with trapped holes
and trapped electrons left behind from the experiments for curves *a* and *c*
respectively, thus reducing the low field free carrier yields. Curves *a* and *b, c*
and *d* actually superpose at high fields, but curves *b* and *d* have been shifted
downwards for clarity of presentation. [After Chance and Braun 1973.]

the experimental results in good agreement with Onsager theory, while for specimens
containing trapped carriers, the free carrier–trapped carrier recombination will reduce
the quantum yield of free carriers as shown in Fig. 6.14. This effect is more serious at low
fields ($< 10^3$ V cm^{-1}) because free carriers can diffuse to the surface and recombine
with trapped charges of opposite sign. This effect may also be responsible for the high
values of S/I reported by Batt *et al.* [1968] and Geacintov and Pope [1971] and also for
the electrode dependence of quantum yield [Geacintov and Pope 1971]. The excitation
with strongly absorbed light would produce a higher density of trapped charges at the
surface than in the bulk. However, Hughes [1971] has reported that his experimental
results on field dependence of free carrier generation in X-ray excited anthracene fit well
to the Onsager model. Recently, Pai and Enck [1975] have reported that the Onsager
model provides a quantitative interpretation of their results on field dependent

photogeneration in amorphous selenium. The field dependence of quantum yield indicates that the quantum yield of free carriers is controlled by geminate recombination as originally proposed by Kepler and Coppage [1966].

(B) Auto-ionization

Figure 6.3 shows the charge carrier generation processes via intermediate steps involving excitons, and Tables 6.1 and 6.2 summarize various charge carrier generation mechanisms and their macroscopic features observed in anthracene. The nature of AI states differs from another according to the mode of excitation. For example, the photoionization of a triplet exciton produces auto-ionizing states that are primarily triplet in character, whereas the photoionization of a singlet exciton produces auto-ionizing states that are primarily singlet in character. The internal conversion and hence the branching ratio could be very different for these two types of states even though the energies involved could be equal.

Considering that the charge carriers are generated via a two-photon process, the three types of AI processes described by eqs. (6.15), (6.17), and (6.22) will simultaneously contribute to photocarrier generation. Thus in the spectral region, where optical excitation of spin-allowed states proceeds by the two-photon absorption, the charge carrier concentration n at a given excitation energy hv of light intensity i can be described by [Johnston and Lyons 1970]

$$\frac{dn}{dt} = \gamma_{Si}[S]^2 + \sigma_{Sp}[S]i + K_c i^2 \qquad (6.143)$$

where γ_{Si} is the bimolecular rate constant for carrier generation by exciton–exciton collision, σ_{Sp} is the cross-section for photoionization of singlet excitons, and K_c is the two-photon absorption coefficient. For strongly absorbed light $[S]$ can be approximately expressed as

$$[S] \simeq \varepsilon_S i \tau_S \qquad (6.144)$$

where ε_S and τ_S are, respectively, the absorption coefficient and lifetime of excitons at this particular excitation energy hv. From eqs. (6.143) and (6.144) the ratio of participation of these three types of AI processes is given by

$$\text{Exciton collision/exciton photoionization/exciton AI} = \gamma_{Si}\varepsilon_S^2\tau_S^2/\sigma_{Sp}\varepsilon_S\tau_S/K_c \qquad (6.145)$$

Since all of these parameters are dependent on hv, thus each of these three types can be expected to dominate the carrier generation processes for a particular wavelength region of the incident light. In the following we shall discuss briefly these three types of AI processes.

(a) EXCITON AUTO-IONIZATION

The one-photon direct carrier generation process has just been discussed in Section 6.2.2(A). The quantum yield due to this process is $\eta \simeq 10^{-4}$ for anthracene and it occurs at $hv \geq 4$ eV (in the region 310–200 nm).

TABLE 6.1. *Charge carrier generation mechanisms observed in anthracene* [*after Castro 1971*]

Mechanism	Light intensity dependence	Wavelength dependence	Efficiency	Remarks
Single photon extrinsic	Linear	Reproduces crystal absorption spectrum; $N \propto k$ (extinction depth)	Quantum yield: $\eta(e^+) = 10^{-4}$ to 10^{-2} hole/photon	Observed only for hole production; therefore $\eta(e^+) \neq \eta(e^-)$. (A)
Single photon intrinsic	Linear	4 eV threshold [4.0 4.4 5.6 eV]	$\eta(e^+) = \eta(e^-) = 10^{-4}$ at $298°\mathrm{K}$; $E = 10^4$ V cm^{-1}	No extinction depth dependence, as indicated by lack of a polarization effect. The yield of 3.1 eV excitons is essentially unity.
Direct two-photon	Quadratic	2 eV threshold	Two-photon cross-section for carrier generation: 10^{-30} cm sec	Two-photon cross-section for 3.1 eV exciton generation: 10^{-26} cm sec (B).
Singlet exciton–singlet exciton collision ionization	Quadratic	$N \propto k$	Carrier generation rate constant: 0.9×10^{-12} cm^3 sec^{-1}	Exciton–exciton total annihilation rate constant: 1.5×10^{-8} cm^3 sec^{-1} (C).
Singlet exciton–triplet exciton collision ionization	Quadratic	Fixed wavelength experiment: $h\nu = 1.8$ eV	Carrier generation rate constant: 10^{-12} cm^3 sec^{-1}	Triplet exciton (1.8 eV) produced by intersystem crossing of singlet exciton (3.1 eV). (D).
Single photon ionization of a singlet exciton	Cubic	Fixed wavelength experiment: $h\nu = 1.8$ eV	Ionization cross-section: 2×10^{-19} cm^2 (E); 0.6×10^{-19} cm^2 (F)	Proportional to I^3 instead of I^2 since the singlet excitons are produced by two-photon absorption. (E) (F).
Single photon ionization of a triplet exciton	Quadratic	Fixed wavelength experiments: $h\nu = 2.6$ eV $h\nu = 2.35$eV	Ionization cross-section: 4×10^{-21} cm^2 5×10^{-22} cm^2	(G). (B).

(A) Le Blanc [1967]; (B) Strome [1968]; (C) Braun [1968]; Johnston and Lyons [1968]; (D) Fourny et al. [1968]; (E) Kepler [1967]; (F) Courtens et al. [1967]; (G) Holzman et al. [1967].

TABLE 6.2. *Charge carrier generation via exciton dissociation processes [after Lavrushko and Benderskii 1971]*

Dissociation process	Carrier generator rate	Photoresponse	
		Strong recombination	Weak recombination
Singlet exciton dissociation at electrode–semiconductor interface	$D_s \dfrac{d[S]}{dx}\Big\|_{x=0}$	$\left(\dfrac{\eta_{sd}}{\gamma}\right)^{1/2}\dfrac{\phi_0^{1/2}}{t_u^{1/2}}$	$(\eta_{sd})\Phi_0$
Singlet–singlet annihilation	$\gamma_{si}[S]^2$	$\left(\dfrac{\gamma_{si}\tau_s^2}{\gamma}\right)^{1/2}\dfrac{\phi_0}{t_u}$	$\left(\dfrac{\gamma_{si}\varepsilon_0\tau_s^2}{2}\right)\dfrac{\phi_0^2}{t_u}$
Singlet exciton photoionization	$\sigma_{sp}[S]\dfrac{\phi_0}{t_u}$	$\left(\dfrac{\sigma_{sp}\tau_s}{\varepsilon_0\gamma}\right)^{1/2}\dfrac{\phi_0}{t_u}$	$\left(\dfrac{\sigma_{sp}\tau_s}{2}\right)\dfrac{\phi_0^2}{t_u}$
Singlet–triplet annihilation	$\gamma_{sTi}[S][T]$	$\left(\dfrac{\gamma_{sTi}\tau_s\tau_T}{\gamma}\right)^{1/2}\dfrac{\phi_0}{t_u^{1/2}}$	$\left(\dfrac{\gamma_{sT}\varepsilon_0\tau_s\tau_T}{4}\right)\phi_0^2$
Triplet–triplet annihilation	$\gamma_{Ti}[T]^2$	$\left(\dfrac{\gamma_{Ti}\tau_T^2}{\gamma}\right)^{1/2}\phi_0$	$\left(\dfrac{\gamma_{Ti}\varepsilon_0\tau_T^2}{6}\right)\phi_0^2 t_u$
Triplet exciton photoionization	$\sigma_{Tp}[T]\dfrac{\phi_0}{t_u}$	$\left(\dfrac{\phi_{TP}\tau_T}{\gamma}\right)^{1/2}\dfrac{\phi_0}{t_u^{1/2}}$	$\left(\dfrac{\sigma_{TP}\tau_T}{6}\right)\phi_0^2$

ϕ_0, integrated density of photons incident on the specimen during the illumination pulse.
t_u, illumination time.
ε_0, absorption coefficient of the crystal.
η_{sd}, quantum yield of the singlet exciton dissociation at electrode–semiconductor interface.
γ, carrier recombination constant.
σ_{sp}, σ_{Tp}, cross-sections of photoionization of singlet and triplet excitons, respectively.
τ_s, τ_T, lifetimes of the singlet and triplet excitons, respectively.
Strong and weak carrier recombinations correspond to the cases at low and high applied·fields, respectively.

The two-photon direct carrier generation process becomes predominant if $\varepsilon_S \ll 1$ [cf. eq. (6.145)]. This process involves a direct transition from the valence band to the conduction band or to an auto-ionizing state above E_c. Strome [1968] has used a giant-pulsed ruby laser to generate first and second anti-strokes stimulated Raman lines in liquid nitrogen and liquid oxygen to obtain 40 nsec pulses of light at 597 nm (2.07 eV), 571 nm (2.16 eV), and 525 nm (2.35 eV) for his measurements of photocurrents and prompt fluorescence in anthracene. His results are shown in Fig. 6.15. It can be seen that both the photocurrent and fluorescence have a square-law dependence on light intensity because the prompt fluorescence is directly proportional to the number of singlet excitons. His results are not consistent with singlet photoionization or singlet–singlet collision mechanisms for carrier generation because these two mechanisms would yield photocurrents larger by factors from 15 to 75 depending upon the polarization. He has considered the square-law dependence for his results as evidence for direct two-photon photocarrier generation in anthracene. Similar results have also been reported by Kepler [1971] who used the second harmonic of

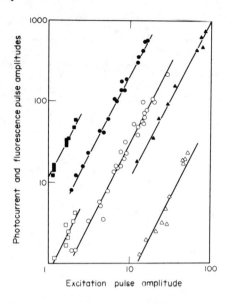

Fɪɢ. 6.15. Photocurrent (open symbols) and fluorescence (solid symbols) pulse amplitude as functions of excitation pulse
amplitude at 597 (triangles), 571 (squares), and 525 ṅm
(circles). One unit of photocurrent is 0.34 nA cm^{-2}; one unit
of fluorescence is 5.7 W cm^{-3}; and one unit of excitation is
3400, 3000, and 2500 W cm^{-2} at 597, 571, and 525 nm
respectively. The anthracene crystal specimen was 1.1 mm
thick and 6.0 mm in diameter, and was biased at 900 V.
[After Strome 1968.]

neodymium light of 530 nm (2.35 eV) as the excitation light. It should be noted that in
the two-photon process the quantum yield is about 10^{-4}, which is of the same order as
that in the one-photon process.

(b) EXCITON PHOTOIONIZATION

There are two possible processes, namely the singlet exciton photoionization and the
triplet exciton photoionization [Choi 1964].

It is extremely important to mention that in order to observe an intrinsic carrier
generation mechanism, it is necessary to prepare carefully an extremely clean surface
and an ultra-pure crystal, and to use high intensity weakly absorbed light to avoid the
trapping and recombination effects at the surface and in the bulk. Using a Q-switched
ruby laser to produce 20 nsec light pulses of photon energy 1.8 eV, Courtens *et al.*
[1967] have measured the photocurrent as a function of light intensity in anthracene.
Their results are shown in Fig. 6.16. The photocarrier density n is proportional to i^3
(light intensity), and n is proportional to $[S]^{3/2}$ and obviously $[S]$ is proportional to i^2.
The specimen they used was in cylindrical shape (3.6 mm in length and 6.2 mm in
diameter) with two silver electrodes on the end faces and was biased with 1500 V for the
photocurrent measurements. Their results lead to the following conclusions: (i) the
charge carriers are generated by a three-photon process; (ii) the dominant mechanism

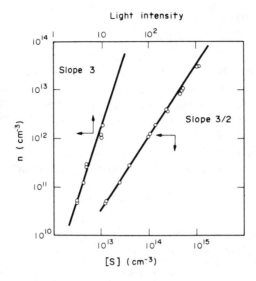

FIG. 6.16. The peak charge carrier density as a function of the peak singlet exciton density [S] and the peak laser light intensity. The light intensity is in arbitrary units with one unit equal to 5×10^{24} photons cm^{-2} sec^{-1}. [After Courtens, Bergman, and Jortner 1967.]

involves two-photon excitations of a singlet exciton followed by one-photon photo-ionization of the singlet exciton according to eq. (6.17), since singlet–triplet exciton collision, which also involves a three-proton process, does not lead to the production of charge carriers in crystalline anthracene [Silver *et al.* 1963, Kepler 1967, Delacote and Schott 1968]; (iii) the cross-section for photoionization of singlet excitons $\sigma_{Sp} \simeq 10^{-19}$ cm^2 at $\lambda = 644$ nm for anthracene, which is in good agreement with those reported by other investigators [Kepler 1967, Fourny *et al.* 1968, Johnston and Lyons 1970], σ_{Sp} is practically independent of the light wavelength; (iv) the bimolecular recombination rate constant is about $(5 \pm 3) \times 10^{-7}$ cm^3 sec^{-1} for charge carrier recombination determined from the decay of photocarrier density versus time results, which is also in reasonable agreement with that reported by Helfrich and Schneider [1966].

For weakly absorbed light (that is $\varepsilon_S \simeq 1$) the rate of free carrier generation by exciton photoionization is approximately 5×10^2 times greater than that by exciton collision ionization [Bergman *et al.* 1967, Johnston and Lyons 1970]. This may explain why Courtens *et al.* [1967] did not observe the fourth-power law dependence of light intensity ($n \propto i^4$).

The triplet exciton photoionization has been studied by several investigators [Holzman *et al.* 1967, Hernandez 1968]. The cross-section for photoionization of triplet excitons in $\tau_{Tp} \simeq 4 \times 10^{-21}$ cm^2 which is much smaller than τ_{Sp}. The values of τ_{Sp} and τ_{Tp} are dependent on applied voltage because of geminate recombination.

(c) EXCITON COLLISION IONIZATION

There are two most probable processes, namely, the singlet exciton–singlet exciton collision ionization and the singlet exciton–triplet exciton collision ionization.

For singlet–singlet collision ionization $\varepsilon_S \gg 1$, so that the singlet exciton density is sufficiently high to make the collision ionization predominant. Many investigators have studied the carrier generation by these processes [Choi 1964, 1965, 1967, Choi and Rice 1962, 1963, Kearns 1963, Silver et al. 1963, Bergman et al. 1966, 1972, 1974, Braun 1968, Jortner 1968, Jortner and Bixon 1969]. If ε_S is so large that the first term is much larger than the sum of the last two terms on the right-hand side of eq. (6.143), then the rate of carrier generation by the bi-excitonic mechanism can be written as

$$\frac{dn(x)}{dt} = \gamma_{Si}[S]^2 \tag{6.146}$$

For the excitation light pulse of width at half-peak height W, the carrier concentration can be approximately expressed as

$$n(x) = W\gamma_{Si}[S]^2 \tag{6.147}$$

where $[S]$ is given by eq. (6.120). Thus the total carriers generated by the singlet–singlet collision ionization is [Braun 1968]

$$N = \int_0^\infty n(x)\,dx$$

$$= \int_0^\infty W\gamma_{Si}\left\{\frac{\varepsilon_S i_S \tau_S}{1-\varepsilon_S^2 D_S \tau_S}\left[\exp(-\varepsilon_S x)-\exp(-x/\sqrt{D_S\tau_S})\right]\right\}^2 dx$$

$$= \frac{W\gamma_{Si}\tau_S^2\varepsilon_S i_S^2}{2(1+\varepsilon_S\sqrt{D_S\tau_S})^3} \tag{6.148}$$

Using 0.3×10^{-6} sec pulses of approximately monochromatic, unpolarized light at wavelengths between 315 and 410 nm, Braun [1968] has measured electron carriers generated in anthracene crystals at an applied field of 10^4 V cm^{-1}. His results in Fig. 6.17 show a nearly quadratic dependence of the number of carriers generated on the incident light intensity. This is a two-photon process involving singlet–singlet collision ionization according to eq. (6.22). Since from eq. (6.148) the parameter $B = N(1+\varepsilon_S\sqrt{D_S\tau_S})^3/\varepsilon_S i_S^2 = W\gamma_{Si}\tau_S^2$ should be independent of wavelength, Braun has plotted B as a function of wavelength and found that for wavelengths between 390 and 340 nm, B is practically independent of wavelength, in good agreement with the prediction of eq. (6.148). Obviously, at wavelengths beyond this range other mechanisms for carrier generation become important.

Taking γ_S as the rate constant for singlet exciton disappearance via exciton–exciton annihilation (cf. Table 7.1 of Chapter 7) and assuming that all the highly excited states, formed by the annihilation, decay via the first singlet exciton state, and that the diffusion term is neglected, then the steady state concentration of singlet excitons can be determined from the following equation [Braun 1968]:

$$\frac{\partial[S]}{\partial t} = \varepsilon_S i_S \exp(-\varepsilon_S x)-\frac{[S]}{\tau_S}-\frac{\gamma_S}{2}[S] \tag{6.149}$$

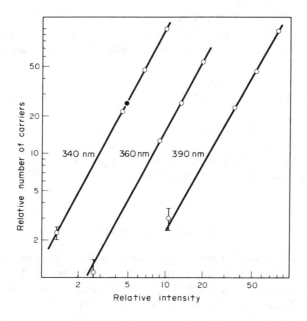

FIG. 6.17. Number of photogenerated electron carriers as a function of
light intensity. The units are arbitrary and data for each
wavelength have been arbitrarily displaced for clarity. For
340 nm, slope = 1.87; for 360 nm, slope = 1.85; and for
390 nm, slope = 1.77. The fully blackened point in the
340 nm curve corresponds to $i = 2.5 \times 10^{18}$ photons
$cm^{-2} sec^{-1}$ and $N = 3.75 \times 10^6 cm^{-3}$. [After Braun 1968.]

Using the expression for $[S]$ from eq. (6.149) and following eqs. (6.147) and (6.148), it
can be easily shown that

$$N = W\gamma_{Si}F(b)/\tau_S^2\varepsilon\gamma_S^2 \qquad (6.150)$$

where

$$F(b) = b - 4[(1+b)^{1/2} - 1] - 2\ln\left\{\frac{4[(1+b)^{1/2} - 1]}{b[(1+b)^{1/2} + 1]}\right\} \qquad (6.151)$$

and

$$b = 2\gamma_S \tau_S^2 \varepsilon_S i_S \qquad (6.152)$$

For $N = 3.75 \times 10^6 cm^{-3}$, $i = 2.5 \times 10^{18}$ photons $cm^{-2} sec^{-1}$ at 340 nm, $\varepsilon_S \simeq 2$
$\times 10^4 cm^{-1}$; Braun [1968] has calculated the rate constant for carrier generation by
singlet–singlet collision ionization which is $\gamma_{Si} = 0.9 \times 10^{(-12\pm0.4)} cm^3 sec^{-1}$ for
anthracene. This value agrees well with that reported by Johnston and Lyons [1970].

   Using a Q-switched ruby laser to produce a light beam of about 50 nsec duration
with photon energy of 1.79 eV, Hasegawa and Yoshimura [1966] have observed a
fourth power law dependence of the photocurrent in anthracene crystal on light
intensity in the low intensity region, and a quadratic power law dependence in the high

intensity region. The fourth power law dependence can be explained as due to the exciton–exciton interaction for generation of carriers and a two-photon process for production of excitons. They attributed the quadratic power law dependence to recombination effects.

Autoionization occurs not only in anthracene, but also in other organic solids, such as in naphthalene [Braun and Dobbs 1970], in benzene, biphenyl, and pyrene [Castro 1971].

Fourny *et al.* [1968] have reported that a quadratic power law for light intensity dependence of photocurrent has been observed in anthracene using excitation light of photon energy of 1.8 eV. They have attributed the photocarrier generation to singlet exciton–triplet exciton collision ionization following eq. (6.25).

6.3. PHOTOEMISSION

Photoemission is generally referred to as emission of electrons into vacuum from a solid or injection of electrons or holes from a solid into another solid, resulting from an interaction between photons and a solid. The photoemission of carriers from a solid into vacuum is referred to as the external photoemission, while that from a solid into another solid such as from a contact electrode to a semiconductor is referred to as the internal photoemission. Photoemission has been extensively studied in recent years as a tool to study energy levels, and particularly ionization energies of molecular crystals [Gutmann and Lyons 1967, Kochi *et al.* 1970], and to obtain information about the injection contact potential barriers, surface states and electronic structures of solids [Williams and Dresner 1967, Baessler and Vaubel 1968, Vaubel and Baessler 1968, Caywood 1970]. In the following we shall discuss (i) photoinjection from a contacting electrode to a crystalline solid, and (ii) photoemission from a crystalline solid to vacuum.

6.3.1. Photoinjection from electrical contacts

(A) Metallic contacts

An extensive compilation of experimental data on this subject for inorganic semiconductors has been given by Williams [1970] and Ruppel [1970] and for organic semiconductors by Caywood [1970]. Metallic contacts can be roughly classified into two groups, namely ohmic contacts and blocking contacts [Rose 1963]. An ohmic contact can be considered as a reservoir of carriers which is always ready to supply as many carriers as needed. Usually, at a given field, the ohmic contact can supply more carriers than the bulk material can carry. Thus the current is bulk limited. Further increase in current injection by photoexcitation does not affect the current, and therefore no photoemission current can be observed.

A blocking contact can inject only very few carriers which are much less than what the bulk material can carry. Thus the current is contact limited. If a light of energy $hv \geq \phi_B$ (where ϕ_B is the potential barrier height—the difference between the Fermi level of the metal and the conduction band edge of the semiconductor as shown in Fig. 2.4) is used to illuminate the metal contact, photoinjection from the contact will take place and the photoemission without taking into account the effects of scattering and relaxation can be written, according to the Fowler theory [Fowler 1931, Hughes and Du Bridge 1932], as

$$J_{\text{ph}} = C(h\nu - \phi_B)^2 \quad \text{for} \quad h\nu \geq \phi_B \tag{6.153}$$

This equation is valid only for $\phi_B \geq 0.5$ eV if $h\nu - \phi_B$ is greater than some multiple of $kT (\geq 6kT)$ and $h\nu \leq 1.5\phi_B$ because beyond this range the assumptions for the simple Fowler theory is no longer valid [Kadlec and Gundlach 1976]. Furthermore, eq. (6.153) is valid only for photoinjection from a metal to a semiconductor having wide energy bands (with bandwidths larger than 0.5 eV). Most metals, inorganic semiconductors, and insulators can satisfy these conditions. For materials and conditions for which eq. (6.153) is valid, the measurements of J_{ph} as a function of $h\nu$ and then the extrapolation of the plot of $J_{\text{ph}}^{1/2}$ versus $h\nu$ to $J_{\text{ph}} = 0$ yield ϕ_B. However, eq. (6.153) cannot be applied to narrow energy band materials, such as in most low mobility organic semiconductors (e.g. anthracene). In narrow energy band materials an electron injected into the narrow band can diffuse only a few angstroms (about 5 Å) and would then be captured in a bound state of the Coulomb image potential (cf. Figs. 2.12 and 2.13). Therefore an injected electron must have a sufficient momentum perpendicular to the surface before it can escape from the image force to enter the conduction band.

(a) FROM A METAL INTO WIDE BANDWIDTH SEMICONDUCTORS

This is the most common case and it has been treated in detail by some investigators [Hughes and Du Bridge 1932, Abeles 1972, Kadlec and Gundlach 1976]. As has been mentioned before, in this case the photoemission measurements would directly yield the barrier height of the blocking contact ϕ_B. For semiconductors for which the image force effect is negligible, the photoemission method is reasonably accurate for determining ϕ_B. A typical example is given in Fig. 6.18 for a gold contact to a CdS crystal specimen in which $\phi_B = 0.7$ eV is obtained for this contact potential barrier height. A similar agreement to the Fowler theory has been reported for silicon [Spitzer *et al.* 1962] and for other inorganic semiconductors [Mead and Spitzer 1963, 1964]. By measuring the threshold energy $h\nu_0 = \phi_B$ for electron emission and that for hole emission, the sum of these two threshold energies gives the energy band gap of the semiconductor. For example, the threshold energy for electron emission from gold to *n*-GaP is 1.30 eV [Cowley and Heffner 1964] and that for hole emission from gold to *p*-GaP is 0.72 eV [White and Logan 1963]. The sum of these is 2.02 eV, which is about the energy band gap of GaP, which is 2.18 eV [Tietjen and Amick 1966]. It should be noted that the barrier heights determined by means of photoemission measurements depend on the specimen surface conditions. The effects of surface states have been discussed by several investigators [Mead and Spitzer 1963, 1964, Goodman 1964].

For large band gap materials the Fowler plot at $h\nu$ close to $h\nu_0$ is usually not linear, that means $(J_{\text{ph}})^{1/2}$ is not proportional to $(h\nu - \phi_B)$, or deviates from the linear Fowler plot. Figure 6.19 shows a typical $(J_{\text{ph}})^{1/2}$ versus $h\nu$ curve for an MIM (metal–insulator–metal) structure. The asymmetry of the $Al-Al_2O_3-Al$ $(\phi_1 > \phi_2)$ shown in the inset of Fig. 6.19 is probably due to the different pre-history of the two

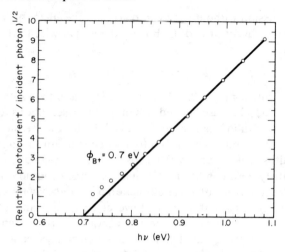

FIG. 6.18. (Relative photocurrent per incident photon)$^{1/2}$ as a function of photon quantum energy for an evaporated gold contact to a cadmium sulphide crystal specimen. [After Goodman 1964.]

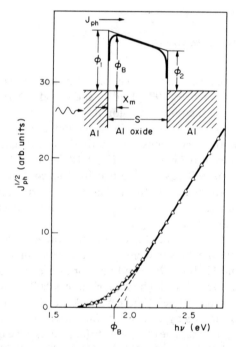

FIG. 6.19. (Photoemission current)$^{1/2}$ as a function of incident photon quantum energy for an Al (1000 Å)–Al$_2$O$_3$(30 Å)–Al (200 Å) structure with the 200 Å thick top electrode illuminated. The corners of the barriers are rounded off due to the image force as shown in the inset. [After Kadlec and Gundlach 1975.]

interfaces [Pollack and Morris 1964]. This non-linear behaviour or deviation has been observed by many investigators [Gundlach and Kadlec 1975, Jourdain and Despujols 1973, Lewicki et al. 1972], and this makes it difficult to determine accurately the value of ϕ_B. Several factors may be responsible for this deviation and they are [Kadlec and Gundlach 1976]: (i) the electron distribution may be smeared out about the Fermi level but the deviation is too large to be explained by this effect; (ii) the scattering of electrons in the conduction band may play a role but it is likely that it can be important for thick specimens and not for films thinner than 50 Å; (iii) the quantum mechanical transmission coefficient $T(E_x)$, which affects the photocurrent, is not equal to zero for $E_x < \phi_B$ because some of the electrons can tunnel through the potential barrier, and is not equal to 1 for $E_x > \phi_B$ because some of the electrons are reflected, resulting in a smearing of the step-like form of $T(E_x)$ and thus causing the deviation of the $J_{ph}^{1/2}$ versus hv curve, where $E_x = m v_x^2/2$ and v_x is the normal component of the electron velocity; (iv) it is possible that the barrier height is not uniform over the whole area of the interface, some regions may have $\phi_B = 1.9$ eV and some regions may have $\phi_B = 1.7$ eV.

It should be noted that the barrier height may depend on specimen thickness and light intensity, possibly due to space charge effects [Gundlach and Kadlec 1972]. Obviously, because of the image force effect the barrier height is field dependent. Berglund and Powell [1971] have derived the expression for the photoemission current versus applied voltage characteristics for an MIM structure shown in Fig. 6.20, and it is given by

$$J_{ph} \sim \left\{ hv - \phi_B - \left[\frac{q}{4\pi\varepsilon S}(qV + \phi_1 - \phi_2) \right]^{1/2} \right\}^p \exp\left(-\frac{x_m}{\lambda} \right) \qquad (6.154)$$

where S is the thickness of the specimen, x_m is the distance from the illuminated electrode to the location of the maximum barrier height, λ is the mean free path, and p is a parameter depending on the kind of electron excitation and lies $1 \le p \le 3$. It can be seen in Fig. 6.20 that the experimental results agree well with eq. (6.154) for the Si–SiO$_2$–Au structures. If the insulator is too thin, the incident light illuminating the first electrode can also reach the second electrode (opposite to the first) and excite carriers there. Consequently, the photoemission current is composed of two types of currents (electrons and holes) under an applied field. Unless the component from the opposite electrode is negligibly small, the photoemission consisting of two types of carriers would influence the thickness and hv dependence of J_{ph}. Furthermore, the electron–electron and electron–phonon interactions also play an important role in this dependence. A comprehensive review of all these effects on the internal photoemission has recently been given by Kadlec [1976].

The application of various internal photoemission experiments for the determination of the barrier height, the potential barrier shape, the charge distribution in the insulator, the topographical distribution of the barrier height, the trap distribution in the barrier, and also the mean free paths and energy losses of the carriers has recently been critically reviewed by Kadlec and Gundlach [1976].

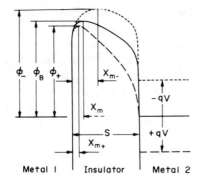
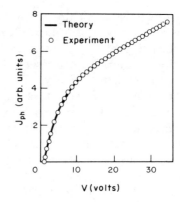

FIG. 6.20. Energy diagram of an MIM structure (Si–SiO$_2$–Au) and voltage dependence of photoemission current from Si into SiO$_2$ for $hv = 5$ eV (Si acts as metal 1 in this case). Thickness of SiO$_2$ = 2450 A. [After Berglund and Powell 1971.]

(b) FROM A NARROW BANDWIDTH EMITTER INTO WIDE BANDWIDTH SEMICONDUCTORS

A degenerate n-type or p-type silicon or similar semiconductor can act as a narrow bandwidth emitter. Goodman [1966, 1968] has studied experimentally the photoemission from such a narrow bandwidth emitter into an insulator. Photons with energies hv_0 or hv_1 can excite electrons from the conduction band and those with energy hv_2 can excite electrons from the valence band of a degenerate n-type semiconductor into the conduction band of an insulator as shown in Fig. 6.21. For a given photon energy hv, the excited electrons are distributed over only a narrow band of the width within a few kT of the bottom of the conduction band, which are far narrower than those excited from the metal. Thus within the range of photon energies $\phi_B \leq hv \leq \phi_B + E_g$, the photoemission current (or the quantum yield for photoemission) as a function of photon energy is simply

$$J_{ph} = C(hv - hv_0) \tag{6.155}$$

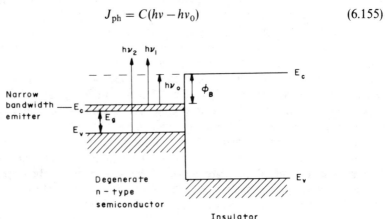

FIG. 6.21. Illustrating various transitions possible in photoemission of electrons from a degenerate n-type semiconductor into an insulator.

For the Si–SiO$_2$ system the experimental results fit eq. (6.155) if the photon energies hv is less than $\phi_B + E_g$, and then fit eq. (6.153) rather than eq. (6.155) when $hv > \phi_B + E_g$ because at higher photon energies the electrons excited from the valence band become dominant as shown in Fig. 6.21.

(c) FROM A METAL INTO NARROW BANDWIDTH CRYSTALS

The narrow band can be considered as a delta function. This means that within the band the energy distribution of excited holes (or excited electrons) is uniform, and implies that the quantum yield for photon energies above the threshold hv_0 is independent of photons energy as shown in Fig. 6.22(a). In organic crystals there are usually several narrow bands separated by roughly equally spaced levels due to molecular vibrations. For example, in anthracene, the molecular vibration has frequencies corresponding to energies of approximately 0.2 eV but the electronic bandwidth is about 0.02 eV. For clarity only two of such narrow valence bands are shown in Fig. 6.22(b). Each band gives rise to a step function of photoemission current starting at different photon energies. The combination of these step functions gives a staircase quantum yield Y, and the derivative $dY/d(hv)$ gives a clear picture of energy level splitting by molecular vibration as shown in Fig. 6.22(b).

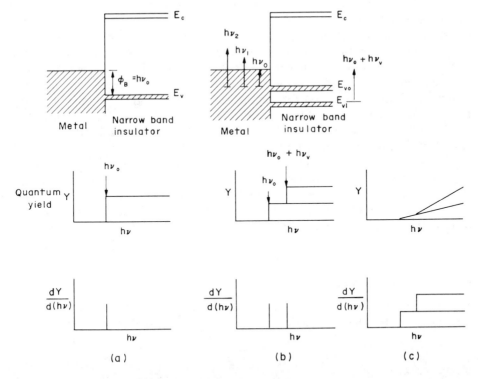

FIG. 6.22. Schematic diagrams illustrating photoemission from metal into narrow band crystals: (a) with one narrow band when the acceptance angle is not considered, (b) with two narrow bands when the acceptance angle is not considered, and (c) with two narrow bands when the acceptance angle is taken into account.

Figure 6.22(b) gives only the basic concept about the behaviour of narrow bands. In fact the quantum yield of the ith narrow band should be proportional to $(hv - \phi_{Bi})$ since the number of excited carriers which can surmount the potential barrier ϕ_{Bi} is proportional to $(hv - \phi_{Bi})$ as shown in Fig. 6.22(c) [cf. also the top energy diagram of Fig. 6.22(b)]. The total quantum yield Y from all the vibrationally split levels should be the sum of the linear ramps as shown in Fig. 6.22(c). In this case the derivative $dY/d(hv)$ gives a series of step functions starting at different photon energies. For photoemission from a metallic contact to an organic crystal, the quantum yield Y should follow the relation schematically shown in Fig. 6.22(c) and $Y \propto (hv - \phi_B)^2$ rather than those shown in Fig. 6.22(b) with $Y \propto (hv - \phi_B)$. Williams and Dresner [1967] have measured the photoemission quantum yield as a function of photon energy for anthracene with various metallic contacts. The experimental arrangement used by them for these measurements is shown in Fig. 6.23(a). Light enters through the transparent electrode (deionized water) and the organic crystal specimen, and is then absorbed by the metallic electrodes. The metallic electrode was vacuum deposited on the organic crystal specimen at liquid nitrogen temperature in order to ensure good adherence. Some of their results are shown in Fig. 6.23(b) in which it can be seen that Y is proportional to $(hv - \phi_B)^2$. Extrapolation of these curves yields the value of ϕ_B for a given metallic electrode. The values of ϕ_B (the potential barrier or the threshold energy for photoemission) for Au, Ag, Pb, Al, and Mg electrodes are 1.17, 1.20, 1.60, 1.86, and 1.97 eV, respectively. The roughly periodic departures of the points from the straight line in Fig. 6.23(b) can be considered as the indication of the vibrational splitting. The derivative $dY/d(hv)$ of the curves shown in Fig. 6.23(c) gives even a much clearer picture of the vibrational splitting. The peaks of $dY/d(hv)$ versus hv curves for various metal electrodes occur at about the same positions of the peaks of the anthracene absorption spectrum. For details about the photoemission measurements for the determination of the nature of metallic contacts on organic crystals and the conduction states in organic crystals, the reader is referred to some original papers [Williams and Dresner 1967, Dresner 1968, 1970, Williams 1970, Caywood 1970, Baessler and Vaubel 1968, Baessler et al. 1968, Vaubel and Baessler 1968].

The above arguments for photoemission of holes from metal to narrow band crystals apply equally well to cases for photoemission of electrons from metal to narrow band crystals. It should also be noted that from the arguments given in (b) and (c) above, it is not difficult to estimate the photoemission spectrum for the cases from a narrow bandwidth emitter into crystals with a series of narrow bands.

(B) Electrolytic contacts

Photoemission can also occur from an electrolytic contact into an organic crystal. In general, an electrolytic electrode would inject holes if the optical excitation energy

$$hv > E_I - \phi_{EL} \qquad (6.156)$$

or inject electrons if

$$hv > \phi_{EL} - \chi \qquad (6.157)$$

where E_I and χ are, respectively, the ionization energy and the electron affinity of the

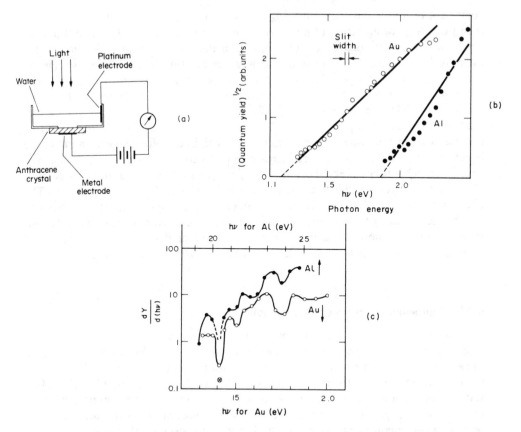

FIG. 6.23. (a) Electrode arrangement for photoemission measurements, (b) quantum yield as a function of photon energy for photoemission from Au and Al electrodes into anthracene crystal, and (c) derivative of the quantum yield with respect to photon energy. The point ⊗ belongs to the aluminium curve. [After Williams and Dresner 1967.]

organic crystal; and ϕ_{EL} is the work function of the electrolytic electrode [Pope *et al.* 1962, Kallmann and Krasnansky 1967]. For example, an electrolytic solution containing iodine acts as a good hole-injecting contact into anthracene if the contact is illuminated by light of suitable wavelength and biased at positive polarity [Kallmann and Pope 1959, 1960, 1962]. Using electrolyte $NaI + I_2$ as electrode for anthracene, Kallmann and Pope [1960] have observed that the photoemission current at light wavelength of 436 nm is much higher than that caused by the absorption of the anthracene. They have attributed the hole injection to the following reaction processes [cf. also Mehl *et al.* 1969]:

$$I_2 + h\nu \rightarrow 2I \qquad (6.158)$$

$$I + e \text{ (from anthracene)} \rightarrow I^- + h \text{ (to anthracene)} \qquad (6.159)$$

However, Mehl *et al.* [1969] have proposed another possible mechanism responsible for the hole injection, that is the donor–acceptor reaction between excited iodine molecules and anthracene excitons. They have also reported that the diffusion coefficient of iodine molecules in CCl_4 determined from the photocurrent measurements due to photoinjection of holes from the solution of $CCl_4 + I_2$ into anthracene agrees well with the theoretical value based on the donor–acceptor reaction at the interface. For more information about such charge reaction between redox systems in electrolytes and organic crystals with and without optical excitation, the reader is referred to the work of Mehl and his coworkers [Mehl 1971, Hale and Mehl 1971], Gerischer [1973], and Willing et al. [1974].

Dyes or electron acceptors dissolved in electrolytes can form good photoinjecting electrodes. Such electrodes under illumination by light of suitable wavelength would inject holes and thus cause a large increase in photocurrent, and this photoinjection phenomenon has been observed in anthracene, perylene, and dibenzanthracene crystals [Meier 1974]. Dyes can shift the spectral sensitivity of organic photoconductors. The photocurrents induced in organic photoconductors by means of excitation of dyes can also be used to study the photochemical behaviour of excited molecules, the relations between organic photoconductance, photochemistry of dyes, photodegradation of polymers, photobiological problems, and energy transfer [Meier 1974].

6.3.2. Photoemission from crystalline solids

In general, photoemission from crystalline solids means volume photoemission resulting from optical absorption in the bulk, which should be distinguished from surface photoemission resulting from optical absorption at the surface. The contribution from excitation of surface states is small because the total number of surface states is small compared to the number of states in the bulk which can participate in photoemission processes. Therefore the photoemission is mostly a volume effect. The photoemission quantum yield increases with increasing specimen thickness and reaches a saturation value when the specimen thickness exceeds the penetration (or absorption) depth or the escape depth. If the specimen thickness is smaller than the penetration depth, the photoemission may exhibit a roughly periodic nature of its variation with specimen thickness, going through a minimum at thicknesses equal to an odd number of quarter-wavelengths and through a maximum at thicknesses equal to an integral number of half-wavelengths of the incident light [Ramberg 1967]. To simplify matters, such an interference effect is ignored in the following discussion and the photoemission is assumed to be a bulk phenomenon.

The photoemission consists of three steps: (1) an electron excited to a high energy state by the absorption of a photon, (2) the scattering of the excited electron on its way moving to the vacuum–solid interfaces, and (3) the escape of the electron over the potential barrier at the surface of the solid. Surface states may affect volume photoemission indirectly through band bending. However, for the sake of simplicity and clarity, we neglect the band-bending effects. Figure 6.24 illustrates the three steps and the band diagram and the relative energy levels for photoemission from an organic crystal into vacuum. The light intensity of the excitation with photon energy hv at the same distance x from the vacuum–solid interface is

$$I(v, x) = I_0(1 - R) \exp(-\varepsilon x) \qquad (6.160)$$

where I_0 is the light intensity in vacuum, R and ε are, respectively, the reflection and the

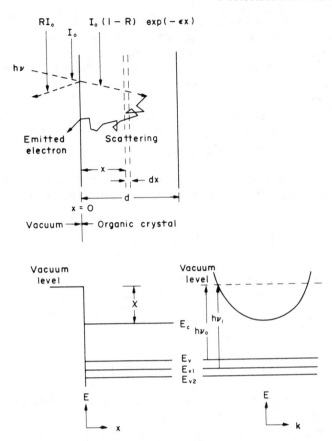

FIG. 6.24. Schematic diagrams illustrating excitation, scattering, and escape of an excited electron, the hole band E_r, the vibrationally excited hole bands E_{r1}, E_{r2}, and so on. The conduction band is assumed to be parabolic and the valence bands are assumed to be almost independent of the **k** vector. Localized states in the energy gap due to imperfections are not shown for the sake of clarity.

absorption coefficients of the crystal, which are dependent on light wavelength. $I(v, x)$ can be defined as the number of photons of energy hv arriving at x per second. Before going to organic crystals, we take a simple inorganic semiconductor as an example to illustrate the procedures of deriving an expression for the photoemission yield Y. Supposing that the energy of incident photons hv is larger than $E_g + \chi$, then the photocarriers generated within dx at x is

$$dn = \alpha \varepsilon I(v, x) dx$$

$$= \alpha \varepsilon I_0 (1 - R) \exp(-\varepsilon x) dx \qquad (6.161)$$

where α is the probability that a photon absorbed will excite an electron to a high energy state. The energy $hv > (E_g + \chi)$ may excite some electrons to a high energy state above the vacuum level, but it may also excite some to levels below the vacuum level if the electrons being excited are located much lower than E_v in the valence band. Clearly, only those above the vacuum level can lead to photoemission. Furthermore, even though some excited electrons at x may have energies above the vacuum level, they will lose part of their energies by the scattering processes during their motion towards the vacuum–solid interface. Thus by the time they reach the interface some of them may have lost so much energy that their energies become lower than the vacuum level and they will not be able to escape over the potential barrier. The probability for an excited electron at x to escape to contribute to photoemission can be written as [Spicer 1968]

$$P_{esc}(v, x) = B(v) \exp(-x/L_{esc}) \tag{6.162}$$

where $B(v)$ is a constant and L_{esc} is the escape depth. Both $B(v)$ and L_{esc} are dependent on the electron energy. The total photocarriers emitted from the crystal specimen is then

$$n = \int_0^\infty \alpha \varepsilon I_0 (1 - R) \exp(-\varepsilon x) B \exp(-x/L_{esc}) dx$$

$$= \frac{\alpha \varepsilon B I_0 (1 - R)}{\varepsilon + 1/L_{esc}} \tag{6.163}$$

The quantum yield is defined as

$$Y = \frac{n}{I_0 (1 - R)} = \frac{\alpha B}{1 + 1/\varepsilon L_{esc}} \tag{6.164}$$

Since B and L_{esc} are functions of electron energy, they must be directly related to scattering mechanisms and absorption processes (direct transition without involving a change in momentum or indirect transition involving a change in momentum). Kane [1962] has analysed theoretically the quantum yield versus photon energy characteristics near the threshold for a general band structure and for a variety of photocarrier generation and scattering mechanisms involving volume and surface states (volume- and surface-scattering processes) in semiconductors. The initial fast-rising portion of the $Y - hv$ curve can be expressed in the form

$$Y = A(hv - E_{th})^S \tag{6.165}$$

where A and S are constants and E_{th} is the threshold energy required for photoemission. For intrinsic and lightly doped semiconductors, $E_{th} = \chi + E_g$. Depending on the photocarrier generation and scattering mechanisms, S varies from $S = 1$ to $S = 5/2$ [Kane 1962]. For the cases without involving surface scattering the values of S are as follows:

$S = 1$ for direct transition without volume scattering.
$S = 2$ for direct transition involving elastic scattering.
$S = 5/2$ for indirect transition with or without elastic scattering.

In practical cases the photocarriers suffer both volume and surface scattering. However, $S = 1$ in the high energy region and $S = 3$ in the region near the threshold have been experimentally observed in III–V compounds such as InSb, GaSb, InAs, and GaAs [Gobeli and Allen 1966]. However, Kane's theory depends on the shape of the edge of the valence band and, therefore, cannot be applied to organic crystals with narrow valence bands.

In organic crystals excitons are intermediate states in photocarrier generation processes. If this involves a one-quantum process, the carriers are generated by AI of high energy excitons (AI excitons) as has been discussed in Section 6.2.2. However, for photoemission from an organic crystal to vacuum, the AI excitons must at the energy states be higher than the vacuum level. In organic crystals the molecular vibrations split each hole band into many well-separated bands with an energy separation of one vibrational quantum hv_v as shown in Figs. 6.22 and 6.24. The electron escape probability P_{esc} involves both the scattering probability and the surface transmission probability. Near the threshold, P_{esc} may be represented only by the surface transmission probability $T(E_K)$ because the scattering probability varies slowly in this energy region [Kochi *et al.* 1970]. $T(E_K)$ depends on the surface potential barrier profile and is difficult to determine. According to Kochi *et al.*, $T(E_K)$ can be approximately expressed as

$$T(E_K) \simeq \begin{cases} 0 & \text{for} \quad E_K < 0 \\ (E_K)^l & \text{for} \quad E_K \geq 0 \end{cases} \tag{6.166}$$

where E_K is the kinetic energy of excited electron just outside the crystal surface, and l is a constant with values $\frac{1}{2} \leq l \leq 1$. Neglecting energy losses, E_K can be expressed as (cf. Fig. 6.24)

$$E_K = hv - hv_n = hv - hv_0 - n\hbar\omega \tag{6.167}$$

where hv_0 denotes the threshold energy for ionization (corresponding to the ionization potential E_{th}), $\hbar\omega$ is the molecular vibrational energy quantum, and n is the quantum number. Using $P_{esc} \simeq T(E_K)$ and n to replace Condon's overlap integral $\langle n|0 \rangle$, Kochi *et al.* [1970] have derived a relation between the quantum yield and the photon energy near the threshold, and it is given by

$$Y \propto (hv - E_{th})^S \tag{6.168}$$

with $\frac{5}{2} \leq S \leq 3$.

Kochi *et al.* [1970] have also measured Y as a function of hv for anthracene, naphthacene, pentacene, perylene, indanthrone, and tetrathionaphthacene, and found that near the threshold their experimental results are in good agreement with the empirical cube law

$$Y \propto (hv - E_{th})^3 \tag{6.169}$$

By plotting $Y^{1/3}$ versus hv as shown in Fig. 6.25, the extrapolation of the curves to $Y = 0$ yields E_{th}. The values of E_{th} are in good agreement with those reported by other investigators [Lyons and Morris 1960, Terenin and Vilesov 1961, Harada and Inokuchi 1966, Batley and Lyons 1968, Marchetti and Kearns 1970, Sworakowski 1972].

Sworakowski [1972, 1974], using an approach similar to that used by Kochi *et al.* [1970] and Marchetti and Kearns [1970], has derived the following expression for the photoemission quantum yield:

$$Y = \beta(v) \int_{x=0}^{\infty} \int_{\Delta E = \Delta E_l}^{\Delta E_\mu} \int_{-\pi/2}^{\pi/2} n(\Delta E) T(E'_K) \exp\left(-\frac{x}{L_{esc} \cos\theta} - \varepsilon x \right) d\theta \, d\Delta E \, dx \tag{6.170}$$

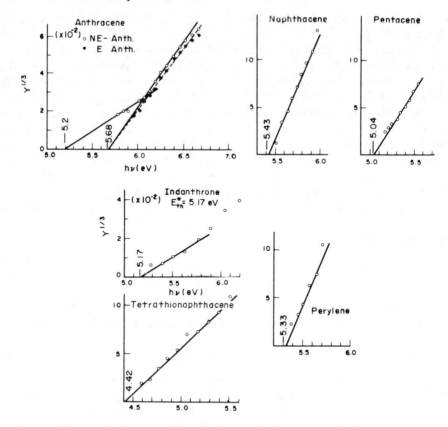

F IG. 6.25. The plot of $Y^{1/3}$ versus photon energy $h\nu$ near the threshold for six organic crystals. NE-Anth denotes non-evaporated anthracene, and E-Anth denotes the anthracene specimen after more than 5 h of evacuation at a pressure of 10^{-3} Torr. [After Kochi, Harada, Hirooka, and Inokuchi 1970.]

where $\beta(\nu)$ is a parameter depending only on the photon energy and including the overall efficiency of carrier generation and the surface reflection coefficient, θ is the angle between the direction of electron motion and the normal to the surface, L_{esc} is the escape depth which is assumed to be of the order of a few intermolecular distances, E'_K is slightly different from E_K defined by eq. (6.167), and it is

$$E'_K = h\nu + E'_v(x) + \Delta E \qquad (6.171)$$

in which ΔE denotes changes in the energy of a valence level of a molecule due to thermal vibration or any other types of defects continuously distributed in energy and E'_v is the energy of the valence level measured from the vacuum level $[E'_v(x) < 0]$, and $n(\Delta E)$ is the distribution function for the density of states, which is assumed to be Gaussian

$$n(\Delta E) = \frac{N_0}{\sigma \sqrt{2\pi}} \exp\left[-\frac{(\Delta E)^2}{2\sigma^2}\right] \qquad (6.172)$$

in which N_0 is the total density of states, σ is the dispersion (of the order of a few kT), and ΔE is measured with respect to E_v'. It is obvious that the solution of eq. (6.170) for all over the spectral region must be of numerical computation. However, an approximate solution of eq. (6.170) gives

$$Y \propto (hv - E_{th})^{1/m} \tag{6.173}$$

which is valid in the region sufficiently far above the threshold, where $m > 1$ is a constant. The most probable value of m is 2.

The approach of Sworakowski has been commented on by Aihara [1973] on two main points: (i) the influence of intramolecular vibrations has been neglected in deriving eq. (6.170), which may affect the $Y - hv$ relation near the threshold, and (ii) the influence of scattering of electrons has been neglected in deriving eq. (6.170), which may affect the photoemission energy distribution bands. Later, Sworakowski [1974] has given a detailed discussion about these effects. Regarding point (i) the Gaussian distribution function $n(\Delta E)$ should be a good approximation to represent the distribution of the density of states of low energy bands at about the valence level. Regarding point (ii), the major scattering causing substantial energy losses is the electron–hole pair production by electron–electron scattering (Auger effect), and this would occur only when $hv > 2E_g$, that means hv is well above the threshold [Kochi et al. 1970]. In organic crystals, the scattering does not affect the quantum yield for photon energies close to the threshold. Other scattering mechanisms, such as the electron–phonon scattering, which gives rise to energy losses of the order of 10^{-3} eV per collision, have negligible effects. The change in photoionization cross-section for hv close to the threshold is negligibly small [Druger 1972, Sworakowski 1974]. However, Sworakowski [1972] has used the relation

$$Y \propto (hv - E_{th})^{1/2} \tag{6.174}$$

and plotted the Y^2 versus hv curve. The extrapolation of this curve to $Y^2 = 0$ yields the value of E_{th}, which is close to that determined from eq. (6.169), indicating that eq. (6.174) is valid for the above-threshold region. The slight difference in the value of E_{th} may be due to the fact that the square-root extrapolation yields the E_{th} corresponding to the first maximum in the photoelectron energy distribution spectrum (from the most probable position of the first valence band to the vacuum level) while the cubic law extrapolation gives the E_{th} corresponding to the onset of photoemission. Sworakowski [1974] has also shown that by means of suitable approximation eq. (6.170) can be reduced to eq. (6.168) valid only near the threshold.

(A) Effects of surface conditions

In the above analysis the energy bands are assumed to be flat up to the surface. In fact, there are always surface states present at the surface which induce space charge near it causing the band to bend up or down as shown in Fig. 6.26. Gases such as oxygen adsorbed on the surface may result in a change of the order of 1 eV in E_{th}. If the space charge causes the band to bend up by ΔE, the threshold energy for photo-emission $E_{th} = |E_{th}|_{\text{flat band}} + \Delta E$. Conversely, if the band is bent down by ΔE, then $E_{th} = |E_{th}|_{\text{flat band}} - \Delta E$. To include the effects of surface conditions we have to know

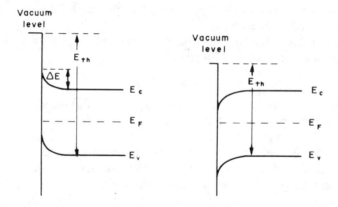

FIG. 6.26. Effects of band bending on the threshold for photoemission.

the variation of E_v with x measured from the vacuum level. However, for this case the analysis for the $Y-hv$ relation will be very much mathematically involved.

(B) Effects of imperfections in crystals

The photoemission from defects has been observed in the case of oxygen acceptor levels in tetracene [Spicer 1967] and in magnesium phthalocyanine [Thornber 1969] and in potassium–anthracene complex [Marchetti and Kearns 1966]. Obviously, local changes in the polarization energy and impurities with lower ionization energies present in the crystal lattice would cause changes in shape of photoemission curves. Sworakowski [1972] has studied the effects of imperfections. He has suggested that eq. (6.170) can still be used to calculate Y including the effects of imperfections provided that

$$n(\Delta E) \quad \text{is changed to} \quad n(\Delta E)h(E_t)$$

and

$$(6.175)$$

$$E'_K \quad \text{is changed to} \quad E'_K = hv + E_v + E_t + \Delta E$$

where $h(E_t)$ is the energy distribution function of the defects in the crystal and E_t is the dominant defect level. He has also computed Y as a function of hv for the case with traps confined in a single discrete level and that with traps distributed exponentially and Gaussianly. In general, the defects tend to broaden the low energy tails, leading to a decrease in the threshold. In the studies on photoemission of such complexes, or of organic crystals in general, there are two intricate problems to be solved—the method of determining the threshold energy and the method of identifying the origin of the emitted electrons. By doping the anthracene crystals with alkali metals—caesium, potassium and sodium—Yoshimura [1970] has studied the effects of dopants. The threshold energy is dependent on the amount of the metal doped. When the dopant concentration is low, there are two peaks on the $Y-hv$ curves irrespective of the kind of

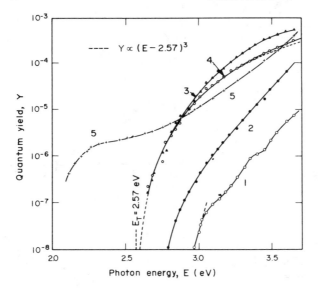

Fig. 6.27. Photoemission quantum yield as a function of photon
energy for anthracene crystals doped with various amounts
of caesium. The amount of caesium increases as the number
cited increases. [After Yoshimura 1970.]

alkali metal doped as shown in Fig. 6.27. The separation of the two peaks is about
0.2 eV, which corresponds to the energy of molecular vibrations. As the first peak
occurs at 3.14 eV, which corresponds to the first singlet absorption peak in a crystalline
anthracene, Yoshimura suggested that the photoelectrons are excited via the first
singlet exciton state and are emitted with the aid of the surface double layer formed
between alkali metal and anthracene. When the dopant concentration is high, Y
follows a cubic law dependence upon the incident photon energy. The threshold energy
deduced from the $Y^{1/3} - h\nu$ plot in this case is interpreted as the energy level of the
surface states below the vacuum level.

(C) Electron energy distribution and energy loss relation

Figure 6.28(b) shows the relative number of the emitted electrons with the kinetic
energy E_k relative to the vacuum level as a function of E_k, and these curves are
abbreviated to EDC. Figure 6.28(c) shows the relative number of electrons with the
energy loss $E_p = E_m - E_k$ (E_m is the maximum kinetic energy) as a function of E_p, and
these curves are abbreviated to ELC. It can be seen from the EDC that there are two
groups of photoelectrons. One group is formed by slow electrons which are almost
stationary in the same kinetic energy range (0.3–0.5 eV), and practically independent of
the increase in photon energy though the height of their peak (the first peak in the EDC)
increases with increasing photon energy. The other group is formed by fast electrons
which are shifted towards higher kinetic energy as the photon energy is increased. The
behaviour of fast electrons can be seen more clearly in the ELC. At $h\nu = 7.75$ eV, the

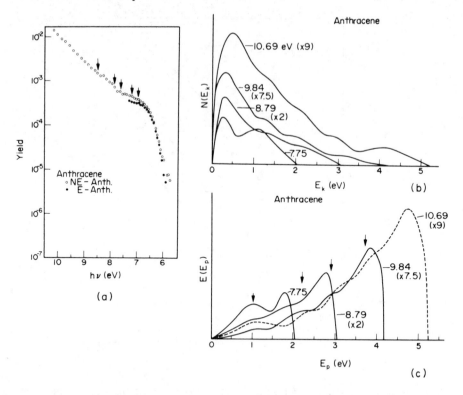

Fig. 6.28. (a) Spectral distribution of the photoemission of quantum yield, (b) electron kinetic energy distribution curves, and (c) electron energy–loss curves for anthracene crystals. [After Kochi, Harada, and Inokuchi 1970.]

first peak occurs at $E_p = 1.0$ eV; at $hv = 8.79$ eV, the second shoulder occurs at $E_p \simeq 2.2$ eV, the third at $E_p = 2.9$ eV; and at $hv = 9.84$ eV the fourth shoulder occurs at $E_p \simeq 3.7$ eV, and so on. The positions of such peaks and shoulders are practically independent of the photon energy. This means that there are energy losses of 1.2, 1.9, and 2.7 eV, respectively, relative to the position of the first peak. The energy structures which do not shift in the ELC and which do shift in the EDC linearly with the photon energy can be explained in terms of non-direct transition [Berglund and Spicer 1964, Blodgett and Spicer 1966]. In Fig. 6.28(a) the inflection points marked by arrows are located at 6.93, 7.18, 7.61, 7.83, and 8.48 eV, implying that energies higher by 1.28, 1.53, 1.96, 2.18 and 2.83 eV, respectively, compared to the threshold energy $E_{th} \simeq 5.65$ eV. By summarizing the results as follows:

$\Delta E_{\Delta Y}$ (eV)	ΔE_p (eV)
0	0
1.28 ⎫ 1.53 ⎬	1.2
1.96 ⎫ 2.18 ⎬	1.9
2.83	2.7

it can be concluded that the correspondence between $\Delta E_{\Delta Y}$ and ΔE_p can be considered as the indication of the existence of additional valence bands [Kochi *et al.* 1970].

(D) Multi-quantum processes

Two-photon photoemission processes have been observed by Pope *et al.* [1965]. If the rate of photoemission is proportional to the light intensity, the excitation is a one-photon process; and if it varies quadratically with the light intensity, the excitation is a two-photon process. The mechanisms which may contribute to this double quantum process are: direct two-photon excitation, exciton photoionization, exciton–exciton collision ionization, exciton–conduction electron ionization, and conduction electron ionization. The excitons involved could be neutral singlet or triplet excitons or ionic excitons [Gutmann and Lyons 1967].

6.4. PHOTOVOLTAIC EFFECTS

Photovoltaic effects in solids may be caused by (a) bulk photovoltaic effects—a Dember photovoltage arises due to the diffusion of non-equilibrium photocarriers with different electron and hole mobilities in the bulk of the solid, (b) surface photovoltaic effects—a photovoltage arises due to the potential barrier (Schottky barrier) near the interface between a metal and a semiconductor or an insulator and due to the presence of surface states there, (c) depletion-layer photovoltaic effects—a photovoltage arises due to the built-in field at the junction to drive the two types of charge carriers in opposite directions such as *p–n* homo-junction or hetero-junction devices, and (d) anomalous photovoltaic effects—a photovoltage can also arise due to a combination of several mechanisms such as the Dember effect in micro-regions, the photovoltaic effects at *p–n* junctions, Schottky barriers, or strains at grain boundaries, etc. Any photovoltaic effects taking place in a solid involve (i) light absorption in the solid, (ii) mobile charge carriers generated by the absorption, (iii) an internal discontinuity or non-uniform distributions of impurities or defects giving rise to an internal electric field to separate the two types of carriers, and (iv) electrical contacts. In the following we shall discuss briefly various photovoltaic effects.

6.4.1. Bulk photovoltaic effects

As early as 1931 Dember has observed a potential developed across a cuprous oxide specimen when subjected to illumination with a strong light in the absorption region of this material, and later this phenomenon is referred to as the Dember effect. This phenomenon was soon found also in diamond and zinc sulphide [Robertson *et al.* 1932], in germanium and silicon [Moss *et al.* 1953, 1959, Tauc 1957], and in CdS [Kallmann *et al.* 1960]. In organic crystals Kallmann and Pope [1959] were the first to report that when an anthracene crystal was illuminated with a strong light of the fundamental absorption region of the anthracene ($\lambda = 3650$ Å), a photovoltage of magnitude up to 0.2 V was produced with the negative polarity on the illuminated surface. This phenomenon is attributed to the diffusion of more photogenerated holes than electrons from the illuminated to the non-illuminated surface since the hole mobility is larger than the electron mobility in anthracene crystals. Since then, many investigators [Plotnikov and Matalygina 1961, Nakada and Kojima 1964, Geacintov *et al.* 1966, Vladimirov *et al.* 1969, Tavares 1970, Lyons and Newman 1971, Vertsimakha *et al.* 1973] have reported the observation of the photovoltaic effects in organic crystals. Vladimirov *et al.* [1969] and Vertsimakha *et al.* [1973] have reported that if the anthracene crystal was illuminated with the light of the wavelengths unimportant to absorption (e.g. $\lambda > 4000$ Å), the polarity of the photovoltage produced was reversed to that observed by Kallmann and Pope [1959]. They have attributed this phenomenon to the photorelease of the charge carriers from the traps in the crystal and to the formation of anti-barrier bending of the bands at the illuminated surface. It has also been reported that the spectral distribution of the photovoltage amplitude correlates closely to the absorption spectrum, and the change of light intensity changes only the photovoltage amplitude but does not affect this correlation; and that the photovoltage is sensitive to the surface conditions [Nakada and Kojima 1964, Vladimirov *et al.* 1969, Vertsimakha *et al.* 1973]. Absorption of oxygen on crystal surfaces reduces the photovoltage amplitude.

(A) The Dember effect

On the basis of the following assumptions: (i) there are no traps in the bulk of the specimen; (ii) the specimen thickness is much larger than the diffusion lengths of the carriers; and (iii) there is recombination at the surface, the open-circuit photovoltage or the sometimes called Dember voltage for small illumination light intensities (i.e. for the excess photogenerated carrier concentration less than the equilibrium carrier concentration) can be easily evaluated by solving the diffusion equations based on the small signal theory, and it is given by [Moss 1959]:

$$V_{ph} = \frac{i(D_p S)(\mu_p - \mu_n)}{\mu_n n_0} \left\{ \frac{\sinh(d/L_p) + S(D_p/L_p)[\cosh(d/L_p) - 1]}{[S^2 + (D_p/L_p^2)]\sinh(d/L_p) + 2S(D_p/L_p)\cosh(d/L_p)} \right\} \quad (6.176)$$

For large illumination light intensities (i.e. for the excess photogenerated carrier concentration larger than the equilibrium carrier concentration), V_{ph} can be derived on the basis of the large signal theory, and it is given by [Moss 1959]:

$$V_{ph} = \left(\frac{kT}{q}\right)\left(\frac{\mu_p - \mu_n}{\mu_p + \mu_n}\right)\ln\left[1 + (Sd/2D_p)\left(1 + \frac{\mu_p}{\mu_n}\right)\right] \quad (6.177)$$

In these two equations i is the illumination light intensity, S is the surface recombination velocity, and L_p is the diffusion length of holes. It can be seen that either when $S = 0$ or when $\mu_n = \mu_p$, $V_{ph} = 0$. It should be noted that in general there are always traps in solids and the local charge is not neutral. The effects of traps and local charge non-neutrality have not been taken into account in eqs. (6.176) and (6.177). However, the Dember effect always occurs together with other photovoltaic effects and special precautions are required in order to observe this effect.

In general, the spectral distribution of the photovoltage correlates well with the absorption spectrum. In the strong absorption region the photovoltage may reach a certain maximum limiting value, and its sign indicates the sign of majority carriers, while in the weak absorption region the photovoltage may be reduced to zero or even be changed to the reverse sign. Vladimirov *et al.* [1969] have reported that when an anthracene crystal is illuminated within the strong absorption region ($\lambda < 4000\,\text{Å}$) the photovoltage spectrum correlates well with the absorption spectrum and its sign indicates a *p*-type conductivity; and that when the anthracene crystal is illuminated in the weak absorption region ($\lambda > 4000\,\text{Å}$) its sign indicates an *n*-type conductivity, and this inversion occurs at $\lambda = 4000\,\text{Å}$ at room temperature. Tavares [1970] has observed that the photovoltage in naphthalene illuminated by the red light (7500 Å) is strongly enhanced if the specimen has been previously illuminated by the violet light (4100 Å) or by the green light (5100 Å). This phenomenon has been attributed to the effects of the local charge non-neutrality or the space charge. Lyons and Newman [1971] have also reported that in tetracene the thinner the specimen, the higher is the photovoltage generated in the weak absorption region, and that the photovoltage is almost independent of specimen thickness in the strong absorption region. It may be expected that the positions of the energy levels of traps determine the long wavelength range of the action spectra and that the slow photovoltage component is associated with the trapping and detrapping processes in the specimen.

The photovoltaic effects in organic materials have received little attention in the past because of their low efficiencies (less than 0.1 %). Most photovoltaic effect in organic materials involve the effects of the surface states and the Schottky barrier. We shall come back to this topic in Section 6.4.2.

(B) The photomagnetoelectric effect

When a magnetic field H_z is applied transversely to the diffusion current, the electrons and holes will be deflected in opposite directions, resulting in the creation of a voltage V_y across the specimen. This phenomenon is generally referred to as the photomagnetoelectric (PME) effect [Garreta and Grosvalet 1956, van Roosbroeck 1956, Pankove 1971]. This phenomenon, unlike the Dember effect, occurs even if $\mu_n = \mu_p$. For a given illumination light intensity the PME photovoltage depends on the surface recombination velocity S and the carrier lifetime τ. Thus the measurements of the PME photovoltages as functions of wavelength at various experimental conditions will enable the determination of S, τ, carrier diffusion lengths, and other bulk properties [Moss *et al.* 1953, Bulliard 1954]. If the applied magnetic field is tilted in the *y–z* plane instead of being perpendicular to the *y*-axis, then the H_y component will interact with the circulating open-circuit current in the specimen, creating a torque. This phenomenon is called the "photomechanical effect" [Garreta and Grosvalet 1956]. By measuring the PME torques rather than PME photovoltages, no electrodes are required, and thus the electrode effects can be eliminated, and this is an important advantage when using the PME effect for studying bulk properties.

6.4.2. Surface photovoltaic effects

The effects of surface conditions on photovoltage have been studied by one of the following means: (a) to expose the specimen in various ambient gases, (b) to modify the surface potential barrier by applying high electric fields, and (c) to excite the specimen by increasing the light intensity. The surface photovoltage is due mainly to the bending of the energy band at the surface, and thus its magnitude and sign depend strongly on the surface states of the specimen surface.

(A) Schottky barrier photovoltages

There are three possible photoeffects that can take place at the Schottky barrier: (1) Light absorbed in the metal will raise electrons in the metal to energy levels high enough to surmount the barrier as shown in Fig. 6.29 (process 1). The threshold for this to occur is in fact the measure of the barrier height ϕ_B. Electrons with sufficient energy to enter the solid will acquire negative charges, thus creating a photovoltage across the barrier. (2) Light with $hv \geq E_g$ can generate electron–hole pairs in the depletion region. The high electric field in the depletion region will separate the photogenerated carriers as shown in Fig. 6.29 (process 2), resulting in a photovoltage between the metallic electrode and the bulk of the solid. (3) Light of longer wavelengths is usually absorbed deep in the bulk of the solid and generates electron–hole pairs there, and one type of carriers (minority carriers) diffuse to the junction as shown in Fig. 6.29 (process 3). This also contributes to the total photovoltage across the barrier.

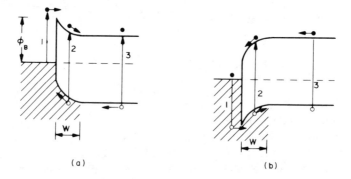

(a)   (b)

FIG. 6.29. Energy band diagrams of electron and hole Schottky barriers and three possible photovoltaic effects at these barriers.

In general, the contribution from process 1 is small [Mead 1966]. The major contribution to the photovoltaic effects is from processes 2 and 3. Process 1 is, in fact, similar to photoemission which has been discussed in Section 6.3. Electrons injected from a metal into a semiconductor experience a force directed away from the contact. Of course, the probability of the occurrence of the photoemission depends on the thickness of the metal film which is the illuminated electrode. If the electron attenuation length is sufficiently long compared with the thickness of the metal film, then the photoresponse is proportional to $(hv - \phi_B)^2$ [Crowell et al. 1962]. Processes 2 and 3 occur when $hv \geq E_g$. A typical curve showing the spectral dependence of the photoresponse is shown in Fig. 6.30. Thus the measurements of the threshold energies at which these processes occur may be used to determine the barrier height ϕ_B and the energy gap of the semiconductor E_g [Mead and Spitzer, 1963, 1964]. When the

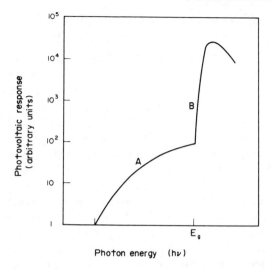

Fig. 6.30. Spectral dependence of photoresponse. (a) Photoemission from metal, and (b) band-to-band excitation.

photon energies are much larger than E_g, then the electron–hole pairs are generated very close to the interface between the illuminated electrode and the semiconductor because of the high absorption there. The decrease of photoresponse at high photon energies is due to the strong surface recombination. At $hv \geq E_g$ and on the assumption that the energy band diagram of the Schottky barrier is as shown in Fig. 6.29(a) without interface states, then the photocurrent due to process 2 (photocarriers generated in the depletion region) for the exciting light of wavelength λ is given by [Hovel 1975]

$$J_{dr} = qT(\lambda)i(\lambda)[1 - \exp(-\varepsilon W)] \tag{6.178}$$

and the photocurrent due to process 3 (photogenerated holes collected from the bulk) is given by [Hovel 1975]

$$J_p = \frac{qi\varepsilon Lp}{(\varepsilon^2 L_p^2 - 1)} T \exp(-\varepsilon w) \left[\varepsilon L_p \right.$$

$$\left. - \frac{(SL_p/D_p)\{\cosh[(d-w)/L_p] - \exp[-\varepsilon(d-w)]\} + \sinh[(d-w)/L_p] + \varepsilon L_p \exp[\varepsilon(d-w)]}{(SL_p/D_p)\sinh[(d-w)/L_p] + \cosh[(d-w)/L_p]} \right] \tag{6.179}$$

where $T(\lambda)$ is the transmission of the metal film into the underlying semiconductor, $i(\lambda)$ is the incident photon flux, ε is the absorption coefficient, S is the surface recombination velocity, d is the semiconductor specimen thickness, and w is the width of the depletion layer and is given by eq. (2.26) [cf. Chapter 2]. The total photocurrent is the sum of J_{dr} and J_p. Several effects which would affect the photocurrent are not included in the derivation of eqs. (6.178) and (6.179), and these effects are (a) the surface states at the metal–semiconductor interface, (b) the reflection from the illuminated electrode surface, (c) the image potential, and (d) the interfacial dielectric layer. By including these effects the analysis becomes not amenable.

(B) Photovoltaic effects in organic semiconductor–metal Schottky barriers

Using a transparent metal film, or other transparent materials as electrode to form a Schottky barrier so that the light can enter the organic semiconductor through such an electrode, many investigators have observed the photovoltaic effects in anthracene [Killesreiter and Baessler 1971], in tetracene (naphthacene) [Geacintov *et al.* 1966, Reucroft *et al.* 1969, Ghosh and Feng 1973, Matsumura *et al.* 1975], in phthalocyanine [Ghosh *et al.* 1974], and in chlorophyll-a [Tang and Albrecht 1975]. Most of the experimental results reported in the literature suggest that electron–hole pairs may be photogenerated directly, or even if excitons are formed first and then separated into individual carriers, they do not have to diffuse to the surface for carrier generation since the excitons may be ionized due to their interactions with impurities [Fedorov and Benderskii 1971] or with phonons [Usov and Benderskii 1970]. In general, carriers can be collected by the Schottky barrier contact only when they are generated at a distance less than $W +$ diffussion length of these carriers from the contact. Using a sandwich configuration with a blocking contact on one side of an organic solid specimen and with an ohmic contact on the other, several investigators [Ghosh *et al.* 1973, 1974, Tang and

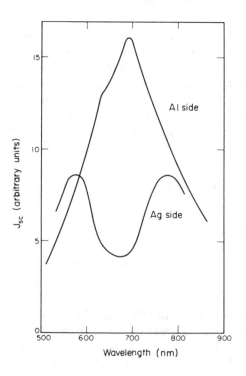

FIG. 6.31. Spectral dependence of short-circuit photo-current in an Al/MgPh/Ag system. Mg-phthalocyanine film thickness: 1500 Å. Al side and Ag side mean, respectively, the light incident on Al side and Ag side. [After Ghosh, Morel, Feng, Shaw, and Rowe 1974.]

Albrecht 1975] have investigated the photovoltaic effects by comparing the spectral response when light is illuminating the barrier contact side with that when light is illuminating the ohmic contact side. For the aluminium/magnesium-phthalocyanine/silver Schottky barrier system, the aluminium forms a barrier contact, while the silver forms an ohmic contact. Figure 6.31 shows the short-circuit photocurrent J_{sc} as a function of light wavelength for a magnesium-phthalocyanine film of 1500 Å in thickness. The spectrum of J_{sc} correlates closely to the absorption spectrum with the peak around 690 nm when light is incident on the aluminium surface. The Schottky barrier acts as a sink for minority carriers [electrons in the present case—cf. Fig. 6.29(b)] and as a barrier to majority carriers (holes). Thus the aluminium side is negative and the silver side is positive. If excitons are formed first under illumination, then excitons will diffuse towards the barriers and dissociate with the electrons to the aluminium side and the holes to the silver side [Ghosh *et al.* 1974].

When light is incident on the silver surface, the main peak disappears, but there are two peaks on the two sides of the main peak, as shown in Fig. 6.31, and these two peaks tend to move towards the regions of lower absorption coefficient in thicker films. The most photosensitive region is the barrier region next to the aluminium contact. Only a very small fraction of photon flux can reach the barrier if the silver surface is illuminated because of the high absorption and the distance $(d - w)$ is larger than the carrier diffusion length. Therefore, in this case most photogenerated carriers are lost due to recombination or trapping. Since light of wavelengths at low absorption regions can penetrate deeper to the material, the photoresponse increases with decreasing absorption coefficient. Of course, after they reach the peaks, they decrease because some photon energies may not be absorbed in the regions where the absorption coefficients are too low. In the aluminium/magnesium-phthalocyanine/silver system Ghosh *et al.* [1974] have reported that the short-circuit photocurrent $J_{sc} \propto i^m$ and the open-circuit $V_{oc} \propto \ln i$ as expected for a Schottky barrier device, where i is the incident light intensity and m is a constant and is equal to about 0.5. The dependence of J_{sc} on the square root of i indicates a high electron–hole recombination rate. At 690 nm and with light incident on the aluminium side (barrier contact surface) the photovoltaic efficiency is about 0.01 % [Ghosh *et al.* 1974].

A similar phenomenon observed in the aluminium/magnesium-phthalocyanine/silver system has also been observed in the aluminium/tetracene/gold system [Lyons and Newman 1971, Ghosh and Feng 1973, Fang 1973]. For light incident on the aluminium electrode, the spectral dependence of short-circuit photocurrent has a one-to-one correspondence with that of optical absorption, implying the presence of a Schottky barrier at the aluminium–tetracene interface. It is interesting to note that the photocurrent–photovoltage characteristics measured by varying the load resistance at a fixed incident light intensity are similar to those for inorganic semiconductor solar cells. A typical one is shown in Fig. 6.32. The maximum power that can be obtained from such a device is at the point for which the largest rectangle can be inscribed in the curve ($I_{ph} V_{ph}$ product is maximum). For white light incident on the aluminium electrode surface, the maximum power conversion efficiency is about 10^{-4} % [Ghosh and Feng 1973]. The open-circuit photovoltages of up to 1.2 V has been observed with aluminium electrode negative with respect to the gold electrode in tetracene films, but both photocurrent and photovoltage measured in vacuum are different from those measured in air, indicating that the adsorbed air (or oxygen) may create traps at the

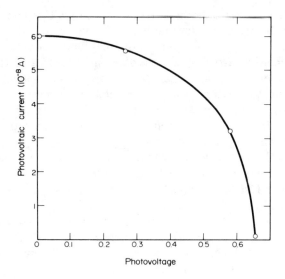

FIG. 6.32. The photocurrent–photovoltage curve for the
Al/tetracene/Au system. Tetracene film thickness:
5×10^{-5} cm. [After Ghosh and Feng 1973.]

interface (surface states), and thus affect the photoresponse [Lyons and Newman 1971].

Matsumura *et al.* [1975] have reported that the photovoltage in the tetracene–gold system consists of two components—one with a short response time due to the electron transfer across the boundary between gold and tetracene, and the other with a long response time due to the diffusion in the tetracene film as shown in Fig. 6.33. When light is incident on Au^1, the photovoltage exhibits a rapid negative response, followed by a slow rise towards the positive value (the polarity at Au^1 electrode with Au^2 at zero voltage as reference). When light is turned off it becomes more positive for a short time and then decays slowly. This photovoltage–time curve can be resolved into two components. One broken-line curve has a negative polarity and a fast response, following almost the same shape as the incident light pulse. The other component shown by broken-line curve has a positive polarity and a slow response. This phenomenon can be observed with the direction of illumination from either side of the gold electrodes. The fast component is associated with the electron transfer to Au^1 due to the interaction of the photogenerated excitons with Au^1 which takes place mainly within 20 Å from the boundary. The slow component is associated with the charge carrier generation in the bulk of tetracene. The holes move towards Au^1 electrode because of the potential gradient in the barrier region, thus making the photovoltage positive at the Au^1 electrode. The slow decay after the cease of illumination indicates a thermal recombination process through traps.

Tang and Albrecht [1975] have reported that there exists a barrier region at the contact between the *p*-type chlorophyll-a (Chl-a) and the metal of low work function which is directly related to the observed rectifying and photovoltaic properties of Cr/Chl-a/Hg (single barrier) and Cr/Chl-a/Hg-In (double barriers) systems. Figure 6.34 shows the current–voltage and photocurrent–photocurrent–photovoltage charac-

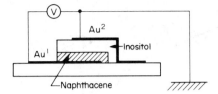

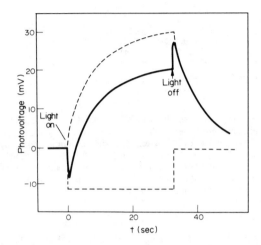

Fɪɢ. 6.33. Electrode arrangement for photovoltaic measurements, and the photovoltage as a function of time when a rectangular light pulse is applied on the Au^1 electrode ($\lambda = 500$ nm). [After Matsumura, Uohashi, Furusawa, Yamamoto, and Tsubomura 1975.]

teristics of the Cr/Chl-a/Hg system in which the dominant Schottky barrier is at the Cr/Chl-a boundary, and the Chl-a/Hg contact may be ohmic or only slightly blocking. The activation energy measurements give the barrier height on the order of 0.4–0.7 eV and the power conversion efficiency is better than 10^{-2} % [Tang and Albrecht 1975].

Schottky barrier solar cells are attractive because of their simplicity and low cost in fabrication, and particularly for the cells using polycrystalline or amorphous films in which the presence of grain boundaries may make the conventional *p–n* junction devices difficult to perform their functions. Organic materials can be competitive to inorganic materials for Schottky barrier solar cells if the former can be improved by increasing the purity (or reducing concentrations of various traps), increasing the quantum efficiencies for charge carrier generation (or lowering the resistivity or increasing the mobility), and providing good ohmic contacts. Such improvements would raise the power conversion efficiency. However, a quantum efficiency of 2.3 % and the conversion efficiency of 0.2 % for Schottky barrier solar cells made from thin films (100–1000 Å) of the organic dye hydroxy squareylium have been reported [Merritt and Hovel 1975]. In organic materials, polymers can be easily fabricated in the

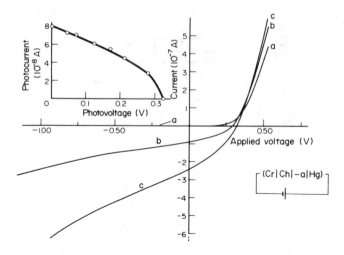

F IG. 6.34. The current–voltage characteristics of the Cr/Chl–a/Hg system: *a*, device
unilluminated; *b*, device illuminated by light of wavelength 745 nm and
intensity of 7.6×10^{-4} W cm^{-2}; *c*, device illuminated by light of wavelength 745 nm and intensity of 2.7×10^{-3} W cm^{-2}. The insert is the
photocurrent–photovoltage characteristics measured by varying the load
resistance while the device is constantly illuminated by light of wavelength
745 nm. [After Tang and Albrecht 1975.]

form of large-area thin films and, therefore, have the potential for solar cells. Using a
model of charge carrier separation at the Schottky barrier between a metal electrode
and a polymer film, Reucroft *et al.* [1974] have calculated the theoretical efficiency for
photovoltaic energy conversion for the polymeric system—Poly (N-vinylcarbazole)
(PVK) and the (1 : 1) complex between PVK and 2,4,7-trinitro-9-fluorenone (TNF). The
theoretical energy conversion efficiency estimated by these investigators may approach
a maximum value of 2% with a 2 eV barrier for this material.

6.4.3. Photovoltaic properties of *p–n* junctions

Since people realized that the conventional energy resources in this world are
limited, many scientists have attempted to search for other unconventional energy
sources, particularly the conversion of the unlimited solar energy source. A great deal of
theoretical and experimental results on photovoltaic properties of *p–n* junctions is now
available in the literature. In organic materials most photovoltaic phenomena are
usually considered to involve Schottky barrier effects rather than *p–n* homojunctions.
In this present section we shall not review the work on the theory of photovoltaic effects
in *p–n* junctions. The reader who is interested in the theory and experimental results of
p–n junction solar cells made of inorganic semiconductors is referred to some excellent
books and articles [Pankove 1971, Hovel 1975, Backus 1976, Moss *et al.* 1973,
Landsberg 1975, Milnes and Feucht 1972, Williams 1978].

Like the organic semiconductor *p–n* junctions, the combination of an *n*-type and a *p*-
type organic semiconductor can form a photovoltaic device. Photocurrents of the

order of 10^{-9}–10^{-8} Å and photovoltages up to 200 mV have been observed in such organic photovoltaic devices. The photocurrent increases with increasing light intensity, and its response time (rise and decay time) is less than 10^{-2} sec [Meier 1968, 1974]. The electrons in the junction move from the *p*-type material to the *n*-type material, irrespectively of the direction of light illumination. This indicates that the carrier behaviour is exactly the same as that in inorganic *p–n* junction solar cells. The following pairs of materials are just to cite a few as examples that can be used to form *p–n* junction photovoltaic devices: dyes [malachite green (*n*-type)] – merocyanine (*p*-type); rhodamine B (*n*-type) – phthalocyanine (*p*-type); crystal violet (*n*-type) –merocyanine FX (*p*-type) [Meier 1974].

Some previous results on photovoltaic effects in organic semiconductors are given in Table 6.3.

6.4.4. Anomalous photovoltaic effects

Some semiconductors in thin film form exhibit a high photovoltage when exposed to intense white light, and in some cases the photovoltage may be much larger than the potential across the energy gap. Such an anomalous photovoltaic effect has been observed in many semiconductors such as in germanium [Kallmann *et al.* 1972, Ma *et al.* 1975], in silicon [Kallmann *et al.* 1961], and in many compound semiconductors [Pankove 1971]. The semiconductor films exhibiting this phenomenon are usually fabricated using an oblique vapour deposition technique (the insulating substrate is inclined with respect to the plane normal to the direction of vapour). Photovoltages as high as 5000 V have been observed in III–V compound semiconductors [Adirovich *et al.* 1966]. Such an anomalous photovoltaic effect is sensitive to the specimen preparation processes; it depends on the angle of oblique evaporation with respect to the substrate, the substrate temperature during oblique deposition, the film thickness, the pressure and composition of the gases in the vacuum chamber in which the film is fabricated, substrate material, and its surface treatment, etc. The adsorption and desorption of oxygen in the film may reverse the polarity of the photovoltage [Pankove 1971]. Furthermore, the photovoltage generally increases with decreasing temperature, but in some materials it is a minimum at a certain temperature, and increases at both lower and higher temperatures [Cheroff 1961].

Several models have been put forward to explain such anomalous phenomena. However, none of them can account for all of the observed phenomena. It is generally believed that the large photoinduced potential difference along the length of the film is due to the addition of photovoltages generated in micro-elements connected in a series arrangement. Such micro-elements could be micro *p–n* junctions, or the Dember effect in micro-crystals separated by grain boundaries, or both. It is also believed that the surface recombination which leads to anomalous photogenerated carrier distribution is responsible for the photovoltage polarity and the spectral sensitivity of the films [Korsunskii and Sominskii 1973, Onishi *et al.* 1974]. Obviously, anomalous photovoltaic effects are interesting and may be important for device applications. However, further work, both experimental and theoretical, is needed to clarify the ambiguities. A more detailed account of these effects has been given in the book of Pankove [1971].

TABLE 6.3. *Summary of previous experimental results of photovoltaic effects in organic crystals*

Organic crystal	Crystal preparation	Specimen thickness (μm)	Contacting electrodes	Polarity of front surface (illuminated surface)	Wavelength range (Å)	Temperature range (°K)	Remarks	References
Anthracene	Single crystals (solution grown)	5.1–10	Electrolyte 0.01 M NaCl solution	Negative	3650	293	Maximum photo-voltage = 200 mV	Kallmann and Pope [1959]
Anthracene	Single crystals and polycrystals	100–300	Conducting quartz-glass and aluminium	Negative	3100–3650	261–328	Pulse-photovoltage	Plotnikov and Matalygina [1961]
Anthracene	Sublimation films	20–50	SnO_2	Negative	–	300	Surface photo-voltage	Nakada and Kojima [1964]
Anthracene	Single crystals (zone refined)	30–50	SnO_2	Negative or positive	3500–4200	300–375	Sign inversion of photovoltage occurs at $\lambda = 4000$ Å	Vladimirov et al. [1969]
Anthracene	Single crystals (zone refined)	40	SnO_2	Negative or positive	3700–4200	300	Dember photo-voltage	Vertsimakha et al. [1973]
Tetracene	Single crystal (sublimation grown)	20	Electrolyte 1 M NaCl solution	Negative	2200–5600	300	Maximum photovoltage = 60 mV	Geacintov et al. [1966]
Tetracene	Single crystal	60–100	Silver	Negative or positive	3300–6900	300–400	Transient photovoltage	Reucroft et al. [1969]
Tetracene	Vacuum evaporated thin film	0.3–1.5	Aluminium and gold	Negative or positive	3400–5700	293	Open-circuit photovoltage = 1.2 V rectification	Lyons and Newman [1971]
Tetracene	Thin film	0.5–1.0	Aluminium and gold	Negative	3400–5700	293	Open-circuit photovoltage = 0.6 ± 0.06 V rectification	Ghosh and Feng [1973]

Metal-free phthalocyanine	Thin film (vacuum sublimation)	1–6	Aluminium	Negative	6250	300	pulse-photovoltage	Usov and Benderskii [1968]
Magnesium-phthalocyanine	Thin film	0.2–0.5	Aluminium and silver	Negative	4000–8500	293–333	p-n junction photovoltage	Fedorov and Benderskii [1971]
Magnesium-phthalocyanine	Thin film	0.15–0.5	Aluminium and silver	Negative	5000–8500	293	Schottky-barrier photovoltage	Ghosh et al. [1974]
Chlorophyll-a	Thin film	0.8–3.0	Aluminium and mercury	Negative or positive	4000–8000	293	Schottky-barrier photovoltage	Tang and Albrecht [1975]
PVK–TNF	Thin film	5–25	SnO_2 and gold	Negative or positive	4250–7750	298	Schottky-barrier photovoltage	Reucroft et al. [1974, 1975]

CHAPTER 7

Luminescence

7.1. INTRODUCTORY REMARKS

Luminescence was known and studied long before the discovery of the quantum picture of atoms by Rutherford and Bohr. In organic semiconductors this subject has been extensively studied for the past three decades [Gutmann and Lyons 1967, Birks 1970]. Studies of luminescence can provide information about the nature, structure, size and shape, and excited states of an atom or a molecule in a solid. Luminescence has been used as an experimental technique by physicists and chemists for studying the properties of materials, and its related phenomena have been technologically employed for various luminescent electron devices such as scintillation counters. The process of luminescence is the de-excitation of excited atoms or molecules (or the annihilation of excitons through recombination) by re-emission of absorbed energy as light quanta. As has been discussed in Chapter 6, singlet and triplet excitons are formed when an organic semiconductor is illuminated with a light absorbed by it. The processes of photocarrier generation resulting from the dissociation of excitons are generally referred to as charge transfer processes, while the processes involving the conversion of the absorbed energy into luminescence photons or phonons, or the migration of excitons in the solid, or the collision of excitons to form excitons of different kinds, are generally referred to as energy transfer processes. The charge transfer processes result in electric conduction and potential difference between two points involving electric charge transport, and these processes have been discussed in some detail in the previous chapters. The energy transfer processes are associated with excitation which separates two types of charge carriers (negatively and positively charged) to such distances within which they are coupled with each other, and with radiative and non-radiative transitions which either bring the excited charge carriers from one energy level to a lower energy level, or lead to recombination of the two types of charge carriers. As these processes do not involve the transport of net electric charges, they can result in luminescence and energy transport (diffusion of excitons—electrons and holes coupling and moving in the same direction). In this chapter we shall confine ourselves to the luminescence and its related phenomena which are relevant to electrical transport in organic semiconductors.

Since an atom or a molecule in a crystal perturbs its neighbours upon absorption or emission of energy, part of the excitation energy is necessarily consumed to increase the atomic or molecular vibration. The increase in atomic or molecular vibration implies the increase in interference with subatomic radiative transition processes and with energy transport in the crystal. Thus the lower the temperature the higher is the efficiency for luminescence because the probability of the competing non-radiative transition, which depends on the coupling between the excited atom or molecule and its neighbours, is smaller. However, it should be noted that for some organic materials, to lower the temperature may involve a change in phase from gas to liquid or from liquid to solid, and therefore in this case the intermolecular coupling and hence the probability of non-radiative radiation are increased.

A molecule may be excited by many different methods and an excited molecule may give up its absorbed energy through many different mechanisms. Figure 7.1 summarizes these methods and mechanisms. To facilitate the discussion it is desirable to define first some important terms generally used in the literature.

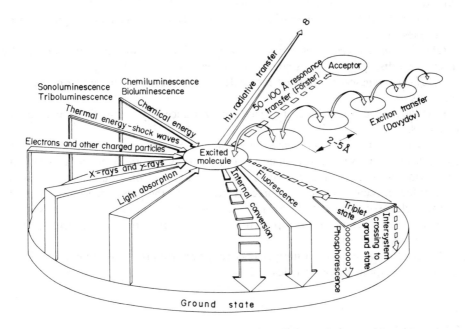

FIG. 7.1. The various possible ways of excitation and the various possible mechanisms of de-excitation of a molecule. [After Windsor 1965.]

(A) *Luminescence.* Part of the energy of absorbed photons or high-energy particles is converted into emitted electromagnetic radiation in excess of thermal radiation. Luminescent materials are sometimes called "luminophors" or "lumophors" or "scintillators".

(B) *Fluorescence.* The states from which the emission originates and terminates have the same multiplicity. The multiplicity of the state is defined by $2S + 1$, where S is the total spin of the state. For example, the radiative transition from the first excited singlet S_1 to the ground state singlet S_0 produces fluorescence with $\Delta S = (S)_{S_1} - (S)_{S_0} = 0$. The natural lifetime (the duration of detectable afterglow) of fluorescence for organic molecules usually lies in the range from 10^{-6} to 10^{-10} sec. Fluorescent materials are sometimes called "fluorophors" or "fluors".

(C) *Phosphorescence.* The states from which the emission originates differ in multiplicity or in spin. For example, the radiative transition from the lowest excited triplet T_1 to the ground state singlet S_0 produces phosphorescence with $\Delta S = (S)_{T_1} - (S)_{S_0} = 1$. Since highly probable transition can occur only for $\Delta S = 0$, phosphorescence which is attributed to a forbidden transition has relatively long lifetimes (10^{-3}–10^2 sec). Materials exhibiting phosphorescence are sometimes called phosphors.

(D) *Photoluminescence.* Excited molecules are produced by absorption of light (low energy photons such as visible or ultraviolet light).

(E) *Roentgenoluminescence.* Excited molecules are produced by absorption of high energy photons, such as X-rays or gamma-rays.

(F) *Cathodoluminescence.* Excited molecules are produced by energetic electrons or cathode rays.

(G) *Ionoluminescence (or radioluminescence).* Excited molecules are produced by α-particles or ions.

(H) *Triboluminescence (or sonoluminescence).* Excited molecules are produced by mechanical disruption such as sound and shock waves, or grinding.

(I) *Chemiluminescence.* Excited molecules are produced by energy from chemical reactions.

(J) *Bioluminescence.* Excited molecules are produced by energy from biochemical reactions.

(K) *Thermoluminescence.* Excited molecules are produced thermally, for example, by collisional activation in which thermal energy (vibrational, rotational, and translational) is converted during a collision to electron energy of excitation. Thermoluminescence is also sometimes referred to as the phenomenon in which the thermal release of trapped carriers with rising temperature gives rise to the emission of light, and this effect has been employed for the determination of the energy levels of traps (cf. Section 5.4).

(L) *Electroluminescence.* Luminescence resulting from the recombination is due to the action of applied electric fields.

7.1.1. The Franck–Condon principle

The concept of the configuration diagram based on the Franck–Condon principle is useful in explaining the non-radiative transition. In the Born–Oppenheimer approximation the total wave function of a vibronic state ψ can be expressed as the product of electronic wave function ϕ_e and vibrational wave function ϕ_v. Thus we may write

$$\psi_{li} = \phi_{le}(q, r)\phi_{li}(r) \tag{7.1}$$

as the total wave function of the li vibronic state, where i indicates the ith vibrational state of the lower electronic state l; and

$$\psi_{mj} = \phi_{me}(q, r)\phi_{mj}(r) \tag{7.2}$$

as the total wave function of the mj vibronic state, where j indicates the jth vibrational state of the higher electronic state m, and q and r are, respectively, the electronic and nuclear coordinates. According to the Franck–Condon principle, there are no changes in the nuclear coordinates during an electronic transition and therefore the most probable vibronic transitions are vertical because the time (less than 3×10^{-14} sec) required for an electronic transition is negligibly small compared with the period of nuclear vibration. The electronic transitions are most probable when the kinetic energies of the nuclei are a minimum (that is, at the endpoints of their vibrations; for example, at $a_2 c_2$ and $a_2^* c_2^*$ in Fig. 7.2). The lowest vibrational levels may be possible exceptions to this principle [Condon 1947]. If the molecule is in its ground state E_l with vibrational energy corresponding to lowest sublevel $a_0 c_0$ as shown in Fig. 7.2, the electron may be excited from this level to state E_m at vibrational level endpoint a_2^* by absorption of a proton with energy $E_A = h\nu_A$. The excited molecule then vibrates between a_2^* and c_2^* and may emit phonons and relax to sublevel $a_0^* c_0^*$. Then it may proceed back to state E_l at sublevel $a_2 c_2$ by emitting a photon of energy $E_B = h\nu_B$. The molecule can relax to its lowest sublevel $a_0 c_0$ by emitting phonons.

The energy-emitted E_B is lower than the energy-absorbed E_A. The difference between these two energies is called the "Franck–Condon shift". In general, the degradation of

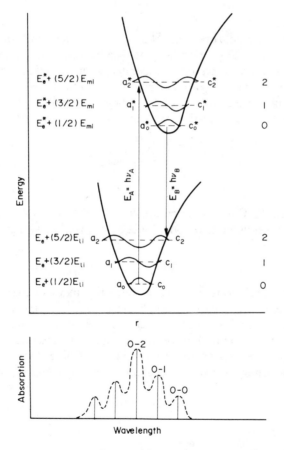

FIG. 7.2. Franck–Condon configuration diagram and absorption spectrum for a diatomic molecule. In the absorption curves the dashed curve is for molecules in solution and the solid lines are for molecules in vapour form. E_A is the absorption at the Franck–Condon maximum and E_B is the emission at the Franck–Condon minimum. r is the nuclear separation.

optical energy is called the "Stokes shift". Thus the Franck–Condon shift is a Stokes shift due to the displacement of the molecule Δr. The absorption spectrum depends on the relative position of the minimum potential energy level and on whether the molecules are in vapour or in solution as shown in Fig. 7.2.

7.1.2. The radiative and non-radiative transition processes

The transition moment (or matrix elements of the electric dipole moment) between two states ψ_{li} and ψ_{mj} given by eqs. (7.1) and (7.2), based on the Franck–Condon principle, can be expressed as

$$M_{mj \to li} = \langle \psi_{mj} | M | \psi_{li} \rangle$$
$$= \langle \phi_{me} | M | \phi_{le} \rangle \langle \phi_{mj} | \phi_{li} \rangle$$
$$= |\bar{M}_{ml}| |S_{mj, \, li}| \tag{7.3}$$

where M is the electric dipole operator. $\bar{M}_{ml}$ is the mean electronic transition moment, and $S_{mj, \, li}$ is the vibrational overlap integral. The Franck–Condon maximum, which represents a vertical transition on the configuration diagram, corresponds to a maximum overlap integral ($|S_{mj, \, li}|$). The probability of the radiative transition from a state corresponding to wave function ψ_{mj} to a state corresponding to wave function ψ_{li} is proportional to the square of the electronic transition moment

$$P^r_{m \to l} \propto |M_{mj \to li}|^2$$
$$= |\bar{M}_{ml}|^2 |S_{mj, \, li}|^2$$
$$= A_{m \to l} F \tag{7.4}$$

where F is the Franck–Condon factor which is equal to $|S_{mj, li}|^2$, $A_{m \to l}$ is Einstein's coefficient of spontaneous emission, which is equal to $|\bar{M}_{ml}|^2$, and is given by [e.g. Birks 1970]

$$A_{m \to l} = (8 \pi h v^3_{mj \to li} \eta^3 c^{-3}) B_{m \to l} \tag{7.5}$$

in which h is the Planck constant, η is the refractive index of the medium where a molecule undergoes a transition, c is the light velocity, $v_{mj \to li}$ is the radiative frequency corresponding to the energy difference between the states mj and li, and $B_{m \to l}$ is Einstein's coefficient for induced absorption or induced emission.

The probability of the non-radiative transition from a state of wave function ψ_{mj} to a state of wave function ψ_{li} is proportional to the electronic factor [Birks 1970]

$$p^{nr}_{m \to l} \propto |H_{mj \to li}|^2$$
$$= |\bar{J}_{ml}|^2 F \tag{7.6}$$

where

$$|H_{mj \to li}| = \langle \phi_{mj} | J_{ml} | \phi_{li} \rangle \tag{7.7}$$

in which

$$|J_{ml}| = \langle \phi_{me} | J_N | \phi_{le} \rangle \tag{7.8}$$

$|\bar{J}_{ml}|^2$ is the electronic factor, and J_N is the nuclear kinetic energy operator.

The non-radiative transition processes always compete with the radiative transition processes, and are therefore very important in energy transfer processes and luminescent phenomena. But, unfortunately, they are poorly understood. Both $|\bar{M}_{ml}|^2$ and $|\bar{J}_{ml}|^2$ involve the wave functions of the initial and final electronic states, so that non-radiative transitions are subject to the same multiplicity, symmetry, and parity selection rules as radiative transitions [Birks 1970]. For example, the electric dipole transition between electronic states of different multiplicity (or different spins) is spin forbidden (e.g. the $T_1 \rightarrow S_0$ transition), while the transition between electronic states having the same multiplicity is spin allowed (e.g. the $S_1 \rightarrow S_0$ transition).

The electric dipole operator M implies that the radiative transition is a vertical transition between the potential energy surfaces corresponding to the different electronic states. The nuclear kinetic energy operator J_N implies that the non-radiative transition is a horizontal transition involving crossing or tunnelling from the potential energy surface of an electronic state to the continuum of iso energetic vibrational levels of the potential energy surface associated with a different electronic state. For example, in the intersystem crossing process $S_1 \rightarrow T_1$ the energy of the initial state S_1 is usually all electronic and that of the final state T_1 is electronic and vibrational (molecular and lattice), but the two states have practically the same total energy.

Since we are interested in the transitions not in a completely isolated molecule but in a solid containing interacting molecules, an analysis of the behaviour of such a system involves a complete Hamiltonian and therefore become formidable. However, Gouterman [1962] and Robins and Frosch [1962, 1963] have put forward theories for non-radiative transitions in terms of molecular vibrational factors. For more details in this field the reader is referred to the aforementioned papers and also the following references: Birks [1970], Becker [1969], Craig and Walmsley [1968], Förster [1960], Kasha [1959], Wolf [1959], McClure [1959], and Wotherspoon and Oster [1960].

In general there are five processes for radiative transitions in solids [Gooch 1973] and they are: (i) band-to-band transitions, (ii) transitions via shallow donor or acceptor levels, (iii) donor–acceptor transitions, (iv) transitions via deep recombination centres, and (v) exciton transitions. In inorganic semiconductors, processes (i)–(iv) may be dominant. However, in organic semiconductors, process (v) is the most important transition process. It is generally accepted [Birks 1970] that singlet excitons radiatively decay produce luminescence. The quantum yield of luminescence depends on the relative probability of radiative and non-radiative transitions and the relative efficiency of producing radiative excitons. There are many possible processes for non-radiative transitions. In inorganic semiconductors multiphonon transition and Auger recombination processes are dominant non-radiative transition processes [Landsberg 1970]. But in organic semiconductors, again the process of exciton transitions leading to non-radiative transitions is the most important non-radiative transition process. Figures 7.1 and 7.3 and Table 7.1 outline all possible energy transfer processes. Before going to Section 7.2 it is desirable to define clearly various energy transfer processes as follows.

(a) *Internal conversion.* This is generally referred to as the conversion of electronic energy into vibrational energy. If the optical absorption by the crystal in its ground state results in an excited state which is vibronic including both electronic and vibrational excitation such as process A in Fig. 7.2, then internal conversion can occur in

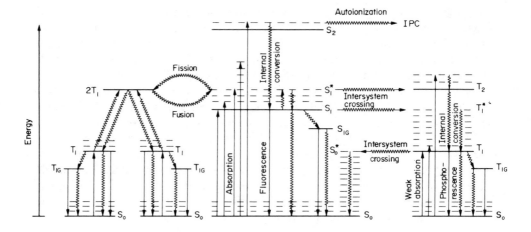

FIG. 7.3. Typical energy levels and energy transfer processes in organic molecular crystals. S_0 is the ground state level; S_1, S_{1G}, S_2, and T_1, T_{1G}, T_2 are, respectively, the first, trapped, and second excited singlet and triplet levels; $2T_1$ is the double triplet exciton pair state level; S_1^* and T_1^* are, respectively, the high vibronic–electronic levels of S_1 and T_1 states.

which thermal motions of the lattice remove the vibrational portion of the exciton such as process B in Fig. 7.2, thus causing the excitation to become only electronic. In general, internal conversion is a non-radiative transition between two states of like multiplicity. If the excited state resulting from process B is strongly coupled to lattice vibrations, the crystal may revert to its ground state through a non-radiative process $c_2 \rightarrow c_0$ (Fig. 7.2) which converts the excitation energy into heat.

(b) *Intersystem crossing.* This term was first used by Kasha [1950] to describe the internal conversion (or non-radiative) process involving states of different multiplicity. This process which depends on spin–orbit coupling commonly occurs between the lowest first excited singlet (S_1) to the lowest first excited triplet ($T_1 \rightsquigarrow S_1$) via a highly vibrating level of T_1 (i.e. T_1^*), and occurs in a time of 10^{-8} sec. The actual cross over from one electronic energy level to another is adiabatic. (This means that no net energy change is involved.) Similarly, the intercrossing from the lowest first excited triplet to S_0 ($T_1 \rightsquigarrow S_0$) via a highly vibrating level of S_0 (i.e. S_0^*) can also occur, but the rate is lower by a factor of 10^4 or larger. Of course, after intersystem crossing phosphorescence may appear due to the $T_1 \rightarrow S_0$ radiative transition process.

(c) *Radiative transfer.* This transfer process involves the emission of a photon by the molecule D (donor) and the subsequent reabsorption of this photon by the molecule A (acceptor). The conditions for such an energy transfer process to occur are that the fluorescence spectrum of the molecule D must overlap the absorption spectrum of the molecule A, and that the medium between the two molecules must be transparent to the wavelength of the radiation involved, but molecules D and A need not be identical.

TABLE 7.1. *Summary of transition processes in organic crystals*

	Process		Kinetics
1.	Absorption	$S_0 + h\nu \xrightarrow{\varepsilon_s} S_1$	
		$S_0 + h\nu \xrightarrow{\varepsilon_T} T_1$	
2.	Radiative transition		
	Fluorescence (strong)	$S_1 \xrightarrow{\alpha_1} S_0 + h\nu$	
	Phosphorescence (weak)	$T_1 \xrightarrow{\beta_1} S_0 + h\nu$	
3.	Non-radiative transition		
	Internal conversion	$S_1 \xrightarrow{\alpha_2} S_0 + \text{phonons}$	
	(Single molecular level)	$S_2 \text{ or } S_n \xrightarrow{\alpha_4} S_1 + \text{phonons}$	No multiplicity change
		$S_1^* \xrightarrow{\alpha_5} S_1 + \text{phonons}$	
		$S_0^* \xrightarrow{\alpha_6} S_0 + \text{phonons}$	
		$T_1 \xrightarrow{\beta_2} S_0 + \text{phonons}$	With multiplicity change
		$T_2 \text{ or } T_n \xrightarrow{\beta_4} T_1 + \text{phonons}$	No multiplicity change
		$T_1^* \xrightarrow{\beta_5} T_1 + \text{phonons}$	
4.	Energy transfer	$S_1 + S_0 \longrightarrow S_0 + S_1$	
		$T_1 + S_0 \longrightarrow S_0 + T_1$	

(continued overleaf)

Table 7.1 (Contd.)

Process	Kinetics
5. Intersystem crossing	

$$S_1 + S_0 \xrightarrow{\gamma_f} T_1 + T_1$$

$$S_{1G} + S_0 \xrightarrow{\gamma_{fG}} T_1 + T_{1G}$$

$\left.\begin{array}{c}\\ \\ \end{array}\right\}$ With multiplicity change

Homofission
Heterofission

Homofusion
(triplet–triplet annihilation)

$$T_1 + T_1 \underset{K_{-1}}{\overset{K_1}{\rightleftharpoons}} (T_1 T_1), \ (T_1 T_1) \xrightarrow{K_2} S_1^*, \ T_2, \text{ or } T_1^*$$

$$T_1 + T_1 \xrightarrow{\gamma_T} \begin{cases} \xrightarrow{\gamma_{Tr}} S_1^* + S_0 \\ \xrightarrow{\gamma_{Tn}} T_2 + S_0 \end{cases}$$

Heterofusion
(triplet–trapped triplet annihilation or heterogeneous annihilation)

$$T_1 + T_{1G} \underset{K_{-1G}}{\overset{K_{1G}}{\rightleftharpoons}} (T_1 T_{1G}), \ (T_1 T_{1G}) \xrightarrow{K_{2G}} S_{1G}^* \text{ or } T_{1G}^*$$

$$T_1 + T_{1G} \xrightarrow{\gamma_G} \begin{cases} \xrightarrow{\gamma_{Gr}} S_{1G}^* + S_0 \\ \xrightarrow{\gamma_{Gn}} T_{1G}^* + S_0 \end{cases}$$

Intersystem crossing without involving fission or fusion processes

$$S_1 \xrightarrow{k_{IS_1}} T_1$$

$$S_1 \xrightarrow{K_{IS_2}} T_1^*$$

$$S_1^* \xrightarrow{K_{IS_3}} T_2$$

$$T_1 \xrightarrow{K_{IT}} S_0^*$$

6. Singlet–singlet annihilation

$$S_1 + S_1 \xrightarrow{\gamma_S}$$

$$\xrightarrow{\gamma_{Sr}} S_1^* + S_0 \to S_1 + S_0 + phonons$$

$$\downarrow \text{fluorescence}$$

$$\xrightarrow{\gamma_{Si}} e + h \quad \text{(auto-ionization, IPC, EPE)}$$

$$\xrightarrow{\gamma_{Sf}} T_1^* + T_1^* \quad \text{(fission)}$$

7. Singlet–triplet annihilation

$$S_1 + T_1 \xrightarrow{\gamma_{ST}}$$

$$\xrightarrow{\gamma_{STe}} T_2^* + S_0$$

$$\xrightarrow{\gamma_{STi}} e + h \quad \text{(ionization, IPC)}$$

8. Singlet–trap interaction

$$S_1 + trap \xrightarrow{\alpha_{HG}} S_{1G} + S_0 + phonon$$

$$\xrightarrow{\alpha_{G1}} \text{radiative transition}$$

$$\xrightarrow{\alpha_{G2}} \text{non-radiative transition}$$

Triplet–trap interaction

$$T_1 + trap \xrightarrow{\beta_{HG}} T_{1G} + S_0 + phonon$$

$$\xrightarrow{\beta_{G1}} \text{radiative transition}$$

$$\xrightarrow{\beta_{G2}} \text{non-radiative transition}$$

9. Singlet exciton–free carrier interaction

$$S_1 + e \text{ (or } h)$$

$$\xrightarrow{K_{SC}} S_0^* + e \text{ (or } h)$$

$$\xrightarrow{K'_{SC}} T^* + e \text{ (or } h)$$

Singlet exciton–trapped carrier interaction

$$S_1 + e_t \text{ (or } h_t)$$

$$\xrightarrow{K_{SCt}} S_0^* + e \text{ (or } h)$$

$$\xrightarrow{K'_{SCt}} T_1^* + e \text{ (or } h)$$

Triplet exciton–free carrier interaction

$$T_1 + e \text{ (or } h) \xrightarrow{K_{TC}} S_0 + e \text{ (or } h)$$

(continued overleaf)

Table 7.1 (Contd.)

Process	Kinetics
Triplet exciton–trapped carrier interaction	$T_1 + e \text{ (or } h_t) \xrightarrow{\quad K_{TCt} \quad} S_0 + e \text{ (or } h)$
10. Exciton–photon interaction	$S_1 + h\nu \xrightarrow{\quad \sigma_{Sp} \quad} e + h$
	$T_1 + h\nu \xrightarrow{\quad \sigma_{Tp} \quad} e + h$
11. Exciton–surface interaction	S_1 + surface (metal or electrolyte electrodes of dye layer) $\longrightarrow e \text{ (or } h)$
	T_1 + surface (metal or electrolyte electrodes or dye layer) $\longrightarrow e \text{ (or } h)$
12. Exciton (energy) migration (e.g. trapping) —multimolecular level	$\left.\begin{array}{l} S_{1(\text{host})} \longrightarrow S_{1(\text{guest})} \\ T_{1(\text{host})} \longrightarrow T_{1(\text{guest})} \end{array}\right\}$ No multiplicity change

(d) *Resonance transfer.* The conditions for resonance transfer are the same as those for radiative transfer. But in resonance transfer no actual emission or absorption of photons takes place. Rather the transition dipole of molecule A produces a field inducing stimulated emission of molecule D. It can also be thought of as a process in which molecule A absorbs the photon before molecule D has finished emitting it [Windsor 1965]. By this process energy may be transferred to an unexcited molecule over distances ranging from 5 to 100 Å [Förster 1959, 1960].

(e) *Exciton transfer.* An excited state is not confined to one molecule but rather interacts with molecules surrounding it. For a pair of molecules such an interaction leads to a splitting of a single molecular energy level into two levels. This type of splitting is generally referred to as Davydov splitting or exciton-type splitting [Davydov and Sheka 1965, Davydov 1971]. If a crystal consists of n molecules, the interaction may split a single excitation level into a band of n levels, and the stronger the interaction the greater is the splitting. This is why the lowest energy level of an excited state of a molecular crystal is lower than the energy level of the corresponding excited state of a free or isolated molecule. The exciton splitting is due to an overlap of neighbouring molecular orbitals, and thus the energy eigenvalues of the wave functions of exciton states are determined by an electronic coupling parameter corresponding to the resonance integral. In other words, excitons can be thought of as mobile quasi-particles which are neutral in charge and whose mobility increases with increasing value of the resonance integral. In pure molecular crystals the coupling energy between an adjacent excited molecule in a singlet state and an unexcited molecule lies typically in the wave number range $100-10,000 \, \text{cm}^{-1}$ ($1 \, \text{eV} = 8067 \, \text{cm}^{-1}$), and the time required for the transfer of the excitation from one molecule to the other is of the order of $10^{-12}-10^{-14}$ sec, which is shorter than the period of a lattice vibration. For this case energy can transport rapidly along a chain of molecules to a site in which it either decays radiatively to produce luminescence or becomes trapped by an impurity molecule with lower energy levels or at some defects in the crystal lattice. For anthracene the mean diffusion length of an exciton is larger than 1000 Å [Agranovich *et al.* 1957, 1959, Vishaevskii and Borisov 1956]. However, if the coupling energy between adjacent molecules is very small, such as the excitation of a triplet state of the molecule for which the coupling energy is of the order of $1 \, \text{cm}^{-1}$, then there is competition between excitation transfer and radiation of the excitation energy as phosphorescence. Therefore, exciton transfer depends on the rate processes of interactions which will be discussed in later sections.

7.2. THE FORMATION AND BEHAVIOUR OF EXCITONS

Absorption and reflectance spectra often show the structure for photon energies below the energy band gap, indicating that there are excited states of energy levels below E_c and above E_v. If the excited states are located at the energy level E_{ex} then the excited electron at this level is electrostatically bound with the hole in the valence band to form an exciton. The binding energy of the exciton is $E_c - E_{\text{ex}}$ referred to a free electron and a free hole, and the absorption or excitation energy is $E_{\text{ex}} - E_v$ which is smaller than $E_c - E_v = E_g$. Excitons are unstable with respect to the ultimate recombination in which the electron drops into the hole producing either a photon or phonons.

Excitons can be formed in almost any insulating crystals although some types of excitons are intrinsically unstable with respect to dissociation into free electrons and holes. Excitons can be generated by direct optical, indirect optical, and carrier injection (double injection) processes. Singlet excitons in anthracene can be generated very easily. Triplet excitons in anthracene can be generated either directly by absorption of red light or indirectly by producing first singlet excitons [by one-photon (blue light) or two-photon (red light) absorption] followed by intersystem crossing. Bombardment with high energy particles or X-rays can also produce excitons. In general, when $\nabla_k E_v = \nabla_k E_c$ (the electron and hole group velocities are equal), the electron and hole may be bound by their Coulombic attraction to form an exciton. Excitons can be classified into two groups based on two different limiting approximations; one group is based on the tight-binding approximation, which was first investigated by Frenkel [1931] and Peierls [1932], and is generally referred to as Frenkel excitons, and the other is based on the weakly-binding approximation, which was first proposed by Wannier [1937] and Mott [1938], and is generally referred to as Wannier excitons. In the former an excited state of a molecule is a state in which an electron has been removed from the filled orbital and occupies a previously empty orbital of higher energy, leaving a hole in the original orbital (ground state), and the excitation is confined within or near the molecule, while in the latter the molecules (or atoms) are packed so closely that interaction among molecules is strong which reduces the Coulombic interaction between the electron and the hole and hence increases their separation as shown schematically in Fig. 7.4.

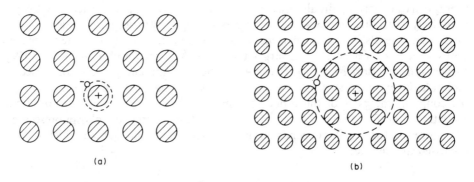

(a)   (b)

FIG. 7.4. Schematic illustration of (a) the Frenkel exciton, and (b) the Wannier exciton.

In molecular crystals the covalent binding of atoms within a molecule is much stronger than the van der Waals binding between molecules. This is indicated by the fact that electronic excitation lines of an isolated molecule appear also in the crystalline solid with the same molecules, with only a slight shift in frequency. At low temperatures the excitation lines are quite sharp although there may be more structures to the lines in the crystal than in the isolated molecule because of the Davydov splitting. On this basis the Frenkel model of excitons has been used extensively to explain luminescent phenomena. However, the Wannier model of excitons for molecular crystals has been studied by Rice and Jortner [1965, 1967], but direct experimental evidence for the existence of Wannier excitons in molecular crystals is still lacking [Pope and Burgos 1966, Pott and Williams 1969]. In the following we shall confine ourselves to the discussion of Frenkel excitons.

7.2.1. Dynamic properties of excitons

There are several possible ways through which an exciton can interact with another exciton or other particles. In this section we shall discuss various exciton interactions which are directly associated with photoluminescence and electroluminescence. Before doing that, we review briefly some basic transition processes.

The various exciton energy levels and transition processes in organic crystals are summarized in Fig. 7.3. The absorption spectra in anthracene and tetracene single crystals corresponding to various energy levels and transition energies are shown in Fig. 7.5. To facilitate the description we take anthracene crystals as an example. The transition from the first excited singlet state S_1 to the ground state S_0 occurs with a lifetime of the order of 10^{-8} sec giving predominantly fluorescence. The transition from the first excited triplet state T_1 to the ground state S_0 is forbidden by spin selection

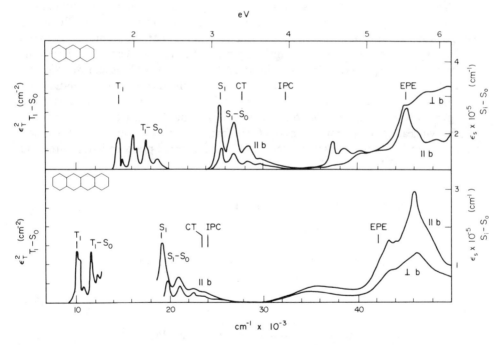

Fig. 7.5. Singlet–singlet (S_0–S_1) and singlet–triplet (S_0–T_1) absorption coefficients $\varepsilon_s\,(\mathrm{cm}^{-1})$ and $\varepsilon_T\,(\mathrm{cm}^{-1})$, respectively, as functions of wave number. *Top*: anthracene; *bottom*: tetracene. S_0–T_1 spectra determined by delayed fluorescence which is proportional to ε_T^2 (ε_T for anthracene at the peak is about $3.4 \times 10^{-4}\,\mathrm{cm}^{-1}$). S_0, ground state level; S_1, lowest excited singlet level; T_1, lowest excited triplet level; CT, charge transfer exciton level; IPC, intrinsic photoconductivity threshold; EPE, external photoemission threshold: $\|\,b$, the electric vector of the light parallel to the b crystal axis; $\perp b$, the electric vector of the light perpendicular to the b crystal axis. [After Swenberg and Geacintov 1973.]

rules, and is therefore difficult to detect. The $S_0 \rightarrow T_1$ absorption spectra shown in Fig. 7.5 were obtained indirectly by measuring the delayed fluorescence produced by the triplet–triplet annihilation process, which is proportional to ε_T^2, where ε_T is the absorption coefficient of the weak $S_0 \rightarrow T_1$ transition. The lifetime of delayed fluorescence is of the order of 10^{-3} sec. There is no absolute "allowedness" and "forbiddenness" in transition, though there is a great difference in degree. The lifetime and hence the diffusion length of triplet excitons are comparatively long and therefore there is a high probability for them to collide with defects resulting in non-radiative transition. However, a weak radiative $T_1 \rightarrow S_0$ transition does occur in a time of the order of 10^{-3} sec, producing phosphorescence. The separation between S_1 and S_0 is about 3.15 eV and that between T_1 and S_0 is about 1.83 eV for anthracene [Avakian and Merrifield 1968]. Although higher excited singlet or triplet excitons may exist, they would rapidly decay non-radiatively to the lowest singlet or triplet exciton states with a very short lifetime ($\sim 10^{-12}$–10^{-13} sec). For this reason we consider only the lowest singlet and triplet exciton bands for the discussion of dynamic properties of excitons.

Exciton states based on the Frenkel model are constructed on the basis of wave functions of neutral molecular excited states. However, exciton states constructed on the basis of wave functions of nearest-neighbour positive and negative ion pairs are generally referred to as charge transfer exciton states or Wannier-type exciton states, in which the electron and hole are separated by one or more lattice constants but are still correlated to each other. Such states were first proposed by Pope *et al.* [Pope and Burgos 1966, Pope *et al.* 1965] and are labelled CT in Fig. 7.5. The separation between CT and S_0 is about 3.4 ± 0.5 eV for anthracene and the lifetime of excited CT is of the order of 10^{-9} sec [Avakian and Merrifield 1968, Swenberg and Geacintov 1973].

As the photon energy (or wave number) increases, the intrinsic photoconductivity threshold (IPC) occurs as shown in Figs. 7.3 and 7.5. This threshold corresponds to the energy gap between S_0 and the lowest conduction band. Further increase in photon energy will lead to photoemission of electrons from the crystal to outside of its surface, and the threshold for its onset is referred to as the external photoemission threshold and is labelled EPE in Fig. 7.5.

(A) Free exciton–free exciton interactions

(i) SINGLET–SINGLET INTERACTIONS

$$S_1 + S_1 \xrightarrow{\gamma_{sr}} S_1^* + S_0 \xrightarrow{\alpha_5} S_1 + S_0 + \text{phonons}$$
$$\downarrow$$
$$\text{fluorescence} \tag{7.9}$$

$$S_1 + S_1 \xrightarrow{\gamma_{si}} e + h \text{ (auto-ionization, IPC, EPE)} \tag{7.10}$$

$$S_1 + S_1 \xrightarrow{\gamma_{sf}} T_1^* + T_1^* \text{ (fission)} \tag{7.11}$$

$$\gamma_s = \gamma_{sr} + \gamma_{si} + \gamma_{sf}$$

where S_1^* and T_1^* are singlet and triplet excitons at high vibronic–electronic levels (hot excitons). Transition processes (7.9)–(7.11) are self-explanatory. Processes (7.9) and (7.10) have been observed in anthracene and naphthalene at high light intensities [Tolstoi and Abramov 1967, Pope 1967, Bergman *et al.* 1967, Johnston and Lyons 1968, Braun 1968, Braun and Dobbs 1971]. The auto-ionization (AI) state rapidly decays non-radiatively to S_1 in a time of 10^{-12}–10^{-13} sec. The generation rate of electrons and holes through the AI process is proportional to the square of the exciting light intensity. However, the fission process (7.11) has not yet been observed experimentally [Swenberg and Geacintov 1973]. Processes (7.9) and (7.10) should be independent of applied magnetic field, but process (7.11) is expected to be slightly dependent on magnetic field.

(ii) TRIPLET–TRIPLET INTERACTIONS

$$T_1 + T_1 \xrightarrow{\gamma_{Tr}} S_1^* + S_0 \xrightarrow{\alpha_5} S_1 + S_0 + \text{phonons} \qquad (7.12)$$

$$\downarrow$$

strong delayed fluorescence or weak phonons

$$T_1 + T_1 \xrightarrow{\gamma_{Tn}} T_2 + S_0 (\text{or } T_1^* + S_0) \xrightarrow{\beta_4 \text{ or } \beta_5} T_1 + S_0 + \text{phonons} \qquad (7.13)$$

$$\downarrow$$

strong phonons or weak phosphorescence

$$\gamma_T = \gamma_{Tr} + \gamma_{Tn}$$

The collision of two triplet excitons to produce a singlet (S_1 or S_1^*) or a higher level triplet (T_2 or T_1^*) is called the fusion, and when they are an identical triplet it is called the homofusion. The experimental evidence for the triplet–triplet annihilation in the form of delayed fluorescence following process (7.12) has been observed as early as in 1958 by Sponer *et al.* and Blake and McClure, although the interpretation was available later [Hochstrasser 1962, Nieman and Robinson 1962, Sternlicht *et al.* 1963, Kepler *et al.* 1963]. This triplet–triplet annihilation phenomenon has been observed in anthracene [e.g. Singh *et al.* 1965], in pyrene [Avakian and Abramson 1965], in naphthalene [Misra and McGlynn 1965, Port and Wolf 1968], and in a number of mixed crystals [Port and Wolf 1968]. The delayed fluorescence, whose intensity is proportional to the square of the exciting light intensity, enables the determination of singlet–triplet absorption spectra and the triplet exciton diffusion coefficient and is usually considered as strong evidence of the migration of triplet excitons. The integrated intensity of emitted delayed fluorescence from the crystal is given by [Avakian and Merrifield 1968]

$$\phi = \frac{1}{2}\gamma_{Tr} \int_{\text{volume}} [T]^2 dV = \frac{1}{2} f \gamma_T \int_{\text{volume}} [T]^2 dV \qquad (7.14)$$

where $[T]$ is the concentration of triplet excitons at position r and at time t, γ_{Tr} is the rate constant of the triplet–triplet annihilation which produces delayed fluorescence, γ_T is the total overall bimolecular annihilation rate constant, and f is the fraction of triplet–triplet annihilation which leads to delayed fluorescence. It should be noted that the bimolecular annihilation process is dependent on the spatial distribution of triplet excitons $[T]$, while the monomolecular decay does not depend on the spatial distribution of $[T]$ because for the former the triplet is captured by or interacted with an impurity or defect.

The triplet–triplet interaction leading to the formation of an ion-pair differs from the singlet–singlet interaction leading to AI in that the rate-determining step for the former case involves the rate of encounter of two triplet excitons, whereas that for the latter case involves the rate of producing a positive ion and a free electron [Choi and Rice 1963]. To evalue γ_T we consider that the annihilation rate of two nearby excitons is comparable in magnitude to the rate of the collision of two excitons, and follow the scheme [Jortner *et al.* 1965, Swenberg 1969]

$$\left.\begin{array}{c} T_1 + T_1 \ \xrightleftharpoons[K_{-1}]{K_1} \ (T_1\,T_1) \\[2em] (T_1\,T_1) \ \xrightarrow{K_2} \ S_1^*, T_2 \text{ or } T_1^* \end{array}\right\} \tag{7.15}$$

where K_1 is the rate of encounter of two triplets to form an intermediate complex state $(T_1\,T_1)$, K_{-1} is the dissociation rate of the complex state to form two separate triplets, and K_2 is the annihilation rate of the complex state to fuse into an excited singlet or an excited triplet [cf. Table 7.1]. The overall annihilation (fusion) rate constant for the triplet–triplet interaction is therefore given by [Merrifield 1968, Johnson and Merrifield 1970]

$$\gamma_T = \frac{K_1}{1 + K_{-1}/K_2} \tag{7.16}$$

γ_T has been measured for anthracene [Avakian and Merrifield 1968], for tetracene [Tomkiewicz *et al.* 1971], and for pyrene [Yarmus *et al.* 1973]. It should be noted that f in eq. (7.14) depends not only on the relative rates of annihilation leading to the fusion of $(T_1\,T_1)$ into excited singlets and excited triplets, but also on the efficiency of producing the zero vibronic fluorescence state S_1 and on the quantum yield for prompt fluorescence. Naturally, the value of f is between 0 and 1.

Obviously, when the Zeeman splitting is large compared with thermal agitation under applied magnetic fields, the magnetic moments of triplet excitons will be partially aligned altering the populations of the spin states, thus causing a change in the $T_1 + T_1$ annihilation rate. It is expected that process (7.12) is magnetic field dependent, but process (7.13) is not. Johnson *et al.* [1967, 1970] have reported that the annihilation rate is strongly dependent on applied magnetic field, and that the delayed fluorescence intensity in anthracene increases in weak magnetic fields up to a peak (about 5 % higher than the value in the absence of the magnetic fields) and then diminishes with increasing magnetic field. At high fields, such a field dependent fluorescence is also dependent on the direction of applied magnetic field. Magnetic field effects have been used as a powerful tool for the investigation of the behaviour of triplet excitons [Swenberg and Geacintov 1973, Yarmus *et al.* 1973, Avakian and Merrifield 1968]. Analysis based on such magnetic field effects has led to the conclusion that one out of about 25 triplet–triplet encounters gives rise to annihilation [Merrifield 1968] rather than one out of 9 predicted by earlier investigators [Jortner *et al.* 1965].

(iii) FISSION OF SINGLETS INTO TWO TRIPLETS

$$S_1 + S_0 \xrightarrow{\gamma_f} T_1 + T_1 \qquad (7.17)$$

This is simply the inverse of process (7.12) and is therefore magnetic field dependent. In anthracene the total energy of two triplets is $E(2T_1) \simeq 3.66$ eV, while the energy of one singlet is $E(S_1) \simeq 3.15$ eV; and in tetracene $E(2T_1) \simeq 2.52$ eV, while $E(S_1) \simeq 2.4$ eV. Thus for homofission to occur S_1 must be excited to S_1^*. This fission process can be suppressed at low temperatures. The low fluorescence efficiency in tetracene and pentacene has been attributed to this homofission process [Swenberg and Stacy 1968, Geacintov *et al.* 1969, Merrifield *et al.* 1969, Singh *et al.* 1965].

(iv) SINGLET–TRIPLET INTERACTIONS

$$S_1 + T_1 \xrightarrow{\gamma_{ST}} \begin{cases} \xrightarrow{\gamma_{STe}} T_2^* \quad \text{or} \quad T_3 + S_0 & (7.18) \\ \xrightarrow{\gamma_{STi}} e + h \text{ (AI, IPC)} & (7.19) \end{cases}$$

Process (7.18) has been observed in anthracene [Babenko *et al.* 1971] and in tetracene [Ern *et al.* 1971] due to the annihilation of S_1 to promote the excitation of T_1 since $T_1 \to T_2^*$ or $T_1 \to T_3$ are allowed transitions. The single–triplet annihilation may also lead to intrinsic photocarrier generation [process (7.19)] because the sum of singlet and triplet energies is higher than the IPC threshold [Fourny *et al.* 1968].

(B) Free exciton–trapped exciton interactions

In real crystals there are always traps capturing excitons. Exciton traps are sites capable of holding the energy that otherwise propagates through the lattice. These traps are generally localized and non-periodic states in the crystal. Thus the radiative transition rate is determined by the specific electronic structure of the trap site. The presence of traps changes the spectral energy distribution, especially the fluorescence and electroluminescence, and also changes the time dependence of population and depopulation of excitons. In organic semiconductors three types of traps have been identified [Helfrich and Lipsett 1965, Rice and Jortner 1967, Wolf and Benz 1971], and they are: (a) guest molecules such as tetracene doped in anthracene, (b) defects or lattice imperfections–structural defects, and (c) self-trapping due to a lattice relaxation induced by excitons.

(i) FREE SINGLET–TRAPPED SINGLET INTERACTIONS

In general the molecule which transfers the energy is referred to as the sensitizer and the molecule which receives the energy as the activator. Either a host molecule or an impurity (guest) molecule can be the sensitizer. Energy transfer implies the exciton migration through sensitizers, and the energy eventually given to an activator is referred to as the exciton trapping.

$$S_1 + \text{Trap} \xrightarrow{\alpha_{HG}} S_{1G} + S_0 + \text{phonons}$$

$$\begin{array}{l} \downarrow \alpha_{G1} \\ \longrightarrow \text{radiative transition} \\ \alpha_{G2} \\ \longrightarrow \text{non-radiative transition} \end{array} \tag{7.20}$$

where α_{HG} is the energy transfer rate constant. The most famous case is the sensitized fluorescence in the anthracene–tetracene mixed crystals [Winterstein *et al.* 1934]. It is well known that the intensity of fluorescence in tetracene is relatively stronger than that in anthracene, thus the presence of traces of tetracene suppresses the blue–violet fluorescence from anthracene and enhances the green–yellow fluorescence from tetracene. This implies that the cross-section for excitation of tetracene molecules is increased by the incorporation of the tetracene molecules into the anthracene crystal, or, in other words, the tetracene molecules act as traps for the electronic excitation energies which are travelling as excitons in the anthracene crystal. A comprehensive review on singlet exciton energy transfer processes in organic solids has recently been presented by Powell and Soos [1975].

(ii) FREE TRIPLET–TRAPPED TRIPLET INTERACTION

$$T_1 + T_{1G} \xrightarrow{\gamma_G} \begin{array}{l} \xrightarrow{\gamma_{Gr}} S_{1G}^* \to \text{guest fluorescence} \qquad (7.21) \\ \xrightarrow{\gamma_{Gn}} T_{1G}^* \qquad (7.22) \end{array}$$

$$\left. \begin{array}{l} T_1 + T_{1G} \underset{K_{-1G}}{\overset{K_{1G}}{\rightleftarrows}} (T_1\,T_{1G}) \\ \\ (T_1\,T_{1G}) \xrightarrow{K_{2G}} S_{1G}^* \text{ or } T_{1G}^* \end{array} \right\} \tag{7.23}$$

Process (7.23) is called "heterofusion" representing the fusion between a free exciton and a trapped exciton. It has been found that the delayed host fluorescence is entirely due to host–host annihilation, while the delayed guest fluorescence is entirely due to host–guest annihilation, which is magnetic field dependent. The heterofusion may also give rise to a vibrational excited guest triplet (T_{1G}^*) which does not lead to fluorescence [Groff *et al.* 1970, Misra 1969, Yee and El-Sayed 1970].

(iii) FISSION OF TRAPPED (GUEST) SINGLETS INTO TWO TRIPLETS

$$S_{1G} + S_0 \xrightarrow{\gamma_{fG}} T_1 + T_{1G} \qquad (7.24)$$

This is simply the inverse of process (7.21) and is therefore magnetic field dependent. Geacintov *et al.* [1971] have reported that the presence of an excited pentacene singlet as trapped exciton in a tetracene crystal can give rise to fission into one free tetracene triplet and one trapped pentacene triplet. Schwob and Williams [1972] have attributed the anisotropy of the magnetic field dependence of the prompt fluorescence in anthracene to fission of charge transfer excitons.

(C) Exciton–charge carrier interactions

$$S_1 + e \text{ (or } h) \left[\begin{array}{l} \xrightarrow{K_{Sc}} S_0^* + e \text{ (or } h) \qquad (7.25) \\[2ex] \xrightarrow{K'_{Sc}} T_1^* + e \text{ (or } h) \qquad (7.26) \end{array} \right.$$

$$S_1 + e_t \text{ (or } h_t) \left[\begin{array}{l} \xrightarrow{K_{Sct}} S_0^* + e \text{ (or } h) \qquad (7.27) \\[2ex] \xrightarrow{K'_{Sct}} T_1^* + e \text{ (or } h) \qquad (7.28) \end{array} \right.$$

$$T_1 + e \text{ (or } h) \xrightarrow{K_{Tc}} S_0^* + e \text{ (or } h) \qquad (7.29)$$

$$T_1 + e_t \text{ (or } h) \xrightarrow{K_{TCt}} S_0^* + e \text{ (or } h) \qquad (7.30)$$

The interaction of singlet excitons or triplet excitons with free electrons or holes (e or h) or with trapped electrons or holes (e_t or h_t) can lead to their annihilation following processes (7.25)–(7.30). Processes (7.26) and (7.27), (7.28), and (7.30) give rise to detrapping of trapped electrons and trapped holes, thus causing an increase in photoconductivity as has been discussed in Section 6.2. Such a non-radiative destruction of excitons has been observed by many investigators for singlets [Many *et al.* 1969, Pope *et al.* 1971, Wakayama and Williams 1971, 1972, Schott and Berrehar 1973] and for triplets [Simpson *et al.* 1968, Ern and Merrifield 1968, Helfrich 1966, Weisz *et al.* 1970, Levinson *et al.* 1970, Frankevich *et al.* 1972, Geacintov *et al.* 1970, Ern *et al.* 1971]. Obviously, the exciton carrier interactions lead to a decrease in both fluorescence intensity and exciton lifetime. Helfrich [1966] was the first to show that excess electrons introduced into anthracene crystals by carrier injection cause a decrease of triplet lifetime. Wakayama and Williams [1971, 1972] have reported that excitons are quenched by carriers and suggested that the interaction of excitons with charge carriers provides an additional mechanism for non-radiative exciton decay.

The change of quantum yield of fluorescence with carrier density due to exciton carrier interactions may be written as [Wakayama and Williams 1971, 1972]

$$\Delta Q \simeq \frac{\alpha_1}{\alpha_1 + \alpha_2 + K_{sc}} \tag{7.31}$$

where α_1 and α_2 are, respectively, radiative and non-radiative singlet decay constants, and K_{sc} is the singlet carrier interaction rate which may be written as

$$K_{sc} = n_T(1 - \theta)\sigma v + n_T \theta \sigma (u + v^2/3u), \quad u > v \tag{7.32}$$

$$= n_T(1 - \theta)\sigma v + n_T \theta \sigma (v + u^2/3v), \quad u < v \tag{7.33}$$

where σ is the reaction cross-section between a singlet and a carrier, u and v are, respectively, the thermal velocities of singlet excitons and carriers, n_T is the total carrier density, and $n_T \theta$ is the free carrier density which is denoted by n (n_T and n are for total and free electron densities and p_T and p are for total and free hole densities).

The quenching of triplet excitons due to their interaction with carriers is magnetic field dependent and anisotropic in a manner similar to that for the triplet exciton fusion [Ern and Merrifield 1969]. The triplet carrier interaction rate constant decreases with increasing magnetic field. We shall come back to the effects of exciton carrier interactions on the quantum yield of photoluminescence and electroluminescence in Sections 7.3 and 7.4.

It is worth mentioning that the effect of magnetic fields on the quenching of phosphorescence has been studied by means of oxygen quenching of triplet excitons. However, the oxygen quenching of triplet excitons was not observed in aromatic hydrocarbons [Geacintov and Swenberg 1972] but it was observed in polycrystalline 2-chloroanthracene [Johnson *et al.* 1971].

(D) Exciton–surface interactions

There are two processes for quenching the mobile molecular excitons at the boundary between an organic semiconductor and an electrode. They are: (1) *charge transfer*—an exciton can transfer an electron to an adjacent trapping centre at the interface, producing a free hole in an organic semiconductor (e.g. oxygen molecules adsorbed at the surface can act as electron trap centres), and (2) *energy transfer*—an exciton can transfer its energy to the acceptor molecules present at or adjacent to the surface of the organic semiconductor. Usually such an electron transfer is a rather slow process as compared with the energy transfer between an organic semiconductor and an electrode.

A metal on the surface of a molecular crystal can influence the electronic states of the surface molecules in two ways: first, it can modify their energetic position, and, secondly, it can affect the lifetime of excited states. The first effect results from the discontinuity of the dielectric constant across the interface, which leads to a change in the molecular polarization energy. If the polarization is enhanced, surface molecules may act as traps for excitons. The second effect results from non-radiative transitions induced by metallic electrodes.

In general, the exciton-quenching zone at the surface is very narrow (about 20 Å) [Pope *et al.* 1971]. In the case of electroluminescence the excitons are generated from the electron–hole recombination and the recombination zone is of the order of 10^3 Å, and thus for this case the exciton–surface interactions may not be so important as compared with other non-radiative transitions. However, in the case of photocon-duction the dissociation of excitons at the boundary generates free carriers and is therefore one of the important processes for photocarrier generation. The behaviour of excitons in contact with a metal surface has been studied theoretically and experimen-tally in some detail by Agranovich *et al.* [1972], Killesreiter and Baessler [1971, 1972], Chance and Prock [1973], and Michel-Beyerle *et al.* [1974].

(E) Exciton–photon interactions

$$S_1 + hv \xrightarrow{\sigma_{SP}} e + h \tag{7.34}$$

$$T_1 + hv \xrightarrow{\sigma_{Tp}} e + h \tag{7.35}$$

Processes (7.34) and (7.35) represent, respectively, the photoionization of a singlet exciton and that of a triplet exciton. For anthracene the photoionization cross-section of singlet excitons σ_{Sp} has been measured to be about 2×10^{-19} cm^2 for photons of 1.78 eV and about 5×10^{-19} cm^2 for photons of 2.9 eV [Kepler 1967], and that of triplet exciton σ_{Tp} about 4×10^{-21} cm^2 [Holzman *et al.* 1967]. The theoretical values of σ_{Sp} and σ_{Tp} reported by Hernandez [1968] are in good agreement with the experimental values. The exciton–photon interactions may possibly compete with the exciton–exciton interactions in producing charge carriers when the light intensity is sufficiently high [Kepler 1967, Delacote and Schott 1968].

(F) Exciton–phonon interactions

The effects of the interaction between Frenkel excitons and phonons arising from intramolecular vibrations can be summarized as follows [Rice and Jortner 1967]: (i) the main effect is the reduction of the bandwidth relative to the free exciton value, and, in general, the upper half of the band is compressed much more than the lower half; (ii) the effective mass of the exciton increases with increasing exciton–phonon coupling because of the exciton motion being retarded by nuclear displacements; (iii) the wider the free exciton band the less effective is the exciton–phonon coupling; and (iv) the exciton–phonon coupling affects the electronic transition intensity spectrum.

The intermolecular vibrations or lattice vibrations interact with phonons causing a broadening and shift of the exciton energy levels [Rice and Jortner 1967]. The exciton–phonon coupling affects greatly the exciton migration. In extremely pure crystals the energy transfer rate constant increases with increasing temperature, indicating that the exciton–phonon coupling is the dominant path-limiting mechanism. The nature and the effects of the exciton–phonon interactions are not yet fully understood. However, for the details of the present state of knowledge in this area the reader is referred to the work of Rice and Jortner [1967], Sceats and Rice [1975], Davydov [1971], and Craig and Walmsley [1968].

Some rate constants for exciton interactions in anthracene and tetracene crystals are given in Table 7.2.

TABLE 7.2. *Rate constants for exciton interactions in crystalline anthracene and tetracene [after Geacintov et al. 1975]*

Process	Tetracene	Anthracene	Magnetic field dependent
(1) Fission: $S_0 + S_1 \xrightarrow{\gamma_f} T_1 + T_1$	$\gamma_f \approx 15 \times 10^{-12}$ cm³ sec⁻¹	Negligible	Yes
(2) Fusion: $T_1 + T_1 \xrightarrow{\gamma_{Tr}} S_1 + S_0$	$\gamma_{Tr} \approx 10^{-9}$ cm³ sec⁻¹	$\gamma_{Tr} \simeq 10^{-11}$ cm³ sec⁻¹	Yes
(3) Singlet–triplet annihilation: $S_1 + T_1 \xrightarrow{\gamma_{STe}} S_0 + T_1$	$\gamma_{STe} \approx 2 \times 10^{-7}$ cm³ sec⁻¹	$\gamma_{STe} \simeq 10^{-8}$ cm³ sec⁻¹	No
(4) Singlet–singlet annihilation: $S_1 + S_1 \xrightarrow{\gamma_{Sr}} S_1 + S_0$	—	$\gamma_{Sr} \simeq 10^{-8}$ cm³ sec⁻¹	No
(5) Annihilation of S_1 by free carriers: $S_1 + e$ (or h) $\xrightarrow{K_{Sc}} S_0^* + e$ (or h)	—	$K_{Sc} \simeq 10^{-8}$ cm³ sec⁻¹	No
(6) Annihilation of S_1 by trapped carriers: $S_1 + e_t$ (or h_t) $\xrightarrow{K_{Sct}} S_0^* + e$ (or h)	—	$K_{Sct} \simeq 10^{-8}$ cm³ sec⁻¹	?
(7) Annihilation of triplets by free carriers: $T_1 + e$ (or h) $\xrightarrow{K_{Tc}} S_0 + e$ (or h)	—	$K_{Tc} \simeq 10^{-9}$ cm³ sec⁻¹	Yes
(8) Annihilation of triplets by trapped carriers: $T_1 + e_t$ (or h_t) $\xrightarrow{K_{Tct}} S_0 + e$ (or h)	$K_{Tct} = (5 \pm 1) \times 10^{-9}$ cm³ sec⁻¹	$K_{Tct} \simeq 10^{-11}$ cm³ sec⁻¹	Yes
(9) Singlet lifetime (no quenchers)	$T_s \simeq 2 \times 10^{-10}$ sec (at 298° K)	$T_S \simeq 2 \times 10^{-3}$ sec	{ Yes (tetracene) / No (anthracene)
(10) Triplet lifetime (no quenchers)	$T_T = (50 - 200) \times 10^{-6}$ sec	$T_T > 20 \times 10^{-3}$ sec	No

7.2.2. Exciton transport processes

One of the most important properties of excitons is their ability of transporting energy without involving the migration of net electric charge. We have mentioned earlier that when the electron and hole group velocities are equal, the Coulombic attraction between the electron and hole may form an exciton. If both the electron and hole concentrations are high the electron–electron and hole–hole Coulombic repulsions are large, which tend to reduce the attractive Coulombic interaction. The internal field created by the potential fluctuations of the band edges due to high concentrations of both types of carriers tends to separate the electrons and the holes causing the dissociation of excitons. However, if the internal field is due to a deformation potential, the direction of the force acting on the electron will be the same as that acting on the hole, thus causing the exciton to move as an entity from a large energy gap region to a low energy gap region. The exciton–exciton annihilation phenomenon provides strong evidence, though indirect, that singlet and triplet excitons are mobile. The spatial and temporal exciton distributions have been studied theoretically and experimentally, and the exciton motion is correctly described by diffusion equations [Merrifield and Avakian 1968, Powell and Kepler 1969]. Generally, there are three different basic mechanisms of exciton transport of energy, and they are briefly described as follows.

(i) *Electromagnetic wave packet transport*. The energy is transported by polariton, which is an intimate mixture of a photon and an exciton. When a photon contributes one quantum of excitation to the electromagnetic flux, it will travel as a wave packet inside a crystal [Knox 1973, Dexter and Knox 1965].

(ii) *Hopping transport*. If the exciton is self-trapped it might jump to another site of the perfect lattice along a chain of molecules (e.g. anthracene molecules) until it falls into a trap (e.g. tetracene molecule) such as in sensitized luminescence [Trlifaj 1958, Munn and Siebrand 1970].

(iii) *Long range resonance transfer*. This transfer process is based on the dipole–dipole coupling mechanism [Förster 1959, 1960, Kallmann 1960, Powell and Soos 1972] in which it is not necessary to have a chain of molecules to carry the energy. Supposing that an acceptor molecule A possesses absorptive transitions which coincide in energy with the luminescent transitions of a donor molecule D, then both may occur together without involving radiative emission. This implies that the excitation energy is transferred from D to A due to coupling between two systems, or, in other words, the photon is absorbed by A before D has finished emitting. The dipole–dipole interaction energy is proportional to r^{-3} and the probability of energy transfer is proportional to the square of the interaction energy and thus proportional to r^{-6}, where r is the separation distance between A and D. For the quadrupole–dipole coupling, the probability of energy transfer is proportional to r^{-8} [Dexter 1953].

(A) Coherent transport

In organic semiconductors the overlap between molecules is small and the electrons are highly localized. This implies that the intermolecular interaction is small and carriers or excitons do not move easily through the solid. However, there are two limiting cases generally used for analysing the exciton transport problems which are (i) coherent transport and (ii) incoherent transport.

For the coherent transport the time required for an exciton to transfer from one molecule to another τ_{et} is much shorter than the displacement time required for molecules to change from the old to new equilibrium positions τ_{md} under a change in the force of interaction between neighbouring molecules upon excitation of one molecule in the solid. This means that the motion of an exciton is not accompanied by a local lattice deformation, and therefore can be described using a band model because its mean free path is greater than the lattice spacing and it can move coherently over several lattice spacings before being scattered. In general, coherent exciton motion may be expected to occur only in ultrapure organic crystals at low temperatures at which both impurities and phonons are ineffective in limiting the exciton mean free path.

The conduction for the coherent exciton motion depends on the three inter-actions which are the vibrational coupling between neighbouring molecules

$$\sum_r \tfrac{1}{4} m_R \omega_1^2 X_r (X_{r+1} + X_{r-1}),$$ the electronic coupling between adjacent molecules J, and

the exciton–phonon coupling $\tfrac{1}{2} m_R \omega_2^2 X_R$. The Schrödinger equation for the excited electron wave function ψ can be written in the tight-binding approximation as a linear combination of electron wave function ϕ localized on individual molecules. Such an equation has been given in eq. (1.29). Since $|a_n\{X_r\}|^2$ is the probability of finding the excited electron at molecule n, then a_n is analogous to a wave function. By substituting eq. (1.29) into the time dependent Schrödinger equation, a_n can be expressed in terms of the aforementioned three interactions [Munn and Siebrand 1970, Holstein 1959]. This equation has been given in eq. (1.30). In the slow exciton limit ($J \ll \hbar\omega_1$) and the slow phonon limit ($J \gg \hbar\omega_1$), the quadratic exciton–phonon coupling is assumed to be more substantial than the intermolecular electronic and intermolecular vibrational coup-lings. In the quasi-free exciton limit, the intermolecular electron coupling is the most substantial of the three interactions.

In the slow exciton limit for coherent transport, the slow exciton is followed adiabatically by the deformation it induces on the zero-point vibration, while in the slow phonon limit for coherent transport the slow motion of the deformation is followed adiabatically by the exciton. In the slow exciton limit, the exciton velocity W_i in a state i is given by

$$W_i = (a/\hbar)(\partial E_i/\partial q_i) \tag{7.36}$$

where E_i is the energy of the exciton state i and q_i is the corresponding wave vector. With this velocity, a mean free time τ_i can be defined as the time within which the exciton travels before encountering a site with a different number of phonons. Thus the exciton diffusion coefficient for coherent transport is

$$D_{ce} = \langle \tau_i W_i \rangle$$
$$= (4Ja^2/\pi\hbar)[(1-\xi)^2(1+\xi)/\xi] \tag{7.37}$$

where

$$\xi = \exp(-\hbar\omega_0/kT) \tag{7.38}$$

In the slow phonon limit, W_i is the phonon velocity. The exciton diffusion coefficient for coherent transport is

$$D_{cp} = (Ba^2/\pi\hbar)[(1-\xi)^2(1+\xi)/\xi] \tag{7.39}$$

where B is the phonon bandwidth which is

$$B = \hbar\omega_1^2/\omega_0 \tag{7.40}$$

It can be seen that D_{ce} or D_{cp} decreases rapidly with increasing temperature. It is generally agreed that coherent transport occurs at low temperatures and incoherent hopping transport is dominant at high temperatures. There is a narrow transition region, centred about the transition temperature T_t, at which the coherent and incoherent hopping contributions to the total exciton diffusion coefficient are equal. Munn and Siebrand [1970] have pointed out that theoretically the transport is coherent at temperatures up to $T \simeq \hbar\omega_0/K$, which is of the order of the melting point of the crystal, but other factors may restrict the transport to incoherent hopping, such as the scattering from acoustic lattice modes. The observed triplet exciton diffusion in the c' direction of anthracene crystals at low temperatures has been considered as a coherent transport.

(B) Incoherent transport

The wavelength of the light which is absorbed to create excitons in a crystal is much larger than both the absorption length and the intermolecular separation in most cases. A typical example is $\lambda = 5000$ Å, $\alpha^{-1} = 100$ Å, and $a = 10$ Å. Therefore it would be expected that some coherence in all excitons should exist at the time of their creation. It probably does exist for a very brief instant, but in most situations the initial coherence associated with the photon's spatial extent is washed out by vibrational broadening.

In the aforementioned interactions the electron–phonon interaction is taken to be the strongest of the three. For the incoherent hopping transport in the slow exciton limit the exciton exchange energies are small compared with phonon dispersion energies, and the transfer of the electron between adjacent molecules is rate determining. In the slow phonon limit the electron exchanges are large compared with phonon dispersion energies, and the transfer of the electron between adjacent molecules is limited by the rate of phonon transfer. In the slow exciton limit $J \ll \hbar\omega_1$ the weak intermolecular electronic coupling causes the exciton to transfer to neighbouring sites, while the stronger intermolecular vibrational coupling provides a final density of states, so that the transfer develops uniformly in time. Since the energy conversion requires that the vibrational occupation numbers of two adjacent molecules interchange, the exciton transfer can occur only between two adjacent sites separated by a with the same initial vibrational quantum numbers. Thus the rate of transfer from site n to site $n+1$ is equal to that from site n to site $n-1$, and hence the diffusion coefficient is

$$D_{he} = 2d^2W_T(n \to n+1) \tag{7.41}$$

where $W_T(n \to n+1)$ is the thermal average hopping rate which is given by [Munn and Siebrand 1970]

$$W_T(n \to n+1) = 2\pi\omega_0(J/\hbar\omega_1)[\xi/(1+\xi)] \tag{7.42}$$

The diffusion coefficient decreases with decreasing temperature and approaches zero at low temperatures. However, at low temperatures coherent diffusion becomes important.

In the slow phonon limit $J \gg \hbar\omega_1$ the intermolecular electronic coupling provides the density of states, while the relatively weaker intermolecular vibrational coupling determines the rate of exciton transfer. Under this condition the exciton diffusion coefficient is [Munn and Siebrand 1970]

$$D_{hp} = (\pi a^2 B^2 / qhJ)(1 + \xi)^{-2} \qquad (7.43)$$

In this case the density of states is independent of temperature; thus the temperature dependence of D_{hp} is that of the rate of phonon transfer. The difference between D_{he} (slow exciton limit) and D_{hp} (slow phonon limit) lies in their difference in temperature dependence. Figure 7.6 shows the temperature dependence of D_{he} and μ_{he} (slow exciton limit) and D_{hp} and μ_{hp} (slow phonon limit).

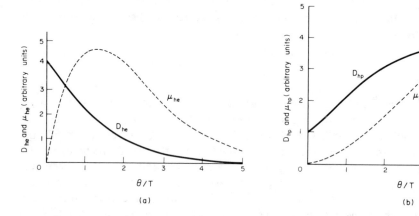

FIG. 7.6. Temperature dependence of hopping diffusion coefficients (solid lines) and mobilities (broken lines) in (a) the slow exciton limit, and (b) the slow phonon limit. $\theta = \hbar\omega_0/k$. [After Munn and Siebrand 1970.]

The total diffusion coefficient is the sum of the coherent and incoherent hopping contributions discussed above. By defining the transition T_t as the temperature at which the two contributions are equal, it is easy to plot T_t as a function of the ratio of the free exciton and free phonon bandwidth $\zeta = 4J/B$ from eqs. (7.37), (7.39), (7.41), and (7.43), and the result is shown in Fig. 7.7. In the following we shall discuss the methods of determining exciton diffusion coefficients.

7.2.3. Diffusion of excitons

One of the most important properties of excitons is their capability of transporting energy without the transport of net charge. There is a great deal of experimental evidence on diffusion of triplet and singlet excitons in anthracene and other organic crystals [Wolf 1967, Avakian and Merrifield 1968, Powell and Soos 1975]. In the following we shall review briefly this important transport process.

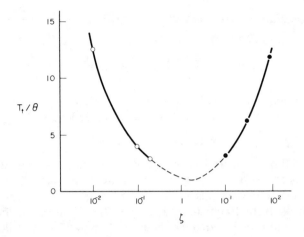

Fig. 7.7. Transition temperature between coherent and hopping diffusion as a function of $\zeta \cdot \zeta = 4J/B$ is the ratio of the free exciton and free phonon bandwidths. Open circles represent the slow exciton limit and filled circles the slow phonon limit. The curve in the intermediate region is roughly indicated by the broken line. [After Munn and Siebrand 1970.]

(A) Diffusion of triplet excitons

The well-known experimental evidence on the motion of triplet excitons is the delayed fluorescence in anthracene resulting from triplet–triplet fusion (or annihilation) as has been mentioned in the previous section. According to eq. (7.14) the diffusion coefficient can be determined by measuring the dependence of the integrated delayed fluorescence intensity on the spatial distribution of excitons created by spatially distributed excitation. Triplet excitons generated by irradiating the specimen with light of sufficient energy and very small beam cross-section would diffuse out of the illuminated regions following the rate equation

$$\frac{\partial[T]}{\partial t} = G_T - (\beta_1 + \beta_2)[T] - \beta_{HG}[T] - (K_{Tc} + K_{Tct})[T]$$

$$- \gamma_T[T]^2 - \gamma_G[T][T_G] + \sum_{ij} D_{Tij} \frac{\partial^2[T]}{\partial x_i \partial x_j}$$

$$= G_T - \beta_T[T] - \gamma_T[T]^2 - \gamma_G[T][T_G] + D_T \nabla^2[T] \qquad (7.44)$$

where G_T is the triplet exciton generation rate, β_{HG} is the rate of excitons being trapped, D_{Tij} is the triplet diffusion coefficient in crystalline directions i and j, and D_T is the average triplet diffusion coefficient. Using a continuous helium–neon laser light to excite an anthracene specimen, Avakian and Merrifield [1964] have measured the luminescence intensity as a function of ruling period (a Ronchi ruling is a grating with alternating opaque and transparent strips) under low excitation intensity. Under this condition

$$\beta_T[T] \gg \gamma_T[T]^2 + \gamma_G[T][T_G] \qquad (7.45)$$

Thus for the one-dimensional case and under the steady state conditions, eq. (7.44) becomes

$$D_T \frac{d^2[T]}{dx^2} - \beta_T[T] = 0 \tag{7.46}$$

Solving this equation in the usual way with the boundary condition $[T] \to [T]_0$ as $x \to 0$ and $[T] \to 0$ as $x \to \infty$, we can find the diffusion length

$$L_T = (D_T/\beta_T)^{1/2} = (D_T \tau_T)^{1/2} \tag{7.47}$$

where τ_T is the lifetime of the triplet excitons. By assuming the spread of the excitation as a diffusion process, Avakian and Merrifield [1964] have found the triplet diffusion length L_T for anthracene crystals to be $10 \pm 5 \, \mu m$ in the *ab* plane for specimen thicknesses ranging from $10 \, \mu m$ to $250 \, \mu m$, and the diffusion coefficient D_T corresponding to this diffusion length to be about $10^{-4} \, cm^2 \, sec^{-1}$. It should be noted that there is an uncertainty of this direct measurement method because the presence of excitations in the unilluminated regions, which have been considered as evidence of exciton diffusion, could also be caused by scattered or diffracted light rather than by diffused excitons.

However, Ern *et al.* [1966] have extended the method of Avakian and Merrifield [1964] by using a low frequency chopper to switch on and off the laser beam enabling the dynamic measurements of the time required for the delayed fluorescence to build up and to decay. This method can provide a means of distinguishing the light scattering or diffraction phenomena from the exciton diffusion because the latter is a rather slow process. By normalizing the integrated delayed fluorescence intensity as

$$\phi_N(t) = \phi(t)/\phi_{\text{steady state}} \tag{7.48}$$

the buildup and the decay of $\phi(t)$ after the switch on and off of the illuminating light are shown in Fig. 7.8. The measurements were carried out with the Ronchi rulings of different period x_0 and different window to period ratio r, and a simplified approximation was made for stray light in the shadow region as shown in Fig. 7.8. Their measurements yield a value of $2 \times 10^{-4} \, cm^2 \, sec^{-1}$ for the triplet exciton diffusion coefficient in the *ab* plane of anthracene crystals at room temperature. A number of investigators [Levine *et al.* 1966] have reported a similar value of D_T for anthracene.

On the basis of the assumption that triplet excitons are quenched at the crystal surfaces, Kepler and Switendick [1965] have studied the dependence of triplet lifetime on specimen thickness and, from their results, determined the triplet diffusion coefficient of anthracene to be $0.4–2.0 \times 10^{-2} \, cm^2 \, sec^{-1}$ along the *c*-axis perpendicular to the *ab* plane. By generating triplet excitons through the intersystem crossing process from the blue-light-generated excitons, and by controlling the depth of the crystal specimen within which triplet excitons are generated by varying the wavelength of the incident light, Kepler and Switendick [1965] have obtained the excitation spectrum for delayed fluorescence in anthracene, which is shown in Fig. 7.9. In the short wavelength region their results are in contrast to the theoretical prediction from eq. (7.14). However, they have attributed the decrease of delayed fluorescence intensity with the decrease of wavelength in the short wavelength region to the diffusion of triplet excitons to the crystal specimen surface and their quenching there, and from these results they have obtained the value of D_T equal to $10^{-2} \, cm^2 \, sec^{-1}$ for the triplet

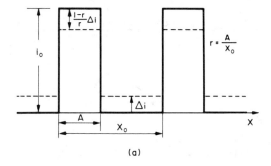

(a)

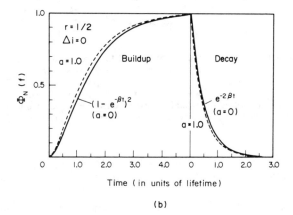

(b)

FIG. 7.8. (a) The exciting light intensity distribution in the Ronchi ruling method for observing triplet exciton diffusion. The solid line represents the ideal geometrical shadow of the ruling, and the dashed line the assumed intensity distribution with stray light intensity Δi in the shadow regions. (b) The time dependence of buildup and decay of normalized delayed fluorescence $\phi_N(t)$ due to exciton diffusion. The parameter $a = 2\pi \sqrt{D\tau}/x_0$, where x_0 is the ruling period and τ the triplet lifetime; and the parameter r gives the ruling window to period ratio. The solid curves correspond to the absence of diffusion ($D = 0$, $a = 0$) and the dashed curves are for $a = 1.0$. [After Ern, Avakian, and Merrifield 1966.]

motion perpendicular to the *ab* plane of the crystal. Similar experiments have also been performed by Williams *et al.* [1966, 1967]. Their mean values determined for anthracene triplet excitons are $(0.2–1.0) \times 10^{-2}$ cm^2 sec^{-1} for the diffusion coefficient and 70–150 μm for the diffusion length in the direction perpendicular to the *ab* plane.

Using high energy ionizing radiation to produce tracks of the excitons and then studying the time dependence of the delayed fluorescence intensity, King and Voltz [1966] and Voltz *et al.* [1966] have obtained the value of $(3–6) \times 10^{-6}$ cm^2 sec^{-1} for D_T based on α-particle and electron excitations, while Perkins [1968] has obtained the value of $(0.5–1.5) \times 10^{-4}$ cm^2 sec^{-1} for D_T based on a pulsed 600 kV X-ray excitation in anthracene.

On the basis of the exclusive generation of triplet excitons at the anthracene crystal surface by energy transfer from an absorbed dye, and on the comparison of prompt and

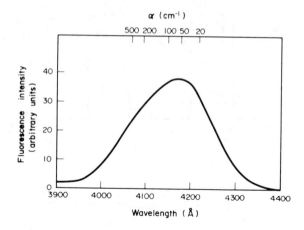

FIG. 7.9. Excitation spectrum for delayed fluorescence in anthracene crystals at room temperature with the wavelengths of the exciting light in the region of the singlet absorption tail. The incident light was polarized with its electric vector parallel to the *a* crystal axis, and its wave vector was perpendicular to the *ab* plane of the crystal. [After Kepler and Switendick 1965.]

sensitized delayed fluorescence, Nickel and Maxdorf [1971] have obtained the value of $(5 \pm 1) \times 10^{-7}$ cm^2 sec^{-1} for D_T in the c' direction (perpendicular to the *ab* plane).

On the basis of the field dependence of the effectiveness of triplet excitons in reducing the spin lattice relaxation time of protons in an anthracene crystal, Maier *et al.* [1967] have obtained a value of $(5 \pm 1) \times 10^{-12}$ sec for the average time spent by the triplet exciton on any molecule, and estimated the value of 2.5×10^{-4} cm^2 sec^{-1} for D_T.

It can be seen that the value of D_T is dependent on the method used for its determination. D_T is determined by the average exciton velocity and the average exciton scattering time (or the relaxation time), thus D_T is strongly dependent on the properties of the material and its surface conditions. By means of a spectroscopic approach to analyse the absorption spectrum for generation of triplet excitons in polarized light, Avakian *et al.* [1968] have measured the Davydov splitting and optical line width for the 0–0 transition of the singlet–triplet excitation spectrum for delayed fluorescence in anthracene crystals and their results are shown in Fig. 7.10. From these results they have deduced the value of 0.5×10^{-4} cm^2 sec^{-1} for D_T along the *a* axis in the *ab* plane, which is in reasonable agreement with that measured by the direct method. They have suggested that the spectroscopic approach is less accurate than the direct method in determining D_T, but it could provide other information concerning the exciton motion. From the scattering time and exciton velocity determined from the spectroscopic approach, they have obtained a value of about 0.1 Å for the exciton scattering length, which is much smaller than the nearest-neighbour intermolecular distance. Such a small scattering length indicates that the exciton motion is a hopping process with a strong interaction between the exciton and the lattice rather than a nearly free propagation with infrequent scattering events.

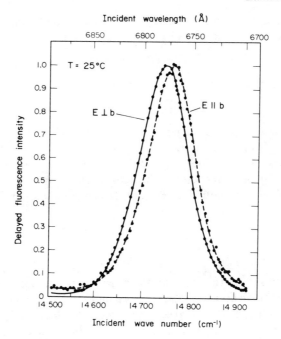

Incident wavelength (Å)

Incident wave number (cm⁻¹)

FIG. 7.10. Davydov splitting in the first "line" (0–0 transition) in the normalized singlet–triplet excitation spectrum for delayed fluorescence in anthracene crystals. The wave vector of the incident polarized light was perpendicular to the *ab* plane of the crystal. The true height of the "line" with $E \parallel b$ crystal axis is about 7 times smaller than that with $E \perp b$ axis. [After Avakian, Ern, Merrifield, and Suna 1968.]

(B) Diffusion of singlet excitons

The well-known experimental evidence on the motion of singlet excitons is the diffusion of an excitation to an activator as has been mentioned in the previous section. The diffusion of singlet excitons is governed by the following kinetic equation:

$$\frac{\partial [S]}{\partial t} = G_S - (\alpha_1 + \alpha_2)[S] - \alpha_{HG}[S] - (K_{sc} + K_{sct})[S]$$

$$- \gamma_s [S]^2 + \sum D_{Sij} \frac{\partial^2 [S]}{\partial x_i \partial x_j}$$

$$= G_S - \alpha_S [S] - \gamma_s [S]^2 + D_S \nabla^2 [S] \qquad (7.49)$$

where G_S is the singlet exciton generation rate, α_{HG} is the rate of excitons being trapped, D_{Sij} is the singlet diffusion coefficient in crystalline directions i and j, and D_s is the average singlet diffusion coefficient. For low concentration of singlet excitons

$$\alpha_S [S] \gg \gamma_s [S]^2 \qquad (7.50)$$

then eq. (7.49) reduces to

$$\frac{\partial [S]}{\partial t} = G_s - \alpha_s [S] + D_s \nabla^2 [S] \qquad (7.51)$$

Under steady state conditions and using the boundary conditions $[S] = [S]_0$ as $x \to 0$ and $[S] \to 0$ as $x \to \infty$, the solution of eq. (7.51) gives

$$[S] = [S]_0 \exp(-x/L_S) \tag{7.52}$$

where L_s is the singlet exciton diffusion length

$$L_S = (D_s/\alpha_s)^{1/2} = (D_s\tau_s)^{1/2} \tag{7.53}$$

and τ_s is the lifetime of the singlet excitons. It can also be easily shown that the concentration of singlet excitons will decay exponentially when the excitation is switched off. To study the diffusion of singlet excitons, the following four methods are generally used: (a) bulk quenching, (b) surface quenching, (c) bimolecular recombination, and (d) photoconduction. Each of these methods is now briefly described as follows.

(a) BULK QUENCHING

The host-sensitized energy transfer (the sensitizer is a molecule of the host crystal such as anthracene) in a doped crystal (a doped molecule such as a tetracene molecule in anthracene acts as the activator to receive the energy) has been studied by measuring the relative fluorescence intensity or decay time as a function of activator concentration. In a doped crystal (e.g. tetracene-doped anthracene), if the host molecules are excited, their energy will be transferred to randomly distributed activators. If the concentration of host molecules (sensitizers) N_S is much larger than that of doped impurity molecules (activators) N_A, then the ratio of the integrated fluorescence intensity of the sensitizers in a pure specimen $I_s(0)$ to that in a doped specimen $I_s(N_A)$ varies linearly with increasing doped activator concentration N_A following the relation [Wolf 1967]

$$I_S(0)/I_S(N_A) = 1 + K_{\text{eff}}N_A \tag{7.54}$$

where K_{eff} is an effective energy transfer parameter which is related to the singlet exciton energy transfer rate $K(t)$ in the following simple equations [Powell and Kepler 1969, 1970]

$$\frac{d[S]}{dt} = G_s - \alpha_s[S] - K(t)[S] \tag{7.55}$$

$$\frac{dn_A}{dt} = K(t)[S] - \alpha_A n_A \tag{7.56}$$

where n_A is the concentration of excited activators; α_s and α_A are, respectively, the fluorescence decay rates for the sensitizers and activators. Since $K(t)$ depends on specific assumptions about exciton motion and trapping, thus by solving eqs. (7.55) and (7.56) for a δ-function excitation at time $t = 0$ and for the case $N_S \gg N_A$, the average value of $K(t)$ over the excitation lifetime can be obtained [Powell and Soos 1975], and it is

$$K_{\text{eff}} = (a/\alpha_s)\{1 + (b/a)[\pi(\alpha_s + N_A a)]^{1/2}\} \tag{7.57}$$

where a is a constant determined from the quenching of the host fluorescence lifetime at long times and b is a constant determined by fitting the experimental time dependence of $K(t)$ to the expression

$$K(t) = N_A(a + bt^{-1/2}) \qquad (7.58)$$

For the case of purely exponential decay, then $b \to 0$ and eq. (7.57) reduces to

$$K_{\text{eff}} = a/\alpha_s \qquad (7.59)$$

From an experimental consistency point of view, it is not desirable to use two different specimens, one undoped (pure) and the other doped, for obtaining $I_s(0)$ and $I_S(N_A)$, respectively, since the background imperfection conditions of the two specimens are generally not identical. To avoid this difficulty, the measurement is generally made on the integrated fluorescence intensity of the sensitizers $I_s(N_A)$ and that of the activators $I_A(N_A)$ separately in the same doped specimen. Their ratio can be written as

$$I_A(N_A)/I_S(N_S) = K_{\text{eff}} N_A \qquad (7.60)$$

if the sensitizer and activator quantum efficiencies are approximately the same.

To determine the diffusion coefficient it is necessary to know the fluorescence lifetime of the sensitizer τ_s, which is generally found to vary linearly with N_A following the relation [Powell and Soos 1975]

$$\tau_S(0)/\tau_S(N_A) = 1 + K_{\text{eff}} N_A \qquad (7.61)$$

In general $\tau_s(0) = \alpha_s^{-1}$ is the lifetime of singlet excitons in the pure crystal and larger than $\tau_s(N_A)$ which is the lifetime of singlet excitons in the activator-doped crystal. From the bulk-quenching experiment the apparent diffusion coefficient D_s^{app} is given by

$$D_s^{\text{app}} = \tfrac{1}{6}(C^2/t_H^{\text{app}}) \qquad (7.62)$$

and the average diffusion length

$$L_S = [D_s \tau_s(0)]^{1/2} \qquad (7.63)$$

where t_H^{app} is defined as the apparent hopping time given by

$$t_H^{\text{app}} = \tau_s(0)/K_{\text{eff}} \qquad (7.64)$$

and C is the lattice constant. Since the lattices in organic crystals are not simple cubic, C has to be estimated by an equivalent cubic unit cell volume. The diffusion parameters for some organic crystals determined by this method are given in Table 7.3.

The two most common energy transfer mechanisms are the exciton diffusion and the long range resonance transfer (LRRT). The former has a constant transfer rate, while the latter has an energy transfer rate which varies with $t^{-1/2}$. To distinguish between these two mechanisms, it is convenient to use the time-resolved spectroscopy (TRS) method [Powell 1973, Powell and Kepler 1970]. In this method the total time evolution of the fluorescence of the sensitizers and activators is observed and yields the time dependence of the energy transfer rate $K(t)$. From a careful analysis, the constants a and b of eq. (7.58) can be determined and the contribution of these two energy transfer mechanisms can be estimated. Furthermore, this method permits a more accurate determination of D_s, L_s, and t_H. For details of this method the reader is referred to the original papers of Powell and his coworkers.

(b) SURFACE QUENCHING

In the surface-quenching experiment one side of a thin specimen is excited and the opposite side is deposited with a sensitive fluorescence detector layer as shown in Fig. 7.11. By measuring the fluorescence intensity through the detector as a function of specimen thickness d and excitation wavelength λ, it is easy to determine $L_S = [D_s\tau_s]^{1/2}$. The kinetic equation governing this quenching process [Simpson 1956] is given by

$$\frac{d[S]}{dt} = \varepsilon_s I_S \exp(-\varepsilon_s x) - [S]/\tau_s + D_s \frac{d^2[S]}{dx^2} \tag{7.65}$$

where ε_s is the absorption coefficient of the specimen, which depends on the excitation wavelength. Obviously, under steady state conditions $\partial[S]/\partial t = 0$, $D_s\tau_s$ can be determined if $[S]_{x=d}$ as a function of d and the boundary condition for $[S]_{x=0}$ are known. The value of $(D_s\tau_s)^{1/2}$ for the assumption of pure reflection $\partial[S]/\partial x = 0$ at $x = 0$ is about half as large as that based on the assumption of pure absorption $[S]_{x=0} = 0$. In most cases the values of $(D_s\tau_s)^{1/2}$ lie between these two extreme cases. Furthermore, the surface condition at $x = d$ and the trapping effect for $x > d$, and also the reabsorption, play a very important role in the diffusion for energy transfer. The values of L_S^{app} and D_S^{app} for the same organic solids determined by this method are given in Table 7.3.

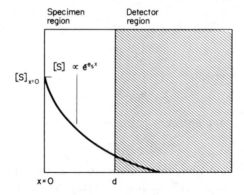

FIG. 7.11. Schematic illustration of the surface-quenching experiment showing the initial exciton concentration distribution ($[S]_x$ at $t = 0$).

(c) BIMOLECULAR RECOMBINATION

When two singlet excitons approach each other and interact, one molecule will return to its ground state through bimolecular recombination and the other will then be raised to a higher excited state, resulting in a decrease in fluorescence intensity. The kinetic equation describing this process is

$$\frac{d[S]}{dt} = G_s - \alpha_s[S] - \gamma_s[S]^2 \tag{7.66}$$

TABLE 7.3. *Singlet-exciton diffusion parameters for some organic solids*

Method	Crystal	K_{eff}	t_H^{app} (nsec)	$\tau_S(0)$ (nsec)	L_S^{app} (Å)	D_S^{app} (cm² sec⁻¹)	Remarks	Ref.
Bulk quenching	Naphthalene (host) Anthracene (activator)	5.0×10^4	1.7×10^{-3}	85		3.6×10^{-4}	By X-ray excitation	A
		4.1×10^3	2.6×10^{-2}	85		2.3×10^{-5}	Thin-film or powder samples	A
	Naphthalene (host) Tetracene (activator)	2.8×10^3	3.8×10^{-2}	85		1.6×10^{-5}	Thin-film or powder samples	A
	Anthracene (host) Tetracene (activator)	5.0×10^4	4.0×10^{-4}	20		1.5×10^{-3}	Thin-film or powder samples	B
Surface quenching	Naphthalene			78	500	9.6×10^{-4}	Polycrystalline samples	C
	Anthracene			10	1300	8.4×10^{-3}	Tetracene on back surface	D
				10	3800	7.2×10^{-2}	Perylene on back surface	B, E
				10	$420 \perp ab$ plane	8.7×10^{-4}	Energy transfer to silver surface	F
	Tetracene			0.4	120	1.7×10^{-3}	Tetraquinone or capriblue on back surface	G
Bimolecular quenching	Anthracene			10	894	4.0×10^{-3}	$r_S = 1.0 \times 10^{-8}$ cm³ sec⁻¹	H, I
	Pyrene			112	3	3.6×10^{-9}	$r_S = 9.0 \times 10^{-15}$ cm³ sec⁻¹	J, K
	Perylene			12	2	3.2×10^{-8}	$r_S = 8.0 \times 10^{-14}$ cm³ sec⁻¹	K

Table 7.3 (Contd.)

Method	Crystal	K_{eff}	t_H^{app} (nsec)	$\tau_S(0)$ (nsec)	L_S^{app} (Å)	D_S^{app} (cm² sec⁻¹)	Remarks	Ref.
Photo-conductivity	Anthracene			10	730	2.7×10^{-3}	Dependence of	M
				10	780∥b	3.0×10^{-3}	photoconductivity	D
					1050∥a	5.4×10^{-3}	or the	D
	Tetracene			0.4	290'	1.1×10^{-2}	wavelength	N
	Perylene			12	100	4.2×10^{-5}	of excitation	N

A, Uchida and Tomura [1974].
B, Takahashi and Tomura [1971].
C, Gallus and Wolf [1968].
D, Kurik [1972].
E, Takahashi and Tomura [1970].
F, Kallmann et al. [1971].
G, Vaubel and Baessler [1970].

H, Babenko et al. [1971].
I, Fourny et al. [1973].
J, Vol et al. [1971].
K, Inoue et al. [1972].
M, Benderskii and Lavrushko [1972].
N, Mulder [1968].

which contains a bimolecular annihilation term $\gamma_s[S]^2$. The rate constant γ_s for the singlet–singlet annihilation can be determined by analysing the fluorescence intensity or decay time as a function of excitation intensity or exciton generation rate G_s. The bimolecular quenching rate coefficient γ_s is composed of two rates; namely, the rate of encounter of two excitons and the rate of interaction after encounter. On the basis of the framework of the strong-scattering random walk model, γ_s can be expressed as [Powell and Soos 1975]

$$\gamma_s = 8\pi D_s \langle R_e \rangle \qquad (7.67)$$

where $\langle R_e \rangle$ is the effective interaction distance for annihilation. The measured values of γ_s are discrepant among different investigators. Powell and Soos [1975] have given a critical review about the bimolecular quenching experiment and the experimental conditions for obtaining γ_s. Several additional mechanisms do play a role in the experiment to affect the values of γ_s, such as the generation of an excited triplet exciton from the excited singlet exciton, and its subsequent decay to a triplet exciton through a non-radiative transition process, the density of singlet excitons, and the excitation intensity, etc. The values of τ_s, D_s^{app}, and L_s^{app} for some organic solids determined by this method are given in Table 7.3.

(d) PHOTOCONDUCTION

The values of L_S determined from the conductivity measurements are mainly based on the exciton–surface interaction which produces free carriers due to exciton dissociation. These carriers contribute to photocurrent and hence photoconductivity. Thus L_S can be obtained by measuring the photoconductivity as a function of excitation wavelength, which permits the measurement of the distance from the illuminated surface of the specimen at which the excitons are produced due to light absorption [Steketee and De Jonge 1962]. When the excitons diffuse back to the surface, then due to exciton–surface interaction, they dissociate into free carriers, thus inducing photocurrents. If the illuminated surface is now treated as an absorber, eq. (7.65) can also be applied to the present situation, but for this case $[S]_{x=0} = 0$. The values of L_s^{app} and D_s^{app} for some organic solids determined by the photoconduction method are given in Table 7.3.

For the details of the above four methods and the diffusion parameters the reader is referred to the original papers of Powell and his co-workers [1970, 1972, 1973, 1975].

7.3. PHOTOLUMINESCENCE

In organic crystals, absorption of light will generate singlet and triplet excitons, and it is these singlet excitons that radiatively decay, producing fluorescence, and these triplet excitons that weakly radiatively decay, producing phosphorescence.

7.3.1. Fluorescence

Singlet excitons can be generated not only directly by absorption of light but also indirectly by triplet–triplet annihilation as has been discussed in the previous section. Since there is a great difference in lifetime between singlet and triplet excitons (e.g. they are about 10^{-8} and 10^{-2} sec, respectively, for anthracene), fluorescence produced by singlet excitons generated directly by absorption of light or other external means is referred to as the prompt fluorescence, and that produced indirectly through triplet–triplet annihilation as the delayed fluorescence.

(A) Prompt fluorescence—theory

CASE I: PERFECT CRYSTALS IN THE ABSENCE OF EXTERNAL ELECTRIC FIELDS AND REABSORPTION

For this case the rate equation for free singlet excitons generated directly by the absorption of light with the intersystem crossing into the triplet states neglected may be written as

$$\frac{d[S]}{dt} = G_S - (\alpha_1 + \alpha_2)[S] \tag{7.68}$$

where α_1 and α_2 are the radiative and non-radiative transition rate constants, respectively, for the monomolecular decay of singlet excitons; G_S is the rate of singlet exciton generation, which can be expressed as

$$G_S = \varepsilon_s(\lambda) i_s(x, t) \tag{7.69}$$

in which $\varepsilon_s(\lambda)$ is the absorption coefficient for the light intensity $i_s(x, t)$ for singlet exciton generation. In the steady state $d[S]/dt = 0$ and the rate of fluorescence emission is $\alpha_1[S]$. Thus the fluorescence quantum efficiency η_s can be written as

$$\eta_s = \frac{\alpha_1[S]}{\varepsilon_s i_s} = \frac{\alpha_1}{\alpha_1 + \alpha_2} \tag{7.70}$$

Of course, η_s can also be defined as the ratio of the number of fluorescence photons emitted to the number of molecules excited into singlet states (the number of photons absorbed to produce singlet excitons). The fluorescence spectrum $\psi(v)$ is defined as the relative fluorescence intensity as a function of frequency v. If the relative fluorescence intensity is normalized to the absorbed light intensity, then η_s can be expressed as

$$\eta_s = \int_0^\infty \psi(v) dv \tag{7.71}$$

Equation (7.68) can be solved for transient conditions. Supposing that a δ-function light flash of negligible duration illuminates the crystal at $t = 0$ and produces singlet excitons of initial concentration $[S]_0$, then the fluorescence response function defined as the relative fluorescence intensity $I_F(t)$ at any time t after such a δ-function excitation can be obtained by solving eq. (7.68), and it is

$$I_F(t) = \alpha_1 \exp(-t/\tau_s) \tag{7.72}$$

where τ_s is the singlet lifetime or the molecular fluorescence lifetime which is

$$\tau_s = (\alpha_1 + \alpha_2)^{-1} \tag{7.73}$$

Obviously, η_s can also be determined by the following expression

$$\eta_s = \int_0^\infty I_F(t)dt = \alpha_1/(\alpha_1 + \alpha_2)$$

$$= \tau_s/\tau_{s1} \tag{7.74}$$

where τ_{s1} is the radiative singlet lifetime which is

$$\tau_{S1} = \alpha_1^{-1} \tag{7.75}$$

It should be noted that α_1 is in fact Einstein's A coefficient for spontaneous emission summed over the complete fluorescence spectrum [Birks 1970].

CASE II: PERFECT CRYSTALS IN THE ABSENCE OF EXTERNAL
ELECTRIC FIELDS BUT INCLUDING REABSORPTION

If reabsorption due to the overlap of the absorption and fluorescence spectra occurs in a perfectly pure crystal with a probability a, then the fluorescence emission probability is reduced to $(1-a)$. This implies that reabsorption decreases the rate of decay of $[S]$. Thus, for this case, eq. (7.68) is changed to

$$\frac{d[S]}{dt} = G_S - \left[(1-a)\alpha_1 + \alpha_2\right][S]$$

$$= \varepsilon_s i_s - [S]/\tau_{\text{obs}} \tag{7.75a}$$

where τ_{obs} is the observed (or technical) fluorescence lifetime

$$\tau_{\text{obs}} = \left[(1-a)\alpha_1 + \alpha_2\right]^{-1}$$

$$= \tau_s(1-a\eta_s)^{-1} \tag{7.76}$$

The observed quantum efficiency of the fluorescence emission is then

$$\eta_{\text{obs}} = \eta_s(1-a)\left[1 + a\eta_s + a^2\eta_s^2 + \ldots\right]$$

$$= \eta_s(1-a)/(1-a\eta_s) \tag{7.77}$$

The successive terms in eq. (7.77) correspond to photon escapes after 1, 2, 3, ..., emissions. The reabsorption depends on the overlap of the absorption spectrum $\varepsilon_s(v)$ and the fluorescence spectrum $\psi(v)$ on the specimen thickness d through which the fluorescence photons escape, and on $[S]$. These parameters are related in the following equation:

$$a\eta_s = \int_0^\infty \psi(v)(1 - \exp\{-\varepsilon_s(v)[S]d\})dv \tag{7.78}$$

CASE III: REAL CRYSTALS IN THE ABSENCE OF EXTERNAL
ELECTRIC FIELDS AND REABSORPTION

Real crystals contain both chemical (impurity) and structural defects which create traps to trap singlet excitons. Therefore a real crystal is a multicomponent fluorescence system which consists of the host fluorescence spectrum and the guest fluorescence spectra. The rate equations for free singlet excitons $[S]$ and trapped singlet excitons $[S_G]$ with the intersystem crossing into the triplet states neglected may be written as

$$\frac{d[S]}{dt} = \varepsilon_s i_s - (\alpha_1 + \alpha_2)[S] - \alpha_{HG}[S] \tag{7.79}$$

$$\frac{d[S_G]}{dt} = \alpha_{HG}[S] - (\alpha_{G1} + \alpha_{G2})[S_G] \tag{7.80}$$

where α_{HG} is the rate constant for the free singlet excitons to relax into traps (or called the energy-quenching or energy transfer rate constant for singlet excitons), and α_{G1} and α_{G2} are, respectively, the rate constants for the radiative and non-radiative transition for the thermally released trapped excitons. In the steady state $d[S]/dt = d[S_G]/dt = 0$, and thus from eqs. (7.79) and (7.80) it can be shown that the fluorescence quantum efficiency for real crystals is given by

$$\eta_{sr} = \frac{\alpha_1[S]}{\varepsilon_s i_s} = \frac{\alpha_1}{\alpha_1 + \alpha_2 + [\alpha_{HG}^2/(\alpha_{G1} + \alpha_{G2})]}$$

$$= \frac{\eta_s}{1 + \tau_s[\alpha_{HG}^2/(\alpha_{G1} + \alpha_{G2})]} \tag{7.81}$$

and the fluorescence lifetime is given by

$$\tau_{sr} = \{\alpha_1 + \alpha_2 + [\alpha_{HG}^2/(\alpha_{G1} + \alpha_{G2})]\}^{-1}$$

$$= \frac{\tau_s}{1 + \tau_s[\alpha_{HG}^2/(\alpha_{G1} + \alpha_{G2})]} \tag{7.82}$$

The ratio of τ_{sr} to η_{sr} is

$$\tau_{sr}/\eta_{sr} = \tau_s/\eta_s = \alpha_1^{-1} = \tau_{s1} \tag{7.83}$$

which is the same as that for the cases of perfect crystals.

CASE IV: REAL CRYSTALS IN THE ABSENCE OF EXTERNAL ELECTRIC FIELDS BUT INCLUDING REABSORPTION

This case is similar to Case II but with traps. Thus the rate equations for $[S]$ and $[S_G]$ are

$$\frac{d[S]}{dt} = \varepsilon_s i_s - [(1-a)\alpha_1 + \alpha_2][S] - \alpha_{HG}[S] \tag{7.84}$$

$$\frac{d[S_G]}{dt} = \alpha_{HG}[S] - [(1-a_G)\alpha_{G1} + \alpha_{G2}][S_G] \tag{7.85}$$

where a_G is the probability of reabsorption of the guest fluorescence. Following the same methods given above, it can be easily shown that the observed (or technical) fluorescence lifetime τ_{obsr} in real crystals is given by

$$\tau_{obsr} = \tau_s\{(1 - a\eta_s) + \tau_s[\alpha_{HG}^2/(\alpha_{G1} + \alpha_{G2})](1 - a_G\eta_{Gs})^{-1}\}^{-1} \tag{7.86}$$

and the observed quantum efficiency of the fluorescence emission in real crystals is given by

$$\eta_{obsr} = \eta_s\{1 - a - [\alpha_{HG}^2/(\alpha_{G1} + \alpha_{G2})](1 - a_G\eta_{Gs})^{-1}\}$$

$$\times \left\{ 1 + \left[a + \left(\frac{\alpha_{HG}^2}{\alpha_{G1} + \alpha_{G2}}\right)\left(\frac{1}{1 - a_G\eta_{GS}}\right)\right]\eta_s \right.$$

$$\left. + \left[a + \left(\frac{\alpha_{HG}^2}{\alpha_{G1} + \alpha_{G2}}\right)\left(\frac{1}{1 - a_G\eta_{GS}}\right)\right]^2\eta_s^2 + \ldots \right\}$$

$$= \frac{\eta_s\{1 - a - [\alpha_{HG}^2/(\alpha_{G1} + \alpha_{G2})](1 - a_G\eta_{Gs})^{-1}\}}{1 - \{a + [\alpha_{HG}^2/(\alpha_{G1} + \alpha_{G2})][1/(1 - a_G\eta_{Gs})]\}\eta_s} \qquad (7.87)$$

where

$$\eta_{Gs} = \frac{\alpha_{G1}}{\alpha_{G1} + \alpha_{G2}} \qquad (7.88)$$

CASE V: REAL CRYSTALS IN THE PRESENCE OF
EXTERNAL ELECTRIC FIELDS

Several investigators [Wotherspoon *et al.* 1970, Pope *et al.* 1971, Kalinowski *et al.* 1974, 1976] have reported the modulation of the prompt fluorescence by charge carriers in the presence of electric fields. If charge carriers are injected into a crystal and occupy a portion of traps, then the trapped carriers act as centres to quench prompt fluorescence due to the reduction of free singlet excitons, and to decrease the effective concentration of trapped excitons and hence the trapped excitons fluorescence.

To manifest the physical picture and to avoid complicated mathematical involvement, it is assumed that traps are confined in a single discrete energy level E_t, with the concentration N_t, and they trap either free singlet excitons or free charge carriers (e.g. injected holes). The rate equation for $[S]$ for this case may be written as [Kalinowski *et al.* 1976]

$$\frac{d[S]}{dt} = \varepsilon_s i_s - (\alpha_1 + \alpha_2')[S] - K_{sti} N_t[S]$$

$$- (K_{sct} - K_{sti})p_t[S] \qquad (7.89)$$

where α_2' is the monomolecular non-radiative decay rate constant including exciton quenching at the crystal–electrode interface, K_{sti} is the second order rate constant governing the singlet-exciton trap interaction, K_{sct} is the second order rate constant for the exciton-trapped carrier interaction, and p_t is the concentration of trapped carriers (for this case, trapped holes). Since the ratio of free to trapped carrier concentration is generally very small (e.g. less than 10^{-4}), the singlet exciton free carrier interaction can be ignored. If the incident photon flux is low, the singlet–singlet annihilation can also be ignored.

On the assumption that the traps are deep traps so that thermally assisted detrapping can be neglected, the rate equation for trapped singlet excitons $[S_G]$ may be written as [Kalinowski *et al.* 1976]

$$\frac{d[S_G]}{dt} = K_{sti}(N_t - p_t)[S] - (\alpha_{G1} + \alpha_{G2})[S_G] \qquad (7.90)$$

The total fluorescence is the sum of the contributions from the free and trapped singlet excitons and is a function of emitted fluorescence wavelength λ. Thus

$$\phi_{\text{total}} = \phi + \phi_t = \alpha_1[S]\psi(\lambda) + \alpha[S_G]\psi_t(\lambda) \qquad (7.91)$$

where ϕ and ϕ_t are, respectively, the free and trapped exciton fluorescence, and $\psi(\lambda)$ and $\psi_t(\lambda)$ are, respectively, the fluorescence spectrum due to free and trapped excitons. In the steady state and from eqs. (7.89)–(7.91) it can be shown that

$$\phi_{\text{total}} = \frac{\varepsilon_s i_s[\eta_s(\alpha_1 + \alpha'_2)\psi(\lambda) + K_{sti}(N_t - p_t)\eta_{Gs}\psi_t(\lambda)]}{(\alpha_1 + \alpha'_2) + (K_{sct} - K_{sti})p_t} \qquad (7.92)$$

It can easily be seen that if the effect of trapped carriers is neglected, the terms involving p_t should be removed in eqs. (7.89)–(7.92).

(B) Prompt fluorescence—experiments

The measurements of prompt fluorescence spectra and decay time in organic materials, and particularly in anthracene, have long been the subject of extensive investigation, because such results would help the studies of chemical and structural defects and their effects on material properties. A great deal of experimental fluorescence results are available in the literature. In the following only some typical experimental results are given.

(i) GENERAL SPECTRUM OF PURE ANTHRACENE CRYSTALS

In the fluorescence spectrum of anthracene crystals, the broad bands observed at room temperature can be resolved into a number of narrow well-separated bands at low temperatures. Figure 7.12 shows the typical b-polarized fluorescence spectrum of ultra-pure anthracene single crystals at 5.6°K. Several narrow bands in the spectrum have the widths at half-height of 3 cm^{-1} or less. The band 25,103 cm^{-1} is generally considered as the fluorescence from the host band edge E. Most narrow bands are associated with 2-methylanthracene impurity plus phonons. The band 23,699 cm^{-1} may be assigned as E-1404 vibration, and the band 23,308 cm^{-1} as E-391-1404. Lyons and Warren [1972] have given a detailed discussion about the origins of all bands. They observed all levels involving phonons or molecular vibrations, or both, except that the fluorescence from the electronic origin 25,103 cm^{-1} was either absent or very weak. This may imply that coupling with such vibrations is necessary for emission.

The fluorescence vibronic band maxima positions change slightly with changing temperature as shown in Fig. 7.13. This indicates that the density of states near the minimum energy of the exciton band may be considerably greater than that at higher energies. An increase in temperature will cause an increase in reabsorption and hence cause an apparent shift in the position of maximum intensity. Of course, the change in

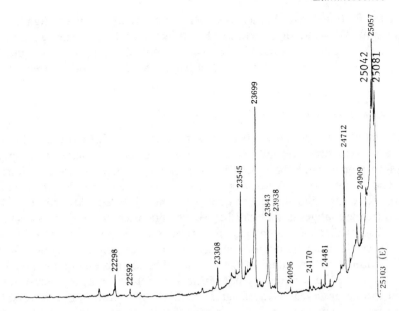

FIG. 7.12. Fluorescence spectrum of ultrapure anthracene single crystals in ||b polarization. Specimen thickness: 20 μm; temperature: 5.6 °K. [After Lyons and Warren 1972.]

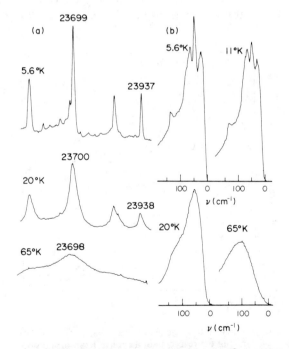

FIG. 7.13. Temperature dependence of anthracene fluorescence (a) vibronic bands, and (b) phonon structure near the effective origin E. [After Lyons and Warren 1972.]

phonon distribution due to the increase in temperature may also play a significant role in such a shift. The width of the vibronic bands increases with increasing temperature. However, the width of the narrow impurity bands changes little with increasing temperature over the temperature range of 5.6–65°K [Lyons and Warren 1972].

(ii) EFFECTS OF REABSORPTION

When the absorption and emission spectra overlap, both the observed emission spectra and decay times depend upon the measuring geometry and crystal specimen size. This phenomenon has been considered as an indication of the reabsorption effects [Birks and Munro 1967, Munro *et al.* 1972]. To determine the probability for reabsorption *a* the following techniques are generally used [Munro *et al.* 1972]: (a) Measure the changes in size and shape of the fluorescence band shapes, which are generally assumed to be Lorentzian, as functions of temperature. (b) Observe the effects of crystal specimen thickness on the emission spectra and decay rate. Since the excited portion of the crystal is not the same at all temperatures and the crystal is not perfectly uniform, this technique is not satisfactory. (c) Since the fraction of the total emission spectra observed at wavelengths shorter than a certain value, beyond which the reabsorption is unimportant, decreases as the reabsorption increases as shown in Fig. 7.14, the measurement of the ratio of the spectral areas will enable the determination of *a*.

For anthracene dissolved in polymethylmethacrylate (PMM) to form a dilute anthracene solution, reabsorption is negligible because of low concentration of

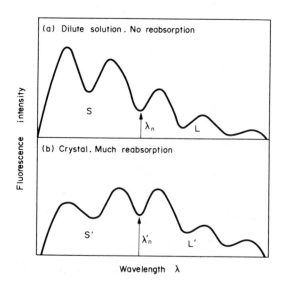

FIG. 7.14. Illustrating the use of spectral areas for calculating the reabsorption factor *a*. Reabsorption occurs for $\lambda < \lambda_n$ and it is assumed that $\lambda_n = \lambda'_n$. After normalization $L + S = 1 = L' + S'$, and $a = 1 - L/L'$. [After Munro, Logan, Blair, Lipsett, and Williams 1972.]

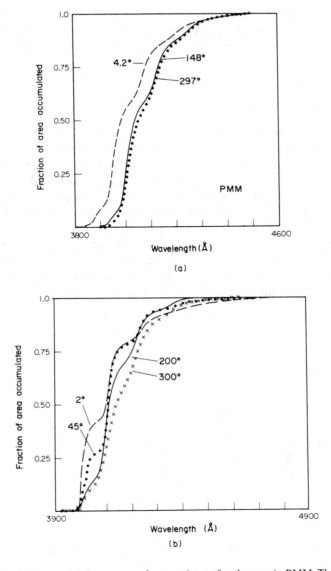

FIG. 7.15. (a) Accumulated area curves for a specimen of anthracene in PMM. There is little change in the curves with temperature indicating that reabsorption in the specimen is negligible. (b) Accumulated area curves for sublimation flake containing few defects (nearly defect-free anthracene crystal). The abrupt changes in slope as maxima are approached or passed and the curves for 2 and 45 °K cross the other curves. The crossing is thought to be due to the presence of some defect fluorescence at low temperatures. There is a great change in the curves with temperature indicating the effects of reabsorption. [After Munro, Logan, Blair, Lipsett, and Williams 1972.]

anthracene. This can be seen in the accumulated area curves in Fig. 7.15(a) which show practically no change with temperature. However, the slopes of such curves change as the temperature is changed for a nearly defect-free crystal as shown in Fig. 7.15(b) indicating the effects of reabsorption. Munro *et al.* [1972] have used these accumulated area curves to calculate a and reported that the fluorescence lifetime after correcting the reabsorption is almost independent of temperature for anthracene crystal over the temperature range from 350°K to 2°K.

(iii) EFFECTS OF IMPERFECTIONS

By three-point bending highly purified anthracene crystals at 373°K under N_2 to introduce an excess concentration of well-characterized basal dislocations, Williams *et al.* [1976] have measured the prompt fluorescence spectrum and lifetimes. The spectrum following excitation using a tunable dye laser at 24,700 cm^{-1} at 4°K is shown in Fig. 7.16. It can be seen that the spectrum retains the characteristic feature of the anthracene spectrum. Both shallow and deep traps introduced into anthracene crystals as a result of plastic deformation are located at 300 cm^{-1} and 1500 cm^{-1} below the S_1 level, and they may thus involve displaced anthracene molecules in the vicinity of dislocations. Lifetimes at 24,780 cm^{-1} peak and 23,535 cm^{-1} peak are, respectively, 6.0 nsec and 7.5 nsec at 4°K, and they indicate that the emitting species responsible for both peaks are single molecules.

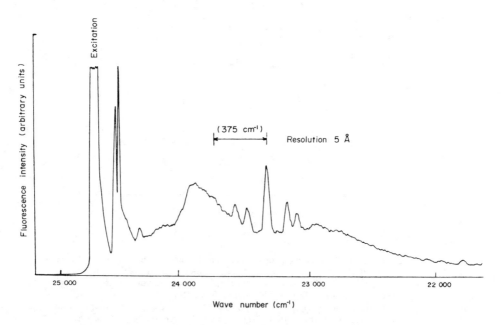

FIG. 7.16. Prompt fluorescence spectrum of a deformed anthracene single crystal at 4° K following "defect" excitation at 24700 cm^{-1}. [After Williams, Clarke, Thomas, and Shaw 1976.]

Even in most highly purified crystals presently available, there are always a number of bands which are not part of the main spectrum, indicating that there are always various defects in the crystals. Thus by doping single crystals with various dopants, it is possible to identify the origins for some of those bands. Lyons and Warren [1972] have studied the bands arising from doping with ten different organic molecules. Some of their results for anthracene single crystals doped with 2-methylanthracene, naphthacene, and pyrene are shown in Figs. 7.17–7.19.

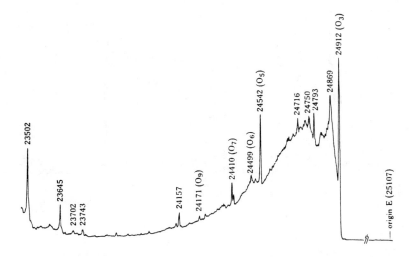

FIG. 7.17. Fluorescence from anthracene single crystal doped with 2-methylanthracene of 300 μmol/mol at 5.6° K. Specimen thickness: 10 μm. [After Lyons and Warren 1972.]

Some of the most likely chemical impurities in anthracene are methylanthracenes, and, in particular, those with similar melting points to anthracene. In Fig. 7.17 the narrow band at 24,912 cm^{-1} is due to the fluorescence origin for 2-methylanthracene, which coincides with the strongest impurity band found in zone-refined anthracene (O_3). The vibrational frequencies of 2-methylanthracene are close to those of anthracene, except for the 370 and 500 vibrations. Theoretically, the host exciton band should move towards the impurity level, that is, to the lower energy level for 2-methylanthracene.

Figure 7.18 shows that for the low concentration of naphthacene, 0.07 μmol/mol, a violet fluorescence can be observed. As the concentration is increased to 12 μmol/mol, a green fluorescence can also be observed together with some host fluorescence. When the concentration is increased to 700 μmol/mol no host fluorescence can be detected even with long exposures to excitation.

The spectrum for the anthracene doped with pyrene in shown in Fig. 7.19. At the concentration of 580 μmol/mol a band at E-62 is dominant and this band is termed an X-band (X_3). A band at E-23 (X_1) does not appear in a-polarization ($\perp b$).

Except for the dopant of 2-methylanthracene there is a correlation of the number and intensity of impurity bands with the size and shape of the guest molecule as shown in Fig. 7.20.

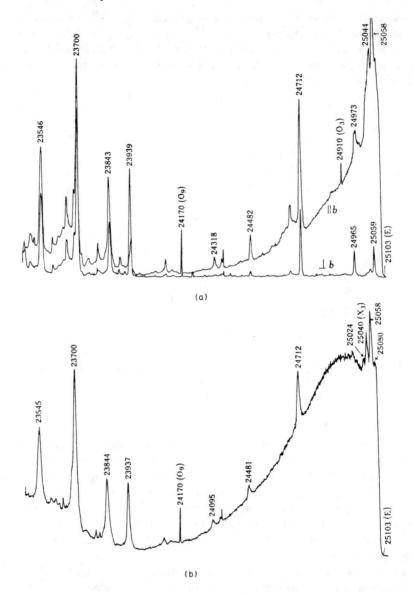

FIG. 7.18. Fluorescence from anthracene single crystal doped with naphthacene:
(a) 0.07 μmol/mol at 5.6° K and specimen thickness of 25 μm;
(b) 12 μmol/mol at 5.6° K and specimen thickness of 9 μm. [After Lyons
and Warren 1972.]

(iv) EFFECTS OF EXTERNAL ELECTRIC FIELDS

Using distilled water as electrodes and the experimental arrangement shown in Fig.
7.21, Kalinowski and Godlewski [1974] have measured the prompt fluorescence
intensity from sublimation-grown anthracene crystals of thicknesses 34 and 75 μm as a

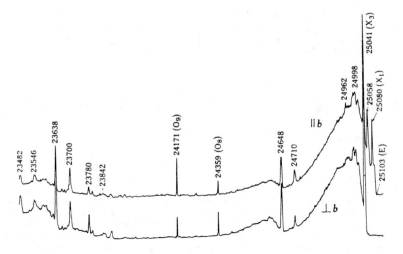

FIG. 7.19. Fluorescence from anthracene single crystal doped with pyrene of 580 μmol/mol at 5.6° K. Specimen thickness: 15 μm. [After Lyons and Warren 1972.]

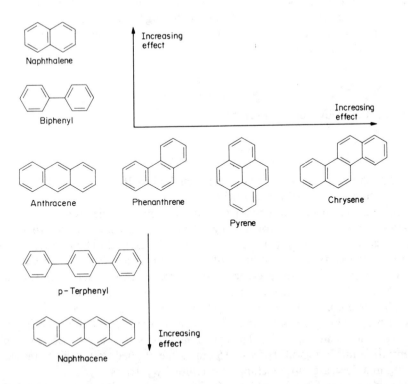

FIG. 7.20. Relative size of guest molecules and their relative effects on the host fluorescence of anthracene. [After Lyons and Warren 1972.]

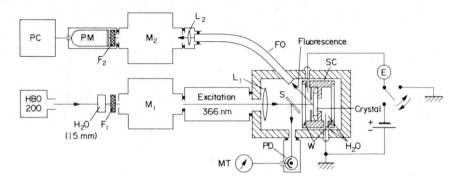

FIG. 7.21. Schematic diagram of the experimental arrangement. The cell is made of cross-linked polyethylene and placed in the black-coated brass housing. The crystal and the suprasil window *W* are attached with paraffin wax; the observed front fluorescence is transmitted through the filter optic (*FO*) and focused with the lens L_2 onto the M_2 monochromator slit; F_2 Corning glass filter 3-74; *PM* photomultiplier tube (EMI 6256 S); *PC* photon counter; *HBO* 200 Hg 200 W lamp; F_1 Corning glass filter 7-37; M_1 and M_2 SPM-1 Zeiss monochromators; *S* suprasil slide (thickness 150 μm) beam splitter; *PD* MVS (Zeiss) vacuum photodiode; *MT* nano-ammeter; *SC* shutter to make the illuminated area of the crystal equal to its effective area under the rear electrical water contact; *E* electrometer. [After Kalinowski and Godlewski 1974.]

function of applied electric field. In Chapter 3 we have discussed the behaviour of injected charge carriers under electric fields. According to Mott and Gurney [1940], the concentration of carriers (say, holes) $p(x)$ at any point x from the carrier injecting contact at $x = 0$ can be expressed as

$$p(x) = p(0)\left(\frac{x_0}{x_0 + x}\right)^2 \tag{7.93}$$

where x_0 is the Debye length given by

$$x_0 = \left[\frac{2\varepsilon kT}{q^2 p(0)}\right]^{1/2} \tag{7.94}$$

Figure 7.22 shows the distribution of the injected holes. If $p(0) = 7 \times 10^{14}$ cm^{-3}, then $x_0 = 0.12$ μm and $p(x) \to 0$ for $x \geqslant 0.50$ μm.

When a sufficiently high and positive voltage is applied to the illuminated contact, a saturation photocurrent will flow through the crystal specimen leading to a uniform space charge distribution [$p_v(x) = $ constant]. For the excitation light of $\lambda_{\mathrm{ex}} = 366$ nm, $l_a \simeq 0.3$ μm, the average concentration of holes in the region $0 < x < 0.5$ μm is higher in the absence than that in the presence of an external electric field. For applied voltages below the threshold voltage for the saturation photocurrent to occur ($V < 120$ V in Fig. 7.23) the fluorescence intensities plotted in $\delta = [\phi(V) - \phi(0)]/\phi(0)$ for two emission wavelengths $\lambda_e = 495$ nm and $\lambda_e = 565$ nm increase with increasing applied voltage as shown in Fig. 7.23. This phenomenon is due to the decrease in hole concentration in the region $0 < x < 0.5$ μm as the applied voltage is increased, thus resulting in a decrease in quenching of excitons by carriers.

The value of $\delta = [\phi(V) - \phi(0)]/\phi(0)$ for a fixed applied voltage (500 V across specimens of thicknesses of 34 μm and 75 μm) is a function of emission wavelength λ_e

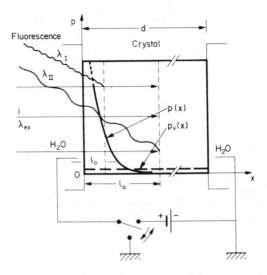

FIG. 7.22. Illustrating the experimental situation. l_a is the penetration depth of the excitation light i of wavelength λ_{ex}, $[i = i_0 \exp(-x/l_a)]$; l_0 is the reverse of the absorption coefficient of the fluorescent light and is a function of fluorescent light wavelength. [After Kalinowski, Godlewski, and Chance 1976.]

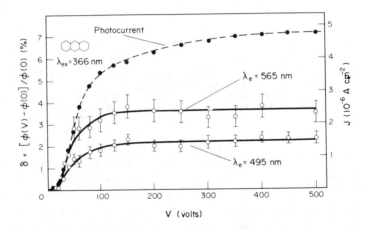

FIG. 7.23. The effect of a steady voltage on the intensity of anthracene fluorescence at room temperature for two emission wavelengths λ_e. The applied electric field is perpendicular to the *ab* plane of the crystal of thickness of 34 μm. The excitation intensity is $(4 \pm 2) \times 10^{15}$ photons/cm^2 sec. [After Kalinowski, Godlewski, and Chance 1976.]

and it exhibits several prominent maxima. Three of these maxima occur at λ_e of 465 nm (2.67 eV), 495 nm (2.51 eV), and 527 nm (2.35 eV), and are reproducible from crystal to crystal and do not depend on the history of the crystal. Kalinowski *et al.* [1976] have attributed these maxima to the emission from trapped singlet excitons.

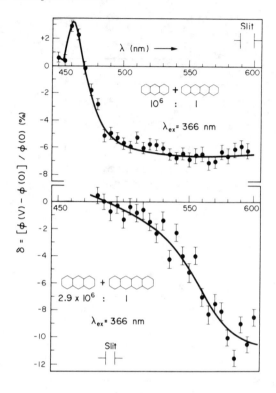

FIG. 7.24. Variation of the charge-induced change in fluorescence with emission wavelength for two tetracene-doped anthracene crystals: *top*: sublimation flake, $d = 23\ \mu m$, the modulation voltage $V = +200$ V. *Bottom*: melt-grown crystal, $d = 100\ \mu m$, the modulation voltage $V = +400$ V. In both cases the exciting photon flux is $(4 \pm 2) \times 10^{15}$ photons/cm² sec. [After Kalinowski, Godlewski, and Chance 1976.]

The injected holes also influence the green fluorescence of tetracene singlets produced by trapping of free anthracene singlet excitons. Figure 7.24 shows such an influence for tetracene-doped anthracene crystals. For a fixed applied voltage, the fractional change in fluorescence intensity δ first increases with positive values, reaches a peak, and then decreases to become negative with increasing fluorescence wavelength. This phenomenon may be explained by the fact that the excitation light generates holes and electrons, and tetracene molecules form shallow traps for holes and deep traps for electrons [Kalinowski *et al.* 1976].

(C) Delayed fluorescence—theory

CASE I. PERFECT CRYSTALS IN THE ABSENCE OF EXTERNAL
ELECTRIC FIELDS AND EXCLUDING INTERSYSTEM CROSSING

For this case the rate equations for free triplet excitons and for free singlet excitons created indirectly by triplet–triplet annihilation are given by

$$\frac{d[T]}{dt} = G_T - (\beta_1 + \beta_2)[T] - \gamma_T[T]^2 \tag{7.95}$$

$$\frac{d[S]_{\text{ind}}}{dt} = \tfrac{1}{2}f\gamma_T[T]^2 - (\alpha_1 + \alpha_2)[S]_{\text{ind}} \tag{7.96}$$

where β_1 and β_2 are the monomolecular radiative and non-radiative transition rate constants, respectively, γ_T is the total bimolecular triplet–triplet annihilation rate constant, f is the fraction of triplet–triplet annihilation which creates indirectly singlet excitons $[S]_{\text{ind}}$ and the value of f is approximately 0.4 for anthracene, and G_T is the rate of triplet exciton generation, which can be expressed as

$$G_T = \varepsilon_T(\lambda)i_T(x, t) \tag{7.97}$$

in which $\varepsilon_T(\lambda)$ is the absorption coefficient for the light intensity $i_T(x, t)$ for triplet exciton generation. Under steady state conditions and for low exciton concentrations, the solution of eqs. (7.95) and (7.96) yields the intensity of emitted delayed fluorescence per unit volume of crystal ϕ_{DF}. It is given by [Avakian and Merrifield 1968]

$$\phi_{DF} = \alpha_1[S]_{\text{ind}} = \left(\frac{\alpha_1}{\alpha_1 + \alpha_2}\right)\left\{\tfrac{1}{2}f\gamma_T[T]^2\right\}$$

$$= \left(\frac{\alpha_1}{\alpha_1 + \alpha_2}\right)\left\{\tfrac{1}{2}f\gamma_T\left(\frac{\varepsilon_i i_T}{\beta_1 + \beta_2}\right)^2\right\} \tag{7.98}$$

It should be noted that in eq. (7.98) the reabsorption effect has been excluded. To include this effect we need only to replace all α_1 with $(1 - a)\alpha_1$. Since the radiative triplet transition is weak, the reabsorption from this source can be ignored.

CASE II: PERFECT CRYSTALS IN THE ABSENCE OF EXTERNAL ELECTRIC FIELDS BUT INCLUDING INTERSYSTEM CROSSING

Triplet exciton fusion due to triplet–triplet annihilation and singlet exciton fission (the inverse of triplet–triplet annihilation) always coexist following the reaction [Johnson and Merrifield 1970, Groff *et al.* 1970]

$$T_1 + T_1 \underset{\gamma_f}{\overset{\gamma_{Tr}}{\rightleftarrows}} S_1 + S_0 \tag{7.99}$$

The fission and fusion rate constants γ_f and τ_{Tr} are related by [Merrifield *et al.* 1969]

$$\gamma_f/\gamma_{Tr} = \frac{9}{2}\exp[-(2E_T - E_s)/kT] = C(T) \tag{7.100}$$

where E_T is the energy of a triplet exciton and E_s is the energy of a singlet exciton.

On the assumption that the excitation is uniform throughout the crystal specimen (i.e. the extinction depth for the absorbed light is much greater than the exciton diffusion length), the rate equations for the concentrations of triplet and singlet excitons are [Groff *et al.* 1970]:

$$\frac{d[T]}{dt} = \varepsilon_T i_T - (\beta_1 + \beta_2)[T] - \gamma_T[T]^2 + [a_T(\alpha_1 + \alpha_2) + 2\alpha'][S] \quad (7.101)$$

$$\frac{d[S]}{dt} = \varepsilon_s i_s + \frac{1}{2} f\gamma_T[T]^2 - (\alpha_1 + \alpha_2)[S] - \alpha'[S] \quad (7.102)$$

where a_T is the fraction of singlet excitons leading to triplet excitons via intersystem crossing but not involving fission, α' is the rate constant for singlet fission which is [Groff *et al.* 1970]

$$\alpha' = [S_0]\gamma_f = [S_0]C\gamma_{Tr} \quad (7.103)$$

and

$$\gamma_{Tr} = f\gamma_T.$$

Under steady state conditions with low exciton concentrations (low excitation levels) the solution of the above rate equations with $i_T = 0$ leading to the prompt fluorescence intensity

$$\phi = \alpha_1[S] = \frac{\alpha_1(\varepsilon_s i_s)}{\alpha_1 + \alpha_2 + \alpha'} \quad (7.104)$$

and that with $i_s = 0$ leading to the delayed fluorescence intensity

$$\phi_{DF} = \alpha_1[S]_{ind} = \frac{\alpha_1\left[\frac{1}{2}\gamma_{Tr}\varepsilon_T^2 i_T^2/(\beta_1 + \beta_2)^2\right]}{\alpha_1 + \alpha_2 + \alpha'} \quad (7.105)$$

CASE III: REAL CRYSTALS IN THE ABSENCE OF EXTERNAL
ELECTRIC FIELDS AND EXCLUDING FISSION OF SINGLETS

For this case, supposing that the concentration of triplet excitons trapped at the level E_t lower in energy is $[T_G]$ with the rate constant β_{HG}, the rate equations may be written as [Siebrand 1965]:

$$\frac{d[T]}{dt} = \varepsilon_T i_T - (\beta_1 + \beta_2)[T] - \gamma_T[T]^2 + p_G[T_G] - \{\beta_{HG} + \gamma_G[T_G]\}[T] \quad (7.106)$$

$$\frac{d[T_G]}{dt} = \varepsilon_G i_G + \beta_{HG}[T] - (\beta_{G1} + \beta_{G2})[T_G] - \{p_G + \gamma_G[T]\}[T_G] \quad (7.107)$$

$$\frac{d[S]_{ind}}{dt} = \frac{1}{2}\{f\gamma_T[T]^2 + f_G\gamma_G[T][T_G]\} - (\alpha_1 + \alpha_2)[S]_{ind} \quad (7.108)$$

where $\varepsilon_T i_T$ and $\varepsilon_G i_G$ are, respectively, the generation rates of free triplet excitons and trapped triplet excitons from their ground state; β_{HG} is the rate constant for the free triplet excitons to relax into traps; p_G is the rate constant for detrapping the trapped triplet excitons from traps, which is generally temperature dependent; γ_G is the bimolecular triplet–trapped triplet annihilation constant and f_G is the fraction of triplet–trapped triplet annihilation which creates indirectly singlet excitons; β_{G1} and β_{G2} are, respectively, the rate constants for the radiative and non-radiative transition for the thermally released trapped triplet excitons. On the assumption that (a) the direct absorption by traps is unimportant ($\varepsilon_G i_G = 0$), (b) only a negligibly small portion of the free and trapped triplet excitons decay by triplet–triplet or triplet–trapped triplet annihilation ($\gamma_T[T]^2$ and $\gamma_G[T][T_G]$ are small), and (c) the trapping is not saturated (weak excitation or large trap concentrations), the solution of eqs. (7.106)–(7.108) under steady state conditions yields the intensity of emitted delayed fluorescence, and for low excitation levels it is given by [Siebrand 1965]

$$
\phi_{DF} = \left(\frac{\alpha_1}{\alpha_1 + \alpha_2}\right)\left[\frac{1}{2}f\gamma_T\left(\frac{\varepsilon_T i_T}{\beta_1 + \beta_2}\right)^2\right]
$$
$$
\times \left[1 + \frac{f_G}{f}\left(\frac{\beta_{HG}}{\beta_{G1} + \beta_{G2} + p_G}\right)\right] \bigg/ \left[1 + \left(\frac{\beta_{G1} + \beta_{G2}}{\beta_1 + \beta_2}\right)\left(\frac{\beta_{HG}}{\beta_{G1} + \beta_{G2} + p_G}\right)\right]^2 \quad (7.109)
$$

Again, if the reabsorption is included, α_1 in the above equations has to be replaced with $(1 - a)\alpha_1$.

If there are many types of traps in the crystal p_G, β_{HG}, β_{G1}, β_{G2}, and $\gamma_G[T_G]$ have to be replaced with

$$
p_G \to \sum_i p_{Gi}, \qquad \beta_{G2} \to \sum_i \beta_{G2i}
$$

$$
\beta_{HG} \to \sum_i \beta_{HGi}, \qquad \gamma_G[T_G] \to \sum_i \gamma_{Gi}[T_{Gi}]
$$

$$
\beta_{G1} \to \sum_i \beta_{G1i},
$$

and for the trap type i at the i trapping energy level, β_{HGi} and p_{Gi} are given by [Arnold *et al.* 1970]:

$$
\beta_{HGi} = Z_i N_{ti} \quad (7.110)
$$

$$
p_{Gi} = p_{G0} \exp(- E_{ti}/kT) \quad (7.111)
$$

where N_{ti} and E_{ti} are, respectively, the density and the depth of the traps of type i, Z_i is the triplet exciton velocity times the trap cross-section, and p_{G0} is the pre-exponential factor in the trap release rate. Then eq. (7.109) becomes

$$
\phi_{DF} = \left(\frac{\alpha_1}{\alpha_1 + \alpha_2}\right)\left[\frac{1}{2}f\gamma_T\left(\frac{\varepsilon_T i_T}{\beta_1 + \beta_2}\right)^2\right]\left[1 + \sum_i \frac{f_{Gi}}{f}A_i\right] \bigg/ \left[1 + \sum_i A_i B_i\right]^2 \quad (7.112)
$$

where

$$
A_i = \frac{\beta_{HGi}}{\beta_{G1i} + \beta_{G2i} + p_{Gi}} \quad (7.113)
$$

$$
B_i = \frac{\beta_{G1i} + \beta_{G2i}}{\beta_1 + \beta_2} \quad (7.114)
$$

Siebrand [1965] has used eq. (7.112) to interpret the experimental results of Singh and Lipsett [1964] based on the assumption that there are three types of traps in the anthracene crystal.

CASE IV: REAL CRYSTALS IN THE PRESENCE OF
EXTERNAL ELECTRIC FIELDS BUT EXCLUDING FISSION
OF SINGLETS

The interaction between excitons and charge carriers may result in quenching of delayed fluorescence [Helfrich 1966, Levinson *et al.* 1970, Ern *et al.* 1971, Wakayama and Williams 1971 and 1972, Frankevich *et al.* 1972, Pope and Selsby 1972] and in photoenhancement of SCL currents [Helfrich 1967, Many *et al.* 1969, Geacintov *et al.* 1970]. When an electric field is applied to an organic crystal specimen, carriers are injected into an organic crystal and they can act as paramagnetic impurities to interact with both triplet and singlet excitons, thus increasing the rate of non-radiative decay. Therefore, for this case a term $-K_T[T]$ has to be added on the right-hand side of eq. (7.95) or eq. (7.106), and a term $-K_s[S]_{ind}$ to be added on the right-hand side of eq. (7.96) or eq. (7.108), where K_T and K_s are, respectively, the rate constants for triplet exciton and singlet exciton deactivation due to their interaction with charged carriers of total concentration N, including both free and trapped charge carriers.

For low excitation levels, the change in intensity of emitted delayed fluorescence due to triplet exciton carrier interactions may be estimated by

$$\frac{\phi_{DF} \text{ (with } K_T[T] \text{ and } K_s[S]_{ind} \text{ terms)}}{\phi_{DF} \text{ (without these terms)}} \propto \left(\frac{\beta_1 + \beta_2}{\beta_1 + \beta_2 + K_T}\right)^2 \qquad (7.115)$$

For high excitation levels, eqs. (7.95) or (7.106) can be approximately simplified to

$$\frac{d[T]}{dt} \simeq \varepsilon_T i_T - \gamma_T[T]^2 \qquad (7.116)$$

It can be seen that under steady state conditions the delayed fluorescence intensity is proportional to the excitation intensity and is practically not dependent upon the triplet exciton lifetime since for this case the bimolecular decay is dominant $(\beta_1 + \beta_2 + K_T)[T] \ll \gamma_T[T]^2$ and $(\beta_{G1} + \beta_{G2} + K_T) [T_G] \ll \gamma_G[T] [T_G]$, and also $\gamma_G[T] [T_G] \ll \gamma_T[T]^2$. However, eq. (7.116) is an approximation equation. The injected carriers do influence the annihilation process of triplet excitons and the quantum efficiency of singlet transitions [Wakayama and Williams 1972].

CASE V: REAL CRYSTALS IN THE PRESENCE OF EXTERNAL MAGNETIC
FIELDS AND INCLUDING INTERSYSTEM CROSSING

The magnetic moments of triplet excitons will be partially aligned if the applied magnetic field and temperature are of such values that the Zeeman splitting is large compared to thermal energies. Since the annihilation rate is expected to follow some kind of spin selection rule, the alignment of the magnetic moments of triplet excitons due to the applied magnetic field should affect the population of the spin states and hence affect the annihilation rate.

On the assumption that only γ_{Tr} in eqs. (7.100)–(7.105) is magnetic field dependent and that the crystals are sufficiently pure so that the trapping effects can be ignored, then by using eq. (7.104) for prompt fluorescence and eq. (7.105) for delayed fluorescence for magnetic field $H = 0$, the total singlet decay rate constant due to the presence of the external magnetic field H is

$$\alpha_{\text{eff}}(H) = \alpha_{\text{eff}}(0)Y/(1+Y) \tag{7.117}$$

where

$$\alpha_{\text{eff}}(0) = \alpha_1 + \alpha_2 + \alpha' \tag{7.118}$$

$$Y = -\left\{\frac{[\phi(H)-\phi(0)]/\phi(0)}{[\phi_{DF}(H)-\phi_{DF}(0)]/\phi_{DF}(0)}\right\} \tag{7.119}$$

where (H) and (0) denote the parameters at magnetic field H and at zero magnetic field. $\phi(H)$, $\phi(0)$, $\phi_{DF}(H)$, and $\phi_{DF}(0)$ can be easily measured experimentally.

(D) Delayed fluorescence—experiments

One of the most sensitive methods for detecting traps is that of sensitized delayed fluorescence [Pope and Selsby 1972]. The triplet exciton lifetime measurement has been well accepted as a monitor to determine the purity of organic crystals [Lupien and Williams 1968]. Temperature dependence of delayed fluorescence intensity and triplet decay rate also provides further information about traps. In the following only some typical experimental results are given.

(i) UNDOPED SINGLE CRYSTALS

The triplet exciton trapping characteristics and the nature of traps have been studied by analysing the experimental results on the temperature dependence of delayed fluorescence intensity and triplet decay rate [Avakian *et al.* 1963, Singh and Lipsett 1964, Siebrand 1965, Arnold *et al.* 1970]. Using the experimental system shown in Fig. 7.25, Arnold *et al.* [1970] have measured the triplet decay rate and delayed fluorescence in highly purified anthracene crystals as functions of temperature and studied the effects of dislocations introduced by controlled plastic bending. They used a rotating shutter to enable the crystal to be optically pumped for 30 msec, followed by a dead time of 2 msec, and then a measurement of the delayed fluorescence for the next 30 msec. They also used direct excitation to generate triplet excitons to avoid involving an intersystem-crossing process which is temperature dependent. Figure 7.26 shows the delayed fluorescence intensity as a function of temperature for an anthracene crystal before and after plastic bending to a radius of curvature of 1.4 cm. The curve for the unbent is similar to those reported by other investigators [Singh and Lipsett 1964, Goode and Lipsett 1969]. In the unbent crystal there are three maxima in the fluorescence intensity at $10°$, $45°$, and $175°$ K. In the bent crystal these three maxima are displaced downward. Siebrand [1965] has interpreted the temperature dependence results of Singh and Lipsett on the basis of a triplet exciton trapping model with three trapping levels. Each trapping level with a trapped triplet lifetime, long compared to the lifetime of a free triplet exciton, gives rise to such a maximum. He has also determined the depth and concentration of the traps from the results of Singh and Lipsett. The essential differences between the results obtained before and after plastic bending of the crystal are: (i) the bending causes a reduction in the steady state fluorescence intensity below $390°$ K as shown in Fig. 7.26; (ii) the bending causes an increase in the triplet

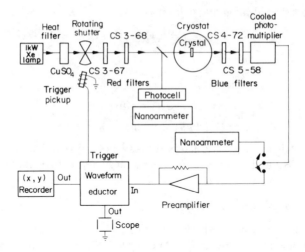

F<small>IG</small>. 7.25. Experimental arrangement for the measurements of triplet decay constants and delayed fluorescence in organic crystals as functions of temperature. [After Arnold, Whitten, and Damask 1970.]

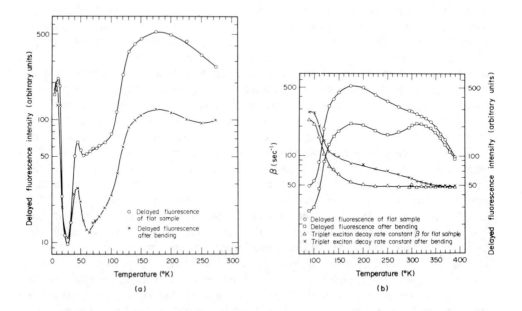

F<small>IG</small>. 7.26. Delayed fluorescence intensity and triplet exciton decay rate constant as functions of temperature for an anthracene crystal before and after bending to a 1.4 cm radius of curvature in the temperature region (a) 6–273° K, and (b) 90–390° K. [After Arnold, Whitten, and Damask 1970.]

decay rate constant below $390°$ K as shown in Fig. 7.26; and (iii) the shape of the curve for the bent crystal is quite different from that of the unbent one for temperatures above $200°$ K. Following Siebrand's trapping model, Arnold *et al.* [1970] have attributed these differences to the presence of traps due to dislocations created by plastic bending.

From eqs. (7.106)–(7.108) the triplet lifetime can be calculated if only one type of traps is effective since these equations are linearized for the assumptions made by Siebrand. Since the singlet excitons decay so rapidly as compared to the triplet excitons [Birks *et al.* 1962], it is reasonable to assume that $d[S]_{ind}/dt = 0$ shortly after the shutter is closed. Thus from eq. (7.108) the delayed fluorescence intensity can be written as

$$\phi_{DF} = \alpha_1 [S]_{ind}$$

$$= \frac{1}{2}\left(\frac{\alpha_1}{\alpha_1 + \alpha_2}\right)\{f\gamma_T[T]^2 + f_G\gamma_G[T][T'_G]\} \tag{7.120}$$

The solutions of eqs. (7.106) and (7.107) for the case without the bimolecular terms and with only one discrete trapping level are [Arnold *et al.* 1970]

$$[T] = D_1 \exp(-\beta_+ t) + D_2 \exp(-\beta_- t) \tag{7.121}$$

$$[T'_G] = D_3 \exp(-\beta_+ t) + D_4 \exp(-\beta_- t) \tag{7.122}$$

where

$$\beta_\pm = \tfrac{1}{2}(\beta'_{HG} + \beta_1 + \beta_2 + p'_G + \beta'_{G1} + \beta'_{G2})$$
$$\pm \tfrac{1}{2}[(p'_G + \beta'_{G1} + \beta'_{G2} - \beta'_{HG} - \beta_1 - \beta_2)^2 + 4p'_G\beta'_{HG}]^{1/2} \tag{7.123}$$

$$D_1 = \{[T]_0/(\beta - \beta_+)\}\{\beta_- - \beta_1 - \beta_2 - \beta'_{HG} + p'_G\beta'_{HG}/(\beta_{G1} + \beta_{G2} + p'_G)\} \tag{7.124}$$

$$D_2 = [T]_0 - D_1 \tag{7.125}$$

$$D_3 = D_1(\beta_1 + \beta_2 + \beta'_{HG} - \beta_+)/p'_G \tag{7.126}$$

$$D_4 = D_2(\beta_1 + \beta_2 + \beta'_{HG} - \beta_-)/p'_G \tag{7.127}$$

and

$$[T]_0 = \frac{\varepsilon_T i_T}{\beta_1 + \beta_2 + [p'_G\beta'_{HG}/(\beta'_{G1} + \beta'_{G2} + p'_G)]} \tag{7.128}$$

Arnold *et al.* [1970] have used the above analysis to explain their experimental results shown in Fig. 7.26. If it is assumed that no traps are effective in the unbent crystal at temperatures above $200°$ K since its delayed fluorescence does not indicate detrapping above $200°$ K, then the decrease in delayed fluorescence intensity and triplet lifetime after bending above this temperature must be due to new traps introduced by bending. In Fig. 7.27 is shown the theoretical intensity ratio of the delayed fluorescence of the bent crystal to that before bending (dashed curve) obtained from eq. (7.112). The calculation for the unbent crystal is based on three different types of traps, while that for the bent crystal is based on four different types of traps. However, for temperatures above $200°$ K the same dashed curve can be obtained if the calculation for the bent crystal is based on one major type of traps following eq. (7.120). It can be seen that the experimental results are in good agreement with the theoretical model. The fluores-

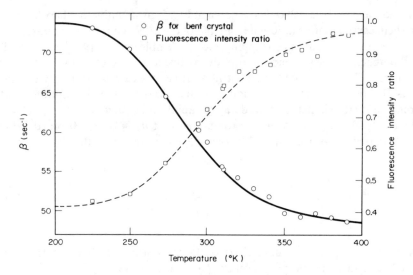

FıG. 7.27. Triplet decay rate constant (open circles: experimental, solid line: theoretical) and fluorescence intensity ratio (ratio of the fluorescence intensity of the bent crystal to that before bending; open squares: experimental, dashed line: theoretical) as functions of temperature. $E'_t = 0.3\,\text{eV}$; $\beta_{HG} = 26\,\text{sec}^{-1}$; $\beta_{G1} + \beta_{G2} = 4 \times 10^6\,\text{sec}^{-1}$; $\beta_1 + \beta_2 = 48\,\text{sec}^{-1}$; $Z' = 2 \times 10^{-10}\,\text{cm}^3\,\text{sec}^{-1}$; $p_{G0} = 8 \times 10^{11}\,\text{sec}^{-1}$; where $\beta_{HG} = Z'N'_t$; $p_G = p_{G0}\exp(-E'_t/kT)$; N'_t and E'_t are, respectively, the density and the depth of the traps. for a particular type of traps; Z' is the triplet exciton velocity times the trap cross-section. [After Arnold, Whitten, and Damask 1970.]

cence intensity ratio is chosen as the parameter for the comparison purpose because it is independent of crystal reabsorption and other processes not related to triplet trapping. In Fig. 7.27 is also shown the theoretical triplet decay rate constant (solid curve) calculated from eqs. (7.121) and (7.123). Again, the experimental results are in good agreement with the theoretical model. The constants given in Fig. 7.27 are those chosen for the best fit to the two sets of results. It should be noted that β_+ is much larger than β_- at any temperature. Thus eq. (7.120) can be written as

$$\phi_{DF} = \phi_{DF0}\exp[-2\beta_- t] = \phi_{DF0}\exp[-2\beta t] \tag{7.129}$$

where β is the triplet decay rate constant and ϕ_{DF0} is the delay fluorescence intensity at the start of the measurements (at $t = 0$). The triplet lifetime is β^{-1}, which is smaller than 24 msec for anthracene even for the purest anthracene crystals so far reported in the literature [Arnold *et al.* 1970].

(ii) DOPED SINGLE CRYSTALS

It is well known that the trapping processes are responsible for the sensitized fluorescence phenomenon—the doped impurity molecules of low concentration convert the host fluorescence into guest fluorescence due to doped impurity molecules. The delayed fluorescence is even more sensitive to doped impurity molecules because of the effectiveness of their energy transfer processes. In fact the measurement of sensitized delayed fluorescence is a useful method for determining the concentration of

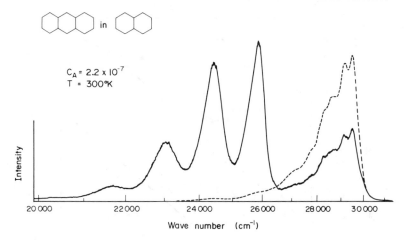

F IG. 7.28. Spectra of prompt (broken line) and delayed (solid line) fluorescence in naphthalene containing 2×10^{-7} part of anthracene at room temperature. Anthracene emission is below $26{,}500 \, \text{cm}^{-1}$. [After Benz, Häcker, and Wolf 1970.]

impurity molecules or traps and their energy levels. The sensitivity of this method is 100 times higher than the method based on sensitized prompt fluorescence. In Fig. 7.28 are shown the prompt (dashed curve) and the delayed (solid curve) fluorescence spectra for naphthalene containing 2×10^{-7} part of anthracene at room temperature. The anthracene spectrum is mainly in the wave number below $26{,}500 \, \text{cm}^{-1}$. It can be seen that for prompt fluorescence the anthracene concentration is too low to convert much of the naphthalene emission into anthracene emission, thus anthracene emission is almost absent as indicated by the dashed curve. However, in the delayed fluorescence spectrum the anthracene emission is much more intense than that of naphthalene as indicated by the solid curve. The energy transfer can be determined by measuring the fluorescence quantum efficiency ratio η_A/η_N as a function of guest molecule concentration, where η_A is quantum efficiency of guest molecules (anthracene), and η_N is that of host molecules (naphthalene). The results for such ratios are shown in Fig. 7.29. Again, this ratio is about 100 times higher in the delayed fluorescence than in the prompt fluorescence [Benz *et al.* 1970].

(iii) EFFECTS OF EXTERNAL ELECTRIC FIELDS

To study electrical transport problems it is very useful to know the field and temperature dependence of the spatial distribution of charge carriers inside the specimen, and, therefore, a method of obtaining this information is important. Pope and Weston [1974] have used the method of measuring the delayed fluorescence lifetime as a function of applied voltage to determine the average excess charge (hole) density in anthracene crystal provided with a hole-injecting contact. The fractional change in the triplet exciton decay rate is given by [Helfrich 1967]

$$\frac{\Delta \beta}{\beta(0)} = \frac{\beta(V) - \beta(0)}{\beta(0)} = K_{Tct} \langle p_t \rangle \tau_T \qquad (7.130)$$

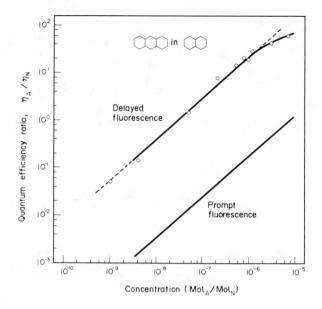

FIG. 7.29. Fluorescence quantum efficiency ratio η_A/η_N for prompt
and delayed fluorescence in naphthalene crystals as a
function of anthracene concentration in the crystals. The
anthracene concentration was measured by absorption.
[After Benz, Häcker, and Wolf 1970.]

where $\beta(0)$ and $\beta(V)$ are, respectively, the triplet exciton decay rates in the absence and
in the presence of applied voltage V; $\langle p_t \rangle$ is the average trapped hole density;
$\tau_T = 1/\beta(0)$; and K_{Tct} is the bimolecular triplet-trapped carrier interaction rate
constant. Equation (7.130) holds for the interaction of triplet excitons with free holes if
K_{Tct} and $\langle p_t \rangle$ are replaced with K_{Tc} and $\langle p \rangle$. But when both free and trapped holes
are present, then eq. (7.130) has to be modified to include both contributions. By
assuming trapped carriers to be dominant, the measurement of $\Delta\beta/\beta(0)$ as a function of
V will allow the determination of $\langle p_1 \rangle$. The measured $\Delta\beta/\beta(0)$ as a function of V for
anthracene is shown in Fig. 7.30, and the corresponding J–V characteristics in Fig. 7.31.
At $V \simeq 60$ V the current saturation commences and $\Delta\beta/\beta(0)$ reaches its maximum
value. At this point $a\{\Delta\beta/\beta(0)\}/dV = 0$, which implies that $p_{\text{Total}} = $ constant. In the
quasi-saturation region ($V > 60$ V) the current density can be written as [Bonham and
Lyons 1973]

$$J = J_s + J_0 \exp(\beta_{sc} d^{-1/2} V^{1/2}/kT) \tag{7.131}$$

where J_s is the saturation current density, J_0 is a constant, and β_{sc} is the Schottky
coefficient [cf. eq. (2.24)]. Equation (7.131) implies that at $V = 60$ V the carrier
reservoir becomes depleted and the field dependent Schottky emission begins. Under
this condition the field throughout most of the specimen can also be assumed to be
approximated by V/d. Thus J can also be approximately expressed as

$$J = q\mu_p \langle p \rangle V/d \tag{7.132}$$

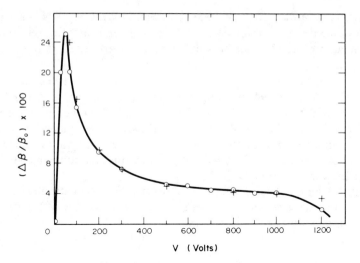

FIG. 7.30. Voltage dependence of average excess total charge density, $p_T \propto \Delta\beta/\beta_0$, in anthracene crystals of thickness of $230\,\mu m$. Electrodes: $0.1\,M\,Ce^{4+}$ in $7.5\,M\,H_2SO_4$ (anode) and H_2O (cathode). o experimental; + theoretical using eq. (7.132) to determine $\langle p \rangle$ and normalizing at $V = 300\,V$ with $\beta = 0.0083\,cm^{1/2}\,V^{-1/2}$. [After Pope and Weston 1974.]

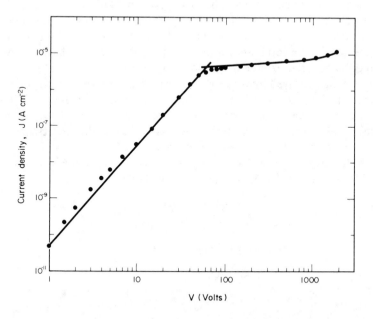

FIG. 7.31. Voltage dependence of hole current density through an anthracene crystal of thickness of $230\,\mu m$. Electrode: $0.1\,M\,Ce^{4+}$ in $7.5\,M\,H_2SO_4$ (anode) and H_2O (cathode). Electrode area: $1.6 \times 10^{-2}\,cm^2$. [After Pope and Weston 1974.]

Assuming $p/p_t = $ constant for $V < 1000$ V, Pope and Weston [1974] have calculated $\Delta\beta/\beta(0)$ as a function of V using eqs. (7.130)–(7.132), and their theoretical results are also shown in Fig. 7.30. It can be seen that the experimental results are in good agreement with the theoretical ones.

They have also measured $J–V$ and $\Delta\beta/\beta(0)–V$ characteristics for anthracene specimens of thickness of 94 μm using a hole-injecting contact of Au and a neutral contact of Hg and found that $J \propto V^3$ for $V < 300$ V, followed by a steeply rising current ($J \propto V^{18}$) for $V > 300$ V, but $\Delta\beta/\beta(0)$ is zero over the entire voltage. This phenomenon indicates that the steeply rising at $V > 300$ V may not be due to a traps-filled limit [Lampert and Mark 1970] but possibly due to double injection or other high-field effects (cf. Chapter 5).

Since $\Delta\beta/\beta(0)$ is an optical signal and the triplet exciton is unaffected by the electric field, changes in charge carrier density can be measured precisely without reference to the internal and external fields [Pope and Weston 1974].

(iv) EFFECTS OF EXTERNAL MAGNETIC FIELDS

The magnetic field experiments provide a powerful tool for the studies of triplet–triplet annihilation processes, and the spin states of triplet excitons. Several investigators [Johnson *et al.* 1967, Groff *et al.* 1970] have reported that the delayed fluorescence intensity due to triplet–triplet annihilation depends on the magnitude of the external magnetic field and its direction relative to the crystallographic axes of the crystal specimen. Figure 7.32 shows the magnetic field dependence of the prompt and delayed fluorescence intensities for tetracene crystals at 300°K, 238°K, and 192°K. The decrease in magnetic field dependence at higher temperatures is consistent with the theoretical prediction from eqs. (7.100) and (7.105). The trend of magnetic field dependence for the prompt fluorescence is just contrary to that for the delayed fluorescence as expected from eqs. (7.104), (7.105), and (7.117)–(7.119).

The prompt and delayed fluorescence intensities also depend on the orientation of the magnetic field as shown in Fig. 7.33. The orientation angles for the level-crossing resonances to occur in the prompt fluorescence coincide with those in the delayed fluorescence, indicating that singlet fission is involved in the fluorescence emission processes. Taking the concentration of tetracene molecules $[S_0] = 3.37 \times 10^{21}$ cm^{-3} at $T = 293$°K, Groff *et al.* [1974] have calculated α' and γ_f, γ_{Tr}, and $\Delta E = 2E_T - E_s$ based on eqs. (7.100), (7.104), (7.105), and (7.117) and they are

$$\alpha' = (6.3 \pm 0.7) \times 10^8 \text{ sec}^{-1}$$

$$\gamma_f = (1.9 \pm 0.2) \times 10^{-13} \text{ cm}^3 \text{ sec}^{-1}$$

$$\gamma_{Tr} = 9 \times 10^{-10} \text{ cm}^3 \text{ sec}^{-1}$$

$$\Delta E = 1900 \pm 200 \text{ cm}^{-1}$$

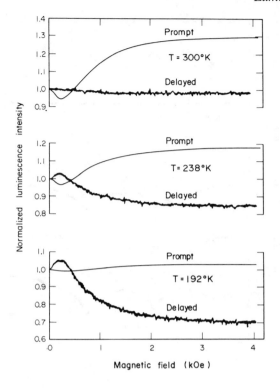

F<small>IG</small>. 7.32. Magnetic field dependence of prompt and delayed fluorescence intensities in a tetracene crystal at various temperatures. The field was oriented at $-20°$ with respect to the b axis in the ab plane of the crystal. [After Groff, Avakian, and Merrifield 1970.]

7.3.2 Phosphorescence

(A) Theory

Phosphorescence is due to the weak radiative transition from the triplet state T_1 to the ground state S_0, and this phenomenon is associated with the long-lived triplet excitons. In the following, several important cases are considered.

C<small>ASE</small> I: PERFECT CRYSTALS IN THE ABSENCE OF EXTERNAL
ELECTRIC FIELDS

For steady excitation to produce singlet excitons at a low exciton level i_s, the triplet excitons will be generated through the intersystem crossing following the rate equations

$$\frac{d[S]}{dt} = \varepsilon_s i_s - (\alpha_1 + \alpha_2)[S] - \alpha'[S] \tag{7.133}$$

$$\frac{d[T]}{dt} = K_{IST}[S] - (\beta_1 + \beta_2)[T] \tag{7.134}$$

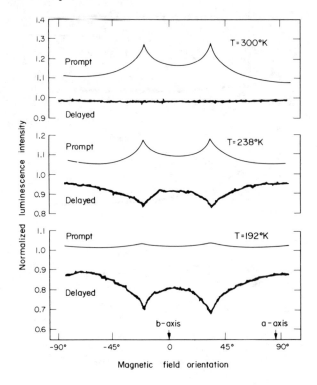

Fɪɢ. 7.33. Dependence of prompt and delayed fluorescence intensity on the orientation of a 4 kOe magnetic field in the *ab* plane of a tetracene crystal. [After Groff, Avakian, and Merrifield 1970.]

where K_{IST} is the rate constant for intersystem crossing, which is given by

$$K_{IST} = [a_T(\alpha_1 + \alpha_2) + 2\alpha'] \qquad (7.135)$$

and α' and a_T have been defined in eqs. (7.101)–(7.103). Under steady state conditions $d[S]/dt = d[T]/dt = 0$, the solution of eqs. (7.133) and (7.134) gives the rate of phosphorescence emission $\beta_1[T]$. Thus the phosphorescence quantum efficiency which is defined as the ratio of the number of phosphorescence photons emitted to the number of molecules excited into triplet states can be written as

$$\eta_T = \frac{\beta_1}{\beta_1 + \beta_2} \qquad (7.136)$$

Of course, η_T can also be expressed in terms of the phosphorescence spectrum $\psi_T(v)$ which is defined as the relative phosphorescence intensity as a function of frequency v. It is

$$\eta_T = \int_0^\infty \psi_T(v)\,dv \qquad (7.137)$$

The triplet quantum yield ϕ_T is defined as

$$\phi_T = \frac{\text{number of triplet excitons generated}}{\text{number of quanta absorbed into a singlet state}}$$

$$= \frac{K_{IST}[S]}{\varepsilon_s i_s} = \frac{K_{IST}}{\alpha_1 + \alpha_2 + \alpha'}$$

$$= \frac{(\beta_1 + \beta_2)[T]}{\varepsilon_s i_s} = \frac{(\beta_1 + \beta_2)[T]}{(\alpha_1 + \alpha_2 + \alpha')[S]} \tag{7.138}$$

The phosphorescence quantum yield ϕ_p is defined as

$$\phi_p = \frac{\text{number of phosphorescence photons emitted}}{\text{number of quanta absorbed into a singlet state}}$$

$$= \frac{\beta_1[T]}{\varepsilon_s i_s} = \eta_T \phi_T \tag{7.139}$$

Equations (7.133) and (7.134) can be solved for transient conditions. Supposing that a δ-function light-flash of negligible duration illuminates the crystal at $t = 0$ and produces singlet excitons of initial concentration $[S]_0$, then the solution of eqs. (7.133) and (7.134) under the initial condition $[T] = 0$ at $t = 0$ yields

$$[T] = \frac{K_{IST}[S]_0}{[(\alpha_1 + \alpha_2 + \alpha') - (\beta_1 + \beta_2)]} \{\exp[-(\beta_1 + \beta_2)t] - \exp[-(\alpha_1 + \alpha_2 + \alpha')t]\} \tag{7.140}$$

Thus the phosphorescence quantum yield can also be expressed as

$$\phi_p = \frac{\int_0^\infty \beta_1[T]\,dt}{[S]_0} = \frac{\beta_1 K_{IST}}{(\alpha_1 + \alpha_2 + \alpha')(\beta_1 + \beta_2)} = \eta_T \phi_T \tag{7.141}$$

Since $(\alpha_1 + \alpha_2 + \alpha') \gg \beta_1 + \beta_2$ at $t > (\alpha_1 + \alpha_2 + \alpha')^{-1}$, eq. (7.140) can be approximated to

$$[T] = \left(\frac{K_{IST}[S]_0}{\alpha_1 + \alpha_2 + \alpha'}\right) \exp[-(\beta_1 + \beta_2)t] \tag{7.142}$$

The triplet lifetime is given by

$$\tau_T = (\beta_1 + \beta_2)^{-1} \tag{7.143}$$

and the radiative triplet lifetime is

$$\tau_{Tl} = \beta_1^{-1} \tag{7.144}$$

CASE II: REAL CRYSTALS IN THE ABSENCE OF EXTERNAL ELECTRIC FIELDS AND EXCLUDING INTERSYSTEM CROSSING

Following eqs. (7.106)–(7.108) and the same assumptions made for Case III in

Section 7.3.1(C), the rate equations can be written as [Smith 1968]

$$\frac{d[T]}{dt} = \varepsilon_T i_T - (\beta_1 + \beta_2)[T] - \beta_{HG}[T] + p_G[T_G] \qquad (7.145)$$

$$\frac{d[T_G]}{dt} = \beta_{HG}[T] - (\beta_{G1} + \beta_{G2})[T_G] - p_G[T_G] \qquad (7.146)$$

In the steady state $d[T]/dt = d[T_G]/dt = 0$, the solution of eqs. (7.145) and (7.146) yields

$$[T] = \frac{\varepsilon_T i_T}{(\beta_1 + \beta_2)} \left[1 + \left(\frac{\beta_{G1} + \beta_{G2}}{\beta_1 + \beta_2} \right) \left(\frac{\beta_{HG}}{\beta_{G1} + \beta_{G2} + p_G} \right) \right]^{-1} \qquad (7.147)$$

and

$$[T_G] = \frac{\beta_{HG}[T]}{\beta_{G1} + \beta_{G2} + p_G} \qquad (7.148)$$

If the radiative transition probability is assumed to be independent of trapping, then the rate of phosphorescence emission (or the phosphorescence intensity) is

$$\phi_{ph} = \beta_1 \{[T] + [T_G]\} \qquad (7.149)$$

CASE III: REAL CRYSTALS IN THE PRESENCE OF EXTERNAL
ELECTRIC FIELDS BUT EXCLUDING INTERSYSTEM CROSSING

Several investigators have reported the effects of injected charge carriers due to the application of electric fields on the rate of triplet decay and on the phosphorescence through the interaction of triplet excitons with charge carriers [Pope *et al.* 1970, Wakayama and Williams 1972]. To include the effect of triplet and charge carrier interaction, a term $-K_T[T]$ has to be added on the right-hand side of eq. (7.145) and a term $+K_T[T]$ added on the right-hand side of eq. (7.146). For the direct triplet exciton quenching by a short range interaction to occur, it is necessary that either the triplet excitons or the charge carriers must be mobile [Singh *et al.* 1965]. As the thermal velocity of charge carriers is much greater than that of triplet excitons, the concentrations of free or trapped carriers inside the crystal will play the major role in the change of ϕ_{ph} if only triplet quenching is involved. Thus, following the same approach given in Section 7.3.1(C)—Case IV, the change in intensity of emitted phosphorescence due to triplet exciton carrier interactions may be estimated by

$$\frac{\phi_{ph} (\text{with } K_T[T] \text{ term})}{\phi_{ph} (\text{without the } K_T[T] \text{ term})} \propto \frac{\beta_1 + \beta_2}{\beta_1 + \beta_2 + K_T} \qquad (7.150)$$

(B) Experiments

The measurements of phosphorescence intensity will allow the determination of the concentration and the lifetime of triplet excitons and the measurements of delayed fluorescence will allow the study of the interaction of triplet excitons with other particles or traps, particularly with these measurements made at various temperatures.

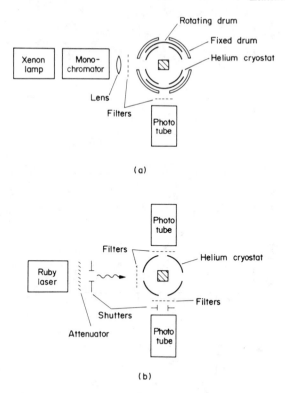

FIG. 7.34. Experimental arrangements: (a) chopped light with
d.c. xenon lamp excitation; (b) Q-switched laser
experiments. The upper phototube monitored the
prompt fluorescence and the lower phototube monitored the phosphorescence. [After Smith 1968.]

To measure both the phosphorescence and the delayed fluorescence intensities in
anthracene crystals, Smith [1968] has used the experimental arrangements shown in
Fig. 7.34. The triplet excitons can be generated in anthracene by either (i) directly using
red light from an xenon lamp, or (ii) using two-photon absorption of Q-switched ruby
laser light to yield singlet excitons first and then to produce indirectly triplet excitons
through inter-system crossing. These two methods are clearly shown in Fig. 7.34. For
the detection of weak red and near infrared phosphorescence, Smith used cooled
photomultipliers such as EMI 9558, S-20, and RCA 7102, S-1 photomultipliers.

Using the chopped light method, Smith [1968] has measured the relative phosphorescence spectrum for undoped anthracene crystals as a function of temperature and the
results are shown in Fig. 7.35. When these curves are folded about the 0–0 line at
14,740 cm^{-1} there is approximately mirror symmetry of the vibrational lines. At 300°K
there is no measurable shift of the 0–0 line between absorption and emission, implying
that pure crystal electron states and free excitons are dominant at room temperature.
When the temperature is lowered to 20°K, the 0–0 line at 14,740 cm^{-1} is almost absent
and the most prominent line appears at about 13,860 cm^{-1}, corresponding to a shift of
880 cm^{-1}, which is consistent with the triplet exciton trapping model. The two

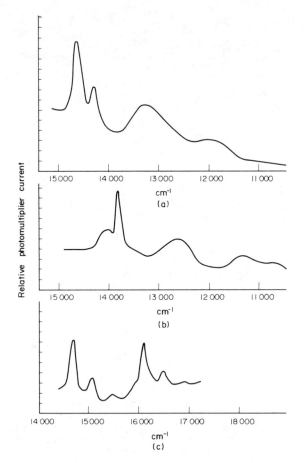

FIG. 7.35. Triplet phosphorescence spectrum of anthracene crystals at (a) 300° K, and (b) at 18° K. Phosphorescence excitation spectrum at 33° K shown in (c). The phosphorescence energy scales are reflected about the 0–0 line. Resolution was 50 Å with no correction applied for small spectral variations of detector sensitivity and excitation intensity. [After Smith 1968.]

prominent low temperature phosphorescence peaks in Fig. 7.35 can be considered to correspond to two trapping levels, one located at 880 cm^{-1} and the other located at 2020 cm^{-1} from the 0–0 line. In general, at low temperatures the interaction between free and shallowly trapped excitons is more efficient in yielding the delayed blue fluorescence than is the free–free exciton bimolecular interaction which is dominant at room temperature.

Smith [1968] has also studied the trapping effects in a variety of nominally pure anthracene crystals of similar size by measuring the relative delayed (blue) fluorescence intensity and the relative delayed (red) phosphorescence using a laser pulse in the 10–40 msec time range. A log–log plot of the results in Fig. 7.36 shows that the delayed blue intensity varies over several orders of magnitude for a given red emission rate, implying

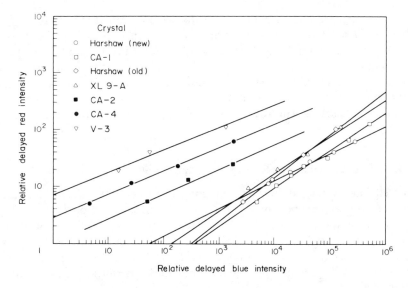

F IG. 7.36. Red versus blue intensity for various anthracene crystals following laser excitation at room temperature. [After Smith 1968.]

that different crystals contain different types and concentrations of traps (various impurities and structural imperfections) and these traps play an important role in this phenomenon. However, it is apparent that a given rate of delayed blue emission does not directly give information about the total concentration of triplet excitons in anthracene. It is important to note that with only 0.25 ppm tetracene doped in the anthracene crystal, the delayed blue fluorescence intensity is weaker by a factor of about 10^6, while the red phosphorescence emission is comparable to that of undoped anthracene. This implies that the localized or trapped triplet excitons avoid bimolecular interaction and do not change greatly their radiative and non-radiative lifetimes.

7.4. ELECTROLUMINESCENCE

Electroluminescence has been observed in undoped anthracene crystals [Helfrich and Schneider 1965, 1966, Williams and Schadt 1970, Schwob and Williams 1973, 1974, Hwang and Kao 1973, 1974, Kunkel and Kao 1976], in tetracene-doped anthracene crystals [Dresner and Goodman 1970, Kawabe *et al.* 1971, Zschokke-Gränacher *et al.* 1967, Schwob *et al.* 1970, 1971], in naphthalene crystals [Lohmann and Mehl 1969], in tetracene [Kalinowski and Godlewski 1975], in pyrene [Basurto and Burshtein 1975], and in pentacene [Wakayama *et al.* 1974], either with double injection or with single injection electrodes. For double injection the recombination of the injected holes with the injected electrons will yield singlet and triplet excitons, and it is these singlet excitons that radiatively decay producing electroluminescence. For single injection (say, hole injection) the other type of charge carriers (say, electrons) must be supplied from the counter electrode (say, electrons from the cathode), but in this case the supply of minority carriers occurs only at relatively high fields [Hwang and Kao 1973]. Thus the so-called single injection would become double injection at the threshold voltage for the onset of electroluminescence. In fact, many materials have been used as electrodes [Mehl and Funk 1967, Dresner 1969, Williams *et al.* 1972] to produce electroluminescence in anthracene (cf. Tables 2.1 and 2.2). That different materials used for electrodes result in different values of threshold voltages indicates that there are no perfect ohmic contacts and that different electrode materials in contact with a molecular crystal surface will form different potential barriers for carrier injection.

7.4.1. Electroluminescence under time-invariant fields

It is well known that at high current injection level I the electroluminescence brightness B is linearly proportional to I, and at low current injection level $B \propto I^2$. The location of the luminous zone inside the specimen depends on the concentration and distribution of injected electrons and holes, and the width of the zone is therefore also the width of the carrier recombination zone [Helfrich and Schneider 1965]. In the following we shall present the theory and the experimental results for d.c. electroluminescence.

(A) Theory

In Section 5.1 we have discussed the theory of filamentary double injection in solids. It is likely that multiple current filaments may simultaneously exist between two parallel plane electrodes. For such a case we can always consider that within a domain of radius r_d is enclosed only one current filament and that the total current between the plane electrodes may be expressed as

$$I_T = I_{\text{domain } 1} + I_{\text{domain } 2} + \ldots = \sum_n I_n \simeq HI \qquad (7.151)$$

This means that the total current can be represented by the current in one domain I multiplied by a constant H. We can also assume that the current is uniformly distributed within the area of πr_d^2 provided that r_d is chosen small enough to satisfy this condition (cf. Section 5.5).

In molecular crystals, for example, in undoped and doped anthracene, both the electron and hole mobilities are generally small [Kepler 1960, Le Blanc 1967, Munn and Siebrand 1970] and the recombination rate constant is large [Helfrich and Schneider 1966, Silver and Sharma 1967, Morris and Silver 1969], resulting in a small space charge overlap. Thus the simultaneous injection of electrons and holes from the contacting electrodes will produce two-carrier space-charge-limited currents within the filament, and lead to electroluminescence when two types of carriers meet and recombine radiatively.

After the onset of electroluminescence in a molecular crystal with a fixed emission spectrum, the electroluminescence brightness is governed by the external quantum efficiency [Bergh and Dean 1972] η_q given by

$$\eta_q = \eta_i \eta_g \eta_e = \eta_{\text{int}} \, \eta_e \qquad (7.152)$$

where η_i is the carrier injection efficiency which is the ratio of the current due to minority carriers to the total current; and if it is assumed that J_n is the current due to minority carriers, then

$$\eta_i = J_n/(J_n + J_p) \tag{7.153}$$

η_g is the light generation efficiency; $\eta_{int} = \eta_i \eta_g$ is the internal quantum efficiency which is a function of the total current density and temperature of the electroluminescence specimen; and η_e is the light extraction efficiency which is defined as the ratio of power loss due to the light transmission within the electroluminescence specimen to the total power losses which consist of both the losses in the bulk and on the surface, and can be considered to be fixed for a given specimen.

For double injection the recombination of the injected electrons with the injected holes in the organic crystal specimens will yield singlet and triplet excitons. It is generally accepted that the singlet excitons producing fluorescence are partly generated directly by electron–hole recombination and partly generated indirectly by triplet–triplet recombination in pairs according to the following relation [Helfrich and Schneider 1965, 1966]:

$$20(e+h) \rightarrow 5[S]_{dir} + 15[T]$$
$$\rightarrow 5[S]_{dir} + 3[S]_{ind} \tag{7.154}$$

where e and h represent, respectively, the electron and hole, $[S]_{dir}$ and $[S]_{ind}$ represent, respectively, the singlet excitons produced by the direct and the indirect processes, and $[T]$ represents the triplet excitons. It is the efficiency of generating $[S]_{dir}$ and $[S]_{ind}$ and their subsequent population in the crystal which govern the electroluminescent intensity, but the threshold voltage is mainly governed by local field effects on the electrode surfaces. Since there is a great difference in lifetime between the singlet and the triplet excitons in molecular crystals (e.g., they are 10^{-8} and 10^{-2} sec, respectively, in anthracene), the total electroluminescence consists of prompt electroluminescence due to $[S]_{dir}$ and delayed electroluminescence due to $[S]_{ind}$ and exhibits time constants corresponding to both of these decays. But in the steady state the electroluminescence is the combination of these two.

To develop a time dependent equation for excitons and hence for electroluminescence, the following assumptions are made:

(i) Singlet and triplet excitons generated due to the recombination of injected electrons and holes have the generation rates G_s and G_T, respectively.

(ii) Singlet and triplet excitons have, respectively, the rate constants α_1 and β_1 for the radiative transition with the emission of photons and α_2 and β_2 for the non-radiative transition with the emission of phonons.

(iii) Singlet and triplet excitons may relax into traps at the levels ΔE_t lower in energy with the rate constants α_{HG} and β_{GH}, respectively.

(iv) Trapped excitons may be thermally detrapped, and then decay radiatively with the rate constants α_{G1} and β_{G1}, and non-radiatively with the rate constants α_{G2} and β_{G2} for singlet and triplet excitons, respectively.

(v) The effective rate constant for triplet–triplet annihilation is γ_T, and that for triplet–trapped triplet annihilation is γ_G.

(vi) Excitons may be depopulated by the interactions between excitons and charge carriers or between excitons and surface states. But in the following analysis it is assumed that the effect of these interactions is small [Pope and Selsby 1972] and can be neglected.

(1) Prompt electroluminescence

(a) *Undoped crystals*. On the assumption that in the deep traps $\Delta E_t \gg kT$ and that the inter-system crossing into the triplet states can be ignored, the rate equations for singlet excitons generated directly by the electron–hole recombination may be written as [Avakian and Merrifield 1968, Wolf and Benz 1971]:

$$d[S]_{\text{dir}}/dt = G_s - (\alpha_1 + \alpha_2)[S]_{\text{dir}} - \alpha_{HG}[S]_{\text{dir}} - K_s[S]_{\text{dir}} \qquad (7.155)$$

$$d[S_G]_{\text{dir}}/dt = \alpha_{HG}[S]_{\text{dir}} - (\alpha_{G1} + \alpha_{G2})[S_G]_{\text{dir}} \qquad (7.156)$$

where $[S]$ and $[S_G]$ represent, respectively, the free and trapped singlet excitons, and K_s is the carrier–singlet exciton reaction rate. In the deep trap case α_{HG} is larger than K_s so that the last term in eq. (7.155) may be neglected. In the steady state $d[S_G]_{\text{dir}}/dt = d[S]_{\text{dir}}/dt = 0$, and thus from eqs. (7.155) and (7.156) we obtain

$$[S]_{\text{dir}} = \frac{G_s}{\alpha_1 + \alpha_2 + \alpha_{HG}^2/\alpha_G} \qquad (7.157)$$

where $\alpha_G = \alpha_{G1} + \alpha_{G2}$, and G_s is the number of singlet excitons per unit volume generated per unit time, and can be written as [Helfrich and Schneider 1965, 1966]

$$G_s = g_s J_z(r)/qL_s \qquad (7.158)$$

in which L_s is the diffusion length of the singlet exciton, g_s is the fraction of electron–hole pairs that produces the singlet excitons immediately after recombination, approximately equal to 1/4 for anthracene [Helfrich and Schneider 1966], and $J_z(r)$ is the current density in the filament. Thus the internal quantum efficiency can be written as

$$\eta_{\text{int}} = \alpha_1 \int_0^{2\pi} \int_0^d \int_0^{r_d} [S]_{\text{dir}}\, rd\theta\, dz\, dr$$

$$= \{\alpha_1/[\alpha_1 + \alpha_2 + K(N_s)]\}(2\pi g_s d/q\lambda_s)I \qquad (7.159)$$

where

$$K(N_s) = \alpha_{HG}^2/\alpha_G \qquad (7.160)$$

The brightness of the prompt electroluminescence B is proportional to η_q. If it is assumed that η_e is a constant, then B is linearly proportional to the total current I.

$$B \propto \eta_{\text{int}}\eta_e = b_1 I \qquad (7.161)$$

where b_1 is a constant.

(b) *Doped crystals*. In general the doped guest molecules tend to quench the host molecular fluorescence and to emit the guest molecular fluorescence. Of course, the quantum yield of guest molecular fluorescence depends on the dopant concentration. For example, an anthracene crystal doped with 1 ppm of tetracene will emit green light from tetracene instead of blue light from anthracene. The rate equations (7.155) and (7.156) can be used for doped crystals. Thus in the steady state we have

$$[S_G]_{\text{dir}} = \left(\frac{\alpha_{HG}}{\alpha_{G1} + \alpha_{G2}} \right) \frac{G_s}{\alpha_1 + \alpha_2 + \alpha_{HG}^2/\alpha_G} \tag{7.162}$$

In this case α_{HG} based on the hopping model can be defined as [Wolf and Benz, 1971, Suna 1970]

$$\alpha_{HG} = C_G/t_{hs} \tag{7.163}$$

where C_G is the dopant (guest molecule) concentration and t_{hs} is the singlet exciton hopping time. Following the same procedure, the internal quantum efficiency for doped crystal can be written as

$$\eta_{\text{int}}^d = (K_{es}C_G) \frac{\alpha_1}{\alpha_1 + \alpha_2 + K(N_s)} \frac{2\pi g_s d}{qL_s} I \tag{7.164}$$

where K_{es} is defined as the energy transfer constant for singlet excitons [Wolf and Benz 1971] and is given by

$$K_{es} = \alpha_{G1}/\alpha_1 (\alpha_{G1} + \alpha_{G2})t_{hs} \tag{7.165}$$

Thus, the brightness of the prompt electroluminescence is also linearly proportional to I,

$$B^d \propto \eta_{\text{int}}^d \eta_e = b_2 I \tag{7.166}$$

where b_2 is a constant.

(2) Delayed electroluminescence

The rate equations for the free triplet $[T]$ and trapped triplet $[T_G]$ excitons and those for free singlet $[S]_{\text{ind}}$ and trapped singlet $[S_G]_{\text{ind}}$ created indirectly by triplet–triplet annihilation are given by

$$d[T]/dt = G_T - (\beta_1 + \beta_2)[T] - \beta_{HG}[T] - \gamma_G[T][T_G] - \gamma_T[T]^2 - K_T[T] \tag{7.167}$$

$$d[T_G]/dt = \beta_{HG}[T] - (\beta_{G1} + \beta_{G2})[T_G] - \gamma_G[T][T_G] \tag{7.168}$$

$$d[S]_{\text{ind}}/dt = \tfrac{1}{2}f\gamma_T[T]^2 - (\alpha_1 + \alpha_2)[S]_{\text{ind}} - K_s[S]_{\text{ind}} \tag{7.169}$$

$$d[S_G]_{\text{ind}}/dt = \tfrac{1}{2}f'\gamma_G[T][T_G] - (\alpha_{G1} + \alpha_{G2})[S]_{\text{ind}} \tag{7.170}$$

where f and f' are, respectively, the fractions of triplet–triplet and triplet–trapped triplet annihilations, which create singlet excitons, and the value of f and f' is approximately 0.4 for anthracene, K_T is the carrier–triplet excitons reaction rate, and G_T is the number of triplet excitons per unit volume generated per unit time and it is given by [Helfrich and Schneider 1966]

$$G_T = g_T J_z(r)/qL_T \tag{7.171}$$

in which L_T is the diffusion length of the triplet exciton and g_T is the fraction of electron–hole pairs that produces the triplet excitons after recombination, approximately equal to 3/4 for anthracene. It should be noted that in eqs. (7.167) and (7.169) γ_T is the effective overall rate constant for bimolecular triplet–triplet annihilation, which includes the probability of producing $[T]$ from this process [Wakayama and Williams 1972]. In the deep trap case the last terms of eqs. (7.167) and (7.169) are very small and can be neglected, and in the steady state

$$d[T]/dt = d[T_G]/dt = d[S]_{ind}/dt = d[S_G]_{ind}/dt = 0$$

However, to solve the coupling equations (7.167)–(7.170) we have to make some approximations. It is therefore convenient to treat this problem for two cases as follows:

(a) *Low injection (or low current) case.* In this case it can be assumed that the monomolecular decay is dominant and therefore $(\beta_1 + \beta_2)[T] \gg \gamma_T[T]^2$ and $(\beta_{G1} + \beta_{G2})[T_G] \gg \gamma_G[T][T_G]$.

(i) Undoped crystals. From eqs. (7.167) and (7.169) we have

$$[S]_{ind} = \tfrac{1}{2}f\gamma_T[1/(\alpha_1 + \alpha_2)][T]^2 \tag{7.172}$$

and

$$[T] = \frac{g_T/qL_T}{\beta_1 + \beta_2 + \beta_{HG}} J_z(r) \tag{7.173}$$

Thus the internal quantum efficiency is

$$\eta_{int} = \alpha_1 \int_0^{2\pi} \int_0^d \int_0^{r_d} [S]_{ind}\, r\, \theta\, dz\, dr$$

$$= \tfrac{1}{2}f\gamma_T[\alpha_1/(\alpha_1 + \alpha_2)](\beta_1 + \beta_2 + \beta_{HG})^{-2}(g_T/qL_T)^2\, 2\pi d$$

$$\times \int_0^{r_d} J_z^2(r)dr \tag{7.174}$$

If we assume η_e is a constant, then the brightness of the delayed electroluminescence can be written as

$$B \propto \eta_{int}\, \eta_e = b_3 \int_0^{r_d} J_z^2(r)dr \tag{7.175}$$

where b_3 is a constant.

(ii) Doped crystals. In this case the guest molecules tend to quench the host molecular fluorescence. From eqs. (7.168) and (7.170) we have

$$[S_G]_{ind} = \tfrac{1}{2}f'\gamma_G[T][T_G]/(\alpha_{G1} + \alpha_{G2}) \tag{7.176}$$

and

$$[T_G] = [\beta_{HG}/(\beta_{G1} + \beta_{G2})][T] \tag{7.177}$$

Based on the hopping model β_{HG} is given by [Wolf and Benz 1971, Suna 1970]

$$\beta_{HG} = C_G/t_{hT} \tag{7.178}$$

where t_{hT} is the triplet exciton hopping time. Thus from eqs. (7.173), (7.176), and (7.177) the internal quantum efficiency for the doped crystals can be written as

$$\eta_{int}^{d} = (K_{eT}C_{G})\frac{1}{2}f\gamma_{T}\left(\frac{\alpha_{1}}{\alpha_{1}+\alpha_{2}}\right)\left(\frac{1}{\beta_{T}+\beta_{HG}}\right)^{2}\left(\frac{g_{T}}{qL_{T}}\right)^{2}2\pi d \int_{0}^{r_{d}} J_{z}^{2}(r)dr \quad (7.179)$$

where K_{eT} is the energy transfer constant for triplet excitons [Wolf and Benz 1971] and is given by

$$K_{eT} = \frac{f'\gamma_{G}}{f\gamma_{T}}\frac{\alpha_{G1}(\alpha_{1}+\alpha_{2})}{\alpha_{1}(\alpha_{G1}+\alpha_{G2})}\frac{1}{t_{hT}} \quad (7.180)$$

Therefore the brightness of the delayed electroluminescence can be written as

$$B^{d} \propto \eta_{int}^{d}\eta_{e} = b_{4}\int_{0}^{r_{d}} J_{z}^{2}(r)dr \quad (7.181)$$

where b_{4} is a constant.

(b) *High injection (or high current) case.* In this case it can be assumed that the bimolecular decay is dominant and therefore $(\beta_{1}+\beta_{2})[T] \ll \gamma_{T}[T]^{2}$ and $(\beta_{G1}+\beta_{G2})[T_{G}] \ll \gamma_{G}[T][T_{G}]$.

(i) Undoped crystals. By assuming that $[S_{G}]_{ind}$ is negligibly small, then from eqs. (7.167)–(7.170) and in the steady state, we obtain

$$[S]_{ind} = \frac{1}{2}f\gamma_{T}[1/(\alpha_{1}+\alpha_{2})][T]^{2} \quad (7.182)$$

and

$$[T]^{2} = \frac{1}{2}[g_{T}J_{z}(r)/qL_{T}](1+2\theta+\sqrt{1+4\theta})/2\theta \quad (7.183)$$

where

$$\theta = [g_{T}J_{z}(r)r/qL_{T}](t_{hT}/C_{G})^{2} \quad (7.184)$$

Since $\theta \gg 1$ for undoped crystals, we have

$$[T]^{2} = \frac{1}{2}g_{T}J_{z}(r)/qL_{T} \quad (7.185)$$

From eqs. (7.182) and (7.183) we obtain

$$\eta_{int} = \alpha_{1}\int_{0}^{2\pi}\int_{0}^{d}\int_{0}^{r_{d}}[S]_{ind}\,r\,d\theta\,dz\,dr$$
$$= \frac{1}{2}f\gamma_{T}[\alpha_{1}/(\alpha_{1}+\alpha_{2})](\frac{1}{2}g_{T}/gL_{T})2\pi d\,I \quad (7.186)$$

and thus the brightness of the delayed electroluminescence becomes

$$B \propto \eta_{int}\eta_{e} = b_{5}I \quad (7.187)$$

where b_{5} is a constant.

(ii) Doped crystals. From eqs. (7.167)–(7.170) and in the steady state we have

$$[S_{G}]_{ind} = \frac{1}{2}f'[\beta_{HG}/(\alpha_{G1}+\alpha_{G2})][T] \quad (7.188)$$

and

$$[T] = \frac{1}{2}\beta_{HG}[1+\sqrt{1+4\theta}] \quad (7.189)$$

Since $\theta \ll 1$ for doped crystals, we have

$$[T] = \tfrac{1}{2}\sqrt{2}(\gamma_T/\beta_{HG}^2)^{1/2}[g_T J_z(r)/qL_T] \tag{7.190}$$

From eqs. (7.188) and (7.190) we have

$$\eta_{\text{int}}^d = \frac{f'\gamma_T^{1/2}}{2\sqrt{2}} \frac{\alpha_{G1}}{(\alpha_{G1}+\alpha_{G2})^2} \frac{g_T}{qL_T} 2\pi d\, I \tag{7.191}$$

and thus the brightness of the delayed electroluminescence is

$$B^d \propto \eta_{\text{int}}^d \eta_e = b_6 I \tag{7.192}$$

where b_6 is a constant.

In the steady state the brightness of electroluminescence is the sum of the brightness of prompt and delayed electroluminescence. Thus the brightness as a function of current can be deduced as follows.

(a) *Low injection case*: For undoped crystals

$$B_T = b_1 I + b_3 \int_0^{r_d} J_z^2(r)dr \tag{7.193}$$

for doped crystals

$$B_T = b_2 I + b_4 \int_0^{r_d} J_z^2(r)dr \tag{7.194}$$

(b) *High injection case*: For undoped crystals

$$B_T = b_1 I + b_5 I \tag{7.195}$$

for doped crystals

$$B_T = b_2 I + b_6 I \tag{7.196}$$

(II) VOLTAGE DEPENDENCE OF ELECTROLUMINESCENT INTENSITY

It has been analysed that under high injection conditions the electroluminescent intensity is proportional to the current level I. If we know the I–V characteristics we can easily deduce the relationship of electroluminescent brightness with the applied voltage.

According to Kao and Elsharkawi [1976] the current density inside a current filament can be expressed as

$$J_z(r) = \overline{W} J_{z0} \tag{7.197}$$

where J_{z0} is the current density at the centre of the filament $r = 0$ and $\overline{W}$ is dependent on the distribution of traps and recombination centres. For electron and hole traps confined in two separate single discrete energy levels, $\overline{W}$ can be approximately given by

$$\overline{W} = \left[1 + \left(\frac{1}{6}\frac{Ad J_{z0}}{V}\right)^{1/2} r\right]^{-2} \tag{7.198}$$

where A is

$$A = q^{-1} \left[\frac{\mu_n D_p + \mu_p D_n}{D_n D_p (\mu_n \Omega_n + \mu_p \Omega_p)^2} \right] \langle v\sigma_R \rangle \Omega_n \Omega_p \qquad (7.199)$$

$$\Omega_n = 1 + n_t/n \qquad (7.200)$$

$$\Omega_p = 1 + p_t/p \qquad (7.201)$$

If the organic crystal has one deep trapping level at which the traps act as recombination centres and one shallow hole-trapping level at which the traps act as traps, the double injection current is given by [Hwang and Kao 1978]

$$J = \left(\frac{9}{8} \right) \mu_n \mu_n \tau (P_{TR0} + p_0 - n_0) (V^2/d^3) \qquad (7.202)$$

where τ is the carrier lifetime; P_{TR0}, p_0 and n_0 are the densities of occupied recombination centres (traps), free holes and electrons, respectively, under thermal equilibrium conditions. By using J as J_{z0} and eqs. (7.197)–(7.202), the electroluminescent brightness can be expressed as

$$B_T \simeq b_T I$$

$$= b_T \int_0^{2\pi} \int_0^{r_d} J_z(r) r \, dr \, d\theta$$

$$\simeq A_a (V^2/d^3)(1 + a_1 V/d^{3/2} + a_2 V^2/d^3) \qquad (7.203)$$

where A_a, a_1, a_2, and b_T are constants.

Similarly, if the traps are Gaussian distributed, the double injection current–voltage relation would be of the form

$$J \propto V^{m+1}/d^{2m+1} \qquad (7.204)$$

In this case the electroluminescent brightness may become

$$B_T \simeq A_b \frac{V^{m+1}}{d^{2m+1}} \left[1 + \frac{b_1 V^{(m+1)/2m}}{d^{(2m+1)/2m}} + \frac{b_2 V^{(m+1)/m}}{d^{(2m+1)/m}} \right] \qquad (7.205)$$

where A_b, m, b_1, and b_2 are constants.

It is most likely that prior to the onset of observable electroluminescence, the current is dominated by one type of charge carriers from one electrode, the other being a neutral or blocking contact. But when the applied voltage $V \geq V_{th}$ the space charge built up near the blocking contact becomes sufficient to turn on the carrier injection from it, and therefore the current increases sharply to another regime as shown in Fig. 5.15. In this regime both types of carriers, holes and electrons, play almost equally important roles in the conduction current, and both electrodes under such conditions can be assumed to be ohmic contacts. Equations (7.203) and (7.205) are derived on the basis of ohmic contacts, but the effects of field-enhanced detrapping and reabsorption of electroluminescence are excluded. If these effects are taken into account, it would be expected that I and hence B_T would increase more rapidly with V than would be predicted from eqs. (7.203) and (7.205).

(III) TEMPERATURE DEPENDENCE OF ELECTROLUMINESCENT
INTENSITY

The temperature dependent phenomenon may be explained in terms of three processes: (a) exciton–trapped exciton interactions, (b) exciton–carrier interactions and (c) exciton–surface state interactions, which control the electroluminescent intensity and are temperature dependent. The surface state at the interface between the contacting electrode and the anthracene crystal quenches singlet excitons and the quenching rate decreases with increasing temperature. However, the effect of surface states may be very small [Pope and Selsby 1972] as compared with those of processes (a) and (b), and therefore process (c) for the present discussions is ignored.

If the temperature for the peak electroluminescent brightness is defined as the brightness characteristic temperature T_b, then it is possible that for temperatures lower than T_b process (a) is dominant and for temperatures higher than T_b process (b) becomes important. The physical meaning of T_b can be thought of as the characteristic temperature of these processes, at which the singlet exciton-attempt-escape frequency is equal to the carrier–singlet exciton reaction rate

$$v \exp(-E_{ti}/kT_b) = K_{sc} = ZN_T$$

or

$$T_b = \frac{E_{ti}/k}{\ln(v/ZN_T)} \tag{7.206}$$

where E_{ti} is the trapped singlet exciton energy measured from the singlet exciton energy level, v is the singlet exciton-attempt-escape frequency factor, and K_{sc} and Z are, respectively, the rate and the rate constant of carrier–singlet exciton reactions. In anthracene it is generally accepted that the thermal velocity of carriers v is larger than that of singlet excitons u. If this is the case, K_{sc} may be written as [Wakayama and Williams 1971]

$$K_{sc} = N_T(1-\theta_t)\sigma v + N_T\theta_t\sigma(v + u^2/3v) = ZN_T \tag{7.207}$$

where N_T is the total carrier density (electrons and holes), θ_t is the ratio of free carrier density to the total carrier density which includes both free and trapped carriers, and σ is the reaction cross-section between a carrier and an exciton.

For temperatures lower than T_b the brightness increases with increasing current. Since B_T is proportional to I for the high injection case, the temperature dependence of B_T can be explained in terms of the temperature dependence of I. By assuming $I \propto \mu_{\text{eff}}$ and μ_{eff}, the effective carrier mobility, is temperature dependent following the relation

$$\mu_{\text{eff}} \propto \exp(-E_{ti}/kT) \tag{7.208}$$

then the electroluminescent brightness is

$$B_T \propto I \propto \exp(-E_{ti}/kT) \tag{7.209}$$

Equation (7.209) shows B_T increases with increasing T for $T < T_b$.

For temperatures higher than T_b the brightness decreases though the current still increases with increasing temperature, and the electroluminescence disappears at a certain temperature depending on the applied voltage. It has been experimentally observed that the interaction of singlet excitons [Wakayama and Williams 1971] or of triplet excitons [Wakayama and Williams 1972] with charge carriers quenches the fluorescence. The change of temperature may not affect very much the carrier injection from the electrodes but it would affect the value of θ_t. For undoped anthracene crystals the brightness as a function of temperature for $T > T_b$ can be written as

$$B_T[T] = \frac{[b_1 I]_{T_b}}{1 + K_{sc}/[\alpha_1 + \alpha_2 + K(N_s)]}$$

$$+ \frac{\left[b_3 \int_0^{r_d} J_z^2(r)dr\right]_{T_b}}{[1 + K_{sc}(\alpha_1 + \alpha_2)][1 + K_T/(\beta_1 + \beta_2 + \beta_{HG})]} \qquad (7.210)$$

for low injection, and

$$B_T[T] = \frac{[b_1 I]_{T_b}}{1 + K_{sc}/[\alpha_1 + \alpha_2 + K(N_s)]} + \frac{[b_5 I]_{T_b}}{1 + K_{sc}/(\alpha_1 + \alpha_2)} \qquad (7.211)$$

for high injection, where $[b_1 I]_{T_b}$, $\left[b_3 \int_0^{r_d} J_z^2 (r)dr\right]_{T_b}$, and $[b_5 I]_{T_b}$ are defined in eqs. (7.161), (7.175), and (7.187), but at temperature T_b. The values of $K_{sc}/[\alpha_1 + \alpha_2 + K(N_s)]$, $K_{sc}/(\alpha_1 + \alpha_2)$, and $K_T/(\beta_1 + \beta_2 + \beta_{HG})$ generally increase with increasing temperature.

It should be noted that the nature of exciton–trapped exciton interactions and carrier–exciton interactions is still not fully understood. However, the above argument serves, at least, to explain qualitatively the temperature dependence phenomena.

(B) Experiments

In general, for the high injection case the electroluminescent brightness is indirectly proportional to current according to eqs. (7.195) and (7.196), and this theoretical prediction agrees well with all presently available experimental results as shown in Fig. 7.37. For the low injection case, the B_T vs I relationship becomes non-linear according to eqs. (7.193) and (7.194). If $b_3 > b_1$ or $b_4 > b_2$, B_T becomes proportional to I^2. Some experimental results following this square law are also shown in Fig. 7.37. For doped crystals the presence of guest molecules quenches the host molecular fluorescence and exhibits the guest molecular fluorescence. But it should be noted that the electroluminescence yield from host and guest molecules changes with current and that the transfer of excitation energy from host to guest molecules decreases with increasing current [Zschokke-Gränacher *et al.* 1967]. Schwob *et al.* [1970] have reported that under the low injection condition the tetracene fluorescence in tetracene-doped anthracene is dominant and the electroluminescent brightness is a function of dopant concentration. This is expected on the basis of eqs. (7.164), (7.179), and (7.194). It is possible that the brightness increases with increasing dopant concentration because under such a condition most recombinations occur in traps. However, as the current increases the probability of carriers being trapped by the guest decreases, and so does

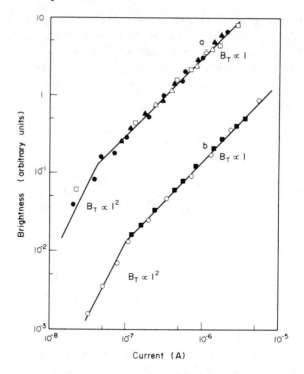

FIG. 7.37. Electroluminescent brightness as a function of current for (a) undoped anthracene crystals, and (b) anthracene crystals doped with tetracene. Solid curves are based on the theory, and experimental results are after Williams and Schadt [1970] ▲, Mehl and Funk [1967] □, Schwob and Zschokke-Gränacher [1971] ○, Kawabe, Masuda, and Namba [1971] △, ■, and Hwang and Kao ●. [After Hwang and Kao 1974.]

the yield of guest fluorescence. Therefore, for the high injection case the host fluorescence of the tetracene-doped anthracene becomes dominant and its brightness increases approximately linearly with current as expected from eqs. (7.164), (7.191), and (7.196).

Using undoped anthracene crystals cleared along the *ab* plane with a pair of double injection electrodes (silver as anode and sodium and anthracene in tetrahydrofuran as cathode) of 3 mm diameter, we have measured the electroluminescent brightness as a function of current, applied voltage, and temperature. Figure 7.38 shows that B_T increases very rapidly with increasing applied voltages as expected [cf. Section 7.4.1(A)].

It should be noted that similar results can also be obtained using silver for both anode and cathode [Hwang and Kao 1973]. This indicates that sodium + anthracene in tetrahydrofuran is a good electron-injecting contact but silver is a relatively poor hole-injecting contact. Electroluminescence can occur only when field-enhanced injection of both types of carriers can take place [Sworakowski *et al.* 1974, Kunkel and Kao 1976]. Sworakowski *et al.* [1974] have used a sodium–potassium alloy for the electron-injecting contact, and silver, gold, and aluminium for the positively biased electrode for

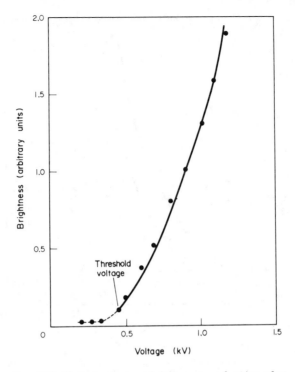

F<small>IG</small>. 7.38. Electroluminescent brightness as a function of applied voltage for undoped anthracene crystals at 20° C. Solid curve is based on the theory and ● experimental results. [After Hwang and Kao 1974.]

their studies of electroluminescence in anthracene. They have proposed that the injected electrons captured in traps located near the positive electrode may enhance the field there (may build up a field as large as 10^7 V cm^{-1}), thereby enabling holes to tunnel from the electrode or from surface states into the bulk crystal. Thus the total current is completely determined by the electron–hole recombination processes and, of course, both the current and electroluminescent intensity increase with increasing concentration of holes tunnelling through the barrier.

Figure 7.39 shows that the electroluminescent brightness of undoped anthracene increases with increasing temperature. This phenomenon has also been observed using a pair of silver electrodes [Hwang and Kao 1973]. This phenomenon may be explained in terms of three processes discussed in Section 7.4.1(A). The semilogarithmic plot of B_T vs $1/T$ from the data given in Fig. 7.39 for the temperature range from -20–$40°$C gives an activation energy E_{ti} of 0.21 eV. Using this value for E_{ti}, the value of 10^7 sec^{-1} for v [Mulder 1968], and the value of 10^{-8} cm^3 sec^{-1} for K_{sc} [Schott and Berrehar 1973], and assuming N_T to be 3×10^{11} cm^{-3} (this value is of the same order of that used by other investigators [Wakayama and Williams 1971, 1972]), T_b has been estimated to be about 313°K (or 40°C), which is in good agreement with experiment. It should be noted that the average value of 313°K for T_b was calculated using the average value of N_T which was determined and used by other investigators [Wakayama and Williams 1971,

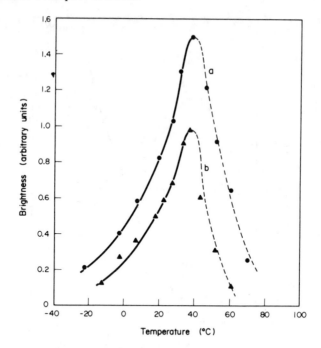

FIG. 7.39. Electroluminescent brightness as a function of temperature for undoped anthracene crystals of about 1 mm in thickness for (a) 1.2 kV applied voltage, and (b) 1.0 kV applied voltage. Solid curves are based on the theory and ●, ▲, and dashed curves are experimental results. [After Hwang and Kao 1974.]

1972] assuming an average carrier distribution over the complete electrode area. This value of T_b should correspond to our measured peak value of B_T because we measured only the total electroluminescent brightness over the complete electrode area. Since J_{z0} in the filament is much larger than $\overline{J_z}(r)$ at $r > 0$, we can expect that N_T inside the filament at $r = 0$ may be several orders of magnitude larger than the average value of N_T, and also T_b at $r = 0$ would be much larger than the average value of T_b. The results shown in Fig. 7.39 can be explained qualitatively on the basis of eqs. (7.210) and (7.211) for $T > T_b$.

It is interesting to note that the overall electroluminescent intensity is proportional to current under high injection conditions, and this relation is practically independent of temperature for $T < T_b$. Typical results for anthracene for the temperature range from 128°K to 323°K are shown in Fig. 7.40.

In general, the spectral distribution of the electroluminescence is independent of electrode material, but does depend on crystal preparation and temperature. Figure 7.41 shows the electroluminescence and optically excited fluorescence spectra of anthracene crystals melt-grown under an argon atmosphere. It can be seen that there is a considerable portion of defect emission in the electroluminescence spectra, particularly at low temperatures. For the anthracene doped with tetracene the electroluminescence will be a mixture of anthracene and tetracene emission, the intensity ratio depending

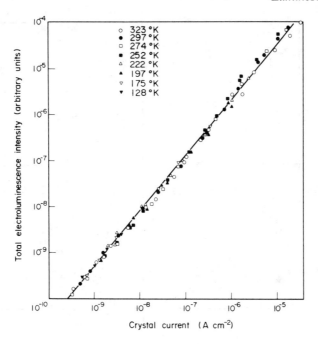

Fɪɢ. 7.40. Electroluminescent intensity as a function of current passing through anthracene crystals of thickness of 2 mm at various temperatures. [After Williams and Schadt 1970.]

upon the concentration of the dopant [Dresner and Goodman 1970, Kawabe *et al.* 1971, Schwob *et al.* 1970, 1971, Zschokke-Gränacher *et al.* 1967]. By adding a suitable amount of tetracene, the colour of the electroluminescence of anthracene can be controlled from blue to green. The anthracene–tetracene system involves the transfer of excitation energy from the host crystal to the guest molecules. At room temperature the fluorescence efficiency is approximately 10^5 times higher for the guest (tetracene) than for the host (anthracene).

$$\eta_T/\eta_A \simeq 10^5 C_T \qquad (7.212)$$

where η_T and η_A are, respectively, the quantum yield for the fluorescence of tetracene and anthracene, and C_T is the mole fraction of tetracene in anthracene. If $C_T = 10^{-5}$ then both partners have the same intensity of fluorescence. Fluorescence spectra for $C_T = 5 \times 10^{-6}$ and electroluminescence spectra for $C_T = 10^{-3}$ are shown in Fig. 7.42.

Some experimental results on electroluminescence in anthracene are summarized in Table 7.4.

7.4.2. Electroluminescence under time-varying fields

(A) Pulsed electroluminescence

Pulsed electroluminescence has been observed by many investigators [Helfrich and

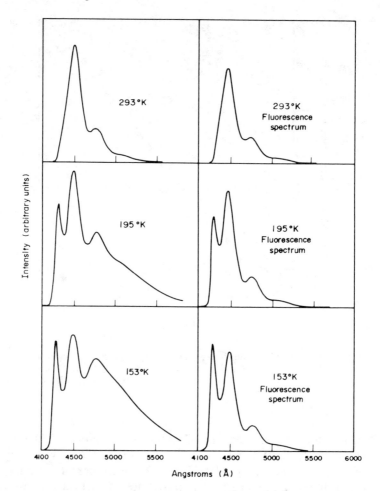

FIG. 7.41. Electroluminescence and optically excited fluorescence spectra of anthracene crystals melt-grown under an argon atmosphere. Electroluminescence (on the left side) were all taken at a current level of $10^{-6}\,\mathrm{A\,cm^{-2}}$. [After Williams and Schadt 1970.]

Schneider 1966, Williams and Schadt 1970, Zvyagintsev 1970]. Electroluminescence consists of two components, namely, the fast component, generally referred to as the "prompt electroluminescence", and the slow component, generally referred to as the "delayed electroluminescence". The "fast" light transient marks the time when the two leading carrier fronts meet in the specimen, while the "slow" light transient involves triplet–triplet annihilation and electron and hole detrapping processes. The time dependence of the "fast" current enables the determination of the carrier recombination rate constant, and the "slow" current transient can be used to monitor any change of exciton generation rate which may arise from a decrease of mobile carriers due to trapping.

The steady state (or d.c.) electroluminescence spectrum is independent of the electrode material, but depends on temperature and crystal preparation. Under pulse

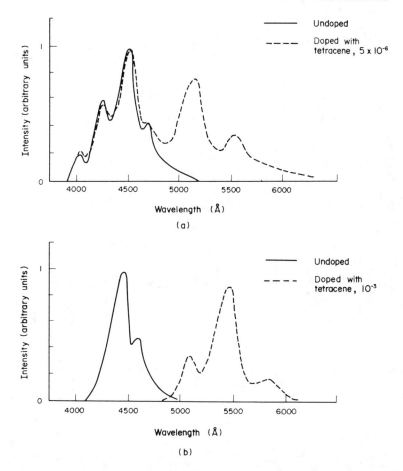

F IG. 7.42. The spectra of (a) fluorescence, and (b) electroluminescence for undoped
anthracene and tetracene-doped anthracene crystals. [After Zschokke-
Gränacher, Schwob, and Baldinger 1967.]

voltage conditions, however, a carrier injection mechanism of a normally blocking
contact becomes apparent. Electroluminescence first appears at the time, after the
application of a voltage pulse, corresponding to the transit time of $d^2/(\mu_n + \mu_p)V$ for the
cases with both electron-injecting and hole-injecting contacts. But for the cases with
one electron-injecting contact and one normally hole-blocking contact, then electroluminescence appears at the time corresponding to the transit time of electrons
alone; that is, on the arrival of the electron space charge front at the anode to enhance
the hole injection. For the cases with two carrier-injecting contacts the electroluminescent intensity is proportional to the current irrespective of the current level,
while for the cases with only one carrier-injecting contact and one normally blocking
contact the relationship between the electroluminescent intensity and the
current depends on the current level as shown in Fig. 7.43 [Williams and Schadt
1970].

TABLE 7.4. Summary of some previous experimental results of electroluminescence in anthracene crystals

Specimen preparation	Specimen thickness	Form of contacts	Light emission peak wavelength (Å)	Threshold field (V cm^{-1})	Current–voltage dependence $J \propto V^n$	Temperature range (°K)	Remarks	References
Single crystals (solution grown)	10-20 μ	Liquid contacts, single injection	4210 4440	4×10^5	$n > 10$	293	D.C. EL Pulsed EL	A, B
Single crystals (melt grown)	1-5 mm	Liquid contacts, double injection	4300	Unspecified	$n > 2$	293	A.C. EL D.C. EL Pulsed EL	C, D
Single crystals doped with tetracene	1.5 mm	Liquid contacts, double injection	4230 4450 4740 4792 5300	Unspecified	$n > 2$	293	A.C. EL D.C. EL	E
Films (vacuum evaporated)	10 μ	Solid contacts, double injection	4100–5400	Unspecified	$n > 3$	293	D.C. EL	F, G
Single crystals (zone refined)	2 mm	Solid contacts, double injection	4500	350	$n > 2$	100–350	D.C. EL	H
Single crystals doped with tetracene	Unspecified	Liquid contacts, double injection	4340 5480		$n > 3$	293	D.C. EL	I, J
Single crystals doped with tetracene	Unspecified	Solid contacts, double injection	4250 4950 5250	7000	$n > 6$	293	D.C. EL	K
Single crystals (zone refined)	2.2-5 mm	Solid contacts, double injection	4300	6×10^3	$6 < n < 12$	293	D.C. EL	L
Single crystals	50 μ	Solid contacts, double injection	4450	1×10^5		293	A.C. EL	M

A, Pope et al. [1963].
B, Sano et al. [1965].
C, Helfrich and Schneider [1965].
D, Helfrich and Schneider [1966].
E, Zschokke-Gränacher et al. [1967].
F, Dresner [1969].
G, Dresner and Goodman [1970].
H, Williams and Schadt [1970].
I, Schwob et al. [1970].
J, Schwob and Zschokke-Gränacher [1971].
K, Kawabe et al. [1971].
L, Williams et al. [1972].
M, Brodzeli et al. [1970].

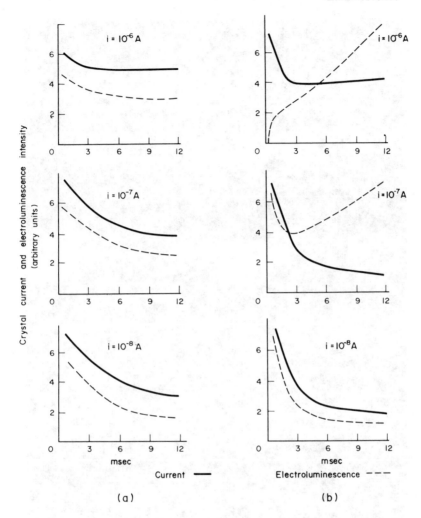

FIG. 7.43.   Time dependence of the electroluminescent intensity and the current passing through anthracene crystals provided with (a) one electron-injecting and one hole-injecting contact, and (b) one electron-injecting and one silver paste contact. [After Williams and Schadt 1970.]

(B) Sinusoidal a.c. electroluminescence

Electroluminescence under sinusoidal a.c. fields has been studied by several investigators [Brodzeli *et al.* 1970, Kunkel and Kao 1976]. In the following is summarized mainly the recent work of Kunkel and Kao [1976].

Using pure undoped anthracene crystals of thickness of about 0.4 mm, a sodium electrode as the electron-injecting contact and a silver electrode as the hole-injecting contact (in fact a silver electrode is a hole-blocking contact at low fields, and can become a hole-injecting contact only at high fields), Kunkel and Kao have studied the electroluminescence under continuously sinusoidal a.c. and half-wave rectified

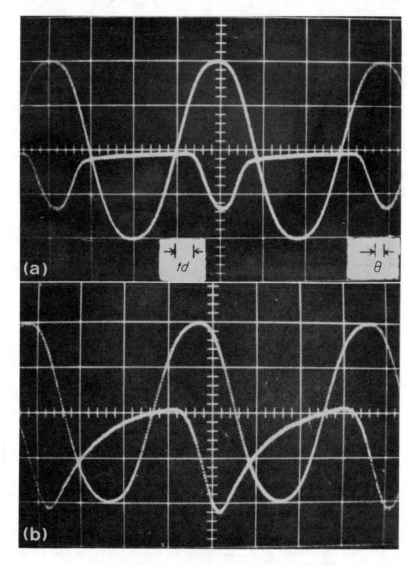

F<small>IG</small>. 7.44. Oscillograms illustrating the wave form of a.c. voltages and that of corresponding light output which appears only when the silver electrode is at the positive
voltage half-cycles. t_d is the time delay and θ is the phase shift. (a) Frequency:
500 Hz. Temperature: 40°C. Vert. scale: voltage, 400 V/div; brightness,
0.5 V/div. Horiz. scale: time, 0.5 msec/div. (b) Frequency: 5000 Hz.
Temperature: 20°C. Vert. scale: voltage, 400 V/div; brightness, 0.1 V/div. Horiz.
scale: time, 0.05 msec/div. [After Kunkel and Kao 1976.]

sinusoidal a.c. fields at various temperatures. The typical wave forms of a.c. voltage
applied across the specimen and the electroluminescence produced in it are shown in
Fig. 7.44. It can be seen that electroluminescence appears only when the silver electrode
is at the positive voltage half-cycles, indicating that neither does the sodium electrode
inject holes, nor does the silver electrode inject electrons.

The peak brightness as a function of frequency for sinusoidal and half-wave rectified sinusoidal a.c. fields are shown in Fig. 7.45. Both decrease monotonically with increasing frequency. A similar feature is also found for integrated light output as a function of frequency as shown in Fig. 7.46 although the rate of decrease with increasing frequency for the peak brightness is much greater than that for the integrated light output. Both the peak brightness and the integrated light output at a given peak to zero voltage under a sinusoidal a.c. field are much smaller than those

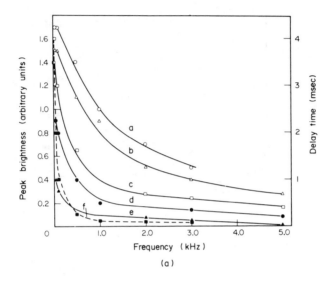

(a)

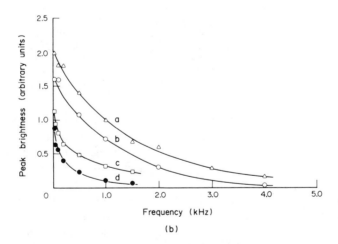

(b)

Fɪɢ. 7.45. (a) Peak electroluminescent brightness (*a–e*) and delay time (*f*) as functions of frequency at an applied continuously sinusoidal a.c. voltage of 1600 V peak to peak (or 800 V peak to zero) and at various temperatures. *a*, 20°C; *b*, 30°C; *c*, 40°C; *d*, 0°C; *e*, −20°C; and *f*, 40°C. (b) Peak electroluminescent brightness as a function of frequency at an applied half-wave rectified a.c. voltage of 800 V peak and at various temperatures. *a*, 30°C; *b*, 40°C; *c*, 20°C; and *d*, 0°C. [After Kunkel and Kao 1976.]

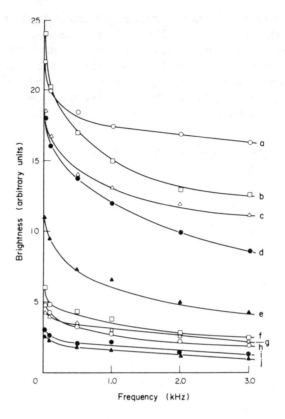

Brightness (arbitrary units)

Frequency (kHz)

F IG. 7.46. Integrated light output as a function of frequency
and temperature at an applied sinusoidal a.c.
voltage of 1600 V peak to peak ($f-j$) and at an
applied half-wave rectified sinusoidal a.c. voltage
of 800 V peak ($a-e$). a, 31°C; b, 38°C; c, 19°C; d,
47°C; e, 56°C; f, 15°C; g, 31°C; h, 7°C; i, 0°C; and j,
56°C. [After Kunkel and Kao 1976.]

under a half-wave rectified sinusoidal a.c. field at any given frequency and temperature,
although in both cases electroluminescence appears only on the one half-wave in which
the sodium electrode is at the negative and the silver electrode at the positive polarity.
All these results are consistent for applied voltages up to 2400 V peak to peak for a.c.
and up to 1200 V peak for half-wave rectified a.c. voltages. The temperature
dependence of electroluminescent brightness is given in Fig. 7.47. They show that for
both sinusoidal a.c. and half-wave rectified sinusoidal a.c. voltages the brightness
increases with increasing temperature, reaches a peak at a certain critical temperature,
and then decreases with increasing temperature in a manner similar to that for d.c.
electroluminescence reported earlier by Hwang and Kao [1973, 1974], but the critical
temperature for the peak to occur is frequency dependent for half-wave rectified
sinusoidal a.c. voltages, and is practically independent of frequency for sinusoidal a.c.
voltages.

In Fig. 7.44 it can be seen that there is a time delay (t_d) between the time when the

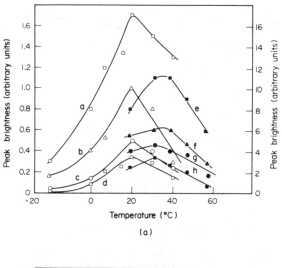

(a)

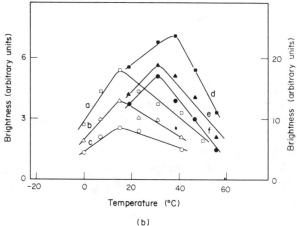

(b)

FIG. 7.47. (a) Peak electroluminescent brightness as a function of temperature at various
frequencies. Applied voltage: 1600 V peak to peak sinusoidal a.c. voltage of
frequencies: *a*, 100 Hz; *b*, 1000 Hz; *c*, 3000 Hz; *d*, 5000 Hz. Applied voltage:
800 V peak half-wave rectified sinusoidal a.c. voltage of frequencies; *e*, 20 Hz; *f*,
500 Hz; *g*, 1000 Hz; *h*, 2000 Hz. (b) Integrated light output as a function of
temperature at various frequencies. Applied voltage: 1600 V peak sinusoidal
a.c. voltage of frequencies; *a*, 20 Hz; *b*, 500 Hz; *c*, 2000 Hz. Applied voltage:
800 V peak half-wave rectified sinusoidal a.c. voltage of frequencies; *d*, 20 Hz; *e*,
500 Hz; *f*, 2000 Hz. [After Kunkel and Kao 1976.]

voltage is applied and the time when the electroluminescence appears. The frequency
dependence of the delay time is also shown in Fig. 7.45 for sinusoidal a.c. voltages. A
similar frequency dependence of the delay time has also been observed for half-wave
rectified sinusoidal a.c. voltages. By measuring the actual voltage when the electroluminescence appears (turn-on voltage) and then when it disappears (turn-off
voltage) during the half-wave rectified cycle, they have also plotted "turn-on" voltage

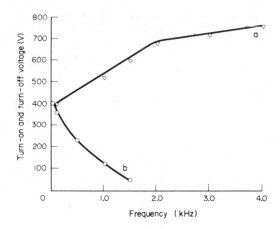

FIG. 7.48. Turn-on voltage (a) and turn-off voltage (b) as
functions of frequency at an applied half-wave
rectified sinusoidal a.c. voltage of 800 V peak and
at the temperature of 40°C. [After Kunkel and
Kao 1976.]

and "turn-off" voltage as functions of frequency, and they are shown in Fig. 7.48.

In Fig. 7.44 it can also be seen that there is a phase shift θ between the peak of applied voltage and the peak of electroluminescent brightness. This phase shift increases with increasing frequency; it is not noticeable at 20 Hz, but becomes significant at 5000 Hz. The leading portion of the wave form of electroluminescence is similar in shape to that of applied voltage for all frequencies, but the tailing portion is different. There is a significant exponential tail developed on the trailing edge at high frequencies as shown in Fig. 7.44(b). The "turn off" is defined as the voltage at the point where the electroluminescence wave form becomes exponential since beyond this point the light output results predominantly from delayed fluorescence.

In anthracene crystals the electron and hole mobilities are small and the carrier recombination rate is large. This implies that the recombination or luminescence zone is narrow in this material. If it is assumed that the sodium electrode (cathode) provides a good electron-injecting contact and the silver electrode (anode) is practically a blocking contact for hole injection, then electroluminescence would appear only when the electrons arriving at the anode from the injecting cathode can build up a negative space charge density sufficiently high to modify the potential barrier and hence to enhance hole injection at the anode. The delay time required for the onset of electroluminescence can be thought of as the time required for the electrons to travel from the cathode to the anode (transit time) and the time taken to build up a space charge density sufficient to modify the barrier at the anode. Thus the delay time would be expected to depend on the amplitude of the applied voltage, the frequency, and its direction (unidirectional or alternating). Hwang and Kao [1973] have found that the threshold voltage V_{th} for the onset of d.c. electroluminescence is affected by the on-and-off time. If the voltage source is switched off and then immediately switched on again, V_{th} is reduced, but if the switch-off time is increased, V_{th} remains practically unchanged. This suggests that there is a relaxation time involved in the build-up of negative space charge near the anode, possibly due to electron traps. Under a sinusoidal a.c. field, if the positive half-cycle helps to build up the space charge and hence to produce electroluminescence, the negative half-cycle tends to drive the space charge off. But some trapped electrons are still in traps to contribute to the enhancement of hole injection from the anode. Thus the higher the frequency the more are electrons trapped in the traps. This explains why the delay time decreases with increasing frequency as shown in Fig. 7.45, and why the "turn-on" and "turn-off" voltages are frequency dependent as shown in Fig. 7.48. As the frequency is further increased, the delay time gradually saturates at a value of 40 μsec, which is closely equal to the electron transit time. The electron transit time measured by an 800 V rectangular pulse is about 30 μsec [Kunkel and Kao 1976].

Based on this model, the electroluminescence can be observed only when one-half of a period of the applied voltage is greater than the transit time t_t. This implies that the critical frequency f_c beyond which the electroluminescence disappears can be expressed as

$$f_c \simeq \frac{1}{2t_t} \qquad (7.213)$$

By assuming that the time taken to build up the space charge and to modify the potential barrier is small compared to the electron transit time, and that the transit time across the specimen of thickness of 0.4 mm at an 800 V peak-to-zero voltage is 30 μsec, Kunkel and Kao [1976] have estimated f_c to be 12 kHz. This agrees fairly well with the experimental result as electroluminescence was still detectable at frequencies as high as 10 kHz though the signal was very small. As the frequency increases, the number of electrons injected and transported to the anode per cycle decreases. Thus the peak electroluminescent brightness and the light output per cycle decreases with increasing frequency. The integrated light output depends upon both the light output per cycle and the repetition rate, and therefore it decreases more slowly with frequency than the peak brightness. In the case of a half-wave rectified a.c. excitation the negative space charge is not driven off after the elapse of each positive half-cycle, and consequently $f_c > (1/2t_t)$ and both the peak brightness and the integrated light output are much higher than those excited by a sinusoidal a.c. field.

The exponential tail in the electroluminescence wave form as shown in Fig. 7.44 may be caused by the delayed fluorescence generated by triplet–triplet annihilation. The lifetime of the triplet exciton is of the order of milliseconds [Williams and Schadt 1970] and thus after the excitation ceases the fluorescence does not cease immediately but decays exponentially with a time constant related to the triplet lifetime. In fact the exponential tail exists on the electroluminescence wave form at all frequencies but becomes much more pronounced at high frequencies [Kunkel and Kao 1976].

The temperature dependence phenomenon in d.c. electroluminescence has been explained [Hwang and Kao 1974] in terms of the exciton–carrier interactions and the detrapping of trapped carriers. It is expected that the similar temperature dependence phenomenon in a.c. electroluminescence shown in Fig. 7.47 is also caused by these two processes. It is possible that the change in temperature will affect the trapped carrier concentration much more than the carrier injection from the electrodes. The increase in free carrier density with increasing temperature results in the initial increase in brightness. The space charge formed near the anode, and hence the onset of electroluminescence, are mainly due to the free carriers produced by the following processes: (a) the electron injection from the cathode, and (b) the detrapping process. The peak in the brightness–temperature curves shown in Fig. 7.47 occurs when the efficiency of both processes (a) and (b) reaches an optimum point. In the case of a half-wave rectified sinusoidal a.c. excitation, process (a) is more dominant than process (b) at low frequencies; thus the peak in the brightness–temperature curves is shifted to higher temperatures for lower frequencies, while in the case of a sinusoidal a.c. excitation, free carriers will be driven off in every alternate half-cycle, so that process (a) does not dominate process (b) and therefore the critical temperature for such a peak to occur is not sensitive to frequency. At high temperatures the brightness decreases with increasing temperature because the process of exciton–carrier interactions becomes predominant [Hwang and Kao 1974, Kunkel and Kao 1976].

7.4.3. Electroluminescence under magnetic fields

In the initial state of electroluminescence the electron–hole pairs are separated and uncorrelated, but in the final state they are Frenkel excitons, that means each electron–hole pair is on one molecule. Therefore an intermediate Wannier or charge transfer state (an electron–hole pair on nearest-neighbour sites) must exist [Morris and Silver 1969]. Electroluminescence modulation by a magnetic field is a sensitive technique for detecting the presence of such charge transfer excitons.

An electron and a hole tend to come close together under their Coulombic field, and when they are on nearest-neighbour molecules they will form a charge transfer exciton [CT]. Such a [CT] can have either a singlet $[CT]_S$ or a triplet $[CT]_T$ character, and may decay into an excited singlet or an excited triplet exciton, which then decays to the ground state emitting one photon or undergoing a non-radiative transition as shown in Fig. 7.49(a). Of course, triplet excitons may also undergo triplet–triplet annihilation to form either one singlet or one triplet exciton via a [CT] intermediate as shown in Fig. 7.49(b). The fission of a [CT] exciton into two triplet excitons as shown in Fig. 7.49(c) can also occur in organic crystals. From Fig. 7.49(c) it can be seen that when the applied magnetic field at a certain orientation decreases (γ_1 and γ_{-1}), the delayed electroluminescent intensity will be decreased but the prompt electroluminescent intensity will be increased because of the competition between the two decay channels (γ_{-1} and K_3) of the $[CT]_S$ excitons. The effects of magnetic field strength and orientation on delayed and prompt electroluminescent intensities are shown in Fig. 7.50. The prompt electroluminescence exhibits the same magnetic field strength and orientation dependence as the delayed electroluminescence but differs in sign and magnitude [Schwob and Williams 1972, 1973]. A similar phenomenon has also been observed by several investigators [Johnson and Merrifield 1970, Groff *et al.* 1970] (cf. Figs. 7.32 and 7.33). The intensities of the prompt and delayed electroluminescence B and B^d, respectively, with the magnetic field effects can be compared through the following relation [Schwob and Williams 1973]:

$$(B - B_0)/B_0 = -C(B^d - B_0^d)/B_0^d + \Delta \qquad (7.214)$$

where the subscript 0 denotes the intensities at zero magnetic field, C is a constant depending on the conditions of carrier injecting contacts, and Δ is the small isotropic

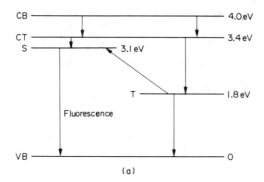

(a)

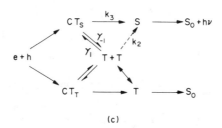

(b)

(c)

FIG. 7.49. (a) Energy levels of an anthracene crystal and decay channels
for recombining carriers. *CB*, conduction band; *CT*, charge
transfer exciton state; *VB*, valence band; *S*, singlet exciton
state; *T*, triplet exciton state. (b) Schematic illustration of
T–T annihilation. *k*'s represent the rate constants involved.
(c) Schematic illustration of the decay channels for carrier
recombination. *k*'s and *γ*'s represent the rate constants
involved. [After Schwob and Williams 1973.]

modulation depending on the imperfections of the crystals. The relative change of the
delayed electroluminescence in the presence of a magnetic field may be written as

$$(B^d - B_0^d) = (\gamma_1 - \gamma_{1,0})/\gamma_{1,0} \tag{7.215}$$

and the relative change of the prompt electroluminescent intensity as

$$(B - B_0)/B_0 = -(\gamma_1 - \gamma_{-1,0})/k_3 \tag{7.216}$$

where γ's and k's are the rate constants involved in the decay channels as shown in Fig.

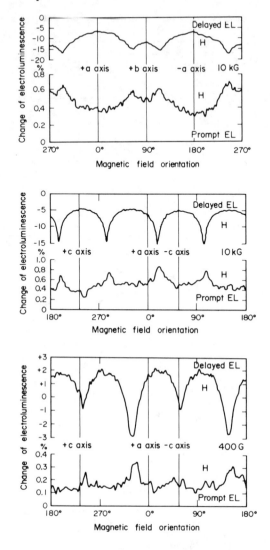

F<small>IG.</small> 7.50. Variation of the delayed and prompt electroluminescent intensities (in per cent change) with magnetic
field rotating in different plane of an anthracene
crystal. *Top*: 10 kG in *ab* plane; middle: 10 kG in *ac*
plane; bottom: 400 G in *ac* plane. [After Schwob and
Williams 1973.]

7.49. Schwob and Williams [1973] have also observed that the magnetic field effects increase as temperature increases in the temperature range of 20–80°C. Their results lead to the following conclusions: (1) Fission of a charge transfer exciton of singlet character $[CT]_S$ causes an increase in prompt electroluminescent intensity with an increase in magnetic field. (2) The rate constant for the $[CT]_S$ fission process is about $10^7 \ \text{sec}^{-1}$ at room temperature and increases with increasing temperature, giving an activation energy of 0.2 eV. (3) The lifetime of the charge transfer excitons in anthracene is about 10^{-9} sec and the energy of the [CT] state is about 3.45 eV above the singlet ground state. (4) Fission of a charge transfer exciton with overall triplet spin

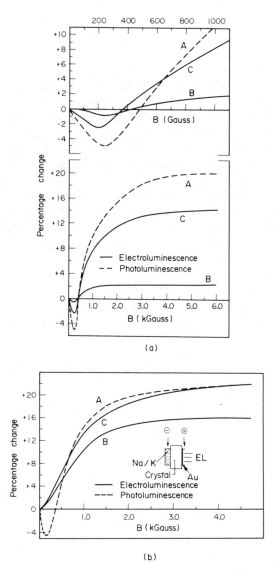

(a)

(b)

FIG. 7.51. (a) Relative change of electroluminescent and photoluminescent intensities as functions of magnetic field strength in a high field resonance direction with a low delayed component of EL (electroluminescence). Curve *B* was measured for the red edge and curve *C* for the short wavelength and maximum emission. (b) The same as (a) but with a large delayed component of EL. Curve *B* was measured for the red edge and curve *C* for the total EL. Photoluminescence was obtained at low excitation levels ($\lambda_{ex} = 366$ nm). There is no difference in the change of photoluminescent intensity measured for the red edge or for the short wavelengths as shown in curve *A*. Tetracene crystal thickness: 118 μm. [After Kalinowski and Godlewski 1975.]

will not result in a detectable magnetic field effect. (5) In anthracene charge transfer excitons are an important intermediate in the production of singlet excitons from the annihilation of two charge carriers—these charge transfer excitons may then undergo fission into two triplet excitons.

The triplet exciton–charge carrier interactions also influence the magnetic field effects on electroluminescence. The presence of charge carriers tends to quench triplet excitons. However, the magnetic field tends to decrease the triplet exciton–charge carrier interaction rate constant. This is demonstrated in Fig. 7.51; the low field dip in the magnetic field dependence curve decreases as the delayed component of the electroluminescence is increased, and it disappears when the delayed component becomes the dominant mechanism of electroluminescence as shown in Fig. 7.51(b). There is a similarity between the magnetic field dependence for the red edge electroluminescence in pure tetracene crystals [curve B in Fig. 7.51(a)] and that for total electroluminescence in pentacene-doped tetracene crystals shown in Fig. 7.52, indicating that at the red edge the electroluminescence, at least partly, originates from trapped states (S_{1G}), which can fission into two inequivalent triplet excitons (T_1, T_{1G})—heterofission [Geacintov *et al.* 1971, Kalinowski and Godlewski 1975]. The variation of total electroluminescence with magnetic field is determined by the interplay of three effects: (a) the change of the rate for free singlet exciton fission, (b) the change of the rate for heterofission, and (c) the decrease of the rate of triplet exciton–charge carrier interaction due to the presence of the magnetic fields. At low excitation levels, the prompt fluorescence is predominant and its magnetic field dependence is mainly due to (a). The delayed component of electroluminescence is determined, to a large extent, by the nature and concentration of traps, and thus its magnetic field dependence is mainly controlled by (b) and (c).

It should also be noted that the change of triplet exciton concentration in the magnetic field does not cause any observable change in the prompt electrolumi-

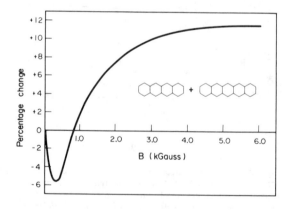

Fig. 7.52. The change in electroluminescent intensity as a function of magnetic field strength for a tetracene crystal doped with pentacene in one of the high field resonance directions. Crystal thickness $d = 36\,\mu m$; pentacene concentration: 2×10^3 ppm. [After Kalinowski and Godlewski 1975.]

nescence. In a 10^{-5} mol/mol tetracene-doped anthracene crystal, the triplet lifetime is reduced by about two orders of magnitude as compared to a pure anthracene crystal. Of course, in this case no delayed electroluminescence could be detected, but the magnetic field effects on the prompt electroluminescence was exactly that shown in Fig. 7.50 [Schwob and Williams 1973]. It should also be noted that the effect of magnetic field on the emission originated from the triplet–triplet annihilation is negligible [Groff *et al.* 1970, Kalinowski and Godlewski 1975].

Bibliography

BOOKS

ABELES, F. (ed.): *Optical Properties of Solids*, North-Holland (Amsterdam and New York) 1972.

BACKUS, C. E. (ed.): *Solar Cells*, IEEE Press, Institute of Electrical and Electronics Engineers Inc. (New York) 1976.

BECKER, R. S.: *Theory and Interpretation of Fluorescence and Phosphorescence*, Wiley Interscience (New York) 1969.

BEER, A. C.: *Galvanomagnetic Effects in Semiconductors*, Academic Press (New York) 1963.

BERGH, A. A. and DEAN, P. J.: *Light Emitting Diodes*, Clarendon Press (Oxford) 1976.

BIRKS, J. B.: *Scintillation Counters*, McGraw-Hill (New York) 1953.

BIRKS, J. B.: *Photophysics of Aromatic Molecules*, Wiley Interscience (New York) 1970.

BIRKS, J. B. (ed.): *Organic Molecular Photophysics*, Vols. 1 and 2, Wiley (New York) 1973 and 1975.

BLAKEMORE, J. S.: *Semiconductor Statistics*, Pergamon Press (London) 1962.

BOGUSLAVSKII, L. I. and VANNIKOV, A. V.: *Organic Semiconductors and Biopolymers*, Plenum Press (New York) 1970.

BOLTAKS, B. I.: *Diffusion in Semiconductors*, Academic Press (New York) 1963.

BRILLOUIN, L.: *Horizons in Biochemistry*, Academic Press (New York) 1962.

BROPHY, J. J. and BUTLREY, J. W.: *Organic Semiconductor*, Macmillan (London) 1962.

BUBE, R. H.: *Photoconductivity of Solids*, Wiley (New York) 1960.

BUBE, R. H.: *Electronic Properties of Crystalline Solids—An Introduction to Fundamentals*, Academic Press (New York) 1974.

BULLMANN, W.: *Crystal Defects and Crystalline Interfaces*, Springer-Verlag (Berlin, New York) 1970.

BURSTEIN, E. and LUNDQVIST, S.: *Tunneling Phenomena in Solids*, Plenum Press (New York) 1969.

CHOPRA, K. L.: *Thin Film Phenomena*, McGraw-Hill (New York) 1969.

COHEN, E. and THIRRING, W. (eds.): *The Boltzmann Equation—Theory and Application*, Springer-Verlag (New York) 1973.

CONWAY, B. E.: *Theory and Principles of Electrode Processes*, Ronald Press (New York) 1965.

CONWELL, E. M.: *High Field Transport in Semiconductors*, Academic Press (New York) 1967.

COTTRELL, A. H.: *Theory of Crystal Dislocations*, Blackie & Son (London) 1962.

COULSON, C. A.: *Valence*, Oxford University Press (Oxford) 1961.

CRAIG, D. P. and WALMSLEY, S. H.: *Excitons in Molecular Crystals—Theory and Applications*, Benjamin (New York) 1968.

CRAWFORD, JR., J. H. and SLIFKIN, L. M.: *Point Defects in Solids*: Vol. 1, *General and Ionic Crystals*; Vol. 2, *Semiconductors and Molecular Crystals*; Vol. 3, *Defects in Metals*, Plenum Press (New York) 1975.

DANDEL, R., LE FEBURE, R., and MOSER, C.: *Quantum Chemistry*, Wiley Interscience (New York) 1959.

DAVYDOV, A. S.: *Theory of Molecular Excitons*, Nauka, Moscow (1968), in Russian (English translation by S. B. Dresner, Plenum Press, New York) 1971.

DEVREESE, J. T.: *Polarons in Ionic Crystals and Polar Semiconductors*, Antwerp Advanced Study Institute 1971 on Fröhlich Polarons and Electron–Phonon Interactions in Polar Semiconductors, North-Holland (New York) 1972.

DEXTER, D. L. and KNOX, R. S.: *Excitons*, Wiley Interscience (New York) 1965.

DUKE, C. B.: *Tunneling in Solids*, Solid Physics Supplement 10, Academic Press (New York) 1969.

ELLIOTT, R. J. and GIBSON, N. F.: *Solid State Physics*, Macmillan (London) 1974.

FOX, D., LABES, M. M., and WEISSBERG, A. (eds.): *Physics and Chemistry of the Organic Solid State*, Vol. 1, Wiley Interscience (New York) 1961.

FOX, D., LABES, M. M., and WEISSBERG, A. (eds.): *Physics and Chemistry of the Organic Solid State*, Vol. 2, Wiley Interscience (New York) 1965.

FOX, D., LABES, M. M., and WEISSBERG, A. (eds.): *Physics and Chemistry of the Organic Solid State*, Vol. 3, Wiley Interscience (New York) 1967.

FRANKL, D. R.: *Electrical Properties of Semiconductor Surfaces*, Pergamon Press (New York) 1967.

GOOCH, C. H.: *Injection Electroluminescent Devices*, Wiley (New York) 1973.

GOSSICK, B. R.: *Potential Barriers in Semiconductors*, Academic Press (New York) 1964

GRÜNBAUM, F. A.: *The Boltzmann Equation*, New York University (New York) 1972.

GUTMANN, F. and LYONS, L. E.: *Organic Semiconductors*, Wiley (New York) 1967.

HANNAY, N. B. (ed.): *Semiconductors*, Reinhold (New York) 1959.

HARPER, W. R.: *Contact and Frictional Electrification*, Oxford University Press (Oxford) 1967.

HARRIS, S.: *An Introduction to the Theory of the Boltzmann Equation*, Holt, Rinehart & Winston (New York) 1971.

HEAVENS, O. S.: *Optical Properties of Thin Solid Films*, Dover Publications (New York) 1965.

HENISCH, H. K.: *Rectifying Semiconductor Contacts*, Oxford University Press (Oxford) 1957.

HENISCH, H. K.: *Electroluminescence*, Macmillan (London) 1962.

HOCHTIN, A. R. (ed.): *Kinetic Processes in Gases and Plasmas*, Academic Press (New York) 1969.

HODGKIN, A.: *Conduction of Nervous Impulse*, Thomas (London) 1964.

HOVEL, H. J.: *Solar Cells, Semiconductors and Semimetals*, Vol. 11 (eds. R. K. Willardson and A. C. Beer), Academic Press (New York) 1975.

HUGHES, A. L. and DU BRIDGE, L. A.: *Photoelectric Phenomena*, McGraw-Hill (New York) 1932.

IOFFE, A. F.: *Semiconductor Elements and Thermoelectric Cooling*, Infosearch (London) 1957.

IOFFE, A. F.: *Physics of Semiconductors*, Academic Press (New York) 1960.

IVEY, H. F.: *Electroluminescence and Related Effects*, Supplement 1, *Advances in Electronics and Electron Physics* (ed. L. Marton), Academic Press (New York) 1963.

KALLMANN, H. and SILVER, M.: *Symposium on Electrical Conductivity in Organic Solids*, Wiley Interscience (New York) 1961.

KALLMANN, H. and SPRUCH, G.: *Luminescence of Organic and Inorganic Materials*, Wiley (New York) 1962.

KIRKWOOD, J. G.: *Selected Topics in Statistical Mechanics*, Gordon & Breach (New York) 1967.

KLUG, H. P. and ALEXANDER, I. E.: *X-ray Diffraction Procedures*, Wiley (New York) 1954.

KNOX, R. S.: *Theory of Excitons, Solid State Physics*, Supplement 5, Academic Press (New York) 1963.

KORSUNSKII, M. I.: *Anomalous Photoconductivity*, Jerusalem, Israel, Israel Program for Sci. Translations, 1973.

KUPER, C. G. and WHITEFIELD, G. D.: *Polarons and Excitons*, Oliver & Boyd (London) 1962.

LAMB, D. R.: *Electrical Conduction Mechanisms in Thin Insulating Films*, Methuen (London) 1967.

LAMPERT, M. A. and MARK, P.: *Current Injection in Solids*, Academic Press (New York) 1970.

LANDSBERG, P. T.: *Solid State Theory—Methods and Applications*, Wiley Interscience (New York) 1969.

LE COMBER, P. G. and MORT, J.: *Electronic and Structural Properties of Amorphous Semiconductors*, Academic Press (New York) 1973.

LEIBOVIC, K. N.: *Nervous System Theory—An Introductory Study*, Academic Press (New York) 1972.

LEVERENZ, H. W.: *An Introduction to Luminescence of Solids*, Dover Publications (New York) 1968.

LIM, E. C.: *Excited States*, Vols. 1 and 2, Academic Press (New York) 1974.

MANY, A., GOLDSTEIN, Y., and GROVER, N. B.: *Semiconductor Surfaces*, North-Holland (Amsterdam) 1965.

MASUDA, K. and SILVER, M.: *Energy and Charge Transfer in Organic Semiconductors*, Plenum Press (New York) 1974.

MATARE, H. F.: *Defect Electronics in Semiconductors*, Wiley (New York) 1971.

MAZO, R. M.: *Statistical Mechanics of Transport Processes*, Pergamon Press (London) 1967.

MEHL, W.: *Reactions of Molecules at Electrodes* (ed. N. S. Hush), Wiley (New York) 1971.

MEIER, H.: *Spectral Sensitization*, Focal Press (London and New York) 1968.

MEIER, H.: *Organic Semiconductors—Dark and Photoconductivity of Organic Solids*, Verlag-Chemie (Germany) 1974.

MILNES, A. G.: *Deep Impurities in Semiconductors*, Wiley (New York) 1973.

MILNES, A. G. and FEUCHT, D. L.: *Heterojunctions and Metal–Semiconductor Junctions*, Academic Press (New York) 1972.

MOORE, W. J.: *Seven Solid States*, Benjamin (New York) 1967.

MORT, J. and PAI, D. M.: *Photoconductivity and Related Phenomena*, Elsevier (Amsterdam) 1976.

MOSS, T. S.: *Optical Properties of Semiconductors*, Academic Press (New York) 1959.

MOSS, T. S., BURRELL, G. J., and ELLIS, B.: *Semiconductor Opt-electronics*, Butterworths (London) 1973.

MOTT, N. F. and DAVIS, E. A.: *Electronic Processes in Non-crystalline Materials*, Clarendon Press (Oxford) 1971.

MOTT, N. F. and GURNEY, R. W.: *Electronic Processes in Ionic Crystals*, Dover Publications (New York) 1940.

MURRELL, J. N., KETTLE, S. F. A., and TEDDER, J. M.: *Valence Theory*, Wiley (New York) 1965.

NAG, B. R.: *Theory of Electrical Transport in Semiconductors*, Pergamon Press (Oxford) 1972.

O'DWYER, J. J.: *The Theory of Electrical Conduction and Breakdown in Solid Dielectrics*, Clarendon Press (Oxford) 1973.

OKAMOTO, Y. and BRENNER, W.: *Organic Semiconductors*, Reinhold (New York) 1964.

PANKOVE, J. I.: *Optical Processes in Semiconductors*, Prentice-Hall (Englewood Cliffs) 1971.
PATTERSON, J. D.: *Introduction to the Theory of Solid State Physics*, Addison-Wesley (Massachusetts) 1971.
PEPPER, M. (hon. ed.): *Metal–Semiconductor Contacts*, Conference Series Number 22, Institute of Physics (London) 1974.
PICK, R. M.: *Computational Solid State Physics* (eds. F. Herman, N. W. Dalton, and T. R. Koehler), Plenum Press (New York) 1972.
PULLMANN, B. (ed.): *Quantum Mechanics of Molecular Conformations*, Wiley Interscience (New York) 1976.
READ, W. T.: *Dislocations in Crystals*, McGraw-Hill (New York) 1953.
REMBAUM, A. and LANDEL, R. (eds.): *Electrical Conduction Properties of Polymers, J. of Polym. Sci.*, Part C, Polymer Symposia No. 17, Wiley Interscience (New York) 1967.
REXER, E.: *Organische Halbleiter*, Akademie-Verlag (Berlin) 1966.
RHODERICK, E. H.: *Metal–Semiconductor Contacts*, Oxford University Press (Oxford) 1978.
ROSE, A.: *Concepts in Photoconductivity and Allied Problems*, Wiley Interscience (New York) 1963.
RYWKIN, S. M.: *Photoelectric Effects in Semiconductors*, Consultants Bureau (New York) 1964.
SALEM, L.: *The Molecular Orbital Theory of Conjugated Systems*, Benjamin (New York) 1966.
SCHRIEFFER, J. R.: *Theory of Superconductivity*, Benjamin (New York) 1964.
SEITZ, F.: *Modern Theory of Solids*, McGraw-Hill (New York) 1940.
SHEWMON, P. G.: *Diffusion in Solids*, McGraw-Hill (New York) 1963.
SHOCKLEY, W.: *Electrons and Holes in Semiconductors*, van Nostrand (New York) 1950.
SIMMONS, J. G.: *DC Conduction in Thin Films*, Mills & Boon (London) 1971.
SMITH, A. C., JANAK, J. F., and ALDER, R. B.: *Electronic Conduction in Solids*, McGraw-Hill (New York) 1967.
SMITH, R. A.: *Semiconductors*, Cambridge University Press (Cambridge) 1964.
SOLYMAR, L. and WALSH, D.: *Lectures on the Electrical Properties of Materials*, Clarendon Press (Oxford) 1970.
STEPANOV, B. I. and GRIBKOVSKII, V. P. (translated by Scripta Technica Limited and edited by S. Chomet): *Theory of Luminescence*, Iliffe Books Ltd. (London) 1968.
STREITWIESER, A.: *Molecular Orbital Theory for Organic Chemists*, Wiley (New York) 1961.
SZE, S. M.: *Physics of Semiconductor Devices*, Wiley (New York) 1969.
TAUC, J.: *Amorphous and Liquid Semiconductors*, Plenum Press (New York) 1974.
THORNTON, P. R.: *The Physics of Electroluminescent Devices*, E. and F. N. Spon Ltd. (London) 1967.
TOLANSKY, S.: *Surface Microtopography*, Wiley (New York) 1960.
TREDGOLD, R. H.: *Space Charge Conduction in Solids*, Elsevier (Amsterdam) 1966.
VAN DER ZIEL, A.: *Solid State Physical Electronics*, Prentice-Hall (Englewood Cliffs) 1968.
VIJH, A. K.: *Electrochemistry of Metals and Semiconductors—The Application of Solid State Science to Electrochemical Phenomenon*, Marcel Dekker (New York) 1973.
VON HIPPEL, A. R.: *Dielectrics and Waves*, Wiley (New York) 1954.
WILLIAMS, E. W.: *Solar Cells*, IEE Special Publication, Institution of Electrical Engineers (UK) 1978.
WILSON, A. H.: *The Theory of Metals*, Cambridge University Press (Cambridge) 1953.
WINCHELL, A. N.: *The Optical Properties of Organic Compounds*, Academic Press (New York) 1954.
WINDER, H. H.: *Intermetallic Semiconducting Films*, Pergamon Press (New York) 1970.
WUNDERLICH, B.: *Macromolecular Physics*, Vols. 1 and 2, Academic Press (New York) 1976.
ZIMAN, J. M.: *Electrons and Phonons*, Clarendon Press (Oxford) 1967.
ZWORYKIN, V. K. and RAMBERG, E. G.: *Photoelectricity and its Applications*, Wiley (New York) 1949.

ARTICLES

ABBATI, I. and BRAICOVICH, L.: "Concepts in ultraviolet photoelectron spectroscopy of solids", *Riv. Nuovo Cimento* **4**, 293–322 (1974).
ABBATI, I., BRAICOVICH, L., and DE MICHELIS, B.: "Notes on the method of ultraviolet photoelectron spectroscopy of solids", *Riv. Nuovo Cimento* **4**, 323–34 (1974).
ABDELMALIK, T. G. and COX, G. A.: "Charge transport in nickel phthalocyanine crystals. I: Ohmic and space-charge-limited currents in vacuum ambient", *J. Phys.* C **10**, 63–74 (1977).
ABDUVAKHIDOV, KH. M., VOLKOV, A. S., and GALAVANOV, V. V.: "Lifetime of non-equilibrium current carriers in indium antimonide at 78°K", *Soviet Phys.—Semicond.* **1**, 788–90 (1967).
ABKOWITZ, M., LAKATOS, A., and SCHER, H.: "AC conductivity and AC photoconductivity in amorphous and crystalline insulators", *Phys. Rev.* B **9**, 1813–22 (1974).
ADAMEC, V. and CALDERWOOD, J. H.: "Electrical conduction in dielectrics at high fields", *J. Phys.* D: *Appl. Phys. (London)* **8**, 551–60 (1975).
ADAMEC, V. and CALDERWOOD, J. H.: "Electric-field-enhanced conductivity in dielectrics", *J. Phys.* D: *Appl. Phys. (London)* **10**, L79–81 (1977).

ADIROVICH, E. I.: "Electric fields and currents in dielectrics", *Soviet Phys.—Solid State* **2**, 1282 (1961).

ADIROVICH, E. I., MIRZAMAKHMUDOV, T., RUBINOV, V. M., and YUABOV, YU. M.: "Narrow-band-gap semiconductor films which develop photovoltages up to 5000 V", *Soviet Phys.—Solid State* **7**, 2946–8 (1966).

ADIROVICH, E. I., ROSLYARKOVA, V. F. and YUABOV, YU. M.: "Anomalous photo-voltage effect in gallium phosphide", *Soviet Phys.—Semicond.* **2**, 848–9 (1969).

ADIROVICH, E. I., RUBINOV, V. M., and YUABOV, YU. M.: "Phenomenon of anomalously high photovoltages in GaAs films", *Soviet Phys.—Solid State* **6**, 2540–1 (1965).

ADLER, D.: "Insulating and metallic states in transition metal oxides", *Solid State Phys.* **21**, 1–113 (1968).

ADOLPH, J.: "Behavior of space-charge-limited current in anthracene single crystals", *Helv. Phys. Acta* **38**, 409–30 (1965).

ADOLPH, J.: "Diffusion length of triplet excitons in anthracene crystals", *J. Chem. Phys.* **46**, 4252–4 (1967).

ADOLPH, J., BALDINGER, E., CZAJA, W., and GRÄNACHER, I.: "Energy distribution of trapping centers in anthracene", *Phys. Lett.* **6**, 137–9 (1963).

ADOLPH, J., BALDINGER, E., and GRÄNACHER, I.: "Influence of the optical activation of trapped carriers on SCL current in anthracene", *Phys. Lett.* **8**, 224–5 (1964).

ADOLPH, J., BALDINGER, E., GRÄNACHER, I., and SCHADT, M.: "Temperature dependence of space-charge-limited current in anthracene crystals", *Z. Angew. Math. Phys.* **17**, 326–9 (1966).

ADOLPH, J. and WILLIAMS, D. F.: "Temperature dependence of singlet–triplet intersystem crossing in anthracene crystals", *J. Chem. Phys.* **46**, 4248–51 (1967).

ADVANI, G., GOTTLING, N., and OSMAN, T.: "Thin film triode research", *Proc. IRE* **50**, 1530–1 (1962).

AFANASEVA, G. K., ALEKSANDROV, K. S., and KITAIGORODSKII, A. I.: "Elastic constants of anthracene", *Phys. Stat. Sol.* **24**, K61–63 (1967).

AGARWAL, S. C.: "Analysis of the thermally stimulated capacitor-discharge method for characterizing localized states in amorphous semiconductors", *Phys. Rev.* **B 10**, 4340–9 (1974).

AGARWAL, S. C. and FRITZSCHE, H.: "Attempts to measure thermally stimulated currents in chalcogenide glasses", *Phys. Rev.* **B 10**, 4351–7 (1974).

AGARWAL, V. K.: "Breakdown conduction in thin dielectric films, a bibliographical survey", *Thin Solid Films* **24**, 55–70 (1974).

AGRANOVICH, V. M., FAIDYSH, A. N., and KUCHEROV, I. YA.: "The length of the diffusion displacement of excitons in anthracene crystals", *Ukr. Fiz. Zh.* **2**, 61–67 (1957).

AGRANOVICH, V. M., and KONOBEEV, YU. V.: "On the exciton free path in a molecular crystal", *Opt. Spectrosc.* **6**, 155–9 (1959).

AGRANOVICH, V. M., MAL'SHUKOV, A. G., and MEKHTIEV, M. A.: "Surface excitons and electrostatic image forces at the metal–dielectric boundary with allowance for field penetration into the metal", *Zh. Eksp. Teor. Fiz.* **63**, 2247–87 (1972).

AIHARA, J.: "On the threshold law for the photoemission from the organic solid", *Phys. Stat. Sol.* (a), **17**, K37–40 (1973).

AIHARA, J. and INOKUCHI, H.: "Photoionization of naphthalene", *Bull. Chem. Soc. Japan* **43**, 1265–7 (1970).

AIHARA, J. and INOKUCHI, H.: "The photoionization of molecular and crystalline anthracene", *Bull. Chem. Soc. Japan* **47**, 2631–3 (1974).

AIHARA, J., TSUDA, H., and INOKUCHI, H.: "Direct determination of charge transfer energy", *Bull. Chem. Soc. Japan* **43**, 2439–41 (1970).

AKAMUTU, H. and INOKUCHI, H.: "On the electrical conductivity of violanthrone, iso-violanthrone and pyranthrone", *J. Chem. Phys.* **18**, 810–11 (1950).

AKAMUTU, H. and INOKUCHI, H.: "Photoconductivity of violanthrone", *J. Chem. Phys.* **20**, 1481–4 (1952).

AKAMUTU, H., INOKUCHI, H., and MATSUNAGA, Y.: "Electrical conductivity of the perylene–bromine complex", *Nature* **173**, 168–9 (1954).

AKAMUTU, H., INOKUCHI, H., and MATSUNAGA, Y.: "Organic semiconductors with high conductivity. I: Complexes between polycyclic aromatic hydrocarbons and halogens", *Bull. Chem. Soc. Japan* **29**, 213 (1956).

AKASAKA, Y., MURAKAMI, K., MASUDA, K., and NAMBRA, S.: "Optical absorption bands of the radical in irradiated anthracene single crystals", *Mol. Cryst. Liq. Cryst.* **13**, 377–80 (1971).

AKIMOV, I. A. and MESHKOV, A. M.: "Determination of the sign of photo-current carrier charge by the capacitor method", *Soviet Phys.—Doklady* **10**, 439–42 (1965).

AKON, C. D. and CRAIG, D. P.: "Polarization ratio (PR) of anthracene fluorescence", *Trans. Faraday Soc.* **62**, 1673–7 (1966).

AKON, C. D. and CRAIG, D. P.: "Ultraviolet polarization ratios in anthracene–tetracene mixed crystals", *Trans. Faraday Soc.* **63**, 56–60 (1967).

AKOPYAN, A. A. and GRIBNIKOV, Z. S.: "Injection effects in structures with ohmic contacts and weak heating of carriers", *Soviet Phys.—Semicond.* **9**, 981–5 (1976).

ALADEKOMO, J. B., ARNOLD, S., and POPE, M.: "Triplet exciton diffusion and double photon absorption in tetracene", *Phys. Stat. Sol.* (b) **80**, 333–40 (1977).

ALEKSANDROV, V. V., BELKIND, A., and SUMROV, V. V.: "Photoelectron emission with tetracene–alkali metal charge transfer", *Izv. Akad. Nauk Latv. SSR Sec. fiz.-tekhn. nauk* **3**, 58–65 (1974).

ALEXANDER, P., LACEY, A. R., and LYONS, L. E.: "Absorption and luminescence in anthracene crystals", *J. Chem. Phys.* **34**, 2200–1 (1961).

ALFANO, F. R., SHAPIRO, S. L., and POPE, M.: "Fission rate of singlet excitons in tetracene crystal measured with picosecond laser pulses", *Opt. Commun.* **9**, 388–91 (1973).

ALLCOCK, G. R.: "On the polaron rest energy and effective mass", *Adv. Phys.* **5**, 412–51 (1956).

ALLEN, T. H.: "Ellipsometry measurements on thin organic films", 1972 Annual Meeting Optical Soc. America, Abs. TUG14, also *J. Opt. Soc. Am.* **62**, 1375 (1972).

ALLEN, T. H.: "The study of molecular contamination kinetics using ellipsometry", *7th Space Simulation Conference, November 12–14* (1973).

ALMAZOV, A. B., KULIKÒVA, E. V., and STAFEEV, V. I.: "Investigation of the ambipolar equation describing the distribution of non-equilibrium carriers in semiconductors", *Soviet Phys.—Semicond.* **7**, 228–32 (1973).

ALMELEH, N. and HARRISON, S. E.: "Trapping effects in the organic semi-conductor triphenylene", *J. Phys. Chem. Solids* **27**, 893–901 (1966).

ANCKER-JOHNSON, B.: "Some plasma effects in semiconductors", *IEEE Trans. Nucl. Sci.* NS-14, No. 6, 27–39 (December 1967).

ANCKER-JOHNSON, B., ROBBINS, W. P., and CHANG, D. B.: "Transient high-density injection in a semiconductor with traps", *Appl. Phys. Lett.* **16**, 377–80 (1970).

ANDERSEN, T. N., WOOD, D. W., LIVINGSTON, R., and EYRING, H.: "Electrical properties of some charge-transfer complexes under high pressure", *J. Chem. Phys.* **44**, 1259–63 (1966).

ANDERSON, J. C. and NORIAN, K. H.: "A quasi-equilibrium thermally stimulated current process", *Solid-State Electron.* **20**, 335–42 (1977).

ANDERSON, P. W.: "Absence of diffusion in certain random lattices", *Phys. Rev.* **109**, 1492–1505 (1958).

ANDERSON, P. W.: "Localized magnetic states and Fermi-surface anomalies in tunneling", *Phys. Rev. Lett.* **17**, 95–97 (1966).

ANDERSON, W. A., DELAHOY, A. E., and MILANO, R. A.: "An 8 % efficient layered Schottky-barrier solar cell", *J. Appl. Phys.* **45**, 3913–15 (1974).

ANDREESHCHEW, E. A., VIKTOROVA, V. S., KILIN, S. F., and ROZMAN, I. M.: "Quenching of anthracene fluorescence by carbon tetrachloride", *Opt. Spectrosc.* **34**, 530–2 (1973).

ANDRIESCH, A. M., SHUTOV, S. D., ABASHKIN, V. G., and CHARNII, M. R.: "Investigation of the thermally stimulated depolarization current in amorphous films of As_2S_3", *Soviet Phys.—Semicond.* **8**, 1254–7 (1975).

ANGUS, J. G. and MORRIS, G. C.: "Ionization potential of the anthracene molecule from Rydberg absorption bands", *J. Mol. Spectrosc.* **21**, 310–24 (1966).

ANTHEUNIS, D. A., SCHMIDT, J., VAN DER WAALS, J. H.: "Spin-forbidden radiationless processes in isoelectronic molecules: anthracene, acridine, and phenazine: a study by microwave induced delayed phosphorescence", *Mol. Phys.* **27**, 1521–41 (1974).

ANTULA, J.: "Hot-electron concept for Poole–Frenkel conduction in amorphous dielectric solids", *J. Appl. Phys.* **43**, 4663–8 (1972).

ANZAI, H., ITOH, U., TAKAKUSA, M., and NAKAJIMA, T.: "Organic semi-conductor", *Cir. Electrotech. Lab. (Japan)* **157**, 1–158 (1966).

AOYAGI, Y., MASUDA, K., and NAMBA, S.: "Explanation of light-intensity dependence of photoconductivity in zinc–phthalocyanine", *J. Appl. Phys.* 249–51 (1972).

AOYAGI, Y., MASUDA, K., and NAMBA, S.: "Drift mobility measurements in copper phthalocyanine single crystals", *Mol. Cryst. Liq. Cryst.* **22**, 301–8 (1973).

APPEL, J.: "Polarons", *Solid State Phys.*, **21**, 193–391 (1968).

APPELBAUM, J. A.: "S-d exchange model of zero-bias tunneling anomalies", *Phys. Rev. Lett.* **17**, 91–95 (1966).

APPELBAUM, J. A.: "Exchange model of zero-bias tunneling anomalies", *Phys. Rev.* **154**, 633–43 (1967).

APRILEST, G., GAROFANO, T., NAVA, F., and SANTANGELO, M.: "On the nature of trapping levels in anthracene crystals from the study of thermally stimulated current", *Nuovo Cimento* B **1**, 85–99 (1971).

APSLEY, N. and HUGHES, H. P.: "Temperature- and field-dependence of hopping conduction in disordered systems", *Phil. Mag.* **30**, 963–72 (1974).

ARCHER, R. J.: "Determination of the properties of films on silicon by the method of ellipsometry", *J. Opt. Soc. Am.* **52**, 970–7 (1962).

ARDEN, W., KOTANI, M., and PETER, L. M.: "Triplet exciton decay processes in crystalline tetracene", *Phys. Stat. Sol.* (b) **75**, 621–31 (1976).

ARDEN, W., PETER, L. M., and VAUBEL, G., "Excimer emission from glassy films of some aromatic hydrocarbons", *J. Lumin.* **9**, 257–66 (1974).

ARIS, F. C., BRODRIBB, J. D., HUGHES, D. M., and LEWIS, T. J.: "A comparison of trapping phenomena in anthracene determined by photon induced current, spectroscopy and 'time-of-flight' analysis", *Sci. Papers, Inst. Org. Phys. Chem. Wroclaw Tech. Univ.* **7**, 182–7 (1974).

ARIS, F. C. and LEWIS, T. J.: "Charge trapping phenomena in anthracene crystals and the influence of crystal deformation", *J. Chem. Soc. Faraday Trans.* (I), **69**, 1872–86 (1973).

ARIS, F. C., LEWIS, T. J., THOMAS, J. M., WILLIAMS, J. O., and WILLIAMS, D. F.: "Influence of deformation on the mobility and lifetimes of charge-carriers in anthracene crystals", *Solid State Commun.* **12**, 913–17 (1973).

ARISTOV, A. V., KUZIN, V. A., and CHERKASOV, A. S.: "Stimulated lasing by solutions of anthracene derivatives", *Opt. Spectrosc.* **35**, 192–4 (1973).

ARNETT, P. C.: "Transient conduction in insulators at high fields", *J. Appl. Phys.* **46**, 5236–43 (1975).

ARNETT, P. C. and KLEIN, N.: "Poole–Frenkel conduction and the neutral trap", *J. Appl. Phys.* **46**, 1399–40 (1975).

ARNOLD, S.: "Optically induced fission spectrum in anthracene", *J. Chem. Phys.* **61**, 431–2 (1974).

ARNOLD, S., ALFANO, R. R., POPE, M., YU, W., HO, P., SELSBY, R., THARRATS, J., and SWENBERG, C. E.: "Triplet exciton caging in two dimensions", *J. Chem. Phys.* **64**, 5104–20 (1976).

ARNOLD, S., HU, H. T., and POPE, M.: "Exciton sounding of α-particle induced radiation defects in anthracene", *Mol. Cryst. Liq. Cryst.* **36**, 179–92 (1976).

ARNOLD, S., SWENBERG, C. E., and POPE, M.: "Triplet exciton caging in two dimensions, magnetic field effects", *J. Chem. Phys.* **64**, 5115–20 (1976).

ARNOLD, S. and WHITTEN, W. B.: "Triplet excitation spectrum of phenanthrene crystals", *Mol. Cryst. Liq. Cryst.* **18**, 83–86 (1972).

ARNOLD, S., WHITTEN, W. B., and DAMASK, A. C.: "Triplet exciton trapping by dislocations in anthracene", *J. Chem. Phys.* **53**, 2878–84 (1970).

ARNOLD, S., WHITTEN, W. B., and DAMASK, A. C.: "Davydov splitting and band structure for triplet excitons in pyrene", *Phys. Rev.* B **3**, 3452–7 (1971).

ARNOV, D. A., KOTOV, E. P., and KOTOV, YA. P.: "The influence of carrier capture in deep traps on small-signal characteristics of double-injection currents in semiconductors", *Phys. Stat. Sol.* (a) **30**, 97–105 (1975).

ASADA, T., SUGIHARA, K., KIMURA, Y., and OWAKI, S.: "Spectral dependence of photocurrent in anthracene", *Mem. Inst. Sci. Ind. Res., Osaka University*, **22**, 89–91 (1965).

ASAF, U. and STEINBERGER, I. T.: "Photoconductivity and electron transport parameters in liquid and solid xenon", *Phys. Rev.* B **10**, 4466–8 (1974).

ASHLEY, K. L., BAILEY, R. L., and BUTLER, J. K.: "GaAs:Mn double injection devices", *Solid-State Electron.* **16**, 1125–31 (1973).

ASHLEY, K. L. and MILNES, A. G.: "Double-injection in deep-lying impurity semiconductors", *J. Appl. Phys.* **35**, 369–74 (1964).

ASLANGUL, C. and KOLTIS, P.: "Density operator description of excitons in molecular aggregates: optical absorption and motion. I: Dimer Problem", *Phys. Rev.* B **10**, 4364–82 (1974).

ASPNES, D. E. and BOTTKA, N.: "Electrical-field effects on dielectric function of semiconductors and insulators" in *Semiconductor and Semimetals* (eds. R. K. Willardson and A. C. Beer) **9**, 457–543 (1972).

ATTIX, F. H.: "Luminescence degration", *Nucleonics* **17**, 60–61 (1959).

ATTIX, F. H.: "High-level dosimetry by luminescence degration", *Nucleonics* **17**, 142–8 (1959).

AUST, R. B., BENTLEY, W., and DRICKAMER, H.: "Behaviour of fused ring aromatic hydrocarbons at very high pressures", *J. Chem. Phys.* **41**, 1356–64 (1964).

AUST, R. B., SAMARA, G. A., and DRICKAMER, H.: "Effect of pressure on the properties of TCNQ and its complexes", *J. Chem. Phys.* **41**, 2003–6 (1964).

AUSTIN, I. G. and GARBETT, E. S.: "Far infra-red vibrational spectra of crystalline and amorphous As_2Se_3", *Phil. Mag.* **23**, 17–28 (1971).

AUSTIN, I. G. and MOTT, N. F.: "Polarons in crystalline and non-crystalline materials", *Adv. Phys.* **18**, 41–102 (1969).

AVAKIAN, P. and ABRAMSON, E.: "Singlet–triplet absorption spectra in naphthalene and pyrene crystals", *J. Chem. Phys.* **43**, 821–2 (1965).

AVAKIAN, P., ABRAMSON, E., KEPLER, R. G., and CARIS, J. C.: "Indirect observation of singlet–triplet absorption in anthracene crystals", *J. Chem. Phys.* **39**, 1127–8 (1963).

AVAKIAN, P., ERN, V., MERRIFIELD, R. E., and SUNA, A.: "Spectroscopic approach to triplet exciton dynamics in anthracene", *Phys. Rev.* **165**, 974–80 (1968).

AVAKIAN, P. and MERRIFIELD, R. E.: "Experimental determination of the diffusion length of triplet excitons in anthracene crystals", *Phys. Rev. Lett.* **13**, 541–3 (1964).

AVAKIAN, P. and MERRIFIELD, R. E.: "Triplet excitons in anthracene crystals—a review", *Mol. Cryst.* **5**, 37–77 (1968).

AVAKIAN, P. and WOLF, H. C.: "The temperature dependence of the energy transport in anthracene–tetracene mixed crystals", *Z. Phys.*, **165**, 439–44 (1961).

AVANESYAN, H. S., BENDERSKII, V. A., BRIKENSTEIN, V. KH., BROUDE, V. L., KORSHUNOV, L. I., LAVRUSHKO, A. G., and TARTAKOVSKII, I. I.: "Anthracene crystals under intensive optical pumping", *Mol. Cryst. Liq. Cryst.* **29**, 165–74 (1974).

AVANESYAN, H. S., BENDERSKII, V. A., BRIKENSTEIN, V. KH., BROUDE, V. L., and LAVRUSHKO, A. G.: "Stimulated radiation of tetracene impurity-containing anthracene crystals", *Phys. Stat. Sol.* (a), **19**, K121–3 (1973).

AVANESYAN, H. S., BENDERSKII, V. A., BRIKENSTEIN, V. KH., LAVRUSHKO, A. C., and FILLIPPOV, P. G.: "Stimulated emission of mixed molecular crystals", *Phys. Stat. Sol.* (a) **30**, 781–9 (1975).

AVERY, J. S. and MASON, R.: "Exciton states in polymers", *J. Phys. Chem.* **69**, 784–7 (1965).

AZZAM, R. M. A. and BASHARA, N. M.: "Generalized ellipsometry for surfaces with directional preference: Applications to diffraction gratings", *J. Opt. Soc. Am.* **62**, 1521–3 (1972).

BABAEV, A. S., KUZNETSOV, V. V., and STAFEEV, V.: "Current–voltage characteristics of metal–organic film–metal structure", *Soviet Phys.—Semicond.* **7**, 692–4 (1973).

BABENKO, S. D., BENDERSKII, V. A., GOLDANSKII, V. J., LAVRUSHKO, A. G., and TYCHINSKII, V. P.: "Annihilation of excited singlet states in anthracene solutions", *Chem. Phys. Lett.* **8**, 598–600 (1971).

BACCARANI, G.: "Current transport in Schottky-barrier diodes", *J. Appl. Phys.* **47**, 4122–6 (1976).

BAESSLER, H., BECKER, G., and RIEHL, N.: "Energy distribution of hole traps in anthracene crystals", *Phys. Stat. Sol.* **15**, 347–54 (1966).

BAESSLER, H., HERRMANN, G., RIEHL, N., and VAUBEL, G.: "Space-charge-limited currents in tetracene single-crystals", *J. Phys. Chem. Solids* **30**, 1579–85 (1969).

BAESSLER, H. and KILLESREITER, H.: "Hot carrier injection into molecular crystals and its relevance to the field dependence of photocurrents", *Phys. Stat. Sol.* (b), **53**, 183–92 (1972).

BAESSLER, H. and KILLESREITER, H.: "Band gap-determination from auto-ionization data in molecular crystals", *Mol. Cryst. Liq. Cryst.* **24**, 21–31 (1973).

BAESSLER, H., KILLESREITER, H., and VAUBEL, G.: "Exciton-induced photo-currents in molecular crystals", *Discuss. Faraday Soc.* **51**, 48–53 (1971).

BAESSLER, H., RIEHL, N., and VAUBEL, G.: "Calculation of the width of the second conduction band in anthracene crystals from photoemission data", *Phys. Stat. Sol.* **26**, 607–10 (1968).

BAESSLER, H., RIEHL, N., and VAUBEL, G.: "Photoemission of electrons from alkali and alkaline earth metals into anthracene crystals", *Mol. Cryst. Liq. Cryst.* **9**, 249–64 (1969).

BAESSLER, H. and VAUBEL, G.: "Surface states of anthracene crystals", *Phys. Rev. Lett.* **21**, 615–17 (1968).

BAESSLER, H. and VAUBEL, G.: "Photoemission of electrons from alkali metals into anthracene", *Solid State Commun.* **6**, 97–99 (1968).

BAESSLER, H. and VAUBEL, G.: "Photoinjection of electrons into anthracene crystals", *Solid State Commun.* **6**, 631–3 (1968).

BAESSLER, H. and VAUBEL, G.: "Detection of photoemission of electrons into anthracene crystals by studying recombination radiation", *Mol. Cryst. Liq. Cryst.* **11**, 25–34 (1970).

BAGLEY, B. G.: "The field dependent mobility of localized electron carriers", *Solid State Commun.* **8**, 345–8 (1970).

BALDWIN, B. A. and OFFEN, H. W.: "Effects of high pressures on the phosphorescence of aromatic hydrocarbons", *J. Chem. Phys.* **46**, 4509–14 (1967).

BALLARD, W. P. and CHRISTY, R. W.: "Switching effects in election-beam deposited polymer films", *J. Non-Cryst. Solids* **17**, 81–88 (1975).

BALSEV, I. and HAM, J. M.: "Drift of electron-hole drops in exciton density gradients", *Phys. Stat. Sol.* B **65**, 531–6 (1974).

BARANTELY, W. A., LARIMOR, O. G., DAPKUS, P. D., HASZKO, S. E., and SAIL, R. H.: "Effect of dislocations on green electroluminescence efficiency GaP grown by liquid phase epitaxy", *J. Appl. Phys.* **46**, 2629–37 (1975).

BARBE, D. F. and MCRAE, H. F.: "Current transport mechanisms in semi-insulators having an injecting contact", *J. Vacuum Sci. Tech.* **9**, 70–73 (1971).

BARBE, D. F. and WESTGATE, C. R.: "Bulk trapping states in β-phthalocyanine single crystals", *J. Chem. Phys.* **52**, 4046–54 (1970).

BARDEEN, J.: "Theory of the work function", *Phys. Rev.* **49**, 653–63 (1936).

BARDEEN, J.: "Surface states and rectification at a metal-semiconductor contact", *Phys. Rev.* **71**, 717–27 (1947).

BARDEEN, J.: "Superconducting fluctuations in one-dimensioned organic solids", *Solid State Commun.* **13**, 357–9 (1973).

BARNETT, A. M.: "Current filaments in semiconductors", *IBM Jl. Res. Dev.* **13**, 522–8 (1969).

BARNETT, A. M.: "Current filament formation", in *Semiconductors and Semimetals* (eds. R. K. Willardson and A. C. Beer), Academic Press (New York) **6**, 141–200 (1970).

BARNETT, A. M. and JENSEN, H. A.: "Observation of current filaments in semi-insulating GaAs", *Appl. Phys. Lett.* **12**, 341–2 (1968).

BARNETT, A. M. and MILNES, A. G.: "Filamentary conduction in two carrier space charge limited devices", *IEEE Trans. Electron Devices* ED **13**, 816 (1966).

BARNETT, A. M. and MILNES, A. G.: "Filamentary injection in semi-insulating silicon", *J. Appl. Phys.* **37**, 4215–23 (1966).

BARON, R.: "Effects of diffusion on double injection in insulators", *Phys. Rev.* **137**, A272–85 (1965).

BARON, R.: "Effects of diffusion and thermal generation on double injection in semiconductors", *J. Appl. Phys.* **39**, 1435–46 (1968).

BARON, R., MARSH, O. J., and MAYER, J. W.: "Transient response of double injection in a semiconductor of finite cross-section", *J. Appl. Phys.* **37**, 2614–26 (1966).

BARON, R. and MAYER, J. W.: "Double injection in semiconductors", in *Semiconductors and Semimetals* (eds. R. K. Willardson and A. C. Beer), Academic Press (New York) **6**, 201–314 (1970).

BARON, R., NICOLET, M. A., and RODRIGUEZ, V.: "Differential step response of unipolar space-charge-limited current in solids", *J. Appl. Phys.* **37**, 4156–8 (1966).

BARTLING, J.: "Charge, current and electric field development in a film of insulating materials with trap", *J. Phys. D* **9**, 1619–37 (1976).

BARU, V. G., and TEMNITSKII, YU. N.: "Transient space-charge-limited and contact-limited photoinjection currents in an insulator", *Soviet Phys.—Semicond.* **5**, 1862–8 (1972).

BARU, V. G. and TEMNITSKII, YU. N.: "Negative differential resistance of a semiconductor controlled by photoinjection from contact", *Soviet Phys.—Semicond.* **7**, 211–16 (1973).

BASHMAKOVA, M. I.: "Change of electrical conductivity of organic semiconductors in the melting region", *Izv. Vyssh. Ucheb. Zaved Fiz. (USSR)* No. 4, 55–9 (1973); English translation in *Soviet Phys. J.* **16**, 486–9 (1973).

BASURTO, J. G. and BURSHTEIN, Z.: "Electroluminescence studies in pyrene single crystals", *Mol. Cryst. Liq. Cryst.* **31**, 211–19 (1975).

BASURTO, J. G., BURSHTEIN, Z. and LEVINSON, J.: "Studies of band-structure exciton processes and trap distribution in pyrene", *Mol. Cryst. Liq. Cryst.* **31**, 131–44 (1975).

BATEMAN, R. J., CHANCE, R. R., and HORNIG, J. F.: "Fluorescence reabsorption in anthracene single crystals: lifetime variations with emission wave-length and temperature", *Chem. Phys.* **4**, 402–8 (1974).

BATLEY, M. and LYONS, L. E.: "Electrical conductivity in organic solids at high pressures", *Aust. J. Chem.* **19**, 345–50 (1966).

BATLEY, M. and LYONS, L. E.: "Photo-electric emission from donor–acceptor solids and donor molecules", *Mol. Cryst.* **3**, 357–74 (1968).

BATRA, I. P., KANAZAWA, K. K., SCHECHTMAN, B. H., and SEKI, H.: "Charge-carrier dynamics following pulsed photoinjection", *J. Appl. Phys.* **42**, 1124–30 (1971).

BATRA, I. P., KANAZAWA, K. K., and SEKI, H.: "Discharge characteristics of photoconducting insulators", *J. Appl. Phys.* **41**, 3416–22 (1970).

BATRA, I. P. and SCHECHTMAN, B. H.: "Pulse induced photocurrents and field distributions in photoconductor–dielectric structure", *J. Phys. Chem. Solids* **32**, 769–86 (1971).

BATRA, I. P., SCHECHTMAN, B. H., and SEKI, H.: "Transient space-charge-limited currents in photoconductor–dielectric structures", *Phys. Rev. B* **2**, 1592–1603 (1970).

BATRA, I. P., SCHECHTMAN, B. H., and SEKI, H.: "Transient space-charge-limited currents in photoconductor–dielectric structures: a comment", *Phys. Rev. B* **3**, 3571–2 (1971).

BATRA, I. P. and SEKI, H.: "Photocurrent due to pulse illumination in the presence of trapping", *J. Appl. Phys.* **41**, 3409–15 (1970).

BATT, R. H., BRAUN, C. L., and HORNIG, J. F.: "Electric-field and temperature dependence of photoconductivity", *J. Chem. Phys.* **49**, 1967–8 (1968).

BAUSER, H.: "Dark and photocurrents in anthracene crystals with injecting electrodes", *Third Organic Crystal Symposium, Chicago, Illinois, USA*, paper 15 (1965).

BAUSER, H.: "Injection of holes into anthracene crystals in the presence of iodine vapors", *Z. Phys.* **188**, 423–43 (1965).

BAUSER, H. and PERNISZ, U.: "Field effect in anthracene crystal", *Chem. Phys. Lett.* **11**, 213–15 (1971).

BAUSER, H., PERNISZ, U., and SCHUMACHER, K.: "Singlet-oxygen-induced hole current in anthracene crystals", *Chem. Phys. Lett.* **17**, 47–48 (1972).

BAUSER, H. and RUF, H. H.: "Excitation spectrum of photocurrent in anthracene crystals in singlet absorption region", *Phys. Stat. Sol.* **32**, 135–49 (1969).

BAVERSTOCK, K. F. and DYNE, P. J.: "Photoconductivity of γ-irradiated methyltetrahydrofuran glass at 77°K", *Can. J. Chem.* **48**, 2182–91 (1970).

BAYLISS, N. L. and RIVIERE, J. C.: "Photoconductivity of naphthalene and anthracene", *Nature* **163**, 765 (1949).

BAZAKUTSA, V. A. and MOKHOV, G. D.: "Comments on the nature of the 'anomalous' photoconductivity", *Soviet Phys.—Semicond.* **9**, 447–9 (1975).

BAZHINA, I. N., VERZILOV, V. S., GRECHISHKIN, V. S., GRECHISHKINA, R. V., and GUSAROV, V. M.: "ESR and the Hall effect in organic charge transfer complex", *J. Struct. Chem.* **14**, 872–4 (1973).

BEATTIE, A. R. and CUNNINGHAM, R. W.: "Large-signal photomagneto electric effect", *Phys. Rev.* **125**, 533–40 (1962).

BEATTIE, A. R. and CUNNINGHAM, R. W.: "Large-signal photoconductivity effect", *J. Appl. Phys.* **35**, 353–9 (1964).

BEATTIE, A. R. and LANDSBERG, P. T.: "Auger effect in semiconductors", *Proc. Roy. Soc. (London)* A **249**, 16–29 (1959).

BEATTIE, A. R. and LANDSBERG, P. T.: "One-dimensional overlap functions and their application to Auger recombination in semiconductors", *Proc. Roy. Soc. (London)* A **258**, 486–95 (1960).

BEBELAAR, D.: "Time resolved molecular spectroscopy using high power solid state lasers in pulse transmission mode: a re-examination of the $S_n \leftarrow S_1$ spectra of naphthalene and anthracene", *Chem. Phys.* **3**, 205–16 (1974).

BECKER, R. S. and CHEN, E.: "Extension of electron affinities and ionization potentials of aromatic hydrocarbons", *J. Chem. Phys.* **45**, 2403–10 (1966).

BECKER, G., RIEHL, N., and BAESSLER, H.: "Injection-determined dark-current in anthracene single crystals", *Phys. Lett.* **20**, 221–2 (1966).

BECKMANN, K. H. and MEMMING, R.: "Photoexcitation and luminescence in redox processes on gallium phosphide electrodes", *J. Electrochem. Soc.* **116**, 368–73 (1969).

BELKIND, A. I. and ALEKSANDROV, S. R.: "Photoemission and photoionization of molecular crystals", *Phys. Stat. Sol.* (a) **9**, 105–12 (1972).

BELKIND, A. I. and BOK, J.: "Photoelectron emission from molecular crystals near the threshold", *Phys. Stat. Sol.* (a) **22**, K37–40 (1974).

BELKIND, A. I. and GRECHOV, V. V.: "Energy levels of polyacene crystals", *Phys. Stat. Sol.* (a) **26**, 377–84 (1974).

BELKIND, A. I. and KALENDAREV, R. I.: "Photoionization of quasicontinuous traps in tetracene crystals", *Phys. Stat. Sol.* (a) **14**, 681–8 (1972).

BENDERSKII, V. A., BELKIND, A. I., FEDOROV, M. I., and ALEXANDROV, S. B.: "Impurity-level energies in magnesium phthalocyanine films", *Fiz. Tverd. Tela* **14**, 790–4 (1972).

BENDERSKII, V. A., BRIKENSHTEIN, V. KH., and LAVRUSHKO, A. G.: "Nonlinear quenching of the fluorescence of anthracene crystals at low temperatures", *Soviet Phys.—Solid State* **15**, 189–91 (1973).

BENDERSKII, V. A. and LAVRUSHKO, A. G.: "Reabsorption of light and photoconductivity of anthracene crystals", *Soviet Phys.—Solid State* **13**, 1072–6 (1971).

BENDERSKII, V. A. and LAVRUSHKO, A. G.: "Investigations of pulsed photocurrents in anthracene crystals", *Soviet Phys.—Solid State* **13**, 1280–3 (1971).

BENDERSKII, V. A. and LAVRUSHKO, A. G.: "Fluorescence quenching and photoeffect in anthracene crystals", *Opt. Spectrosc.* **32**, 390–2 (1972).

BENDERSKII, V. A. and RUDENKO, T. S.: "Sensitized photoconductivity of anthracene crystals", *Soviet Electrochem.* **3**, 618–22 (1967).

BENHAM, W. E.: "Theory of the internal action of thermionic systems at moderately high frequencies", *Phil. Mag.* **5**, 641–62 (1928).

BENNELT, R. G. and McCARTIN, P. J.: "Radiationless deactivation of the fluorescent state of substituted anthracene", *J. Chem. Phys.* **44**, 1969–72 (1966).

BENNETT, A. J.: "Some electric properties of solid surfaces", *Crit. Rev. Solid State Sci.* **4**, 261–77 (1974).

BENSON, R. and GEACINTOV, N. E.: "Deuterium effect on the quenching of aromatic hydrocarbon triplet excited states by oxygen", *J. Chem. Phys.* **60**, 3251–7 (1974).

BENTLEY, W. and DRICKAMER, H. G.: "Behaviour of thirteen charge transfer complexes at high pressures", *J. Chem. Phys.* **42**, 1573–87 (1965).

BENZ, K. W., HÄCKER, W., and WOLF, H. C.: "Das Spectrum der verzögerten in Antracen- und Naphthalin-Kristallen", *Z. Naturforsch.* **25a**, 657–64 (1970).

BENZ, K. W. and WOLF, H. E.: "Fluorescence and energy transfer in phenanthrene crystals", *Z. Naturforsch.* **19a**, 381–6 (1964).

BEPLER, W.: "Photoströme und Ladungsträgerbeweglichkeiten in Pyreneinkristallen", *Z. Phys.* **185**, 507–17 (1965).

BEREZIN, G. N., ROMANETS, A. N., SONDAEVSKII, V. P., and UZDOVSKII, V. V.: "Investigation of current filaments with the aid of mirror electron microscope", *Soviet Phys.—Semicond.* **7**, 510–11 (1973).

BERG, J. I. and MILLER, N. C.: "Trap-perturbed emission-limited current in amorphous selenium films", *J. Appl. Phys.* **46**, 4812–18 (1975).

BERGH, A. A. and DEAN, J. P.: "Light-emitting diodes", *Proc. IEEE* **60**, 156–223 (1972).

BERGLUND, C. N. and POWELL, R. J.: "Photoinjection into SiO_2: electron scattering in the image force potential well", *J. Appl. Phys.* **42**, 573–9 (1971).

BERGLUND, C. N. and SPICER, W. E.: "Photoemission studies of copper and silver: theory", *Phys. Rev.* **136**, A1030–44 (1964).

BERGMAN, A., BERGMAN, D., and JORTNER, J.: "Singlet exciton collisions in crystalline anthracene", *Israel J. Chem.* **10**, 471–83 (1972).

BERGMAN, A. and JORTNER, J.: "Photoconductivity of crystalline anthracene induced by tunable dye lasers", *Phys. Rev.* B **9**, 4560–74 (1974).

BERGMAN, A., LEVINE, M., and JORTNER, J.: "Collision ionization of singlet excitons in molecular crystals", *Phys. Rev. Lett.* **18**, 593–6 (1967).

BERLINSKY, A. J., CAROLAN, J. F., and WEILER, L.: "Band structure parameters for solid TTF-TCNQ", *Solid State Commun.* **15**, 795–801 (1974).

BERMAN, M., KERCHNER, H. R., and ERGUN, S.: "Determination of the optical properties of absorbing uniaxial crystals from reflectance at oblique incidence", *J. Opt. Soc. Am.* **60**, 646–8 (1970).

BERREHAR, J., DELANNOY, P., and SCHOTT, M.: "Drift mobility of holes in crystalline tetracene", *Phys. Stat. Sol.* (b) **77**, K119–22 (1976).

BERREHAR, J., SCHOTT, M., and DELANNOY, P.: "Optical absorption of crystalline tetracene in the low-energy tail of the $S_0 \rightarrow S_1$ transition", *Phys. Stat. Sol.* (a) **32**, K 37–39 (1975).

BERRY, R. S., JORTNER, J., MACKIE, J. C., PYSH, E. S., and RICE, S. A.: "Search for a charge-transfer state in crystalline anthracene", *J. Chem. Phys.* **42**, 1535–40 (1965).

BEST, A. P., GARFORTH, F. M., INGOLD, C. K., POOLE, H. G., and WILSON, C. L.: "Excited states of benzene", *J. Chem. Soc.* Part I–Part XII, 406–516 (1948).

BETHE, H. A.: "Theory of the boundary layer of crystal rectifiers", MIT Radiation Laboratory Report 43–12 (1942).

BETTERIDGE, D. and THOMPSON, M.: "Interpretation of ultraviolet photoelectron spectra by simplified methods", *J. Mol. Struct.* **4**, 341–71 (1974).

BIEN, V. C. and LYONS, L. E.: "An electron spin response study of pure anthracene doped with alkali metals", *Aust. J. Chem.* **23**, 261–7 (1970).

BIERMAN, A.: "Acceleration of molecular excitons by an electric field", *Phys. Rev.* **130**, 2266–70 (1963).

BILGER, H. R., LEE, D. H., NICOLET, M. A., and McCARTER, E. R.: "Noise and equivalent circuit of double injection", *J. Appl. Phys.* **39**, 5913–18 (1968).

BIRKS, J. B.: "Scintillation from naphthalene–anthracene crystals", *Proc. Phys. Soc.* A **63**, 1044–6 (1950).

BIRKS, J. B.: "Temperature dependence of organic scintillation materials", *Phys. Rev.* **95**, 277 (1954).

BIRKS, J. B.: "The fluorescence and scintillation decay times of crystalline anthracene", *Proc. Phys. Soc. (London)* **79**, 494–6 (1962).

BIRKS, J. B.: "Quenching of excited singlet and triplet states in aromatic hydrocarbons by oxygen and nitric oxide", *J. Lumin.* **1**, **2**, 154–65 (1970).

BIRKS, J. B.: "Energy transfer in organic system X. Pure and mixed crystals", *J. Phys.* B **3**, N12, 1704–14 (1970).

BIRKS, J. B.: "The influence of reabsorption and defects on anthracene crystal fluorescence", *Mol. Cryst. Liq. Cryst.* **28**, 117–29 (1974).

BIRKS, J. B. and BLACK, F. A.: "Deterioration of anthracene under α-particle irradiation", *Proc. Phys. Soc. (London)* A **64**, 511 (1951).

BIRKS, J. B. and BROOKS, F. O.: "Scintillation response of anthracene to 6–30 kev photoelectrons", *Proc. Phys. Soc.* B **69**, 721–30 (1956).

BIRKS, J. B., KING, T. A., and MUNRO, I. H.: "The photoluminescence decay of organic crystals", *Proc. Phys. Soc. (London)* **80**, 355–61 (1962).

BIRKS, J. B. and MUNRO, I. H.: "The fluorescence lifetimes of aromatic molecules", *Progr. React. Kinet.* **4**, 239–303 (1967).

BLACK, F. A.: "The decay in fluorescence efficiency of organic materials on irradiation by particles and photons", *Phil. Mag.* **44**, 263–7 (1953).

BLAGODAROV, A. N. and LUTSENKO, E. L.: "The contact and bulk conductivity in organic semiconductors", *Phys. Stat. Sol.* (a) **21**, K115–18 (1974).

BLAGODAROV, A. N., LUTSENKO, E. L., and ROSENSHTEIN, L. D.: "Low-frequency conductivity of phthalocyanine films with unipolar injection", *Soviet Phys.—Solid State* **11**, 2747–8 (1970).

BLAIR, B. and TREDGOLD, R. H.: "Conductivity and photoconductivity in benzene and pyridine", *J. Phys.* D **7**, 1995–2002 (1974).

BLAKE, A. E., CHARLESBY, A., and RANDLE, K. J.: "Simultaneous thermoluminescence and thermally stimulated current in polyethylene", *J. Phys.* D **7**, 759–70 (1974).

BLAKE, N. W. and McCLURE, D. S.: "Delayed singlet–singlet emission from molecular crystals", *J. Chem. Phys.* **29**, 722–4 (1958).

BLAKNEY, R. M. and GRUNWALD, H. P.: "Small-signal current transient in insulators with traps", *Phys. Rev.* **159**, 658–64 (1967).

BLASQUEZ, G. and VAN VLIET, K. M.: "Noise in single injection diodes of spherical and cylindrical geometry", *J. Appl. Phys.* **47**, 762–7 (1976).

BLINOV, L. M. and KIRICHENKO, N. A.: "Stark effect and charge-transfer states in organic semiconductors", *Soviet Phys.—Solid State* **14**, 2163–4 (1973).

BLOCH, A. N.: "Systematics of one-dimensional organic conductors", *Solid State Commun.* **14**, 99 (1974).

BLOCH, A. N., FERRARIS, J. P., and COWAN, D. O.: "Temperature dependent X-ray observation of TTF–TCNQ", *Phys. Lett.* **47A**, 313–14 (1974).

BLOCH, A. N., FERRARIS, J. P., COWAN, D. O., and POEHLER, T. O.: "Microwave conductivities of the organic conductors TTF–TCNQ and ATTF–TCNQ", *Solid State Commun.* **13**, 753–7 (1973).

BLODGETT, A. J. and SPICER, W. E.: "Experimental determination of the density of states in nickel", *Phys. Rev.* **146**, 390–402 (1966).

BLOSSEY, D. F.: "Wannier exciton in an electric field I. optical absorption by bound and continuum states", *Phys. Rev.* B **2**, 3976–90 (1970).

BLOSSEY, D. F.: "Wannier exciton in an electric field II electroabsorption in direct-band-gap in solids", *Phys. Rev.* B **3**, 1382–91 (1971).

BLOSSEY, D. F.: "One-dimensional Onsager theory for carrier injection in metal–insulator systems", *Phys. Rev.* B **9**, 5183–7 (1974).

BLOSSEY, D. F. and HANDLER, P.: "Electroabsorption" in *Semiconductors and Semimetals* (eds. R. K. Willardson and A. C. Beer), Academic Press (New York), **9**, 257–315 (1973).

BLOSSEY, D. F. and ZALLEN, R.: "Surface and bulk photoresponse of crystalline As_2S_3", *Phys. Rev.* B **9**, 4306–13 (1974).

BLUM, H., MATTERN, P. L., ARNDT, R. A., and DAMASK, A. C.: "Gamma radiation induced color centers in anthracene", *Mol. Cryst.* **3**, 269–80 (1967).

BODY, R. G. and ROSS, I. G.: "Spectra of impurities in molecular crystals", *Organic Crystal Symposium, Chicago* (1965).

BOE, W. and FRIZSCH, L.: "Poole–Frenkel effect on centres with screened coulomb potential", *Phys. Stat. Sol.* B **61**, K35–39 (1974).

BÖER, K. W., OBERLANDER, S., and VOIGT, J.: "Theory of glow curves", *Z. Naturforsch.* **13a**, 544–7 (1958).

BÖER, K. W. and ROTHWARF, A.: "Materials for solar photovoltaic energy conversion", *Ann. Rev. Material Sci.* **6**, 303–33 (1976).

BOGOMOLOV, V. N., KUDINOV, E. K., and FIRSOV, YU. A.: "Polaron nature of the current carriers in rutile (TiO_2)", *Soviet Phys.—Solid State* **9**, 2502–13 (1968).

BOGUS, C.: "Measurement of photoconductivity in anthracene crystals", *Z. Phys.* **184**, 219–28 (1965).

BOGUS, C.: "Charge-carrier lifetime measurements on photoconductor anthracene", *Z. Naturforsch.* **21a**, 667–8 (1966).

BOGUS, C.: "Pulsed photoconductivity in anthracene crystals", *Z. Phys.* **207**, 281–93 (1967).

BOGUSLAVSKII, L. and LOZHKIN, B. T.: "Electrochemical reactions with participation of excitons at anthracene electrodes", *Surf. Sci.* **38**, 413–32 (1973).

BOGUSLAVSKII, L. I., LOZHKIN, B. T., and MARGULIS, V. B.: "Surface state and limiting currents on an anthracene electrode", *Soviet Electrochem.* **4**, 62–64 (1968).

BOHUN, A.: "Thermoemission and photoemission of sodium chloride", *Czech. J. Phys.* **4**, 91–93 (1954).

BOLOTNIKOVA, T. N. and NAUMOVA, T. M.: "Concentration dependence of the intensity of triplet–triplet absorption of some aromatic hydrocarbons", *Opt. Spectrosc.* **23**, 206–8 (1967).

BONCH-BRUWVICH, V. L. and LANDSBERG, E. G.: "Recombination mechanisms", *Phys. Stat. Sol.* **29**, 9–43 (1968).

BONHAM, J. S.: "SCLC theory for a Gaussian trap distribution", *Aust. J. Chem.* **26**, 927–39 (1973).

BONHAM, J. S. and LYONS, L. E.: "Field effects in the saturation region of space charge limited current in molecular crystals", *Aust. J. Chem.* **26**, 489–97 (1973).

BONHAM, J. S., LYONS, L. E., and WILLIAMS, D. F.: "Concerning the interpretation of thermal activation energies for hole injection in anthracene", *J. Chem. Phys.* **56**, 1782–3 (1972).

BONSCH, G.: "Directional and temperature dependence of light pulses from anthracene crystals excited by alpha particles", *Z. Phys.* **212**, 497 (1968).

BONSCH, G., KLEINERT, G., and FLAMMERSFELD, A.: "Scintillation light yield of thin films of anthracene", *Z. Phys.* **222**, N4, 363–71 (1969).

BOON, M. R.: "Space charge limited currents with coulombic trapping", *Thin Solid Films* **9**, 457–8 (1972).

BOON, M. R.: "The effect of image forces on space charge limited conduction in thin films", *Thin Solid Films* **9**, 461–3 (1972).

BOON, M. R.: "Space charge limited currents in samples containing fully ionised donors: an analytical non-parametric treatment", *Thin Solid Films* **9**, 469–70 (1972).

BOOTH, A. H.: "Calculation of electron trap depth from thermoluminescence", *Can. J. Chem.* **32**, 214–15 (1954).

BORKMAN, R. F., TSIVOGLOV, S., and VAN DULLEN, R.: "Absence of a host-crystal deuterium effect on the phosphorescence lifetime of naphthalene", *J. Chem. Phys.* **59**, 4589–92 (1973).

BOROFFKA, H.: "The photoconductivity of anthracene", *Z. Phys.* **160**, 93–108 (1960).

BOSCHI, R., CLAR, E., and SCHMIDT, W.: "Photoelectron spectra of polynuclear aromatics. III: The effect of nonplanarity in sterically overcrowded aromatic hydrocarbons", *J. Chem. Phys.* **60**, 4406–18 (1974).

BOSMAN, A. J. and CREVECOEUR, C.: "Mechanism of the electrical conduction in Li-doped NiO", *Phys. Rev.* **144**, 763–70 (1966).

BOSMAN, A. J. and VAN DAAL, H. J.: "Small-polaron versus band conduction in some transition-metal oxides", *Adv. Phys.* **19**, 1–117 (1970).

BOUCHRIHA, H., DELACOTE, G., DELANNOY, P., and SCHOTT, M.: "Interaction of triplet excitons with trapped and free holes in crystalline anthracene. Exciton quenching, current enhancement and magnetic fields", *J. Phys.* **35**, 577–87 (1974).

BOUCHRIHA, H., SCHOTT, M., BISCEGLIA, M., and DELACOTE, G.: "Quenching of triplet excitons by trapped holes in crystalline pyrene", *Chem. Phys. Lett.* **23**, 183–6 (1973).

BOUDRY, M. R.: "Computed transient solution for the M–S–M diode with traps", *Electron. Lett.* **4**, 193–5 (1968).

BOWEN, E. J.: "Fluorescence spectra of naphthacene molecules in solid solution of anthracene with the variation of wavelengths", *Nature* **153**, 653 (1944).

BOWEN, E. J.: "The fluorescence of naphthacene in anthracene", *J. Chem. Phys.* **13**, 306 (1945).

BOWEN, E. J., MICKIEWICZ, E., and SMITH, F.: "Resonance transfer of electronic energy in organic crystals", *Proc. Phys. Soc. (London)* A **62**, 26–31 (1949).

BOWEN, E. J. and WEST, K.: "Solvent quenching of the fluorescence of anthracene", *J. Chem. Soc.* 4394 (1955).

BOWLT, C.: "Space-charge-limited currents in amorphous arsenic trisulphide", *Proc. Phys. Soc. (London)* **80**, 810–11 (1962).

BRADDLEY, L. L., SCHWOB, H. P., WEITZ, D., and WILLIAMS, D. F.: "Delayed electroluminescence quenching in anthracene", *Mol. Cryst. Liq. Cryst.* **23**, 271–82 (1973).

BRADDY, J. J. and MOORE, W. H.: "Actinoelectric effects in tartaric acid crystals", *Phys. Rev.* **55**, 308–11 (1939).

BRADDY, J. J. and TOOD, J. D.: "Photoconductivity of tartaric acid crystals", *Phys. Rev.* **56**, 859 (1939).

BRADLEY, A. and HAMMES, J. P.: "Photoconductivity in thin organic films", *J. Electrochem. Soc.* **110**, 543–8 (1963).

BRAGG, W. H.: "The structure of organic crystals", *Proc. Phys. Soc. (London)* **34**, 33–50 (1921).

BRAGG, W. H.: "The crystalline structure of anthracene", *Proc. Phys. Soc. (London)* **35**, 167–9 (1922).

BRANDER, R. W., GERKIN, R. E., and HUTCHISON, JR, C. A.: "Electron magnetic resonance of triplet states and the detection of energy transfer in crystals", *J. Chem. Phys.* **37**, 447–8 (1962).

BRATTAIN, W. H. and BARDEEN, J.: "Surface properties of germanium", *Bell System Tech. J.* **32**, 1–41 (1953).

BRATTAIN, W. H. and GARRETT, C. G. B.: "Surface states", in *Methods of Experimental Physics* (eds. K. Lark-Horovitz and V. A. Johnson), Vol. 6, Solid State Physics, Academic Press (New York), 136–44 (1959).

BRAU, A. and FARGES, J. P.: "Further aspects of the electrical conduction in the highly anisotropic organic semiconductor TEA(TCNQ)$_2$", *Phys. Stat. Sol.* B **61**, 257–65 (1974).

BRAUN, C. L.: "Singlet exciton–exciton interaction ionization in anthracene", *Phys. Rev. Lett.* **21**, 215–19 (1968).

BRAUN, C. L. and CHANCE, R. R.: "Germinate charge-pair recombination in molecular crystals", in *Energy and Charge Transfer in Organic Semiconductors* (eds. K. Masuda and M. Silver), Plenum (New York), 17–24 (1974).

BRAUN, C. L. and DOBBS, G. M.: "Intrinsic photoconductivity in naphthalene single crystals", *J. Chem. Phys.* **53**, 2718–25 (1970).

BRAUN, H. P. and MICHEL-BEYERLE, M. E.: "Triplet exciton stimulated current pulses in anthracene crystals", *Chem. Phys. Lett.* **21**, 76–79 (1973).

BRAUN, H. P. and MICHEL-BEYERLE, M. E.: "Current and fluorescence oscillations in anthracene crystals", *Phys. Stat. Sol.* (a) **37**, 73–92 (1976).

BRÄUNLICH, P.: "Light effects on space-charge controlled currents in CdS", *Phys. Stat. Sol.* **21**, 383–91 (1967).

BRÄUNLICH, P. and KELLY, P.: "II. Correlations between thermoluminescence and thermally stimulated conductivity", *Phys. Rev.* B **1**, 1596–1606 (1970).

BRAUNSTEIN, A. I., BRAUNSTEIN, M., and PICUS, G. S.: "Voltage dependence of the barrier heights in Al$_2$O$_3$ tunneling junctions", *Appl. Phys. Lett.* **8**, 95–98 (1966).

BRAUNSTEIN, A. I., BRAUNSTEIN, M., PICUS, G. S., and MEAD, C. A.: "Photoemissive determination of barrier shape in tunnel junctions", *Phys. Rev. Lett.* **14**, 219–21 (1965).

BREE, A., CARSWELL, D. J., and LYONS, L. E.: "Photo- and semi-conductance of organic crystals. I: Photo-effects in tetracene and anthracene", *J. Chem. Soc.* 1728–33 (1955).

BREE, A. and KATAGIRI, S.: "Absorption and fluorescence of anthracene", *J. Mol. Spectrosc.* **17**, 24–47 (1965).

BREE, A., KATAGIRI, S., and SUART, S. R.: "Vibrational assignment of anthracene-d_{10} from the fluorescence and absorption spectra", *J. Chem. Phys.* **44**, 1788–92 (1966).

BREE, A. and KYDD, R. A.: "Hole traps at the surface of anthracene crystals", *J. Chem. Phys.* **40**, 1775–6 (1964).

BREE, A. and KYDD, R. A.: "Infrared spectrum of anthracene crystals", *J. Chem. Phys.* **48**, 5319–45 (1968).

BREE, A. and LYONS, L. E.: "The polarized ultraviolet spectrum of tetracene single crystals by a photoconductance method", *J. Chem. Phys.* **22**, 1630 (1954).

BREE, A. and LYONS, L. E.: "Effects of the gaseous atmosphere on the photo- and semiconduction aromatic hydrocarbon crystals", *J. Chem. Phys.* **25**, 384 (1956).

BREE, A. and LYONS, L. E.: "Effect of heat upon anthracene photoconductance in air", *J. Chem. Phys.* **25**, 1284–5 (1956).

BREE, A. and LYONS, L. E.: "Measurement of thickness of thin crystals by interferometry", *J. Chem. Soc. (London)* 2658–61 (1956).

BREE, A. and LYONS, L. E.: "Absorption and reflection by anthracene of plane-polarised light", *J. Chem. Soc. (London)* 2662–70 (1956).

BREE, A. and LYONS, L. E.: "Photo- and semiconductance of organic crystals. Part III: Effect of oxygen on the surface photo-current and some photo-chemical properties of solid anthracene", *J. Chem. Soc.* 5179 (1960).

BREE, A., REUCROFT, P. J., and SCHNEIDER, W.G.: "Trapping centers and electronic conduction processes in anthracene and 9-10,dichloro-anthracene", in *Electrical Conductivity in Organic Solids* (eds. H. Kallmann and M. Silver), Wiley Interscience (New York), 113–25 (1961).

BRENIG, W., DÖHLER, G. H., and HEYSZENAU, H.: "$T^{-1/3}$- hopping in organic charge transfer crystals", *Phys. Lett.* A **39**, 175–6 (1972).

BRENIG, W., DÖHLER, G. H., and HEYSZENAU, H.: "Conduction by thermally assisted hopping in organic charge-transfer crystals", *Proceedings of the International Conference on Semiconductors* **11**, 490–4 (1972).

BRENNAN, W. D., BROPHY, J. J., and SCHONHORN, H.: "Electrical conductivity in pyrolyzed polyacrylonitrile", in *Organic Semiconductors* (eds. J. J. Brophy and J. W. Buttrey), Macmillan (New York), 159–68 (1962).

BRENNER, H. C.: "Electron spin polarized triplet state energy transfer in mixed organic single crystals at lower temperatures", *J. Chem. Phys.* **59**, 6362–79 (1973).

BRIAN, M. T.: "High frequency small signal characteristics of the SCL heterojunction transistor", *Solid–State Electron* **17**, 47–59 (1974).

BRIDGE, N. J. and VINCENT, D.: "Fluorescence and Raman spectra of pure and doped anthracene crystals at 4°K", *J. Chem. Soc. Faraday Trans.* II, **68**, 1522–35 (1972).

BRIGHT, A. A., CHAIKIN, P. M., and McGHIE, A. R.: "Photoconductivity and small-polaron effects in tetracyanoquinodimethan", *Phys. Rev.* B **10**, 3560–8 (1974).

BRIGODIOT, M. and LEBAS, J. M.: "Vibration frequencies of anthracene", *C. R. Acad. Sci. Paris*, **268B**, 51–54 (1969).

BRILLANTE, A., CRAIG, D. P., MAU, A. W. H., and RAJIKAN, J.: "Impurity induced phosphorescence in anthracene crystal", *Chem. Phys. Lett.* **30**, 5–10 (1975).

BRINEN, J. S. and ARLOFF, M. K.: "Zero-field splitting in phosphorescent triplet states of aromatic hydrocarbons", *J. Chem. Phys.* **45**, 4747–50 (1966).

BRODIN, M. S. and MARISOVA, S. V.: "Dependence of the refractive index of anthracene crystals at 20° K on their thickness", *Ukr. Fiz. Zh.* **6**, 742–4 (1961).

BRODIN, M. S. and MARISOVA, S. V.: "Dependence of the spectral distribution of transparency of anthracene single crystals on their thickness", *Ukr. Fiz. Zh.* **6**, 745 (1961).

BRODIN, M. S. and PEKAR, S. I.: "On an experimental determination of the existence of additional anomalous light waves in a crystal in the exciton absorption region", *Soviet Phys.—JETP* **1**, 55–60 (1960).

BRODIN, M. S. and PRIKHOTJKO, A. F.: "Influence of the thickness of anthracene crystals on their absorption curves at 20° K", *Opt. Spektrosk.* **12**, 82–83 (1959).

BRODIN, M. S. and PRIKHOTJKO, A. F.: "Additional abnormal light waves in anthracene in the region of exciton absorption", *Soviet Phys.—JETP* **11**, 1373–5 (1961).

BRODOVIL, V. A., PEKA, G. P., and SMOLYAR, A. N.: "Role of deep trapping levels information of current-voltage characteristics of S-type diode", *Soviet Phys.—Semicond.* **11**, 162–4 (1977).

BRODRIBB, J. D., O'COLMAIN, D., and HUGHES, D. M.: "The theory of photon-simulated current spectroscopy", *J. Phys.* D **8**, 856–62 (1975).

BRODRIBB, J. D., O'COLMAIN, D., and HUGHES, D. M.: "Photon-stimulated current analysis of trap parameters in anthracene", *J. Phys.* D **9**, 253–63 (1976).

BRODZELI, M. I., ELIGULASHVILI, I. A., KERTSMAN, E. L., NAKASHIDZE, G. A., and ROZENSHTEIN, L. D.: "Injection electroluminescence of anthracene", *Soviet Phys.—Semicond.* **4**, 811–12 (1970).

Brojdo, S.: "Characteristics of the dielectric diode and triode at very high frequencies", *Solid State Electron.* **6**, 611–29 (1963).

Bromberg, A., Tang, C. W., and Albrecht, A. C.: "Photoconductivity of chlorophyll-a induced by a tunable dye laser", *J. Chem. Phys.* **60**, 4058–62 (1974).

Broude, V. L., Dolganov, V. K., Slobadskoi, F. B., and Sheka, F. F.: "Exciton–phonon interaction and energy transfer in benzene and isotropically impure deuterobenzene crystals", *J. Lumin.* **8**, 367–74 (1974).

Browen, E. J.: "The fluorescence of naphthacene in anthracene", *J. Chem. Phys.* **13**, 306 (1945).

Brown, F. C.: "Temperature dependence of electron mobility in AgCl", *Phys. Rev.* **97**, 355–62 (1955).

Brown, J. M. and Jordan, A. G.: "Injection and transportation of added carriers in silicon at liquid helium temperatures", *J. Appl. Phys.* **37**, 337–46 (1966).

Bryant, F. J., Bree, A., Fielding, P. E., and Schneider, W. D.: "Trapping centres in anthracene crystals", *Discuss. Faraday Soc.* **28**, 48–53 (1959).

Bube, R. H.: "Photoconductivity and crystal imperfections in CdS crystals. Pt. II. Determination of characteristic photoconductivity quantities", *J. Chem. Phys.* **23**, 18–25 (1955).

Bube, R. H.: "Trap density determination by space-charge-limited currents", *J. Appl. Phys.* **33**, 1733–7 (1962).

Bube, R. H.: "Comparison of solid-state photoelectronic radiation detectors", *Trans. AIME* **239**, 291–300 (1967).

Bube, R. H.: "Photoconductivity in amorphous semiconductors", *RCA Rev.* **36**, 467–84 (1975).

Bube, R. H., Dussel, G. A., Ho, C. T., and Miller, L. D.: "Determination of electron trapping parameters", *J. Appl. Phys.* **37**, 21–31 (1966).

Bube, R. H. and Thomsen, M.: "Photoconductivity and crystal imperfection in CdS crystals. Pt. I: Effect of impurities", *J. Chem. Phys.* **23**, 15–17 (1955).

Bubnov, L. Ya. and Frankevich, E. L.: "Secondary electron emission from some aromatic hydrocarbons", *Phys. Stat. Sol.* (b) **62**, 281–90 (1974).

Bubnov, L. Ya. and Frankevich, E. L.: "Comparison of spectra characteristics loss of slow electrons in anthracene and phthalocyanine of copper with optic absorption spectra", *Fiz. Tverd. Tela.* **16**, 1533–4 (1974).

Bucker, W.: "ESR, Seebeck and Hall measurements on an amorphous system with hopping conduction", *J. Non-Cryst. Solids* **18**, 11–19 (1975).

Buckley, W. D. and Holmberg, S. H.: "Evidence for critical-field switching in amorphous semiconductor materials", *Phys. Rev. Lett.* **32**, 1429–32 (1974).

Buckley, W. D. and Holmberg, S. H.: "Electrical characteristics and threshold switching in amorphous semiconductors", *Solid-State Electron.* **18**, 127–47 (1975).

Buechner, W. and Mehl, W.: "Injection of a negative space charge into anthracene single crystals", *Z. Phys. Chem. (Frankfurt)* **44**, 376–9 (1965).

Buimistrov, V. M., Gorban, A. P., and Litovchenko, V. G.: "Photo-voltage induced by capture of photocarriers by surface traps", *Surf. Sci.* **3**, 445–60 (1965).

Bulliard, H.: "Photomagnetoelectric effect in germanium and silicon", *Phys. Rev.* **94**, 1564–6 (1954).

Bullis, W. M.: "Measurement of carrier lifetime in semiconductors: an annotated bibliography covering the period 1949–1967", Document AD674627 National Technical Information Service, Springfield, Virginia, USA (1968).

Buravov, L. I., Khidekel, M. L., Shchegolev, I. F. and Yagubskii, E. B.: "Superconductivity and dielectric constant of highly conductive complexes of TCQM", *Soviet Phys.—JETP Lett.* **12**, 99–100 (1970).

Burgess, R. E.: "Band conduction and fluctuation in polymeric semi-conductors", *J. Polymer Sci.* C **17**, 51–72 (1967).

Burland, D. M.: "Cyclotron resonance in a molecular crystal-anthracene", *Phys. Rev. Lett.* **33**, 835–6 (1974).

Burland, D. M. and Konzelmann, U.: "Cyclotron resonance and carrier scattering processes in anthracene crystals", *J. Chem. Phys.* **67**, 319–31 (1977).

Burns, G. and Sherwood, J. N.: "Self-diffusion in phenanthrene single crystals", *Mol. Cryst. Liq. Cryst.* **18**, 91–94 (1972).

Burns, G. and Sherwood, J. N.: "Self- and impurity diffusion in anthracene single crystals", *J. Chem. Soc. Faraday Trans.* **68**, 1036–40 (1972).

Burshtein, Z., Levinson, J., and Many, A.: "Determination of the singlet–triplet absorption coefficient in anthracene", *Mol. Cryst. Liq. Cryst.* **25**, 45–52 (1974).

Burshtein, Z. and Many, A.: "Studies of hole-trap distribution and valence-band structure in anthracene", *Mol. Cryst. Liq. Cryst.* **25**, 31–44 (1974).

Burshtein, Z. and Williams, D. F.: "Charge carrier mobility in durene", *J. Chem. Phys.* **66**, 2746–8 (1977).

Burshtein, Z. and Williams, D. F.: "Temperature dependence and anisotropy of electron drift mobilities in anthraquinone", *Mol. Cryst. Liq. Cryst.* **43**, 1–11 (1977).

BURSHTEIN, Z. and WILLIAMS, D. F.: "Temperature dependence of electron transport and generation in biphenyl", *J. Chem. Phys.* **67**, 3392–8 (1977).

BURSHTEIN, Z. and WILLIAMS, D. F.: "Temperature dependence of carrier generation, and transport in para-terphenyl above and below the 180°K phase transition", *J. Chem. Phys.* **68**, 983–8 (1978).

BURSTEIN, E., PICUS, G., and SCLAR, N.: "Optical and photoconductive properties of silicon and germanium", in *Proceedings of the Conference on Photoconductivity, Atlantic City, New Jersey 1954* (ed. R. G. Breckenbridge), Wiley (New York) 353–413 (1956).

BUTCHER, P. N.: "On the rate equation formulation of the hopping conductivity problem", *J. Phys. C, Solid State Phys.* **5**, 1817–29 (1972).

BUTCHER, P. N.: "Stochastic interpretation of the rate equation formulation of hopping transport theory", *J. Phys. C* **7**, 879–92 (1974).

BUTCHER, P. N.: "Stochastic interpretation of the rate equation formulation of hopping transport theory II—the DC limit", *J. Phys. C* **7**, 2645–54 (1974).

BUTLER, W. H. and FROMHOLD, A. T.: "Difference equation solutions for non-steady-state diffusion of charged particle in solids", *J. Phys. Chem. Solids* **35**, 1099–1113 (1974).

BYKOVSKII, YU. A., ELESIN, V. F., and ZUEV, V. V.: "Negative differential conductivity in a semiconductor containing attractive impurity centers", *Soviet Phys.—Semicond.* **3**, 1442–3 (1970).

BYKOVSKII, YU. A., VINOGRADOV, K. N., and ZUEV, V. V.: "Negative resistance of gold-doped $p-n_i-n$ silicon diodes", *Fiz. Tekhn. Poluprov.* **1**, 1559 (1968); *Soviet Phys.—Semicond.* **1**, 1295 (1968).

BYKOVSKII, YU. A., VINOGRADOV, K. N., ZUEV, V. V., and KOZYREV, YU. P.: "Negative photoconductivity of gold-doped silicon", *Soviet Phys.—Semicond.* **3**, 933–5 (1970).

CAMPOS, M.: "Space charge limited current in naphthalene single crystals", *Mol. Cryst. Liq. Cryst.* **18**, 105–15 (1972).

CANALI, C., NICOLET, M. A., and MAYER, J. W.: "Transient and steady state space-charge-limited in CdTe", *Solid–State Electron.* **18**, 871–4 (1975).

CANALI, C., OTTAVIANI, G., and ALBERIGI QUARANTA, A.: "Drift velocity of electrons and holes and associated anisotropic effects in silicon", *J. Phys. Chem. Solids* **32**, 1707–20 (1971).

CARCHANO, H., LACOSTE, R., and SEGNI, Y.: "Bistable electrical switching in polymer thin films", *Appl. Phys. Lett.* **19**, 414–15 (1971).

CARD, H. C.: "One-dimensional analysis of heat-treated aluminum silicon junctions", in *Metal–Semiconductor Contacts* (hon. ed. M. Pepper), Institute of Physics (London) Conference Series No. 22, 129–31 (1974).

CARD, H. C.: "Potential barriers to electron tunneling in ultra-thin films of SiO_2", *Solid State Commun.* **14**, 1011–14 (1974).

CARD, H. C.: "Factors affecting avalanche injection into insulating layers from a semiconductor surface", *Solid-State Electron.* **17**, 501–2 (1974).

CARD, H. C.: "On the direct currents through interface states in metal-semiconductor contacts", *Solid-State Electron.* **18**, 881–3 (1975).

CARD, H. C.: "Collection velocity of excess minority carriers at metal-semiconductor contacts in solar cells", *J. Appl. Phys.* **47**, 4964–7 (1976).

CARD, H. C.: "Photovoltaic properties of MIS–Schottky barriers", *Solid-State Electron.* **20**, 971–6 (1977).

CARD, H. C. and RHODERICK, E. H.: "Studies of tunnel MOS diodes. I: Interface effects in silicon Schottky diodes", *J. Phys. D* **4**, 1589–1601 (1971).

CARD, H. C. and RHODERICK, E. H.: "Studies of tunnel MOS diodes. II: Thermal equilibrium considerations", *J. Phys. D* **4**, 1602–11 (1971).

CARD, H. C. and RHODERICK, E. H.: "Conductance associated with interface states in MOS tunnel structures", *Solid-State Electron.* **15**, 993–8 (1972).

CARD, H. C. and RHODERICK, E. H.: "The effect of an interfacial layer on the minority carrier injection in forward-biased silicon Schottky diodes", *Solid-State Electron.* **16**, 365–74 (1973).

CARD, H. C. and SMITH, B. L.: "Green injection luminescence from forward biased Au-GaP Schottky barriers", *J. Appl. Phys.* **42**, 5863–5 (1971).

CARD, H. C. and YANG, E. S.: "MIS–Schottky theory under conditions of optical carrier generation in solar cells", *Appl. Phys. Lett.* **29**, 51–53 (1976).

CARD, H. C. and YANG, E. S.: "Electronic processes at grain boundaries in polycrystalline semiconductors under optical illumination", *IEEE Trans.* ED **24**, 397–402 (1977).

CARDEW, M. H. and ELEY, D. D.: "The semiconductivity of the organic substances Part 3 – Haemoglobin and some amino acids", *Discuss. Faraday Soc.* **27**, 115–28 (1959).

CARLES, D., VAUTIER, C., and TOURAINE, A.: "Caractéristiques courant-tension en régimes de charge d'espace en présence d'une distribution uniforme de pièges. 1: Calcul de la fonction théorique", *Thin Solid Films* **9**, 219–28 (1972).

CARSWELL, D. J.: "Correction between photoconductance and ultraviolet spectrum of anthracene crystal", *J. Chem. Phys.* **21**, 1890 (1953).

CARSWELL, D. J., FERGUSON, J., and LYONS, L. E.: "Photo- and semi-conductance in molecular single crystals", *Nature* **173**, 736 (1954).

CARSWELL, D. J. and LYONS, L. E.: "Photo- and semi-conductance of organic crystals. II: Spectral dependence, quantum efficiency and relation between semi- and photo-effects in anthracene", *J. Chem. Soc.* 1734–40 (1955).

CARUSO, A., SPIRITO, P., and VITALE, G.: "Negative resistance induced by avalanche injection in bulk semiconductors", *IEEE Trans. Electron. Devices* ED **21**, 578–86 (1974).

CARUSOTTO, S., POLACCO, E., and VASELLI, M.: "Experiment on two-photon absorption and coherence of light", *Lettere al Nuovo Cimento*, Serie 1, **2**, 628–30 (1969).

CASALINI, G., CORAZZARI, T., and GAROFANO, T.: "Further investigation of thermally stimulated currents in anthracene crystals", *IL Nuovo Cimento* B **38**, 141–8 (1977).

CASALINI, G., CORAZZARI, T., and GAROFANO, T.: "Evidence for surface states in anthracene crystals", *IL Nuovo Cimento* B **41**, 273–83 (1977).

CASERTA, G., RISPOLI, B., and SERRA, A.: "Space charge-limited current and band structure in amorphous organic films", *Phys. Stat. Sol.* B **35**, 237–248 (1969).

CASEY JR., H. C. and STERN, F.: "Concentration-dependent absorption and spontaneous emission of heavily doped GaAs", *J. Appl. Phys.* **47**, 631–43 (1976).

CASTRO, G.: "On Wannier excitons in anthracene", *J. Chem. Phys.* **46**, 4997–8 (1967).

CASTRO, G.: "Photoconduction in aromatic hydrocarbons", *IBM Jl Res. Dev.* **15**, 27–33 (1971).

CASTRO, G. and HORNIG, J. F.: "Multiple charge-carrier generation processes in anthracene", *J. Chem. Phys.* **42**, 1459 (1965).

CASTRO, G. and ROBINSON, G. W.: "Singlet–triplet absorption of crystalline naphthalene by high-resolution photoexcitation spectroscopy", *J. Chem. Phys.* **50**, 1159–64 (1969).

CAYWOOD, J. M.: "Photoemission from metal contacts into anthracene crystals: a critical review", *Mol. Cryst. and Liq. Cryst.* **12**, 1–26 (1970).

CERRINA, F., MARGARITONDO, G., MIGLIORATO, P., PERFETTI, P., and SALUSTI, E.: "Carrier injection in semiconductors: pre-breakdown electrical conductivity theory", *Nuovo Cimento* **27**, 229–39 (1975).

CHABR, M., FÜNFSCHILLING, J., and ZSCHOKKE-GRÄNACHER, I.: "Magnetic field effects on exciton annihilation processes in tetracene doped anthracene crystals", *Chem. Phys. Lett.* **25**, 387–9 (1974).

CHABR, M. and WILLIAMS, D. F.: "Fission of singlet excitons into triplet-exciton pairs in molecular crystals", *Phys. Rev.* B **16**, 1685–93 (1977).

CHABR, M. and ZSCHOKKE-GRÄNACHER, I.: "Magnetic effects on delayed fluorescence in doped naphthalene and anthracene crystals", *J. Chem. Phys.* **64**, 3903–10 (1976).

CHADWICK, A. V. and SHERWOOD, J. N.: "Point defects in molecular solids", in *Points Defects in Solids* (eds. J. H. Crawford Jr. and L. M. Slifkin), Plenum Press (New York), Vol. 2, Chapter 6, 441–75 (1975).

CHAI, Y. G. and ADERSON, W. W.: "Semiconductor–electrolyte photovoltaic cell energy conversion efficiency", *Appl. Phys. Lett.* **27**, 183–4 (1975).

CHAIKEN, R. F. and KEARNS, D. R.: "Intrinsic photoconduction in anthracene crystals", *J. Chem. Phys.* **45**, 3966–76 (1966).

CHAIKIN, P. M., GARITO, A. F., and HEEGER, A. F.: "Excitonic polarons in molecular crystals", *Phys. Rev.* B **5**, 4966–9 (1972).

CHAIKIN, P. M., GREENE, R. L., ETEMAD, S., and ENGLER, E.: "Thermopower of an isostructural series of organic conductors", *Phys. Rev.* B **13**, 1627–32 (1976).

CHAN, Y. and JAYADEVAIAH, T. S.: "Screening effects on Poole–Frenkel conduction in amorphous solids", *Phys. Stat. Sol.* (b) **49**, K129–33 (1972).

CHANCE, R. R. and BRAUN, C. L.: "Intrinsic photoconduction in anthracene single crystals: electric field dependence of hole and electron quantum yields", *J. Chem. Phys.* **59**, 2269–72 (1973).

CHANCE, R. R. and BRAUN, C. L.: "Temperature dependence of intrinsic carrier generation in anthracene single crystals", *J. Chem. Phys.* **64**, 3573–81 (1976).

CHANCE, R. R. and PROCK, A.: "Role of singlet and triplet excitons in extrinsic photocurrent production in anthracene–gold system", *Phys. Stat. Sol.* (b) **57**, 597–609 (1973).

CHANDROSS, E. A. and FERGUSON, J.: "Absorption and excimer fluorescence spectra of sandwich dimers of substituted anthracene", *J. Chem. Phys.* **45**, 3554–63 (1966).

CHANDROSS, E. A. and FERGUSON, J.: "Photodimerization of crystalline anthracene – the photolytic dissociation of crystalline dianthracene", *J. Chem. Phys.* **45**, 3564–7 (1966).

CHANDROSS, E. A., FERGUSON, J., and McRAE, E. G.: "Absorption and emission spectra of anthracene dimers", *J. Chem. Phys.* **45**, 3546–53 (1966).

CHANG, J. J.: "Nonvolatile semiconductor memory devices", *Proc. IEEE* **60**, 1039–59 (1976).

CHANG, L. L., STILES, P. J., and ESAKI, L.: "Electron barriers in Al–Al$_2$O$_3$–ZnTe and Al–Al$_2$O$_3$–GeTe tunnel junctions", *IBM Jl Res. Dev.* **10**, 484–6 (1966).

CHANG, L. L., STILES, P. J., and ESAKI, L.: "Electron tunneling between a metal and a semiconductor: characteristics of Al–Al$_2$O$_3$–SnTe and GeTe junctions", *J. Appl. Phys.* **38**, 4440–5 (1967).

CHANNON, M. A., MALTBY, J. R., REED, C. E., and SCOTT, C. G.: "The spectral dependence of the surface photovoltage in CdS", *J. Phys.* D **8**, L39–41 (1975).

CHAR, M., FÜNFSCHILLING, J., TSCHARNER, V. V., and ZSCHOKKE-GRÄNACHER, I.: "The effect of a magnetic field on the interaction of triplet excitons in doped anthracene crystal", *Helv. Phys. Acta* **47**, 25 (1974).

CHATTOPADHYAYA, S. K. and MATHUR, V. K.: "Photomagnetoelectric effect in graded band-gap semiconductors", *Phys. Rev.* B **3**, 3390–9 (1971).

CHAUDHURY, N. K. and EL-SAYED, M. A.: "Host-crystal effects on the mechanism of the phosphorescence process of aromatic hydrocarbons", *J. Chem. Phys.* **43**, 1423–4 (1965).

CHAUDHURY, N. K. and GANGULY, S. C.: "The absorption and fluorescence spectra of naphthacene molecules in anthracene crystals", *Proc. Roy. Soc. (London)* A **259**, 419 (1960).

CHEE, K. T. and WEICHMAN, F. L.: "Pt–Cu$_2$O–Cu double injection diodes", *Can. J. Phys.* **55**, 727–34 (1977).

CHEN, I.: "Xerographic discharge characteristics of photoreceptors II", *J. Appl. Phys.* **43**, 1137–43 (1972).

CHEN, I.: "Theory of thermally stimulated current in hopping systems", *J. Appl. Phys.* **47**, 2988–94 (1976).

CHEN, I.: "Effects of bimolecular recombination on photocurrent and photoinduced discharge", *J. Appl. Phys.* **49**, 1162–72 (1978).

CHEN, I. and KAO, C. C.: "Xerographic discharge characteristics of photoreceptors with bulk generation. II Time-varying exposure", *J. Appl. Phys.* **44**, 2718–23 (1973).

CHEN, I. and MORT, J.: "Xerographic discharge characteristics of photoreceptors I", *J. Appl. Phys.* **43**, 1164–70 (1972).

CHEN, I., MORT, J., and TABAK, M. D.: "Photoinduced discharge characteristics of xerographic photoreceptors", *IEEE Trans.* ED **19**, 413–21 (1972).

CHEN, R.: "On the calculation of activation energies and frequency factors from glow curves", *J. Appl. Phys.* **40**, 570–85 (1969).

CHEN, R.: "Simultaneous measurement of thermally stimulated conductivity and thermoluminescence", *J. Appl. Phys.* **42**, 5899–5901 (1971).

CHEN, R.: "On the methods for determining trap depth from glow curves", *J. Mater. Sci.* **9**, 345–9 (1974).

CHEROFF, G.: "Temperature dependence of anomalous photovoltages in ZnS", *Bull. Am. Phys. Soc.* **6**, 110 (1961).

CHILDS, G. E., ERICKS, L. J., and POWELL, R. L.: "Thermal conductivity of solids at room temperature and below—a review and compilation of the literature", Report NBS–Mono–131, Nat. Bur. Stand., Washington DC, USA 1–608 (1973).

CHOI, S. I.: "Ionization of molecular excitons", *J. Chem. Phys.* **40**, 1691–2 (1964).

CHOI, S. I.: "Mutual interaction and ionization of excitons in aromatic molecular crystals", *J. Chem. Phys.* **43**, 1818–25 (1965).

CHOI, S. I.: "Collision annihilation of singlet excitons in molecular crystals", *Phys. Rev. Lett.* **19**, 358–60 (1967).

CHOI, S. I., JORTNER, J., RICE, S. A., and SILBEY, R.: "Charge-transfer exciton states in aromatic molecular crystals", *J. Chem. Phys.* **41**, 3294–3306 (1964).

CHOI, S. I. and RICE, S. A.: "Exciton–exciton interactions and photoconductivity in crystalline anthracene", *Phys. Rev. Lett.* **8**, 410–12 (1962).

CHOI, S. I. and RICE, S. A.: "Exciton–exciton and photoconductivity in crystalline anthracene", *J. Chem. Phys.* **38**, 366–73 (1963).

CHOJNACKI, H.: "Temperature dependence of current carriers mobilities in anthracene", *Mol. Cryst.* **3**, 375–83 (1968).

CHOJNACKI, H.: "On the tight binding calculations for anthracene crystal", *Mol. Cryst.* **5**, 313–15 (1969).

CHOJNACKI, H., LORENZ, K., and PIGON, K.: "Combined hopping and tunneling mechanism of electron transport and doping effects in chloraniltrimethylamine charge-transfer complex", *Phys. Stat. Sol.* (a) **20**, 211–16 (1973).

CHONG, T. and ITOH, N.: "Interaction between excitons and radiation-induced radicals in naphthalene single crystals", *Phys. Stat. Sol.* (b) **74**, 111–20 (1976).

CHONG, T. and ITOH, N.: "Radiation-induced radicals in anthracene single crystals", *Mol. Cryst. Liq. Cryst.* **36**, 99–113 (1976).

CHOW, C. K.: "Square-mean-root approximation for evaluating asymmetric tunneling characteristics", *J. Appl. Phys.* **36**, 559–63 (1965).

CHOWDHURY, M. and BASU, S.: "Some notes on molecular complexes between iodine and polynuclear aromatic hydrocarbons", *J. Chem. Phys.* **32**, 1450–2 (1960).

CHOWDRY, A. and WESTGATE, C. R.: "The role of bulk traps in metal–insulator contact charging", *J. Phys.* D: *Appl. Phys.* **7**, 713–25 (1974).

CHU, C. W., GEBALLE, T. H., HARPER, J. M. E., and GREENE, R. L.: "Pressure dependence of the metal–insulator transition in tetrathiofulvalinium tetracyanoquinodimethane (TTF–TCNQ)", *Phys. Rev. Lett.* **31**, 1491–4 (1973).

CHU, J. L., PERSKY, G. and SZE, M.: "Thermionic injection and space-charge-limited current in reach-through P^+np^+ structures", *J. Appl. Phys.* **43**, 3510–15 (1972).

CHUNG, K. L.: "An anomalous crystal photo-effect in d-tartaric acid single crystals", *Phys. Rev.* **60**, 529–31 (1941).

CHYNOWETH, A. G.: "The effect of oxygen on the photoconductivity of anthracene II", *J. Chem. Phys.* **22**, 1029–32 (1954).

CHYNOWETH, A. G. and SCHNEIDER, W. G.: "Photoconductivity of anthracene I", *J. Chem. Phys.* **22**, 1021–8 (1954).

CISNEROS, G. and MARK, P.: "Unipolar current transport through metal–insulator–metal (MIM) structures supplied with barrier contacts, in the diffusion limit", *Solid-State Electron.* **18**, 563–8 (1975).

CLAESER, R. M. and BERRY, R. S.: "Mobilities of electrons and holes in organic molecular solids: comparison of band and hopping models", *J. Chem. Phys.* **44**, 3797–810 (1966).

CLARK, L. B.: "Dependence of the electronic spectra of anthracene crystals on propagation direction", *J. Chem. Phys.* **51**, 5719–20 (1969).

CLARKE, B. P., THOMAS, J. M., and WILLIAMS, J. O.: "The relationship between crystallographic structure and luminescence in single crystal of acridines", *Chem. Phys. Lett.* **35**, 251–4 (1975).

CLARKE, H. B., NORTHROP, D. C., and SIMPSON, O.: "The scintillation phenomena in anthracene. I: Radiation damage. II: Scintillation pulse shape", *Proc. Phys. Soc.* **79**, 366–72, 372–82 (1962).

CLARKE, R. A., GREEN, M. A., and SHEWCHUN, J.: "Contact area dependence of minority-carrier injection in Schottky barrier diodes", *J. Appl. Phys.* **45**, 1442–3 (1974).

CLARKE, R. H.: "Phosphorescence and Zeeman spectra of anthracene in a mixed crystal", *J. Chem. Phys.* **52**, 2328–32 (1970).

CLARKE, R. H. and HOCHSTRASSER, R. M.: "Electronic Zeeman effect in anthracene", *J. Chem. Phys.* **46**, 4532–3 (1967).

CLEMENTI, E.: "Study of the electronic structure of molecular crystals. I: Molecular wavefunctions and their analysis", *J. Chem. Phys.* **46**, 3842–50 (1967).

CLEMENTI, E., ROOTHAAN, C. C. J., and YOSHIMINE, M.: "Accurate analytical self-consistent field functions for atoms. II: Lowest configurations of the neutral first row atoms", *Phys. Rev.* **127**, 1618–20 (1962).

COBAS, A., CHALVERUS, J., SZMANT, H. H., TRESTER, S., and WEISZ, S. Z.: "Method for growing thin anthracene crystals of large area", *Rev. Sci. Instrum.* **37**, 232 (1966).

COBAS, A., RICHARDSON, P. E., TRESTER, S., and WEISZ, S. Z.: "Measurement of the characteristic of radiation induced hole traps in anthracene crystals", *International Symposium on Luminescence, Munich, 1965*, 168–72 (1965).

COELHO, R.: "Theory of the space-charge build-up and current transient in a weakly conducting layer with an induced conductivity gradient", *Phys. Stat. Sol.* (a), **31**, 563–77 (1975).

COHEN, M. D., KLEIN, E., LUDMER, Z., and YAKHOT, V.: "The absorption and fluorescence properties of pyrene crystal: a theoretical approach. I: Harmonic approximation", *Chem. Phys.* **5**, 15–26 (1974).

COHEN, M. E. and LANDSBERG, P. T.: "Impact ionization theory for traps in semiconductors", *Phys. Stat. Sol.* B **64**, 39–48 (1974).

COHEN, M. H., FRITZSCHE, H., and OVSHINSKY, S. R.: "Simple band model for amorphous semiconductor alloys", *Phys. Rev. Lett.* **22**, 1065–8 (1969).

COHEN, M. J., COLEMAN, L. B., GARITO, A. F., and HEEGER, A. J.: "Electrical conductivity of tetrathiofulvalinium tetracyanoquinodimethan (TTF) (TCNQ)", *Phys. Rev.* B **10**, 1298–1307 (1974).

COLEMAN, L. B., COHEN, J. A., CARITO, A. F., and HEEGER, A. J.: "Conductivity studies on high-purity N-methylphenazinium tetracyanoquinodimethan", *Phys. Rev.* B **7**, 2122–7 (1973).

COLEMAN, L. B., COHEN, M. J., SANDMAN, D. J., YAMAGISHI, F. G., GARITO, A. F., and HEEGER, A. J.: "Superconducting fluctuations and the Peierls instability in an organic solid", *Solid State Commun.* **12**, 1125–32 (1973).

COMPTON, D. M. J., SCHNEIDER, W. G., and WADDINGTON, T. C.: "Photoconductivity of anthracene. III", *J. Chem. Phys.* **27**, 160–72 (1957).

COMPTON, D. M. J., SCHNEIDER, W. G., and WADDINGTON, T. C.: "Photoconductivity of anthracene: the effect of neutron bombardment", *J. Chem. Phys.* **28**, 741–2 (1958).

COMPTON, D. M. J. and WADDINGTON, T. C.: "Photoconductivity and semiconductivity of anthracene: effects of NO_2 and Cl_2", *J. Chem. Phys.* **25**, 1075 (1956).

CONDON, E. U.: "The Franck–Condon principle and related topics", *Am. J. Phys.* **15**, 365–75 (1947).

CONNELL, G. A. N., CAMPHAUSEN, D. L., and PAUL, W.: "Theory of Poole–Frenkel conduction in low-mobility semiconductors", *Phil. Mag.* **26**, 541–51 (1972).

CONWELL, E. M.: "Impurity band conduction in germanium and silicon", *Phys. Rev.* **103**, 51–61 (1956).

CONWELL, E. M.: "Mobility in tetrathiafulvalene-tetracyanoquinodimethane (TTF-TCNQ)", *Phys. Rev. Lett.* **39**, 777–80 (1977).

Cook, B. E. and Le Comber, P. G.: "The optical properties of anthracene crystals in the vacuum ultraviolet", *J. Phys. Chem. Solids* **32**, 1321–9 (1971).

Coppage, F. N. and Kepler, R. G.: "Generation of electrons and holes in anthracene by ruby laser light", *Mol. Cryst.* **2**, 231–9 (1967).

Corazzari, T. and Garofano, T.: "Trapping effects on the mobility of anthracene crystals", *IL Nuovo Cimento* B **33**, 719–31 (1976).

Corazzari, T. and Garofano, T.: "Shallow trapping centres in anthracene crystals", *IL Nuovo Cimento* B **33**, 732–9 (1976).

Coret, A. and Fort, A.: "Enhanced photocurrent in anthracene crystals at 77° K", *Solid State Commun.* **15**, 145–7 (1974).

Coret, A. and Fort, A.: "Photoconductivity by polaritons in anthracene crystals", *IL Nuovo Cimento* B **39**, 544–9 (1977).

Coret, A., Nikitine, S., Zielinger, J. P., and Zouaghi, M.: "Formation and dissociation processes of excitons—influence of localized electric fields", in *Proceedings of the 3rd International Conference on Photoconductivity* (ed. E. M. Pell), Pergamon Press (Oxford) 81–85 (1971).

Corke, N. T., Sherwood, J. N., and Jarnagin, R. C.: "Growth and perfection of organic crystals. 2: Distribution of line defects in pure and doped single crystals of anthracene", In *Proceedings of the 2nd International Conference on Crystal Growth, Birmingham, 1968* (eds. F. C. Frank, J. B. Mullin and H. S. Peiser) North-Holland, 766–70 (1968).

Corker, G. A. and Lundström, I.: "Trapped-electron doping of photovoltaic sandwich cells containing microcrystalline chlorophyll-a", *J. Appl. Phys.* **49**, 686–700 (1978).

Courtens, E., Bergman, A., and Jortner, J.: "Photo-ionization of two-photon-excited singlet excitons in anthracene", *Phys. Rev.* **156**, 948–50 (1967).

Covington, D. W. and Ray, D. C.: "Effect of surface regions on the electrical conductivity of extrinsic semiconductor films", *J. Appl. Phys.* **45**, 2616–20 (1974).

Cowley, A. M. and Heffner, H.: "Gallium-phosphide–gold surface barrier", *J. Appl. Phys.* **35**, 255–60 (1964).

Cox, G. A. and Knight, P. C.: "The temperature dependence of charge carrier drift mobility in single crystals of metal-free phthalocyanine in the β-phase", *Phys. Stat. Sol.* B **50**, K135–7 (1972).

Cox, G. A. and Knight, P. C.: "Electrical conduction mechanism and carrier trapping in β-metal free phthalocyanine single crystals", *J. Phys. Chem. Solids* **34**, 1655–9 (1973).

Cox, G. A. and Knight, P. C.: "Drift mobility, trapping and photogeneration of charge carriers in β-metal free phthalocyanine single crystals", *J. Phys. C* **7**, 146–56 (1974).

Craig, D. P.: "The polarized spectrum of anthracene. II: Weak transitions and second-order crystal field perturbation", *J. Chem. Soc.* 2302–8 (1955).

Craig, D. P.: "Model calculations in the theory of mixed-crystal spectra", *Adv. Chem. Phys.* **8**, 27–46 (1965).

Craig, D. P.: "The neutral exciton model for triplet factor-group splittings in aromatic crystals", *Chem. Phys. Lett.* **22**, 239–42 (1973).

Craig, D. P. and Hobbins, P. C.: "Polarized UV absorption of anthracene", *Nature* **171**, 566 (1953).

Craig, D. P. and Hobbins, P. C.: "The polarized spectrum of anthracene. I: Assignment of the intense short wave length system", *J. Chem. Soc.* 539–48 (1955).

Craig, D. P. and Hobbins, P. C.: "The polarized spectrum of anthracene. III: The system at 3800 Å", *J. Chem. Soc.* 2309–19 (1955).

Craig, D. P. and Philpott, M. R.: "The electronic states of mixed molecular crystals. I: Theory for shallow traps", *Proc. Roy. Soc. (London)* A **290**, 583–601 (1965).

Craig, D. P. and Philpott, M. R.: "The electronic states of fixed molecular crystals. II. Application of the theory to the mixed crystals of naphthalene and deuteronaphthalenes", *Proc. Roy. Soc. (London)* A **290**, 602–12 (1965).

Craig, D. P. and Philpott, M. R.: "The electronic states of mixed molecular crystals. III: The general trapping problems", *Proc. Roy. Soc. (London)* A **293**, 213–34 (1966).

Craig, D. P. and Ross, I. G.: "The triplet–triplet absorption spectra of some aromatic hydrocarbons and related substances", *J. Chem. Soc.* 1589–1606 (1954).

Craig, D. P. and Thirunamachandran, T.: "The electronic spectra of mixed crystals", *Proc. Roy. Soc. (London)* A **271**, 207–17 (1963).

Craig, D. P. and Walmsley, S. H.: "Visible and ultraviolet absorption by molecular crystals", in *Physics and Chemistry of the Organic Solid State* (eds. I. Fox, M. M. Labes, and A. Weissberg), Wiley (New York) **1**, 585–655 (1963).

Crowell, C. R.: "The Richardson constant for thermionic emission in Schottky barrier diodes", *Solid-State Electron.* **8**, 395–9 (1965).

Crowell, C. R.: "Richardson constant and tunneling effective mass for thermionic-field emission in Schottky barrier diodes", *Solid-State Electron.* **12**, 55–59 (1969).

CROWELL, C. R., SPITZER, W. G., HOWARTH, L. E., and LABATE, E. E.: "Attenuation length measurements of hot electrons in metal films", *Phys. Rev.* **127**, 2006–11 (1962).

CROWELL, C. R. and SZE, S. M.: "Current transport in metal–semiconductor barriers", *Solid-State Electron.* **9**, 1035–48 (1966).

CROWELL, C. R. and SZE, S. M.: "Quantum mechanical reflection at metal–semiconductor barriers", *J. Appl. Phys.* **37**, 2683–9 (1966).

CRUICKSHANK, D. W. J.: "A detailed refinement of the crystal and molecular structure of naphthalene", *Acta Cryst.* **10**, 504–8 (1957).

CULP, C. H., ECKELS, D. E., and SIDLES, P. H.: "Threshold switching in melanin", *J. Appl. Phys.* **46**, 3658–60 (1975).

CUMMINS, P. G. and DUNMUR, D. A.: "The electric permittivity of crystalline anthracene", *J. Phys. D* **7**, 451–4 (1974).

CUMMINS, P. G., DUNMUR, D. A., and MUNN, R. W.: "The effective molecular polarizability and local electric field in molecular crystals", *Chem. Phys. Lett.* **22**, 519–22 (1973).

CUNNINGHAM, K., MORRIS, J. M., FÜNFSCHILLING, J., and WILLIAMS, D. F.: "Site selection spectroscopy—luminescence of solutions with laser excitation", *Chem. Phys. Lett.* **32**, 581–5 (1975).

CUNNINGHAM, K. and WILLAMS, D. F.: "Singlet–triplet absorption spectra of anthracene: Franck–Condon factors", *J. Chem. Phys.* **59**, 2150–1 (1973).

CURTIS, JR., O. L.: "Steady-state photoconductivity in the presence of traps", *Phys. Rev.* **172**, 773–8 (1968).

CURTIS, JR., O. L. and SROUR, J. R.: "Steady-state recombination in semiconductors containing two or more trapping levels", *J. Appl. Phys.* **45**, 792–4 (1974).

CUTLER, M.: "A molecular bonding theory for liquid thallium–tellurium alloys. I: Electrical behaviour", *Phil. Mag.* **24**, 381–416 (1971).

CUTLER, M.: "The thermoelectric behavior of disordered systems", *Phil. Mag.* **25**, 173–5 (1972).

CUTLER, M. and MOTT, N. F.: "Observation of Anderson localization in an electron gas", *Phys. Rev.* **181**, 1336–40 (1969).

DAHIYA, R. P. and MATHUR, V. K.: "Space-charge-limited currents in insulators under illumination", *J. Phys. D* **7**, 1512–17 (1974).

DAHIYA, R. P. and MATHUR, V. K.: "Current injection in insulators containing traps under spherical and cylindrical flow", *J. Appl. Phys.* **47**, 3240–7 (1976).

DAHLKE, W. E. and SZE, S. M.: "Tunneling in metal–oxide–silicon structures", *Solid-State Electron.* **10**, 865–73 (1967).

DALAL, V. L., KRESSEL, H., and ROBINSON, P. H.: "Epitaxial silicon solar cell", *J. Appl. Phys.* **46**, 1283–5 (1975).

DALAL, V. L. and LAMPERT, M. A.: "Transient space-charge phenomena in semiconductors at high electric fields", *Solid-State Electron.* **16**, 689–99 (1973).

DAMASK, A. C.: "Charge transport in organic crystals", *U.S. Govt. Res. Dev. Rep. 1969*, **69** (11), 76 (1969) and also in *Solid State Phys.* **11**, 64–67 (1969).

DANNO, T. and INOKUCHI, H.: "Dynamic mechanical behavior of organic molecular crystals. 2: Elastic constants of single-crystal anthracene", *Bull. Chem. Soc. Japan* **41**, 1783–7 (1968).

DASCALU, D.: "Analysis of injection-controlled one-carrier flow in semiconductors", *Rev. Rom. Phys. (Romania)* **19**, 177–90 (1974).

DASCALU, D. and BREZEANU, GH.: "Theory of VHF detection and frequency multiplication with space-charge-limited current (SCLC) silicon diodes", *Solid-State Electron.* **18**, 437–48 (1975).

DATT, S. C., VERMA, J. K. D., and NAG, B. D.: "Electrical conductivity in organic semiconductors", *J. Sci. Indust. Res. (India)* **26** (2) 57–75 (1967).

DAVYDOV, A. S.: "Theory of absorption spectra of molecular crystals" (in Russian), *Zh. Eksperim. Theor. Fiz.* **18**, 210 (1948).

DAVYDOV, A. S.: "Excitons in thin crystals", *Soviet Phys.-JETP* **18**, 496–9 (1963).

DAVYDOV, A. S. and SHEKA, E. F.: "Structure of exciton bands in crystalline anthracene", *Phys. Stat. Sol.* **11**, 877–90 (1965).

DAY, P. and WILLIAMS, R. J. P.: "Spectra and photoconduction of phthalocyanine complexes (I)", *J. Chem. Phys.* **37**, 567–70 (1962).

DAY, P. and WILLIAMS, R. J. P.: "Photoconductivity of copper phthalocyanine in the near infrared", *J. Chem. Phys.* **42**, 4049–50 (1965).

DEAL, B. E., SNOW, E. H., and MEAD, C. A.: "Barrier energies in metal–silicon dioxide–silicon structures", *J. Phys. Chem. Solids* **27**, 1873–9 (1966).

DEAN, P. J.: "Inter-impurity recombinations in semiconductors", *Prog. in Solid State Chemistry* **8**, 1–126 (1973).

DEAN, R. H.: "Transient double injection in germanium", *Appl. Phys. Lett.* **13**, 164–6 (1968).

DEAN, R. H.: "Transient double injection in trap-free semiconductors", *J. Appl. Phys.* **40**, 585–95 (1969).

DEAN, R. H.: "Transient double injection in semiconductors with traps", *J. Appl. Phys.* **40**, 596–602 (1969).

DEARNALEY, G., STONEHAM, A. M., and MORGAN, D. V.: "Electrical phenomena in amorphous oxide films", *Rep. Prog. Phys.* **33**, 1129–91 (1970).

DELACOTE, G.: "Temperature dependence of transient photocurrents in anthracene crystals", *J. Chem. Phys.* **40**, 4315–16 (1965).

DELACOTE, G.: "Direct recombination of photoinjected electrons and holes in an anthracene crystal", *Compt. Rend. Ser.* **262B**, 958–61 (1966).

DELACOTE, G. and CHOJNACKI, H.: "Temperature dependence of current carrier mobilities in anthracene", *Mol. Cryst.* **5**, 309–11 (1969).

DELACOTE, G., FILLARD, J. P., and SCHOTT, M.: "Transient transport properties of optically injected carriers in copper phthalocyanine thin films", *Solid State Commun.* **4**, 137–40 (1966).

DELACOTE, G., QUÉDEC, P., and SCHOTT, M.: "Volume recombination of electrons with trapped holes in an anthracene crystal", *J. Phys. (Paris)* **29**, 1024–34 (1968).

DELACOTE, G. and SCHOTT, M.: "Hall mobility of holes and valence bandwidth in anthracene", *Solid State Commun.* **4**, 177–9 (1966).

DELACOTE, G. and SCHOTT, M.: "On charge carrier generation by exciton–exciton and exciton–photon interactions in crystalline anthracene", *Mol. Cryst.* **3**, 385–91 (1968).

DELANNOY, P. and SCHOTT, M.: "High-field anisotropies in the effect of magnetic field on D.C. photoconductivity of anthracene and tetracene", *Phys. Lett.* **30A**, 357–8 (1969).

DELANNOY, P. and SCHOTT, M.: "Ruby-laser-excited delayed fluorescence of crystalline tetracene. Combined effects of trap saturation and bimolecular processes", *Phys. Stat. Sol.* (b) **70**, 119–31 (1975).

DELANNOY, P., SCHOTT, M., and BERREHAR, J.: "Comment on the origin of photoenhanced currents in organic insulators under inhomogeneous excitation", *Phys. Stat. Sol.* (a) **32**, 577–84 (1975).

DELANY, J. L. and HIRSCH, J.: "Conductivity induced by electron bombardment anthracene", *Solid State Commun.* **4**, 107–9 (1966).

DELANY, J. L. and HIRSCH, J.: "Carrier generation and transport in anthracene under electron bombardment", *J. Chem. Phys.* **48**, 4717–24 (1968).

DEMBER, H.: "Photoelectromotive force in cuprous oxide crystals", *Z. Phys.* **32**, 554, 856 (1931), ibid. **33**, 207 (1932).

DEN ENGELSEN, D.: "Ellipsometry of anisotropic films", *J. Opt. Soc. Am.* **61**, 1460–6 (1971).

DENTON, G. A., FRIEDMAN, G. M., and SCHETZINA, J. F.: "Switching effects in polycrystalline cadmium telluride films", *J. Appl. Phys.* **46**, 3044–8 (1975).

DE SMET, D. J.: "Ellipsometry of anisotropic thin films", *J. Opt. Soc. Am.* **64**, 631–8 (1974).

DEULING, H. J.: "Double injection currents in long p–i–n diodes with one trapping level", *J. Appl. Phys.* **41**, 2179–84 (1970).

DEVAUX, P. and QUÉDEC, P.: "Electron drift mobility in copper phthalocyanine single crystals", *Phys. Lett.* **28A**, 537–8 (1969).

DEVAUX, P. and SCHOTT, M.: "Thermally stimulated currents without optical excitation. Application to copper phthalocyanine", *Phys. Stat. Sol.* **20**, 301–9 (1967).

DE WIT, H. J.: "Some numerical calculations on Hostein's small-polaron theory", *Philips Res. Repts.* **23**, 449–60 (1968).

DEXTER, D. L.: "A theory of sensitized luminescence in solids", *J. Chem. Phys.* **21**, 836–50 (1953).

DHARIWAL, S. R., KOTHARI, L. S., and JAIN, S. C.: "Theory of transient photovoltaic effects used for measurement of lifetime of carriers in solar cells", *Solid-State Electron.* **20**, 297–304 (1977).

DIAMOND, J. B.: "Computational method for generalized susceptibility", in *Computational Methods in Band Theory* (eds. P. M. Marcus, J. F. Janak, and A. R. Williams), Plenum Press (New York) 347–54 (1971).

DIENES, G. J.: "On the fluorescence decay curve of anthracene", *Mol. Crys.* **3**, 293–6 (1967).

DIETRICH, G., BAUSER, H., and PICK, H.: "Charge-carrier generation at photooxidized surface of anthracene crystals. I: Enhancement of hole generation of surface traps", *Phys. Stat. Sol.* (a) **32**, 113–21 (1975).

DIETRICH, G., BAUSER, H., and PICK, H.: "Charge-carrier generation at photooxidized surfaces of anthracene crystals. II: Correlation between enhanced hole generation and fluorescence quenching", *Phys. Stat. Sol.* (a) **32**, 403–11 (1975).

DIMARCO, P. G. and GIRO, F.: "Electric field and temperature dependence of hole quantum yield in dibenzothiophene single crystals", *Phys. Stat. Sol.* (a) **32**, 263–8 (1975).

DIMARIA, D. J. and FEIGE, F. J.: "Photocurrent spectroscopy in thin-film insulators: voltage dependence of external-circuit current", *Phys. Rev.* B **9**, 1874–83 (1974).

DIMTRIEV, A. P., STEFANOVICH, A. E., and TSENDIN, L. D.: "Formation of a carrier density and electric field discontinuity in double injection of hot carriers", *Soviet. Phys.—Semicond.* **9**, 894–7 (1976).

DISMUKES, G. C., HAGER, S. L., MORINE, G. H., and WILLARD, J. E.: "Light intensity dependence of photocurrents and bleaching rates for trapped electrons in γ-irradiated organic glasses", *J. Chem. Phys.* **61**, 426–7 (1974).

DITTMER, G.: "Electron conduction, electron emission and electroluminescence of MIM sandwich structures with A1$_2$O$_3$ insulating layers", *Thin Solid Films* **9**, 141–72 (1972).

DIX, G., HELBERG, H. W., and WARTENBERG, B.: "Mikrowellenapparatur zur Messung anisotroper Dielektrizitätskonstanten—Beispiel: Anthrazen", *Phys. Stat. Sol.* (a) **5**, 633–6 (1971).

DODELET, J. P. and FREEMEN, G. R.: "Ionization in liquid hydrocarbons: γ radiolysis and electron range distribution functions", *J. Chem. Phys.* **60**, 4657–9 (1974).

DÖHLER, G. H. and HEYSZENAU, H.: "Conduction in relaxation-case semiconductors", *Phys. Rev. Lett.* **30**, 1200–2 (1973).

DOLEZALEK, F. K. and SPEAR, W. E.: "Electronic transport properties of some low mobility solids under high pressures", *J. Non-Cryst. Solids* **4**, 97–106 (1970).

DOLOCAN, V.: "On the double injection negative resistance in magnetic field", *Solid-State Electron.* **18**, 229–34 (1975).

DONATI, D., GUARINI, G. G. T., and SARTI-FANTONI, P.: "Photoreactions and fluorescence ageing in crystalline anthracene", *Mol. Cryst. Liq. Cryst.* **21**, 289–97 (1973).

DONNINI, J. M.: "Effect of a magnetic field on the photoconductivity of anthracene", *CR Acad. Sci. Paris* **266B**, 626–9 (1968).

DONNINI, J. M.: "Photoconduction activation energy and conduction band determinations in pure and doped aromatic crystals", *J. Phys. Soc.* (*Japan*) **32**, 455–61 (1972).

DONNINI, J. M. and ABETINO, F.: "Internal polarization in aromatic crystals. Study as a function of temperature and applied electric field", *CR Acad. Sci. Paris* **228B**, No. 2, 185–8 (1969).

DONNINI, J. M. and ABETINO, F.: "Photoemission of metal electrons in anthracene crystals", *CR Acad. Sci. Paris* **269A** and **B**, No. 1, 34–7 (1969).

DONNINI, J. M. and PESTEIL, P.: "Influence of a magnetic field on photoconductivity of anthracene", *CR Acad. Sci. Paris* **266B**, 724–7 (1968).

DONNINI, J. M. and PESTEIL, P.: "Organic excitons and electron–hole pairs", *CR Acad. Sci. Paris* **268B**, 82–84 (1969).

DONOVAN, T. M. and SPICER, W. E.: "Changes in the density of states of germanium on disordering as observed by photoemission", *Phys. Rev. Lett.* **21**, 1572–5 (1968).

DOUSMANIS, G. C. and DUNCAN, R. C. JR.: "Calculations of the shape and extent of space-charge regions in semiconductor surfaces", *J. Appl. Phys.* **29**, 1627–9 (1958).

DOW, J. D. and REDFIELD, D.: "Electroabsorption in semiconductors: The excitonic absorption edge", *Phys. Rev.* **B1**, 3358–71 (1970).

DREESKAMP, H. and ZANDER, M.: "Delayed fluorescence of benzophenone-aromatic hydrocarbon mixed crystals", *Z. Naturforsch.* **28A**, 45–50 (1973).

DRESNER, J.: "Photo-Hall effect in anthracene", *Phys. Rev.* **143**, 558–63 (1966).

DRESNER, J.: "Conduction band structure in anthracene determined by photoemission", *Phys. Rev. Lett.* **21**, 356–8 (1968).

DRESNER, J.: "Double injection electroluminescence in anthracene", *RCA Rev.* **30**, 322–34 (1969).

DRESNER, J.: "Photoemission of holes and valence band structure in anthracene", *Mol. Cryst.* **11**, 305–8 (1970).

DRESNER, J.: "Volume generation and Hall mobility of holes in anthracene", *J. Chem. Phys.* **52**, 6343–7 (1970).

DRESNER, J., CAMPOS, M., and MORENO, R. A.: "Limitation on deep trapping of injected space charge in naphthalene and CdS monocrystals", *J. Appl. Phys.* **44**, 3708–12 (1973).

DRESNER, J. and GOODMAN, A. M.: "Anthracene electroluminescent cells with tunnel-injection cathodes", *Proc. IEEE* **58**, 1868–9 (1970).

DRESNER, J. and SHALLCROSS, F. V.: "Rectification and space charge limited currents in CdS films", *Solid-State Electron.* **5**, 205–10 (1962).

DRICKAMER, H. G.: "Pi electron systems at high pressures", *Science* **156**, 1183–9 (1967).

DRIEDONKS, F. and ZIJLSTRA, R. J. J.: "Double injection and noise in germanium diode", in *Physical Aspects of Noise in Electronic Devices* (ed. P. Peregrinus) 95–98 (1968).

DRIVER, M. C. and WRIGHT, G. T.: "Thermal release of trapped space-charge in solids", *Proc. Phys. Soc. (London)* **81**, 141–7 (1963).

DRUGER, S. D.: "Photoionization and photogeneration of carriers in anthracene", *Chem. Phys. Lett.* **17**, 603–7 (1972).

DUBININ, N. V. and LUTSENKO, E. L.: "Relaxation of stimulated conductance and photocurrent in an organic semiconductor acriflavine", *Soviet Phys.—Semicond.* **7**, 443–6 (1973).

DUBOVSKII, O. A. and KONOBEEV, Y. V.: "Capture of free excitons in molecular crystals by shallow traps", *Soviet Phys.—Solid State* **6**, 2071–8 (1965).

DU-CHATENIER, F. J.: "Space-charge-limited photocurrent in vapour-deposited layers of red lead monoxide", *Philips Res. Repts.* **23**, 142–50 (1968).

DUKE, C. B.: "Polymers and molecular solids: new frontiers in surface", *Surf. Sci.* **70**, 674–91 (1978).

DUNBAR, P. M. and HAUSER, J. R.: "A study of efficiency in low resistivity silicon solar cells", *Solid-State Electron.* **19**, 95–102 (1976).

DUROCHER, D. F. and WILLIAMS, D. F.: "Temperature dependence of triplet diffusion in anthracene", *J. Chem. Phys.* **51**, 1675–6 (1969).

DUSSEL, G. A. and BÖER, K. W.: "Field-enhanced ionization" and "Field quenching as mechanism of negative differential conductivity in photo-conducting CdS", *Phys. Stat. Sol.* B **39**, 375–389 (1970) and B **39**, 391–402 (1970).

DUSSEL, G. A. and BUBE, R. H.: "Further considerations on a theory of superlinearity in CdS and related materials", *J. Appl. Phys.* **37**, 13–21 (1966).

DUSSEL, G. A. and BUBE, R. H.: "Validity of steady-state photoconductivity lifetime under nonsteady-state conditions", *J. Appl. Phys.* **37**, 934–5 (1966).

DUSSEL, G. A. and BUBE, R. H.: "Electric field effects in trapping processes", *J. Appl. Phys.* **37**, 2797–804 (1966).

DUSSEL, G. A. and BUBE, R. H.: "Theory of thermally stimulated conductivity in a previously photoexcited crystal", *Phys. Rev.* **155**, 764–79 (1967).

DYNE, P. J. and MILLER, O. A.: "Photochemistry of electrons trapped in glasses of methyltetrahydrofuran at 77°K", *Can. J. Chem.* **43**, 2696–2707 (1965).

EASTMAN, D. E.: "Photoemission studies of the electronic structure of transition metals", *J. Appl. Phys.* **40**, 1387–94 (1969).

EASTMAN, D. E.: "Photoelectric work functions of transition, rare-earth and noble metals", *Phys. Rev.* B **2**, 1–2 (1970).

EDMOND, J. T.: "Electronic conduction in As_2Se_3, As_2Se_3Te and similar materials", *Br. J. Appl. Phys.* **17**, 979–89 (1966).

EDWARDS, J., KAWABE, K., STEVENS, G., and TREDGOLD, R. H.: "Space-charge conduction and electrical behavior of aluminum nitride single crystals", *Solid State Commun.* **3**, 99–100 (1965).

EGGERT, E. and HÄNSEL, H.: "Characteristic Debye temperature of anthracene", *Phys. Stat. Sol.* (b) **47**, K69–73 (1971).

EIERMANN, R., HOFBERGER, W., and BÄSSLER, H.: "Localized valence states in the forbidden gap of non-crystalline tetracene", *J. Non-Cryst. Solids* **28**, 415–28 (1978).

ELEY, D. D.: "Semiconductivity in biological molecules", in *Horizons in Biochemistry* (eds. M. Kasha and B. Pullman), Academic Press, 341–80 (1962).

ELEY, D. D.: "Energy gap and pre-exponential factor in dark conduction by organic semiconductors", *J. Polymer Sci.* C **17**, 73–91 (1967).

ELEY, D. D., FAWCETT, A. S., and WILLIS, M. R.: "Semiconductivity of organic substances. Electrode injection of charge carriers into crystals of small aromatic molecules", *Trans. Faraday Soc.* **64**, 1513–27 (1968).

ELEY, D. D., INOKUCHI, H., and WILLIS, M. R.: "The semiconductivity of organic substances. Part 4: Semi-quinone type molecular complexes", *Discuss. Faraday Soc.* **28**, 54–63 (1959).

ELEY, D. D. and LESLIE, R. B.: "Conduction in nucleic acid components", *Nature* **147**, 898 (1963).

ELEY, D. D., METCALFE, E., and WHITE, M. P.: "Semiconductivity of organic substances. Part 17: Effects of ultraviolet and visible light on the conductivity of the sodium salt of deoxyribonucleic acid", *J. Chem. Soc. Faraday Trans.* I **71**, 955–60 (1975).

ELEY, D. D. and NEWMAN, O. M. G.: "Semiconductivity of organic substances. Part 15: Compounds related to stilbene", *Trans. Faraday Soc.* **66**, 1106–12 (1970).

ELEY, D. D. and PARFITT, G. D.: "The semiconductivity of organic substances. Part 2", *Trans. Faraday Soc.* **51**, 1529–39 (1955).

ELEY, D. D., PARFITT, G. D., PERRY, M. J., and TAYSUM, D. H.: "The semi-conductivity of organic substances. Part 1", *Trans. Faraday Soc.* **49**, 78–86 (1953).

ELEY, D. D., and PETHIG, R.: "Microwave Hall effect measurements on organic and biological semiconductors", in *2nd International Conference on Conduction in Low Mobility Materials, Fliat, Israel*, Taylor & Francis (London) 397–402 (1971).

ELEY, D. D. and SPIVEY, D. J.: "The semiconductivity of organic substances. Part 6: A range of proteins", *Trans. Faraday Soc.* **56**, 1432–42 (1960).

ELEY, D. D. and SPIVEY, D. J.: "The semiconductivity of organic substances. Part 7: Polyamides; Part 8: Porphyrins and dipyromethenes, *Trans. Faraday Soc.* **57**, 1 (1961) and Part 8, ibid. **58**, 405 (1962).

ELEY, D. D. and SPIVEY, D. J.: "Semiconductivity of organic substances. Part 9: Nucleic acid in the dry state", *Trans. Faraday Soc.* **58**, 411–15 (1962).

ELEY, D. D. and THOMAS, P. W.: "Semiconductivity of organic substances. Part 4: Electrode effects in proteins", *Trans. Faraday Soc.* **64**, 2459–62 (1968).

ELEY, D. D. and WILLIS, M. R.: "The electrical conductivity of solid free radicals and the electron tunneling mechanism", in *Symposium on Electrical Conductivity in Organic Solids* (eds. H. Kallmann and M. Silver), Wiley Interscience (New York) 257–76 (1961).

ELLIS, D. M.: "Deuteron nuclear magnetic resonance in anthracene crystals", *J. Chem. Phys.* **46**, 4460–3 (1967).

EL-SAYED, M. A. and SIEGEL, 3.: "Method of magnetophoto-selection of the lowest excited triplet states of aromatic molecules", *J. Chem. Phys.* **44**, 1416–23 (1966).

EL-SAYED, M. A., WAUK, M. T., and ROBINSON, G. W.: "Retardation of singlet and triplet excitation migration in organic crystals by isotropic dilution", *Mol. Phys.* **5**, 205–8 (1962).

ELSHARKAWI, A. R. and KAO, K. C.: "Effects of temperature on current instabilities caused by recombination centers in semiconductors", *Solid-State Electron.* **16**, 1355–62 (1973).

ELSHARKAWI, A. R. and KAO, K. C.: "Study of the optical properties of anthracene thin films by ellipsometry", *J. Opt. Soc. Am.* **65**, 1269–73 (1975).

ELSHARKAWI, A. R. and KAO, K. C.: "Space-charge created by an ohmic injecting contact, and its effects in an electrolyte–insulator system with traps distributed non-uniformly in energy and in space", *Solid-State Electron.* **19**, 939–47 (1976).

ELSHARKAWI, A. R. and KAO, K. C.: "Switching and memory phenomena in anthracene thin films", *J. Phys. Chem. Sol.* **38**, 95–96 (1977).

EMERALD, R. L. and MORT, J.: "Transient photoinjection of electrons from amorphous selenium into trinitrofluorenone", *J. Appl. Phys.* **45**, 3943–5 (1974).

EMIN, D.: "Correlated small-polaron hopping motion", *Phys. Rev. Lett.* **25**, 1751–5 (1970).

EMIN, D.: "Vibrational dispersion and small-polaron motion: enhanced diffusion", *Phys. Rev. B* **3**, 1321–37 (1971).

EMIN, D.: "Lattice relaxation and small-polaron hopping motion", *Phys. Rev. B* **4**, 3639–51 (1971).

EMIN, D.: "Energy spectrum of an electron in a periodic deformable lattice", *Phys. Rev. Lett.* **28**, 604–7 (1973).

EMIN, D.: "On the existence of free and self-trapped carriers in insulators: an abrupt temperature dependent conductivity transition", *Adv. Phys.* **22**, 57–116 (1973).

EMIN, D.: "Aspects of the theory of small polarons in disordered materials", in *Electronic and Structural Properties of Amorphous Semiconductors* (eds. P. G. Le Comber and J. Mort), Academic Press (New York) 261–328 (1973).

EMIN, D. and HOLSTEIN, T.: "Studies of small-polaron motion. IV: Adiabatic theory of the Hall effect", *Ann. Phys.* **53**, 439–520 (1969).

EMTAGE, P. R.: "Enhancement of metal to insulator tunneling by optical phonons", *J. Appl. Phys.* **38**, 1820–5 (1967).

EMTAGE, P. R.: "Conduction by small polarons in large electric fields", *Phys. Rev. B* **3**, 2685–9 (1971).

EMTAGE, P. R. and O'DWYER, J. J.: "Richardson–Schottky effect in insulators", *Phys. Rev. Lett.* **16**, 356–8 (1966).

ENCK, R. C.: "Two-photon photogeneration in amorphous selenium", *Phys. Rev. Lett.* **31**, 220–3 (1973).

EREMENKO, V. V. and MEDVEDEV, V. S.: "Dependence of the photoconductivity and the intensity of luminescence of anthracene crystals on the excitation wavelength", *Soviet Phys.—Solid State* **2**, 1426–8 (1960).

ERN, V.: "Anisotropy of triplet exciton diffusion in anthracene", *Phys. Rev. Lett.* **22**, 343–5 (1969).

ERN, V.: "Delayed fluorescence wave form from free-deep trapped triplet exciton annihilation", *Mol. Cryst. Liq. Cryst.* **18**, 1–16 (1972).

ERN, V., AVAKIAN, P., and MERRIFIELD, R. E.: "Diffusion of triplet excitons in anthracene crystals", *Phys. Rev.* **148**, 862–7 (1966).

ERN, V., BOUCHRIHA, H., BISCEGLIA, M., ARNOLD, S., and SCHOTT, M.: "Total and radiative triplet–triplet exciton annihilation rate constant in pyrene crystals", *Phys. Rev. B* **8**, 6038–42 (1973).

ERN, V., BOUCHRIHA, H., FOURNY, J., and DELACOTE, G.: "Triplet exciton–trapped hole interaction in anthracene crystals", *Solid State Commun.* **9**, 1201–3 (1971).

ERN, V. and McGHIE, A. R.: "Quenching of triplet excitons in anthracene crystals by internal beta-irradiation", *Mol. Cryst. Liq. Cryst.* **15**, 277–82 (1971).

ERN, V. and MERRIFIELD, R. E.: "Magnetic field effect on triplet exciton quenching in organic crystals", *Phys. Rev. Lett.* **21**, 609–11 (1968).

ERN, V., SAINT-CLAIR, J. L., SCHOTT, M., and DELACOTE, G.: "Effects of exciton interactions on the fluorescence yield of crystalline tetracene", *Chem. Phys. Lett.* **10**, 287–90 (1971).

ERN, V., SUNA, A., and MERRIFIELD, R. E.: "Dependence of delayed fluorescence on the velocity of a. moving pattern of light", *J. Appl. Phys.* **42**, 2770–3 (1971).

ERN, V., SUNA, A., TOMKIEWICZ, Y., AVAKIAN, P., and GROFF, R. P.: "Temperature dependence of triplet-exciton dynamics in anthracene crystals", *Phys. Rev. B* **5**, 3222–34 (1972).

ESAKI, L.: "New horizons in semimetal alloys", *IEEE Spectrum* **3**, 74–86 (1966).

ESAKI, L.: "Tunneling studies on the group V semimetals and the IV–VI semiconductors", *J. Phys. Soc. Japan*, Suppl. **21**, 589–97 (1966).

Esaki, L. and Chang, L. L.: "New transport phenomena in a semiconductor superlattice", *Phys. Rev. Lett.* **33**, 495–8 (1974).

Esaki, L., Chang, L. L., Stiles, P. J., O'Kane, D. F., and Wiser, N.: "Phonon-assisted tunneling in bismuth tunnel junction", *Phys. Rev.* **167**, 637–9 (1968).

Esaki, L. and Stiles, P. J.: "Study of electronic band structures by tunneling spectroscopy: bismuth", *Phys. Rev. Lett.* **14**, 902–4 (1965).

Esaki, L. and Stiles, P. J.: "New phenomena in semimetals and semiconductors", *Phys. Rev. Lett.* **15**, 152–4 (1965).

Esaki, L. and Stiles, P. J.: "BiSb alloy tunnel junctions", *Phys. Rev. Lett.* **16**, 574–6 (1966).

Esaki, L. and Stiles, P. J.: "New type of negative resistance in barrier tunneling", *Phys. Rev. Lett.* **16**, 1108–11 (1966).

Evans, M. G. and Gergely, J.: "A discussion of the possibility of bands of energy levels in proteins. Electronic interaction in non-bonded systems", *Biochem. Biophys. Acta* **3**, 188–97 (1949).

Ewald, M. and Durocher, G.: "Photoionization and recombination delayed fluorescence in anthracene single crystals", *Chem. Phys. Lett.* **12**, 385–8 (1971).

Fabre, E., Mautref, M., and Mircea, A.: "Trap saturation in silicon solar cells", *Appl. Phys. Lett.* **27**, 239–41 (1975).

Faidysh, A. N.: "Luminescence and photoconductivity of anthracene crystals", *Bull. Acad. Sci. USSR Phys. Sec.* **24**, 562–6 (1960).

Faidysh, A. N.: "Length of diffusion displacement of excitons in anthracene crystals", *Visn. Kyyiv. Univ.* **3**, *Ser. Fiz. ta Khim.* **1**, 53 (1960).

Faidysh, A. N. and Gaevsky, A. S.: "Influence of triplet exciton annihilation on quantum efficiency and phosphorescence decay law in benzo-phenone crystals", *Mol. Cryst. Liq. Cryst.* **19**, 13–24 (1972).

Faidysh, A. N. and Kucherov, I. Ya.: "Migration and transfer of the energy of electrical excitation in anthracene and naphthalene crystals", *Fiz. Sb. L'vovsk. Gos. Univ.* **3**, 40 (1957).

Fang, P. H.: "Transient photovoltaic current in tetracene", *Japan J. Appl. Phys.* **12**, 536–9 (1973).

Fang, P. H.: "Reinvestigation of photovoltaic current anomaly in naphthacene", *Japan. J. Appl. Phys.* **13**, 1232–4 (1974).

Fang, P. H., Golubovic, A., and Diamond, N. A.: "Photovoltaic current anomaly in naphthacene", *Japan J. Appl. Phys.* **11**, 1298–1302 (1972).

Fano, U.: "Effects of configuration interaction on intensities and phase shifts", *Phys. Rev.* **124**, 1866–78 (1961).

Farges, J. P. and Brau, A.: "Study at 10^8 and 2×10^8 Hz of the electrical conductivity and dielectric constant of the high anisotropic organic semiconductor TEA (TCNQ)$_2$", *Phys. Stat. Sol.* B **61**, 669–73 (1974).

Farges, J. P. and Brau, A.: "Dependence of the thermopower on temperature and crystal direction in the highly anisotropic organic semiconductor TEA (TCNQ)$_2$", *Phys. Stat. Sol.* B **64**, 269–75 (1974).

Farnsworth, H. E., Schlier, R. E., George, T. H., and Burger, R. M.: "Ion bombardment-cleaning of germanium and titanium as determined by low-energy electron diffraction", *J. Appl. Phys.* **26**, 252–3 (1955).

Fedorov, M. I. and Benderskii, V. A.: "Characteristics of magnesium phthalocyanine thin-film photocells", *Soviet Phys.—Semicond.* **4**, 1198–9 (1971).

Fedorov, M. I. and Benderskii, V. A.: "Formation of *p–n* junctions by doping magnesium phthalocyanine films", *Soviet Phys.—Semicond.* **4**, 1720–1 (1971).

Feilchenfeld, H. and Many, A.: "The measurement of the dark conductivity of anthracene by use of pulse techniques", *Abstr. Organic Crystal Symposium, NRC Ottawa, Canada* **99** (1962).

Ferguson, J. and Schneider, W. G.: "Polarization of anthracene crystal fluorescence", *J. Chem. Phys.* **25**, 780 (1956).

Ferguson, J. and Schneider, W. G.: "The polarized fluorescence of very thin anthracene crystals", *Can. J. Chem.* **36**, 1070–80 (1958).

Ferguson, J. and Schneider, W. G.: "On the spectral response of photo-conduction in thin single crystals of anthracene", *Can. J. Chem.* **36**, 1633–9 (1958).

Feynman, R. P.: "Slow electrons in a polar crystal", *Phys. Rev.* **97**, 660–5 (1955).

Feynman, R. P., Hellwarth, R. W., Iddings, C. K., and Platzman, P. M.: "Mobility of slow electrons in a polar crystal", *Phys. Rev.* **127**, 1004–17 (1962).

Fielding, P. E. and Jaragin, R. C.: "Excimer and defect structure for anthracene and some derivatives in crystals, thin films, and other rigid matrices", *J. Chem. Phys.* **47**, 247–52 (1967).

Fields, D. E. and Moran, P. R.: "Analytical and experimental check of a model for correlated thermoluminescence and thermally stimulated conductivity", *Phys. Rev.* B **9**, 1836–41 (1974).

Fillard, J. P. and Schott, M.: "Variation of the mobility of charge carriers with the temperature in thin copper phthalocyanine layers", *CR Acad. Sci. Paris* **262B**, 1287–90 (1966).

FISCHER, D.: "Oriented growth of anthracene crystals in an electric field", *Mater. Res. Bull.* **3**, 759–64 (1968).

FISCHER, J. E. and ASPNES, D. E.: "Electroreflectance: a status report", *Phys. Stat. Sol.* (b) **55**, 9–32 (1973).

FISHER, J. C. and GIAEVER, I.: "Tunneling through thin insulating layers", *J. Appl. Phys.* **32**, 172–7 (1961).

FISHER, R.: "Absorption and electroabsorption of trigonal selenium near the fundamental absorption edge", *Phys. Rev.* B **5**, 3087–94 (1972).

FLANDROIS, S., LIBERT, P., and DUPUIS, P.: "Effect of the crystallization conditions and trapped solvent on the physical properties of TCNQ salts", *Phys. Stat. Sol.* (a) **28**, 411–15 (1975).

FLEET, B., KIRKBRIGHT, G. F., and PICKFORD, C. J.: "Application of electro-luminescence technique to determination of aromatic hydrocarbons", *Talanta* **15**, 566–70 (1968).

FLEMING, R. J.: "Transient photoconductivity current-time profiles in low carrier mobility solids", *J. Appl. Phys.* **45**, 4944–9 (1974).

FLEMINS, P. D.: "Hydrodynamic behavior of triplet excitons", *J. Chem. Phys.* **59**, 3199–206 (1973).

FONASH, S. J.: "The role of the interfacial layer in metal–semiconductor solar cells", *J. Appl. Phys.* **46**, 1286–9 (1975).

FONASH, S. J.: "Outline and comparison of the possible effects present in a metal–thin-film–insulator–semiconductor solar cell", *J. Appl. Phys.* **47**, 3597–602 (1976).

FONASH, S. J.: "Metal–insulator–semiconductor solar cells: theory and experimental results", *Thin Solid Films* **36**, 387–92 (1976).

FORMM, B. and TEUBNER, W.: "Screening effects in amorphous semiconductors influencing the Poole–Frenkel conduction", *Phys. Stat. Sol.* (a) **40**, 63–68 (1977).

FÖRSTER, T.: "Excitation transfer", in *Comparative Effects of Radiation* (eds. M. Burton, J. S. Kirby-Smith, and J. L. Magee), Wiley (New York) (1960).

FÖRSTER, T.: "Transfer mechanisms of electronic excitation energy", *Discuss. Faraday Soc.* No. 27, 7–17 (1959), also *Radiation Res.* Suppl. **2**, 326–36 (1960).

FORT, A. and CORET, A.: "Photogeneration of free carriers near 4000 Å at 77°K in anthracene crystals", *J. Chem. Phys.* **62**, 3269–70 (1975).

FOURNY, J. and DELACOTE, G.: "High-temperature dependence of electron and hole mobilities in anthracene crystal", *J. Chem. Phys.* **50**, 1028–9 (1969).

FOURNY, J., DELACOTE, G., and SCHOTT, M.: "Singlet–triplet interactions in crystalline anthracene", *Phys. Rev. Lett.* **21**, 1085–8 (1968).

FOURNY, J., SCHOTT, M., and DELACOTE, G.: "Simultaneous observation of fluorescence quenching by singlet–singlet and by singlet–triplet exciton interactions in crystalline anthracene", *Chem. Phys. Lett.* **20**, 559–62 (1973).

FOWLER, R. H.: "The analysis of photoelectric sensitivity curves for clean metals at various temperatures", *Phys. Rev.* **38**, 45–56 (1931).

FOWLER, R. H. and NORDHEIM, L.: "Electron emission in intense electric fields", *Proc. Roy. Soc. (London)* **119A**, 173–81 (1928).

FOX, D.: "Electronic states of aromatic solids", in *Electrical Conductivity of Organic Solids* (eds. H. Kallmann and M. Silver), Wiley Interscience (New York) 239–45 (1961).

FOX, S. J.: "Decay of surface potential in electrophotograph: single-carrier case", *J. Appl. Phys.* **45**, 610–17 (1974).

FRANK, R. I. and SIMMONS, J. G.: "Space charge effects on emission-limited current flow in insulators", *J. Appl. Phys.* **38**, 832–40 (1967).

FRANKEVICH, E. L.: "The nature of a new effect of variation of photo-conductivity of organic semiconductors in a magnetic field", *Soviet Phys.—JETP* **23**, 814–19 (1966).

FRANKEVICH, E. L. and BALABANOV, E. I.: "New effect of increasing the photoconductivity of organic semiconductors in a weak magnetic field", *JETP Lett.* **1**, 169–71 (1965).

FRANKEVICH, E. L. and BALABANOV, E. I.: "A study of carrier mobility in organic materials", *Soviet Phys.–Solid State* **7**, 570–5 (1965).

FRANKEVICH, E. L. and BALABANOV, E. I.: "Changes in the photoconductivity of an anthracene single crystal in magnetic field", *Soviet Phys.–Solid State* **8**, 682–4 (1966).

FRANKEVICH, E. L. and BALABANOV, E. I.: "Investigation of current-carrier drift in organic substances (tetracene)", *Phys. Stat. Sol.* **14**, 523–9 (1966).

FRANKEVICH, E. L., BALABANOV, E. I., and VSELYUKSKAYA, G. V.: "Investigation of the change in the photoconductivity of organic semiconductors in a magnetic field", *Soviet Phys.—Solid State* **8**, 1567–8 (1966).

FRANKEVICH, E. L. and RUMYANTSEV, B. M.: "Quenching of anthracene luminescence by a magnetic field", *Zh. Eksp. Teor. Fiz.* **6**, 553–6 (1967).

FRANKEVICH, E. L. and RUMYANTSEV, B. M.: "Recombination luminescence of anthracene in a magnetic field", *Zh. Eskp. Teor. Fiz.* **53**, 1942–54 (1967).

FRANKEVICH, E. L. and RUMYANTSEV, B. M.: "Charge-transfer excitons and delayed fluorescence of anthracene", *Phys. Stat. Sol.* **30**, 329–40 (1968).

FRANKEVICH, E. L. and RUMYANTSEV, B. M.: "Transfer of excitation energy in anthracene by a new type of exciton", *Zh. Eksp. Teor. Fiz.* **9**, 424–8 (1969).

FRANKEVICH, E. L. and RUMYANTSEV, B. M.: "New type of magnetically sensitive fluorescence produced by excitation of tetracene on the surface of an anthracene crystal", *Soviet Phys.–JETP* **36**, 1064–8 (1973).

FRANKEVICH, E. L. and SOKOLIK, I. A.: "Formation of current carriers from excitons with charge transfer in anthracene", *Fiz. Tverd. Tela* **9**, 3441–7 (1967).

FRANKEVICH, E. L. and SOKOLIK, I. A.: "Photoconductivity and recombination of triplet excitons in anthracene", *Soviet Phys.—Solid State* **9**, 1532–5 (1967).

FRANKEVICH, E. L. and SOKOLIK, I. A.: "Investigation of the mechanism of current carrier production in anthracene by application of the effect of photoconductivity variation in a magnetic field", *Soviet Phys.—JETP* **25**, 790–4 (1967).

FRANKEVICH, E. L. and SOKOLIK, I. A.: "On the mechanism of the magnetic field effect of anthracene photoconductivity", *Solid State Commun.* **8**, 251–3 (1970).

FRANKEVICH, E. L., SOKOLIK, I. A., and LUKIN, L. V.: "Triplet exciton–charge carrier interaction in anthracene", *Phys. Stat. Sol.* (b) **54**, 61–65 (1972).

FRANKL, D. R.: "Conditions for quasi-equilibrium in a semiconductor surface space charge layer", *Surf. Sci.* **3**, 101–8 (1965).

FRANKL, D. R.: "Reanalysis of the temperature dependence of surface conductance in clean germanium", *Surf. Sci.* **4**, 201–4 (1966).

FRANZ, W.: "Elektronische Leitung in kristallinen Isolatoren", *Z. Phys.* **132**, 285–311 (1952).

FRANZ, W.: "Dielektrischer Durchschlag", in *Handbuch der Physik* (ed. S. Flügge), Springer-Verlag (Berlin), **17**, 155–263 (1956).

FRASS, M. L., GRINBERG, J., BLEHA, W. P., and JACOBSON, A. D.: "Novel charge-storage-diode structure for use with light activated displays", *J. Appl. Phys.* **47**, 576–83 (1976).

FREEMAN, J. R., KALLMANN, H. P., and SILVER, M.: "Persistent internal polarization", *Rev. Mod. Phys.* **33**, 553–73 (1961).

FREEMAN, L. B. and DAHLKE, W. E.: "Theory of tunneling into interface states", *Solid-State Electron.* **13**, 1483–1503 (1970).

FRENKEL, J.: "On the electrical resistance of contacts between solid conductors", *Phys. Rev.* **36**, 1604–18 (1930).

FRENKEL, J.: "Some remarks on the theory of the photo-electric effect", *Phys. Rev.* **38**, 309–20 (1931).

FRENKEL, J.: "On the transformation of light into heat in solids", Part I, *Phys. Rev.* **31**, 17–44 (1931); Part II, *Phys. Rev.* **31**, 1276–94 (1931).

FRENKEL, J.: "On the absorption of light and trapping of electrons and holes in crystalline dielectrics", *Phys. Z. Sowjetunion* **9**, 136 and 158 (1936).

FRENKEL, J.: "On pre-breakdown phenomena in insulator and electronic semiconductors", *Phys. Rev.* **54**, 647–8 (1938) [Also *Tech. Phys. USSR* **5**, 685–6 (1938)].

FRIDKIN, V. M. and ABDULGAMIDOV, S. A.: "The distribution of volumetric changes in single crystals of anthracene polarized by photoconductivity", *J. Phys. Chem. Solids* **27**, 220–2 (1966).

FRIEDMAN, L.: "Hall effect in the polaron-band regime", *Phys. Rev.* **131**, 2445–6 (1963).

FRIEDMAN, L.: "Transport properties of organic semiconductor", *Phys. Rev. A* **133**, 1668–78 (1964).

FRIEDMAN, L.: "Electron–phonon interaction in organic molecular crystals", *Phys. Rev.* **140A**, 1649–67 (1965).

FRIEDMAN, L.: "Hall conductivity of amorphous semiconductors in the random phase model", *J. Non-Cryst. Solids* **6**, 329–41 (1971).

FRIEDMAN, L.: "Theory of the Hall effect of amorphous semiconductors in the random phase model", in *Electronic and Structural Properties of Amorphous Semiconductors* (eds. P. G. Le Comber and J. Mort), Academic Press (New York), 363–72 (1973).

FRIEDMAN, L. and HOLSTEIN, T.: "Studies of polaron motion. Pt. III: The Hall mobility of the small polarons", *Ann. Phys.* **21**, 494–549 (1963).

FRITZSCHE, H.: "Resistivity and Hall coefficient of antimony doped germanium at low temperatures", *J. Phys. Chem. Solids* **6**, 69–80 (1958).

FRITZSCHE, H.: "Piezoresistance of *n*-type germanium", *Phys. Rev.* **115**, 336–45 (1959).

FRITZSCHE, H.: "Effect of shear on impurity conduction in *n*-type germanium", *Phys. Rev.* **119**, 1899–1900 (1960).

FRITZSCHE, H.: "A review of some electronic properties of amorphous substances", in *Electronic and Structural Properties of Amorphous Semiconductors* (eds. P. G. Le Comber and J. Mort), Academic Press (New York) 55–125 (1973).

FRITZSCHE, H.: "Switching and memory in amorphous semiconductors", in *Amorphous and Liquid Semiconductors* (ed. J. Tauc), Plenum Press (New York) 313–60 (1974).

FRITZSCHE, H. and CUEVAS, M.: "Impurity conduction in transmutation-doped *p*-type germanium", *Phys. Rev.* **119**, 1238–45 (1960).

FROEHLICH, D. and MAHR, H.: "Two-photon spectroscopy in anthracene", *Phys. Rev. Lett.* **16**, 895–7 (1966).

FRÖHLICH, H.: "On the theory of dielectric breakdown in solids", *Proc. Roy. Soc. (London)* A **188**, 521–32 (1947).

FRÖHLICH, H.: "Electron in lattice fields", *Adv. Phys.* **3**, 325–61 (1954).

FRÖHLICH, H. and PARANJAPE, B. V.: "Dielectric breakdown in solids", *Proc. Phys. Soc. (London)* B **69**, 21–32 (1956).

FRÖHLICH, H. and SEITZ, F.: "Notes on the theory of dielectric breakdown in ionic crystals", *Phys. Rev.* **79**, 526–7 (1950).

FRÖHLICH, H. and SEWELL, G. L.: "Electric conduction in semiconductors", *Proc. Phys. Soc* **74**, 643–7 (1959).

FÜNFSCHILLING, J. and WILLIAMS, D. F.: "Interaction of triplet excitons with charge carriers in crystalline anthracene", *Chem. Phys. Lett.* **31**, 551–4 (1975).

FÜNFSCHILLING, J. and WILLIAMS, D. F.: "Low temperature magnetic field modulation of delayed fluorescence in anthracene and naphthalene", *Chem. Phys. Lett.* **38**, 329–33 (1976).

FÜNFSCHILLING, J. and ZSCHOKKE-GRÄNACHER, I.: "Triplet–triplet exciton annihilation in tetracene-doped anthracene crystals", *Helv. phys. Acta* **43**, 768–70 (1970).

FÜNFSCHILLING, J. and ZSCHOKKE-GRÄNACHER, I.: "Interaction of triplet excitons in doped anthracene crystals", *Helv. Phys. Acta* **46**, 14 (1973).

FURST, M., KALLMANN, H., and KRAMER, B.: "Absolute light emission efficiency of crystal anthracene for gamma-ray excitation", *Phys. Rev.* **89**, 416–17 (1953).

GADZUK, J. W.: "Resonance tunneling through impurity states in metal–insulator–metal junctions", *J. Appl. Phys.* **41**, 286–91 (1970).

GAEHRUS, H. J. and WILLING, F.: "Contact-dependent electron transfer quenching of singlet and triplet existons at the surface of organic crystals", *Phys. Stat. Sol.* (b) **27**, 355–64 (1975).

GAFUROV, KH. M., MULIKOV, V. F., and GACHKOVSKII, V. F.: "Effect of the local paramagnetic centers on optical and photoelectric properties of anthracene", *J. Structural Chem.* **6**, 621–3 (1965).

GALANIN, M. D., DEMCHUK, M. I., KHAN-MAGOMETOV, SH. D., CHERNYAVSKII, A. F., and CHIZHIKOVA, Z. A.: "Attenuation time of exciton luminescence in an anthracene crystal at 4.2°K", *Zh. Eksp. Teor. Fiz.* **20**, 260–4 (1974).

GALANIN, M. D., KHAN-MAGOMETOV, SH. D., and CHIZHIKOVA, Z. A.: "Polarization of superradiance in anthracene crystals", *Kratk. Sobshch. Fiz.* **7**, 21–24 (1973).

GALKIN, G. N., KHARAKHORIN, F. F., and SHATKOVSKII, E. V.: "Recombination of nonequilibrium carriers in indium arsenide at high excitation levels", *Soviet Phys.—Semicond.* **5**, 387–91 (1971).

GALLO, G. F.: "Optical xerography", *IEEE Trans.* IA **8**, 372–82 (1972).

GALLUS, G. and WOLF, H. C.: 'Exciton diffusion in vapor deposited naphthalene films", *Z. Naturforsch.* **23** A, 1333–9 (1968).

GALOLUBOV, S. I. and KNOBEEV, YU. V.: "Energy transfer by excitons in tetracene-doped anthracene crystals", *Phys. Stat. Sol.* (b) **79**, 79–84 (1977).

GAMMIL, L. S. and POWELL, R. C.: "Energy transfer in perylene doped anthracene crystals", *Mol. Cryst. Liq. Cryst.* **25**, 123–30 (1974).

GAMOUDI, M., ROSENBERG, N., GUILLAUD, G., MAITROT, M., and MESNARD, G.: "Analysis of deep and shallow trapping of holes in anthracene", *J. Phys. C* **7**, 1149–59 (1974).

GANGULY, B. N.: "Transient space-charge-limited currents in photoconductor–dielectric structure: a field-dependent mobility case", *Phys. Rev. B* **12**, 1275–84 (1975).

GANGULY, S. C.: "Fluorescence of anthracene in presence of naphthacene", *Nature* **151**, 673 (1943).

GANGULY, S. C.: "On the fluorescence spectra of naphthacene in solid solution of anthracene for different exciting wavelengths", *J. Chem. Phys.* **13**, 128–30 (1945).

GANGULY, S. C. and CHAUDHURY, N. K.: "Energy transport in organic phosphors", *Rev. Mod. Phys.* **31**, 990–1017 (1959).

GAPOZZI, V., MARIOTT, G., MONTAGNA GINGOLANIC, A. and MINAFRA, A.: "Space-charge-limited currents in GaSe at different temperatures", *Phys. Stat. Sol.* (a) **40**, 93–100 (1977).

GARABER, N. and ZSCHOKKE-GRÄNACHER, I.: "Delayed fluorescence excitations spectra in doped anthracene crystals", *Phys. Stat. Sol.* (b) **85**, 505–10 (1978).

GARBEN, B.: "The effect of pressure on the electric conduction in dried liquid benzene", *Phil. Mag.* **29**, 1281–8 (1974).

GARLICK, G. F. J. and GIBSON, A. F.: "The electron trap mechanism of luminescence in sulphide and silicate phosphors", *Proc. Phys. Soc. (London)* **60**, 574–90 (1948).

GARLICK, G. F. J. and WILKINS, M. H. F.: "Short period phosphorescence and electron traps", *Proc. Roy. Soc. (London)* **184A**, 408–33 (1945).

GAROFANO, T.: 'Trapping effects in pulsed photoconductivity of anthracene crystals", *Nuovo Cimento* **21B**, 376–94 (1974).

GAROFANO, T. and CORAZZARI, T.: "Isothermal current decay in anthracene crystals", *Nuovo Cimento* **26**, 23–29 (1975).

GAROFANO, T., CORAZZARI, T., and CASALINI, G.: "Thermally stimulated currents in anthracene crystals", *Il Nuovo Cimento* B **38**, 133–40 (1977).

GAROFANO, T. and MORELLI, S.: "Trapping levels in anthracene crystals by thermally stimulated currents", *Nuovo Cimento* **13B**, 174–84 (1973).

GARRETA, O. and GROSVALET, J.: "Photo-magneto-electric effect in semiconductors", *Progr. Semicon.* **1**, 165–94 (1956).

GARRETT, C. G. B.: "Organic semiconductors", *Radiation Research Supplement* **2**, 340–8 (1960), also in *Semiconductors* (ed. N. B. Hannay), Reinhold (New York) 634–46 (1959).

GARRETT, C. G. B. and BRATTAIN, W. H.: "Physical theory of semiconductor surfaces", *Phys. Rev.* **99**, 376–87 (1955).

GARRETT, S. G. E., PETHIG, R., and SONI, V.: "Switching and other high field effects in organic films", *J. Chem. Soc. Faraday Trans.* II, **70**, 1732–40 (1974).

GÄRTNER, W.: "Spectral distribution of the photo-magnetoelectric effect in semiconductors: theory", *Phys. Rev.* **105**, 823–9 (1957).

GAZSO, J.: "Electrical behaviour of thin layer Au–polyethylene–Al sandwiches", *Hung. Acad. Sci. (Budapest)* Report KFKI-7354 (1973).

GEACINTOV, N. E., BINDER, M., SWENBERG, C. E., and POPE, M.: "Exciton dynamics in α-particle tracks in organic crystals—magnetic field study of scintillation in tetracene crystals", *Phys. Rev.* B **12**, 4113–34 (1975).

GEACINTOV, N. E., BURGOS, J., POPE, M., and STROM, C.: "Heterofission of pentacene excited singlets in pentacene-doped tetracene crystals", *Chem. Phys. Lett.* **11**, 504–8 (1971).

GEACINTOV, N. E. and POPE, M.: "Photogeneration of charge carriers in anthracene", *J. Chem. Phys.* **45**, 3884–5 (1966).

GEACINTOV, N. E. and POPE, M.: "Generation of charge carriers in anthracene with polarized light", *J. Chem. Phys.* **47**, 1194–5 (1967).

GEACINTOV, N. E. and POPE, M.: "Low-lying valence band states and intrinsic photoconductivity in crystalline anthracene and tetracene", *J. Chem. Phys.* **50**, 814–22 (1969).

GEACINTOV, N. E. and POPE, M.: "Intrinsic photoconductivity in organic crystals", in *Proceedings of the 3rd International Conference on Photoconductivity* (ed. E. M. Pell), Pergamon Press (Oxford) 289–95 (1971).

GEACINTOV, N. E., POPE, M., and FOX, S.: "Magnetic field effects on photoenhanced currents in organic crystals", *J. Phys. Chem. Solids* **31**, 1375–9 (1970).

GEACINTOV, N. E., POPE, M., and KALLMANN, H.: "Photogeneration of charge carriers in tetracene", *J. Chem. Phys.* **45**, 2639–49 (1966).

GEACINTOV, N. E., POPE, M., and VOGEL, F.: "Effect of magnetic field on the fluorescence of tetracene crystals: exciton fission", *Phys. Rev. Lett.* **22**, 593–6 (1969).

GEACINTOV, N. E. and SWENBERG, C. E.: "Possible effects of magnetic fields on the quenching of triplet excited states of polynuclear hydrocarbons by oxygen", *J. Chem. Phys.* **57**, 378–89 (1972).

GELMONT, B. L. and SHUR, M. S.: "Motion of current filaments in transverse electric and magnetic fields. Linear theory", *Soviet Phys.—Semicond.* **7**, 1311–14 (1974).

GEPPERT, D. V.: "Space charge limited tunnel emission into an insulating film", *J. Appl. Phys.* **33**, 2993–5 (1962).

GEPPERT, D. V.: "Theoretical shape of metal–insulator–metal potential barriers", *J. Appl. Phys.* **34**, 490–3 (1963).

GERHOLD, G. A.: "Wannier excitons in anthracene", *J. Chem. Phys.* **45**, 1889–92 (1966).

GERHOLD, G. A.: "Reply to comments on Wannier excitons in anthracene", *J. Chem. Phys.* **46**, 4998 (1967).

GERISCHER, H.: "Semiconductor electrode reactions", in *Advances in Electrochemistry and Electrochemical Engineering* (ed. P. Delahay), Wiley Interscience **1**, 139–232 (1961).

GERISCHER, H.: "Kinetics of reduction–oxidation reaction on metals and semiconductors. III: Semiconductor electrodes", *Z. Phys. Chem.* **27**, 48–79 (1961).

GERISCHER, H.: "Semiconductor electrochemistry", in *Physical Chemistry: An Advanced Treatise* (ed. H. Eyring), Vol. IXA. Academic Press (New York) 463–542 (1970).

GERISCHER, H.: "Elektrodenreaktionen mit angeregten elektronischen Zuständen", *Ber. Bunsenges. Phys. Chem.* **77**, 771–82 (1973).

GERLACH, E.: "Surface states of anthracene crystals", *Phys. Rev.* **183**, 807–8 (1969).

GHEORGHITA-OANCEA, G.: "On the structure and semiconductor properties of thin anthracene layers", *Rev. Phys. Acad. Rep. Populaire Roumaine* **8**, 361–75 (1964), and **9**, 23–53 (1964).

GHEORGHITA-OANCEA, G.: "Electrical dark- and photoconductivity of thin anthracene layers in vacuum", *Br. J. Appl. Phys.* **12**, 579–80 (1961).

GHOSH, A. K. and FENG, T.: "Rectification, space-charge-limited current, photovoltaic and photoconductive properties of Al/tetracene/Au sandwich cell", *J. Appl. Phys.* **44**, 2781–8 (1973).

GHOSH, A. K., MOREL, D. L., FENG, T., SHAW, R. F., and ROWE, C. A., JR.: "Photovoltaic and rectification properties of Al/Mg-phthalocyanine/Ag Schottky barrier cells", *J. Appl. Phys.* **45**, 230–6 (1974).

GIAEVER, I.: "Energy gap in superconductors measured by electron tunneling", *Phys. Rev. Lett.* **5**, 147–8 (1960).

GIAEVER, I.: "Electron tunneling between two superconductors", *Phys. Rev. Lett.* **5**, 464–6 (1960).

GIBBONS, D. J. and SPEAR, W. E.: "Electron hopping transport and trapping phenomena in orthorhombic sulphur crystals", *J. Phys. Chem. Solids* **27**, 1917–25 (1966).

GILL, W. D. and BATRA, I. P.: "Transient optical double injection in insulators", *J. Appl. Phys.* **42**, 2067–73 (1971).

GILL, W. D. and KANAZAWA, K. K.: "Transient photocurrent for field-dependent mobilities", *J. Appl. Phys.* **43**, 529–34 (1972).

GINZBURG, V. L.: "High temperature superconductivity", *J. Polymer Sci.* Part C, **29**, 3–16 (1970).

GLAESER, R. and BERRY, R. S.: "Band structure and charge mobility in crystals of polarizable molecules", in *Third Organic Crystal Symposium, Chicago, USA*, paper 13 (1965).

GLAESER, R. and BERRY, R. S.: "Mobilities of electrons and holes in organic molecular solids—comparison of band and hopping models", *J. Chem. Phys.* **44**, 3797–810 (1966).

GLARUM, S. H.: "Electron mobilities in organic semiconductors", *J. Phys. Chem. Solids*, **24**, 1577–83 (1963).

GLOCKNER, E. and WOLF, H. C.: "The fluorescence spectrum of anthracene crystals", *Z. Naturforsch.* **24A**, 943–51 (1969).

GLOCKNER, E. and WOLF, H. C.: "Fluorescence spectra of the mixed crystal system in anthracene-perdeuteroanthracene mixed exciton band", *Chem. Phys. Lett.* **27**, 161–6 (1974).

GOBELI, G. W. and ALLEN, F. G.: "Direct and indirect excitation processes in photo electric emission from silicon", *Phys. Rev.* **127**, 140–6 (1962).

GOBELI, G. W. and ALLEN, F. G.: "Photoelectric threshold and work function", in *Semiconductors and Semimetals* (eds. R. K. Willardson and A. C. Beer), Academic Press (New York) **2**, 263–300 (1966).

GOBRECHT, H. and HOFLANN, D.: "Spectroscopy of traps by fractional glow technique", *J. Phys. Chem. Solids* **27**, 509–22 (1966).

GODLEWSKI, J. and KALINOWSKI, J.: "Photosensitive space charge controlled currents in tetracene crystal", *Phys. Stat. Sol.* (a) **32**, K173–80 (1975).

GOEPPERT-MAYER, M. and SKLATER, A. L.: "Calculations of the lowest excited levels of benzene", *J. Chem. Phys.* **6**, 645–52 (1938).

GOLDER, J., NICOLET, M. A., and SHUMKA, A.: "Noise of space-charge-limited currents in solids is thermal", *Solid-State Electron.* **16**, 1151–8 (1973).

GOLDSMITH, G. J.: "Photoconductivity of single crystals of anthracene", *Phys. Rev.* **93**, 929 (1954).

GOLUBOV, S. I. and KONOBEEV, YU. V.: "Energy transfer by excitons in a tetracene-doped anthracene crystal", *Phys. Stat. Sol.* (b) **79**, 79–84 (1977).

GOOD, R. H. and MULLER, W.: "Field emission", in *Handbuch der Physik*, Vol. 21, Springer-Verlag (Berlin) 176–231 (1956).

GOODE, D. H.: "Effect of triplet exciton trap saturation and bimolecular decay on the delayed fluorescence of anthracene", in *Molec. Lumin. Int. Conf. 1968*, Benjamin (New York) 751–63 (1969).

GOODE, D. H. and LIPSETT, F. R.: "Delayed fluorescence of anthracene at low temperatures: failure of square dependence on exciting light", *J. Chem. Phys.* **51**, 1222–7 (1969).

GOODE, D. H., LUPIEN, Y., SIEBRAND, W., WILLIAMS, D. F., THOMAS, J. M., and WILLIAMS, J. O.: "Triplet excitons as probes for structural imperfections in crystalline anthracene", *Chem. Phys. Lett.* **25**, 308–11 (1974).

GOODINGS, E. P.: "Polymeric conductors and superconductors", *Endeavour*, ICI Publications (UK), XXXIV, No. 123, 123–30 (1975).

GOODMAN, A. M.: "Evaporated metallic contacts to conducting cadmium sulfide single crystals", *J. Appl. Phys.* **35**, 573–80 (1964).

GOODMAN, A. M.: "Photoemission of electrons from silicon and gold into silicon dioxide", *Phys. Rev.* **144**, 588–93 (1966).

GOODMAN, A. M.: "Photoemission of holes from silicon into silicon dioxide", *Phys. Rev.* **152**, 780–6 (1966).

GOODMAN, A. M.: "Photoemission of electrons and holes into silicon nitride", *Appl. Phys. Lett.* **13**, 275–7 (1968).

GOODMAN, A. M.: "Electron energy-band diagrams of insulators determined from internal photoemission measurements", in *Proceedings of the 3rd International Conference on Photoconductivity* (ed. E. M. Pell), Pergamon Press (Oxford) 69–74 (1971).

GOODMAN, A. M. and ROSE, A.: "Double extraction of uniformly generated electron–hole pairs from insulators with non-injecting contacts", *J. Appl. Phys.* **42**, 2823–30 (1971).

GOODMAN, A. M. and WARFIELD, G.: "Dember effect in silver chloride", *Phys. Rev.* **120**, 1142–8 (1960).

GÖRLICH, P.: "Problems of photoconductivity", *Advance in Electronics and Electronic Physics*, Academic Press, XIV (c) 37–84 (1961).

GORODETSKII, S. M., ZHDANOVICH, N. S., and RAVICH, YU. I.: "Anomalous photomagnetic effect in silicon", *Soviet Phys.–Semicond.* **7**, 853–8 (1973).

GOSAR, P. and CHOI, S. I.: "Linear-response theory of the electron mobility in molecular crystals", *Phys. Rev.* **150**, 529–38 (1966).

GOSAR, P. and CHOI, S. I.: "Properties of narrow band and small polaron propagators", in *Excitons, Magnons, and Phonons in Molecular Crystals*, Beirut, Lebanon (Cambridge, England) 175–84 (1968).

GOSAR, P. and PRELOVSEK, P.: "Electron mobility in a disordered narrow band solid", *Z. Phys.* **266**, 299–304 (1974).

GOSAR, P. and VILFAN, I.: "Phonon-assisted current in organic molecular crystals", *Mol. Phys.* **18**, 49–61 (1970).

GOSSICK, B. R.: "Electrical characteristics of a metal–semiconductor contact. III", *Surf. Sci.* **28**, 469–88 (1971).

GOUTERMAN, M.: "Radiation less transitions: a semiclassical model", *J. Chem. Phys.* **36**, 2846–53 (1962).

GRÄNACHER, I.: "Dependence of the space-charge-limited current in anthracene on crystal thickness", *Solid State Commun.* **3**, 331–3 (1965).

GRANT, A. J. and DAVIS, E. A.: "Hopping conduction in amorphous semiconductors", *Solid State Commun.* **15**, 291–4 (1974).

GRANT, W. J. C.: "Role of rate equations in the theory of luminescent energy transfer", *Phys. Rev.* B **4**, 648–63 (1971).

GRAY, P. V.: "Observation of surface and impurity states in silicon by oxide layer tunneling", *Phys. Rev. Lett.* **9**, 302–5 (1962).

GRAY, P. V.: "Tunneling from metal to semiconductors", *Phys. Rev.* **140**, A179–86 (1965).

GREEN, M. A.: "Enhancement of Schottky solar cell efficiency above its semiempirical limit", *Appl. Phys. Lett.* **27**, 287–8 (1975).

GREEN, M. A.: "The open-circuit voltage of vertical junction solar cells", *J. Phys.* D **9**, L57–59 (1976).

GREEN, M. A.: "The depletion layer collection efficiency for *p–n* junction, Schottky diode, and surface insulator solar cells", *J. Appl. Phys.* **47**, 547–54 (1976).

GREEN, M. A. and GODFREY, R. B.: "MIS solar cell—general theory and new experimental results for silicon", *Appl. Phys. Lett.* **29**, 610–12 (1976).

GREEN, M. A., KING, F. D., and SHEWCHUN, J.: "Minority carrier MIS tunnel diodes and their application to electron and photovoltaic energy conversion. I: Theory", *Solid-State Electron.* **17**, 551–61 (1974).

GREEN, M. E.: "Pressure theory of the thermoelectric and photovoltaic effects", *J. Appl. Phys.* **35**, 2689–94 (1964).

GREEN, M. E.: "A proposed model for electron injection into some organic semiconductor in the dark", *J. Phys. Chem.* **69**, 3510–13 (1965).

GREEN, M. E.: "Charge-carrier injection and pre-exponential factor in semiconducting organic substances", *J. Chem. Phys.* **51**, 3279–80 (1969).

GREENWOOD, D. A.: "The Boltzmann equation in the theory of electrical conduction in metals", *Proc. Phys. Soc.* **71**, 585–96 (1958).

GREER, W. L.: "Exciton transport in substitutionally impure molecular crystals", *J. Chem. Phys.* **60**, 744–53 (1974).

GREGOR, L. V.: "Polymer dielectric films", *IBM Jl Res. Dev.* **12**, 140–62 (1968).

GRENBEL, W., WOLFF, V., and KRÜGER, H.: "Electric field induced texture changes in certain nematic/cholesteric liquid crystal mixtures", *Mol. Cryst. Liq. Cryst.* **24**, 103–11 (1973).

GRENET, J., VAUTIER, C., CARLES, D., and CHABRIER, J. J.: "Space charge conduction in presence of Gaussian distribution of localized states", *Phil. Mag.* **28**, 1265–77 (1973).

GRIBNIKOV, Z. S.: "Anisotropic Dember photo-emf of a semiconductor with a periodic doping profile", *Soviet Phys.–Semicond.* **7**, 823–5 (1973).

GRIBNIKOV, Z. S.: "Injection filament in a semiconduction with deep traps", *Soviet Phys.—Semicond.* **11**, 184–6 (1977).

GROFF, R. P., AVAKIAN, P., and MERRIFIELD, R. E.: "Coexistence of exciton fission and fusion in tetracene crystals", *Phys. Rev.* B **1**, 815–17 (1970).

GROFF, R. P., AVAKIAN, P., and MERRIFIELD, R. E.: "Magnetic field dependence of delayed fluorescence from tetracene crystals", *J. Lumin.* **1-2**, 218–23 (1970).

GROFF, R. P., MERRIFIELD, R. E., and AVAKIAN, P.: "Singlet and triplet channels for triplet-exciton fusion in anthracene crystals", *Chem. Phys. Lett.* **5**, 168–70 (1970).

GROFF, R. P., MERRIFIELD, R. E., SUNA, A., and AVAKIAN, P.: "Magnetic hyperfine modulation of dye-sensitized delayed fluorescence in an organic crystal", *Phys. Rev. Lett.* **29**, 429–31 (1972).

GROFF, P. R., SUNA, A., and MERRIFIELD, R. E.: "Temperature dependence of conductivity of tetrathiafulvalene–tetracyanoquinodimethane (TTF–TCNQ) single crystals", *Phys. Rev. Lett.* **33**, 418–21 (1974).

GROSS, B., SESSLER, G. M., and WEST, J. E.: "TSC studies of carrier trapping in electron and γ-irradiated teflon", *J. Appl. Phys.* **47**, 968–75 (1976).

GU, J, KAWABE, M., MASUDA, K., and NAMBA, S.: "Electroluminescence of anthracene with powdered graphite electrodes and ambient gas effects on the electrode", *J. Appl. Phys.* **48**, 249–94 (1977).

GUDKOV, I. D., KOZLENKOVA, N. I., and KURLOVA, G. A.: "Conductivity of amorphous films of copper naphthalocyanine", *Soviet Phys.—Solid State* **11**, 1203–6 (1969).

GUGESHASHVILI, M. I., ELIGHLASHVILI, I. A., NAKASHIDZE, G. A., ROZENSHTEIN, L. D., and CHAVCHANIDZE, V. V.: "Injection-induced long-wavelength photoconductivity of anthracene", *Soviet Phys.—Semicond.* **2**, 125–7 (1968).

GUGESHASHVILI, M. I., ELIGHLASHVILI, I. A., NAKASHIDZE, G. A., ROZENSHTEIN, L. D., and HATIASHVILI, A. A.: "Distribution of naphthacene levels in anthracene", *Soviet Electrochem.* **4**, 533–4 (1968).

GUNDLACH, K. H. and HELDMAN, G.: "The effect of the $E(k)$-relation on tunneling through asymmetric barriers", *Phys. Stat. Sol.* **21**, 575–9 (1967).

GUNDLACH, K. H. and KADLEC, J.: "Space-charge dependence of the barrier height on insulator thickness in Al–(Al-oxide)–Al sandwich", *Appl. Phys. Lett.* **20**, 445–6 (1972).

GUNDLACH, K. H. and KADLEC, J.: "Spectral dependence of the photoresponse in MIM (metal–insulator–metal) structures: influence of the electrode thickness", *Thin Solid Films* **28**, 107–17 (1975).

GUNDLACH, K. H. and KADLEC, J.: "Interfacial barrier height measurements from voltage dependence of the photocurrent", *J. Appl. Phys.* **46**, 5286–7 (1975).

GUPTA, H. M.: "Theory of isothermal current–time characteristics determination of trap parameters", *J. Appl. Phys.* **48**, 3448–54 (1977).

GUPTA, H. M. and VAN OVERSTRAETEN, R. J.: "Theory of thermal dielectric relaxation and direct determination of trap parameters", *J. Phys. C* **7**, 3560–72 (1974).

GUPTA, H. M. and VAN OVERSTRAETEN, R. J.: "Theory of isothermal dielectric relaxation and direct determination of trap parameters", *J. Appl. Phys.* **47**, 1003–9 (1976).

GURARI, M.: "Self energy of slow electrons in polar materials", *Phil. Mag.* **44**, 329–36 (1953).

GUREVICH, L. E. and GASYMOV, T. M.: "Thermal EMF of a semiconductor in a strong electric field", *Soviet Phys.—Solid State* **9**, 2752–6 (1968).

GUTMANN, F.: "Some comments on carrier generation and transport in short range order systems", *J. Polymer Sci. C* **17**, 41–49 (1967).

GUTMANN, F. and KEYZER, H.: "Electrical conduction in chlorpromazine", *Nature* **205**, 1102–3 (1965).

HAARER, D. and CASTRO, G.: "Exciton induced photoemission in anthracene", *Chem. Phys. Lett.* **12**, 277–80 (1971).

HAARER, D., SCHMID, D., and WOLF, H. C.: "Electron spin resonance of triplet excitons in anthracene", *Phys. Stat. Sol.* **23**, 633–8 (1967).

HABERKORN, R.: "Injection of electrons into *p*-chloranil crystals via exciton decay at metal and aqueous electrodes", *Ber. Bunsenges. Phys. Chem.* **77**, 928–38 (1973).

HABERKORN, R. and MICHEL-BEYERLE, M. E.: "Onsager's ion recombination model in one dimension", *Chem. Phys. Lett.* **23**, 128–230 (1973).

HADEK, V. and ULBERT, K.: "Metal chamber for the measurement of the temperature dependence of conductivity and the thermoelectric power of organic semiconductors", *Rev. Sci. Instrum.* **38**, 991–2 (1967).

HAGENLOCHER, A. K. and CHEN, C. T.: "Space-charge-limited current instabilities in $n^+ - \pi - n^+$ silicon diodes", *IBM Jl Res. Dev.* **13**, 533–6 (1969).

HAHN, T. S. and SCHETZINA, J. F.: "Photovoltaic effect in uniaxially stressed germanium", *Phys. Rev. B* **7**, 729–32 (1973).

HALE, J. M. and BAESSLER, H.: "Extrinsic photoconductivity in aromatic hydrocarbons", *Trans. Faraday Soc.* **64**, 2452–8 (1968).

HALE, J. M. and MEHL, W.: "Approximate solution to the one carrier problem of a semiconductor–electrolyte system", *Surf. Sci.* **4**, 221–3 (1966).

HALE, J. M. and MEHL, W.: "Photo-generation of carriers in molecular crystals through electrochemical electrode processes", *Electrochim. Acta* **31**, 1483–95 (1968).

HALE, J. M. and MEHL, W.: "Quenching of excited states at the interface, molecular crystal/electrolyte", *Discuss. Faraday Soc.* **51**, 54–60 (1971).

HALL, J. L., JENNINGS, P. A., and McCLINTOCK, R. M.: "Study of anthracene fluorescence excited by the ruby-generated giant pulse laser", *Phys. Rev. Lett.* **11**, 364–6 (1963).

HALL, R. N.: "Power rectifiers and transistors", *Proc. IRE* **40**, 1512–18 (1952).

HALL, R. N., RACETTE, J. H. and EHRENREICH, H.: "Direct observation of polarons and phonons during tunneling in group 3–5 semiconductor junctions", *Phys. Rev. Lett.* **4**, 456–8 (1960).

HAMANN, C.: "On the electric and thermoelectric properties of copper phthalocyanine single crystals", *Phys. Stat. Sol.* **20**, 481–91 (1967).

HAMANN, C.: "On the trap distribution in thin film of copper phthalocyanine", *Phys. Stat. Sol.* **26**, 311–18 (1968).

HAMANN, C.: "A model of the electrical transport of phenomena in imperfect crystals of copper phthalocyanine. I: Basic ideas and treatment of bulk defects", *Phys. Stat. Sol.* (b) **55**, 585–94 (1973).

HAMANN, C. and LEHMANN, G.: "A model of the electrical transport phenomena in imperfect crystals of copper phthalocyanine. II: Surface states of phthalocyanine single crystals", *Phys. Stat. Sol.* (b) **60**, 407–13 (1973).

HAMANN, C., STARKE, M., and WAGNER, H.: "The influence of an impurity on the energetic trap distribution of copper phthalocyanine", *Phys. Stat. Sol.* (a) **16**, 463–8 (1973).

HAMPTON, E. M., SHAH, B. S., and SHERWOOD, J. N.: "The growth and perfection of orthorhombic (α) sulphur single crystals", *J. Crystal Growth* **22**, 22–28 (1974).

HANAI, T., HAYDON, D. A., and TAYLOR, J.: "Polar group orientation and the electrical properties of lecithin biomolecular leaflets", *J. Theor. Biol.* **9**, 278–96 (1965).

HANDY, R. M.: "Electrode effects on aluminum oxide tunnel junctions", *Phys. Rev.* **126**, 1968–73 (1962).

HANEMANN, D.: "Structure and adsorption characteristics of clean (111) and (111) surfaces of gallium antimonide", in *International Conference on Semiconductor Physics in 1960, Prague*, Czechoslovak Academy of Science, 540–3 (1961).

HÄNSEL, H.: "Über den Leitungsmechanismus des organischen Festkörpers", *Ann. Phys.* **24**, 147–54 (1970).

HARADA, Y. and INOKUCHI, H.: "Photoemission from polycyclic aromatic crystals in the vacuum ultraviolet region", *Bull. Chem. Soc. Japan* **39**, 1443–8 (1966).

HARADA, Y., MARUYAMA, Y., SHEROTANI, I., and INOKUCHI, H.: "Electrical conductivity of organic semiconductors at high pressures", *Bull. Chem. Soc. Japan* **37**, 1378–80 (1964).

HARGREAVES, A. and RIZVI, S. H.: "The crystal and molecular structure of biphenyl", *Acta Cryst.* **15**, 365–73 (1962).

HARRACH, L. A. and HUGHES, R. C.: "X- and gamma-radiation damage to single crystal anthracene", *Mol. Cryst.* **5**, 141–3 (1968).

HARRIS, R. A.: "Pi-electron Hamiltonian", *J. Chem. Phys.* **47**, 3967–71 (1967).

HARRIS, R. A.: "Generalized time-dependent Hartree theory and the coupling between sigma and pi-electrons", *J. Chem. Phys.* **47**, 3972–6 (1967).

HARRISON, S. W., FISCHER, C. R., and ARNOLD, S.: "Calculation of Davydov splitting of the first triplet state in pyrene", *J. Chem. Phys.* **57**, 1102–5 (1972).

HARRISON, W. A.: "Tunneling from an independent-particle point of view", *Phys. Rev.* **123**, 85–89 (1961).

HARTKE, J. L.: "The three-dimensional Poole–Frenkel effect", *J. Appl. Phys.* **39**, 4871–3 (1968).

HARTMAN, T. E.: "Tunneling through asymmetric barriers", *J. Appl. Phys.* **35**, 3283–94 (1964).

HARTMAN, T. E., BLAIR, J. C., and BAUER, R.: "Electrical conduction through SiO films", *J. Appl. Phys.* **37**, 2468–74 (1966).

HARTMAN, T. E. and CHIVIAN, J. S.: "Electron tunneling through thin aluminium oxide films", *Phys. Rev.* **134**, A1094–1101 (1964).

HARTMANN, G. C. and LIPARI, N. O.: "Transient space-charge-perturbed currents following time-dependent injection", *J. Appl. Phys.* **44**, 1676–81 (1973).

HARTMANN, G. C. and LIPARI, N. O.: "Investigation of transient space-charge-perturbed currents", *J. Appl. Phys.* **46**, 2821–7 (1975).

HARTMANN, G. C. and NOOLANDI, J.: "Charge transfer at photoconductor liquid interfaces", *J. Chem. Phys.* **66**, 3498–508 (1977).

HARTMANN, G. C. and SCHMIDLIN, F. W.: "Transient photostimulated charge transfer from a photoconductor to an insulating fluid", *J. Appl. Phys.* **46**, 266–78 (1975).

HASEGAWA, K.: "Dark resistivity of anthracene measured by potentiometric method", *Japan J. Appl. Phys.* **3**, 633–6 (1964).

HASEGAWA, K. and SCHNEIDER, W. G.: "Ruby-laser excited photocurrents in anthracene single crystals", *J. Chem. Phys.* **39**, 1346–7 (1963).

HASEGAWA, K. and SCHNEIDER, W. G.: "Ruby-laser excited photocurrents in anthracene", *J. Chem. Phys.* **40**, 2533–7 (1964).

HASEGAWA, K. and YOSHIMURA, S.: "Two photon-excited currents in anthracene crystals", *J. Phys. Soc. Japan* **20**, 460–1 (1965).

HASEGAWA, K. and YOSHIMURA, S.: "Photo carrier generation in anthracene due to exciton interaction of two photon excited singlets", *Phys. Rev. Lett.* **14**, 689–90 (1965).

HASEGAWA, K. and YOSHIMURA, S.: "Bulk generation of photocarriers via two photon absorption in anthracene single crystal", *J. Phys. Soc. Japan* **21**, 2626–33 (1966).

HAYWARD, D. and PETHIG, R.: "Frequency dependence of the conductivity of molecular solids", *Phys. Stat. Sol.* (a) **32**, K177–80 (1975).

HEEGER, A. J.: "(TTE) (TCNQ): a one-dimensional metal", *Solid State Commun.* **14**, 99 (1974).

HEILMEIER, G. H. and HARRISON, S. E.: "Charge transport in copper phthalocyanine single crystals", *Phys. Rev.* **132**, 2010–16 (1963).

HEILMEIER, G. H. and WARFIELD, G.: "Investigation of bulk currents in metal-free phthalocyanine crystals", *J. Chem. Phys.* **38**, 163–8 (1963).

HEILMEIER, G. H., WARFIELD, G., and HARRISON, S. E.: "Applicability of the band model to metal-free phthalocyanine single crystals", *J. Appl. Phys.* **34**, 2278–81 (1963).

HEILMEIER, G. H. and ZANONI, L. A.: "Surface studies of α-Cu phthalocyanine films", *J. Phys. Chem. Solids* **25**, 603–11 (1964).

HELFRICH, W.: "Lichtempfindliche raumladungsbeschränkte Ströme", *Phys. Stat. Sol.* **7**, 863–8 (1964).

HELFRICH, W.: "Destruction of triplet excitons in anthracene by injected electrons", *Phys. Rev. Lett.* **16**, 401–3 (1966).

HELFRICH, W.: "Space-charge-limited and volume-controlled currents in organic solids", in *Physics and Chemistry of the Organic Solid State*, Wiley Interscience (New York) **3**, 1–65 (1967).

HELFRICH, W. and LIPSETT, F. R.: "Fluorescence and defect fluorescence of anthracene at 4.2°K", *J. Chem. Phys.* **43**, 4368–76 (1965).

HELFRICH, W. and MARK, P.: "Space-charge-limited currents in anthracene as a means for determining the hole mobility", *Z. Phys.* **166**, 370–85 (1962).

HELFRICH, W. and MARK, P.: "A proof for the space charge in space-charge-limited hole currents in anthracene", *Z. Phys.* **168**, 495–503 (1962).

HELFRICH, W. and MARK, P.: "Eine Bestimmung der effektiven Zustandsdichte des Bandes für überschüssige Defekteletronen in Anthrazen", *Z. Phys.* **171**, 527–36 (1963).

HELFRICH, W., RIEHL, N., and THOMA, P.: "Optical and electrical measurements of glow-curves in anthracene", *Phys. Lett.* **10**, 31–32 (1964).

HELFRICH, W. and SCHNEIDER, W. G.: "Recombination radiation in anthracene crystals", *Phys. Rev. Lett.* **14**, 229–31 (1965).

HELFRICH, W. and SCHNEIDER, W. G.: "Transients of volume-controlled current and of recombination radiation in anthracene", *J. Chem. Phys.* **44**, 2902–9 (1966).

HELLER, W. R.: "Kinetic-statistical theory of dielectric breakdown in nonpolar crystals", *Phys. Rev.* **84**, 1130–50 (1951).

HENDERSON, H. T. and ASHLEY, K. L.: "A negative resistance diode based upon double injection in thallium doped silicon", *Proc. IEEE* **57**, 1677 (1969).

HENDERSON, H. T. and ASHLEY, K. L.: "Space charge limited current in neutron-irradiated silicon with evidence of the complete Lampert triangle", *Phys. Rev.* **186**, 811–15 (1969).

HENDERSON, H. T., ASHLEY, K. L., and SHEN, M. K. L.: "Third side of the Lampert triangle: evidence of traps-filled-limit single carrier injection", *Phys. Rev. B* **6**, 4079–80 (1972).

HENISCH, H. K. and SMITH, W. R.: "Switching in organic polymer films", *Appl. Phys. Lett.* **24**, 589–91 (1974).

HENISCH, H. K., SMITH, W. R., and WIHL, W.: "Field-dependent photo-response of threshold switching systems", *Amorphous Liq. Semicond.* **1**, 567–70 (1973).

HEPPELL, G. E. and HARDWICK, R.: "Radiation damage study of crystalline anthracene", *Trans. Faraday Soc.* **63**, 2651–5 (1967).

HERLET, A.: "Die Abhängigkeit der Stromdichte eines *p–i–n* Gleichrichters von der Breite seiner Mittelzone", *Z. Phys.* **141**, 335–45 (1955).

HERLET, A. and SPENKE, E.: "Gleichrichter mit *p–i–n* bezw. mit *p–s–n* Struktur unter Gleichstrombelastung", *Z. angew. Phys.* **7**, 99–107 (1955), 149–163 (1955).

HERMANN, A. M.: "Calculation of transient photocurrents in insulators with trapped space charge", *J. Appl. Phys.* **44**, 926–8 (1973).

HERMANN, A. M. and REMBAUM, A.: "Spin resonance, Hall effect, and transport properties of poly (*N*-vinyl carbazole) – iodine complex", *J. Polymer Sci.* Part C, **17**, 107–23 (1967).

HERNANDEZ, J. P.: "Photo-ionization of crystalline anthracene", *Phys. Rev.* **169**, 746–9 (1968).

HERNANDEZ, J. P., and COLD, A.: "Two-photon absorption in anthracene", *Phys. Rev.* **156**, 26–35 (1967).

HERRING, C. and NICHOLS, M. H.: "Thermionic emission", *Rev. Mod. Phys.* **21**, 185–270 (1949).

HICKMOTT, T. W.: "Electron emission electroluminescence and voltage-controlled negative resistance in $Al–Al_2O_3–Au$ diodes", *J. Appl. Phys.* **36**, 1885–96 (1965).

HICKMOTT, T. W.: "Electroluminescence and photoresponse of Ta_2O_5 and TiO_2 diodes", *J. Electrochem. Soc.* **113**, 1223–5 (1966).

HILL, R. M.: "Poole–Frenkel conduction in amorphous solids", *Phil. Mag.* **23**, 59–86 (1971).

HILL, R. M.: "Injection controlled conduction", *Thin Solid Films* **15**, 369–91 (1973).

HINO, S., SEKI, H., and INOKUCHI, H.: "Photoelectron spectra of *p*-terphenyl in gaseous and solid states", *Chem. Phys. Lett.* **36**, 335–9 (1975).

HIRAI, T. and NAKATA, O.: "Formation of thin polyacrylonitrile films and their electrical properties", *Japan J. Appl. Phys.* **7**, 112–21 (1968).

HIROTA, N.: "Use of triplet-state energy transfer in obtaining singlet–triplet absorption in organic crystals", *J. Chem. Phys.* **44**, 2199–200 (1966).

HIRTH, H. and STOCKMAN, F.: "Electron and hole mobilities in solid and liquid benzene C_6H_6", *Phys. Stat. Sol.* (b) **51**, 691–9 (1972).

HJORTENBERG, D., POPOVIC, J., and BRANDT, B.: "Luminescence of anthracene as a possible means of α- and electron-ray dosimetry", *Proceedings of the International Conference on Luminescence, Budapest, 1966*, 1 and 2, Budapest, Akademiai Kiado, 2108–11 (1968).

HOCHSTRASSER, R. M.: "The luminescence of organic molecular crystals", *Rev. Mod. Phys.* **34**, 531–50 (1962).

HOCHSTRASSER, R. M.: "Spectral effects of strong exciton coupling in the lowest electronic transition of perylene", *J. Chem. Phys.* **40**, 2599–64 (1964).

HOCHSTRASSER, R. M. and WESSEL, J. E.: "Time resolved fluorescence of anthracene in mixed crystals at $2°K$", *Chem. Phys.* **6**, 19–33 (1974).

HOESTEREY, D. C.: "Photocarrier generation in anthracene", *J. Chem. Phys.* **36**, 557–8 (1962).

HOESTEREY, D. C.: "High-field photoconduction in anthracene", *Bull. Am. Phys. Soc.* **13**, 479 (1968).

HOESTEREY, D. C. and LETSON, G. M.: "The trapping of photocarriers in anthracene by anthraquinone, anthrone and naphthacene", *J. Phys. Chem. Solids* **24**, 1609–15 (1963).

HOESTEREY, D. C. and ROBISON, G. W.: "On the diffusion coefficient of triplet excitons in anthracene", *J. Chem. Phys.* **54**, 1709–12 (1971).

HOFBERGER, W.: "Structure and optical properties of polycrystalline evaporated tetracene films", *Phys. Stat. Sol.* (a) **30**, 271–8 (1975).

HOFBERGER, W. and BÄESSLER, H.: "Diffusion of triplet excitons in amorphous tetracene", *Phys. Stat. Sol.* (b) **69**, 725–30 (1975).

HOKADO, K. and SCHNEIDER, W. G.: "Thermally stimulated currents and carrier trapping in anthracene crystals", *J. Chem. Phys.* **40**, 2937–45 (1964).

HOLM, R.: "The electric tunnel effect across thin insulator films in contacts", *J. Appl. Phys.* **22**, 569–74 (1951).

HOLMES-WALKER, W. A. and UBBELHODL, A. R.: "Electron transfer in alkali metal–hydrocarbon complexes", *J. Chem. Soc.* 720–8 (1954).

HOLONYAK JR., N.: "Double injection diodes and related DI phenomena in semiconductors", *Proc. IRE* **50**, 2421–8 (1962).

HOLONYAK JR., N. and BEVACQUE, S. E.: "Oscillations in semiconductors due to deep levels", *Appl. Phys. Lett.* **2**, 71–73 (1963).

HOLONYAK JR., N., ING JR., S. W., THOMAS, R. C., and BEVACQUE, S. E.: "Double injection with negative resistance in semi-insulators", *Phys. Rev. Lett.* **8**, 426–8 (1962).

HOLSTEIN, T.: "Polaron motion. I: Molecular crystal model. II: Small polaron", *Ann. Phys. (New York)* **8**, 325–42, 343–89 (1959).

HOLSTEIN, T.: "Sign of the Hall coefficient in hopping-type charge-transport", *Phil. Mag.* **27**, 225–33 (1973).

HOLSTEIN, T. and FRIEDMAN, L.: "Hall mobility of the small polaron. II", *Phys. Rev.* **165**, 1019–31 (1968).

HOLZMAN, P., MORRIS, R., JARNAGIN, R. C., and SILVER, M.: "Photoconductivity in anthracene crystals due to excitation of triplets", *Phys. Rev. Lett.* **19**, 506–8 (1967), errata: ibid. **19**, 940 (1967).

HOMMA, S. and TAKEMOTO, S.: "Temperature dependence of scintillation pulses in anthracene and CsI(Tl)", *Rev. Sci. Instrum.* **32**, 1055–6 (1961).

HOOGENSTRAATEN, W.: "Electron traps in ZnS phosphors", *Philips Res. Rep.* **13**, 515 (1958).

HORI, Y., IWASHIMA, S., and INOKUCHI, H.: "Intrinsic electrical conductivity of violanthrene, $C_{34}H_{18}$", *Bull. Chem. Soc. Japan* **43**, 3294–6 (1970).

HOSHEN, J. and KOPELMAN, R.: "Exciton surface states in molecular crystals", *J. Chem. Phys.* **61**, 330–8 (1974).

HOVEL, H. J.: "Transparency of thin metal films on semiconductor substrates", *J. Appl. Phys.* **47**, 4968–70 (1976).

HOVEL, H. J.: "Solar cells for terrestrial applications", *Solar Energy* **19**, 605–15 (1977).

HOWARTH, D. J. and SONDHEIMER, E. H.: "The theory of electronic conduction in polar semiconductor", *Proc. Roy. Soc.* A **219**, 53–74 (1953).

HÜCKEL, E.: "Quantentheoretische Beiträge zum Problem der aromatischen und ungesättigten Verbindungen. III", *Z. Phys.* **76**, 628–48 (1932).

HUG, G. and BERRY, S.: "Interaction of electrons and holes in molecular crystals", *J. Chem. Phys.* **55**, 2516–21 (1971).

HUGGINS, C. M. and SHARBAUGH, A. H.: "Dielectric properties of some powdered organic semiconductors", *J. Chem. Phys.* **38**, 393–7 (1963).

HUGHES, A. J., HOLLAND, P. A., and LETTINGTON, A. H.: "Control of holding currents in amorphous threshold switches", *J. Non-Cryst. Solids* **17**, 89–99 (1975).

HUGHES, R. C.: "Geminate recombination of X-ray excited electron–hole pairs in anthracene", *J. Chem. Phys.* **55**, 5442–7 (1971).

HUNTINGTON, H. B., GANGOLI, S. G., and MILLS, J. L.: "Ultrasonic measurements of the elastic constants of anthracene", *J. Chem. Phys.* **50**, 3844–9 (1969).

HWANG, W. and KAO, K. C.: "A unified approach to the theory of current injection in solids with traps uniformly and non-uniformly distributed in space and energy, and size effects in anthracene films", *Solid-State Electron.* **15**, 523–9 (1972).

HWANG, W. and KAO, K. C.: "Electroluminescence in anthracene crystals caused by field induced minority carriers at moderate temperatures", *J. Chem. Phys.* **58**, 3521–2 (1973).

HWANG, W. and KAO, K. C.: "On the theory of filamentary double injection and electroluminescence in molecular crystals", *J. Chem. Phys.* **60**, 3845–55 (1974).

HWANG, W. and KAO, K. C.: "Studies of the theory of single and double injections in solids with a Gaussian trap distribution", *Solid-State Electron.* **19**, 1045–7 (1976).

HWANG, W. and KAO, K. C.: "Double injection in organic solids", in *Proceedings of 2nd Conference on Electrical and Related Properties of Organic Solids held on September 18–23, 1978, in Karpacz, Poland*, 33–40 (1978).

HYNES, J. R. and HORNBECK, J. A.: "Trapping of minority carriers in silicon. I. *p*-type silicon", *Phys. Rev.* **97**, 311–21 (1955).

HYNES, J. R. and HORNBECK, J. A.: "Trapping of minority carriers in silicon. II: *n*-type silicon", *Phys. Rev.* **100**, 606–15 (1955).

IANNUZZI, M. and POLACCO, E.: "Double photon excitation of fluorescence in anthracene", *Phys. Rev. Lett.* **13**, 371–2 (1964).

IANNUZZI, M. and POLACCO, E.: "Polarization dependence of laser-induced fluorescence in anthracene", *Phys. Rev.* **138**, 806–8 (1965).

IEDA, M., SAWA, G., and KATO, S.: "A consideration of Poole–Frenkel effect on electric conduction in insulators", *J. Appl. Phys.* **42**, 3737–40 (1971).

IEDA, M., SAWA, G., and SHINEDARA, U.: "A decay process of surface electric charges across polyethylene film", *Japan J. Appl. Phys.* **6**, 793–4 (1967).

IEDA, M., SAWA, G., and TAKEUCHI, R.: "Decay processes of different kinds of surface electric charges across polyethylene film", *Japan J. Appl. Phys.* **8**, 809 (1969).

IEDA, M., TAKAI, Y., and MIZUTANI, T.: "Photoconduction processes in polymers", in *Memoirs of the Faculty of Engineering, Nagoga University, Nagoga, Japan*, **29**, 1–59 (1977).

IEDA, M. and TAKEUCHI, R.: "Effects of temperature, *γ*-ray irradiation and crystallinity on decay process of surface electric charges across polyethylene film", *Japan J. Appl. Phys.* **9**, 727–8 (1970).

IIZUKA, E., KEIRA, T., and WADA, A.: "Light scattering by liquid crystals of poly-*γ*-benzylglutamates in electric field", *Mol. Cryst. Liq. Cryst.* **23**, 13–49 (1973).

ILEGEMS, M. and QUEISSER, H. J.: "Current transport in relaxation case-GaAs", *Phys. Rev.* B **12**, 1443–51 (1975).

INCE, A. N. and OATLEY, C. W.: "The electrical properties of electroluminescent phosphors", *Phil. Mag.* VII-**46**, 1031–1103 (1955).

INOKUCHI, H.: "Semi- and photoconductivity of molecular single crystals, anthracene and pyrene", *Bull. Chem. Soc. Japan* **29**, 131–3 (1956).

INOKUCHI, H.: "Catalytic activity of organic semiconductors and enzymes", *Discuss. Faraday Soc.* **51**, 183–9 (1971).

INOKUCHI, H. and AKAMATU, H.: "Electrical conductivity of organic semiconductors", *Solid State Phys.* **12**, 93–148 (1961).

INOKUCHI, H., HARADA, Y., and MARUYAMA, Y.: "Electrical properties of the single crystal and thin film of α, α′-diphenyl-β picrylhydrazyl", *Bull. Chem. Soc. Japan* **35**, 1559–61 (1962).

INOKUCHI, H., HORI, Y., and MARUYAMA, Y.: "Intrinsic conduction of polycyclic aromatic single crystal, violanthrene A, $C_{34}H_{18}$", *2nd International Conference on Conduction in Low-mobility Materials*, 375–81 (1971).

INOKUCHI, H., KURODA, H., and AKAMATO, H.: "On the electrical conductivity of the organic thin films; perylene, coronene, and violanthrene", *Bull. Chem. Soc. Japan* **34**, 749–53 (1961).

INOKUCHI, H. and NAKAGAKI, M.: "The density of polycyclic aromatic compounds", *Bull. Chem. Soc. Japan* **32**, 65–67 (1959).

INOUE, A. and NAGAKURA, S.: "The bimolecular annihilation of excitons and singlet exciton migration in anthracene crystals", *Mol. Cryst. Liq. Cryst.* **25**, 199–204 (1974).

INOUE, A., YOSHIHARA, K., and NAGAKURA, S.: "Exciton–exciton interaction and exciton migration in anthracene, pyrene and perylene crystals", *Bull. Chem. Soc. Japan* **45**, 1973–6 (1972).

INOUE, T.: "Paramagnetic centers in gamma-irradiated anthracene single crystals", *J. Phys. Soc. Japan* **25**, 914 (1968).

IOFFE, A. E.: "Heat transfer in semiconductors", Proc. Int. Conf. on Electron Transport, *Can. J. Phys.* **34**, 1342–55 (1956).

IOFFE, A. E.: "Properties of various semiconductors", *J. Phys. Chem. Solids* **8**, 6–14 (1959).

ISHIHARA, Y. and NAKADA, I.: "Generation of charge carriers in crystalline anthracene doped with tetracene", *J. Phys. Soc. Japan* **25**, 1512 (1968).

ISHIHARA, Y. and NAKADA, I.: "Energy gap of the crystalline anthracene", *J. Phys. Soc. Japan* **28**, 667–74 (1970).

ITOH, U.: "Enhancement of photoconduction by detrapping in anthracene crystal", *J. Phys. Soc. Japan* **35**, 514–17 (1973).

ITOH, U.: "Photoconduction in anthracene and anthracene-analogue compounds", Electrotechnical Lab., Japan Report No. 752 (1975).

ITOH, U. and ANZAI, H.: "Enhancement of photoconduction by detrapping in phenazine-doped anthracene crystal", *J. Phys. Soc. Japan* **36**, 1491 (1974).

ITOH, U., ANZAI, H., and TAKEISHI, K.: "Electrical conduction of anthracene crystal", *J. Phys. Soc. Japan* **35**, 810–13 (1973).

ITOH, U. and IZUMI, T.: "Ruby-laser induced photoconduction of α- and β-form crystals of 9,10-dichloroanthracene", *Chem. Phys. Lett.* **17**, 522–4 (1972).

ITOH, U. and IZUMI, T.: "Photoconduction associated with triplet state of 9,10-dichloroanthracene single crystal", *J. Phys. Soc. Japan* **34**, 1110 (1973).

ITOH, U. and TAKEISHI, K.: "Various detrapping processes in anthraquinone-doped anthracene crystals", in *Energy and Charge Transfer in Organic Semiconductors* (eds. K. Masuda and M. Silver), Plenum Press (New York) 25–30 (1974).

IVEY, H. F.: "Electroluminescence and semiconductor lasers", *IEEE Jl Quantum Electron.* QE **2**, 713–26 (1966).

IZUMI, T. and ITOH, U.: "Ruby-laser induced photoconduction in 9,10-dichloroanthracene and 9,10-dibromoanthracene", *J. Phys. Soc. Japan* **32**, 214–16 (1972).

JÄGER, J.: "LCAO–HCO calculation of condensed aromatic hydrocarbons. 1: Anthracene", *Phys. Stat. Sol.* **35**, 731–6 (1969).

JAKLEVIC, R. C. and LAMBE, J.: "Molecular vibration spectra by electron tunneling", *Phys. Rev. Lett.* **17**, 1139–40 (1966).

JANSEN, P., HELFRICH, W., and RIEHL, N.: "Die Wirkung von Licht auf raumladungsbeschränkte Defektelektronenströme in Anthrazen", *Phys. Stat. Sol.* **7**, 851–61 (1964).

JARNAGIN, R. C., GILLILAND, J., KIM, J. S., and SILVER, M.: "Physical and chemical effects at the anthracene–electrolytic interface due to photo-electrolysis", *J. Chem. Phys.* **39**, 573–9 (1963).

JOHNSON, E. O.: "Large-signal surface photovoltage studies with germanium", *Phys. Rev.* **111**, 153–66 (1958).

JOHNSON, H. R., WILLIAMS, R. H., and MEE, C. H. B.: "The anomalous photovoltaic effect in cadmium telluride", *J. Phys. D* **8**, 1530–41 (1975).

JOHNSON, P. M.: "Multiphoton ionization spectroscopy: a new state of benzene", *J. Chem. Phys.* **62**, 4562–3 (1975).

JOHNSON, R. C., ERN, V., WILLY, D. W., and MERRIFIELD, R. E.: "Oxygen quenching of delayed fluorescence in the solid state", *Chem. Phys. Lett.* **11**, 188–91 (1971).

JOHNSON, R. C. and MERRIFIELD, R. E.: "Effects of magnetic fields on the mutual annihilation of triplet excitons in anthracene crystals", *Phys. Rev. B* **1**, 896–902 (1970).

JOHNSON, R. C., MERRIFIELD, R. E., AVAKIAN, P., and FLIPPEN, R. B.: "Effects of magnetic fields on the mutual annihilation of triplet excitons in molecular crystals", *Phys. Rev. Lett.* **19**, 285–7 (1967).

JOHNSON, V. A.: "Seebeck effect in semiconductors", in *Progress in Semiconductors*, Heywood (London) **1**, 63–97 (1956).

JOHNSON, V. A. and LARK-HOROWITZ, K.: "Theory of thermoelectric power in semiconductors with applications to germanium", *Phys. Rev.* **92**, 226–32 (1953).

JOHNSTON, G. R.: "Current–voltage response for dark conduction through anthracene crystals", *Chem. Phys. Lett.* **3**, 699–701 (1969).

JOHNSTON, G. R. and LYONS, L. E.: "The effect of surface purity on the photo-generation of charge carriers in anthracene single crystals", *Chem. Phys. Lett.* **2**, 489–92 (1968).

JOHNSTON, G. R. and LYONS, L. E.: "Photocarrier generation mechanisms in anthracene crystals under ultra-high vacuum", *Aust. J. Chem.* **23**, 1571–9 (1970).

JOHNSTON, G. R. and LYONS, L. E.: "Dark conduction through anthracene crystals", *Aust. J. Chem.* **23**, 2187–2204 (1970).

JOHNSTON, G. R. and LYONS, L. E.: "On the compensation effect in electrical conduction through organic crystals", *Phys. Stat. Sol.* **37**, K43–45 (1970).

JONES, H. and ZENER, C.: "The theory of the change in resistance in a magnetic field", *Proc. Roy. Soc. (London)* **145A**, 268–77 (1934).

JONES, J. E.: "Photoelectric spectral response of certain solids", *J. Appl. Phys.* **44**, 96–99 (1973).

JONES, J. H.: "Anthracene and anthracene–tetracene crystals from vapor", *Mol. Cryst.* **3**, 393–6 (1968).

JONES, P. E.: "Spectral shifts and broadening of the fluorescence of anthracene and tetracene in several host crystals at high pressures", *J. Chem. Phys.* **48**, 3448–56 (1968).

JONES, P. E. and NICOL, M: "Excimer fluorescence of crystalline anthracene and naphthalene produced by high pressure", *J. Chem. Phys.* **43**, 3759–60 (1965).

JONES, P. E. and NICOL, M.: "Fluorescence of doped crystals of anthracene, naphthalene, and phenanthrene under high pressures: role of excimers in energy transfer to the guest molecules", *J. Chem. Phys.* **48**, 5457–64 (1968).

JONES, P. E. and NICOL, M.: "Excimer emission of naphthalene, anthracene, and phenanthrene crystals produced by very high pressures", *J. Chem. Phys.* **48**, 5440–7 (1968).

JONES, W., THOMAS, J. M., and WILLIAMS, J. O.: "Electron-induced transformation in an organic molecular crystal", *Mater. Res. Bull.* **10**, 1031–5 (1975).

JONES, W., THOMAS, J. M., and WILLIAMS, J. O.: "Electron and optical microscopic studies of a stress-induced phase transition in 1-8-dichloro-10-methylanthracene", *Phil. Mag.* **32**, 1–11 (1975).

JONES, W., THOMAS, J. M., WILLIAMS, J. O., and HOBBS, L. W.: "Electron microscopic studies of extended defects in organic molecular crystals. Part I: *p*-terphenyl", *J. Chem. Soc. Faraday Trans.* II **71**, 138–43 (1975).

JONSCHER, A. K.: "Electronic properties of amorphous dielectric films", *Thin Solid Films* **1**, 213–34 (1967).

JONSCHER, A. K.: "Frequency-dependence of conductivity in hopping systems", *J. Non-Cryst. Solids* **8–10**, 293–315 (1972).

JONSCHER, A. K.: "The role of contacts in frequency-dependent conduction in disordered solids", *J. Phys. C* **6**, L235–9 (1973).

JONSCHER, A. K.: "AC conductivity and high field effects", in *Electronic and Structural Properties of Amorphous Semiconductors* (eds. P. G. Le Comber and J. Mort), Academic Press (New York) 329–62 (1973).

JONSCHER, A. K.: "Electrical conduction in polymers", in *International Meeting on Electrets, Charge Storage, and Transport* in *Dielectrics, Frankfurt, Germany* (ed. M. M. Perlman), Verlag-Chemie (Weinheim/Bergstrasse, Germany) 29–43 (1974).

JONSCHER, A. K.: "Hopping losses in polarisable dielectric media", *Nature* **250**, 191–3 (1974).

JORTNER, J.: "Collisions of singlet excitons in molecular crystals", *Phys. Rev. Lett.* **20**, 244–7 (1968).

JORTNER, J.: "Radiationless transitions", *Pure Appl. Chem.* **27**, 389–437 (1971).

JORTNER, J. and BIXON, M.: "Comments on electronic relaxation processes in molecular crystals", *Mol. Cryst.* **9**, 213–37 (1969).

JORTNER, J., CHOI, S. I., KATZ, J. L., and RICE, S. A.: "Triplet energy transfer and triplet–triplet interaction in aromatic crystals", *Phys. Rev. Lett.* **11**, 323–6 (1963).

JORTNER, J., RICE, S. A., KATZ, J. L., and CHOI, S. A.: "Triplet excitons in crystals of aromatic molecules", *J. Chem. Phys.* **42**, 309–23 (1965).

JOSHI, N. V. and CASTILLON, M.: "Electric field dependence of photoconductivity in naphthalene crystal", *Chem. Phys. Lett.* **46**, 317–18 (1977).

JOURDIAN, M. and DESPUJOLS, J.: "Internal photoelectric effect in aluminium–silicon monoxide–gold structures", *Thin Solid Films* **16**, 249–56 (1973).

JURGIS, A. and SILINSH, E. A.: "On the interaction of electrons and holes in a molecular crystal", *Phys. Stat. Sol.* (b) **53**, 735–43 (1972).

KABASHIMA, S. and KAWAKUBO, T.: "High frequency conductivity of NiO", *J. Phys. Soc. Japan* **24**, 493–7 (1968).

KADLEC, J.: "Theory of internal photoemission in sandwich structures", *Physics Rep.* **26C**, 69–98 (1976).

KADLEC, J. and GUNDLACH, K. H.: "Dependence of the barrier height on insulator thickness in Al–(Al-oxide) – Al sandwiches", *Solid State Commun.* **16**, 621–3 (1975).

KADLEC, J. and GUNDLACH, K. H.: "Results and problems of internal photo-emission in sandwich structures", *Phys. Stat. Sol.* (a) **37**, 11–28 (1976).

KADLEC, J. and GUNDLACH, K. H.: "Spatial distribution of photoexcited electrons in sandwich structures", *J. Appl. Phys.* **47**, 672–6 (1976).

KAGAN, N. E.: "An approach to biosynthesis of organic superconductors", *J. Polymer Sci.* Part C, **29**, 191–8 (1970).

KAINO, H.: "Transient photocurrents in pyrene–TCNE single crystals", *J. Phys. Soc. Japan* **36**, 1500 (1974).

KAINO, H.: "Transient photocurrent in pyrene–TCNE single crystals. I: Current controlled by surface layers", *J. Phys. Soc. Japan* **39**, 708–14 (1975).

KAJIWARA, T., INOKUCHI, H., and MINOMURA, S.: "Charge mobility of organic semiconductors under high pressures", *Bull. Chem. Soc. Japan* **40**, 1055–8 (1967).

KALINOWSKI, J. and GODLEWSKI, J.: "Triplet exciton–charge carrier interaction in crystalline tetracene", *Phys. Stat. Sol.* (a) **20**, 403–10 (1973).

KALINOWSKI, J. and GODLEWSKI, J.: "Luminescence modulation in organic crystals by exciton–charge carrier interaction", *Acta Phys. Pol.* A **46**, 523–38 (1974).

KALINOWSKI, J. and GODLEWSKI, J.: "Singlet exciton–charge carrier interaction in anthracene crystals", *Phys. Stat. Sol.* (b) **65**, 789–96 (1974).

KALINOWSKI, J. and GODLEWSKI, J.: "Heterofusion of triplet excitons in pentacene-doped tetracene crystals", *Chem. Phys. Lett.* **25**, 499–504 (1974).

KALINOWSKI, J. and GODLEWSKI, J.: "Magnetic field effects on photocurrents in tetracene crystals", *Pr. Nauk Inst. Chem. Org. Fiz. Politech. Wroclaw* **7**, 122–33 (1974).

KALINOWSKI, J. and GODLEWSKI, J.: "Magnetic field effects on recombination radiation in tetracene crystals", *Chem. Phys. Lett.* **36**, 345–8 (1975).

KALINOWSKI, J., GODLEWSKI, J., and CHANCE, R. R.: "Evidence for trapped-exciton fluorescence in anthracene crystals at room temperature", *J. Chem. Phys.* **64**, 2389–94 (1976).

KALINOWSKI, J., GODLEWSKI, J., and JANKOWIAK, R.: "Pressure-induced phase transition in tetracene crystals", *Chem. Phys. Lett.* **43**, 127–9 (1976).

KALINOWSKI, J., GODLEWSKI, J., and SINGNERSKI, R.: "Electroluminescence in tetracene crystals", *Mol. Cryst. Liq. Cryst.* **33**. 247–59 (1976).

KALLMANN, H.: "Energy transfer processes", in *Comparative Effects of Radiation* (eds. M. Burton, J. S. Kirby-Smith, and J. L. Magee), Wiley (New York) 1960.

KALLMANN, H.: "Production and annihilation of triplet excitons in organic materials near absorbing surface", *Z. Naturforsch.* **26a**, 799–802 (1971).

KALLMANN, H., KRAMER, B., HAIDEMENAKIS, E., MCALEER, W. J., BARKEMEYER, H., and POLLAK, P. I.: "Photovoltages in silicon and germanium layers", *J. Electrochem. Soc.* **108**, 247–51 (1961).

KALLMANN, H., KRAMER, B., SHAIN, J., and SPRUCH, G. M.: "Photovoltaic effects in CdS crystals", *Phys. Rev.* **117**, 1482–6 (1960).

KALLMANN, H. and KRASNANSKY, V. J.: "Charge injection in aromatic hydrocarbons", *J. Polymer Sci.* C **17**, 241–2 (1967).

KALLMANN, H. and POPE, M.: "Preparation of thin anthracene single crystals", *Rev. Sci. Instrum.* **29**, 993–4 (1958).

KALLMANN, H. and POPE, M.: "Photovoltaic effect in organic crystals", *J. Chem. Phys.* **30**, 585–6 (1959).

KALLMANN, H. and POPE, M.: "Electrolytic contacts for photoconductivity measurements", *Rev. Sci. Instrum.* **30**, 44–86 (1959).

KALLMANN, H. and POPE, M.: "Positive hole injection in organic crystals", *J. Chem. Phys.* **31**, 300–1 (1960).

KALLMANN, H. and POPE, M.: "Surface controlled bulk conductivity in organic crystals", *Nature* **185**, 753 (1960).

KALLMANN, H. and POPE, M.: "Bulk conductivity in organic crystals", *Nature* **186**, 31–33 (1960).

KALLMANN, H. and POPE, M.: "Theory of the hole injection and conductivity in organic materials", *J. Chem. Phys.* **36**, 2482–5 (1962).

KALLMANN, H. and ROSENBERG, B.: "Persistent internal polarization", *Phys. Rev.* **97**, 1596–1691 (1955).

KALLMANN, H., SPRUCH, G. M., and TRESTER, S.: "Photovoltages larger than the bandgap in thin films of germanium", *J. Appl. Phys.* **43**, 469–75 (1972).

KALLMANN, H., VAUBEL, G., and BAESSLER, H.: "Interaction of singlet excitons in organic materials with an absorbing surface", *Phys. Stat. Sol.* (b) **44**, 813–20 (1971).

KALRA, A. D., SIMMONS, J. G., and NADKARI, G. S.: "Electrical properties of thin-film Al–CeF$_3$–Al capacitors", *J. Appl. Phys.* **46**, 5076–9 (1975).

KAMB, K.: "Some recent developments in the theory of electronic states of anthracene and similar crystals", *Prog. Theor. Phys. (Kyoto)*, Suppl. No. 40, 136–58 (1967).

KAMURA, Y., SEKI, K., and INOKUCHI, H.: "Near and vacuum ultraviolet absorption spectra of polycrystalline and amorphous evaporated films of naphthalacene, pentacene, perylene and coronene", *Chem. Phys. Lett.* **30**, 35–38 (1975).

KANAZAWA, K. K. and BATRA, I. P.: "Deep-trapping kinematics", *J. Appl. Phys.* **43**, 1845–53 (1972).

KANE, E. O.: "Theory of photoelectric emission from semiconductors", *Phys. Rev.* **127**, 131–41 (1962).

KANETO, K., YOSHINO, K., HIKIDA, M., and INUISHI, Y.: "Emission spectra of phthalocyanine", *Technol. Rep. Osaka Univ. Japan* **23**, 493–502 (1973).

KANETO, K., YOSHINO, K., and INUISHI, Y.: "Effects of a magnetic field on triplet–triplet annihilation in a Pt–phthalocyanine single crystal at 4.2°K", *Chem. Phys. Lett.* **40**, 505–7 (1976).

KANETO, K., YOSHINO, K., KAO, K. C., and INUISHI, Y.: "Electroluminescence in polyethylene terephthalate", *Jap. J. Appl. Phys.* **13**, 1023–4 (1974).

KAO, C. C. and CHEN, I.: "Xerographic discharge characteristics of photoreceptors with bulk generation. 1: Flash exposure", *J. Appl. Phys.* **44**, 2708–17 (1973).

KAO, K. C.: "Theory of high field electric conduction and breakdown in dielectric liquids", *IEEE Trans. Electrical Insulation* EI **11**, 121–8 (1976).

KAO, K. C. and ELSHARKAWI, A. R.: "On the theory of double injection in solids with traps non-uniformly distributed in energy and in space", *Int. J. Electron.* **41**, 607–16 (1976).

KAO, K. C., GILES, L. J., and CALDERWOOD, J. H.: "Behaviour of divalent impurities in sodium chloride crystals", *J. Appl. Phys.* **39**, 3955–61 (1968).

KAO, K. C., PHAHLE, A. M., and CALDERWOOD, J. H.: "The electric conduction in *n*-hexane under ultraviolet and gamma radiations", *IEEE Trans. Electrical Insulation* EI **2**, 108–14 (1967).

KAO, K. C. and RASHWAN, M. M.: "Pressure dependence of electroluminescence in dielectric liquids", *Proc. IEEE* **62**, 856–8 (1974).

KAO, K. C. and RASHWAN, M. M.: "On the thermal activation energy for high-field electric conduction in dielectric liquids", *IEEE Trans. Electrical Insulation* EI **13**, 86–93 (1978).

KAO, K. C. and SILL, R. J.: "Space-charge limited currents and negative differential resistance in high-resistivity *p*-type silicon", *Solid State Commun.* **10**, 149–51 (1972).

KAO, K. C., WHITHAM, W., and CALDERWOOD, J. H.: "Time dependent polarization in alkali halides", *J. Phys. Chem. Solids* **31**, 1019–26 (1968).

KAPLAN, T. A. and MAHANTI, S. D.: "Electrical conductivity in biological semiconductors", *J. Chem. Phys.* **62**, 100–7 (1975).

KARAKUSHAN, E. I. and STAFEEV, V. I.: "Magnetodiodes", *Soviet Phys.—Solid State* **3**, 493–8 (1961).

KARL, N.: "Laser emission from an organic molecular crystal", *Phys. Stat. Sol.* (a) **13**, 651–5 (1972).

KARL, N.: "Organic semiconductors", *Adv. Solid State Phys.* **14**, 261–90 (1974).

KARL, N. and PROBST, K. H.: "Studies on the efficiency of purification by various zone refining devices using the system anthracene-phenazine. Suggestion of a critical standard test", *Mol. Cryst. Liq. Cryst.* **11**, 155–71 (1970).

KARL, N., ROHRBACHER, H., and SIEBERT, D.: "Dielectric tensor and relaxation of photoexcited charge carriers in single crystal anthracene in an alternating field without direct contacts", *Phys. Stat. Sol.* (a) **4**, 105–9 (1971).

KARL, N., SCHMID, E., and SEEGER, M.: "Bestimmung der Diffusionskonstante von Leitungselektronen in Anthrazen aus dem Impulsobfall bei drift Experimenten (diffusion constant)", *Z. Naturforsch.* **25**, 382–91 (1970).

KASHA, M.: "Characterization of electronic transition in complex molecules", *Discuss. Faraday Soc.* **9**, 14–19 (1950).

KASHA, M.: "Collisional perturbation of spin-orbited coupling and the mechanism of fluorescence quenching: a visual demonstration of the perturbation", *J. Chem. Phys.* **20**, 71–74 (1952).

KASHA, M.: "Relation between exciton bands and conduction bands in molecular lamellar systems", *Rev. Mod. Phys.* **31**, 162–9 (1959).

KASICA, H., WLODARSKI, W., KURCZEWSKA, H. and SZYMANSKI, A.: "Bipolar injection as a cause of electrical switching phenomena in thin organic films", *Thin Solid Films* **30**, 325–33 (1975).

KASSING, R.: "Calculation of the frequency dependence of the admittance of SCLC diodes", *Phys. Stat. Sol.* (a) **28**, 107–17 (1975).

KASSING, R. and KAHLER, E.: "The small signal behaviour of SCLC-diodes with deep traps", *Solid State Commun.* **15**, 673–6 (1974).

KATAYAMA, Y., YASUNAGA, H., and TAKEYA, K.: "Triplet exciton and photocarrier-generation in lead phthalocyanine", *Rep. Univ. Electro-Commun. (Japan)* **23**, 25–28 (1972).

KATON, J. E.: "Organic semiconductors", US patent 3,267,115 (1966), and *Chem. Abs.* **65**, 13003 f (1966).

KATON, J. E. and WILDI, B. S.: "Semiconducting organic polymers derived from nitriles: thermoelectric power and thermal conductivity measurements", *J. Chem. Phys.* **40**, 2977–82 (1964).

KATUL, J. and ZAHLAN, A. B.: "Evidence for the existence of surface states in tetracene microcrystals", *Phys. Rev. Lett.* **6**, 101–2 (1961).

KATZ, J. L., JORTNER, J., CHOI, S., and RICE, S. A.: "Triplet excitons bands in aromatic crystals", *J. Chem. Phys.* **39**, 1897–9 (1963).

KATZ, J. L., RICE, S. A., CHOI, S. I., and JORTNER, J.: "On the excess electron and hole band structures and carrier mobility in naphthalene, anthracene and several polyphenyls", *J. Chem. Phys.* **39**, 1683–97 (1963).

KAWABE, M., MASUDA, K., and NAMBA, S.: "Electroluminescence of green light region in doped anthracene", *Japan J. Appl. Phys.* **10**, 527–8 (1971).

KAWABE, M., MASUDA, K. and NAMBA, S.: "Penetration of electrons into anthracene crystals", *J. Appl. Phys.* **42**, 501–2 (1971).

KAWABE, M., YAMAGUCHI, J., and AOYAGI, Y.: "Uniaxial stress effects on electrical conductivity of DPPH single crystal", *J. Phys. Soc. Japan* **21**, 394 (1966).

KAWAOKA, K. and KEARNS, D. R.: "Emission properties of tetracene–pyrene and pentacene–pyrene mixed crystals", *J. Chem. Phys.* **41**, 2095–7 (1964).

KAWASAKI, K., KANON, K., and IIZUKA, M.: "Effects of water vapor on the structure and electrical properties of anthracene", *Surf. Sci.* **5**, 263–6 (1966).

KAZZAZ, A. A. and ZAHLAN, A. V.: "Exciton diffusion in naphthalene crystals", *Phys. Rev.* **124**, 90–95 (1961).

KEARNS, D. R.: "Resonance transfer model of electron and hole conduction in anthracene", *J. Chem. Phys.* **35**, 2269–70 (1961).

KEARNS, D. R.: "Generation of charge carriers in organic molecular crystals through multiple exciton interactions", *J. Chem. Phys.* **39**, 2697–703 (1963).

KEARNS, D. R.: "Electronic conduction in organic molecular solids", *Adv. Chem. Phys.* **7**, 282–338 (1964).

KEARNS, D. R.: "Photogeneration of charge carriers in anthracene by a single exciton process", *J. Chem. Phys.* **40**, 1452–3 (1964).

KEATING, P. N.: "Thermally stimulated emission and conductivity peaks in the case of temperature dependent trapping cross sections", *Proc. Phys. Soc. (London)* **78**, 1408–15 (1961).

KEATING, P. N.: "On two carrier injection currents in insulators", *Solid State Commun.* **1**, 210–13 (1963).

KEATING, P. N.: "Effect of shallow trapping and the thermal-equilibrium recombination center occupancy on double-injection currents in insulators", *Phys. Rev.* **135**, A1407–13 (1964).

KEATING, P. N.: "Photovoltaic effect in photoconductors", *J. Appl. Phys.* **36**, 564–70 (1965).

KEISS, H.: "Theoretical considerations concerning saturation of photocurrents", *J. Phys. Chem. Solids* **28**, 1473–83 (1967).

KELDYSH, L. V.: "The effect of a strong electric field on the optical properties of insulating crystals", *Soviet. Phys. JETP* **34**, 788–90 (1958).

KELKER, H.: "History of liquid crystals", *Mol. Cryst. Liq. Cryst.* **21**, 1–48 (1973).

KELLER, R. A.: "Electronic conductivity in molecular crystals: an alternative to the Bloch-function approach", *J. Chem. Phys.* **38**, 1076–83 (1963).

KELLER, R. A. and RAST, H. E.: "Tunneling model for electron transport and its temperature dependence in crystals of low carrier mobility. Example: anthracene", *J. Chem. Phys.* **36**, 2640–3 (1962).

KELLOGG, R. E.: "Second triplet state of anthracene", *J. Chem. Phys.* **44**, 411–12 (1966).

KELLY, P. and BRÄUNLICH, P.: "Phenomenological theory of thermoluminescence", *Phys. Rev.* B **1**, 1587–95 (1970).

KELLY, P., LAUBITZ, M., and BRÄUNLICH, P.: "Exact solutions of the kinetic equations governing thermally stimulated luminescence and conductivity", *Phys. Rev.* B **4**, 1960–8 (1971).

KELNER, S. R. and RUDENKO, A. I.: "Transient current in a high resistivity semiconductor after a photoinjection pulse", *Soviet Phys.—Semicond.* **9**, 1079–80 (1976).

KEMENY, G. and GOKLANY, I. M.: "Polarons and conforms", *J. Theor. Biol.* **40**, 107–23 (1973).

KEMENY, G. and ROSENBERG, B.: "Theory of the pre-exponential factor in organic semiconductors", *J. Chem. Phys.* **52**, 4151–3 (1970).

KEMENY, G. and ROSENBERG, B.: "Small polarons in organic and biological semiconductors", *J. Chem. Phys.* **53**, 3549–51 (1970).

KENKRE, V. M. and KNOX, R. S.: "Generalized-master-equation theory of excitation transfer", *Phys. Rev.* B **9**, 5279–90 (1974).

KEPLER, R. G.: "Charge carrier production and mobility in anthracene crystals", *Phys. Rev.* **119**, 1226–9 (1960).

KEPLER, R. G.: "Charge carrier mobility and production in anthracene", in *Organic Semiconductors* (eds. J. J. Brophy and J. W. Buttrey), Macmillan (London) 1–20 (1962).

KEPLER, R. G.: "Photoconductivity in organic molecular crystals", *IEEE Trans. Nucl. Sci.* **11**, 1–11 (1964).

KEPLER, R. G.: "Energy and charge transport in organic molecular crystals", in *Phonons and Phonon Interaction* (ed. T. A. Bak), Benjamin (New York) 578–640 (1964).

KEPLER, R. G.: "Photoionization of excitons in anthracene", *Phys. Rev. Lett.* **18**, 951–3 (1969).

KEPLER, R. G.: "Electron and hole generation in anthracene crystals", *Pure Appl. Chem.* **27** (3), 515–26 (1971). .

KEPLER, R. G.: *Proceedings of the 2nd International Symposium on Organic Solid Chemistry, Rehovot, Israel*, Butterworths (London), 1971.

KEPLER, R. G., BIERSTEDT, P. E., and MERRIFIELD, R. E.: "Electronic conduction and exchange interaction in a new class of conductive organic solids", *Phys. Rev. Lett.* **5**, 503 (1960).

KEPLER, R. G., CARIS, J. C., AVAKIAN, P., and ABRAMSON, E.: "Triplet excitons and delayed fluorescence in anthracene crystals", *Phys. Rev. Lett.* **10**, 400–2 (1963).

KEPLER, R. G. and COPPAGE, F. N.: "Generation and recombination of holes and electrons in anthracene", *Phys. Rev.* **151**, 610–14 (1966).

KEPLER, R. G. and HOESTEREY, D. C.: "High-field mobility in anthracene crystals", *Phys. Rev.* B 9, 2743–5 (1974).

KEPLER, R. G. and 'MERRIFIELD, R. E.: "Exciton–exciton interaction and photoconductivity in anthracene", *J. Chem. Phys.* **40**, 1173–4 (1964).

KEPLER, R. G. and SWITENDICK, A. C.: "Diffusion of triplet excitons in anthracene", *Phys. Rev. Lett.* **15**, 56–59 (1965).

KEVORKIEN, J., LABES, M. M., LARSON, D. C., and WU, D. C.: "Bistable switching in organic thin films", *Discuss. Faraday Soc.* **51**, 139–43 (1971).

KHAN-MAGOMETOV, S. D.: "Dependence of the mean lifetime of the excited state and photo-luminescence yield of anthracene on the concentration of the impurity produced by irradiation", *Opt. Spectrosc.* **25**, 203–5 (1968).

KHANNO, S. K., BRIGHT, A. A., GARITO, A. F., and HEEGER, A. J.: "Evidence for strong Coulomb interactions in alkali-TCNQ (tetracyanoquinodimethan) salts", *Phys. Rev.* B **10**, 2139–43 (1974).

KHANNO, S. K., EHRENFREUND, E., GARITO, A. F., and HEEGER, A. J.: "Microwave properties of high-purity tetrathiofulvalene–tetracyanoquinodimethan", *Phys. Rev.* B **10**, 2205–20 (1974).

KHO, J. H. T. and POHL, H. A.: "Molecular structure parameters in certain semiconducting polymers", *J. Polymer Sci.* Al, **7**, 139–55 (1969).

KIESS, H.: "Saturated photocurrents in Cd_4CeS_6", *J. Phys. Chem. Solids* **28**, 1465–71 (1967).

KIESS, H.: "The physics of electrical charging and discharging semiconductors", *RCA Rev.* **36**, 667–700 (1975).

KIESS, H.: "Theoretical considerations concerning saturation of photocurrents", *J. Phys. Chem. Solids* **28**, 1473–83 (1976).

KIESS, H. and ROSE, A.: "Can the majority carrier be decreased by injection of minority carriers?", *Helv. Phys. Acta* **46**, 434 (1973).

KIESS, H. and ROSE, A.: "Transport in relaxation semiconductors", *Phys. Rev. Lett.* **31**, 153–4 (1973).

KIKUCHI, M.: "Observation of negative resistance and oscillation phenomena in the forward direction of point contact semiconductor diodes", *Japan J. Appl. Phys.* **2**, 31–46 (1963).

KILLESREITER, H. and BAESSLER, H.: "Exciton reaction at an anthracene/metal interface: charge transfer", *Chem. Phys. Lett.* **11**, 411–14 (1971).

KILLESREITER, H. and BAESSLER, H.: "Dissociation of Frenkel excitons at the interface between a molecular crystal and metal", *Phys. Stat. Sol.* **51**, 657–68 (1972).

KILLESREITER, H. and BAESSLER, H.: "Field dependence of exciton-induced contact-limited photocurrents in molecular crystals", *Phys. Stat. Sol.* (b) **53**, 193–9 (1972).

KILLESREITER, H. and BRAUN, R.: "Transient photocurrents induced by time-dependent diffusion of excitons to the surface of anthracene and p-chloranil single crystals", *Phys. Stat. Sol.* (b) **48**, 201–13 (1971).

KING, T. A. and VOLTZ, R.: "The time dependence of scintillation intensity in aromatic materials", *Proc. Roy. Soc. (London)* A **289**, 424–39 (1966).

KINGSTON, R. H. and NEUSTADTER, S. F.: "Calculation of the space charge electric field and free carrier concentration at the surface of a semiconductor", *J. Appl. Phys.* **26**, 718–20 (1955).

KIROV, K. and ZHELEV, V.: "Study of the effect of an electric field on trap filling in CdS single crystals by the use of thermally stimulated currents", *Phys. Stat. Sol.* **8**, 431–40 (1965).

KITTAKA, S. and MURATA, Y.: "Effect of photo-irradiation on contact charging", *DECHEMA (Dent. Ges. Chem. Apparatus) Monogr.* **72**, 63–72 (1974).

KLEIN, G., VOLTZ, R., and SCHOTT, M.: "Magnetic field effect on prompt fluorescence in anthracene: evidence for singlet exciton fission", *Chem. Phys. Lett.* **16**, 340–4 (1972).

KLEIN, G., VOLTZ, R., and SCHOTT, M.: "On singlet exciton fission in anthracene and tetracene at 77° K", *Chem. Phys. Lett.* **19**, 391–4 (1973).

KLEIN, N.: "A theory of localized electronic breakdown in insulating films", *Adv. Phys.* **21**, 605–45 (1972).

KLEIN, N. and LISAK, Z.: "Extended temperature range for the maximum strength", *Proc. IEEE* **54**, 979–80 (1966).

KLEINERMAN, M., AZARRAGA, L., and MCGLYNN, S. P.: "Emissivity and photoconductivity of organic molecular crystal", *J. Chem. Phys.* **37**, 1825–34 (1962).

KLEINERMAN, M., AZARRAGA, L., and MCGLYNN, S. P.: "The photoconductive and emission spectroscopic properties of organic molecular crystals", *in Luminescence of Organic and Inorganic Materials* (eds. H. Kallmann and G. Spruch), Wiley (New York) 196 (1962).

KLEINERMAN, M. and MCGLYNN, S. P.: "Photoconductivity of organic molecular crystals and the triplet state", *J. Chem. Phys.* **37**, 1369–70 (1962).

KLEINMAN, D. A.: "The forward characteristic of the PIN diode", *Bell System Tech. J.* **35**, 685–706 (1956).

KLIER, K.: "Theory of the boundary layer for the finite crystal of the adsorbent", *Collection Czech. Chem. Commun.* **27**, 920–7 (1962).

KLIMKA, L., BUMELENE, S., and KAL'VENAS, S.: "Double injection and instability of the current in nickel-compensated germanium", *Soviet Phys.—Semicond.* **7**, 1504–5 (1974).

KNIGHTS, J. C. and DAVIS, E. A.: "Photogeneration of charge carriers in amorphous selenium", *J. Phys. Chem. Solids* **35**, 543–54 (1974).

KNONICK, P. L. and LABES, M. M.: "Photoelectric characterization of X-ray damage in anthracene crystals", *Mol. Cryst.* **2**, 293–7 (1967).

KOBAYASHI, T., NAGAKURA, S., IWASHIMA, S., and INOKUCHI, H.: "The spectral dependence of fluorescence decay times observed for the benzo (*g, h, i*) perylene crystal and its mixed crystal with perylene", *J. Mol. Spectrosc.* **41**, 44–53 (1972).

KOCHI, M.: "Photoemission from organic crystals", *Kagaku-No-Ryoiki* **23** (1), 30–34 (1969).

KOCHI, M., HARADA, Y., HIROOKA, I., and INOKUCHI, H.: "Photoemission from organic crystal in vacuum ultraviolet region. IV", *Bull. Chem. Soc. Japan* **43**, 2690–702 (1970).

KOKADO, H. and SCHNEIDER, W. G.: "Thermally simulated currents and carrier trapping in anthracene crystals", *J. Chem. Phys.* **40**, 2937–45 (1964).

KOLOMIETS, B. T. and LEBEDEV, E. A.: "Transient space-charge-limited currents and mobility in chalcogenide glasses", *Soviet Phys.—Semicond.* **7**, 134–5 (1973).

KOLOMIETS, B. T. and NAZAROVA, T. F.: "Hall effect in vitreous materials of the $Tl_2Se.As_2(Se,Te)_3$ system II", *Soviet Phys.—Solid State* **2**, 369–70 (1960).

KOLOMOETS, N. V. and POLNIKV, V. G.: "Characteristics of the transport phenomenon in the space-charge region near the surface of a semiconductor", *Soviet Phys.—Semicond.* **8**, 250–1 (1974).

KOMMANDEUR, J.: "Photoconductivity in organic single crystals", *J. Phys. Chem. Solids* **22**, 339–49 (1961).

KOMMANDEUR, J.: "Conductivity", in *Physics and Chemistry of the Organic Solid State* (eds. D. Fox, M. M. Labes, and A. Weissberger) **2**, 1–66 (1965).

KOMMANDEUR, J., KORINEK, G. J., and SCHNEIDER, W. G.: "The activation energies of photoconduction for some aromatic hydrocarbons", *Can. J. Chem.* **35**, 998–1001 (1957).

KOMMANDEUR, J., KORINEK, G. J., and SCHNEIDER, W. G.: "The change of photoconduction and semiconductor in aromatic hydrocarbons on melting", *Can. J. Chem.* **36**, 513–17 (1958).

KOMMANDEUR, J. and SCHNEIDER, W. G.: "Photoconductivity of anthracene. IV: Bulk photoconduction in single crystals", *J. Chem. Phys.* **28**, 582–9 (1958).

KOMMANDEUR, J. and SCHNEIDER, W. G.: "Photoconductivity of anthracene. V: Effect of imperfections on the bulk photocurrent", *J. Chem. Phys.* **28**, 590–5 (1958).

KONDRASIUK, J. and SZYMANSKI, A.: "Charge carrier mobility in tetracene", *Mol. Cryst. Liq. Cryst.* **18**, 379–82 (1972).

KONDRATENKO, P. A. and KURIK, M. V.: "Determination of the rate of surface annihilation of singlet excitons in anthracene single crystals", *Soviet Phys.—Solid State* **15**, 2062–3 (1974).

KONOROV, P. P.: "Nature of the high-voltage photovoltaic effect in polycrystalline germanium films", *Soviet Phys.—Semicond.* **2**, 1437–40 (1969).

KONTECKY, J.: "On the theory of surface states", *J. Phys. Chem. Solids* **14**, 233–40 (1960).

KORN, A. I.: "Hall effect of holes in anthracene", Ph.D. thesis, City University of New York (1969).

KORN, A. I., ARNDT, R. A., and DAMASK, A. C.: "Hall mobility of holes in anthracene", *Phys. Rev.* **186**, 938–41 (1969).

KORSCH, B., WILLING, F., GAEHRS, H. J., and TESCHE, B.: "Exchange of holes between organic crystals and metal films evaporated in high vacuum", *Phys. Stat. Sol.* (a) **33**, 461–71 (1976).

KORSUNSKII, M. I. and SOMINSKII, M. M.: "Anomalous Dember effect in cadmium telluride films", *Soviet Phys.—Semicond.* **7**, 342–6 (1973).

KORSUNSKII, V. M. and FAIDISH, O. M.: "Transmission of energy in anthracene crystals doped with phenazine and acridine", *Ukr. Fiz. Zh.* **8**, 677–83 (1963).

KORSUNSKII, V. M. and FAIDISH, O. M.: "Luminescence and excitation energy transfer in anthracene crystals with the addition of naphthacene", *Opt. Spektroskopiya Akad. Nauk SSSR Otd. Fiz. Math. Nauk Sb. Statei* **1**, 119–27 (1963).

KOSA-SOMOGYI, I.: "Organic semiconductors", *Fiz. Sz. (Hungary)* **23**, 329–41 (1973).

KOSAKI, M., SUGIYAMA, K., and IEDA, M.: "Ionic jump distance and glass transition of polyvinyl chloride", *J. Appl. Phys.* **42**, 3388–92 (1971).

KOTANI, M., WATANABE, Y., and KATO, T.: "Activation energy for hole injection and analysis of the space-charge layer in anthracene crystals", *Chem. Lett.* **12**, 1459–62 (1974).

KRAJCSOSZKY, J. and VJHELYI, S.: "Preparation of thin anthracene plates", *Soviet Phys.—Crystallography* **4**, 240 (1960).

KRAMER, G. and GHALLA, M.: "X-ray photoconductivity of anthracene", *J. Chem. Phys.* **45**, 1346–51 (1966).

KRISHNA, V. G.: "Delayed fluorescence due to triplet–triplet annihilation theoretical study", *J. Chem. Phys.* **46**, 1735–9 (1967).

KROLL, D. M.: "Theory of electrical instabilities of mixed electronic and thermal origin", *Phys. Rev.* **B 9**, 1669–1706 (1974).

KRONICK, P. L. and LABES, M. M.: "Organic semiconductors. V: Comparison of measurements on single-crystal and compressed microcrystalline molecular complexes", *J. Chem. Phys.* **35**, 2016–19 (1961).

KRYSZEWSKI, M.: "Electrical conductivity of semicrystalline and amorphous polymers and related problems", *J. Polym. Sci. Polym. Symp.* **50**, 359–404 (1975).

KRYSZEWSKI, M., SAPIEHA, S., and SZYMANSKI, A.: "Space charge limited currents in *p*-terphenyl and *p*-quaterphenyl monocrystals", *Acta Phys. Pol.* **33**, 529–39 (1968).

KRYSZEWSKI, M. and SZYMANSKI, A.: "Space charge limited currents in polymers", *J. Polym. Sci. Marcronol. Rev. D* **4**, 245–320 (1970).

KUBO, R.: "Thermal ionization of trapped electrons", *Phys. Rev.* **86**, 929–37 (1952).

KUBO, R.: "Statistical–mechanical theory of irreversible processes. I: General theory and simple applications to magnetic and conduction problems", *J. Phys. Soc. Japan* **12**, 570–86 (1957).

KUBO, R.: "The Boltzmann equation in solid state physics", *Acta Phys. Austriaca*, Suppl. X, 301–40 (1973).

KUHN, H.: "A quantum-mechanical theory of light absorption of organic dyes and similar compounds", *J. Chem. Phys.* **17**, 1198–1212 (1949).

KUHN, H.: "Classical aspects of energy transfer in molecular systems", *J. Chem. Phys.* **53**, 101–8 (1970).

KUHN, H., HUBER, W., HANDSCHIG, A., MARTIN, H., SCHÄFER, F., and BÄR, F.: "Nature of the free electron model: the simple case of the symmetric polymethines", *J. Chem. Phys.* **32**, 467–9 (1960).

KULSHRESHTHA, A. P. and MOOKHERJI, T.: "Electrical conductivity and photoresponse in tetra-cyanoquinodimethan", *Mol. Cryst. Liq. Cryst.* **10**, 75–83 (1970).

KUMAR, B., PARKASH, V., and JASEJA, T. S.: "Multiple photon excitations in molecular crystals by giant pulse ruby laser", *Indian J. Pure Appl. Phys.* **12**, 626–31 (1974).

KUNKEL, H. P. and KAO, K. C.: "Electroluminescence in anthracene crystals under time-varying electric fields", *J. Phys. Chem. Solids* **37**, 863–6 (1976).

KUNSTREICH, S. and OTTO, A.: "Anisotropy of electron energy loss spectra in anthracene single crystals", *Opt. Commun.* **1**, 45–46 (1969).

KURIK, M. V.: "Diffusion length of singlet excitons in anthracene single crystals", *Soviet Phys.—Solid State* **13**, 2421–4 (1972).

KURIK, M. V., PIRYATINSKII, Y. P., POPEL, O. M., and FROLOVA, E. K.: "Temperature dependence of the Davydov splitting in anthracene", *Phys. Stat. Sol.* **37**, 803–6 (1970).

KURIK, M. V. and TSIKORA, L. I.: "Exciton–phonon interaction in crystals of linear polyacenes", *Phys. Stat. Sol. (b)* **66**, 695–702 (1974).

KURODA, H. and FLOOD, E. A.: "Effect of ambient oxygen on the semiconductivities of evaporated films of mesonaphthodianthrene and mesonapthodianthrone", *Can. J. Chem.* **39**, 1475–83 (1961).

KURTIN, S.: "Carrier transport in metal–insulator–metal structures", in *Proceedings of the 3rd International Conference on Photoconductivity* (ed. E. M. Pell), Pergamon (Oxford) 357–60 (1971).

KUVATOV, Z. KH., KAPUSTIN, A. P., and TROFINIOV, A. N.: "Variations in the sign of the anisotropy of the electrical conductivity of a nematic liquid crystal as a function of temperature", *JETP Lett.* **19**, 89–91 (1974).

LABES, M. M.: "Review of the electrical properties of monomeric and polymeric charge-transfer complexes", *J. Polymer Sci. C* **17**, 95–105 (1967).

LABES, M. M., RUDYJ, O. N., and KRONICK, P. L.: "Specific, sensitive electronic detection of iodine via carrier injection into an anthracene crystal", *J. Am. Chem. Soc.* **84**, 499–500 (1962).

LABES, M. M., SEHR, R. A., and BOSE, M.: "Organic semiconductors. I: Some characteristics of the *p*-phenylene dianinechloranil complex", *J. Chem. Phys.* **32**, 1570–92 (1960).

LABES, M. M., SEHR, R. A., and BOSE, M.: "Organic semiconductors. II: The electrical resistivity of organic molecular complexes", *J. Chem. Phys.* **33**, 868–72 (1960).

LABES, M. M., SEHR, R. A., and BOSE, M.: "Semiconductor properties of organic molecular complexes", *Proceedings of the Princeton University Conference on Semiconduction in Molecular Solids*, Princeton, NY (1960).

LACEY, A. R. and LYONS, L. E.: "Zone refining of anthracene", *Rev. Sci. Instrum.* **34**, 309–10 (1963).

LACEY, A. R. and LYONS, L. E.: "Luminescence from zone-refined anthracene at 4°K, and the Davydov splitting", *J. Chem. Soc. (London)* 5393–400 (1964).

LACEY, A. R., LYONS, L. E., and WHITE, J. W.: "The effect of impurities on the fluorescence of anthracene crystal containing tetracene", *J. Chem. Soc.* 3670–5 (1963).

LAKATOS, A. I.: "Steady-state dark and photocurrents in a poly-*n*-vinylcarbazole-trinitrofluorenone photoconductor", *J. Appl. Phys.* **46**, 1744–53 (1975).

LAKATOS, A. I. and MORT, J.: "Transient photocurrents and internal photoemission in metal–insulator–semiconductor systems", in *Proceedings of the 3rd International Conference on Photoconductivity* (ed. E. M. Pell), Pergamon (Oxford) 361–5 (1971).

LAM, Y. W. and RHODERICK, E. H.: "Surface-state density and surface potential in MIS capacitors by surface photovoltage measurements. I", *J. Phys. D* **4**, 1370–5 (1971).

LAM, Y. W. and RHODERICK, E. H.: "Surface-state density and surface potential in MIS capacitors by surface photovoltage measurements. II", *J. Phys. D* **4**, 1376–89 (1971).

LAMBE, J. and JAKLEVIC, R. C.: "Molecular vibration spectra by inelastic electron tunnelling", *Phys. Rev.* **165**, 821–32 (1968).

LAMPERT, M. A.: "Simplified theory of space-charge-limited currents in an insulator with traps", *Phys. Rev.* **103**, 1648–56 (1956).

LAMPERT, M. A.: "Simplified theory of one-carrier currents with field-dependent mobilities", *J. Appl. Phys.* **29**, 1082–90 (1958).

LAMPERT, M. A.: "A simplified theory of two-carrier, space-charge-limited current flow in solids", *RCA Rev.* **20**, 682–701 (1959).

LAMPERT, M. A.: "Injection currents in insulators", *Proc. IRE* **50**, 1781–96 (1962).

LAMPERT, M. A.: "Double injection in insulators", *Phys. Rev.* **125**, 126–41 (1962).

LAMPERT, M. A.: "Volume-controlled current injection in insulators", *Reports on Progress in Physics*, Vol. XXVII, Institute of Physics (London), 329–67 (1964).

LAMPERT, M. A.: "Simplicity in theory: anecdotal account of current injection in solids—in tribute to Albert Rose", *RCA Rev.* **36**, 444–66 (1975).

LAMPERT, M. A. and EDELMAN, F.: "Theory of one-carrier, space-charge-limited currents including diffusion and trapping", *J. Appl. Phys.* **35**, 2971–82 (1964).

LAMPERT, M. A., MANY, A., and MARK, P.: "Space charge limited currents injected from a point contact", *Phys. Rev.* **135**, A1444–53 (1964).

LAMPERT, M. A. and ROSE, A.: "Volume-controlled, two-carrier currents in solids—the injection plasma case", *Phys. Rev.* **121**, 26–37 (1961).

LAMPERT, M. A. and SCHILLING, R. B.: "Current injection in solids: the regional approximation method", in *Semiconductors and Semimetals* (eds. R. K. Willardson and A. C. Beer), **6**, 1–96 (1970).

LAMPERT, M. A. and SUNSHINE, R. A.: "Simplified theory for stable Gunn domains including diffusion", *J. Appl. Phys.* **41**, 4676–91 (1970).

LAND, P. L.: "New methods for determining electron trap parameters from thermoluminescence or conductivity glow curves", *J. Phys. Chem. Solids* **30**, 1681–92 (1969).

LAND, P. L.: "Equations for thermoluminescence and thermally stimulated current as derived from simple models", *J. Phys. Chem. Solids* **30**, 1693–1708 (1969).

LANDAU, L.: "The motion of electrons in a crystal lattice", *Phys. Z. Sowjetunion* **3**, 664–5 (1933).

LANDSBERG, P.T.: "Non-radiative transitions in semiconductors", *Phys. Stat. Sol.* **41**, 457–89 (1970).

LANDSBERG, P. T.: "An introduction to the theory of photovoltaic cells", *Solid-State Electron.* **18**, 1043–52 (1975).

LANDSBERG, P. T. and BEATTIE, A. R.: "Auger effect in semiconductors", *J. Phys. Chem. Solids* **8**, 73–77 (1959).

LANG, D. V.: "Deep-level transient spectroscopy: a new method to characterize traps in semiconductors", *J. Appl. Phys.* **45**, 3023–32 (1974).

LANG, I. G. and FIRSOV, YU. A.: "Kinetic theory of semiconductors with low mobility", *Soviet Phys.—JETP* **16**, 1301–12 (1963).

LANG, I. G. and FIRSOV, YU. A.: "Mobility of small polarons at low temperatures", *Soviet Phys.—Solid State* **5**, 2049–60 (1964).

LANGRETH, D. C.: "Polaron mobility at finite temperatures", *Phys. Rev.* **159**, 717–25 (1967).

LANYON, H. P. D.: "Electrical and optical properties of vitreous selenium", *Phys. Rev.* **130**, 134–43 (1963).

LAVRUSHKO, A. G. and BENDERSKII, V. A.: "Triplet–triplet annihilation and photoconductivity of anthracene crystals", *Soviet Phys.—Solid State* **13**, 225–7 (1971).

LAVRUSHKO, A. G. and BENDERSKII, V. A.: "Efficiency of the various charge-carrier generation processes in anthracene crystals", *Soviet Phys.—Solid State* **13**, 1223–5 (1971).

LAW, H. C. and KAO, K. C.: "Current oscillations caused by recombination centers in semiconductors", *Solid-State Electron.* **13**, 659–69 (1970).

LAX, M.: "Cascade capture of electrons in solids", *Phys. Rev.* **119**, 1502–23 (1960).

LAX, M.: "Present status of semiconductor surface physics", in *International Conference on Semiconductor Physics in 1960, Prague*, Czechoslovak Academy of Science, 484–514 (1961).

LE BLANC JR., O. H.: "Hole and electron drift mobilities in anthracene", *J. Chem. Phys.* **33**, 626 (1960).

LE BLANC JR., O. H.: "Band structure and transport of holes and electrons in anthracene", *J. Chem. Phys.* **35**, 1275–80 (1961); erratum. *J. Chem. Phys.* **36**, 1082 (1962).

LE BLANC JR., O. H.: "Electronic transport in organic solids and liquids", *Abstr. of Organic Crystal Symposium, NRC, Ottawa*, p. 82, 1962; "Electronic transport in crystalline and liquid pyrene", *J. Chem. Phys.* **37**, 916 (1962).

LE BLANC JR., O. H.: "Band theory and the Hall effect in organic crystals", *J. Chem. Phys.* **39**, 2395–7 (1963).

LE BLANC JR., O. H.: "Conductivity", in *Physics and Chemistry of the Organic Solid State* (eds. D. Fox, M. M. Labes, and A. Weissberger), Wiley Interscience (New York) **3**, 133–98 (1967).

LEE, C. H., KEVORKIAN, H. K., REUCROFT, P. J., and LABES, M. M.: "Diffusion in organic crystals. I: Self-diffusion in anthracene", *J. Chem. Phys.* **42**, 1406–9 (1965).

LEE, T. D., LOW, F. E., and PINES, D.: "The motion of slow electrons in a polar crystal", *Phys. Rev.* **90**, 297–302 (1953).

LEE, T. D. and PINES, D.: "Interaction of a nonrelativistic particle with a scalar field with application to slow electrons in polar crystals", *Phys. Rev.* **92**, 883–9 (1953).

LEES, K. J. and WILSON, E. G.: "Intrinsic photoconduction and photoemission in polyethylene", *J. Phys. C* **6**, 3110–20 (1973).

LENZLINGER, M. and SNOW, E. H.: "Fowler–Nordheim tunneling into thermally grown SiO_2", *J. Appl. Phys.* **40**, 278–83 (1969).

LEVINE, J. D. and MARK, P.: "Evaluation of surface-state theories", *Phys. Rev.* **182**, 926–35 (1969).

LEVINE, M., JORTNER, J., and SZÖKE, A.: "Diffusion of triplet excitons in crystalline anthracene", *J. Chem. Phys.* **45**, 1591–1604 (1966).

LEVINSON, J., BURSHTEIN, Z., and MANY, A.: "On the feasibility of observing photoemission from metal contacts into anthracene", *Mol. Cryst. Liq. Cryst.* **26**, 329–47 (1974).

LEVINSON, J., MARREW, J., COBAS, A., and WEISZ, S. Z.: "Annealing of singlet and triplet quenching centres in anthracene", *J. Lumin.* **1, 2**, 726–31 (1970).

LEVINSON, J., WEISZ, S. Z., COBAS, A., and POLON, A.: "Determination of the triplet exciton–trapped electron interaction rate constant in anthracene crystals", *J. Chem. Phys.* **52**, 2794–5 (1970).

LEWICKI, G., MASERJIAN, J., and MEAD, C. A.: "Barrier energies in MIM structures from photoresponse: effect of scattering in the insulating film", *J. Appl. Phys.* **43**, 1764–7 (1972).

LI, H. T. and REGENSBURGER, P. J.: "Photoinduced discharge characteristics of amorphous selenium plates", *J. Appl. Phys.* **34**, 1730–5 (1963).

LI, S. S., LINDHOLM, F. A., and WANG, C. T.: "Quantum yield of metal–semiconductor photodiodes", *J. Appl. Phys.* **43**, 4123–9 (1972).

LIANG, C. Y. and SCALA, E. G.: "Photoconductivity of sodium deoxyribonucleic acid in the dry state", *Nature* **198**, 86–87 (1963).

LILLY, A. C., LOWITZ, D. A., and SCHUG, J. C.: "Space-charge conduction in insulators", *J. Appl. Phys.* **39**, 4360–4 (1968).

LIN, S. F., SPICER, W. E., and SCHECHTMAN, B. H.: "Electron escape depth, surface composition, and charge transfer in tetrathiafulvalene–tetracyanoquinodimethane (TTF–TCNO) and related compounds: photoemission studies", *Phys. Rev. B* **12**, 4184–99 (1975).

LINDHOLM, F. A., NEUGROSCHEL, A., SAH, C. T., GODLEWSKI, M. P. and BARANDHORST, H. W.: "A methodology for experimentally based determination of gap shrinkage and effective lifetime on the emitter and base of p–n junction solar cells and other p–n junction devices", *IEEE Trans.* ED **24**, 402–9 (1977).

LINDHOLM, F. A. and SAH, C. T.: "Fundamental electronic mechanisms limiting the performance of solar cells", *IEEE Trans.* ED **24**, 299–304 (1977).

LINDMAYER, J., REYNOLDS, J., and WRIGLEY, C.: "One carrier space-charge-limited current in solids", *J. Appl. Phys.* **34**, 809–12 (1963).

LINDMAYER, J. and SLOBODSKY, A.: "Two carrier space-change limited currents in solids and the dielectric capacitor", *Solid-State Electron.* **6**, 495–503 (1963).

LINGELBACH, W., STUKE, J., WEISA, G., and TREUSCH, J.: "Temperature-dependent electro-absorption on the indirect edge of trigonal selenium", *Phys. Rev. B* **5**, 243–53 (1972).

LIPSETT, F. R.: "On the production of single crystals of naphthalene and anthracene", *Can. J. Phys.* **35**, 284–98 (1957).

LIPSETT, F. R.: "Furnace for the growth of naphthalene and anthracene crystals", *Rev. Sci. Instrum.* **29**, 423–4 (1958).

LIPSETT, F. R.: "Apparatus for measurements of the luminescence spectra of crystals at NRC, Ottawa", *Bull. Photoelect. Spectrum* **16**, 474–82 (1965).

LIPSETT, F. R.: "The quantum efficiency of luminescence", *Prog. Dielectrics* **7**, 217–319 (1967).

LIPSETT, F. R., COMPTON, D. J., and WADDINGTON, T. C.: "The effect of surface condition on the fluorescence and surface photoconductivity of anthracene", *J. Chem. Phys.* **26**, 1444–5 (1957).

LIPSETT, F. R. and DEKKER, A. J.: "Fluorescent spectra of some organic solid solutions", *Can. J. Phys.* **30**, 165–73 (1952).

LISIAK, K. P. and MILNES, A. G.: "Platinum as a lifetime-control deep impurity in silicon", *J. Appl. Phys.* **46**, 5229–35 (1975).

LISOVENKO, V. A. and SHPAK, M. T.: "Effect of imperfection of crystal lattice on the luminescence properties of anthracene", *Bull Acad. Sci. USSR, Phys. Ser.* **29**, 1308–11 (1966).

LISOVENKO, V. A. and SHPAK, M. T.: "Some peculiarities of luminescence of deformed anthracene crystals", *Phys. Stat. Sol.* **14**, 467–70 (1966).

LITTLE, W. A.: "Superconductivity of organic polymers", *J. Polymer Sci. C* **17**, 3–12 (1967).

LITTLE, W. A.: "The exciton mechanism in superconductivity", *J. Polymer Sci. C* **29**, 17–26 (1970).

LITTLE, W. A. (ed.): *Proceeding of the International Conference of Organic Superconductors, J. Polymer Sci. Part C, Polymer Symposia* **29**, Wiley Interscience (New York) (1970).

LITTON, C. W. and REYNOLDS, D. C.: "Double-carrier injection and negative resistance in CdS", *Phys. Rev.* **133**, A536–41 (1964).

LITVINENKO, V. YU. and FRIDKIN, V. M.: "Optical active electrode injecting holes into anthracene", *Zh. Fiz. Khim.* **40**, 2585–8 (1966).

LITVINENKO, V. YU. and FRIDKIN, V. M.: "Hole injection in anthracene from an electrolyte and space-charge limitation of current", *Soviet Phys.—Solid State* **9**, 2845–6 (1968).

LIU, R. S. H. and KELLOGG, R. E.: "Triplet–triplet energy transfer from the second triplet state of anthracene: spectroscopic methods", *J. Am. Chem. Soc.* **91**, 250–2 (1969).

LLEWELLYN, F. B.: "Operation of ultra-high frequency vacuum tubes", *Bell Syst. Tech. J.* **14**, 632–65 (1935).

LOGAN, L. M., MUNRO, I. H., WILLIAMS, D. F., and LIPSETT, F. R.: "Low-temperature studies of fluorescence emission from anthracene crystals", in *Molec. Lumin. Int. Conf. 1968*, Benjamin (New York), 773–85 (1969).

LOHANICK, A., COOK, J. S., and O'DWYER, J. J.: "The high field conductivity of anthracene", in *1968 Annual Report on Conference on Electrical Insulation and Dielectric Phenomena* (NAS Publication 1705, Washington DC, USA) 160–5 (1969).

LOHMANN, F. and MEHL, W.: "Electrochemical reaction rates at anthracene electrodes. I: Measurements at rotating disk electrodes", *Ber. Bunsenges. Phys. Chem.* **71**, 493–503 (1967).

LOHMANN, F. and MEHL, W.: "Space charge-limited hole currents injected into orthorhombic sulfur from electrochemical contacts", *J. Phys. Chem. Solids* **28**, 1317–21 (1967).

LOHMANN, F. and MEHL, W.: "Rate of electrochemical reactions at the anthracene electrode. II: Pulse measurements", *Electrochem. Acta* **13**, 1469–81 (1968).

LOHMANN, F. and MEHL, W.: "Dark injection and radiative recombination of electrons and holes in naphthalene crystals", *J. Chem. Phys.* **50**, 500–6 (1969).

LOW, F. E. and PINES, D.: "Mobility of slow electrons in polar crystals", *Phys. Rev.* **98**, 414–18 (1955).

LOWER, S. K. and EL-SAYED, M. A.: "Triplet state and molecular electronic processes in organic molecules", *Chem. Rev.* **16**, 199–241 (1966).

LUCAS, I.: "Interpretation of the switching effect in amorphous semiconductors as a recombination instability", *J. Non-Cryst. Solids* **6**, 136–44 (1971).

LUDWIG, W. and ZEISE, U.: "Doppelinjektion in ZnS-Einkristallen", *Phys. Stat. Sol.* **7**, 143–53 (1964).

LUMB, M. D. and STEPHAN, J. A.: "On the interpretation of thermal activation energies for conduction in organic semiconductors", *Chem. Phys. Lett.* **22**, 81–86 (1973).

LUNDSTRÖM, I., CORKER, G. A., and STENBERG, M.: "Charge injection and storage in thin films of chlorophyll-a", *J. Appl. Phys.* **49**, 701–8 (1978).

LUPIEN, Y. and WILLIAMS, D. F.: "Preparation of high-purity anthracene: zone refine and the triplet lifetime", *Mol. Cryst.* **5**, 1–7 (1968).

LUPIEN, Y., WILLIAMS, J. O., and WILLIAMS, D. F.: "Effects of crystal growth environment on defect concentrations in anthracene crystals", *Mol. Cryst. Liq. Cryst.* **18**, 129–41 (1972).

LUTY, T. and PAWLEY, G. S.: "A shell model for molecular crystals", *Phys. Stat. Sol.* B **66**, 309–19 (1974).

LYONS, L. E.: "Electron affinities of some aromatic molecules", *Nature* **166**, 193 (1950).

LYONS, L. E.: "Photoconductance in tetracene-type crystals: theory of spectral dependence", *J. Chem. Phys.* **23**, 220 (1955).

LYONS, L. E.: "Polarized spectra of anthracene crystals and Davydov splitting", *J. Chem. Phys.* **23**, 1973 (1955).

LYONS, L. E.: "Ionized states of molecular crystals", *Aust. J. Chem.* **10**, 365–7 (1957).

LYONS, L. E.: "Ionized states in molecular crystals", *J. Chem. Soc.* 5001–7 (1957).

LYONS, L. E.: "Electron transfer across the boundaries of organic solids", in *Physics and Chemistry of the Organic Solid State* (eds. D. Fox, M. M. Labes, and A. Weissberger), Wiley Interscience (New York), **1**, 746–804 (1963).

LYONS, L. E.: "Organic semiconductors", in *Encyclopedic Dictionary of Physics*, Supplementary **1**, Pergamon (Oxford) 1966.

LYONS, L. E.: "Organic metals? The electrical conductance of organic solids", *Proc. Roy. Soc. (NSW)* **101**, 1–9 (1967).

LYONS, L. E., BREE, A., and CARSWELL, D. J.: "Photo- and semi-conductance of organic crystals. Part I: Photo-effects in tetracene and anthracene", *J. Chem. Soc.* 1728–33 (1955).

LYONS, L. E. and CARSWELL, D. J.: "Photo- and semi-conductance of organic crystals. Part II: Spectral dependence, quantum efficiency and relation between semi- and photo-effects in anthracene", *J. Chem. Soc.* 1734–40 (1955).

LYONS, L. E. and MACKIE, J. C.: "Photo- and semi-conductance in organic crystals. Part VII: Space-charge effect in anthracene", *J. Chem. Soc.* 5186–92 (1960).

LYONS, L. E. and McGREGOR, K. G.: "Photo- and electro-chemical processes at organic crystal electrodes. II: Oxidation of a solute in water by positive holes from an anthracene crystal", *Aust. J. Chem.* **29**, 21–26 (1976).

LYONS, L. E. and McGREGOR, K. G.: "Photo and electro-chemical processes at organic crystal electrodes. III. Hole photo-injection into anthracene single crystals", *Aust. J. Chem.* **29**, 1401–6 (1976).

LYONS, L. E. and MILNE, K. A.: "One-photon intrinsic photogeneration in anthracene crystals", *J. Chem. Phys.* **65**, 1474–84 (1976).

LYONS, L. E. and MORRIS, G. C.: "Photo- and semi-conduction of aromatic hydrocarbon crystals", *Proc. Phys. Soc.* **69**, 1162–4 (1956).

LYONS, L. E. and MORRIS, G. C.: "Photo- and semi-conductance in organic crystals. Part III: Photoeffects in dry air with eleven organic compounds", *J. Chem. Soc.* **8**, 3648–60 (1957).

LYONS, L. E. and MORRIS, G. C.: "Photo- and semi-conductance in organic crystals. Part IV: Spectral dependence of photocurrents in aromatic hydrocarbon crystals in dry air with polarized light", *J. Chem. Soc.* **8**, 3661–8 (1957).

LYONS, L. E. and MORRIS, G. C.: "The intensity of ultraviolet light absorption by monocrystals. Part III: Absorption by anthracene at 295° K, 90° K, and 4° K of plane polarized light of wavelengths 1600–2750 Å", *J. Chem. Soc.* 1551–8 (1959).

LYONS, L. E. and MORRIS, G. C.: "Photo- and semi-conductance in organic crystals. Part VIII: Photoemission of electrons from crystals of aromatic hydrocarbons", *J. Chem. Soc.* 5192–9 (1960).

LYONS, L. E. and MORRIS, G. C.: "Photo- and semi-conductance in organic crystals. Part IX: Photoconductance in anthracene and naphthracene irradiated in the vacuum region", *J. Chem. Soc.* 5200–6 (1960).

LYONS, L. E. and NEWMAN, O. M. G.: "Photovoltages in tetracene films", *Aust. J. Chem.* **24**, 13–23 (1971).

LYONS, L. E. and WARREN, L. J.: "Anthracene fluorescence at low temperatures. I: Purified single crystals", *Aust. J. Chem.* **25**, 1411–25 (1972).

LYONS, L. E. and WARREN, L. J.: "Anthracene fluorescence at low temperatures. II: Doped single crystals", *Aust. J. Chem.* **25**, 1427–41 (1972).

LYONS, L. E. and WHITE, J. W.: "Low temperature quenching of energy transfer", *J. Chem. Phys.* **29**, 447–8 (1958).

LYONS, L. E. and WHITE, J. W.: "Luminescence from anthracene crystals and its temperature dependence", *J. Chem. Soc.* 5213 (1960).

MA, W., ANDERSON, R. M., and HRUSKA, S. J.: "Study on the anomalously high photovoltaic effect in germanium thin films", *J. Appl. Phys.* **46**, 2650–7 (1975).

MACDONALD, J. R.: "Static space charge and capacitance for a single blocking electrode", *J. Chem. Phys.* **29**, 1346–58 (1958).

MACDONALD, J. R.: "Distribution of space charge in homogeneous metal oxide films and semiconductors", *J. Chem. Phys.* **40**, 3735–7 (1964).

MAIER, G., HAEBERLEN, U., and WOLF, H. C.: "Kernspin-Relaxation durch Triplett-Excitonien in Anthracen", *Phys. Lett.* **25A**, 323–4 (1967).

MAKHOTENKO, A. N. and LITVINENKO, V. YU.: "Injection of electrons and electron trapping levels in anthracene single crystals", *Soviet Phys.—Semicond.* **7**, 439 (1973).

MAKHOTENKO, A. N. and LITVINENKO, V. YU.: "Switching effect in anthracene single crystal under electron injection condition", *Soviet Phys. —Semicond.* **7**, 1342–3 (1974).

MAKI, A. H. and ALCACER, L.: "Electrically conducting metal dithiolate–perylene complex", *J. Phys. Chem.* **78**, 215–17 (1974).

MALTBY, J. R., REED, C. E., and SCOTT, C. G.: "Analysis of the surface photo-voltaic effect in photoconductors: CdS", *Surf. Sci.* **51**, 89–108 (1975).

MANFREDOTTI, C., DE BLAST, C., GALASSINI, S., MICOCCI, G., RUGGIERO, L., and TEPORE, A.: "Analysis of SCLC curves by a new direct method", *Phys. Stat. Sol.* (a) **36**, 569–77 (1976).

MANFREDOTTI, C., RIZZO, A., DE BLASI, C., GALASSINE, S., and RUGGIERO, L.: "Hole trapping centers in GaSe", *J. Appl. Phys.* **46**, 4531–6 (1975).

MANFREDOTTI, C., RIZZO, A., VASANELLI, L., GALASSINI, S., and RUGGIERO, L.: "Electron trapping levels in cadmium selenide single crystals", *J. Appl. Phys.* **44**, 5463–9 (1973).

MANIFACIER, J. C. and HENISCH, H. K.: "Minority-carrier injection into semiconductors", *Phys. Rev. B* **17**, 2640–7 (1978).

MANIFACIER, J. C. and HENISCH, H. K.: "Minority-carrier injection into semiconductors containing traps", *Phys. Rev. B* **17**, 2648–54 (1978).

MANSVETOV, G., RUKMAN, G. I., and SAVELÉV, V. A.: "Time characteristics of the photodepolarization process in anthracene", *Soviet Phys.—Solid State* **8**, 1320–2 (1966).

MANY, A.: "High-field effects in photoconducting cadmium sulphide", *J. Phys. Chem. Solids* **26**, 575–85 (1965).

MANY, A.: "Tunneling processes across the CdS–electrolyte interface", *J. Phys. Chem. Solids* **26**, 587–93 (1965).

MANY, A., HARNIK, E., and GERLICH, D.: "On the semiconductivity of crystalline aromatic substances", *J. Chem. Phys.* **23**, 1733–4 (1955).

MANY, A., LEVINSON, J., and TENCHER, I.: "Photoemission of electrons from alkali and alkaline earth metal contacts into anthracene", *Phys. Rev. Lett.* **20**, 1161–3 (1968).

MANY, A., LEVINSON, J., and TENCHER, I.: "Photo-enhanced electron emission from alkali metal contacts into anthracene", *Mol. Cryst.* **5**, 273–94 (1969).

MANY, A. and RAKAVY, G.: "Theory of transient space-charge-limited currents in solids in the presence of trapping", *Phys. Rev.* **126**, 1980–8 (1962).

MANY, A., SIMHONY, M., WEISZ, S. Z., and LEVINSON, J.: "Studies of photoconductivity in iodine single crystals", *J. Phys. Chem. Solids* **22**, 285–92 (1961).

MANY, A., WEISZ, S. Z., and SIMHONY, M.: "Space-charge-limited currents in iodine single crystals", *Phys. Rev.* **126**, 1989–95 (1962).

MAR, H. A. and SIMMONS, J. G.: "Determination of bulk trap parameters using thermal dielectric relaxation techniques", *Solid-State Electron.* **17**, 1181–5 (1974).

MAR, H. A. and SIMMONS, J. G.: "Surface generation statistics and associated thermal currents in metal-oxide–semiconductor structures", *Phys. Rev.* B **11**, 775–83 (1975).

MAR, H. A. and SIMMONS, J. G.: "A review of the techniques used to determine trap parameters in the MNOS structure", *IEEE Trans Electron Devices* ED **24**, 540–6 (1977).

MARAKHONOV, V. M. and SEISYAR, R. P.: "Space-charge-limited current in cadmium sulfide films", *Soviet Phys.—Semicond.* **7**, 769–72 (1973).

MARCHAND, R. L. and SAH, C. T.: "Study of thermally induced deep levels in Al-doped Si", *J. Appl. Phys.* **48**, 336–41 (1977).

MARCHETTI, A. P. and KEARNS, D. R.: "Photoelectric emission from aromatic hydrocarbon-alkali metal films", *J. Chem. Phys.* **44**, 1301–2 (1966).

MARCHETTI, A. P. and KEARNS, D. R.: "Photoelectron emission from aromatic and metalloorganic hydrocarbons", *Mol. Cryst. Liq. Cryst.* **6**, 299–317 (1970).

MARICIC, S., PIFAT, G., and PARANDIC, V.: "Proton conductivity in the solid hydrated haemoglobin", *Biochem. Biophys. Acta* **79**, 293–300 (1964).

MARK, P. and HARTMAN, T. E.: "On the distinguishing between the Schottky and Poole–Frenkel effects in insulators", *J. Appl. Phys.* **39**, 2163–4 (1968).

MARK, P. and HELFRICH, W.: "Space-charge-limited currents in organic crystals", *J. Appl. Phys.* **33**, 205–15 (1962).

MARLOR, G. A. and WOODS, J.: "Space-charge limited currents and electron traps in CdS crystals", *Br. J. Appl. Phys.* **16**, 1449–56 (1965).

MAROWSKY, G.: "Gain-narrowing studies with an organic dye laser", *J. Appl. Phys.* **45**, 2621–3 (1974).

MARSHALL, J. M. and MILLER, G. R.: "Field-dependent carrier transport in non-crystalline semiconductors", *Phil. Mag.* **27**, 1151–68 (1973).

MARUYAMA, Y. and INOKUCHI, H.: "The effect of oxygen on the semiconductivity of quaterylene", *Bull. Chem. Soc. Japan* **39**, 1418–22 (1966).

MARUYAMA, Y. and INOKUCHI, H.: "Charge carrier mobility in anthracene single crystals", *Bull. Chem. Soc. Japan* **40**, 2073–7 (1967).

MARUYAMA, Y. and IWASAKI, N.: "Electrical conduction in amorphous organic films", *J. Non-Cryst. Solids* **16**, 399–406 (1974).

MARUYAMA, Y., KOBAYASHI, T., INOKUCHI, H., and IWASHIMA, S.: "Charge-carrier drift mobility in perylene single crystals", *Mol. Cryst. Liq. Cryst.* **20**, 373–80 (1973).

MASON, R.: "The crystallography of anthracene at $95°K$ and $290°K$", *Acta Cryst.* **17**, 547–55 (1964).

MATHUR, S. C. and KUMAR, B.: "Charge transport calculations in anthraquinone", *Mol. Cryst. Liq. Cryst.* **23**, 85–98 (1973).

MATHUR, V. K. and DAHIYA, R. P.: "Space-charge-limited currents in insulators containing traps distributed in energy", *Solid-State Electron.* **17**, 61–70 (1974).

MATIJEC, R.: "Protons as charge carriers in the electrical conductivity and photoconductivity of anthracene crystals", *Ber. Bunsengesell.* **68**, 964 (1964).

MATSUI, A.: "The polarized absorption edge and the Davydov splitting of anthracene", *J. Phys. Soc. Japan* **21**, 2221–2 (1966).

MATSUI, A. and ISHII, Y.: "Zone refining of anthracene", *Japan J. Appl. Phys.* **6**, 127–34 (1967).

MATSUI, A. and ISHII, Y.: "Optical properties of anthracene single crystals", *J. Phys. Soc. Japan* **23**, 581–90 (1967).

MATSUI, A., ISHII, Y., and HIKITA, T.: "Reflection spectra of anthracene single crystals", *J. Phys. Soc. Japan* **21**, 2091–2 (1966).

MATSUI, A., TOMIOKA, K., OEDA, Y., and TOMOTIKA, T.: "Exciton band structure of crystalline anthracene", *Surf. Sci.* **37**, 849–54 (1973).

MATSUMURA, M., UOHASHI, H., FURUSAWA, M., YAMAMOTO, N., and TSUBOMURA, H.: "The photovoltaic effect in naphthacene – gold layers", *Bull. Chem. Soc. Japan* **48**, 1965–9 (1975).

MAVROYANNIS, C.: "Ground-state energy and excitation spectrum of molecular crystals", *J. Chem. Phys.* **42**, 1772–80 (1965).

MAYER, J. W., BARON, R., and MARSH, O. J.: "Observation of double injection in long silicon *p–i–n* structures", *Phys. Rev.* **137**, A286–95 (1965).

MAYER, J. W., MARSH, O. J., and BARON, R.: "Double injection in long silicon *p–π–n* structures", *J. Appl. Phys.* **39**, 1447–55 (1968).

MAYER, J. W., MARSH, O. J., BARON, R., KIRKUCHI, R., and RICHARDSON, J. M.: "Interpretation of potential-probe measurements in two-carrier structures", *Phys. Rev.* **137**, A295–301 (1965).

MCCARTHY, S. L. and LAMBE, J.: "Enhancement of light emission from metal–insulator–metal tunnel junctions", *Appl. Phys. Lett.* **30**, 427–9 (1977).

MCCLELLAND, B. J.: "Relationship between the lowest triplet excitation energy and electron affinities of an alternant aromatic hydrocarbon", *Trans. Faraday Soc.* **57**, 183–6 (1961).

MCCLURE, D. S.: "Electronic spectra of molecules and ions in crystals. Part I: Molecular crystals", *Solid State Phys.* **8**, 1–47 (1959).

MCCOLL, M. and MEAD, C. A.: "Electron current through thin mica films", *Trans. Met. Soc. AIME* **233**, 502–11 (1965).

MCGARTHY, S. J. and YEE, S. S.: "Space-charge-limited (SCL) effects in thin. Cu compensated CdS structures", *Solid-State Electron.* **17**, 485–9 (1974).

MCGLYNN, S. P., AZUMI, T. and KASHA, M.: "External heavy-atom spin–orbital coupling effect. V: absorption studies of triplet states", *J. Chem. Phys.* **40**, 507–15 (1964).

MCGREGOR, K. G.: "Photo- and electro-chemical processes at organic crystal electrodes. I: Chronophotometry: a technique to study processes at electrodes", *Aust. J. Chem.* **29**, 13–19 (1976).

MCGRIE, A. R., REUCROFT, P. J., and LABES, M. M.: "Dislocation etching of anthracene", *J. Chem Phys.* **45**, 3163 (1966).

MCKEEVER, S. W. S. and HUGHER, D. M.: "Thermally stimulated currents in dielectrics", *J. Phys. D* **8**, 1520–9 (1975).

MCMAHON, D. H. and KESTIGIAN, M.: "Triplet–triplet annihilation in anthracene at low excitation intensities: wavelength and temperature dependence", *J. Chem. Phys.* **46**, 137–42 (1967).

MCMILLAN, W. L. and ROWELL, J. M.: "Tunneling and strong-coupling superconductivity", in *Superconductivity* (ed. R. D. Parks), Marcel Dekker (New York), 561–613 (1969).

MCRAE, E. G. and SIEBRAND, W. G.: "Vibronic coupling criteria in molecular crystals", *J. Chem. Phys.* **41**, 905–6 (1964).

MEAD, C. A.: "Electron transport mechanisms in thin insulating films", *Phys. Rev.* **128**, 2088–93 (1962).

MEAD, C. A.: "Metal–semiconductor surface barrier", *Solid-State Electron.* **9**, 1023–33 (1966).

MEAD, C. A.: "Electron transport in thin insulating films", in *Basic Problems in Thin Film Phys.* (eds. R. Niedermayer and H. Mayer), Proc. Int. Symposium held at Clausthal-Göttingen, 1965, 674–8 (1966).

MEAD, C. A., SNOW, E. H., and DEAL, B. E.: "Barrier lowering and field penetration at metal–dielectric interfaces", *Appl. Phys. Lett.* **9**, 53–58 (1966).

MEAD, C. A. and SPITZER, W. G.: "Photoemission from Au and Cu into CdS", *Appl. Phys. Lett.* **2**, 74–76 (1963).

MEAD, C. A. and SPITZER, W. G.: "Fermi level position at metal–semiconductor interfaces", *Phys. Rev.* **134**, A713–18 (1964).

MEAUDRE, R. and MESNARD, G.: "Transient and alternating currents in insulators or semiconductors: influence of generation and recombination of carriers", *J. Phys. C* **7**, 1271–8 (1974).

MEHENDRU, P. C., PATHAK, N. L., SINGH, S., and METHENDRU, P.: "Electrical conduction in polypropylene thin-films", *Phys. Stat. Sol.* (a) **38**, 355–60 (1976).

MEHL, W.: "Redoxreaktionen an Anthracenelektroden", *Ber. Bunsenges. Phys. Chem.* **69**, 583–9 (1965).

MEHL, W.: "Dark injection of electrons from alkali metals in anthracene", *Solid State Commun.* **6**, 549–51 (1968).

MEHL, W.: "Charge-carrier injection into homo-molecular crystals", *Pure Appl. Chem.* **27**, (3) 499–513 (1971).

MEHL, W.: "Reactions at organic semiconductor electrodes", in *Reactions of Molecules at Electrodes* (ed. N. S. Hush), Wiley Interscience (New York) 305–45 (1971).

MEHL, W. and BÜCHNER, W.: "Durch elektrochemische Doppelinjektion angeregte Elektrolumineszinz in Anthracen-Kristallen", *Z. Phys. Chem.* (Neue Folge) **47**, 76–86 (1965).

MEHL, W., DRURY, J. S., and HALE, J. M.: "Hole injection into anthracene crystals from electronically excited iodine molecules", *Phys. Lett.* **28A**, 205–6 (1968).

MEHL, W. and FUNK, B.: "Dark injection of electrons and holes and radiative recombination in anthracene with metallic contacts", *Phys. Lett.* **25A**, 364–5 (1967).

MEHL, W. and HALE, J. M.: "Insulator electrode reactions", in *Advances in Electrochemistry and Electrochemical Engineering* (ed. P. Delahay), **6**, 399–458 (1967).

MEHL, W. and HALE, J. M.: "Charge transfer processes with electronically excited anthracene molecules", *Discuss. Faraday Soc.* **45**, 30–39 (1968).

MEHL, W., HALE, J. M., and DRURY, J. S.: "Charge transfer processes with electronically excited iodine molecules", *Ber. Bunsenges. Phys. Chem.* **73**, 855–9 (1969).

MEHL, W. and LOHMANN, F.: "The influence of surface oxides of photocurrents at anthracene electrodes", *Surf. Sci.* **44**, 295–376 (1974).

MEHL, W. and WOLF, N. E.: "Photoconductivity in dispersed organic systems", *J. Phys. Chem. Solids* **25**, 1221–31 (1964).

MEIER, H. and ALBRECHT, W.: "Lichtelektrische Untersuchung organischer Halbleiter in Vidicon-Fernsehaufahmeröhren", *Ber. Bunsenges. Phys. Chem.* **73**, 86–94 (1969).

MEIER, H. and ALBRECHT, W.: "Blitzlichtuntersuchungen an organischen Farbstoff-Photohalbleitern", *Z. Naturforsch.* **24a**, 257–66 (1969).

MEIER, H., ALBRECHT, W., and TSCHIRWITZ, U.: "Mechanism of organic semiconductor doping", *Ber. Bunsenges. Phys. Chem.* **73**, 795–805 (1969).

MEIRSSCHANT, S.: "One-carrier space-charge-limited currents in solids containing double level centers", *Solid-State Electron.* **19**, 633–9 (1976).

MELNGAILIS, I. and REDIKER, R. H.: "The madistor-a magnetically controlled semiconductor plasma device", *Proc. IRE* **50**, 2428–35 (1962).

MELNGAILIS, I. and REDIKER, R. H.: "Negative resistance InSb diodes with large magnetic field effects", *J. Appl. Phys.* **33**, 1883–92 (1962).

MELZ, P. J.: "Photogeneration in trinitrofluorenone–poly (*N*-Vinylcarbazole)", *J. Chem. Phys.* **57**, 1694–9 (1972).

MERINSKY, K., BETDO, J., MORIC, M., and KORDOS, B.: "Über die elektrischen Eigenschaften von Ge-Magnetodioden", *Solid-State Electron.* **11**, 187–91 (1968).

MERRIFIELD, R. E.: "Ionized states in a one-dimensional molecular crystal", *J. Chem. Phys.* **34**, 1835–9 (1961).

MERRIFIELD, R. E.: "Vibrational structure of molecular exciton states", *J. Chem. Phys.* **36**, 2519–20 (1962).

MERRIFIELD, R. E.: "Exciton impurity levels in molecular crystals", *J. Chem. Phys.* **38**, 920–4 (1963).

MERRIFIELD, R. E.: "Tight-binding wavefunctions for electrons in molecular crystals in the presence of an electric field", *J. Chem. Phys.* **39**, 3540–1 (1963).

MERRIFIELD, R. E.: "Theory of the vibrational structure of molecular exciton states", *J. Chem. Phys.* **40**, 445–50 (1964).

MERRIFIELD, R. E.: "Moment theorems for vibronic exciton spectra of molecular crystals", *J. Chem. Phys.* **48**, 3693–6 (1968).

MERRIFIELD, R. E.: "Diffusion and mutual annihilation of triplet excitons in organic crystals", *Accounts Chem. Res.* **1**, 129–35 (1968).

MERRIFIELD, R. E.: "Theory of magnetic field effects on the mutual annihilation of triplet excitons", *J. Chem. Phys.* **48**, 4318–19 (1968).

MERRIFIELD, R. E.: "Magnetic effects on triplet exciton interactions", *Pure Appl. Chem.* **27** (3) 481–97 (1971).

MERRIFIELD, R. E., AVAKIAN, P., and GROFF, R. P.: "Fission of singlet excitons into pairs of triplet excitons in tetracene crystals", *Chem. Phys. Lett.* **3**, 155–7 and 386–8 (1969); Errata, ibid. **3**, 728 (1969).

MERRITT, V. Y. and HOVEL, H. J.: "Organic solar cells of hydroxy squarylium", *Appl. Phys. Lett.* **29**, 414–15 (1976).

METTE, H. and PICK, H.: "Electrical conductivity of anthracene monocrystals", *Z. Phys.* **134**, 566–75 (1953).

MEY, W. and HERMANN, M. A.: "Drift mobilities of holes and electrons in naphthalene single crystals", *Phys. Rev. B* **7**, 1652–6 (1973).

MEY, W., SONNONSTINE, J. J., MOREL, D. L., and HERMANN, A. M.: "Drift mobility of holes and electrons in perdeuterated anthracene single crystals", *J. Chem. Phys.* **58**, 2542–6 (1973).

MEYER, W. and NELDEL, H.: "A relation between the energy constant ϵ and the quantity constant a in the conductivity–temperature formula for oxide semiconductors" (data given for TiO_2, ZnO, Fe_2O_3, and U_2O_3), *Z. Tech. Phys.* **18**, 588–93 (1937).

MEYERHOFER, D. and OCHS, S. A.: "Current flow in very thin films of Al_2O_3 and BeO", *J. Appl. Phys.* **34**, 2535–43 (1963).

MEYERHOFER, D. and PASIERB, E. F.: "Light scattering characteristics in liquid crystal storage materials", *Mol. Cryst. Liq. Cryst.* **20**, 279–300 (1973).

MICHEL-BEYERLE, M. E. and HABERKORN, R.: "Temperature-dependent intersystem crossing in anthracene crystals", *Phys. Stat. Sol.* (b) **85**, 473–5 (1978).

MICHEL-BEYERLE, M. E., HABERKORN, R., KINDER, K., and SEIDLITZ, H.: "Direct evidence for the singlet–triplet exciton annihilation in anthracene crystals", *Phys. Stat. Sol.* (b) **85**, 45–49 (1978).

MICHEL-BEYERLE, M. E., HARENGEL, W., and HABERKORN, R.: "The role of image forces at organic crystal/electrode interfaces", *Mol. Crys. Liq. Cryst.* **25**, 323–38 (1974).

MICHEL-BEYERLE, M. E., HARENGEL, W., and KINDER, J.: "Ideal space-charge-limited transients in a Kepler experiment", *Phys. Stat. Sol.* **20**, 563–8 (1973).

MICHEL-BEYERLE, M. E., REBENTROST, F., and WILLIG, F.: "Injection of defect electrons into perylene single crystals by electrolytic contacts", *Solid State Commun.* **7**, 493–5 (1969).

MICHELL, D., ROBINSON, P. M., and SMITH, A. P., "Direct observation of dislocation in anthracene", *Phys. Stat. Sol.* **26**, K93–95 (1968).

MICKANIN, W. and GORDON, N.: "Field-enhanced space-charge-limited hole-currents in thin-oxide MNOS varactors", *IEEE Trans. Electron. Devices* ED **23**, 995–7 (1976).

MIGLIORATO, P., MARGARITONDO, G., and PERFETTI, P.: "Subquadratic *J–V* dependence and double injection in GaP", *Solid State Commun.* **13**, 499–502 (1973).

MIGLIORATO, P., MARGARITONDO, G., and PERFETTI, P.: "Double injection in semiconductors", *J. Appl. Phys.* **47**, 656–63 (1976).

MIN, H. S.: "Green's function approach to thermal noise in space-charge-limited solid state diodes", *Solid-State Electron.* **18**, 908–9 (1975).

MISEK, J., LNCERAL, L., SROBAR, F., and KORTAIN, J.: "Origin of chromatic delay in electroluminescent diodes", *Solid-State Electron.* **20**, 333–4 (1977).

MISRA, T. N. and MCGLYNN, S. P.: "Delayed luminescence of organic mixed crystals", *J. Chem. Phys.* **44**, 3816–28 (1965).

MIYAMOTO, M. and NAKADA, I.: "On the measurement of the internal electrostatic field of anthracene", *J. Phys. Soc. Japan* **33**, 1724 (1972).

MÖHWALD, H., HAARER, D., and CASTRO, G.: "Electron mobility in the 1:1 charge transfer crystal phenanthrene—PMDA", *Chem. Phys. Lett.* **32**, 433–7 (1975).

MOORE, G. F.: "Effect of temperature on the triplet–triplet annihilation rate in anthracene crystals", *Nature* **211**, 1170–1 (1966).

MOORE, G. F.: "Delayed excimer emission from anthracene", *Nature* **212**, 1452–3 (1966).

MOORE, J. S., HOLONYAK JR., N., SIRKIS, M. D., and BLOUKE, M. M.: "Space charge recombination oscillations in silicon", *Appl. Phys. Lett.* **10**, 58–60 (1967).

MOORE, J. S., PENCHINA, C. M., HOLONYAK JR., N., SIRKIS, M. D., and YAMADA, T.: "Electrical oscillations in silicon compensated with deep levels", *J. Appl. Phys.* **37**, 2009–13 (1966).

MOORE, W. and SILVER, M.: "Spatial distribution of trapped electrons in photoconducting anthracene", in *Proceedings of the Conference on Semiconduction in Molecular Solids, Princeton, New Jersey, USA*, **61** (1960).

MOORE, W. and SILVER, M.: "Generation of free carriers in photoconducting anthracene", *J. Chem. Phys.* **33**, 1671–6 (1960).

MOOSER, E. and PEARSON, W. B.: "The chemical bond in semiconductors", in *Progress in Semiconductors*, (ed. A. F. Gibson), Heywood (London) **5**, 105–39 (1960).

MORAN, P. R.: "Band structure limits for anthracene triplet excitons from magnetic resonance", *Phys. Lett.* **28A**, 781–2 (1969).

MORANTZ, D. J. and JAMES, H.: "Structural effects on photoconduction in amorphous films of organic solids", *J. Vacuum Sci. Tech.* **6**, 637–40 (1969).

MOREL, D. L., GHOSH, A. K., FENG, T., STOGRYN, E. L., PURWIN, P. E., SHAW, R. F., and FISHMAN, C.: "High-efficiency organic solar cells", *Appl. Phys. Lett.* **32**, 495–7 (1978).

MOREL, D. L. and HERMANN, A. M.: "Isotope effect for electron mobility in anthracene", *Phys. Lett.* **36A**, 101–2 (1971).

MORGAN, K. and PETHIG, R.: "Increase in d.c. dark conductivity of anthracene in a magnetic field", *Nature* **213**, 900 (1967).

MORGAN, K. and PETHIG, R.: "Search for an electrode dependent effect in the dark conductivity of anthracene", *Nature* **219**, 478–9 (1968).

MORGAN, K. and PETHIG, R.: "Evidence supporting Poole—Frenkel controlled dark conduction in tetracene doped anthracene single crystals", *Nature* **223**, 496–7 (1969).

MORGAN, K. and PETHIG, R.: "Charge carrier mobility determination for organic crystals using a Hall current measurement technique", in *2nd International Conference on Conduction in Low Mobility Materials, Fliat, Israel*, Taylor & Francis (London) 391–6 (1971).

MORRIS, G. C., RICE, S. A., and MARTIN, A. E.: "Study of the reflection spectrum of crystalline anthracene – evidence for the existence of defects", *J. Chem. Phys.* **52**, 5149–58 (1970).

MORRIS, G. C., RICE, S. A., SCEATS, M. G., and MARTIN, A. E.: "Absorption band profile of the origin region of the *b* polarized 4000 Å anthracene crystal transition", *J. Chem. Phys.* **55**, 5610–21 (1971).

MORRIS, G. C. and SCEATS, M. G.: "The 4000 Å transition of crystal anthracene", *Chem. Phys.* **3**, 164–79 (1974).

MORRIS, G. C., SCEATS, M. G., and LUCIA, S. T.: "Surface excitons in crystal anthracene (surface roughness effects)", *Mol. Cryst. Liq. Cryst.* **25**, 339–59 (1974).

MORRIS, H. and YATES, J.: "Calculation of the energy band structure and carrier mobilities in crystalline coronene and ovalene", *Discuss. Faraday Soc.* **51**, 24–36 (1971).

MORRIS, R. and SILVER, M.: "Direct electron–hole recombination in anthracene", *J. Chem. Phys.* **50**, 2969–73 (1969).

MORRIS, S. A. and WILLIAMS, J. O.: "Fluorescence and photoconductivity in crystalline anthracene and some related derivatives", *Mol. Cryst. Liq. Cryst.* **39**, 13–25 (1977).

MORT, J.: "Transient photoinjection of holes from amorphous Se into poly(N-vinylcarbazole)", *Phys. Rev.* **B 5**, 3329–36 (1972).

MORT, J.: "Physics of xerographic photoreceptors", in *Electronic and Structural Properties of Amorphous Semiconductors* (eds. P. G. Le Comber and J. Mort), Academic Press (New York) 589–606 (1973).

MORT, J., CHEN, I., EMERALD, R. L., and SHARP, J. H.: "Space-charge-perturbed xerographic discharge of photoreceptors", *J. Appl. Phys.* **43**, 2285–90 (1972).

MORT, J. and EMERALD, R. L.: "Energy-level structure and carrier transport in the poly(N-vinylcarbazole): trinitrofluorenone charge-transfer complex", *J. Appl. Phys.* **45**, 175–8 (1974).

MORT, J. and LAKATOS, A. I.: "Steady-state and transient photoemission into amorphous insulators", *J. Non-Cryst. Solids* **4**, 119–31 (1970).

MORT, J. and NIELSEN, P.: "Transient accumulation of interfacial charge in photoconductor-dielectric systems", *Phys. Rev.* **B 5**, 3336–42 (1972).

MORT, J., SCHMIDLIN, F. N., and LAKATOS, A. I.: "Transient internal photoemission of carriers in the metal–insulator system", *J. Appl. Phys.* **42**, 5761–9 (1971).

MORTENSEN, O. S., MUNN, R. W., and WILLIAMS, D. F.: "Phenomenological theory of the Hall effect in insulators", *J. Appl. Phys.* **42**, 1192–1203 (1971).

MOSS, T. S., PINCHERLE, L., and WOODWARD, A. M.: "Photoelectromagnetic and photodiffusion effects in germanium", *Proc. Phys. Soc. (Lond.)* **66B**, 743–52 (1953).

MOTT, N. F.: "Conduction in polar crystals. II: The conduction band and ultra-violet absorption of alkali–halide crystals", *Trans. Faraday Soc.* **34**, 500–6 (1938).

MOTT, N. F.: "On the transition to metallic conduction in semiconductors", *Can. J. Phys.* **34**, 1356–68 (1956).

MOTT, N. F.: "Conduction in non-crystalline systems—non-ohmic behaviour and switching", *Phil. Mag.* **24**, 911–34 (1971).

MOTT, N. F.: "The metal–insulator transition in extrinsic semiconductors", *Adv. Phys.* **21**, 785–823 (1972).

MOTT, N. F.: "Conduction in non-crystalline systems. IX: The minimum metallic conductivity", *Phil. Mag.* **26**, 1015–26 (1972).

MOTT, N. F., DAVIS, E. A., and STREET, R. A.: "States in the gap and recombination in amorphous semiconductors", *Phil. Mag.* **32**, 961–96 (1975).

MOTT, N. F. and LITTLETON, M. J.: "Conduction in polar crystals. I: Electrolytic conduction in solid salts", *Trans. Faraday Soc.* **34**, 485–99 (1938).

MOTT, N. F. and TWOSE, W. D.: "The theory of impurity conduction", *Adv. Phys.* **10**, 107–63 (1961).

MOZUMDER, A.: "Effect of an external electric field on the yield of free ions. I: General results from the Onsager theory", *J. Chem. Phys.* **60**, 4300–4 (1974).

MOZUMDER, A.: "Effect of an external electric field on the yield of free ions. II: The initial distribution of ion pairs in liquid hydrocarbons", *J. Chem. Phys.* **60**, 4305–10 (1974).

MULDER, B. J.: "Mean diffusion path of excitons in crystals of anthracene doped with tetracene", *Philips Res. Rept.* **21**, 283–8 (1966).

MULDER, B. J.: "Photoinjection of electrons into anthracene from electrolytic electrodes", *Solid State Commun.* **4**, 615–17 (1966).

MULDER, B. J.: "Anisotropy of light absorption and exciton diffusion in anthracene crystals determined from externally sensitized fluorescence", *Philips Res. Rept.* **22**, 142–9 (1967).

MULDER, B. J.: "Sensitized photoconduction of anthracene", *Philips Res. Rept.* **22**, 553–67 (1967).

MULDER, B. J.: "Symmetry of fluorescence and absorption spectra of anthracene crystals", *J. Phys. Chem. Solids* **29**, 182–4 (1968).

MULDER, B. J.: "Diffusion and surface reactions of singlet excitons in anthracene", *Philips Res. Rept.* Suppl. N4, 1–128 (1968).

MULDER, B. J. and DE JONGE, J.: "Photoconductivity of crystals of anthracene doped with tetracene and acridine", *Rec. Trev. Chim.* **84**, 1503–10 (1965).

MULDER, B. J. and DE JONGE, J.: "Exciton diffusion and the photoconductivity spectrum of anthracene, pyrene and perylene", *Philips Res. Rep.* **21**, 188–95 (1966).

MULDER, B. J. and DE JONGE, J.: "Electronic absorption spectrum of holes in anthracene", *Solid State Commun.* **5**, 203–5 (1967).

MULDER, B. J., DE JONGE, J., and VERMEULEN, G.: "Photocurrent in anthracene crystals under illumination of the negative electrode", *Rec. Trev. Chim.* **B 5**, 31–34 (1966).

MÜLLER, H. P., BÄESSLER, H., and VAUBEL, G.: "Emission spectra of non-crystalline tetracene films", *Chem. Phys. Lett.* **29**, 102–5 (1974).

MÜLLER, H. P., THOMA, P., and VAUBEL, G.: "The phosphorescence of anthracene single crystals and its spectrum", *Phys. State. Sol.* **23**, 253–62 (1967).

MULLER, R. S.: "A unified approach to the theory of space-charge-limited currents in an insulator with traps", *Solid-State Electron.* **6**, 25–32 (1963).

MUNN, R. W.: "Lattice dynamics of a linear chain model of an imperfect molecular crystal", *J. Chem. Phys.* **52**, 64–66 (1970).

MUNN, R. W.: "The local field and charge-carrier mobilities in molecular crystals", *Chem. Phys. Lett.* **16**, 429–31 (1972).

MUNN, R. W.: "Direct calculation of exciton diffusion coefficient in molecular crystals", *J. Chem. Phys.* **58**, 3230–2 (1973).

MUNN, R. W.: "Thermodynamic and physical properties of solids in electric fields", *J. Phys. C* **6**, 3213–32 (1973).

MUNN, R. W.: "Local-field effects on carrier hopping mobilities", *J. Phys. C* **8**, 2721–8 (1975).

MUNN, R. W.: "Causes of carrier trapping asymmetry in aromatic hydrocarbon crystals", *Mol. Cryst. Liq. Cryst.* **31**, 105–13 (1975).

MUNN, R. W.: "Grüneisen parameter of molecular crystals", *Phys. Rev.* **12B**, 3491–3 (1975).

MUNN, R. W., NICHOLSON, J. R., SCHWOB, H. P., WILLIAMS, D. F.: "Dielectric tensor of anthracene as a function of temperature and pressure", *J. Chem. Phys.* **58**, 3828–32 (1973).

MUNN, R. W., NICHOLSON, J. R., SIEBRAND, W., and WILLIAMS, D. F.: "Evidence for an isotope effect on electron drift mobilities in anthracene crystals", *J. Chem. Phys.* **52**, 6442–3 (1970).

MUNN, R. W. and PAWLEY, G. S.: "Debye temperature of anthracene", *Phys. Stat. Sol.* (b) **50**, K11–15 (1972).

MUNN, R. W. and SIEBRAND, W.: "Phonon-limited transport of charge carriers", *Chem. Phys. Lett.* **3**, 655–7 (1969).

MUNN, R. W. and SIEBRAND, W.: "Sign of the Hall effect for hopping transport in molecular crystals", *Phys. Rev. B* **2**, 3435–7 (1970).

MUNN, R. W. and SIEBRAND, W.: "Transport of quasilocalized excitons in molecular crystals", *J. Chem. Phys.* **52**, 47–63 (1970).

MUNN, R. W. and SIEBRAND, W.: "Theory of charge carrier transport in aromatic hydrocarbon crystals", *J. Chem. Phys.* **52**, 6391–406 (1970).

MUNN, R. W. and SIEBRAND, W.: "Theory of the Hall effect in aromatic hydro-carbon crystals", *J. Chem. Phys.* **53**, 3343–57 (1970).

MUNN, R. W. and WILLIAMS, D. F.: "Dielectric tensor and the local electric field in naphthalene", *J. Chem. Phys.* **59**, 1742–6 (1973).

MUNRO, I. H., LOGAN, L. M., BLAIR, F. D., LIPSETT, F. R., and WILLIAMS, D. F.: "Role of defects and reabsorption in the decay of fluorescence of anthracene from 2–350°K", *Mol. Cryst. Liq. Cryst.* **15**, 297–310 (1972).

MURGATROYD, P. N.: "Theory of space-charge-limited current enhanced by Frenkel effect", *J. Phys. D* **3**, 151–6 (1970).

MURGATROYD, P. N.: "Dimensional considerations for space-charge conduction in solids", *J. Phys. D* **3**, 1488–90 (1970).

MURGATROYD, P. N.: "Saturation of reservoir contacts", *Phys. Stat. Sol.* (a) **6**, 217–21 (1971).

MURGATROYD, P. N.: "Theory of space-charge-limited currents with large drift velocities", *Phys. Stat. Sol.* (a) **8**, 259–65 (1971).

MURGATROYD, P. N.: "On the theory of the transition from space-charge-limited to electrode-limited current", *Phys. Stat. Sol.* (a) **11**, 137–43 (1972).

MURGATROYD, P. N.: "On an approximation in space-charge current theory", *Solid-State Electron.* **16**, 287–8 (1973).

MURGATROYD, P. N.: "The general scaling rule for space-charge currents", *Thin Solid Film* **17**, 335–43 (1973).

MURGATROYD, P. N.: "Simple models of injecting metal–insulator contacts", *Am. J. Phys.* **42**, 677–88 (1974).

MURPHY, E. L. and GOOD, R. H.: "Thermionic emission, field emission and the transition region", *Phys. Rev.* **102**, 1464–73 (1956).

MURRELL, J.: "Mobilities of holes and electrons in organic crystals", *Molec. Phys.* **4**, 205–8 (1961).

MURRELL, J. N.: "The photoconductivity of molecular crystals", *Discuss. Faraday Soc.* **28**, 36–47 (1959).

MUSHER, J. I.: "On the nonexistence of ring current in aromatic hydrocarbons", *J. Chem. Phys.* **46**, 1219–21 (1967).

NADKARNI, G. S. and SIMMONS, J. G.: "Determination of the defect nature of MoO_3 films using dielectric-relaxation currents", *J. Appl. Phys.* **43**, 3650–6 (1972).

NADKARNI, G. S. and SIMMONS, J. G.: "Electrical properties and *I–V* characteristics of MoO_3 film under d.c. bias", *J. Appl. Phys.* **43**, 3741–7 (1972).

NADKARNI, G. S. and SIMMONS, J. G.: "Theory and analyses of the a.c. characteristics of defect thin-film insulators", *J. Appl. Phys.* **47**, 114–19 (1976).

NADKARNI, G. S. and SIMMONS, J. G.: "On the theory and analyses of the a.c. characteristics of defect thin-films", *J. Appl. Phys.* **47**, 4223 (1976).

NAITO, M.: "Magnetic field dependence of delayed fluorescence in anthracene crystal", *J. Phys. Soc. Japan* **28**, 796 (1970).

NAKADA, I.: "The optical properties of anthracene single crystals", *J. Phys. Soc. Japan* **17**, 113–18 (1962).

NAKADA, I.: "The optical absorption of anthracene crystal near 4000 Å", *J. Phys. Soc. Japan* **20**, 346–50 (1965).

NAKADA, I.: "The excited state and charge generation in anthracene", in *Third Organic Crystal Symposium, Chicago, Illinois, USA*, paper 35 (1965).

NAKADA, I., ANGA, K., and ICHIMAYA, A.: "The electrical conductivity of anthracene", *J. Phys. Soc. Japan* **19**, 1587–91 (1964).

NAKADA, I. and ISHIHARA, Y.: "The effects of temperature and electric field for the photo-generation of free carriers in anthracene", *J. Phys. Soc. Japan* **19**, 615–701 (1964).

NAKADA, I. and KAIYO, H.: "Effect of gases on the photoconduction of anthracene", *J. Phys. Soc. Japan* **17**, 93–99 (1962).

NAKADA, I. and KOJIMA, T.: "On the surface photo-voltaic effect in anthracene", *J. Phys. Soc. Japan* **19**, 695–701 (1964).

NAKADA, I. and OYAMA, K.: "Oxygen-induced photo-hole injection in anthracene single crystals", *J. Phys. Soc. Japan* **20**, 2299 (1965).

NAKASHIMA, K. and KAO, K. C.: "Conducting filaments and switching phenomena in glassy chalcogenide semiconductors", *J. Non-cryst. Solids* **33**, 189–204 (1979).

NAKAYAMA, T. and ITOH, N.: "Optical excitation of trapped electrons into continuum in naphthalene single crystal", *Solid State Commun.* **16**, 635–8 (1975).

NATHOO, M. H. and JONSCHER, A. K.: "High-field and a.c. properties of stearic acid films", *J. Phys. C* **4**, L301–4 (1971).

NELSON, O. L. and ANDERSON, D. E.: "Hot-electron transfer through thin film Al–Al$_2$O$_3$ triodes", *J. Appl. Phys.* **37**, 66–76 (1966).

NELSON, R. C.: "Contact potential difference between sensitizing dye and substrate", *J. Opt. Soc. Am.* **46**, 1016–19 (1956).

NELSON, R. C.: "Some photoelectric properties of chlorophyll", *J. Chem. Phys.* **27**, 864–7 (1957).

NELSON, R. C.: "Sensitization by dyes", *J. Opt. Soc. Am.* **48**, 1–3 (1958).

NELSON, R. C.: "Photoelectric phenomena in hemoglobin and dyed gelatin", *J. Chem. Phys.* **39**, 112–15 (1963).

NESPUREK, A.: "Space-charge-limited currents in *N, M′*-diphenyl-*p*-phenylenediamine and Gaussian distribution of traps", *Czech. J. Phys. B* **24**, 660–70 (1974).

NESPUREK, A. and SEMEJTEK, P.: "Space-charge limited currents in insulators with the Gaussian distribution of traps", *Czech. J. Phys. B* **22**, 160–75 (1972).

NICHOLAS, K. H. and WOODS, J.: "The evaluation of electron trapping parameters from conductivity glow curves in cadmium sulphide", *Br. J. Appl. Phys.* **15**, 783–95 (1964).

NICKEL, B.: "Sensitized photoconduction and sensitized delayed fluorescence anthracene single crystals", *Mol. Cryst. Liq. Cryst.* **18**, 227–61 (1972).

NICKEL, B.: "Intensity dependence of the sensitized delayed fluorescence of anthracene single crystals", *Mol. Cryst. Liq. Cryst.* **18**, 263–77 (1972).

NICKEL, B.: "Delayed fluorescence from higher excited singlet states of 1,2-benzanthracene and fluoranthene", *Chem. Phys. Lett.* **27**, 84–90 (1974).

NICKEL, B. and MAXDORF, H.: "A new method for the determination of the diffusion coefficient of triplet excitons in the *c′*-direction of crystalline anthracene", *Chem. Phys. Lett.* **9**, 555–8 (1971).

NICOL, M.: "Fluorescence of anthracene in lucite at very high pressures", *J. Opt. Soc. Am.* **55**, 1176–8 (1965).

NICOLET, M. A.: "Unipolar space-charge-limited current in solids with nonuniform spacial distribution of shallow traps", *J. Appl. Phys.* **37**, 4224–35 (1966).

NICOLET, M. A., BILGER, H. R., and ZIJLSTRA, R. J. J.: "Noise in single and double injection currents in solids (I)", *Phys. Stat. Sol.* (b) **70**, 9–45 (1975).

NICOLET, M. A., BILGER, H. R., ZIJLSTRA, R. J. J.: "Noise in single and double injection currents in solids (II)", *Phys. Stat. Sol.* **70**, 415–83 (1975).

NICOLET, M. A., RODRIQUEZ, V., and STOLFA, D.: "Unipolar interface-charge-limited current", *Surf. Sci.* **10**, 146–64 (1968).

NICOLLIAN, E. H. and GOETZBERGER, A.: "Lateral a.c. current flow model for metal–insulator–semiconductor capacitors", *IEEE Trans. Electron. Devices* ED **12**, 108–17 (1965).

NIEMAN, G. C. and ROBINSON, G. W.: "Rapid triplet excitation migration in organic crystals", *J. Chem. Phys.* **37**, 2150–1 (1962).

NORDMAN, J. E. and GREINER, R. A.: "The small-signal inductive effect in a long p–i–n diode", *IEEE Trans. Electron. Devices* ED **10**, 171–7 (1963).

NORDMAN, J. E. and KVINLAUG, H.: "Negative-resistance current–voltage characteristics of an indium antimonide p^+–p–n^+ diode", *J. Appl. Phys.* **39**, 3244–50 (1968).

NORRELL, C. J., POHL, H. A., THOMAS, M., and BERLIN, K. D.: "Electronic conduction and polarization in polyphthalocyanines", *J. Polymer Sci. Polymer Phys. Ed.* **12**, 913–24 (1974).

NORTHROP, D. C.: "Conduction processes in condensed aromatic hydrocarbons", *Proc. Phy. Soc. (London)* **74**, 756–61 (1959).

NORTHROP, D. C. and SIMPSON, O.: "Electrical properties of aromatic hydrocarbons. I: Electrical conductivity", *Proc. Roy. Soc. (London)* A **234**, 124–35 (1956).

NORTHROP, D. C. and SIMPSON, O.: "Electronic properties of aromatic hydrocarbons. II: Fluorescence transfer in solid solutions", *Proc. Roy. Soc. (London)* A **234**, 136–49 (1956).

NORTHROP, D. C. and SIMPSON, O.: "Electronic properties of aromatic hydrocarbons. IV: Photo-electric effects", *Proc. Roy. Soc. (London)* A **244**, 377–89 (1958).

OBRECIMOV, I. V., PRIKHOTJKO, A. F., and RODNIKOVA, O. I.: "Dispersion of anthracene in the visible part of the spectrum", *Zh. Eksp. Teor. Fiz. (Soviet Phys.—JETP)* **18**, 409 (1948).

O'DWYER, J. J.: "The criterion for dielectric breakdown in ionic crystals", *Aust. J. Phys.* **7**, 36–48 (1954).

O'DWYER, J. J.: "The high temperature dielectric breakdown of alkali halides", *Proc. Phys. Soc. (London)* B **70**, 761–8 (1957).

O'DWYER, J. J.: "Current–voltage characteristics of dielectric films", *J. Appl. Phys.* **37**, 599–601 (1966).

O'DWYER, J. J.: "Theory of double charge ejection from a dielectric", *J. Appl. Phys.* **39**, 4356–9 (1968).

O'DWYER, J. J.: "Theory of high field conduction in a dielectric", *J. Appl. Phys.* **40**, 3887–90 (1969).

O'DWYER, J. J.: "Two-carrier model for high field conduction in SiO_2", *J. Appl. Phys.* **44**, 5438–41 (1973).

OFFEN, H. W.: "Absorption spectrum of solid anthracene–TNB complex under pressure", *J. Chem. Phys.* **42**, 430 (1965).

OFFEN, H. W.: "Fluorescence spectra of several aromatic crystals under high pressures", *J. Chem. Phys.* **44**, 699–703 (1966).

OFFEN, H. W. and PHILLIPS, D. T.: "Fluorescence lifetimes of aromatic hydrocarbons under pressures", *J. Chem. Phys.* **49**, 3995–7 (1968).

OHIGASHI, H., SHIROTANI, I., INOKUCHI, H., and MINOMURA, S.: "Pressure effect on energy transfer in anthracene–tetracene mixed crystals", *J. Phys. Soc. Japan* **19**, 1996–7 (1964).

OHIGASHI, H., SHIROTANI, I., INOKUCHI, H., and MINOMURA, S.: "Anomalous fluorescence of pure anthracene crystal under high pressure", *Mol. Cryst.* **1**, 463–6 (1966).

OHKI, K., INOKUCHI, H., and MARUYAMA, Y.: "Charge mobility in pyrene crystals", *Bull. Chem. Soc. Japan* **36**, 1512–15 (1963).

OKAMOTO, K., CHANG, I. Y., and KANTOR, M. A.: "Effect of pressure on the electrical resistance of ferrocene molecule", *J. Chem. Phys.* **41**, 4010–11 (1964).

OKAMOTO, K., ODA, N., ITAYA, A., and KUSABAYSHI, S.: "Magnetic field effect on the photoconductivity of poly-n-vinylcarbazole", *Chem. Phys. Lett.* **35**, 483–6 (1975).

OKUMURA, K.: "Photovoltaic effects at the interface between amorphous selenium and organic polymers", *J. Appl. Phys.* **45**, 5317–23 (1974).

ONEAL, G.: "Hot finger for zone refining anthracene", *Rev. Sci. Instrum.* **33**, 490 (1962).

ONIPKO, A. I. and SUGAKOV, V. I.: "Interaction of excitons and exciton annihilation", *Opt. Spectrosc.* **35**, 108–9 (1973).

ONISHI, H., KUROKAWA, S., and LEYANSU, K.: "Photovoltaic polarity of CdTe films obliquely deposited in vacuum", *J. Appl. Phys.* **45**, 3205–6 (1974).

ONO, T., KIMURA, M., and MIYAMOTO, T.: "Selective epitaxial growth of single crystal anthracene", *J. Appl. Phys.* **48**, 2102–3 (1977).

ONODERA, A., SHIROTANI, I., INOKUCHI, H., and KAWAI, N.: "Conductive tetraselenonaphthacene and iodanil under very high pressures", *Chem. Phys. Lett.* **25**, 296–8 (1974).

ONSAGER, L.: "Deviations from Ohm's law in weak electrolytes", *J. Chem. Phys.* **2**, 599–615 (1934).

ONSAGER, L.: "Initial recombination of ions", *Phys. Rev.* **54**, 554–7 (1938).

O'REILLY, T. J. and DELUCIA, J.: "Injection current flow through thin insulator films", *Solid-State Electron.* **18**, 965–8 (1975).

ORLANDI, G. and SIEBRAND, W. G.: "On the perturbation description of radiationless transitions", *Chem. Phys. Lett.* **14**, 19–22 (1972).

ORMONDROYD, R. F., THOMPSON, M. J., and ALLISON, J.: "Effect of composition and forming parameters on conductance of amorphous chalcogenide threshold switches", *J. Non-Cryst. Solids* **18**, 375–93 (1975).

OSIPOV, V. V. and KHOLODNOV, V. A.: "Current filamentation in diodes due to double injection", *Soviet Phys.—Semicond.* **7**, 604–9 (1973).

Osugi, J. and Hara, K.: "Effect of pressure on the electrical conductivity of organic substances. I: Pyrolyzed polyacrylonitrile", *Rev. Phys. Chem. Japan* **36**, 20–27 (1966).

Osugi, J. and Hara, K.: "Effect of pressure on the electrical conductivity of organic substances. II: α, α'-Diphenyl-β-picryl hydrazyl", *Rev. Phys. Chem. Japan* **36**, 81–87 (1966).

Otani, Y., Matsubara, K., and Nishida, Y.: "Effects of a magnetic field on double-injection negative resistance in long $p^+-\pi-n^+$ structures", *J. Appl. Phys.* **41**, 4711–17 (1970).

Ottaviani, G., Canali, C., Nava, F., and Mayer, J. W.: "Hot drift velocity in high purity Ge between 8°K and 220°K", *J. Appl. Phys.* **44**, 2917–18 (1973).

Owen, A. E. and Robertson, J. M.: "Electronic conduction and switching in chalcogenide glasses", *IEEE Trans. Electron Devices* ED **20**, 105–22 (1973).

Owen, G. P. and Charlesby, A.: "On charge-carrier trapping in insulating solids as investigated by the space-charge-limited current (SCLC) technique: a cautionary note", *J. Phys.* C **7**, L400–2 (1974).

Owen, G. P., Sworakowski, J., Thomas, J. M., Williams, D. F., and Williams, J. O., "Carrier traps in ultra-high purity single crystals of anthracene", *J. Chem. Soc. Trans. Faraday* II, **70**, 853–61 (1974).

Oyama, K. and Nakada, I.: "The trapping of photo-carriers in anthracene crystals by tetracene", *J. Phys. Soc. Japan* **24**, 792–7 (1968).

Oyama, K. and Nakada, I.: "Photoconduction of 9,10-anthraquinone single crystals", *J. Phys. Soc. Japan* **24**, 798–805 (1968).

Padhye, M. R., McGlynn, S. P., and Kasha, M.: "Lowest triplet level of anthracene", *J. Chem. Phys.* **24**, 588–94 (1956).

Padovani, F. A.: "The voltage–current characteristic of metal–semiconductor contacts", in *Semiconductors and Semimetals* (eds. R. K. Willardson and A. C. Beer), Academic Press (New York), Vol. **7A** (1971).

Padovani, F. A. and Stratton, R.: "Field and thermionic-field emission in Schottky barriers", *Solid-State Electron.* **9**, 695–707 (1966).

Page, D. J.: "Some computed and measured characteristics of CdS space-charge-limited diodes", *Solid-State Electron.* **9**, 255–64 (1966).

Pai, D. M.: "Electric-field-enhanced conductivity in solids", *J. Appl. Phys.* **46**, 5122–6 (1975).

Pai, D. M. and Enck, R. C.: "Onsager mechanism of photogeneration in amorphous selenium", *Phys. Rev.* **B11**, 5163–74 (1975).

Paige, E. G. S.: "The electrical conductivity of germanium", in *Progress in Semiconductors* (ed. A. F. Gibson), Wiley (New York), Vol. **8** (1964).

Palatnik, L. S., Petrenko, L. G., and Kopeliovich, A. I.: "Anomalous photovoltaic effect in single films of lead sulfide", *Soviet Phys.—Semicond.* **9**, 559–62 (1975).

Papadakis, A. C.: "Theory of transient space-charge perturbed currents in insulators", *J. Phys. Chem. Solids* **28**, 641–7 (1967).

Paranjape, V. V.: "Low temperature electric field effects in semiconductors", *Proc. Phys. Soc. (London)* **78**, 516–28 (1961).

Pariser, R. and Parr, R. G.: "Theory of the electronic spectra and electronic structure of complex unsaturated molecules. I", *J. Chem. Phys.* **21**, 466–71 (1953).

Pariser, R. and Parr, R. G.: "Theory of the electronic spectra and electronic structure of complex unsaturated molecules. II", *J. Chem. Phys.* **21**, 767–76 (1953).

Paritskii, I. G. and Rozental, A. I.: "Influence of the charge exchange in impurity centers on the space-charge-limited current", *Soviet Phys.—Semicond.* **1**, 210–15 (1967).

Paritskii, I. G. and Rozental, A. I.: "Effects of an inhomogeneous volume distribution of trapping centers on unipolar injection in semiconductors", *Soviet Phys.—Semicond.* **4**, 446–9 (1970).

Parker, C. A. and Hatchard, C. G.: "Delayed fluorescence from solutions of anthracene and phenanthrene", *Proc. Roy. Soc. (London)* A **269**, 574–84 (1960).

Parkinson, G. M., Thomas, J. M., and Williams, J. O.: "Analysis of trap depths in anthracene by thermal and optical release of injected charge carriers", *J. Phys.* C **7**, L310–13 (1974).

Parmenter, R. H. and Ruppel, W.: "Two-carrier space-charge-limited current in a trap-free insulator", *J. Appl. Phys.* **30**, 1548–58 (1959).

Parr, R. G. and Pariser, R.: "On the electronic structure and electronic spectra of ethylene-like molecules", *J. Chem. Phys.* **23**, 711–25 (1955).

Parrott, J. E.: "Reformulation of basic semiconductor transport equations. II", *Solid-State Electron.* **17**, 707–16 (1974).

Parrott, J. E.: "The saturated photovoltage of a *p–n* junction", *IEEE Trans. Electron. Devices* ED **21**, 89–93 (1974).

Patterson, W. R.: "Effects of ohmic contacts on the Dember voltage", *J. Appl. Phys.* **39**, 4034–5 (1968).

Paul, D. E., Lipkin, D., and Weissman, S. I.: "Reactions of sodium metal with aromatic hydrocarbons", *J. Chem. Phys.* **78**, 116 (1956).

PAWLEY, G. S.: "Model for the lattice dynamics of naphthalene and anthracene", *Phys. Stat. Sol.* **20**, 347–60 (1967).

PAWLEY, G. S. and MIKA, K.: "Dynamics of molecular crystals under pressure", *Phys. Stat. Sol.* B **66**, 679–86 (1974).

PEIERLS, R.: "Zur Theorie der Absorptionsspektren fester Körper", *Ann. Phys.* **13**, 905–52 (1932).

PEKAR, S. I.: "The theory of electromagnetic waves in crystal in which excitons are produced", *Soviet Phys.—JETP* **6**, 785–96 (1958).

PEKAR, S. I.: "Dispersion of light in the exciton absorption region of crystals", *Soviet Phys.—JETP* **7**, 813–22 (1958).

PEKAR, S. I.: "On the theory of absorption and dispersion of light in crystals", *Soviet Phys.—JETP* **36**, 314–23 (1959).

PELLEGRINI, B.: "New quantum and electronic theory of metal–semiconductor contacts", *Phys. Rev.* B **7**, 5299–312 (1973).

PELLEGRINI, B.: "A detailed analysis of the metal–semiconductor contact", *Solid-State Electron.* **17**, 217–37 (1974).

PELLEGRINI, B. and SALARDI, G.: "A model of ohmic contacts to semiconductors", *Solid-State Electron.* **18**, 791–8 (1975).

PENCHINA, C. M., MOORE, J. S., and HOLONYAK, N.: "Energy levels and negative photoconductivity in cobalt-doped silicon", *Phys. Rev.* **143**, 634–6 (1966).

PENDER, L. F. and FLEMING, R. J.: "Memory switching in glow discharge polymerized thin films", *J. Appl. Phys.* **46**, 3426–31 (1975).

PERKAMPUS, H. H. and POHL, L.: "On a tempering effect in the fluorescence of thin anthracene films", *Z. Phys. Chem.* **39**, 397–401 (1963).

PERKINS, W. G.: "Prompt and delayed fluorescence of X-ray excited anthracene crystals", *J. Chem. Phys.* **48**, 931–8 (1968).

PERLMAN, M. M., SONNONSTINE, J. J., and PIERRE, J. A.: "Drift mobility determinations using surface-potential decay in insulators", *J. Appl. Phys.* **47**, 5016–21 (1976).

PERNISZ, U.: "Evidence of surface states of an anthracene crystal", in *International Meeting on Electrostatic Charging, Frankfurt, Germany, 29 March 1973 (Weinheim/Bergstrasse)*, Verlag-Chemie (Germany), 53–62 (1974).

PERNISZ, U. and BAUSER, H.: "Field-effect mobility and surface states of the anthracene crystal", *Seventh Molecular Crystal Symposium, Nikko, Japan, 8–12 September*, 133—6 (1975).

PETER, L. and VAUBEL, G.: "Triplet exciton lifetime in crystalline pyrene", *Chem. Phys. Lett.* **18**, 531–4 (1973).

PETER, L. and VAUBEL, G.: "Host–guest singlet–triplet interaction in pyrene-doped fluoranthene", *Chem. Phys. Lett.* **23**, 75–78 (1973).

PETERSEN, K. E. and ADLER, D.: "Electronic nature of amorphous threshold switching", *Appl. Phys. Lett.* **27**, 625–7 (1975).

PETERSON, O. G., WEBB, J. P., and McCOLGIN, W. C.: "Organic dye laser threshold", *J. Appl. Phys.* **42**, 1917–28 (1972).

PETHIG, R. and HAYWARD, D.: "Frequency dependent conductivity of anthracene single crystals in the range 10^{-6} Hz to 32 GHz", *Phys. Stat. Sol.* A **24**, K23–26 (1974).

PETHIG, R. and MORGAN, K.: "D.C. dark Hall mobility measurements on anthracene", *Nature* **214**, 266–7 (1967).

PETHIG, R. and MORGAN, K.: "Hall mobility of electrons in anthracene crystals", *Phys. Stat. Sol.* (b) **43**, K119–21 (1971).

PETICOLAS, W. L., GOLDSBOROUGH, J. P., and RIECKHOFF, K. E.: "Double photon excitation in organic crystals", *Phys. Rev. Lett.* **10**, 43–45 (1963).

PETICOLAS, W. L., NORRIS, R., and RIECKHOFF, K. E.: "Polarization effects in the two-photon excitation of anthracene fluorescence", *J. Chem. Phys.* **42**, 4164–9 (1965).

PETICOLAS, W. L. and RIECKHOFF, K. E.: "Polarization of anthracene fluorescence by one-and two-photon excitation", *Third Organic Crystal Symposium, Chicago (USA)*, paper No. 22 (1965), also *Phys. Lett.* **15**, 230–1 (1965).

PETRILLO, G. A. and KAO, K. C.: "The effects of electrode materials on the switching behaviour of the amorphous semiconductor Si_{12}-Ge_{12}-As_{30}-Te_{48}", *J. Non-Cryst. Solids* **16**, 247–57 (1974).

PETROVA, M. L. and ROZENSHTEIN, L. D.: "Field effect in the organic semiconductor chloranil", *Soviet Phys.—Solid State* **12**, 756–7 (1970).

PETROVA, M. L. and ROZENSHTEIN, L. D.: "Field effect and slow states in thin films of organic semiconductors", *Soviet Phys.—Solid State* **13**, 1948–52 (1972).

PFISTER, G. and WILLIAMS, D. J.: "Photogeneration processes in poly(*N*-vinylcarbazole)", *J. Chem. Phys.* **61**, 2516–26 (1974).

PFISTER, J. C.: "Note on the interpretation of space charge limited currents with traps", *Phys. Stat. Sol.* A **24**, K15–17 (1974).

PHILPOTT, M. R.: "Exciton–photon coupling in crystalline anthracene", *J. Chem. Phys.* **50**, 3925–9 (1969).

PHILPOTT, M. R.: "Exciton transitions in crystalline anthracene with K perpendicular to the (010) plane", *Chem. Phys. Lett.* **17**, 57–62 (1972).

PICCINI, A. and WHITTEN, W. B.: "Annealing behavior of the 7070 Å color center and ESR spectrum in irradiated naphthalene crystals", *Mol. Cryst. Liq. Cryst.* **18**, 333–7 (1972).

PICK, H. and WISSMANN, H.: "Elektronenleitung von Naphthalin-Einkristallen", *Z. Phys.* **138**, 436–40 (1954).

PICKARD, P. S. and PAVIS, M. V.: "Analysis of electron trapping in alumina using thermally stimulated electrical currents", *J. Appl. Phys.* **41**, 2636–43 (1970).

PICKETT, L. W.: "An X-ray study of substituted biphenyls", *J. Am. Chem. Soc.* **58**, 2299–303 (1936).

PIGON, K. and CHOJNACKI, H.: "Ratio and ranges of free carriers in anthracene surface cell", *J. Chem. Phys.* **31**, 272–3 (1959).

PITTELLI, E.: "Tunnel emission into an insulating film with traps", *Solid-State Electron.* **6**, 667–71 (1963).

PITTELLI, E. and RINDNER, W.: "Tunnel and excess currents in stressed Esaki diodes", *Solid-State Electron.* **'10**, 911–16 (1967).

PLOTNIKOV, YU. I.: "The temperature dependence of dark current in anthracene", *Fiz. Tverd. Tela* **4**, 3104–9 (1962).

PLOTNIKOV, YU. I. and MATALYGINA, ZH. I.: "Photoelectromotive force in anthracene", *Soviet Phys.— Solid State* **2**, 2244–51 (1961).

POEHLER, T. O., BLOCK, A. N., FERRARIS, J. P., and COWAN, D. O.: "Far infrared photoconductivity of TFF–CCNQ", *Solid State Commun.* **15**, 337–40 (1974).

POHL, H. A.: "Semiconductor in polymers", in *Modern Aspects of the Vitreous States* (ed. J. D. Mackenzie), Butterworths (London) **2**, 72–113 (1962).

POHL, H. A.: "Theories of electronic behavior in macromolecular solids", *J. Polymer Sci.* C **17**, 13–40 (1967).

POHL, H. A.: "On the possibility of an organic metal", *Phil. Mag.* **26**, 593–600 (1972).

POHL, H. A. and ENGELHANDT, E. H.: "Synthesis and characterization of some highly conjugated semiconducting polymers", *J. Phys. Chem.* **66**, 2085–95 (1962).

POHL, H. A., GOGOS, C. G., and CAPPAS, C.: "Resistivity studies on polymer semiconductors", *J. Polymer Sci.* Part A, **1**, 2207–12 (1963).

POHL, H. A. and OPP, D. A.: "The nature of semiconduction in some acene quinone radical polymers", *J. Phys. Chem.* **66**, 2121–6 (1962).

POHL, H. A., REMBAUM, A., and HENRY, A.: "Effects of high pressure on some organic semiconducting polymers", *J. Am. Chem. Soc.* **84**, 2699–704 (1962).

POLANCO, J. I. and ROBERTS, G. G.: "Thermally assisted tunneling in dielectric films (II)", *Phys. Stat. Sol.* (a) **13**, 603–6 (1972).

POLLACK, S. R.: "Schottky field emission through thin insulating layers", *J. Appl. Phys.* **34**, 877–80 (1963).

POLLACK, S. R. and MORRIS, C. E.: "Electron tunneling through asymmetric films of thermally grown Al_2O_3", *J. Appl. Phys.* **35**, 1503–12 (1964).

POLLACK, S. R. and MORRIS, C. E.: "On Mott's theory of formation of protective oxide films", *Solid State Commun.* **2**, 21–22 (1964).

POLLACK, S. R. and MORRIS, C. E.: "Tunneling through gaseous oxidized films of Al_2O_3", *Trans. Met. Soc. AIME* **233**, 497–501 (1965).

POLLACK, S. M. and SEITCHIK, J. A.: "Electron transport through insulating thin films", in *Applied Solid State Science* (ed. R. Wolfe) **1**, 343–83 (1969).

POLLAK, M.: "Some aspects of non-steady-state conduction in bands and hopping processes", in *Physics of Semiconductors* (Conference at Exeter, Institute of Physics and Physical Society, England) 86–93 (1962).

POLLAK, M.: "Approximations for the a.c. impurity hopping conduction", *Phys. Rev.* **133**, A564–79 (1964).

POLLAK, M.: "Temperature dependence of a.c. hopping conductivity", *Phys. Rev.* **138**, A1822–6 (1965).

POLLAK, M.: "On the frequency dependence of conductivity in amorphous solids", *Phil. Mag.* **23**, 519–42 (1971).

POLLAK, M. and GEBALLE, T. H.: "Low-frequency conductivity due to hopping processes in silicon", *Phys. Rev.* **122**, 1743–53 (1961).

POLLAK, M. and PIKE, G. E.: "A.C. conductivity of glasses", *Phys. Rev. Lett.* **28**, 1449–51 (1972).

POOLE, H. H.: "On the dielectric constant and electrical conductivity of mica in intense fields", *Phil. Mag. (London)* **32**, 112–29 (1916).

POOLE, H. H.: "On the temperature variation of the electrical conductivity of mica", *Phil. Mag. (London)* **34**, 195–204 (1917).

POPE, J. A.: "Molecular orbital theory and crystals", in *Symposium of Electrical Conductivity in Organic Solids* (eds. H. Kallmann and M. Silver), Wiley Interscience (New York) 147 (1961).

POPE, M.: "Electrostatic determination of photo-ionization potentials of solids and liquids", *J. Chem. Phys.* **37**, 1001–3 (1962).

POPE, M.: "Electric currents in organic crystals", *Sci. Am.* **216**, 86–97 (1967).

POPE, M.: "Generation of charge carriers in anthracene with polarized light", *J. Chem. Phys.* **47**, 1194–5 (1967).

POPE, M.: "Autoionization in anthracene", *J. Chem. Phys.* **47**, 2197–8 (1967).

POPE, M.: "Photoelectric effects in thin organic crystals", in *133rd Meeting of the Electrochemical Society, Boston, Mass., USA, 5–9 May, 1968*, Abstracts—Electrochem. Soc. **8** (1968).

POPE, M.: "Charge—transfer exciton state, ionic energy levels, and delayed fluorescence in anthracene", *Mol. Cryst.* **4**, 183–90 (1968).

POPE, M.: "Charge transfer exciton and ionic levels in organic crystals", *J. Polymer Science* C **17**, 233–40 (1969).

POPE, M. and BURGOS, J.: "Charge-transfer excitons state and ionic energy level in anthracene crystal", *Mol. Cryst.* **1**, 395–415 (1966).

POPE, M. and BURGOS, J.: "Autoionization and exciton annihilation in anthracene", *Mol. Cryst.* **3**, 215–26 (1967).

POPE, M., BURGOS, J., and GIACHINO, J.: "Charge-transfer exciton state and energy levels in tetracene crystal", *J. Chem. Phys.* **43**, 3367–71 (1965).

POPE, M., BURGOS, J., and WOTHERSPOON, N.: "Singlet exciton–trapped carrier interaction in anthracene", *Chem. Phys. Lett.* **12**, 140–3 (1971).

POPE, M. and GEACINTOV, N. E.: "Excitement in excitons", *Ind. Res.* **11**, 68–70 (1969).

POPE, M., GEACINTOV, N. E., SAPERSTEIN, D., and VOGEL, F.: "Calculation of the diffusion length, diffusion coefficient and lifetime of triplet excitons in crystalline tetracene", *J. Lumin.* **1–2**, 224–30 (1970).

POPE, M., GEACINTOV, N. E., and VOGEL, F.: "Singlet exciton fission and triplet–triplet exciton fusion in crystalline tetracene", *Mol. Cryst. Liq. Cryst.* **6**, 83–104 (1969).

POPE, M. and KALLMANN, H.: "A.C. and d.c. photoconductivity in anthracene single crystals", in *Symposium on Electrical Conductivity in Organic Solids* (eds. H. Kallmann and M. Silver), Wiley Interscience (New York) 83–104 (1961).

POPE, M. and KALLMANN, H.: "Photoconductivity and semi-conductivity in organic crystals", *Discuss. Faraday Soc.* **51**, 7–16 (1971).

POPE, M., KALLMANN, H., CHEN, A., and GORDON, P.: "Charge injection into organic crystals—influence of electrode on dark and photoconductivity", *J. Chem. Phys.* **36**, 2486–90 (1962).

POPE, M., KALLMANN, H., and GIACHINO, J.: "Double-quantum external photoelectric effect in organic crystals", *J. Chem. Phys.* **42**, 2540–3 (1965).

POPE, M., KALLMANN, H., and MAGNANTE, P.: "Electroluminescence in organic crystals", *J. Chem. Phys.* **38**, 2042–3 (1963).

POPE, M. and SELSBY, R.: "Excitonic sounding: A proposed method for measuring a charge density gradient in anthracene", *Chem. Phys. Lett.* **14**, 226–30 (1972).

POPE, M. and SOLOWIEJCZYK, Y.: "Space charge controlled currents in insulators—detection of virtual anode", *Mol. Cryst. Liq. Cryst.* **30**, 175–200 (1975).

POPE, M. and WESTON, W.: "Voltage dependence of unipolar excess bulk charge density in organic insulators", *Mol. Cryst. Liq. Cryst.* **25**, 205–13, 501–13 (1974).

POPESCU, C.: "The effect of local non-uniformities on thermal switching and high field behaviour of structures with chalcogenide glasses", *Solid-State Electron.* **18**, 671–81 (1975).

POPESCU, C. and HENISCH, H. K.: "Minority carrier injection in relaxation semi-conductors", *Phys. Rev.* B **11**, 1563–8 (1975).

POPESCU, C. and HENISCH, H. K.: "Minority-carrier injection into semi-insulators", *Phys. Rev.* B **14**, 517–25 (1976).

POPESCU, C. and HENISCH, H. K.: "Minority carrier injection in relaxation semiconductors under illumination", *J. Phys. Chem. Solids* **37**, 47–49 (1976).

POPOVIC, Z. D. and SHARP, J. H.: "Flushed photoconductivity action spectra of β-metal free phthalocyanine thin films", *J. Chem. Phys.* **66**, 5076–82 (1977).

PORT, H. and WOLF, H. C.: "Phosphoreszenz und verzögerte Fluoreszenz von Naphthalin in verschiedenen Mischkristallen", *Z. Naturforsch.* **23a**, 315–29 (1968).

POTT, G. T. and WILLIAMS, D. F.: "Electron photoemission from anthracene crystals", *J. Chem. Phys.* **51**, 203–10 (1969).

POTT, G. T. and WILLIAMS, D. F.: "Low temperature electron injection and space-charge-limited transients in anthracene crystals", *J. Chem. Phys.* **51**, 1901–6 (1969).

POWELL, R. C.: "Host-sensitized energy transfer in molecular crystals", *Phys. Rev.* B **2**, 1159–67 (1970).

POWELL, R. C.: "Thermal and sample-size effects on the fluorescence lifetime and energy transfer in tetracene-doped anthracene", *Phys. Rev.* B **2**, 2090–7 (1970).

POWELL, R. C.: "Energy transfer in anthracene and tetracene doped naphthalene crystals", *Phys. Rev.* B **4**, 628–35 (1971).

POWELL, R. C.: "Singlet exciton migration and energy transfer in pyrene doped naphthalene", *J. Chem. Phys.* **58**, 920–5 (1973).

POWELL, R. C.: "Time resolved spectroscopy of anthracene and tetracene doped fluorene crystals", *J. Luminescence* **6**, 285–95 (1973).

POWELL, R. C. and KEPLER, R. G.: "Evidence for long-range exciton-impurity interaction in tetracene-doped anthracene crystals", *Phys. Rev. Lett.* **22**, 636–9 (1969) and Erratum, *Phys. Rev. Lett.* **22**, 1232 (1969).

POWELL, R. C. and KEPLER, R. G.: "Energy transfer in doped organic crystals", *Mol. Cryst. Liq. Cryst.* **11**, 349–60 (1970).

POWELL, R. C. and KEPLER, R. G.: "Comments on diffusion theory of luminescent emission from a doped solid", *Phys. Stat. Sol.* (B) **55**, K89–91 (1973).

POWELL, R. C. and SOOS, Z. G.: "Kinetic models for energy transfer", *Phys. Rev.* B **5**, 1547–56 (1972).

POWELL, R. C. and SOOS, Z. G.: "Singlet exciton energy transfer in organic solids", *J. Luminescence* **11**, 1–45 (1975).

PRICE, P. J. and RADCLIFFE, J. M.: "Esaki tunneling", *IBM Jl Res. Dev.* **3**, 364–71 (1959).

PRIKHOTJO, A. F. and FUGOL, I. YA.: "Luminescence of crystalline anthracene at 20° K", *Opt. i. Spektrosk.* **3**, 335 (1958).

PROBST, K. H. and KARL, N.: "Energy levels of electron and hole traps in the band gap of doped anthracene crystals", *Phys. Stat. Sol.* (a) **27**, 499–508 (1975).

PRUGLO, G. F., BLAGODAROV, A. N., ELIGULASBRILI, I. A., and LUTSENKO, E. L.: "Frequency dependent conductivity of anthracene single crystals under monopolar injection", *Phys. Stat. Sol.* (a) **12**, K47–50 (1972).

PRZYBYLAKA, M.: "Simple cutting device for organic crystals", *Rev. Sci. Instrum.* **34**, 183–4 (1963).

PULFREY, D. L. and McOUAT, R. F.: "Schottky-barrier solar cell calculations", *Appl. Phys. Lett.* **24**, 167–9 (1974).

PUTSEIKO, E. K.: "Application of the condenser method to the determination of the sign of the carrier of the photocurrent", *Dokl. Akad. Nauk SSSR* **67**, 1009 (1949); **59**, 471 (1948).

PUTSEIKO, E. K.: "Kinetics of photoconductivity of phthalocyanines", *Dokl. Akad. Nauk SSSR* **132**, 1299 (1960).

QUEISSER, H. J.: "Sign of the Hall coefficient in relaxation-case semiconductors", *J. Appl. Phys.* **42**, 5567–9 (1971).

QUEISSER, H. J.: "Semiconductors in the relaxation regime", *in Solid State Devices 1972* (hon. ed., P. N. Robson), Conference Series No. 15, Institute of Physics (London) 145–68 (1972).

QUEISSER, H. J.: "On the condition for recombinative space-charge injection", *J. Appl. Phys.* **43**, 3892 (1972).

QUEISSER, H. J.: "Relaxation-case diode in a magnetic field", *Solid-State Electron.* **15**, 358–60 (1972).

QUEISSER, H. J., CASEY, H. C., and VAN ROOSBROECK, W.: "Carrier transport and potential distributions for a semiconductor *p–n* junction in the relaxation regime", *Phys. Rev. Lett.* **26**, 551–4 (1971).

RABIE, S. and RUMIN, N.: "Characterization of trapping kinetics from the lifetime dependence of thermally stimulated conductivity spectra", *Appl. Phys. Lett.* **27**, 29–31 (1975).

RABIE, S. and RUMIN, N.: "Calculations and effects of inhomogeneity in compensated photoconductors", *Solid-State Electron.* **19**, 357–64 (1976).

RACKOVSKY, S.: "Electronic energy transfer in impure solids. II: One impurity interacting with a full lattice", *Mol. Phys.* **26**, 857–72 (1973).

RACKOVSKY, S. and SILBEY, R.: "Electronic energy transfer in impure solids. I: Two molecules embedded in a lattice", *Mol. Phys.* **25**, 61–72 (1973).

RADOMSKA, M.: "Electric conductivity of solid solutions of acridine in anthracene", *Acta Phys. Pol. A (Poland)* A **45**, 413–15 (1974).

RADOMSKA, M. and RADOMSKI, R.: "Phase diagram and growth of single crystals in the anthracene–acridine binary system", *Mol. Cryst. Liq. Cryst.* **18**, 75–81 (1972).

RADOMSKA, M., RADOMSKI, R., SWORAKOWSKI, J., CLARKE, B., THOMAS, J. M., and WILLIAMS, J. Q.: "Phase transitions in acridine and acridine/anthracene mixed crystals", *Seventh Molecular Crystal Symposium, Nikko, Japan [8–12 September 1975]* 65–68 (1975).

RAMAN, R., AZARRAGA, L., and McGLYNN, S. P.: "Photoconductivity of anthracene", *J. Chem. Phys.* **41**, 2516–23 (1964).

RAMAN, R. and McGLYNN, S. P.: "Hole mobility in organic molecular crystals", *J. Chem. Phys.* **40**, 515–18 (1964).

RAMBERG, E. G.: "Optical factors in the photoemission of thin films", *Appl. Optics* **6**, 2163–70 (1967).

RANDALL, J. T. and WILKINS, M. H. F.: "Phosphorescence and electron traps. I: The study of trap distribution. II: The interpretation of long period phosphorescence", *Proc. Roy. Soc. (London)* **184A**, 365–89, 390–407 (1945).

RASHBA, E. I.: "Theory of impurity absorption near the exciton bands in isotopic substitution", *Soviet Phys.—Solid State* **4**, 2417–29 (1963).

RASHBA, E. I.: "Theory of vibronic spectra of molecular crystals", *Soviet Phys.—JETP* **23**, 708–18 (1966).

RAYKERUS, P. A.: "Space-charge-limited currents in insulators with the Frenkel effect taken into consideration", *Radio Eng. Electron. Phys.* **17**, 652–6 (1972).

REEHAL, H. S.: "Poole–Frenkel effect with a screened coulomb potential", *Phys. Stat. Sol.* (b) **80**, K63–65 (1977).

REEIOGLU, H. I. and DeMASSA, T. A.: "Theoretical temperature dependence of thin-film transistors with traps", *J. Appl. Phys.* **47**, 560–4 (1976).

REIMER, B. and BÄESSLER, H.: "Photoconduction in a polydiacetylene crystal", *Phys. Stat. Sol.* (a) **32**, 435–9 (1975).

REIMER, B. and BÄESSLER, H.: "Transient photoconduction in a polydiacetylene single crystal", *Phys. Sol.* (b) **85**, 145–9 (1978).

REMBAUM, A.: "Semiconducting polymers", *J. Polymer Sci.* Part C, No. 29, 157–86 (1970).

REMBAUM, A. and LANDEL, R. (eds.): "Electrical conduction properties of polymers", *J. of Polymer Sci.* Part C, *Polymer Symposia*, No. 17, Wiley Interscience (New York) 1–250 (1969).

REMBAUM, A., MOACANIN, J., and POHL, H. A.: "Polymeric semiconductors", in *Progress in Dielectrics*, (ed. J. B. Birks), Heywood (London) **6**, 43–102 (1965).

REUCROFT, P. J.: "On the nature of electrode effects on the conductivity of anthracene crystals", *J. Chem. Phys.* **36**, 1114–16 (1962).

REUCROFT, P. J. and GHOSH, S. K.: "Carrier generation processes in poly(N-vinylcarbazole) films", *Phys. Rev.* **38**, 803–7 (1973).

REUCROFT, P. J. and GHOSH, S. K.: "The electrical conductivity of poly(divinylbenzene) films", *Thin solid Films* **20**, 363–5 (1974).

REUCROFT, P. J. and GHOSH, S. K.: "Temperature dependence of the dark current in amorphous poly (N-vinylcarbazole) films", *J. Non-Cryst. Solids* **15**, 399–409 (1974).

REUCROFT, P. J., KEVORKIAN, H. K., and LABES, M. M.: "Diffusion in organic crystal. II: lattice and subgrain boundary diffusion", *J. Chem. Phys.* **44**, 4416–20 (1966).

REUCROFT, P. J., KRONICK, P. L., and HILLMAN, E. E.: "Photovoltaic effects in tetracene crystals", *Mol. Cryst. Liq. Cryst.* **6**, 247–54 (1967).

REUCROFT, P. J., KRONICK, P. L., McGHIE, A. R., and LABES, M. M.: "Factors influencing charge-carrier trapping lifetime in organic crystals", *J. Phys. Chem. Solids*, Suppl. No. 1, 105–7 (1967).

REUCROFT, P. J. and MULLINS, F. D.: "Physical basis for exponential carrier trap distributions in molecular solids", *J. Chem. Phys.* **58**, 2918–21 (1973).

REUCROFT, P. J. and MULLINS, F. D.: "Space-charge-limited–trap-limited current in anthracene crystals", *J. Phys. Chem. Solid* **35**, 347–53 (1974).

REUCROFT, P. J., MULLINS, F. D., and HILLMAN, E. E.: "Charge carrier trapping in organic crystals", *Mol. Cryst. Liq. Cryst.* **23**, 179–86 (1973).

REUCROFT, P. J., RUDYJ, O. N., SALOMON, R. E., and LABES, M. M.: "Effect of gases on the conductivity of organic solids. III: Sensitization of bulk photoconductivity in p-chloranil crystals", *J. Phys. Chem.* **69**, 779–83 (1965).

REUCROFT, P. J. and TAKAHASHI, K.: "Charge carrier generation, transport and trapping in amorphous poly(N-vinylcarbazole) films", *J. Non-Cryst. Solids* **17**, 71–80 (1975).

REUCROFT, P. J., TAKAHASHI, K., and ULLAL, H.: "Theoretical efficiency in an organic photovoltaic energy conversion system", *Appl. Phys. Lett.* **25**, 664–5 (1974).

REUCROFT, P. J., TAKAHASHI, K., and ULLAL, H.: "Theoretical and experimental photovoltaic energy conversion in an organic film system", *J. Appl. Phys.* **46**, 5218–23 (1975).

RHODERICK, E. H.: "The physics of Schottky barriers", *J. Phys.* D **3**, 1153–67 (1970).

RHODERICK, E. H.: "Comments on the conduction mechanism in Schottky diodes", *J. Phys.* D **5**, 1920–9 (1972).

RICE, S. A. and GELBART, W. M.: "Relaxation phenomena in excited molecules", *Pure Appl. Chem.*, **27** (3) 361–87 (1971).

RICE, S. A. and JORTNER, J.: "Possible use of high-pressure techniques for the study of the electronic states of molecular crystals", *Proceedings 1st International Conference on Physical Solids at High Pressures, Tucson, Arizona*, 63–168 (1965).

RICE, S. A. and JORTNER, J.: "Comments on the theory of the exciton states of molecular crystals", in

Physics and Chemistry of Organic Solid State (eds. D. Fox, M. M. Labes, and A. Weissberger) **3**, 199– 497 (1967).

RICE, S. A., MORRIS, G. C., and GREER, W. L.: "Conjecture concerning the width of the lowest singlet– singlet transition in crystalline anthracene", *J. Chem. Phys.* **52**, 4279–87 (1970).

RICHMOND, P.: "Electrical forces between particles with arbitrary fixed surface charge distributions in ionic solutions", *J. Chem. Soc. Faraday Trans.* II, **70**, 1066–73 (1974).

RIDLEY, B. K.: "Specific negative resistance in solids", *Proc. Phys. Soc. (London)* **82**, 954–66 (1963).

RIDLEY, B. K. and WATKINS, T. B.: "Negative resistance and high electric field capture rates in semiconductors", *J. Phys. Chem. Solids* **22**, 155–8 (1961).

RIDLEY, B. K. and WATKINS, T. B.: "The dependence of capture rate on electric field and the possibility of negative resistance in semiconductors", *Proc. Phys. Soc. (London)* **78**, 710–15 (1961).

RIEHL, N.: "Observation on aromatic hydrocarbons—electrical conductivity", in *Symposium on Electrical Conductivity in Organic Solids* (eds. H. Kallmann and M. Silver), Wiley Interscience (New York) 61 (1961).

RIEHL, N., BECKER, G., and BAESSLER, H.: "Injection-determined hole current in anthracene crystals", *Phys. Stat. Sol.* **15**, 339–46 (1966).

RIEHL, N. and THOMA, P.: "Delayed fluorescence and thermoluminescence of anthracene single crystals at very low temperatures", *Phys. Stat. Sol.* **16**, 159–69 (1966).

RIETWELD, H. M., MASLEN, E. N., and CLEWS, C. J.: "An X-ray and neutron diffraction refinement of the structure of P-terphenyl", *Acta Cryst.* B **26**, 693–706 (1970).

RINGEL, H., DAMASK, A. C., ARNDT, R. A., and WHITTEN, W. B.: "Annealing of the 6060 Å color center in gamma-irradiated anthracene", *Mol. Cryst. Liq. Cryst.* **14**, 63–70 (1971).

RINGEL, H., WHITTEN, W. B., and DAMASK, A. C.: "Carrier lifetimes in γ-irradiated anthracene", *J. Chem. Phys.* **52**, 1956–9 (1970).

RIZZO, A., MICOCCI, G., and TEPORE, A.: "Space-charge-limited currents in insulators with two sets of traps distributed in energy: theory and experiment", *J. Appl. Phys.* **48**, 3415–24 (1977).

ROBERT, J., LABRUNIE, G., and BOREL, J.: "Static and transient electric field effect on homeotropic thin nematic layer", *Mol. Cryst. Liq. Cryst.* **23**, 197–206 (1973).

ROBERTS, G. G.: "Electron injection into a *p*-type semiconductor", *Phys. Stat. Sol.* **27**, 209–18 (1968).

ROBERTS, G. G.: "Thermally assisted tunnelling and pseudointrinsic conduction: two mechanisms to explain the Meyer–Neldel rule", *J. Phys.* C **4**, 3167–76 (1971).

ROBERTS, G. G.: "Nonextrinsic and space charge limited currents in semiconductors", in *Electronic and Structural Properties of Amorphous Semiconductors* (eds. P. G. Le Comber and J. Mort), Academic Press (New York) 409–24 (1973).

ROBERTS, G. G. and POLANCO, J. I.: "Thermally assisted tunnelling in dielectric films", *Phys. Stat. Sol.* (a) **1**, 409–20 (1970).

ROBERTS, G. G. and SCHMIDLIN, F. W.: "Study of localized levels in semi-insulators by combined measurements of thermally activated ohmic and space-charge-limited conduction", *Phys. Rev.* **180**, 785–94 (1969).

ROBERTS, G. G. and TREDGOLD, R. H.: "Space charge injection into impurity semiconductors. II", *J. Phys. Chem. Solids* **25**, 1349–56 (1964).

ROBERTSON, J. M.: "Structure of naphthalene and anthracene", *Rev. Mod. Phys.* **30**, 155–8 (1958).

ROBERTSON, R., FOX, J. J., and MARTIN, A. M.: "Photo-conductivity of diamonds", *Nature* **129**, 579 (1932).

ROBINSON, G. W.: "Electronic and vibrational excitons in molecular crystals", *Ann. Rev. Phys. Chem.* **21**, 429–74 (1970).

ROBINSON, G. W. and FROSCH, R. P.: "Theory of electronic energy relaxation in the solid phase", *J. Chem. Phys.* **37**, 1962–73 (1962).

ROBINSON, G. W. and FROSCH, R. P.: "Electronic excitation transfer and relaxation", *J. Chem. Phys.* **38**, 1187–1203 (1963).

ROBINSON, P. M.: "Work hardening of anthracene single crystals", *Acta Met.* **16**, 545–51 (1968).

ROBINSON, P. M., RUSSELL, H. J., and SCOTT, H. G.: "Binary phase diagrams of some molecular compounds. I", *Mol. Cryst. Liq. Cryst.* **10**, 61–74 (1970).

ROBINSON, P. M. and SCOTT, H. G.: "The morphology of anthracene crystals", *J. Cryst. Growth* **1**, 187–94 (1967).

ROBINSON, P. M. and SCOTT, H. G.: "Plastic deformation of anthracene single crystals", *Acta Met.* **15**, 1581–90 (1967).

ROBINSON, P. M. and SCOTT, H. G.: "The anthracene–carbazole system", *Mol. Cryst. Liq. Cryst.* **5**, 387– 401 (1969).

ROHRBACHER, H. and KARL, N.: "Optical detrapping of charge carriers in undoped and in tetracene-doped anthracene crystals", *Phys. Stat. Sol.* (a) **29**, 517–27 (1975).

ROOTHAAN, C. C. J.: "New developments in molecular orbital theory", *Rev. Mod. Phys.* **23**, 69–89 (1951).

Rose, A.: "An outline of some photoconductivity processes", *RCA Rev.* **12**, 362–414 (1951).

Rose, A.: "Recombination processes in insulators and semiconductors", *Phys. Rev.* **97**, 322–33 (1955).

Rose, A.: "Space-charge-limited currents in solids", *Phys. Rev.* **97**, 1538–44 (1955).

Rose, A.: "Performance of photoconductors", *Proc. IRE* **43**, 1850–69 (1955).

Rose, A.: "Comparative anatomy of models for double injection of electrons and holes into solids", *J. Appl. Phys.* **35**, 2664–78 (1964).

Rose, A.: "The role of space-charge-limited currents in photoconductivity controlled devices", *IEEE Trans. Electron Devices* ED **19**, 430–3 (1972).

Rosen, G.: "Space-charge-limited currents in non-metallic solids", *Phys. Rev. Lett.* **17**, 692–3 (1966).

Rosen, G.: "Space-charge-limited transient currents in nonmetallic crystals", *Phys. Rev.* **163**, 921–3 (1967).

Rosen, G.: "Diffusion in transient space-charge-limited currents", *Phys. Rev.* B **4**, 667 (1971).

Rosen, R. and Pohl, H. A.: "Some polymers of high dielectric constant", *J. Polymer Sci.* A1, **4**, 1135–49 (1966).

Rosenberg, B.: "Evidence for the triplet state in photoconductivity in anthracene", *Chem. Phys.* **29**, 1108–18 (1958).

Rosenberg, B.: "Photoconduction and *cis–trans* isomerism in β-carotene", *J. Chem. Phys.* **31**, 238–46 (1959).

Rosenberg, B.: "The effect of oxygen absorption of photo- and semiconduction of β-carotene", *J. Chem. Phys.* **34**, 812–19 (1961).

Rosenberg, B.: "Photoconduction and photovoltaic effects in carotenoid pigments", in *Electrical Conductivity in Organic Solids* (eds. H. Kallmann and M. Silver), Wiley Interscience (New York) 291 (1961).

Rosenberg, B.: "Some comments on the photoconductivity of organic molecular crystals and triplet state", *J. Chem. Phys.* **37**, 1371–2 (1962).

Rosenberg, B. B., Harder, H. C., and Postow, E.: "Pre-exponential factor in semiconducting organic substances", *J. Chem. Phys.* **49**, 4108–14 (1968).

Rosenberg, B. and Camiscoli, J. F.: "Photo- and semi-conduction in crystal-line chlorophylls a and b", *J. Chem. Phys.* **35**, 982–91 (1961).

Rosenberg, L. M. and Lampert, M. A.: "Double-injection in the perfect insulator: further analytic results", *J. Appl. Phys.* **41**, 508–21 (1970).

Rosenstock, H. B.: "Luminescent emission from an organic solid with traps", *Phys. Rev.* **187**, 1166–8 (1969).

Rosental, A.: "Time dependence of space-charge conduction", *Phys. Lett.* **46A**, 270–2 (1973).

Rosental, A. and Kalda, A.: "Third side of the Lampert triangle in fitting experimental data", *Phys. Rev.* B **6**, 4077–8 (1972).

Rosental, A. and Lember, L.: "Current transients in the insulator determined by space charge and diffusion", *Phys. Stat. Sol.* **39**, 19–23 (1970).

Rosental, A. and Sapar, A.: "Diffusion effects in one-carrier space-charge limited currents with trapping", *J. Appl. Phys.* **45**, 2787–8 (1974).

Rosseinsky, D. R., Hann, R. A., and Axon, A. J.: "Semiconductor properties of crystalline anthracene—competition of electron and hole photoinjection by redox electrolytes", *J. Chem. Soc. Faraday Trans.* I, **70**, 1982–90 (1974).

Rossiter, E. L. and Warfield, G.: "Transient space-charge-limited currents in amorphous selenium thin films", *J. Appl. Phys.* **42**, 2527–33 (1971).

Roth, L.: "Fluorescence of anthracene excited by high energy excitation", *Phys. Rev.* **75**, 983 (1949).

Rothwarf, A. and Böer, K. W., "Direct conversion of solar energy through photovoltaic cells", *Progress in Solid State Chemistry* **10**, 71–102 (1975).

Rowell, J. M. and Shen, L. Y. L.: "Zero-bias anomalies in normal metal tunnel junctions", *Phys. Rev. Lett.* **17**, 15–19 (1966).

Rozental, A. I. and Paritskii, L. G.: "Space charge limited current in a high-resistivity semiconductor at low voltages", *Soviet Phys.—Semicond.* **5**, 2100–3 (1972).

Rudenko, A. I.: "Transient space-charge-limited currents in a hemispherical semiconductor", *Soviet Phys.—Semicond.* **8**, 1249–53 (1975).

Rudenko, A. I.: "Transient photoinjection currents in semiconductors", *Soviet Phys.—Semicond.* **9**, 608–11 (1975).

Rudenko, A. I.: "Transient double-injection currents in semiconductors", *Soviet Phys.—Semicond.* **9**, 729–33 (1975).

Rudenko, A. I.: "Theory of trap-controlled transient current injection", *J. Non-Cryst. Solids* **22**, 215–18 (1976).

Ruiz-Urbieta, M., Sparrow, E. M., and Eckert, E. R. G.: "Film thickness and refractive indices of dielectric films on dielectric substrate", *J. Opt. Soc. Am.* **61**, 1392–6 (1971).

RUMYANTSEV, B. M. and FRANKEVICH, E. L. "Effect of the electric field on the delayed fluorescence of anthracene", *Opt. Spectrosc.* **25**, 938–42 (1968).

RUMYANTSEV, B. M. and FRANKEVICH, E. L.: "Interaction of charge transfer excitons and triplet molecular excitons in anthracene", *Opt. Spectrosc.* **27**, 427–30 (1969).

RUPPEL, W.: "Photospannungen in Photoleitern. I: Allgemeiner Ausdruck. II: Spezielle Fälle", *Phys. Stat. Sol.* **5**, 657–66, 667–82 (1964).

RUPPEL, W.: "The photoconductor–metal contact", in *Semiconductors and Semimetals* **6**, 315–45 (1970).

RUSIN, B. A. and FRANKEVICH,.E. L.: "Change of the concentration of Wannier excitons in molecular crystals caused by a magnetic field", *Phys. Stat. Sol.* **33**, 885–95 (1969).

RUSIN, B. A., RUMYANTSEV, B. M., ALEXANDROV, I. V., and FRANKEVICH, E. L.: "Anisotropy of magnetic field quenching photoconductivity and delayed fluorescence of anthracene", *Phys. Stat. Sol.* **34**, K103–5 (1969).

RUTKOWSKY, J., DROST, H., and TIMM, U.: "Electrical conductivity behavior of thermally activated polycyclic aromatic hydrocarbons", *Exp. Tech. Phys. (USSR)* **16**, 342–53 (1968).

RYZHENKOV, A. I., KOZHIN, V. M., and MYASNIKOVA, R. M.: "Thermal expansion of molecular crystals: calculation of the expansion tensor of anthracene within a wide temperature range", *Kristallografiya* **13**, 1028–31 (1968).

SABLIKOV, V. A.: "Ambipolar mobility and injection currents in semiconductors with deep traps", *Phys. Stat. Sol.* (a) **15**, 735–45 (1973).

SABLIKOV, V. A.: "Role of space charge in ambipolar transport of carriers in semiconductors", *Soviet Phys.—Semicond.* 1299–1300 (1975).

SACKS, H. K.: "Low frequency oscillations and deep impurities in high resistance GaAs", Ph.D. thesis, Department of Electrical Engineering, Carnegie–Mellon University, Pittsburgh, Pennsylvania (1970).

SADAOKA, Y. and SAKAI, Y.: "Threshold switching of violanthrone in a sandwich-type cell", *Bull. Chem. Soc. Jap.* **49**, 325–6 (1976).

SADAOKA, Y. and SAKAI, Y.: "Switching in copper-phthalocyanine films", *Bull. Chem. Soc. Jap.* **50**, 2222–5 (1977).

SADAOKA, Y., SAKAI, Y., and KISHI, K.: "Threshold switching in anthracene thin films", *Bull. Chem. Soc. Jap.* **50**, 2239–41 (1977).

SAGE, G. W.: "The non-Hecht pulse height-voltage behavior of anthracene", *J. Chem. Phys.* **44**, 3685–7 (1965).

SAGE, M. L.: "Photon-induced electronic transitions in molecular crystals", *J. Chem. Phys.* **38**, 1083–5 (1963).

SAH, C. T.: "Bulk and interface imperfections in semiconductors", *Solid-State Electron.* **19**, 975–90 (1976).

SAH, C. T. and LINDHOLM, F. A.: "Carrier generation, recombination, trapping and transport in semiconductors with position-dependent composition", *IEEE Trans. Electron. Devices* **ED 24**, 358–62 (1977).

SAH, C. T. and SHOCKLEY, W.: "Electron-hole recombination statistics in semiconductors through flaws with many charge conditions", *Phys. Rev.* **109**, 1103–15 (1958).

SAITO, K., HIRAI, T., and NAKADA, O.: "Carrier generation in photoconductivity of anthracene", *J. Phys. Soc. Japan* **22**, 1297–8 (1967).

SAJI, M. and KAO, K. C.: "Some features relevant to switching processes in the amorphous semiconductor $Si_{12}Ge_{10}As_{30}Te_{48}$", *J. Non-Cryst. Solids* **18**, 275–83 (1975).

SAJI, M. and KAO, K. C.: "Thermally stimulated depolarization currents in $Si_{12}Ge_{10}As_{30}Te_{48}$ glass", *Japan J. Appl. Phys.* **15**, 1393–4 (1976).

SAJI, M. and KAO, K. C.: "Direct observation of switching filaments in chalcogenide glasses", *J. Non-Cryst. Solids* **22**, 223–7 (1976).

SAJI, M., LEUNG, C. H., and KAO, K. C.: "Experimental evidence of energy-controlled switching in amorphous semiconductors", *J. Non-Cryst. Solids* **23**, 147–58 (1977).

SAK..., N., SHIROTANI, I., and MINOMURA, S.: "Phase transition of alkali metal cation–TCNQ anion radical simple salts", *Bull. Chem. Soc. Japan* **45**, 3321–5 (1972).

SAKAI, S., YOSHIDA, M., MITSUDO, H., and OOSHIKA, Y.: "Photogeneration of charge carriers in anthracene", *J. Phys. Soc. Japan* **25**, 1513 (1968).

SAKAI, S., YOSHIDA, M., TANAKA, S., MITSUDO, H., and OOSHIKA, Y.: "Extrinsic process of carrier generation in anthracene crystals", *J. Phys. Chem. Solids* **28**, 1913–20 (1967).

SAKAI, Y. and SADAOKA, Y.: "AC conductivity of anthracene in a compressed powder disk", *Japan J. Appl. Phys.* **12**, 1463–4 (1973).

SAKAI, Y., SADAOKA, Y., and YOKOUCHITT, H.: "Electrical properties of evaporated thin films of copper phthalocyanine", *Bull. Chem. Soc. Jap.* **47**, 1886–8 (1974).

SALANECK, W. R. and LIND, E. L.: "Transient photoconductivity effects in an insulator–semiconductor powder system", *J. Appl. Phys.* **43**, 481–9 (1972).

SAMARA, G. A. and DRICKAMER, H. G.: "Effect of pressure on the resistance of fused ring aromatic compounds", *J. Chem. Phys.* **37**, 474 (1962).

SAMOC, A., SAMOC, M., and SWARAKOWSKI, J.: "Interpretation of one carrier thermally stimulated currents and isothermal decay currents. I: Basic concepts", *Phys. Stat. Sol.* (a) **36**, 735–45 (1976).

SAMOC, A., SAMOC, M., and SWARAKOWSKI, J.: "Interpretation of one-carrier thermally stimulated currents and isothermal decay currents. III: Influence of blocking contacts", *Phys. Stat. Sol.* (a) **39**, 337–44 (1977).

SAMOC, A., SAMOC, M., SWARAKOWSKI, J., THOMAS, J. M., and WILLIAMS, J. O.: "Interpretation of one-carrier thermally stimulated currents and isothermal decay currents. II: Application to the evaluation of trap parameters in anthracene", *Phys. Stat. Sol.* (a) **37**, 271–8 (1976).

SAMOC, M.: "Anisotropy of trapping factors in perylene-doped anthracene crystals", *Mol. Cryst. Liq. Cryst.* **34** (letters), 171–6 (1977).

SANGSTER, R. C. and IRVINE, J. W. JR.: "Study of organic scintillators", *J. Chem. Phys.* **24**, 670–715 (1956).

SANO, M. and AKAMATU, H.: "Spectral response of photoconductivity in polycyclic aromatic hydrocarbons", *Bull. Chem. Soc. Japan* **35**, 587 (1962).

SANO, M. and AKAMATU, H.: "Photoconductivity in organic liquid solutions", *Bull. Chem. Soc. Japan* **36**, 480–1 (1963).

SANO, M., POPE, M., and KALLMANN, H.: "Delayed electroluminescence in anthracene crystals", in *Third Organic Crystal Symposium, Chicago, Ill., USA*, Paper 36 (1965).

SANO, M., POPE, M., and KALLMANN, H.: "Electroluminescence and band gap in anthracene", *J. Chem. Phys.* **43**, 2920–1 (1965).

SAPAR, A. and ROSENTAL, A.: "A general method for computing one-carrier injection and extraction current–voltage characteristics in solids", *Toim Eesti NSV Tead Akad. Fuus. Mat. (USSR)* **22**, 276–85 (1973).

SAPOZHNIKOV, M. N.: "Temperature dependence of phononless lines in the spectra of organic crystals", *Soviet Phys.—Solid-State* **15**, 2111–16 (1974).

SASAKI, A. and HAYAKAWA, S.: "Effects of anthracene crystal bending on triplet excitons", *Japan J. Appl. Phys.* **12**, 1806–7 (1973).

SATO, H., ONO, S., and ANDO, S.: "Space-charge-perturbed discharge characteristics of poly-*n*-vinylcarbazole", *J. Appl. Phys.* **45**, 1675–9 (1974).

SAUNDERS, I. J.: "The relationship between thermally stimulated luminescence and thermally stimulated conductivity", *Br. J. Appl. Phys.* **18**, 1219–20 (1967).

SAUNDERS, I. J.: "The thermally stimulated luminescence and conductivity of insulators", *J. Phys. C* **2**, 2181–98 (1969).

SAUNDERS, I. J. and JEWITT, R. H.: "Thin film circuit element", Annual Report on Research Project R7–27, Document AD 480 752, National Technical Information Service, Springfield, Virginia, USA (1965).

SAVOYE, E. D. and ANDERSON, D. E.: "Injection and emission of hot electrons in thin-film tunnel emitters", *J. Appl. Phys.* **38**, 3245–65 (1967).

SAWA, G., IEDA, M., and KITAGAWA, K.: "Relation between pre-exponential factor and activation energy in dark conductivity in polyethylene", *Electron Lett.* **10**, 50–51 (1974).

SAWA, G., KAWADE, M., and IEDA, M.: "Field-assisted trapping in polyethylene", *J. Appl. Phys.* **44**, 5396–7 (1973).

SCALAPINO, D. J. and MARCUS, S. M.: "Theory of inelastic electron-molecule interactions in tunnel junctions", *Phys. Rev. Lett.* **18**, 459–61 (1967).

SCARMOZZIO, R.: "Photovoltaic properties of ZnS crystals and a comparative study with luminescence", *J. Appl. Phys.* **43**, 4652–6 (1972).

SCEATS, M. G.: "Comments on the phonon-induced fluorescence of crystalline anthracene", *Chem. Phys. Lett. (Netherlands)* **29**, 298–304 (1974).

SCEATS, M. G. and MORRIS, G. C.: "Optical parameters from reflectance data— A new and versatile approach", *Phys. Stat. Sol.* (a) **14**, 643–53 (1972).

SCEATS, M. G. and RICE, S. A.: "On the use of Raman scattering to probe exciton–photon coupling in molecular crystals", *J. Chem. Phys.* **62**, 1098–1110 (1975).

SCHADT, M. and WILLIAMS, D. F.: "Low-temperature hole injection and hole trap distribution in anthracene", *J. Chem. Phys.* **50**, 4364–8 (1969).

SCHADT, M. and WILLIAMS, D. F.: "Hall mobility of electrons in anthracene crystals", *Phys. Stat. Sol.* **39**, 223–30 (1970).

SCHADT, M., ZSCHOKKE-GRÄNACHER, I., and BALDINGER, E.: "Optical excitation of adsorption sites in anthracene single crystals", *Helv. Phys. Acta* **41**, 1349–67 (1968).

SCHAFER, D. E., WUDL, F., SCHMIDT, P. H., THOMAS, G. A., FERRARIS, J. P., and COWAN, D. C.: "Some attempts to corroborate existing transport measurements of TTF–TCNQ", *Solid State Commun.* **14**, 99–100 (1974).

SCHAFER, D. E., WUDL, F., THOMAS, G. A., FERRARIS, J. P., and COWAN, D. C.: "Apparent giant conductivity peaks in an anisotropic medium: TTF–TCNQ", *Solid State Commun.* **14**, 347–51 (1974).

SCHAFFERT, R. M.: "A new high-sensitivity organic photoconductor for electrophotography", *IBM Jl Res. Dev.* **15**, 75–89 (1971).

SCHARAGER, C., MULLER, J. C., STUCK, R., and SIFFEST, P.: "Determination of deep levels in semiinsulating cadmium telluride by thermally stimulated current measurements", *Phys. Stat. Sol.* **31**, 249–53 (1975).

SCHARFE, M. E. and TABAK, M. D.: "Bulk space charge and transient photoconductivity in amorphous selenium", *J. Appl. Phys.* **40**, 3230–7 (1969).

SCHECHTMAN, B. H. and SPICER, W. E.: "Experimental determination of the optical density of states for phthalocyanines and porphyrins", *Chem. Phys. Lett.* **2**, 207–9 (1968).

SCHEIN, L. B.: "Electron drift mobilities over wide temperature ranges in anthracene, deuterated anthracene and As_2S_3", *Chem. Phys. Lett.* **48**, 571–4 (1977).

SCHEIN, L. B.: "Temperature independent drift mobility along the molecular direction of As_2S_3", *Phys. Rev. B* **15**, 1024–34 (1977).

SCHEIN, L. B., DUKE, C. B., and MCGHIE, A. R.: "Observation of the band-hopping transition for electrons in naphthalene", *Phys. Rev. Lett.* **40**, 197–200 (1978).

SCHER, H. and MONTROLL, E. W.: "Anomalous transit-time dispersion in amorphous solids", *Phys. Rev. B* **12**, 2455–77 (1975).

SCHETZINA, J. F.: "Steady-state transport in trap-dominated relaxation semiconductors", *Phys. Rev. B* **11**, 4994–8 (1975).

SCHETZINA, J. F.: "Photovoltaic properties of anisotropic relaxation semiconductors", *Phys. Rev. B* **12**, 3339–51 (1975).

SCHILLING, R. B. and LAMPERT, M. A.: "Plasmas injected into solids analytic study of the diffusion corrections", *J. Appl. Phys.* **41**, 1791–8 (1970).

SCHILLING, R. B. and SCHACHTER, H.: "Neglecting diffusion in space-charge limited currents", *J. Appl. Phys.* **38**, 841–4 (1967).

SCHILLING, R. B. and SCHACHTER, H.: "Oscillatory modes associated with one carrier transient space-charge-limited currents", *J. Appl. Phys.* **38**, 1643–6 (1967).

SCHILLING, R. B. and SCHACHTER, H.: "Transient space-charge-limited currents including diffusion", *Solid-State Electron.* **10**, 689–99 (1967).

SCHIOMI, N.: "Radiation damage to optical absorption properties of anthracene crystals", *Bull. Inst. Chem. Res. (Kyoto Univ.)* **46**, 23–25 (1968).

SCHLOTTER, P. and BAESSLER, H.: "Hot carrier motion in crystalline anthracene", *Chem. Phys. Lett.* **24**, 450–2 (1974).

SCHLOTTER, P. and BAESSLER, H.: "Optical detrapping of charge carriers from surface traps in anthracene crystals", *Chem. Phys. Lett.* **30**, 96–98 (1975).

SCHMID, A. P.: "Evidence for the small polaron as the charge carrier in glasses containing transition metal oxides", *J. Appl. Phys.* **39**, 3140–9 (1968).

SCHMIDLIN, F. W.: "Theory of multiple trapping", *Solid State Commun.* **22**, 451–3 (1977).

SCHMIDLIN, F. W. and ROBERTS, G. G.: "Interpretation of thermal activation energies in wide band-gap materials", *Phys. Rev. Lett.* **20**, 1173–6 (1968).

SCHMIDLIN, F. W. and ROBERTS, G. G.: "Study of localized level. II: The meaning of temperature-induced changes in activation on energies for electrical conduction", *Phys. Rev. B* **9**, 1578–90 (1974).

SCHMIDLIN, F. W., ROBERTS, G. G., and LAKATOS, A. I.: "Resistance limited currents in solids with blocking contacts", *Appl. Phys. Lett.* **13**, 353–5 (1968).

SCHMIDT, K. and WEDEL, K.: "Negative photo-effect by double injection into anthracene single crystals", *Phys. Stat. Sol.* **34**, K67–68 (1969).

SCHMIDT, K. and WEDEL, K.: "Negative resistance by double injection into anthracene crystals", *Phys. Stat. Sol.* **35**, K89–91 (1969).

SCHMIDT, P. F.: "On the mechanisms of electrolytic rectification", *J. Electro-Chem. Soc.* **115**, 167–76 (1968).

SCHMILLEN, A. and FALTER, W. W.: "Tetracene as a hole trap in anthracene crystals", *Z. Phys.* **48**, 401–16 (1969).

SCHNEIDER, W. G. and WADDINGTON, T. C.: "Effect of gases on the photoconductivity of anthracene", *J. Chem. Phys.* **26**, 358 (1956).

SCHNEPP, O. and JACOBI, N.: "The lattice vibrations of molecular solids", in *Advances in Chemical Physics* (eds. I. Prigogine and S. A. Rice) **22**, 205–313 (1972).

SCHOTT, M.: "Comment on photo-conduction induced by Q-spoiled lasers in anthracene crystals", *Phys. Lett.* **23**, 92–94 (1966).

SCHOTT, M.: "Remarks on the process of carrier generation in electron-bombarded crystalline anthracene", *Mol. Cryst.* **5**, 229–43 (1969).

SCHOTT, M.: "Yield of carrier generation in organic solids under electron bombardment. II", *Mol. Cryst.* **10**, 399–409 (1970).

SCHOTT, M.: "Magnetic field effects on prompt fluorescence in anthracene. Evidence for singlet exciton fission", *Chem. Phys. Lett.* **16**, 340–4 (1972).

SCHOTT, M. and BERREHAR, J.: "Detrapping of holes by singlet excitons or photons in crystalline anthracene", *Mol. Crys. Liq. Cryst.* **20**, 13–25 (1973).

SCHOTT, M. and BERREHAR, J.: "Charge carrier generation by singlet–singlet excition interaction in crystalline anthracene under weakly absorbed light illumination", *Phys. Stat. Sol.* (b) **59**, 175–86 (1973).

SCHOTT, M. and WOLF, H. C.: "On Wannier excitons in anthracene", *J. Chem. Phys.* **46**, 4996–7 (1967).

SCHUG, J. C., LILLY, A. C., and LOWITZ, D. A.: "Schottky currents in dielectric films", *Phys. Rev.* B **1**, 4811–18 (1970).

SCHULMAN, J. H.: "Application of luminescence changes in organic solids to dosimetry", *J. Appl. Phys.* **28**, 792–5 (1957).

SCHULTZ, T. D.: Tech. Report 9, Solid State and Molecular Theory Group MIT (1956).

SCHWARTZ, L. M. and HORNIG, J. F.: "Photocurrents generated by intense flash illumination", *J. Phys. Chem. Solids* **26**, 1821–4 (1965).

SCHWARTZ, M., DAVIES, H. W., and DOBRIANSKY, B. J.: "Effects of pressure on the resistivity of some organic charge-transfer complexes", *J. Chem. Phys.* **40**, 3257–9 (1964).

SCHWOB, H. P., FÜNFSCHILLING, J., and ZSCHOKKE-GRÄNACHER, I.: "Recombination radiation and fluorescence in doped anthracene crystals", *Mol. Crys. Liq. Cryst.* **10**, 39–45 (1970).

SCHWOB, H. P., SIEBRAND, W. G., WAKAYAMA, N., and WILLIAMS, D. F.: "Delayed fluorescence quenching by charge carriers in anthracene crystals: quartet state model", *Mol. Cryst. Liq. Cryst.* **24**, 259–69 (1973).

SCHWOB, H. P., WEITZ, D., and WILLIAMS, D. F.: "The variation of the carrier recombination region with carrier density in anthracene crystals", *Mol. Cryst. Liq. Cryst.* **24**, 271–82 (1973).

SCHWOB, H. P. and WILLIAMS, D. F.: "Charge transfer exciton fission in anthracene crystals", *Chem. Phys. Lett.* **13**, 581–4 (1972).

SCHWOB, H. P. and WILLIAMS, D. F.: "Charge transfer exciton fission in anthracene crystals", *J. Chem. Phys.* **58**, 1542–7 (1973).

SCHWOB, H. P. and WILLIAMS, D. F.: "Double injection in insulators: theoretical description and analysis of some double injection experiments in anthracene crystals", *J. Appl. Phys.* **45**, 2638–49 (1974).

SCHWOB, H. P. and ZSCHOKKE-GRÄNACHER, I.: "Doppelinjektion und Elektrolumineszenz in dotiertem Anthracenkristallen", *Mol. Cryst. Liq. Cryst.* **13**, 115–36 (1971).

SCHWOB, H. P. and ZSCHOKKE-GRÄNACHER, I.: "A new approach to double injection", *Solid-State Electron.* **15**, 271–6 (1972).

SCOTT, D. R. and MALTENIEKS, O.: "Experimental method for determining the intersystem crossing rate constant from lowest excited singlet to lowest triplet state", *J. Phys. Chem.* **72**, 3354–6 (1968).

SCOTT, J. C., GARITO, A. F., and HEEGER, A. J.: Magnetic susceptibility studies of tetrathiofulvalene–tetracyanoquinodimethan (TTF) (TCNQ) and related organic metals", *Phys. Rev.* B **10**, 3131–9 (1974).

SEAGER, C. H. and EMIN, D.: "High-temperature measurements of the electron Hall mobility in the alkali halides", *Phys. Rev.* B **2**, 3421–31 (1970).

SEARS, G. W.: "Growth of anthracene whiskers by vapor deposition", *J. Chem. Phys.* **39**, 2846–7 (1963).

SEGUI, Y., AI, B., and CARCHANO, H.: "Switching in polystyrene films: transition from on to off state", *J. Appl. Phys.* **47**, 140–3 (1976).

SEIWATZ, R. and GREEN, M.: "Space charge calculation for semiconductors", *J. Appl. Phys.* **29**, 1034–40 (1958).

SEKI, H.: "Field-dependent photoinjection efficiency of carriers in amorphous Se films", *Phys. Rev.* B **2**, 4877–82 (1970).

SEKI, H.: "A study of the initial photocurrent due to pulsed light absorbed in finite thickness", *J. Appl. Phys.* **43**, 1144–50 (1972).

SEKI, H. and BATRA, I. P.: "Photocurrent due to pulse illumination in the presence of trapping II", *J. Appl. Phys.* **42**, 2407–20 (1971).

SEKI, H., BATRA, I. P., GILL, W. D., KANAZAWA, K. K., and SCHECHTMAN, B. H.: "The electrophotographic discharge process", *IBM Jl Res. Dev.* **15**, 213–21 (1971).

SEKI, H. and SCHECHTMAN, B. H.: "Photocurrent and carrier distributions due to steady light absorbed in finite thickness", *J. Appl. Phys.* **43**, 523–8 (1972).

SELIVANEKO, A. S.: "The exciton state of an imperfect molecular crystal", *Sov. Phys. JETP* **5**, 79–83 (1957).

SELSBY, R. G.: "The applicability of diffusion theory to the study of luminescent emission from a doped organic crystal", *Phys. Stat. Sol.* (b) **53**, 169–78 (1972).

SELSBY, R. G.: "Rebuttal to the comment on diffusion theory of luminescent emission from a doped organic solid", *Phys. Stat. Solid.* **55**, K93–96 (1973).

SELSBY, R. G. and SWENBERG, C. E.: "Diffusion theory of luminescent emission from a doped organic solid", *Phys. Stat. Sol.* (b) **50**, 235–9 (1972).

SERVINI, A. and JONSCHER, A. K.: "Electrical conduction in evaporated silicon oxide films", *Thin Solid Films* **3**, 341–65 (1969).

SEWELL, G. L., NETTEL, S. J., and EAGLES, D. M.: "Seminars on localized (small) polarons. 1: Model of thermally activated polaron motion. 2: Interaction of a polarizable KCl crystal with a valence band hole. 3: Optical absorption by small polarons", in *Polarons and Excitons* (eds. C. G. Kuper and G. D. Whitefield), Oliver & Boyd (London) (1962).

SEYKORA, E. J. and KLEIN, R. A.: "An organic 'metal'?", *Nature* **248**, 401–2 (1974).

SHAH, R. M. and SCHETZINA, J. F.: "Excess-carrier transport in anisotropic semiconductors—the photovoltaic effect", *Phys. Rev.* B **5**, 4014–21 (1972).

SHAO, J. and WRIGHT, G. T.: "Characteristics of the space-charge-limited dielectric diode at very high frequencies", *Solid-State Electron.* **3**, 291–303 (1961).

SHARMA, R. D.: "Intrinsic photoconduction in anthracene", *J. Chem. Phys.* **46**, 2841–3 (1967).

SHARMA, R. D.: "Exciton–exciton collision in anthracene crystals", *J. Chem. Phys.* **46**, 3475–8 (1967).

SHARMA, R. D.: "Autoionization in anthracene crystals", *Phys. Rev. Lett.* **18**, 1139–40 (1967).

SHARN, C. F.: "High-energy radiation damage to fluorescent organic solids", *J. Chem. Phys.* **34**, 240–6 (1961).

SHARP, J. H. and LONDON, M.: "Spectroscopic characterization of a new polymorph of metal-free phthalocyanine", *J. Phys. Chem.* **72**, 3230–5 (1968).

SHARP, J. H. and SCHNEIDER, W. G.: "Photoconduction in anthracene induced by triplet excitons", *J. Chem. Phys.* **41**, 3657–8 (1964).

SHAY, J. L., WAGNER, S., BACHMANN, K. J., BUEHLER, E.: "Preparation and properties of InP/CdS solar cells", *J. Appl. Phys.* **47**, 614–18 (1976).

SHEINKMAN, M. K.: "Possibility of Auger recombination in multiply charged centers in germanium and silicon", *Soviet Phys.—Solid State* **7**, 18–21 (1965).

SHEKA, E. F.: "Exciton–phonon interaction and conservation laws in molecular crystal spectra", *11th European Congress in Molecular Spectroscopy*, PA **76**, 68–72 (1973).

SHEN, L. Y. L. and ROWELL, J. M.: "Magnetic field and temperature dependence of the 'zero bias tunneling anomaly' ", *Solid State Commun.* **5**, 189–92 (1967).

SHEN, L. Y. L. and ROWELL, J. M.: "Zero-bias tunneling anomalies—temperature, voltage and magnetic field dependence", *Phys. Rev.* **165**, 566–77 (1968).

SHENG, S. J. and HANSON, D. M.: "Spectroscopic measurement of the space-charge distribution in insulators, semiconductors and photoconductors", *J. Appl. Phys.* **45**, 4954–6 (1974).

SHER, B., LABES, M. M., BOSE, M., UR, H., and WILHELM, F.: "Semiconductor properties of molecular complexes", in *Electrical Conductivity of Organic Solids* (eds. H. Kallmann and M. Silver), Wiley Interscience (New York) (1961).

SHERWOOD, J. N.: "Molecular mobility and crystalline perfection in organic crystals", in *Crystal Growth, Proceedings of the International Conference on Crystal Growth, Boston, 1966, J. Phys. Chem. Solids*, Suppl. 1, 839–42 (1967).

SHERWOOD, J. N.: "Lattice defects in organic crystals", *Mol. Cryst. Liq. Cryst.* **9**, 37–57 (1969).

SHERWOOD, J. N.: "Purification and growth of large anthracene crystals", in *Purification of Inorganic and Organic Materials* (ed. M. Zief), Marcel Dekker (New York) 157–68 (1969).

SHERWOOD, J. N.: "Self-diffusion in organic solids", in *Proceedings of 7th International Symposium on the Reactivity of Solids, Bristol [July 17–21, 1972]*, 252–61 (1972).

SHERWOOD, J. N.: "Lattice defects and the plasticity of organic solids", *Bull. Soc. fr. Mineral. Cristallogr.* **95**, 253–61 (1972).

SHERWOOD, J. N. and THOMSON, S. J.: "Growth of single crystals of anthracene", *J. Sci. Instrum.* **37**, 242–5 (1960).

SHERWOOD, J. N. and THOMSON, S. J.: "The diffusion of anthracene-9-C-14 in single crystals of anthracene", *Trans. Faraday Soc.* **56**, 1442–51 (1960).

SHERWOOD, J. N. and WHITE, D. J.: "Self-diffusion in naphthalene single crystals", *Phil. Mag.* **15**, 745–53 (1967).

SHERWOOD, J. N. and WHITE, D. J.: "Self-diffusion in polycrystalline naphthalene", *Phil. Mag.* **16**, 957–80 (1967).

SHEWCHUN, J., GREEN, M. A., and KING, F. D.: "Minority carrier MIS tunnel diodes and their application to electron and photovoltaic energy conversion. II: Experiment", *Solid-State Electron.* **17**, 563–72 (1974).

SHEWCHUN, J., SINGH, R., and GREEN, M. A.: "Theory of metal–insulator–semiconductor solar cells", *J. Appl. Phys.* **48**, 765–70 (1977).

SHEWCHUN, J., WAXMAN, A., and WARFIELD, G.: "Tunneling in MIS structures. I: Theory", *Solid-State Electron.* **10**, 1165–86 (1967).

SHIMODA, H., SUKIGARA, M., and HONDA, K.: "The sign and the lifetime of charge carriers in auramine under photoconduction", *Mol. Cryst. Liq. Cryst.* **20**, 165–75 (1973).

SHIMOI, N.: "Radiation damage to optical absorption properties of anthracene crystals", *J. Phys. Soc. Japan* **23**, 1177 (1967).

SHINSAKA, K. and FREEMAN, G. R.: "Electric field induced light emission and conductance-loss transients in liquid anthracene", *Can. J. Chem.* **52**, 3559–61 (1954).

SHIROTANI, I., KAWAMURA, H., and IIDA, Y.: "Anomalous electrical behaviors in $[(C_6H_5)_3PCH_3]$ $(TCNQ)_2$ and $[(C_6H_5)_3AsCH_3]-(TCNQ)_2$ at high pressures", *Chem. Lett. Chem. Soc. Japan* **11**, 1053–6 (1972).

SHOCKLEY, W.: "On the surface states associated with a periodic potential", *Phys. Rev.* **56**, 317–23 (1939).

SHOCKLEY, W. and PEARSON, G. L.: "Modulation of conductance of thin films of semiconductors by surface charges", *Phys. Rev.* **74**, 232–3 (1948).

SHOCKLEY, W. and PRIM, R. C.: "Space-charge-limited emission in semiconductor", *Phys. Rev.* **90**, 753–63 (1953).

SHOCKLEY, W. and READ, W. T.: "Statistics of the recombinations of holes and electrons", *Phys. Rev.* **87**, 835–42 (1952).

SHOUSHA, A. H. M.: "Negative differential conductivity due to electrothermal instabilities in thin amorphous films", *J. Appl. Phys.* **42**, 5131–6 (1971).

SHOUSHA, A. H. M.: "Analysis of glow curves for a material containing distributed trap levels", *Thin Solid Films* **20**, 33–41 (1974).

SHPAK, M. T. and SHEREMET, N. I.:"Luminescence of crystalline anthracene", *Opt. Spectrosc.* **17**, 374–9 (1964).

SHUMKA, A. and NICOLET, M. A.: "Impedance of space-charge limited currents with field-dependent mobility", *Solid State Electron.* **7**, 106–7 (1964).

SIDEMAN, J. W.: "Electronic and vibrational states of anthracene", *J. Chem. Phys.* **25**, 115–21 (1956).

SIDEMAN, J. W.: "Polarized absorption and fluorescence spectra of crystalline anthracene at $4°$: spectral evidence for trapped excitons", *Phys. Rev.* **102**, 96–101 (1956).

SIEBRAND, W.: "Polarons in molecular crystals", abstract in Organic Crystal Symposium (NRC, Ottawa, Canada, 1962) 56, and also doctorate thesis, University of Amsterdam (1963).

SIEBRAND, W.: "Vibrational structure of electronic states of molecular aggregates. I and II", *J. Chem. Phys.* **40**, 2223–30, 2231–5 (1964).

SIEBRAND, W.: "Polaron band structure and carrier mobility in crystals of hydrocarbons", *J. Chem. Phys.* **41**, 3574–81 (1964).

SIEBRAND, W.: "Trapping of triplet excitons and the temperature dependence of delayed fluorescence in anthracene crystals", *J. Chem. Phys.* **42**, 3951–4 (1965).

SIEBRAND, W.: "Mechanism of radiationless triplet decay in aromatic hydrocarbons and magnitude of the Franck–Condon factors", *J. Chem. Phys.* **44**, 4055–6 (1966).

SIEBRAND, W.: "Radiationless transitions in polyatomic molecules: calculation of Franck–Condon factors", *J. Chem. Phys.* **46**, 440–7 (1967).

SIEBRAND, W.: "Theoretical evaluation of anharmonic contributions to radiationless transitions in aromatic hydrocarbons", *J. Chem. Phys. Lett.* **2**, 94–95 (1968).

SIEBRAND, W.: "Temperature dependence of radiationless transitions", *J. Chem. Phys.* **50**, 1040–1 (1969).

SIEBRAND, W.: "On the relation between radiative and non-radiative transitions in molecules", *Chem. Phys. Lett.* **9**, 157–9 (1971).

SIEBRAND, W.: "Franck–Condon factors for radiationless transitions", *J. Chem. Phys.* **55**, 5843 (1971).

SIEBRAND, W. and WILLIAMS, D. F.: "Isotope rule for radiationless transitions with an application to triplet decay in aromatic hydrocarbons", *J. Chem. Phys.* **46**, 403–4 (1967).

SIEBRAND, W. and WILLIAMS, D. F.: "Radiationless transitions in polyatomic molecules: anharmonicity, isotope effects, and singlet to ground state transitions in aromatic hydrocarbons", *J. Chem. Phys.* **49**, 1860–71 (1968).

SIEGEL, S. and GOLDSTEIN, L.: "Study of triplet–triplet transfer by method of magnetophotoselection. II: Concentration depolarization", *J. Chem. Phys.* **44**, 2780–5 (1966).

SILBEY, R.: "Radiation damping of exciton states in molecular crystals", *J. Chem. Phys.* **46**, 4029–33 (1967).

SILBEY, R., JORTNER, J., and RICE, S. A.: "The singlet-exciton state of crystalline anthracene", *J. Chem. Phys.* **42**, 1515–34 (1965).

SILBEY, R., JORTNER, J., RICE, S. A., and VALA JR., M. T.: "Exchange effects on the electron and hole mobility in crystalline anthracene and naphthalene", *J. Chem. Phys.* **42**, 733–7 (1965); errata: *J. Chem. Phys.* **43**, 2925–6 (1965).

SILINSH, E. A.: "On the physical nature of traps in molecular crystals", *Phys. Stat. Sol.* (a) **3**, 817–28 (1970).

SILINSH, E. A., BELKIND, A. I., BALODE, D. R., BISENIECE, A. J., GRECHOV, V. V., TAURE, L. F., KURIK, M. V., VERTZYMACHA, J. I., and BOK, I.: "Photoelectrical properties, energy level spectra, and photogeneration mechanisms of pentacene", *Phys. Stat. Sol.* (a) **25**, 339–47 (1974).

SILVER, M.: "Persistent internal polarization effects in anthracene", Ph.D. thesis, New York University (New York, USA) (1959).

SILVER, M.: "Conduction processes in organic crystals", *Proceedings of the 2nd International Conference on*

Conduction in Low Mobility Materials (eds. N. Klein, D. S. Tannhauser, and M. Pollak), Taylor & Francis (London) 347–56 (1971).

SILVER, M.: "Transient space charge limited currents including diffusion", *Solid State Commun.* **15**, 1785–7 (1974).

SILVER, M. and COHEN, L.: "Monte Carlo simulation of anomalous transit-time dispersion of amorphous solids", *Phys. Rev.* B **15**, 3276–8 (1977).

SILVER, M., DELACOTE, G., SCHOTT, M., MENTALCHETA, J., KIM, J. S., and JARNAGIN, R. C.: "Hot-electron spectroscopy in molecular crystals", *Mol. Cryst.* **1**, 195–9 (1966).

SILVER, M., DY, K. S., and HUANG, I. L.: "Monte Carlo calculation of the transient photocurrent in low-carrier-mobility materials", *Phys. Rev. Lett.* **27**, 21–23 (1971).

SILVER, M. and JARNAGIN, R. C.: "Carrier yield in molecular systems due to photo and high energy beta-particle ionization", *Mol. Cryst.* **3**, 461–9 (1968).

SILVER, M., MARK, P., OLNESS, D., HELFRICH, W., and JARNAGIN, R. C.: "On the observation of transient space-charge-limited currents in insulators", *J. Appl. Phys.* **33**, 2988–91 (1962).

SILVER, M. and MOORE, W.: "Spatial distribution of trapped electrons in anthracene", in *Electrical Conductivity in Organic Solids* (eds. H. Kallmann and M. Silver), Wiley Interscience (New York) 105 (1961).

SILVER, M., OLNESS, D., SWICORD, M., and JARNAGIN, R. C.: "Photogeneration of free carriers in organic crystals via exciton–exciton interaction", *Phys. Rev. Lett.* **10**, 12–14 (1963).

SILVER, M., ONN, D. G., and SMEJTEK, P.: "Steady-state and transient currents in organic liquids by injection from a tunnel cathode", *J. Appl. Phys.* **40**, 2222–6 (1969).

SILVER, M., RHO, J. R., OLNESS, D., and JARNAGIN, R. C.: "Drift mobilities of holes and electrons in naphthalene", *J. Chem. Phys.* **38**, 3030–1 (1963).

SILVER, M. and SHARMA, R.: "Carrier generation and recombination in anthracene", *J. Chem. Phys.* **46**, 692–6 (1967).

SILVER, M., SWICORD, M., JARNAGIN, R. C., MANY, A., WEISZ, S. Z., and SIMHONY, M.: "Transient space-charge-limited photocurrents in anthracene", *J. Phys. Chem. Solids* **23**, 419–22 (1962).

SILVER, M., WEISS, S. L., KIM, J. S., and JARNAGIN, R. C.: "Relative contributions of singlet–singlet and singlet–triplet interactions to the photogeneration of carriers in anthracene", *J. Chem. Phys.* **39**, 3163–4 (1963).

SIMHONY, M., and SHAULOF, A.: "Investigation of trapping in iodine single crystals by repeating carrier injection", *Phys. Rev.* **146**, 598–600 (1966).

SIMMONS, J. G.: "Low-voltage current–voltage relationship of tunnel junctions", *J. Appl. Phys.* **34**, 238–9 (1963).

SIMMONS, J. G.: "Generalized formula for the electric tunnel effect between similar electrodes separated by a thin insulating film", *J. Appl. Phys.* **34**, 1793–1803 (1963).

SIMMONS, J. G.: "Electric tunnel effect between dissimilar electrodes separated by a thin insulating film", *J. Appl. Phys.* **34**, 2581–90 (1963).

SIMMONS, J. G.: "Potential barriers and emission-limited current flow between closely spaced parallel metal electrodes", *J. Appl. Phys.* **35**, 2472–81 (1964).

SIMMONS, J. G., "Generalized thermal J–V characteristics for the electric tunnel effect", *J. Appl. Phys.* **35**, 2655–8 (1964).

SIMMONS, J. G.: "Richardson–Schottky effect in solids", *Phys. Rev. Lett.* **15**, 967–8 (1965).

SIMMONS, J. G.: "Note on the barrier heights in Al–Al_2O_3–Al tunnel junctions", *Phys. Rev. Lett.* **17**, 104–5 (1965).

SIMMONS, J. G.: "Poole–Frenkel effect and Schottky effect in metal–insulator–metal systems", *Phys. Rev.* **155**, 657–60 (1967).

SIMMONS, J. G.: "Incorporation of electric-field penetration of the electrodes in the theory of electron tunneling through a dielectric layer", *Br. J. Appl. Phys.* **18**, 269–75 (1967).

SIMMONS, J. G.: "Electrical conduction in thin insulating films", *Endeavour* **27**, 138–43 (1968).

SIMMONS, J. G.: "Transition from electrode-limited to bulk limited conduction processes in metal–insulator–metal systems", *Phys. Rev.* **166**, 912–20 (1968).

SIMMONS, J. G.: "Image force in metal-oxide–metal tunnel junctions", in *Tunneling Phenomena in Solids* (eds. E. Burstein and S. Lundqvist), Plenum Press (New York) 135–48 (1969).

SIMMONS, J. G.: "Electronic conduction through thin insulating films", in *Handbook of Thin Film Technology*, McGraw-Hill (New York), Chapter 14, 14.1–14.50 (1970).

SIMMONS, J. G.: "Theory of metallic contacts on high resistivity solids. I: Shallow traps", *J. Phys. Chem. Solids* **32**, 1987–99 (1971).

SIMMONS, J. G.: "Theory of metallic contacts on high resistivity solids. II: Deep traps", *J. Phys. Chem. Solids* **32**, 2581–91 (1971).

SIMMONS, J. G. and MAR, H. A.: "Thermal bulk emission and generation statistics and associated phenomena in metal–insulator–semiconductor devices under non-steady-state conditions", *Phys. Rev.* B **8**, 3865–74 (1973).

SIMMONS, J. G. and NADKARNI, G. S.: "Stimulated-dielectric-relaxation currents in thin film Al–CeF$_3$–Al samples", *Phys. Rev.* B **6**, 4815–27 (1972).

SIMMONS, J. G. and NADKARNI, G. S.: "Steady-state and non-steady-state current flow in thin film Al–CeF$_3$–Al samples", *Phys. Rev.* B **12**, 5459–64 (1975).

SIMMONS, J. G. and TAM, M. C.: "Theory of isothermal currents and the direct determination of trap parameters in semiconductors and insulators containing arbitrary trap distributions", *Phys. Rev.* B **7**, 3706–13 (1973).

SIMMONS, J. G. and TAYLOR, G. W.: "Non-equilibrium steady-state statistics and associated effects for insulators and semiconductors containing an arbitrary distribution of traps", *Phys. Rev.* B **4**, 502–11 (1971).

SIMMONS, J. G. and TAYLOR, G. W.: "Dielectric-relaxation currents in insulators", *Phys. Rev.* B **5**, 553–6 (1972).

SIMMONS, J. G. and TAYLOR, G. W.: "High-field isothermal currents and thermally stimulated currents in insulators having discrete trapping levels", *Phys. Rev.* B **5**, 1619–29 (1972).

SIMMONS, J. G. and TAYLOR, G. W.: "Dielectric relaxation and its effect on the isothermal electrical characteristics of defect insulators", *Phys. Rev.* B **6**, 4793–803 (1972).

SIMMONS, J. G. and TAYLOR, G. W.: "Dielectric relaxation and its effect on the thermal electric characteristics of insulators", *Phys. Rev.* B **6**, 4804–14 (1972).

SIMMONS, J. G. and TAYLOR, G. W.: "Theory of photoconductivity in amorphous semiconductors containing slowly-varying trap distributions", *J. Phys.* C **6**, 3706–18 (1973).

SIMMONS, J. G. and TAYLOR, G. W.: "Effect of charged-centre scattering on the mobility of photo-excited carriers in defect photoconductors", *Phil. Mag.* **27**, 121–6 (1973).

SIMMONS, J. G. and TAYLOR, G. W.: "Theory of photoconductivity in amorphous semiconductors containing relatively narrow trap bands", *J. Phys.* C **7**, 3051–66 (1974).

SIMMONS, J. G. and TAYLOR, G. W.: "The theory of photoconductivity in defect insulators containing discrete trap levels", *J. Phys.* C **8**, 3353–9 (1975).

SIMMONS, J. G., TAYLOR, W. G., and TAM, M. C.: "Thermally stimulated currents in semiconductors and insulators having arbitrary trap distributions", *Phys. Rev.* B **7**, 3714–19 (1973).

SIMMONS, J. G. and UNTERKOFLER, G. J.: "Potential barrier shape determination in tunnel junctions", *J. Appl. Phys.* **34**, 1828–30 (1963).

SIMMONS, J. G., UNTERKOFLER, G. J., and ALLEN, W. W.: "Temperature characteristics of BeO tunneling structures", *Appl. Phys. Lett.* **2**, 78–80 (1963).

SIMMONS, J. G. and VERDERBER, R. R.: "New thin film resistive memory", *Radio Electron. Eng.* **34**, 81–89 (1967).

SIMMONS, J. G. and VERDERBER, R. R.: "New conduction and reversible memory phenomena in thin insulating films", *Proc. Roy. Soc. (London)* A **301**, 77–102 (1967).

SIMONSEN, M. G. and COLEMAN, R. V.: "Inelastic tunnelling spectra of organic compounds", *Phys. Rev.* B **8**, 5875–87 (1974).

SIMPSON, G. A., CASTEILANOS, J., COBAS, A., and WEISZ, S. Z.: "Radiation-induced paramagnetic centers in anthracene and deuterated anthracene", *Mol. Cryst.* **5**, 165–70 (1968).

SIMPSON, O., "Electrical properties of aromatic hydrocarbons. IV: Diffusion of excitons", *Proc. Roy. Soc. (London)*, A **238**, 402–11 (1956).

SINGH, D. C. and MATHUR, S. C.: "Charge transport in aromatic hydrocarbon crystals", *Mol. Cryst. Liq. Cryst.* **27**, 55–80 (1974).

SINGH, J. and BAESSLER, H.: "Theory of exciton dissociation in molecular crystals at the interface of a metal", *Phys. Stat. Sol.* (b) **62**, 147–52 (1974).

SINGH, R. and SHEWCHUN, J.: "Photovoltaic effect in MIS diodes or Schottky diodes with an interfacial layer", *Appl. Phys. Lett.* **28**, 512–14 (1976).

SINGH, S., JONES, W. J., SIEBRAND, W., STOICHEFF, B. P., and SCHNEIDER, W. G.: "Laser generation of excitons and fluorescence in anthracene", *J. Chem. Phys.* **42**, 330–42 (1965).

SINGH, S. and LIPSETT, F. R.: "Effect of purity and temperature on the fluorescence of anthracene excited by red light", *J. Chem. Phys.* **41**, 1163–4 (1964).

SINGH, S. and STOICHOFF, B. P.: "Double photon excitation of fluorescence in anthracene single crystals", *J. Chem. Phys.* **38**, 2032–3 (1963).

SINHARAY, N. and MELTZER, B.: "Characteristics of insulator diodes determined by space-charge and diffusion", *Solid-State Electron.* **7**, 125–36 (1964).

SJÖLIN, P. G.: "Scintillation decay of anthracene excited by alpha-particles and electrons", *Nucl. Instrum. Methods* **19**, 253–7 (1968).

SKAL, A. S. and SHKLOVSKII, B. I.: "Influence of the impurity concentration on the hopping conduction in semiconductors", *Soviet Phys.—Semicond.* **7**, 1058–9 (1974).

SKAL, A. S., SHKLOVSKII, B. I. and EFROS, A. L.: "Activation energy of hopping conduction", *Soviet Phys.—Solid State* **7**, 316–20 (1975).

SKINNER, S. M.: "Diffusion, static charges and the conduction of electricity in non-metallic solids by a single charge carrier. Pt. I: Electric charges in plastics and insulating materials", *J. Appl. Phys.* **26**, 498–509 (1955).

SLATER, J. C.: "Atomic shielding constants", *Phys. Rev.* **36**, 57–64 (1930).

SLATER, J. C.: "Band theory", *J. Phys. Chem. Solids* **8**, 21–25 (1959).

SLOAN, G. J.: "The purification of anthracene: determination and use of segregation coefficients", *Mol. Cryst.* **1**, 161–4 (1966).

SLOAN, G. J.: "Kinetics of crystallization of anthracene from the vapour", *Mol. Cryst.* **2**, 323–32 (1969).

SLOAN, G. J. and MCGHIE, A. R.: "Purification of tetracene: vapour zone refining and eutectic zone melting", *Mol. Cryst. Liq. Cryst.* **18**, 17–37 (1972).

SLOAN, G. J. and MCGOWAN, N. H.: "Automatic zone refiner for organic compounds", *Rev. Sci. Instrum.* **34**, 60–62 (1963).

SLOAN, G. J., THOMAS, J. M., and WILLIAMS, J. O.: "Basal dislocations in single crystals of anthracene", *Mol. Cryst. Liq. Cryst.* **30**, 167–74 (1975).

SLOBODSKOI, F. B. and SHEKA, E. F.: "Multiparticle character of molecular crystal luminescence", *Soviet Phys.—Solid State* **15**, 860–1 (1973).

SLOBODYANIK, V. V. and FAIDYSH, A. N.: "Influence of oxygen on the photoconductivity of anthracene crystals", *Zh. Fiz. Khim.* **39**, 1041 (1965).

SLOBODYANIK, V. V. and FAIDYSH, A. N.: "Luminescence of anthracene–alkali metal complexes, and their effect on the photoconductivity of anthracene crystals", *Opt. Spectrosc.* **26**, 257–63 (1969).

SLOBODYANIK, V. V. and FAIDYSH, A. N.: "Determination of main carriers sign and mobility in photoconductivity of anthracene crystals", *Ukr. Fiz. Zh. (USSR)* **18**, 1064–8 (1973).

SMITH, G. C.: "Triplet exciton phosphorescence in crystalline anthracene", *Phys. Rev.* **166**, 839–47 (1968).

SMITH, G. C. and HUGHES, R. C.: "Magnetic field effects on triplet-exciton interaction in anthracene", *Phys. Rev. Lett.* **20**, 1358–61 (1968).

SMITH, R. W.: "Properties of ohmic contacts to cadmium sulfide single crystals", *Phys. Rev.* **97**, 1525–30 (1955).

SMITH, R. W. and ROSE, A.: "Space charge limited currents in single crystals of cadmium sulfide", *Phys. Rev.* **97**, 1531–7 (1955).

SOLOMON, P. and KLEIN, N.: "Electroluminescence at high fields in silicon dioxide", *J. Appl. Phys.* **47**, 1023–6 (1976).

SOMMERFELD, A. and BETHE, H.: in *Handbuch der Physik* (eds. H. Geiger and K. Scheel), Vol. XXIV/2, Springer (Berlin) 450 (1933).

SONNONSTINE, T. J. and HERMANN, A. M.: "Drift mobility of holes in phenanthrene single crystals", *J. Chem. Phys.* **60**, 1335–40 (1974).

SONNONSTINE, T. J. and PERLMAN, M. M.: "Transient injection currents in insulators with pre-existing trapped space charge", *Phys. Rev. B* **12**, 4434–42 (1975).

SONNONSTINE, T. J. and PERLMAN, M. M.: "Surface-potential decay in insulators with field-dependent mobility and injection efficiency", *J. Appl. Phys.* **46**, 3975–81 (1975).

SONNONSTINE, T. J., WIGLESWORTH, A., and HERMANN, A. M.: "Transient photoconductivity in azulene single crystals", *J. Chem. Phys.* **59**, 3865–6 (1973).

SOOS, Z. G.: "Zeeman population of optically produced triplet excitons in anthracene", *J. Chem. Phys.* **51**, 2107–12 (1969).

SOOS, Z. G. and POWELL, R. C.: "Generalized random-walk model for singlet-exciton energy transfer", *Phys. Rev. B* **6**, 4035–46 (1972).

SOUKUP, R. J.: "High-voltage vertical multijunction solar cell", *J. Appl. Phys.* **47**, 555–9 (1976).

SPEAR, W. E.: "Transit time measurements of charge carriers in amorphous selenium films", *Proc. Phys. Soc. (London) B* **70**, 669–75 (1957).

SPEAR, W. E.: "The hole mobility in selenium", *Proc. Phys. Soc. (London)* **76**, 826–32 (1960).

SPEAR, W. E.: "Drift mobility techniques for the study of electrical transport properties in insulating solids", *J. Non-Cryst. Solids* **1**, 197–214 (1969).

SPEAR, W. E.: "Electronic transport and localization in low mobility solids and liquids", *Adv. Phys.* **23**, 523–46 (1974).

SPEAR, W. E. and LE COMBER, P. G.: "Substitutional doping of amorphous silicon", *Solid State Commun.* **17**, 1193–6 (1975).

SPEAR, W. E., LE COMBER, P. G., KINMOND, S., and BRODSKY, M. H.: "Amorphous silicon p–n junction", *Appl. Phys. Lett.* **28**, 105–7 (1976).

SPICER, W. E.: "Possible non-one-electron effects in the fundamental optical excitation spectra of certain crystalline solids and their effect on photoemission", *Phys. Rev.* **154**, 385–94 (1967).

SPICER, W. E.: *Proceedings of the 8th International Conference on Phenomena in Ionized Gases*, International Atomic Energy Agency, Vienna 271 (1968).

SPICER, W. E. and DONOVAN, T. M.: "Photoemission and optical studies of amorphous germanium", *J. Non-Cryst. Solids* **2**, 66–80 (1970).

SPIELBERG, D. H., KORN, A. I., and DAMASK, A. C.: " Experimental and theoretical Hall mobilities of holes and electrons in naphthalene", *Phys. Rev.* B 3, 2012–15 (1971).

SPITZER, W. G., CROWELL, C. R., and ATALLA, M. M.: "Mean free path of photoexcited electrons in Au", *Phys. Rev. Lett.* 8, 57–58 (1962).

SPONER, H., KANDA, Y., and BLACKWELL, L. A.: "Delayed fluorescence in naphthalene crystals at $4°K$", *J. Chem. Phys.* 29, 721 (1958).

SPRINGTHORPE, A. J., AUSTIN, J. G., and SMITH, B. A.: "Hopping conduction in $Li_xNi_{1-x}O$ crystals at low temperatures", *Solid State Commun.* 3, 143–6 (1965).

SPROKEL, G. J.: "Conductivity, permittivity, and the electrode space-charge of nematic liquid crystals", *Mol. Cryst. Liq. Cryst.* 26, 45–57 (1974).

SRIVASTAVA, B. B.: "Scaling law for hot-carrier injection in insulators", *Solid-State Electron.* 17, 14–15 (1974).

STACY, W. T. and SWENBERG, C. E.: "Temperature dependence of intersystem crossing in crystalline anthracene", *J. Chem. Phys.* 52, 1962–5 (1970).

STAFEEV, V. I.: "Modulation of diffusion length as a new principle of operation of semiconductor devices", *Soviet Phys.—Solid State* 1, 763–8 (1959).

STAFEEV, V. I.: "Photoconductivity in semiconductor diodes induced by carrier lifetime changes", *Soviet Phys.—Solid State* 3, 1829–33 (1962).

STAFEEV, V. I., KUZNETSOVA, V. V., MALCHANOV, V. P., KARAKASHAN, E. I., AIRAPETYANTS, S. V., and GASANOV, L. S.: "Negative resistance of very thin organic films between metal electrodes", *Soviet Phys.—Semicond.* 2, 642–3 (1968).

STAIGER, E. H.: "High E-field microwave properties of bulk amorphous semiconductors", *J. Non.-Cryst. Solids* 17, 273–80 (1975).

STARKE, M. and HAMANN, C.: "Die electrischen Eigenschaften chlorsubstituierter Kupferphthalocyanine", *Z. Physik. Chem.* 243, 166–76 (1970).

STEBLINA, E. V. and STEBLIN, V. I.: "Frequency-intensity characteristics of the electroluminescence of organic compounds", *Zh. Prikl. Spektrosk. (USSR)* 20, 304–5 (1974).

STEINBERGER, I. T.: "Further experimental evidences on majority carrier injection in CdS single crystals", *J. Phys. Chem. Solids* 15, 354–5 (1960).

STEKETEE, J. W. and DE JONGE, J.: "Photoconductance and spectral absorption of anthracene", *Philips Res. Rept.* 17, 363–81 (1962).

STERN, P. S. and GREEN, M. E.: "Surface states of anthracene and naphthalene", *J. Chem. Phys.* 58, 2507–16 (1973).

STERNLICHT, H., NIEMAN, G. C., and ROBINSON, G. W.: "Triplet–triplet annihilation and delayed fluorescence in molecular aggregates", *J. Chem. Phys.* 38, 1326–35 (1963).

STEVENS, B.: "Effects of molecular orientation of fluorescence emission and energy transfer in crystalline aromatic hydrocarbons", *Spectrochem. Acta* 18, 439 (1962).

STÖCKMANN, F.: "The ranges of charge carriers in photoconductors", in *Proceedings of Conference on Photoconductivity, Atlantic City, New Jersey, 1954* (ed. R. G. Breckenbridge), Wiley (New York) 269–86 (1956).

STÖCKMANN, F.: "Superlinear photoconductivity", *Phys. Stat. Sol.* 34, 351–7 (1969).

STÖCKMANN, F.: "On the dependence of photocurrents on the excitation strength", *Phys. Stat. Sol.* 34, 741–9 (1969).

STÖCKMANN, F.: "Recombination kinetics in photoconductors", in *Proceedings of the 3rd International Conference on Photoconductivity* (ed. E. M. Pell), Pergamon Press (Oxford) 17–22 (1971).

STÖCKMANN, F.: "Photoconductivity—a centennial", *Phys. Stat. Sol.* 15, 381–90 (1973).

STÖCKMANN, F.: "On the classification of traps and recombination centers", *Phys. Stat. Sol.* (a) 20, 217–20 (1973).

STÖCKMANN, F.: "On the concept of 'lifetimes' in photoconductors", *RCA Rev.* 36, 499–507 (1975).

STORBECK, I. and STARKE, M.: "Electric and dielectric properties of poly (tetracyanoethylene)", *Bes. Bunsenges. Phys. Chem.* 89, 343–7 (1965).

STRATTON, R.: "The influence of interelectronic collisions on conduction and breakdown in covalent semiconductors", *Proc. Roy. Soc. (London)* A 242, 355–73 (1957).

STRATTON, R.: "The influence of interelectronic collision on conduction and breakdown in polar crystals", *Proc. Roy. Soc. (London)* A 246, 406–22 (1958).

STRATTON, R.: "The theory of dielectric breakdown in solids", in *Progress in Dielectrics* (Howard, London) 3, 235–92 (1961).

STRATTON, R.: "Theory of field emission from semiconductors", *Phys. Rev.* 125, 67–82 (1962).

STRATTON, R.: "Diffusion of hot and cold electrons in semiconductor barriers", *Phys. Rev.* 126, 2002–14 (1962).

STRATTON, R.: "Volt–current characteristics for tunneling through insulating films", *J. Phys. Chem. Solids* 23, 1177–90 (1962).

STRATTON, R.: "Energy distributions of field emitted electrons", *Phys. Rev.* 135, 794–805 (1964).

STRATTON, R.: Tunneling in Schottky barrier rectifiers", in *Tunneling Phenomena in Solids* (eds. E. Burstein and S. Lundquist), Plenum Press (New York) 105–26 (1969).

STREETMAN, B. G., BLOUKE, M. M., and HOLONYAK, N., JR.: "Current oscillation in Co-doped Si *p–i–n* structures", *Appl. Phys. Lett.* **11**, 200–2 (1967).

STROME JR., F. C.: "Direct, 2-photon photocarrier generation in anthracene", *Phys. Rev. Lett.* **20**, 3–5 (1968).

STROME, F. S. and HAYWOOD, J. S.: "Fluorescence ratio in anthracene—melt and solution with circular and linear polarization of ruby laser excitation", *J. Chem. Phys.* **45**, 4356–7 (1966).

STUART, M.: "Conduction in silicon oxide films", *Br. J. Appl. Phys.* **18**, 1637–40 (1967).

STUART, M.: "Electrode-limited to bulk-limited conduction in silicon oxide films", *Phys. Stat. Sol.* **23**, 595–7 (1967).

STUCK, R., MULLER, J. C., PONPON, J. P., SCHARAGER, C., SCHWAB, C., and STIFFENT, P.: "Study of trapping in mercuric iodide by thermally stimulated current measurements", *J. Appl. Phys.* **47**, 1545–8 (1976).

SUBRAMANIAN, A., GORDON, S. J., and SCHETZINA, J. F.: "Transient photovoltaic effects in anisotropic semiconductors", *Phys. Rev. B* **9**, 536–44 (1974).

SUGAKOV, V. I.: "Theory of electron spectra of deformed molecular crystals", *Soviet Phys.—Solid State* **15**, 1670–1 (1974).

SUKIGARA, M. and NELSON, R. C.: "Band structure and transport of charge carriers in metal-free phthalocyanine", *Mol. Phys.* **17**, 387–96 (1969).

SUN, H. Y., JORTNER, J., and RICE, S. A.: "Resonant transfer of vibrational energy and radiationless transitions in the solid phase: The lifetime of triplets in anthracene", *J. Chem. Phys.* **44**, 2539–40 (1966).

SUNA, A.: "Kinematics of exciton–exciton annihilation in molecular crystals", *Phys. Rev. B* **1**, 1716–39 (1970).

SUSSMAN, A.: "Space-charge-limited currents in copper phthalocyanine thin films", *J. Appl. Phys.* **38**, 2738–48 (1967).

SVENSSON, C. M.: "The conduction mechanism in silicon nitride films", *J. Appl. Phys.* **48**, 392–435 (1977).

SWENBERG, C. E.: "Theory of triplet exciton annihilation in polyacene crystals", *J. Chem. Phys.* **51**, 1753–64 (1969).

SWENBERG, C. E. and GEACINTOV, N. E.: "Exciton interactions in organic solids", in *Organic Molecular Photophysics* (ed. J. B. Birks), Wiley (New York), Vol. 1, Chapter 10, 489–564 (1973).

SWENBERG, C. E., MARKEVICH, D., GEACINTOV, N. E., and POPE, M.: "Room temperature electron drift mobilities in 1,4-dibromonaphthalene", *Phys. Stat. Sol.* (a) **29**, 651–7 (1975).

SWENBERG, C. E., RATNER, M. A., and GEACINTOV, N. E.: "Energy dependence of optically induced exciton fission", *J. Chem. Phys.* **60**, 2152–7 (1974).

SWENBERG, C. E. and STACY, W. T.: "Bimolecular radiationless transition in crystalline tetracene", *Chem. Phys. Lett.* **2**, 327–8 (1968).

SWENBERG, C. E., YARMUS, L., and ROSENTAL, A.: "A non-crossing pair state theorem and its application to impurity determination in molecular crystals", *28th Symposium on Molecular Structure and Spectroscopy* 107 (1973).

SWIATEK, J.: "Electrical conduction mechanism in thin coronene films", *Phys. Stat. Sol.* (a) **38**, 285–91 (1976).

SWORAKOWSKI, J.: "Temperature dependence of saturation currents in the system: anthracene–liquid redox electrode", *Acta Phys. Polon.* **35**, 33–38 (1969).

SWORAKOWSKI, J.: "On the origin of trapping centres in organic molecular crystals", *Mol. Cryst. Liq. Cryst.* **11**, 1–11 (1970).

SWORAKOWSKI, J.: "On the energy gap in anthracene", *J. Phys. Soc. Japan* **29**, 1390 (1970).

SWORAKOWSKI, J.: "Space-charge-limited currents in solids with non-uniform spatial trap distribution", *J. Appl. Phys.* **41**, 292–5 (1970).

SWORAKOWSKI, J.: "Calculation of photoelectron emission curves in organic solids", *Phys. Stat. Sol.* (b), **10**, K89–92 (1972).

SWORAKOWSKI, J.: "Photoemission from molecular crystals", *Phys. Stat. Sol.* (a) **13**, 381–8 (1972).

SWORAKOWSKI, J.: "Effects of impurities and defects on the photoemission from molecular solids", *Phys. Stat. Sol.* (a) **14**, K129–33 (1972).

SWORAKOWSKI, J.: "Trapping states formed by structural defects in molecular crystals", *Mol. Cryst. Liq. Cryst.* **19**, 259–68 (1973).

SWORAKOWSKI, J.: "On the interpretation of photoemission curves in organic solids", *Phys. Stat. Sol.* (a) **22**, K73–77 (1974).

SWORAKOWSKI, J.: "Detection of traps in crystals of anthracene", *Sci. Papers Inst. Org. Phys. Chem. Wroclaw Tech. Univ.* **7**, 191–200 (1974).

SWORAKOWSKI, J.: "Structural disorder as a source of traps in organic crystals", *Mol. Cryst. Liq. Cryst.* **32**, 87–89 (1976).

SWORAKOWSKI, J. and MAGER, J.: "Space-charge-limited currents in single crystals of anthracene doped with perylene", *Acta Phys. Polon.* **36**, 483–6 (1969).

SWORAKOWSKI, J. and PIGON, K.: "Trap distribution and space-charge-limited currents in organic crystals—anthracene", *J. Phys. Chem. Solids*, **30**, 491–6 (1969).

SWORAKOWSKI, J., THOMAS, J. M., WILLIAMS, D. F. and WILLIAMS, J. O.: "The role of defects as charge carrier traps in anthracene single crystals", *Sixth Molecular Crystal Symposium, May 20–25, 1973 (Schloss Elmau, Germany)* 131–4 (1973).

SWORAKOWSKI, J., THOMAS, J. M., WILLIAMS, D. F., and WILLIAMS, J. O.: "Electroluminescence and enhanced double-injection in crystals of anthracene", *J. Chem. Soc. Faraday Trans.* II **70**, 676–84 (1974).

SZE, S. M.: "Current transport and maximum dielectric strength of silicon nitride films", *J. Appl. Phys.* **38**, 2951–6 (1967).

SZE, S. M., CROWELL, C. R., and KAHNG, D.: "Photoelectric determination of the image force dielectric constant for hot electrons in Schottky barriers", *J. Appl. Phys.* **35**, 2534–6 (1964).

SZENT-GYÖRGYI, A.: "The study of energy levels in biochemistry", *Nature* **148**, 157–9 (1941).

SZENT-GYÖRGYI, A.: "Internal photo-electric effect and band spectra in proteins", *Nature* **157**, 875 (1946).

SZYMANSKI, A.: "Space-charge limited currents in polycrystalline *p*-quaterphenyl layers", *Mol. Cryst.* **3**, 339–55 (1968).

SZYMANSKI, A.: "Electrical conductivity of para-polyphenyl and its oligonies", *Acta Phys. Polon.* **34**, 201–15 (1968).

SZYMANSKI, A.: "Determination of trap level distribution from the study space charge limited currents", *Bull. de l'Académie Pol. des Sci.* **16**, 669–75 (1968).

SZYMANSKI, A. and LABES, M. M.: "Charge carrier mobility in tetracene", *J. Chem. Phys.* **50**, 1898–9 (1969).

SZYMANSKI, A., LARSON, D. C. and LABES, M. M.: "A temperature-independent conducting state in tetracene thin film", *Appl. Phys. Lett.* **14**, 88–90 (1969).

TABAK, M. D., ING, S. W., and SCHARFE, M. E.: "Operation and performance of amorphous selenium-based photoreceptors", *IEEE Trans. Electron. Devices* ED **20**, 132–9 (1973).

TABAK, M. D. and SCHARFE, M. E.: "Transition from emission limited to space-charge-limited photoconductivity", *J. Appl. Phys.* **41**, 2114–16 (1970).

TABAK, M. D. and WARTER, P. J.: "Field-controlled photogeneration and free-carrier transport in amorphous selenium films", *Phys. Rev.* **173**, 899–907 (1968).

TAKAHASHI, Y. and TOMURA, M.: "Drift time due to diffusion of singlet excitons in anthracene crystals", *J. Phys. Soc. Japan* **27**, 1369 (1969).

TAKAHASHI, Y. and TOMURA, M.: "The diffusion length of singlet excitons in anthracene layers", *J. Phys. Soc. Japan* **29**, 525 (1970).

TAKAHASHI, Y. and TOMURA, M.: "Diffusion of singlet excitons in anthracene crystals", *J. Phys. Soc. Japan* **31**, 1100–8 (1971).

TAKAI, Y., OSAWA, T., KAO, K. C., MIZUTANI, T., and IDEA, M.: "Effects of electrode materials on photocurrents in polyethylene terephthalate", *Japan J. Appl. Phys.* **14**, 473–9 (1975).

TAMM, I.: "A possible kind of electron binding on crystal surfaces", *Phys. Z. Sowjetunion* **1**, 733–46 (1932) [*Z. Phys.* **76**, 849–50 (1932)].

TANAKA, F. and OSUGI, J.: "The role of Franck–Condon factors in intersystem crossing from the S_1 state of anthracene", *Chem. Phys. Lett.* **27**, 133–7 (1974).

TANAKA, J.: "Electronic spectra of aromatic molecular crystals. II: Crystal structure and spectra of perylene", *Bull. Chem. Soc. Japan* **36**, 1237–49 (1963).

TANAKA, J., KODA, T., SHIONOYA, S., and MIHOMURA, S.: "The effect of pressure on the fluorescence spectra of anthracene, chrysene, and pyrene", *Bull. Chem. Soc. Japan* **38**, 1559–60 (1965).

TANAKA, K. and NIIRA, K.: "The effect of hydrogen on the band structure of naphthalene crystals", *J. Phys. Soc. Japan* **24**, 520–3 (1968).

TANAKA, M. and TANAKA, J.: "Theory of the exciton states of molecular crystals", *Mol. Phys.* **16**, 1–15 (1969).

TANAKA, S. and FAN, H. Y.: "Impurity conduction in *p*-type silicon at microwave frequencies", *Phys. Rev.* **132**, 1516–26 (1963).

TANAKA, T.: "Conduction phenomena in dielectric solids", in *Digest of Literature on Dielectrics* (ed. P. P. Budenstein), Nat. Acad. Sci., Washington DC, USA **35**, 118–68 (1973).

TANDON, J. L., BILGER, H. R., and NICOLET, M. A.: "Thermal equilibrium noise of space charge limited current in silicon for holes with field-dependent mobility", *Solid-State Electron.* **18**, 113–18 (1975).

TANG, C. W. and ALBRECHT, A. C.: "The electrodeposition of films of chlorophyll-a microcrystals and their spectroscopic properties", *Mol. Cryst. Liq. Cryst.* **25**, 53–62 (1974).

TANG, C. W. and ALBRECHT, A. C.: "Chlorophyll-a photovoltaic cells", *Nature* **254**, 507–9 (1975).

TANG, C. W. and ALBRECHT, A. C.: "Photovoltaic effects of metal–chlorophyll-a–metal sandwich cells", *J. Chem. Phys.* **62**, 2139–49 (1975).

TANG, C. W. and ALBRECHT, A. C.: "Transient photovoltaic effects in metal–chlorophyll-a–metal sandwich cells", *J. Chem. Phys.* **63**, 953–61 (1975).

TANGUY, J. and HESTO, P.: "Thermally stimulated currents in organic monomolecular layers", *Thin Solid Films* **21**, 129–43 (1974).

TANTRAPORN, W.: "On the study of metal–semiconductor contacts", *IEEE Trans. on Electron. Devices* ED **19**, 331–8 (1972).

TANTZSCHER, C. and HAMANN, C.: "Current–voltage characteristics and conduction mechanisms in thin-film of copper phthalocyanine with aluminium contacts", *Phys. Stat. Sol.* (a) **26**, 443–9 (1974).

TANTZSCHER, C. and HAMANN, C.: "Current conduction in combined insulating films", *Phys. Stat. Sol.* (a) **27**, 243–6 (1975).

TAUC, J.: "Generation of an e.m.f. in semiconductors with non-equilibrium current carrier concentrations", *Rev. Mod. Phys.* **29**, 308–24 (1959).

TAVARES, A. D.: "Photovoltaic effect in crystals of organic semiconductors as a function of wavelength", *J. Chem. Phys.* **53**, 2520–4 (1970).

TAVARES, A. D.: "Resistivity measurements in organic semiconductors", *Phys. Stat. Sol.* (a) **15**, 599–664 (1973).

TAYLOR, D. M. and LEWIS, T. J.: "Electrical conduction in polyethylene terephthalate and polyethylene films", *J. Phys. D* **4**, 1346–57 (1971).

TAYLOR, G. W. and SIMMONS, J. G.: "Basic equations for statistics, recombination processes and photoconductivity in amorphous insulators and semiconductors", *J. Non-Cryst. Solids* **8–10**, 940–6 (1972).

TAYLOR, G. W. and SIMMONS, J. G.: "Photoconductivity and the determination of trapping parameters in amorphous semiconductors", *J. Phys. C* **7**, 3067–74 (1974).

TAYLOR, G. W. and SIMMONS, J. G.: "Photoconductivity characteristics of defect insulators", *J. Phys. C* **8**, 3360–70 (1975).

TAYLOR, G. W. and SIMMONS, J. G.: "Superlinearity and sensitization in defect photoconductors", *J. Phys. C* **9**, 1013–23 (1976).

TEFFT, W. E.: "Trapping effect in drift mobility experiments", *J. Appl. Phys.* **38**, 5265–72 (1967).

TEICH, M. C. and WOLGA, G. J.: "Two-quantum volume photoelectric effect in sodium", *Phys. Rev.* **171**, 809–14 (1968).

TELL, B., BRIDENBAUGH, P. M., and KASEPER, H. H.: "Photovoltaic properties of Cu_2Se–$AgInSe_2$ heterojunctions", *J. Appl. Phys.* **47**, 619–20 (1976).

TERENIN, A. N.: "Photoelectric properties of semiconducting organic dyes", *Symposium on Electrical Conductivity in Organic Solids* (eds. H. Kallmann and M. Silver), Wiley Interscience (New York) 39–59 (1961).

TERENIN, A. N. and VILESOV, F. Z.: "Photoionization and photo-dissociation of aromatic molecules by vacuum ultraviolet radiation", in *Advances in Photochemistry*, Wiley Interscience (New York) **2**, 385 (1964).

THAXTON, G. D., JARNAGIN, R. C., and SILVER, M.: "Band structure and transport of holes and electrons in homology of anthracene", *J. Phys. Chem.* **66**, 2461–5 (1962).

THOMA, P. and VAUBEL, G.: "Electrical properties of anthracene single crystals at very low temperatures", *Phys. Stat. Sol.* **16**, 663–74 (1966).

THOMAS, J. M. and WILLIAMS, J. O.: "Some electronic and chemical consequences of non-basal dislocations in crystalline anthracene", *Mol. Cryst. Liq. Cryst.* **9**, 59–79 (1969).

THOMAS, J. M. and WILLIAMS, J. O.: "Dislocation and the reactivity of organic solids", *Prog. in Solid State Chem.* **6**, 119–54 (1971).

THOMAS, J. M. and WILLIAMS, J. O.: "Structural imperfections in organic molecular crystals", in *Surface and Defect Properties of Solids*, The Chemical Society (London) **1**, 130–43 (1972).

THOMAS, J. M., WILLIAMS, J. O., and COX, G. A.: "Lattice imperfections in organic solids. Part 3: A study, using the conductivity glow-curve technique, of trapping centers in crystalline anthracene", *Trans. Faraday Soc.* **64**, 2496–504 (1968).

THOMAS, J. M., WILLIAMS, J. O., and TURTON, L. M.: "Lattice imperfections in organic solids. Part 4: A study, using space-charge limited currents, of trapping centres in crystalline anthracene", *Trans. Faraday Soc.* **64**, 2496–504 (1968).

THORNBER, K. K.: "Why we must put atoms back into solids", *Sci. Progr. (London)* **57**, 149–68 (1968).

THORNBER, K. K., MCGILL, T. C., and MEAD, C. A.: "The tunneling time of an electron", *J. Appl. Phys.* **38**, 2384–5 (1967).

THOULESS, D. J.: "Relation between the Kubo–Greenwood formula and Boltzmann equation for electrical conductivity", *Phil. Mag.* **32**, 877–9 (1975).

TIBERGHIEN, A. and DELACOTE, G.: "Calculation of the Davydov splitting in pyrene crystal", *Chem. Phys. Lett.* **14**, 184–9 (1972).

TIBERGHIEN, A., DELACOTE, G., and SCHOTT, M.: "Triplet transfer integrals in aromatic hydrocarbon crystals—influence of nonorthogonality corrections", *J. Chem. Phys.* **59**, 3762–9 (1973).

TIETJEN, J. J. and AMICK, J. A.: "The preparation and properties of vapor-deposited epitaxial GaAs$_{1-x}$P$_x$ using arsine and phosphine", *J. Electrochem. Soc.* **113**, 724–8 (1966).

TIMAN, B. L.: "Relay transport of charge in a metal–insulator–metal system under carrier injection conditions", *Soviet Phys.—Semicond.* **7**, 163–6 (1973).

TIMAN, B. L.: "Recombination amplification of sound in semiconductors", *Soviet Phys.—Semicond.* **7**, 332–6 (1973).

TIMAN, B. L. and FESENKO, V. M.: "Influence of the field ionization of traps and of charge exchange between them on the flow of an injection current", *Soviet Phys.—Semicond.* **7**, 319–21 (1973).

TIMASHEV, S. F.: "Optical generation of carriers by photoionization of deep centers in a strong electric field", *Soviet Phys.—Semicond.* **9**, 67–70 (1975).

TKACH, YU. YA.: "Alternating-current thermally stimulated conductivity method", *Soviet Phys.—Semicond.* **6**, 451–2 (1972).

TOLSTOI, N. A. and ABRAMOV, A. P.: "Interaction of excitons in anthracene", *Fiz. Tverd. Tela* **9**, 340–3 (1967) [*Soviet Phys.—Solid State* **9**, 255–7 (1967)].

TOMKIEWICZ, Y., GROFF, R. P., and AVAKIAN, P.: "Spectroscopic approach to energetics of exciton fission and fusion in tetracene crystals", *J. Chem. Phys.* **54**, 4504–7 (1971).

TOMURA, M. and TAKAHASHI, Y.: "A note on measurements of the diffusion coefficient of singlet excitons in anthracene crystals", *J. Phys. Soc. Japan* **26**, 1325 (1969).

TOMURA, M., TAKAHASHI, Y., MATSUI, A., and ISHII, Y.: "Decay times of emissions and diffusions of singlet excitons in anthracene crystals", *J. Phys. Soc. Japan* **25**, 647 (1968).

TOURAINE, A., VAUTIER, C., and CARLES, D.: "Caractéristiques courant-tension en régimes de charge d'espace en présence d'une distribution uniforme de pièges. II: Application aux résultats expérimentaux des couches mine de sélénium amorphe", *Thin Solid Films* **9**, 229–39 (1972).

TOVE, P. A. and ANDERSON, L. G.: "Transient space-charge-limited currents in light-pulse excited silicon", *Solid-State Electron.* **16**, 961–72 (1973).

TOYOZAWA, Y.: "Self-trapping of an electron by the acoustical mode of lattice vibration. I", *Prog. Theor. Phys.* **26**, 29–44 (1961).

TREDGOLD, R. H.: "On very low mobility carriers", *Proc. Phys. Soc. (London)* **80**, 807–10 (1962).

TRESTER, S., COBAS, A., and WEISZ, S. Z.: "On the difficulty of determining surface lifetime from pulsed space charge limited current measurements", *J. Phys. Chem. Solids* **27**, 1701–3 (1966).

TRLIFAJ, M.: "A contribution to the theory of the resonance transfer of excitation energy in solids", *Czech. J. Phys.* **5**, 463–79 (1955).

TRLIFAJ, M.: "The theory of diffusion of localized excitons in solids", *Czech. J. Phys.* **6**, 533–50 (1956), and "The diffusion of the excitation energy in molecular crystals", *Czech. J. Phys.* **8**, 510–20 (1958).

TRLIFAJ, M.: "Nonradiative destruction of triplet excitons by excess electrons in organic crystals", *Czech. J. Phys. B* **23**, 558–66 (1973).

TRLIFAJ, M.: "Exciton–exciton interactions in molecular crystals", *Pure Appl. Chem.* **37**, 197–210 (1974).

TSARENKOV, G. V.: "Theory of exciton photoelectric effect in metal–semiconductor structures", *Soviet Phys.—Semicond.* **7**, 1016–20 (1974).

UCHIDA, K. and TOMURA, M.: "Lifetimes and diffusion of singlet excitons in naphthalene crystals", *J. Phys. Soc. Japan* **36**, 1358–64 (1974).

UKEI, K., TAKAMOTO, K., and KANODA, E.: "Lead phthalocyanine: a one-dimensional conductor", *Phys. Lett. A* **45**, 345–6 (1973).

UNGER, K.: "Bestimmung von Hafttermspektren mit Hilfe von Glow-Kurven", *Phys. Stat. Sol.* **2**, 1279–98 (1962).

URBACH, F.: "Luminescence of the alkali halides. I: Introduction and orientating observations. II: Methods of measurement—first results—theory of thermoluminescence", *Sitzb. Akad. Wiss. Wien,* Abt IIa, **139**, 353–62, 363–72 (1930).

USOV, N. N. and BENDERSKII, V. A.: "Barrier effect in phthalocyanine films", *Soviet Phys.—Semicond.* **2**, 580–6 (1968).

USOV, N. N. and BENDERSKII, V. A.: "Photoeffect in metal-free phthalocyanine crystals", *Phys. Stat. Sol.* **37**, 535–43 (1970).

USS, V. G. and SIDARAVICHYUS, I. B.: "Transient photoconductivity of poly-N-vinyl carbonzole-amorphous selenium double-layer electrophotographic systems", *Soviet Phys.—Semicond.* **9**, 1091–4 (1976).

VAISNYS, J. R. and KIRK, R. S.: "Electrical properties of copper phthalocyanine at high pressure", *Phys. Rev.* **141**, 641–8 (1966).

VAN DER ZIEL, A.: "Normalized characteristics of $n-v-n$ devices", *Solid-State Electron.* **10**, 267–8 (1967).

VAN DER ZIEL, A.: "A.C. admittance of double-injection space-charge-limited diodes", *Electron. Lett.* **5**, 298–9 (1969).

VAN DER ZIEL, A.: "Alternating current admittance of double injection diodes in the ohmic and the ohmic relaxation regime", *Solid-State Electron.* **13**, 191–3 (1970).

VAN DER ZIEL, A.: "Effect of the displacement current on the alternating current admittance of a double injection diode", *Solid-State Electron.* **13**, 195–7 (1970).

VAN KEER, R. and PHARISEAU, P.: "On the equivalent of the Kubo formalism and a transport equation approach", *Phys. Stat. Sol.* B **61**, 303–10 (1974).

VAN ROOSBROECK, W.: "The transport of added current carriers in a homogeneous semiconductor", *Phys. Rev.* **91**, 282–6 (1953).

VAN ROOSBROECK, W.: "Theory of the photomagnetoelectric effect in semiconductors", *Phys. Rev.* **101**, 1713–16 (1956).

VAN ROOSBROECK, W.: "Theory of current-carrier transport and photoconductivity in semiconductors with trapping", *Bell Syst. Tech. J.* **39**, 515–613 (1960).

VAN ROOSBROECK, W.: "Current-carrier transport and photoconductivity in semiconductors with trapping", *Phys. Rev.* **119**, 636–52 (1960).

VAN ROOSBROECK, W.: "Current-carrier transport with space charge in semiconductors", *Phys. Rev.* **123**, 474–90 (1961).

VAN ROOSBROECK, W.: "Electronic basis of switching in amorphous semiconductor alloys", *Phys. Rev. Lett.* **28**, 1120–3 (1972).

VAN ROOSBROECK, W.: "Principles of electrical behaviour of amorphous semiconductor alloys", *J. Non-Cryst. Solids* **12**, 232–62 (1973).

VAN ROOSBROECK, W. and CASEY, H. C.: "A new regime of semiconductor behavior: carrier transport when dielectric relaxation time exceeds lifetime", *Proceedings of the 10th International Conference on the Physics of Semiconductors* (eds. S. P. Keller, J. C. Hensel, and F. Stern) (USAEC, Springfield, Va) 832–8 (1970).

VAN ROOSBROECK, W. and CASEY, H. C.: "Transport in relaxation semiconductors", *Phys. Rev.* B **5**, 2154–75 (1972).

VAN ROOSBROECK, W. and SHOCKLEY, W.: "Photon-radiative recombination of electrons and holes in germanium", *Phys. Rev.* **94**, 1558–60 (1954).

VAN VLIET, K. M. and BLASQUEZ, G.: "Connection between sources formalism and correlation formalism for transport noise: application to injection diodes", *J. Appl. Phys.* **47**, 768–71 (1976).

VAN VLIET, K. M., FRIEDMANN, A., ZIJLSTRA, R. J. J., GISOLF, A., and VAN DER ZIEL, A.: "Noise in single injection diodes. I: A survey of methods", *J. Appl. Phys.* **46**, 1804–13 (1975).

VAN VLIET, K. M., FRIEDMANN, A., ZIJLSTRA, R. J. J., GISOLF, A., and VAN DER ZIEL, A.: "Noise in single injection diodes. II: Applications", *J. Appl. Phys.* **46**, 1814–23 (1975).

VANNIKOV, A. V., BOGUSLAVSKII, L. I., and MARGULIS, V. B.: "Charge carrier movement in anthracene single crystals", *Fiz. Tekh. Poluprov.* **1**, 935–7 (1967).

VANNIKOV, A. V., LOZHKIN, B. T., and BOGUSLAVSKII, L. I.: "Study of the surface state of anthracene single crystals using low-energy electron pulses", *Soviet Phys.—Solid State* **12**, 426–9 (1970).

VARSHNI, Y. P.: "Band-to-band radiative recombination in groups IV, VI and III–V semiconductors (I)", *Phys. Stat. Sol.* **19**, 459–514 (1967).

VARSHNI, Y. P.: "Band-to-band radiative recombination in groups IV, VI and III–V semiconductors (II)", *Phys. Stat. Sol.* **20**, 9–36 (1967).

VARTANYAN, A. T.: "On the photoconductivity of anthracene", *Dokl. Akad. Nauk SSSR* **71**, 641 (1950) (NRC Technical Translation TT-160).

VARTANYAN, A. T.: "Temperature dependence of the electric conductivity of organic semiconductors", *Bull. Acad. Sci. USSR, Phys. Ser.* **20**, 1412–18 (1956).

VASI, J. and WESTGATE, C. R.: "Diffusion effects in the double injection negative-resistance problem", *Solid-State Electron.* **16**, 269–75 (1973).

VAUBEL, G.: "Recombination radiation in tetracene single crystals", *Phys. Stat. Sol.* **35**, K67–69 (1969).

VAUBEL, G.: "Charge carrier injection by excitons at tetracene (anthracene)–electrolyte interface", *Phys. Stat. Sol.* **38**, 217–21 (1970).

VAUBEL, G.: "Intersystem crossing between guest and host molecules in anthracene crystals", *Chem. Phys. Lett.* **9**, 51–53 (1971).

VAUBEL, G. and BAESSLER, H.: "Determination of the energy of conduction states in anthracene crystals by photoemission of electrons from sodium", *Phys. Stat. Sol.* **26**, 599–606 (1968).

VAUBEL, G. and BAESSLER, H.: "Determination of band-gap in anthracene", *Phys. Lett.* **27A**, 328–9 (1968).

VAUBEL, G. and BAESSLER, H.: "Diffusion of singlet excitons in tetracene crystals", *Mol. Cryst. Liq. Cryst.* **12**, 47–56 (1970).

VAUBEL, G. and BAESSLER, H.: "Excitation spectrum of crystalline tetracene fluorescence: probe for optically induced singlet-exciton fission", *Mol. Cryst. Liq. Cryst.* **15**, 15–25 (1971).

VAUBEL, G. and BAESSLER, H.: "Franck–Condon factors in intermolecular inter-system crossing", *Chem. Phys. Lett.* **11**, 613–16 (1971).

VAUBEL, G., BAESSLER, H., and MÖBIUS, D.: "Reaction of singlet exciton and anthracene/metal interface: energy transfer", *Chem. Phys. Lett.* **10**, 334–6 (1971).

VAUBEL, G. and KALLMANN, H.: "Diffusion length and lifetime of triplet excitons and crystal absorption coefficient in tetracene determined from photocurrent measurements", *Phys. Stat. Sol.* **35**, 789–92 (1969).

VAUBEL, G. and PETER, L.: "Intermolecular triplet energy transfer in naphthalene–pentacene mixed crystals via higher excited triplet states", *Phys. Stat. Sol.* B **51**, 155–9 (1972).

VAUBEL, G., THOMA, P., RIEHL, N., and LENZ, P.: "Stimulation of anthracene by infrared light at varied temperatures and the observation of phosphorescence", in *Proceedings of the International Conference on Luminescence, Budapest, 1966*, Budapest, Akademiai Kiado, **1** and **2**, 387–90 (1968).

VERDERBER, R. R., SIMMONS, J. G., and EALES, B.: "Forming processes in evaporated SiO thin films", *Phil. Mag.* **16** 1049–61 (1967).

VERNON, S. M. and ANDERSON, W. A.: "Temperature effects in Schottky-barrier silicon solar cells", *Appl. Phys. Lett.* **26**, 707–9 (1975).

VERTSIMAKHA, YA. I., KURIK, M. V., LOPATKIN, YU. M., and FSIKORA, L. I.: "Photoconductivity of pentacene crystals", *Soviet Phys.—Solid State* **15**, 666–7 (1973).

VERTSIMAKHA, YA. I., KURIK, M. V., and PIRYATINSKII, YU. P.: "Temperature dependence of the photoconductivity spectrum of anthracene", *Soviet Phys.—Solid State* **14**, 1241–4 (1972).

VERTSIMAKHA, YA. I., KURIK, M. V., and PIRYATINSKII, YU. P.: "Photovoltaic effect in anthracene crystals", *Soviet Phys.—Solid State* **14**, 2035–9 (1973).

VEZZOLI, G. C., WALSH, P. J., and DOREMUS, L. W.: "Threshold switching and the on-state in non-crystalline chalcogenide semiconductors: an interpretation of threshold-switching research", *J. Non-Cryst. Solids* **18**, 333–73 (1975).

VIDADI, Y. A. and ROZENSHTEIN, L. D.: "Photocapacitance and photocurrent memory of thin organic dye films", *Soviet Phys.—Semicond.* **2**, 231–2 (1968).

VIDADI, Y. A., ROZENSHTEIN, L. D., and CHISTYAKOV, E. A.: "Jump and band conductivity in organic semiconductors", *Fiz. Tverd. Tela* **11**, 219–22 (1969) [*Soviet Phys.—Solid State* **11**, 173–6 (1969)].

VILFAN, I.: "Small polaron model of the electron motion in organic molecular crystals", *Phys. Stat. Sol.* (b) **59**, 351–60 (1973).

VINCENT, G., BOIS, D., and PINARD, P.: "Conductance and capacitance studies in GaP Schottky barriers", *J. Appl. Phys.* **46**, 5173–8 (1975).

VINCENT, V. M. and WRIGHT, J. D.: "Photoconductivity and crystal structure of organic molecular complex", *J. Chem. Soc. Faraday Trans.* I **70**, 58–71 (1974).

VISCAKAS, J., MACKUS, P., and SMILGN, A.: "Space-charge-limited currents and high electric field effects in vitreous selenium films", *Phys. Stat. Sol.* **25**, 331–5 (1968).

VISHAEVSKII, V. N. and BORISOV, M. D.: "Determination of the length of the diffusion displacement of excitons in molecular crystals", *Ukr. Fiz. Zh.* **1**, 294, 371 (1956).

VITYUK, N. V., FEDCHUK, A. P., and MIKHO, V. V.: "Anomalous switching effect and polarization memory mechanism in an anthracene–aluminum oxide film structure", *Soviet Phys.—Solid State* **17**, 612–13 75).

VITYUK, N. V. and MIKHO, V. V.: "Electroluminescence of anthracene excited by π-shaped voltage pulses", *Soviet Phys.—Semicond.* **6**, 1479–99 (1973).

VITYUK, N. V. and MIKHO, V. V.: "On asymmetry of electroluminescence for thin anthracene films", *Izv. Vuz. Fiz.* **12**, 11–15 (1973).

VITYUK, N. V. and MIKHO, V. V.: "Temperature dependence of the quenching of anthracene electroluminescence", *Ukr. Fiz. Zh.* **19**, 1041–3 (1974).

VLADIMIROV, V. V., KURIK, M. V., and PIRYATINSKII, YU. P.: "Sign inversion of the photo-e.m.f. in anthracene", *Soviet Phys.—Doklady* **13**, 789–90 (1969).

VODENICHAROVA, M.: "On the mechanism of electrical conductivity in thin polymer organic semiconductor films", *Phys. Stat. Sol.* (a) **28**, 263–8 (1975).

VODYANOI, Y. YA., IVCHENKO, E. L., and FEDOROVICH, N. A.: "Activation mechanism of contact and bulk conductivities in thin iodine-modified organic films", *Soviet Phys.—Solid State* **15**, 1455–6 (1974).

VOGEL, F., GEACINTOV, N. E., and POPE, M.: "Photoinjection of holes into anthracene crystals using aqueous I_3^- solutions: magnetic field effects", *J. Chem. Soc. Trans. Faraday* II, 1208–22 (1973).

VOL, E. D., GOLOYADOV, V. A., KUKUSHKIN, L. S., NABOYKIN, YU. V., and SILAEVA, N. B.: "Exciton–exciton annihilation in pyrene crystals", *Phys. Stat. Sol.* (b) **47**, 685–92 (1971).

VOLKENSTEIN, M. V.: "The conformons", *J. Theor. Biol.* **34**, 193–5 (1972).

VOLLMANN, W.: "Poole–Frenkel conduction in insulators of large impurity densities", *Phys. Stat. Sol.* (A) **22**, 195–203 (1974).

VOLTZ, R., DUPONT, H., and KING, T. A.: "Diffusion of triplet excitons produced by ionizing radiation in aromatic materials", *Nature* **211**, 405–6 (1966).

VOZZHENNIKOV, V. M., ZVONKOVA, Z. V., SHEYAKHINA, L. P., and BEREZKIN, V. V.: "Electrophysical properties of some charge transfer complexes", *Russ. J. Phys. Chem.* **43**, 1285–7 (1969).

VUL, B. M., ZAVARITSKAYA, E. I., and ZAVARITSKII, N. V.: "Tunnel effect in gallium arsenide diodes at low temperatures", *Soviet Phys.—Solid State* **8**, 710–14 (1966).

WADATI, M. and ISIHARA, A.: "Theory of liquid crystals", *Mol. Cryst. Liq. Cryst.* **17**, 95–108 (1972).

WADDINGTON, T. C. and SCHNEIDER, W. G.: "The effect of the ambient gases on the surface photocurrent in anthracene", *Can. J. Chem.* **36**, 789–92 (1958).

WAGENER, J. L. and MILNES, A. G.: "Post-breakdown conduction in forward-biased p–i–n silicon diodes", *Appl. Phys. Lett.* **5**, 186–8 (1964).

WAGENER, J. L. and MILNES, A. G.: "Double-injection experiments in semi-insulating silicon diodes", *Solid-State Electron.* **8**, 495–507 (1965).

WAKAYAMA, N. I., WAKAYAMA, N., and WILLIAMS, D. F.: "Pulsed and steady-state electroluminescence of pentacene doped anthracene crystal", *Mol. Cryst. Liq. Cryst.* **26**, 275–80 (1974).

WAKAYAMA, N. and WILLIAMS, D. F.: "Singlet exciton–charge carrier interaction in anthracene", *Chem. Phys. Lett.* **9**, 45–47 (1971).

WAKAYAMA, N. and WILLIAMS, D. F.: "Carrier–exciton interactions in crystalline anthracene", *J. Chem. Phys.* **57**, 1770–9 (1972).

WALDEN, R. H.: "A method for the determination of high-field conduction law in insulating films in the presence of charge trapping", *J. Appl. Phys.* **43**, 1178–86 (1972).

WALKER, L. G. and PRATT, G. W.: "Low-voltage tunnel-injection blue electroluminescence in ZnS MIS diodes", *J. Appl. Phys.* **47**, 2129–33 (1976).

WANNIER, G. H.: "The structure of electronic excitation levels in insulating crystals", *Phys. Rev.* **52**, 191–7 (1937).

WARMACK, R. J., CALLCOTT, T. A., and SCHWEINLER, H. C.: "Calculations of apparent conductivity for anisotropic medium (TTF) (TCNQ)", *Appl. Phys. Lett.* **24**, 635–7 (1974).

WARTER, P. J.: "Models for field-controlled photogeneration in molecular solids", in *Proceedings of the 3rd International Conference on Photoconductivity* (ed. E. M. Pell), Pergamon Press (Oxford) 311–16 (1971).

WATANABE, Y. and INUISHI, Y.: "Photoconductivity in γ-irradiated anthracene crystals", Tech. Rept., Osaka University, Japan **20**, 910 (1970).

WATANABE, Y., SAITO, N., and INUISHI, Y.: "X-ray induced photoconductivity in anthracene", *Japan J. Appl. Phys.* **7**, 854–61 (1968).

WATANABE, Y., SAITO, N., and INUISHI, Y.: "Carrier production and transport in anthracene", Tech. Rept., Osaka University, Japan **18**, 143–54 (1968).

WAXMAN, A. and LAMPERT, M. A.: "Double injection in insulators. II: Further analytic results with negative resistance", *Phys. Rev.* B **1**, 2735–47 (1970).

WAXMAN, A., SHEWCHUN, J., and WARFIELD, G.: "Tunneling in MIS structures. II: Experimental results on M–SiO$_2$–Si", *Solid-State Electron.* **10**, 1187–98 (1967).

WEBER, W. H.: "Double injection in long p–i–n diodes with deep double-acceptor impurities", *Appl. Phys. Lett.* **16**, 396–9 (1970).

WEBER, W. H., ELLIOTT, R. S., and CEDERQUIST, A. L.: "Small-signal transient double injection in semiconductors heavily doped with deep traps", *J. Appl. Phys.* **42**, 2497–501 (1971).

WEBER, W. H. and FORD, G. W.: "Double injection in semiconductors heavily doped with deep two-level traps", *Solid-State Electron.* **13**, 1333–56 (1970).

WEBMAN, I. and JORTNER, J.: "Energy dependence of two photon-adsorption cross sections in anthracene", *J. Chem. Phys.* **50**, 2706–16 (1969).

WEI, L. S. and SIMMONS, J. G.: "Transient emission and generation currents in metal–insulator–semiconductor capacitors", *Solid-State Electron.* **18**, 853–7 (1975).

WEIGL, J. W.: "Spectroscopic properties of organic photoconductors: absorption spectra of cationic dye films", *J. Chem. Phys.* **24**, 364–70 (1956).

WEISER, K. and LEVITT, R. S.: "Electroluminescent gallium arsenide diodes with negative resistance", *J. Appl. Phys.* **35**, 2431–8 (1964).

WEISZ, S. Z.: "Utilization of radiation damage in the study of exciton impurity interaction", in *1972 Annual Conference on Electrical Insulation and Dielectric Phenomena (NAS Washington, DC, USA)* 346–8 (1973).

Weisz, S. Z., Cobas, A., Richardson, P. E., Szmant, H. H., and Trester, S.: "Detection of radiation damage in anthracene crystals by space-charge-limited current measurements", *J. Chem. Phys.* **44**, 1365–8 (1966).

Weisz, S. Z., Cobas, A., Trester, S., and Many, A.: "Electrode-limited and space-charge-limited transient current in insulators", *J. Appl. Phys.* **39**, 2296–302 (1968).

Weisz, S. Z., Jarnagin, R. C., and Silver, M.: "Carrier trapping on photoconducting anthracene", *J. Chem. Phys.* **40**, 3365–9 (1964).

Weisz, S. Z., Levinson, J., and Cobas, A.: "Interaction of triplet excitons with trapped electrons in anthracene crystals", in *Proceedings of the Third Photo-conductivity Conference, Stanford, August 12–15, 1969* (ed. E. M. Pell), Pergamon (Oxford) 1971; also, *J. Chem. Phys.* **52**, 2794–5 (1970).

Weisz, S. Z., Levinson, J., and Cobas, A.: "A method for obtaining microsecond risetime step function high intensity light pulse", *Rev. Sci. Instrum.* **42**, 261–2 (1971).

Weisz, S. Z., Menendez, J. M., Rojas, L. F., and Dellonte, S.: "The involvement of triplet excitons in the surface carrier generation in anthracene", *Mol. Cryst. Liq. Cryst.* **24**, 45–52 (1973).

Weisz, S. Z., Richardson, P. E., and Cobas, A.: "Triplet sampled radiation damage", *Mol. Cryst.* **4**, 277–92 (1968) and also *Mol. Cryst.* **3**, 168 (1967).

Weisz, S. Z. and Whitlen, W. B.: "Triplet exciton–free hole interaction in pyrene", *Chem. Phys. Lett.* **23**, 187–9 (1973).

Weisz, S. Z., Zahlan, A. B., Gilreath, J., Jarnagin, R. C., and Silver, M.: "Two-photon absorption in crystalline anthracene and naphthalene excited with a xenon flash", *J. Chem. Phys.* **41**, 3491–5 (1964).

Weisz, S. Z., Zahlan, A. B., Silver, M., and Jarnagin, R. C.: "Radiationless transition rate constant determined from delayed fluorescence", *Phys. Rev. Lett.* **12**, 71–73 (1964).

Welder, G.: "Adsorptions", in *Chemische Taschenbücher* (eds. W. Foerst and H. Grünewald), Verlag-Chemie (Germany) **9**, 29 (1970).

Wentworth, W. E., Chen, E., and Lovelock, J. E.: "The pulse-sampling technique for the study of electron attachment phenomena", *J. Phys. Chem.* **70**, 445–8 (1966).

Wertheim, G. K.: "Transient recombination of excess carriers in semiconductors", *Phys. Rev.* **109**, 1086–91 (1958).

West, W.: "Spectroscopy and spectrophotometry", in *Physical Methods of Organic Chemistry*, 3rd edn (ed. A. Weissberger), Wiley (New York), **1**, Part III, 1799–1958 (1960).

Westgate, C. R. and Warfield, G.: "Drift-mobility measurements in metal-free and lead-phthalocyanine", *J. Chem. Phys.* **46**, 94–97 (1967).

Westgate, C. R. and Warfield, G.: "Thermally stimulated trap emptying measurements in phthalocyanine", *J. Chem. Phys.* **46**, 537–8 (1967).

Weston, W., Mey, W., and Pope, M.: "Effects of polishing anthracene single crystals on the triplet exciton lifetime", *J. Appl. Phys.* **44**, 5615–16 (1973).

Wey, H. Y.: "Surface of amorphous semiconductors and their contacts with metals", *Phys. Rev. B* **13**, 3495–505 (1976).

White, H. G. and Logan, R. A.: "GaP surface-barrier diodes", *J. Appl. Phys.* **34**, 1990–6 (1963).

Whitten, W. B. and Arnold, S.: "Pressure modulation of exciton fission in tetracene", *Phys. Stat. Sol.* (b) **74**, 401–7 (1976).

Whitten, W. B., Arnold, S., and Swenberg, C. E.: "Hot band transitions to the triplet state in anthracene and pyrene crystals", *J. Chem. Phys.* **60**, 4219–22 (1974).

Wigner, E. and Bardeen, J.: "Theory of the work function of monovalent metals", *Phys. Rev.* **48**, 84–95 (1935).

Wihksne, K. and Newkirk, A. E.: "Electrical conductivities of α- and β- phthalocyanine", *J. Chem. Phys.* **34**, 2184–5 (1961).

Wildi, B. S.: "Thermoelectric generators", US Patent 3259628 (Monsento Company) (1966).

Wilk, M.: "Semiconductor properties of aromatic hydrocarbons", *Z. Elektrochem.* **64**, 930–6 (1960).

Williams, D. F.: "Phosphorescence spectrum of anthracene crystal", *J. Chem. Phys.* **411**, 344–5 (1967).

Williams, D. F. and Adolph, J.: "Diffusion length of triplet excitons in anthracene crystals", *J. Chem. Phys.* **46**, 4252–4 (1967).

Williams, D. F., Adolph, J., and Schneider, W. G.: "Diffusion of triplet exciton in anthracene crystals", *J. Chem. Phys.* **45**, 575–7 (1966).

Williams, D. F. and Schadt, M.: "A simple organic electroluminescent diode", *Proc. IEEE* **58**, 475 (1970).

Williams, D. F. and Schadt, M.: "DC and pulsed electroluminescence in anthracene and doped anthracene crystals", *J. Chem. Phys.* **53**, 3480–7 (1970).

Williams, D. F. and Schadt, M.: "Temperature dependence of d.c. and pulsed electroluminescence in anthracene crystals", in *Proceedings of the International Conference on Photoconductivity* (ed. E. M. Pell), Pergamon (Oxford) 303–9 (1971).

WILLIAMS, D. F. and SCHNEIDER, W. G.: "Phosphorescence emission from anthracene single crystals", *J. Chem. Phys.* **45**, 4756–7 (1966).

WILLIAMS, J. O.: "Thermal and optical release of injected charge carriers in molecular crystals", *Sci. papers, Inst. Organic and Phys. Chem. Wroclaw Univ.*, **7**, 174–81 (1974).

WILLIAMS, J. O., CLARKE, B. P., THOMAS, J. M., and SHAW, M. J.: "Structural imperfections and singlet exciton traps in crystalline anthracene", *Chem. Phys. Lett.* **38**, 41–46 (1976).

WILLIAMS, J. O., COX, G. A., and THOMAS, J. M.: "Electrical conductivity of plastic crystals", *J. Phys Chem.* **71**, 1542–3 (1967).

WILLIAMS, J. O. and THOMAS, J. M.: "Lattice imperfections in organic solids. Part I: Anthracene", *Trans. Faraday Soc.* **63**, 1720–9 (1967).

WILLIAMS, J. O. and THOMAS, J. M.: "Photochemical reactions inside the electron microscope: preferred dimerization of anthracene at dislocations", *Mol. Cryst. Liq. Cryst.* **16**, 371–5 (1972).

WILLIAMS, R.: "Injection by internal photoemission", in *Semiconductors and Semimetals* (eds. R. K. Willardson and A. Beer) **6**, 97–139 (1970).

WILLIAMS, R. and CAMPOS, M.: "Effect of electric charge on the evaporation of naphthalene single crystals", *J. Appl. Phys.* **41**, 4138–40 (1970).

WILLIAMS, R. and DRESNER, J.: "Photoemission of holes from metal into anthracene", *J. Chem. Phys.* **46**, 2133–8 (1967).

WILLIAMS, R. H. and JOHNSON, H. R.: "A mechanism for the anomalous photovoltaic effect in cadmium telluride", *Solid State Commun.* **16**, 873–5 (1975).

WILLIAMS, W. G., SPONG, P. L., and GIBBONS, D. J.: "Double injection electroluminescence in anthracene and carrier injection properties of carbon fibres", *J. Phys. Chem. Solids* **33**, 1879–84 (1972).

WILLING, F.: "Escape of holes from the surface of organic crystals with electrolytic contacts", *Chem. Phys. Lett.* **40**, 331–5 (1976).

WILLING, F., SCHERER, G., and ROTHÄMEL, W.: "Experimental proof for electrochemical non-equilibrium saturation currents in organic crystals", *Z. Naturforsch.* **29a**, 131–40 (1974).

WINDSOR, M. W.: "Luminescence and energy transfer", in *Physics and Chemistry of the Organic Solid State* (eds. D. Fox, M. M. Labes, and A. Weissberger) **2**, 343–432, 898–908 (1965).

WINTERSTEIN, A. *et al.*: "Fractionation and purification of organic substances by the principle of chromatographic absorption analysis." I: Winterstein, A. and Stein, G. Z., *Physiol. Chem.* **220**, 247–63 (1933); II: ibid. **220**, 263–77 (1933); III; Winterstein, A. and Schön, K. Z., ibid. **230**, 139–45 (1934); IV: ibid. **230**, 146–58 (1934); V: Winterstein, A., Schön, K., and Vetter, H., ibid. **230**, 158–69 (1934); IV: Winterstein, A. and Vetter, H., ibid. **230**, 169–74 (1934).

WINTLE, H. J.: "Surface-charge decay in insulators with nonconstant mobility and with deep trapping", *J. Appl. Phys.* **43**, 2927–30 (1972).

WINTLE, H. J.: "Contact charging of polymers", *J. Phys. D* **7**, L128–31 (1974).

WINTLE, H. J.: "Space charge limited currents in graded films", *Thin Solid Films* **21**, 83–90 (1974).

WINTLE, H. J.: "Absorption currents and steady currents in polymer dielectrics", *J. Non-Cryst. Solids* **15**, 471–86 (1974).

WINTLE, H. J.: "Transient charging currents in insulators", *Solid-State Electron.* **18**, 1039–42 (1975).

WITTMER, M. and ZSCHOKKE-GRÄNACHER, I.: "Electroluminescence of doped anthracene crystals", *Hel. Phys. Acta (Switzerland)* **46**, 405 (1973).

WITTMER, M. and ZSCHOKKE-GRÄNACHER, I.: "Exciton–charge carrier interactions in electroluminescence of crystalline anthracene", *J. Chem. Phys.* **63**, 4187–94 (1975).

WOLF, H. C.: "The electronic spectra of aromatic molecular crystals", *Solid-State Phys.* **9**, 1–81 (1959).

WOLF, H. C.: "Energy transfer in organic molecular crystals: a survey of experiments", in *Advances in Atomic and Molecular Physics.* (eds. D. R. Bates and I. Estermann), Academic Press (New York) **3**, 119–42 (1967).

WOLF, H. C. and BENZ, K. W.: "Energy trapping processes in aromatic crystals", *Pure Appl. Chem.* **27**, 439–55 (1971).

WOOD, D. W., ANDERSEN, T. N., and EYRING, H.: "Electrical properties of some porphyrins under high pressure", *J. Phys. Chem.* **70**, 360–6 (1966).

WOTHERSPOON, N. and OSTER, G.: "Fluorescence and phosphorescence", in *Physical Methods of Organic Chemistry,* 3rd edn. (ed. A Weisberger), Wiley (New York) **1**, Part III, 2063–105 (1960).

WOTHERSPOON, N., POPE, N., and BURGOS, J.: "Fluorescence modulation in anthracene by singlet-trapped hole interactions", *Chem. Phys. Lett.* **5**, 453–5 (1970).

WRIGHT, G. T.: "Scintillation response of organic phosphors", *Phys. Rev.* **91**, 1282–3 (1953).

WRIGHT, G. T.: "Scintillation response of phosphors at low particle energies", *Phys. Rev.* **96**, 569–70 (1954).

WRIGHT, G. T.: "Fluorescence excitation spectrum of anthracene", *Phys. Rev.* **100**, 587–8 (1955).

WRIGHT, G. T.: "Scintillation response of anthracene crystals to short range electrons", *Phys. Rev.* **100**, 588–90 (1955).

WRIGHT, G. T.: "Absolute quantum efficiency of photofluorescence of anthracene crystals", *Proc. Phys. Soc. (London)* B **68**, 241–8 (1955).

WRIGHT, G. T.: "Absolute scintillation efficiency of anthracene crystals", *Proc. Phys. Soc.* B **68**, 929–37 (1955).

WRIGHT, G. T.: "Fluorescence excitation spectra and quantum efficiencies of organic crystals", *Proc. Phys. Soc.* B **68**, 701–11 (1955).

WRIGHT, G. T.: "Scintillation decay times of organic crystals", *Proc. Phys. Soc.* B **69**, 358–72 (1956).

WRIGHT, G. T.: "Mechanisms of space-charge-limited current in solids", *Solid-State Electron.* **2**, 165–89 (1961).

WRIGHT, G. T.: "Transit time effects in the space-charge-limited silicon microwave diode", *Solid-State Electron.* **9**, 1–6 (1966).

WRIGHT, G. T.: "Small-signal characteristics of semiconductor punch-through injection and transit-time diodes", *Solid-State Electron.* **16**, 903–12 (1973).

WRIGHT, H. C. and ALLEN, G. A.: "Thermally stimulated current analysis", *Br. J. Appl. Phys.* **17**, 1181–5 (1968).

WRIGHT, W. H.: "Variations in the efficiency of fluorescence excitation in anthracene", *J. Chem. Phys.* **45**, 874–7 (1966).

WRIGHT, W. H.: "Ultraviolet optical constant of anthracene", *J. Chem. Phys.* **46**, 2951–3 (1967).

WRONSKI, C. R.: "Electronic properties of amorphous silicon in solar cell operation", *IEEE Trans. Electron.* ED **24**, 351–7 (1977).

WRONSKI, C. R., CARLSON, D. E., and DANIEL, R. E.: "Schottky-barrier characteristics of metal–amorphous-silicon diodes", *Appl. Phys. Lett.* **29**, 602–4 (1976).

WU, C. H. and BUBE, R. H.: "Thermoelectric and photothermoelectric effects in semiconductors: cadmium sulfide films", *J. Appl. Phys.* **45**, 648–60 (1974).

WYATT, A. F. G.: "Anomalous densities of states in normal tantalum and niobium", *Phys. Rev. Lett.* **13**, 401–5 (1964).

WYHOF, J. R. and POHL, H. A.: "Hyperelectronic polarization, a physical property of EKA–conjugated polymers", *J. Polymer Sci.* A2 **8**, 1741–54 (1970).

WYSOCKI, J. J., RAPPAPORT, P., DAVISON, E., HAND, R., and LOFERSKI, J. J.: "Lithium-doped, radiation-resistant silicon solar cells", *Appl. Phys. Lett.* **9**, 44–46 (1966).

YAKOVLEV, B. S. and FRANKEVICH, E. L.: "Photogeneration of holes in crystalline tetracene", *Fiz. Tverd. Tela* **11**, 1975–7 (1969).

YAMAGISHI, A. and SOMA, M.: "Temperature dependence of hole injection into anthracene crystals", *Bull. Chem. Soc. Japan* **43**, 3741–5 (1970).

YAMAGUCHI, S.: "Electron diffraction analysis of the dielectric structure of anthracene", *J. Chem. Phys.* **43**, 833–4 (1965).

YANG, C. C. and HARTMANN, G. C.: "Electrostatic separation of a charged particle layer between electrodes", *IEEE Trans. Electron. Devices* ED **23**, 308–12 (1976).

YANG, E.: "Excitonic effects on dielectric properties of solids in a uniform electric field", *Phys. Rev.* B **4**, 2046–53 (1971).

YANG, E. and BUCKMAN, A. B.: "Electroreflectance in a non-uniform field in the small-wave-number approximation and its measurement by ellipsometry", *Phys. Rev.* B **5**, 2242–50 (1972).

YARMUS, L., SWENBERG, C., ROSENTHAL, J., and ARNOLD, S.: "Magnetic field effect on delayed fluorescence from pyrene crystals", *Phys. Lett.* **43A**, 103–4 (1973).

YASUNAGA, H., KOJIMA, K., YOHDA, H., and TAKEYA, H.: "Effect of oxygen on electrical properties of lead phthalocyanine", *J. Phys. Soc. Japan* **37**, 1024–30 (1974).

YEARGAN, J. R. and TAYLOR, H. L.: "The Poole–Frenkel effect with compensation present", *J. Appl. Phys.* **39**, 5600–4 (1968).

YEE, E. M. and EL-SAYED, M. A.: "Effects of traps on migration and annihilation of triplet excitation in phenanthrene crystals", *J. Chem. Phys.* **52**, 3075–90 (1970).

YEH, Y. C. M., ERNEST, F. P., and STIRN, R. J.: "Practical antireflection coating for metal–semiconductor solar cells", *J. Appl. Phys.* **47**, 4107–12 (1976).

YOMOSA, S.: " π -electronic structure of the peptide group and hydrogen-bonded polypeptides", in *Quantum Aspects of Polypeptides and Polynucleotides* (ed. M. Weissbluth), Wiley Interscience (New York) 1–33 (1964).

YOSHIDA, M., SAKAI, S., MITSUDO, H., and OOSHIKA, Y.: "Photogeneration of holes and electrons in anthracene crystals", *J. Phys. Soc. Japan* **25**, 638 (1968).

YOSHIMURA, S.: "Photoemission from an anthracene single crystal doped with alkali-metals", *J. Phys. Soc. Japan* **28**, 701–5 (1970).

ZAHN, M.: "Transient drift-dominated unipolar conduction between concentric cylinders and spheres", *IEEE Trans. Elect. Insulation* EI **11**, 150–7 (1976).

ZAHN, M. and CHATELON, H.: "Charge injection between concentric cylindrical electrodes", *J. Appl. Phys.* **48**, 1797–1805 (1977).

ZAHN, M., TSANG, C. F., and PAO, S. C.: "Transient electric fields and space-charge behavior for unipolar ion conduction", *J. Appl. Phys.* **45**, 2432–40 (1974).

ZANIO, K. R., AKUTAGAWA, W. M., and KIKUCHI, R.: "Transient currents in semiinsulating CdTe characteristics of deep traps", *J. Appl. Phys.* **39**, 2818–28 (1968).

ZGIERSKI, M. Z.: "Vibronic coupling and problem of bulk photoconductivity in molecular crystals", *Phys. Stat. Sol.* (b) **55**, 451–6 (1973).

ZGIERSKI, M. Z.: "Resonance interband interaction and vibronic coupling in molecular crystals", *Phys. Stat. Sol.* (b) **57**, 405–13 (1973).

ZGIERSKI, M. Z.: "Influence of vibronic coupling on the dissociation of Frenkel excitons", *Phys. Stat. Sol.* (b) **59**, 589–93 (1973).

ZGIERSKI, M. Z.: "Herzberg–Teller interaction and vibronic coupling in molecular crystals", *Phys. Stat. Sol.* (b) **65**, 797–800 (1974).

ZHADKO, I. P. and ROMANOV, V. A.: "On the origin of high photovoltages in semiconductors", *Phys. Stat. Sol.* **28**, 797–800 (1968).

ZIMA, V. L. and FAIDISH, O. M.: "Temperature dependence of the quantum yields of luminescence in pure and impurity-admixed crystals of anthracene and naphthalene", *Opt. Spectrosc.* **19**, 203–6 (1965).

ZIMA, V. L. and FAIDISH, O. M.: " Diffusion of excitons in naphthalene and anthracene crystals", *Opt. Spectrosc.* **20**, 566–9 (1966).

ZIMA, V. L., KORSUNSKII, V. M., and FAIDISH, O. M.: "Spectra and transfer condition for electronic exciton energy in pure and doped anthracene crystals", *Izv. Akad. Nauk SSSR Ser. Fiz.* **27**, 519 (1963).

ZIMMERMAN, E. J.: "Scintillation response of anthracene to low-energy protons and helium ions", *Phys. Rev.* **99**, 1199–1202 (1955).

ZSCHOKKE-GRÄNACHER, I., SCHWOB, H. P., and BALDINGER, E.: "Recombination radiation in anthracene doped with tetracene", *Solid State Commun.* **5**, 825–8 (1967).

ZVYAGINTSEV, A. M.: "Preflashover electroluminescence of anthracene", *Zh. Prikl. Spektrosk.* **13**, 165–7 (1970).

ZVYAGINTSEV, A. M. and STELIN, V. I.: "Mechanism of electroluminescence of anthracene", *Fiz. Tverd. Tela* **15**, 1923–5 (1973).

ZVYAGINTSEV, M. V.: "Mechanism of the electroluminescence of anthracene", *Soviet Phys.—Solid State* **15**, 1287–8 (1973).

ZWICKER, H. R., STREETMAN, B. G., HOLONYAK, N. JR., and ANDREWS, A. M.: "Double injection in semiconductors with multivalent trapping centers", *J. Appl. Phys.* **41**, 4697–709 (1970).

Index

Page numbers in **bold figures** indicate items of importance